GW00374653

Directory of Solvents

DIRECTORY OF SOLVENTS

Edited by

B. P. Whim, B.Sc., Ph.D.
formerly of ICI Chemicals & Polymers Ltd
Runcorn, Cheshire

and

P. G. Johnson
European Chlorinated Solvent Association
European Chemical Industry Council (CEFIC)
Brussels

BLACKIE ACADEMIC & PROFESSIONAL
An Imprint of Chapman & Hall
London · Weinheim · New York · Tokyo · Melbourne · Madras

Published by Blackie Academic & Professional, an imprint of Chapman & Hall, 2–6 Boundary Row, London SE1 8HN, UK

Chapman & Hall, 2–6 Boundary Row, London SE1 8HN, UK

Chapman & Hall GmbH, Pappelallee 3, 69469 Weinheim, Germany

Chapman & Hall USA, 115 Fifth Avenue, New York, NY 10003, USA

Chapman & Hall Japan, ITP-Japan, Kyowa Building, 3F, 2-2-1 Hirakawacho, Chiyoda-ku, Tokyo 102, Japan

DA Book (Aust.) Pty Ltd, 648 Whitehorse Road, Mitcham 3132, Victoria, Australia

Chapman & Hall India, R. Seshadri, 32 Second Main Road, CIT East, Madras 600 035, India

---

First edition 1996

Typeset in 10/12pt Times by Acorn Bookwork, Salisbury

Printed in Great Britain at the University Press, Cambridge

ISBN 0 7514 0245 1

A catalogue record for this book is available from the British Library
Library of Congress Catalog Card Number: 95-80924

∞ Printed on acid-free text paper, manufactured in accordance with ANSI/NISO Z39.48-1992 (Permanence of Paper).

Preface

Organic solvents represent an important class of compounds, whose utility is central to industrial and academic chemistry. The impact of solvents, directly or indirectly, in our everyday products, such as paints, surface coatings, adhesives, pharmaceuticals and cleaning products, is enormous, and the interest of industrial and academic researchers in the use of solvents is therefore considerable.

This volume is divided into two parts. Part 1 provides a detailed and authoritative review of the science and technology of solvents and issues relating to solvents, written by respected figures from within the solvents industry. The topics covered are solvency and its measurement, flammability, health and toxicology, environmental issues, legislative information, transport, storage, recovery and disposal, and a review of solvent applications.

Part 2 is intended as an accessible source of reliable and up-to-date data, based principally on information provided by manufacturers and suppliers. It consists of a database of over 350 solvent products, subdivided by Solvent Group. In addition, the data are presented in Key Parameter tables, covering:

Boiling points
Melting points
Evaporation information
Vapour pressure
Flash points
Solubility parameters
Auto ignition temperatures
Names and addresses of manufacturers, with trade names

Users can therefore select potential solvents from the Key Parameter tables, consult the basic data in the individual solvent table, and find the name and address of the manufacturer.

Information on the properties of solvents is widespread in the literature, and there are specialist textbooks dealing with solvents and solvency and listing physico-chemical data of pure compounds. However, there is no readily available source of information collating data for solvents commonly available from industry, as opposed to solvents as materials for research. In recent years there has been increased interest in health and safety, environmental issues and aspects of the legislative control of chemicals, including solvents, and the choice of a given solvent has therefore become more complex. This volume aims to provide in one place a broad spread of physico-chemical data, together with transport, safety, environmental and classification information provided by major European and USA suppliers and manufacturers of industrial organic solvents.

The book is aimed not only at the practitioner who is knowledgeable in the use of solvents and requires a one-stop source of reliable and up-to-date information, but also at the newcomer, for whom it provides an accessible review of major topics and a databank of key information, from which an initial selection of solvent can be made. The volume will be of interest to those engaged in solvents research, technical service and sales, and to operational chemists and engineers. Companies involved in formulating products which use or are based on solvents will find this an invaluable guide to current issues and products. The academic researcher will find much useful information in the range of solvents available and in the reviews of key issues.

The book has been edited by B.P. WHIM, with P.G. JOHNSON acting as co-editor for Part 2. The editors acknowledge with thanks the contributions of the various authors and the solvent suppliers who provided information on their products. They also wish to thank the staff of ICI CHLOR CHEMICALS for their help and advice—particularly Geoff Weemes and Graham Howe.

Finally, we have made every reasonable attempt to provide comprehensive coverage of solvents and their suppliers in Part 2 and shall be happy to receive, via the publishers, additional relevant information for inclusion in the next edition.

B.P.W.
P.G.J.

Contributors

W.C. Aten	Flevolaan 22, 1411 KD Naarden, The Netherlands
P. Davison	BP Chemicals Limited, Hull Research & Technology Centre, Salt End, Hull HU12 8DS, UK
I.D. Dobson	BP Chemicals, Brittanic Tower, Moor Lane, London EC2Y 9BU, UK
G.P. Howe	ICI Chemicals and Polymers Limited, Cleaning Technology and Chlormethanes Business, PO Box 14, The Heath, Runcorn, Cheshire WA7 4QG, UK
P.G. Johnson	CEFIC, Avenue E Van Nieuwenhuyse 4, B-1160, Brussels, Belgium
J. Kelsey	BP Chemicals Limited, Hull Research & Technology Centre, Salt End, Hull HU12 8DS, UK
D.A. King	BP Chemicals Limited, Hull Research & Technology Centre, Salt End, Hull HU12 8DS, UK
A.S. McCormick	BP Chemicals Limited, Hull Research & Technology Centre, Salt End, Hull HU12 8DS, UK
A.M. Moses	475 Chester Road, Hartford, Northwich, Cheshire CW8 2AG, UK
R.L. Rogers	INBUREX GmbH, Wilhelmstraße 2, D-59067 Hamm, Germany
B.P. Whim	Hawthorn Cottage, Dark Lane, Kingsley, Warrington WA6 8BQ, UK

Contents

Colour plates appear between pages 150–151

Part One—Solvents

1 Introduction to solvents

P. DAVISON

1.1 Introduction

Organic solvents have been around for a very long time but their uses have grown to such an extent that few people realise how important they are in our everyday lives. Solvents play a major part in the creation of many consumer products such as furniture, white goods, cars and their components, foodstuffs and packaging, electronic items, books, clothing (including cleaning them) and cosmetics to name a few. They are important in health care since they are key components in pharmaceuticals manufacture as well as in cleaning products and disinfectants. In industry, they are important for metal cleaning, crop protection and arguably, above all, paint production. Paints deserve a special mention as they represent the single largest use for solvents. They greatly enhance the appearance and extend the life of many products through their protective properties, whether it be a simple wooden window frame or a major civil engineering product.

The first use of solvents is shrouded in the unrecorded events of history. The first organic solvents used were almost certainly hydrocarbons, such as turpentines, derived from wood sources and ethanol from wine spirits. Ancient civilisations discovered that by fermenting vegetable matter, such as grapes or sugar cane, they could produce ethanol. In many ways it looked and behaved like water; in other ways it was very different and proved invaluable for dissolving oils and resins so that they could be worn on the skin. With industrialisation, the mass market for cosmetics was born. Other large scale uses started at the same time, initially with coal and eventually oil-derived products. The use of synthesised solvents, cost effectively and consistently produced, has, in the past 100 years, made possible the manufacture of a huge variety of goods which we nowadays take for granted, but without which living standards would be sharply reduced.

The literature contains various definitions of 'solvent'. The Oxford English Dictionary defines solvents as '. . . having the power of dissolving or forming solution with something' (the 'something' is technically referred to as the solute). The substance being dissolved should not be chemically changed by the solvent. Both the solute and solvent can, in principle, be either solid, liquid or gaseous but these definitions need qualification when referring to industrial solvents. Overwhelmingly these are organic liquids used to dissolve, suspend or change the physical properties of other materials and by doing so add value to the final product. Needless to say, industrial solvents also have to be available in commercially useful quantities. Generally, the term encompasses relatively few classes of liquid compounds; either aromatic or aliphatic hydrocarbons, esters, ketones, alcohols, ethers, glycols, aldehydes, amines, glycol ethers, alkyl or aromatic halides that boil within the range of ambient temperature to 250°C but the rule is flexible. (The boiling range definition helps to distinguish solvents from plasticisers which would otherwise fall within this scope.) This is the definition used in this directory and therefore the products which fall within this category are included. This book does not cover inorganic solvents such as supercritical carbon dioxide or ammonia which only have specialised uses or water for which the uses (and importance!) are rather too broad to cover in one volume.

Industrial chemicals generally fulfil three basic functions: they may serve as chemical intermediates, as processing agents or as ingredients in a formulation in applications that depend on certain desirable physical or mechanical properties. Solvents are no exception and bring benefits in each of these categories. It is only, however, in the role of processing aids that true 'solvent like' properties are required. In this use, solvents are a transient ingredient, and although present to fulfil a vital role during a product's life cycle (e.g. to aid application of a coating, facilitate a reaction or recrystallisation of a drug), their separation from the main functional part of the product is required at some point. There are, for most uses, two major technical performance properties that must be considered when selecting the correct solvent for a use: solvency and evaporation characteristics. Solvency is important in all applications because, obviously, the solvent must be capable of dissolving the required substance. Evaporation rate is important in any coating application to control film formation and rheology at various stages of drying; similarly, it is important in determining the cooling effect of a cosmetic product. Sometimes the two factors are interrelated; in a paint it is necessary that good solvency is maintained whilst the solvent evaporates to prevent poor film formation and maintain product quality. This is one reason why most paints comprise a carefully balanced mix of solvents, each bringing special solvency and evaporation characteristics to the final package. There are of course other considerations such as reaction specificity and ease of removal from the process, important in pharmaceutical manufacture, or odour as well as health, safety and environmental considerations such as flammability, toxicity and environmental impact, which will influence selection.

Assuming that the basic technical requirements above are met, a good industrial solvent generally has the following additional properties:

- clear and colourless
- sufficiently volatile to be removable without leaving an unacceptable residue
- no chemical reaction with the substance it is dissolving
- an acceptable odour
- constant physical properties
- low human/animal and environmental toxicity
- biologically degradable

In a commercial environment, solvent users are demanding more and more exacting requirements from the solvents they use. Ever increasing, and consistent, product quality with lower and lower impurity levels are needed for demanding current and new uses. (For example: polyurethane resin users require very low water contents and trace metal concentrations in the solvents they use; odour requirements for inks in food printing applications are increasingly stringent.) Solvents must also be readily and reliably available; security of supply is a key requirement for 'just in time' production processes.

1.2 Solvent classification

Solvents can be classified by one of four basic methods: by solvent power (solubility properties/parameters), evaporation rate/boiling point, chemical structure and hazard classification. Within the latter, classification can be by physical hazard (e.g. flash point), labelling classification, toxicity etc. Hazard classification is complex and is covered in more depth in later chapters.

The groups found when classifying by chemical structure have already been mentioned; the three main classes used are hydrocarbons, oxygenated and chlorinated solvents. Others, such as nitrogen and sulphur-containing solvents are speciality products and not the main focus of this book.

1.2.1 Classification by solvent power

The theory and application of solubility parameters are covered in a later chapter. At a practical level, solvents can be quantified according to three intermolecular forces between solvent molecules: dispersion, polar (dipole–dipole) and hydrogen bonding. Solvents can be grouped into one of four categories according to these forces as shown in Table 1.1.

This is a classification which works generally but not exclusively. There are some exceptions such as symmetric chlorinated solvents which are closer to aromatic solvents in solvency behaviour. Aminated solvents would also be included with hydroxylated materials.

This classification by solubility parameter leads on to three widely used terms: true solvent, non-solvent and latent solvent. For any given solute, a true solvent is one which will dissolve it at room temperature. Conversely, a non-solvent will be unable to dissolve it. A latent solvent cannot dissolve the solute by itself but can be activated in the presence of a true or non-solvent. Latent solvents are normally those that are strongly hydrogen bonded such as alcohols, where the addition of a second solvent strongly disrupts this hydrogen bonding. It should be borne in mind that a true solvent for one solute may be a non-solvent for another.

Table 1.1 Solvent classification by solvent power

Solvent	Dispersion forces	Polar forces	Hydrogen bonding
Oxygenated solvents with OH functionality	Moderate	Mod/High	Donors
Other oxygenated solvents	Mod/High	High	Strong acceptors
Aliphatic hydrocarbons	Low	None	None
Aromatic hydrocarbons	High	Low	Weak acceptors
Chlorinated solvents	Mod/High	High	Strong acceptors

1.2.2 Classification by evaporation rate

This is a simpler classification by relative speed of evaporation which is usually expressed relative to either butyl acetate or diethyl ether. There is no rigid distinction but fast solvents or low boilers typically evaporate more than twice as fast as butyl acetate; slow solvents or high boilers typically evaporate less than half as fast as butyl acetate. Medium boilers fit between these two limits.

1.3 Solvent manufacture

Most solvents are mainly derived from either crude oil or natural gas (methane). Hydrocarbons are all derived from the former with the exception of small amounts of wood derived materials (e.g. turpentine, dipentene). Aliphatic hydrocarbons are produced by refining and distilling various cuts and fractions; aromatics are produced either by cracking (important in Europe) or reforming naphtha streams. With the exception of toluene, hydrocarbon solvents are usually mixtures of different molecules; at least different isomers and often different compounds altogether. (For example, solvent grade xylene is a mixture of meta, ortho and para xylene along with ethyl benzene. The proportion can vary greatly but meta xylene is usually the largest percentage.) Characterisation is primarily by boiling point/evaporation rate and composition (percentage of aromatics, naphthenes (cyclic aliphatics) and paraffins (straight chain aliphatics)).

Oxygenated solvents are usually single molecules and, with a few exceptions at the higher molecular weight end, are available as relatively pure, single isomers. Methane, ethylene, propylene, butylene, air and water are the main building blocks to most of the important ones. The main routes to the primary oxygenated solvents are shown in Table 1.2. Table 1.3 shows the second line derivatives.

Table 1.2 Primary production routes and feedstocks for most important oxygenated solvents

Solvent	Main production routes
Methanol	From methane (via hydrogen, CO and CO_2)
Ethanol	Direct hydration of ethylene
	Fermentation
Propanol	*Isopropanol*: direct and indirect hydration of propylene.
	n-*Propanol*: Oxo reaction of ethylene with subsequent hydrogenation
Butanol	n-*Butanol/isobutanol*: via hydrogenation of respective aldehydes made from oxo reaction of propylene
	2-Butanol: indirect hydration of butene (via sulphate ester)
Acetone	As a byproduct of the cumene hydroperoxide reaction for phenol production (most important route)
	Direct oxidation of hydrocarbons (e.g. naphtha)

Table 1.3 Secondary production routes and feedstocks for other important oxygenated solvents

Solvent	Main production route
Methyl ethyl ketone	Dehydrogenation of 2-butanol
Acetone (secondary route)	Dehydrogenation of isopropanol (most expensive route)
Diacetone alcohol	Condensation of acetone (cool, basic conditions)
Isophorone	Aldol condensation of acetone (hot, basic conditions)
Methyl isobutyl ketone	Acid catalysed dehydration of diacetone alcohol to form mesityl oxide followed by selective hydrogenation
Acetate esters	Acid catalysed esterification of relevant alcohol (methanol, ethanol, propanols, butanols) with acetic acid
	Tischenko reaction via acetaldehyde is also used for ethyl acetate
Glycol ethers	Reaction of either ethylene oxide or propylene oxide with relevant alcohol (methanol, ethanol and butanol commonly used)
Glycol ether esters	Esterification of glycol ethers, usually with acetic acid

Chlorinated hydrocarbon solvents are also pure materials but the routes to most of the important ones are fewer. Most are derived from methane, ethylene, hydrogen chloride and chlorine. The more important routes are shown in Table 1.4.

1.4 Solvent consumption

Estimated 1994 demand figures for industrial solvents are shown in Table 1.5. These exclude use as chemical reaction intermediates.

It can be seen that the total demand for industrial solvents is almost identical in Western Europe and the

Table 1.4 Main production routes and feedstocks for chlorinated solvents

Solvent	Main production route
Methylene chloride	Via methyl chloride chlorination. Methyl chloride is made by hydrochlorination of methanol or direct chlorination of methane
1,1,1-Trichlorethane	Via thermal or photochemical chlorination of 1,1-dichloroethane. The latter is produced through a route which starts from ethylene chlorination to produce 1,2-dichloroethane, followed by dehydrochlorination to vinyl chloride followed finally by hydrochlorination to produce 1,1-dichloroethane
Perchlorethylene (tetrachlorethylene)	Ethylene dichloride chlorination or by oxochlorination of 1,2-dichloroethane or from 1,2-dichloroethane by chlorination to tetrachloroethane followed by dehydrogenation

Table 1.5 Estimated 1994 demand figures for industrial solvents

Solvent	Europe (ktpa)	USA (ktpa)
Aliphatic hydrocarbons	1250	1500
Aromatic hydrocarbons	800	590
Alcohols	900	1000
Ketones	550	600
Esters	470	180
Glycol ethers	380	390
Chlorinated solvents	320	380
TOTAL	4670	4640

USA. The figure for the rest of the world, including Japan, is similar. Total demand over the last 10 years has declined by around 10% despite substantially increased economic growth, clear evidence of the increasingly efficient way in which solvents are being managed and used (Figure 1.1). This will continue into the future sustained by the trend in the changing mix of solvents used. Industry estimates are for a total demand of around 4100 ktpa in the USA and 3800 ktpa in Europe by 2004. Emissions will fall even more dramatically as abatement control techniques are widely adopted (Figure 1.2).

Surface coatings (paints, inks and adhesives) account for approximately 50% of solvent consumption; paint and resin manufacture dominate. The next largest use is in metal cleaning and degreasing followed by pharmaceutical manufacture. Other important end markets include use in rubber processing, agricultural products, detergents, and cosmetics. Minor end uses include explosives, and flavours and fragrances. There are some important end uses for individual solvents such as food processing (mainly aliphatic hydrocarbons), textiles (acetone and chlorinates), and de-icers (isopropanol and ethanol).

Some solvent molecules also happen to be used as chemical feedstocks. These uses rely on molecular struc-

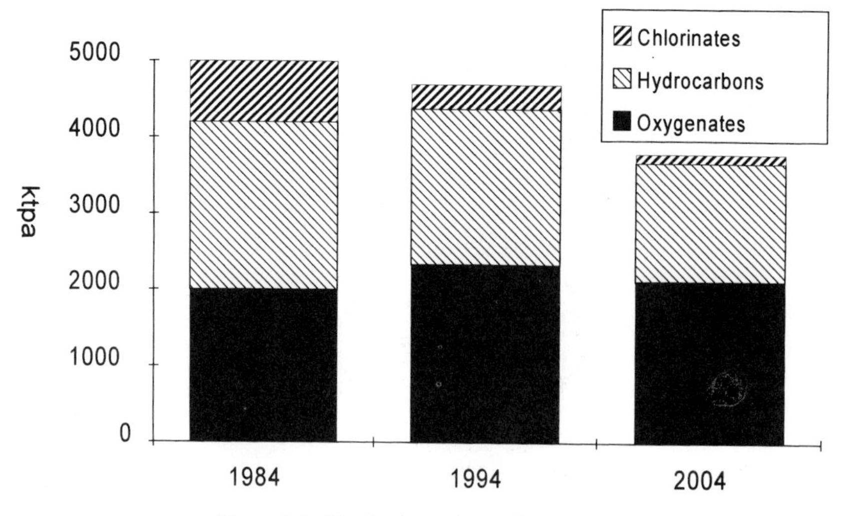

Figure 1.1 Total solvent demand 1984–2004.

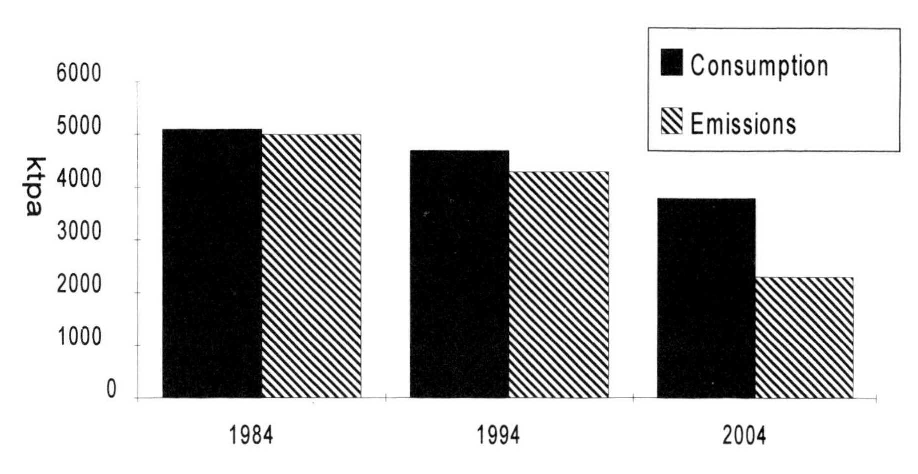

Figure 1.2 Consumption versus emissions 1994–2004.

ture and chemical reactivity characteristics rather than specific solvent-like properties and therefore these chemical intermediate applications lie outside the scope of this book.

1.5 Environmental factors

Environmental legislation is the single biggest influence on the current and future use of many chemicals and solvents are no exception. Organic emissions to air, be they natural in origin (as 90% of the worldwide total are) or man made, are collectively classed as Volatile Organic Compounds (VOC) emissions. Solvents contribute to the VOC emissions inventory (Figure 1.3). VOCs can react in the troposphere with nitrogen oxides (NO_x) to produce ozone. Ozone contributes to the occasional episodes of poor air quality in certain cities or regions. In the USA, however, this issue runs parallel to control of a specific list of chemical emissions (Hazardous Air Pollutants) some of which are organic solvents (Table 1.6). There are other specific issues, such as biodegradability and environmental persistence, which are focused on particular materials rather than solvents in general.

The first environmental legislation on solvents started in California with Rule 66. This state law restricted the use of certain paint formulations containing photochemically active solvents. With hindsight, some of the consequences now seem bizarre in that it allowed unrestricted use of other solvents, and made no attempt to encourage overall reductions in VOC emissions. Rule 66 was eventually superseded by more wide ranging legislation.

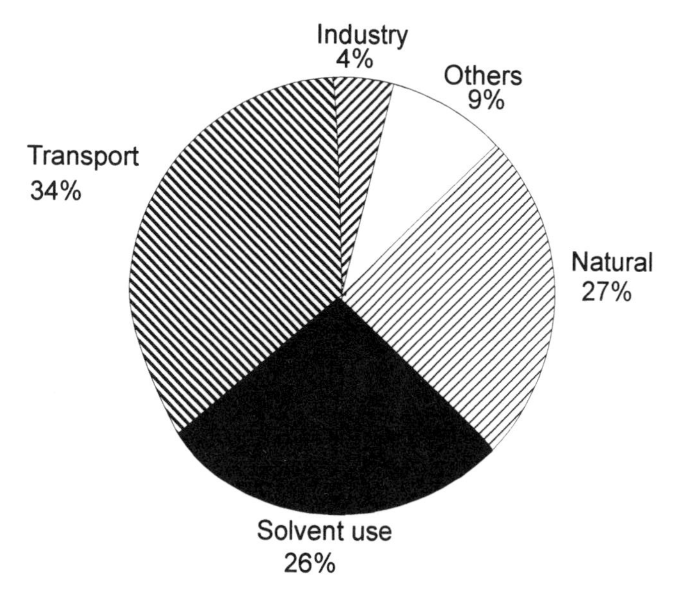

Figure 1.3 West European VOC emissions (excluding methane). Total 14 Mtpa. (Source: Envi-Con).

Table 1.6 Regulatory status of common solvents in the USA

Solvent	Regulated VOC	HAP listed
Butanol (all isomers)	•	
Butyl acetate (both isomers)	•	
Diacetone alcohol	•	
Ethanol	•	
Ethyl acetate	•	
E series glycol ethers	•	•
P series glycol ethers	•	
n-Hexane	•	•
Isophorone	•	•
Methyl ethyl ketone	•	•
Methyl isobutyl ketone	•	•
Propyl acetate (both isomers)	•	
Propanol (both isomers)	•	
Toluene	•	•
Xylene	•	•

Europe historically has lagged behind the USA in legislating to control solvent emissions, principally because the problems of photochemical smogs were less. The legislation gap has shrunk rapidly in the past few years and proposed legislation will be broadly equivalent. Both are working towards the same United Nations Economic Commission (Europe) target of a 30% reduction of VOC emissions as agreed in the Geneva VOC protocol.

Some halogenated solvents (carbon tetrachloride, 1,1,1-trichloroethane), along with chlorofluorocarbons (CFCs) and hydrofluorocarbons (HCFCs) are very stable in the lower atmosphere and can reach the stratosphere where they contribute to the problem of ozone depletion. Almost worldwide agreement has been reached to eliminate the manufacture and use of these materials.

1.6 Health and safety issues

Many people enjoy drinking ethanol in wine, beer and spirits. It is amongst the most widely consumed organic molecules. Methanol, in contrast, can cause damage to the optic nerve (blindness) if drunk. The public recognise and appreciate this toxicological difference. For less familiar molecules, similar small differences in structure often get ignored, even in the face of clear scientific evidence; misleading generalisations have frequently been made in the past. Because of their widespread use, solvents have been extensively tested for their human toxicology over many years and much epidemiological information is also available. This depth of experience and knowledge has resulted in a good understanding of solvent health and safety profiles, particularly in comparison to many newer materials commonly used. This knowledge, coupled with the inherent flexibility of many solvent based applications, has allowed solvent users to move away from those substances used in the past which show adverse toxicological profiles; those remaining in use have good health and safety track records.

It is important not to confuse human toxicity and environmental impact. Whilst, clearly, solvents do vary in their toxicological profiles, there is no evidence to suggest that solvents, at typical ambient air concentrations outside the workplace, pose any health threat to the population.

1.7 The future for solvents

As part of society's increasing drive to minimise the impact it has upon the environment, we are now starting to see solvents managed responsibly by their users, with the consequence that emissions to the environment will be minimised. A result of this will be a decrease in the overall consumption of solvents over the next decade. Behind this simple trend lies an extremely dynamic situation: in order to achieve the desired controls, some solvents will play an increasingly important role in state of the art, efficient systems whilst others will decline. Engineering controls (abatement systems) will be a part of this process of emission reduction along with reformulation to more efficient solvents and the development of new, high efficiency solvents suitable for use in low VOC materials.

The solvents that will be most affected in percentage terms are many of the chlorinates which, because of ozone depletion and toxicity concerns and improved handling are expected to show a marked decline in use over the next decade. Likewise, the use of aromatic and some aliphatic hydrocarbons will show a significant decline whilst others will be flat and some aliphatics may show growth. In the domestic coatings market this will be because of the switch from white spirit to water or iso-paraffin-based decorative paints; in the industrial coatings market, materials such as toluene and xylene will be replaced by higher performance oxygenates blended with aliphatic hydrocarbons. In cleaning too the trend is away from chlorinated materials towards oxygenates and aliphatic hydrocarbons. While some uses of oxygenated solvents will also diminish, this will be offset by the uptake of newer technologies in which they play a fundamental role by virtue of their high solvency power and wide compatibility, even with water.

For what a few years ago was considered a mature market, the solvents industry is currently remarkably dynamic and is responding responsibly and effectively to the needs of a changing world. The key issues for solvents have been highlighted briefly in this introduction and are explored in depth later in the first part of this book. Throughout this century, solvents have had immense impact in enhancing and improving our living standards, often without us realising it. Solvent-using industries are adapting to change, not just to retain what has been achieved so far but to continue to build on the quality of our lives in the future.

2 Solvent action and measurement*

W.C. ATEN

2.1 Basic solvent characteristics

The value of a solvent for industrial use is largely determined by a limited number of properties. Apart from price, these are solvent power, volatility, stability and ease of handling. Other physical properties such as density, refractive index etc. are usually of secondary importance, with viscosity taking an intermediate position. The viscosity of solvents and their blends is in general low and as such of limited importance. Solution viscosity, however, usually plays a major role in solvent selection, but is mainly determined by the type and quantity of solute.

These basic characteristics often depend on a combination of interrelated properties, which individually only describe one aspect of the solvent's performance. Volatility, for instance, is related to boiling point, vapour pressure, relative evaporation rate and heat of vaporisation. For solvents which are flammable, the flashpoint is also related to volatility. For this reason, the solvent power, volatility and viscosity aspects of solvents will be dealt with in more detail.

The ease of solvent handling also depends on odour, flammability, toxicity and environmental factors, some of which are again interlinked. The main factors governing the ease of handling are discussed separately in the chapters dealing with solvent flammability, safety, health and environmental considerations. The same applies to a certain extent to solvent stability. The industrial solvents discussed in this book are, as such, rather stable products. More reactive liquids such as olefins, acids, epichlorohydrin, many amines/amides etc. are generally seen as 'reactants' rather than solvents and are therefore not discussed further.

This chapter emphasises the dependence of basic physical solvent properties on molecular weight and chemical structure. The structure governs the physico-chemical interactions and therefore, to a large extent determines their properties. The physico-chemical properties of a homologous series of solvents are in general related, whereas appreciable differences may exist between members of different homologous series. Within a series, unknown properties of a compound can thus be predicted relatively easily on the basis of molecular weight and the relevant data of neighbouring homologues.

2.1.1 Molecular weight and molecular interactions

The first classification of solvents, after having established their molecular weight, is therefore by their intermolecular interactions. These interactions are divided into three categories:

(i) Dispersion forces as described by van der Waals and London [1]. These are relatively weak interactions between fluctuating dipoles such as those existing, for example, in the molecules of paraffinic hydrocarbon solvents.

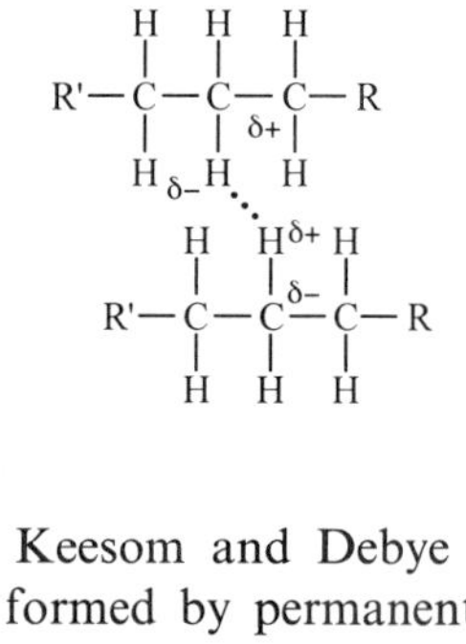

```
       H   H   H
       |   |   |
  R'—C—C—C—R
       |   | δ+|
      Hδ– H   H
           ·.
        H   Hδ+ H
        |   |δ– |
   R'—C—C—C—R
        |   |   |
        H   H   H
```

(ii) Dipole interactions, as described by Keesom and Debye [2]. These are medium strength interactions between oppositely charged parts as formed by permanent or induced dipoles in the molecules of solvents such as ketones.

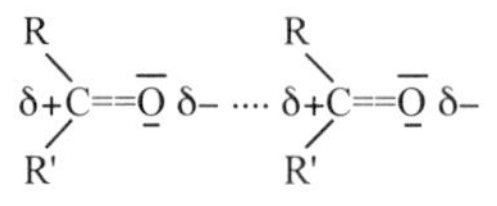

```
   R                    R
    \                    \
  δ+C=O δ– ···· δ+C=O δ–
    /                    /
   R'                   R'
```

* Chemical abbreviations used throughout this chapter are explained at the end of Chapter 3 (p. 69).

(iii) Hydrogen bonds. These relatively strong interactive forces between H-atoms and electronegative atoms such as oxygen, nitrogen etc., can be of inter- and sometimes of intra-molecular origin. Water and alcohols are well known for their hydrogen bonding tendency.

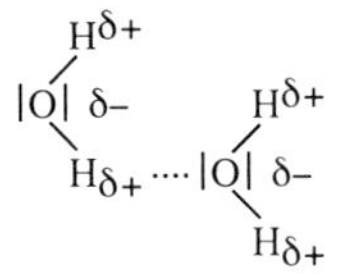

$H^{\delta+}$
|O| $\delta-$ $H^{\delta+}$
$H_{\delta+}$ ···· |O| $\delta-$
$H_{\delta+}$

inter-molecular hydrogen bond between two water molecules

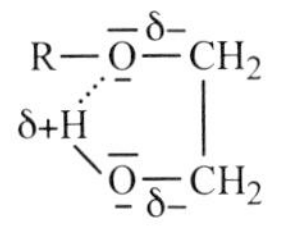

R—O($\delta-$)—CH_2
δ+H |
O($\delta-$)—CH_2

inter-molecular hydrogen bond in glycol ether molecule

An example of the influence of molecular weight on the physical state of three products within a homologous range can be seen in Table 2.1 where intermolecular forces are simply dispersion forces.

The volatility of the alcohols in Table 2.2 decreases with increasing molecular weight and as their boiling points increases, clearly their vapour pressure decreases and they evaporate more slowly. The importance of physical interactions between solvent molecules is shown in Table 2.3, where the effect of the increased molecular weight on volatility is more than compensated for by a decrease in physical interactions. It should be noted, that products have been chosen, which are representative of the three types of interactive forces.

Differences in the volatility of solvents with similar molecular weights will thus be a reflection of the strength of their intermolecular forces. This is highlighted by solvents in Table 2.4 which have been ranked according to increasing boiling point, indicative of stronger molecular interactions.

These data suggest that molecular interactions increase in the order:

alkanes < ethers < esters < ketones < alcohols < glycols

Table 2.1

Product	Formula	Molecular weight	Physical form at room temp.
Methane	CH_4	16	gas
n-Hexane	C_6H_{14}	86	liquid
n-Hexadecane	$C_{16}H_{34}$	226	solid

Table 2.2

Product	Formula	Mol. wt.	Boiling point °C at 101.3 kPa	Vapour pressure kPa at 20°C	R.E.R.[a]
Methanol	CH_3–OH	32	64.6	13.09	3.2
Ethanol	C_2H_5–OH	46	78.3	6.09	2.4
n-Propanol	C_3H_7–OH	60	97.2	1.94	0.94
n-Butanol	C_4H_9–OH	74	117.7	0.57	0.47
n-Pentanol	C_5H_{11}–OH	88	138.5	0.22	

[a] Relative Evaporation Rate (n-buac = 1).

Table 2.3

Product	Formula	Mol. wt.	Boiling point °C at 101.3 kPa	Vapour pressure kPa at 20°C	R.E.R.
Water	H_2O	18	100.0	2.3	0.27
Acetone	CH_3–CO–CH_3	58	56.2	24.9	7.8
n-Pentane	C_5H_{12}	72	36.1	56.5	10.9

Table 2.4

Product	Mol. wt.	Boiling point °C at 101.3 kPa	Vapour pressure kPa at 20°C	Heat of vaporisation at boiling point kJ/kg
Isopentane	72	27.8	76.6	342.4
Diethylether	74	34.5	58.7	360.4
n-Pentane	72	36.1	56.5	357.5
Carbon disulphide	76	46.2	39.5	351.7
1-Chloropropane	78	46.6	37.45	350.8
Methyl acetate	74	56.9	23.24	406.1
Methyl ethyl ketone	72	79.6	10.04	433.1
Benzene	78	80.1	10.00	314.1
Sec. butanol	74	99.5	1.73	550.7
Isobutanol	74	107.9	0.96	568.9
Nitroethane	75	114.0	2.07	468.5
n-Butanol	74	117.7	0.57	582.3
Methyl glycol	76	124.6	0.95	518.9
1,2-Propyleneglycol	76	187.3	0.011	712.0

and in solvents with the same isomeric structure the interactions decrease with increasing degree of branching of the molecules:

n-pentane > isopentane; *n*-butanol > isobutanol > sec. butanol

Furthermore it is evident that the boiling point of a solvent, at a given molecular weight, is higher when the energy required to separate the molecules and bring them in the vapour phase is higher. This energy is the heat of vaporisation. Trouton [3] studied the effects of variation of the molecular weight and the heat of vaporisation of a solvent on its boiling point. Starting from the Clapeyron–Clausius law he calculated that a linear relationship exists between the boiling point and the product of the two other quantities:

$$\text{heat of evapn.} \times \text{mol. wt.} = \text{constant} \times T_{\text{boiling}}$$

When the heat of vaporisation is expressed in kJ/kg and T_{boiling} in Kelvin, the constant is 96. Trouton's law

Table 2.5

Product	Mol. wt.	Boiling point °C at 101.3 kPa	Heat of vaporisation at boiling point kJ/kg	Constant[a]
Water	18	100	2256.9	108.9
Methanol	32	64.6	1101.0	104.4
Acetone	58	56.2	524.6	92.4
Ethylene glycol	62	197.4	846.7	111.6
Isopentane	72	27.8	342.4	81.9
n-Pentane	72	36.1	357.5	83.3
Methyl ethyl ketone	72	79.6	433.1	88.4
Diethyl ether	74	34.5	360.4	86.7
Methylacetate	74	56.9	406.1	91.1
Sec. butanol	74	99.5	550.7	109.4
n-Butanol	74	117.7	582.3	110.3
Nitroethane	75	114.0	468.5	90.8
Carbondisulphide	76	46.2	351.7	83.7
Methyl glycol	76	124.6	518.9	99.2
Benzene	78	80.1	394.1	87.1
Methylene dichloride	85	39.8	329.8	89.6
n-Hexane	86	68.7	335.0	84.3
n-Pentanol	88	138.0	503.5	107.8
Toluene	92	110.6	366.8	88.0
Glycerol	92	290.4	663.3	108.3
Methylisobutyl ketone	100	116.0	364.2	93.6
Diisopropyl ether	102	68.3	290.9	86.9
p-Xylene	106	144.4	339.8	86.3
n-Butyl acetate	116	126.0	309.1	89.9
Methylglycol acetate	118	145.1	368.4	104.0
Butyl glycol	118	170.2	364.7	97.1
Butylglycol acetate	160	191.6	269.8	92.9

[a] Constant $= \dfrac{\Delta H \times \text{M.W.}}{T_{\text{bp}}}$

is only valid for substances with a molecular weight around 100 and with a boiling point which is not extremely high or low in relation to its molecular weight. It does not apply, for example, to highly associated (water, alcohols) or non-associated (isopentane) compounds, as is shown in Table 2.5. It is nevertheless striking that for this wide spread group of solvents both with regard to molecular weight and intermolecular forces, the value of the constant is only 96 ± 15.

As will be discussed later, the intermolecular physical interactions, with their dispersion forces, dipole and hydrogen bond contributions, also play an essential role in describing the solvent power of a liquid. Modern solubility parameter concepts are based on these contributions.

The oldest, most widely used parameter, which gives a very good measure of solvent power, is the Hildebrand solubility parameter [4]. This parameter, δ, is defined as the square root of the cohesive energy density of a compound. The latter, which is the energy required to break the attractive forces between the molecules of $1\,\text{cm}^3$ of material at a certain temperature, is related to the molar heat of vaporisation (ΔH_m) at this temperature, the work needed to expand the volume of the system from the liquid to the vapour phase (RT) and the molar volume of the solvent (V_m), as follows:

$$\delta = \sqrt{\frac{\Delta H_\text{m} - \text{R}T}{V_\text{m}}}$$

in which: R = gas constant, T = temperature in K.

When neglecting the RT term, which is only 5–10% of the value of ΔH_m at 20°C and substituting the heat of vaporisation at 20°C, $\Delta H \times$ mol. weight (M_w) for ΔH_m and M_w/density (d) for V_m, the solubility parameter and hence the solvent power of a solvent is directly related to the square root of the heat of vaporisation:

$$\delta = \sqrt{\frac{\Delta H \times M_\text{w} - \text{R}T}{\frac{M_\text{w}}{d}}} \qquad \delta \approx \sqrt{\Delta H \times d}$$

The relationship between the Hildebrand solubility parameter and the heat of vaporisation at 20°C is shown in Table 2.6, which contains the same solvents as Table 2.5, but now rearranged in order of decreasing heat of vaporisation.

It is clear that the solubility parameter decreases with decreasing heat of vaporisation. The influence of the density can be demonstrated by considering two solvents with approximately the same heat of vaporisation,

Table 2.6

Product	Heat of vaporisation at 20°C kJ/kg	Hildebrand solubility parameter δ (cal/cm^3)
Water	2454	23.4
Methanol	1170	14.5
Ethylene glycol	1030	15.7
Glycerol	998	16.5
n-Butanol	706	11.4
Sec. butanol	646	10.8
Methyl glycol	622	10.8
n-Pentanol	620	10.9
Acetone	560	10.0
Nitroethane	540	11.1
Butylglycol	484	9.1
Methyl ethyl ketone	482	9.3
Benzene	439	9.1
Methyl acetate	436	9.6
Methylisobutyl ketone	431	8.4
Toluene	428	8.9
p-Xylene	404	8.75
n-Butyl acetate	375	8.55
n-Hexane	375	7.25
Diethyl ether	374	7.4
n-Pentane	371	7.05
Carbondisulphide	364	9.9
Butylglycol acetate	363	8.2
Isopentane	348	6.75
Methylene dichloride	343	9.7
Diisopropyl	322	6.9

Table 2.7

Product	Density at 20°C g/ml	Dynamic viscosity at 20°C mPa.s	Refractive index, n^d_{20}	Surf. tension at 20°C, mN/m	Dielectric constant at 20°C, 10 kHz
Methanol	0.7934	0.585	1.3288	23	33.6
Ethanol	0.7894	1.19	1.3619	22	26.0
n-Propanol	0.8039	2.18	1.3854	24	20.1
n-Butanol	0.8089	2.94	1.3993	25	17.8
n-Pentanol	0.8150	3.98	1.4103	26	5.1

Table 2.8

Product	Density at 20°C g/ml	Dynamic viscosity at 20°C mPa.s	Refractive Index, n^d_{20}	Surf. tension at 20°C, mN/m	Dielectric constant at 20°C, 10 kHz
Isopentane	0.6196	0.225	1.3537	15	1.84
Diethyl ether	0.7138	0.230	1.356	17	4.4
n-Pentane	0.626	0.226	1.3575	16	1.84
Carbondisulphide	1.263	0.367	1.6319	32	2.64
1-Chloropropane	0.8952	0.353	1.3879	22	7.70
Methyl acetate	0.9342	0.385	1.3614	26	7.4
Methyl ethyl ketone	0.8050	0.419	1.3788	25	18.5
Benzene	0.8791	0.648	1.501	29	2.3
Sec. butanol	0.8076	4.21	1.3950	23	16.56
Isobutanol	0.8008	2.96	1.3939	23	17.93
Nitroethane	1.050	0.677	1.3916	33	28.04
n-Butanol	0.8089	2.94	1.3993	25	17.8
Methyl glycol	0.9646	1.72	1.4021	31	17.3
1,2-Propyleneglycol	1.0362	59.2	1.4329	36	32.0

isopentane (density at 20°C = 1.325 g/cm^3) and methylene dichloride (density at 20°C = 1.325 g/cm^3). Isopentane, with its low density also has the lowest solubility parameter.

Having discussed the influence of molecular weight and heat of vaporisation at boiling point on volatility and heat of vaporisation at ambient temperature on solvent power, their relationship with some other physical properties will also be reviewed. Examination of the data in Table 2.7 for the normal alcohols homologous series, shows a regular effect of molecular weight on certain physical properties.

However, differing chemical structures greatly influence properties. For example the density of the alcohols increases only slowly with increasing molecular weight, but fluctuations between 0.62 and 1.26 can be observed for the pentanes and carbon disulfide in Table 2.8 due to such differences.

A clear influence of molecular weight and hydrogen bond formation on viscosity can be observed. In the series of the hydrogen bond forming alcohol homologues the viscosity increases from a relatively high value of approximately 0.6 for methanol to 4 for *n*-pentanol. In Table 2.8, the viscosities are much lower, ranging from about 0.23 for the pentanes and diethylether to 0.68 (nitroethane). In these cases, the differences are due to increased intermolecular interactions of the dispersion and dipole type. When hydrogen bond formation takes place due to the presence of OH-groups, the viscosities can be as high as 59.2, as found for propylene glycol!

For the refractive index and the surface tension the relationships are less clear. Low values can, in general,

Table 2.9

Product	Dielectric constant at 20°C, 10 kHz	Solubility parameter δ, cal/cm^3	Fractional polarity
n-Pentane	1.8	7.1	0
Benzene	2.3	9.15	0
Diethylether	4.4	7.4	0.033
n-Butanol	17.8	11.4	0.096
Methyl ethyl ketone	18.5	9.3	0.510
Nitroethane	28.04	11.1	0.710
1,2-Propylene glycol	32.0	12.6	0.468
Water	78.54	23.5	0.819

be observed for solvents with a low density (pentanes, diethylether) and high values for those with a high density (carbon disulfide, benzene, nitroethane, methylglycol and propylene glycol).

The dielectric constant is governed by the ease of formation of dipoles under the influence of electric forces in the solvent. The molecular weight is therefore not expected to have an effect on the dielectric constant but the polarity of the product, as expressed by its Hildebrand solubility parameter and fractional polarity (for more detail see chapter on solvent power), have a pronounced influence. This is demonstrated by the homologous series of *n*-alcohols, where the solubility parameter and fractional polarity decrease from 14.5 and 0.388 respectively for methanol to 10.9 and 0.074 for *n*-pentanol. A few examples taken from the second series are shown in Table 2.9.

2.1.2 *Effect of temperature*

So far, the influences of molecular weight and intermolecular interactions have been discussed on the basic solvent properties at ambient temperature. The effects of temperature changes on these properties will also be reviewed briefly.

According to the first law of thermodynamics the energy added to a system by a temperature increase will cause a reduction of its 'internal' energy. In practical terms this means that a temperature increase will reduce molecular interactions. The effect on the heat of vaporisation is given below:

$$\Delta H_T = \Delta H_{20}(1 - \alpha \Delta T)$$

in which: ΔH_T = heat of vaporisation at temperature T (°C)
ΔH_{20} = heat of vaporisation at 20°C
$\Delta T = T - 20$
α = constant

For most solvents this constant α is around 16×10^{-4} (ranging from approximately 10 for polar to 22 for apolar types).

As a result of the reduction of the heat of vaporisation, solvent volatility increases with temperature. The Antoine equation:

$$\log P = A - \frac{B}{T + C}$$

in which: P = vapour pressure at temperature T, in mm Hg
T = temperature, °C
A, B and C are constants for individual solvents

gives a useful approximation of the change of vapour pressure with temperature. For many solvents constants A, B and C are known to have a validity within a temperature range (for example, those shown in Table 2.10).

The density of a solvent will be lowered due to the volume expansion resulting from a temperature increase according to the expression:

$$V_t = V_{20}(1 + \beta \Delta T)$$

in which: V_t = volume at temp. T
V_{20} = volume at 20°C
$\Delta T = T - 20$
β = expansion coefficient

The expansion coefficient is in the order of 10 to 15×10^{-4}°C and is lower for solvents with a low heat of vaporisation (see Table 2.11).

Table 2.10

Product	Temperature range, °C	Antoine constants		
		A	*B*	*C*
n-Pentane	−4.4/68	6.864	1070.6	232.69
Acetone	−13/55	7.11714	1210.585	229.664
n-Propanol	60/104	7.74416	1437.686	198.463

Table 2.11

Solvent	$\beta \times 10^{-4}$/°C	Heat of vaporisation at boiling point kJ/kg
Acetone	14.3	524.6
Methyl ethyl ketone	13.1	443.1
Methyl isobutyl ketone	11.5	364.2
Ethyl amyl ketone	10.0	240.3

The solubility parameter, which is proportional to both the heat of vaporisation and the density, will thus also decrease with increasing temperature. Nevertheless, a solvent will dissolve most products better at higher temperatures because the intermolecular forces within the solute are also reduced.

Finally the effect of temperature on solvent viscosity can be calculated using the Andrade equation:

$$\log \eta = A + \frac{B}{T}$$

in which: η = kinematic viscosity
T = temperature in Kelvin
A = constant

From this equation it can be derived that the viscosity will be lower at higher temperatures. This conclusion is in full agreement with the relation previously found between the intermolecular interactions and the viscosity of a solvent.

2.1.3 Intramolecular forces

In summary, intermolecular interactions are of great influence on solvent properties, as indeed are the effects of the intramolecular forces. The strength of the (covalent) bonds which bind the atoms in the molecules together thus determines to a large extent the stability of the products (Table 2.12).

Table 2.12 Examples of bond energies (kcal/mole)

Bond	Energy
C—H	98.2
C—F	102
C—C	78–80
C—C[a]	124
C—O[b]	78.3
C=O[c]	185.6
O—H	109.4

[a] In benzene. [b] In methanol. [c] In acetone.

The thermal stability for instance influences the autoignition temperature of inflammable solvents. This autoignition temperature is e.g. higher for the more stable aromatic hydrocarbons like toluene (535°C), than for similar aliphatic ones like heptane (215°C). Branched aliphatics are more stable than the *n*-paraffins (isohexane = 260°C, *n*-hexane = 240°C) and in the series of *n*-paraffins the autoignition temperature decreases with increasing chain length (*n*-pentane = 285°C, *n*-heptane = 215°C). For this reason, when a stable, high-boiling aliphatic solvent is required for a certain application, usually the branched types like SHELLSOL T®, with a boiling range 182–212°C and an autoignition temperature 354°C, are selected. For the combustion in diesel engines, on the contrary, those with a low autoignition temperature, thus the *n*-paraffins, are preferred.

Photochemical stability also can play an important role in solvent selection. Ethers are for instance easily attacked by UV light, to form in the presence of oxygen explosive peroxides; a well-known industrial hazard. Fully halogenated solvents are on the other hand so stable, that, once evaporated, they remain a long time in the atmosphere without being degraded by sunlight. In this way they indirectly, finally affect the stratosphere ozone layer around the earth.

Finally there are many cases where chemical reactivity makes a solvent unsuitable for applications where materials are present which can react with them via their reactive function(s). Certainly water, but also alcohols should for instance be avoided in connection with isocyanate, functions containing coating materials such as 2-component urethane lacquers. Ketones and/or aldehydes cannot be used where there is a danger of oxime formation.

2.2 Solvent power

The solvency of a liquid, or its ability to dissolve a certain solute, is often named its solvent power. This expression should, however, be used with care. There is, for instance, a tendency to rank solvents in order of increasing solvent power, but this is only possible for a specific solute. In former days this was usually a widely used binder for paints. For other solutes the order may well decrease rather than increase. The term solvent power, therefore, may lead to confusion and one should always bear in mind 'for what' when it is said that a solvent has a good solvency. A typical example can be given for water and toluene. Water is an excellent solvent for sugar, but not for fat, whereas the reverse is true for toluene.

2.2.1 Historic concepts

Nevertheless, since the earliest days of solvent technology attempts have been made to systematise solvency and to quantify differences in interaction of solvents with other solvents and solutes. One of the oldest rules in use for this purpose is that a solvent will be miscible with other solvents or dissolve solutes which are similar in chemical character, or: 'alike dissolves alike'. The first attempts to provide a system tabulated miscibility. However, with the growing understanding of solvent action, this approach is hardly used today with the exception of miscibility with water, which is still an important factor both from an end-use and a safety point of view. Water immiscible solvents, for instance, often float on water and this can, when they are flammable, easily cause spread of flames in case of fire. Upon spillage, dilution of toxic solvents is more easy when they are water miscible.

Water miscibility can rank from complete, for the more polar, hydrogen bonding solvent types (glycols, glycolethers, lower alcohols) via intermediate (higher alcohols, ketones and esters) to hardly miscible as in the case of most apolar hydrocarbon, chlorinated hydrocarbon and ether solvent types (Table 2.13).

Higher molecular weight homologues are less water miscible than the lower molecular weight ones. Compare diisopropyl with diethylether, *n*-butyl- with ethylacetate, *n*-butanol with *n*-propanol and methyl-ethylketone with acetone.

Also the paint industry, formerly the main end-user of solvents, attempted to produce a quantitative solvent power data system [5]. This related solvency to certain standard solutes, used in their industry. These could either be a well-known natural (Kauri-resin) or later a synthetic (nitrocellulose) paint binder. The result was the introduction of the Kauri–Butanol number, which applies to hydrocarbon solvents only and the NC-dilution ratio which is used for oxygenated solvents. Another test, used in conjunction with hydrocarbon solvents, is based on the fact that aniline is hardly miscible with aliphatic hydrocarbons but mixes very well with aromatics. The Kauri–Butanol (KB) number as defined in ASTM D 1133 is a measure of the tolerance of a standard solution of Kauri resin in *n*-butanol to hydrocarbon diluent. Standard hydrocarbon solvents used to calibrate the Kauri solution are toluene (KB-number 105) and a 75% v *n*-heptane/25% v toluene blend (KB-number 40). The KB-value increases from approx. 20 to over 100 in the order:

Table 2.13 Water miscibility of solvents

Product	Water miscibility at 20°C, %W	
	Solvent in water	Water in solvent
n-Pentane	<0.01	<0.01
Benzene	0.08	0.06
Diisopropyl ether	0.9	0.6
n-Butylacetate	1.00	1.32
Dichloromethane	2.00	0.17
Nitroethane	4.5	0.9
Diethylether	6.9	1.3
Ethylacetate	7.9	3.0
n-Butanol	7.7	20.1
Methylethylketone	27.1	12.5
Acetone	completely miscible	
n-Propanol	completely miscible	
Methylglycol	completely miscible	
2-Propylene glycol	completely miscible	

isoparaffins < *n*-paraffins < naphthenics < aromatics

Within such a group the KB-number decreases with increasing molecular size. *n*-Hexane has, for instance a KB-number of 28, whereas the value for *n*-hexadecane is only 22. As hydrocarbon solvents are in general based on a mixture of different paraffinic and aromatic constituents, an empirical relationship can be useful for solvents boiling in the range of 150–200°C:

$$\text{KB-number} = 21.5 + 0.205\ (\%\text{w naphthenes}) + 0.723\ (\%\text{w aromatics})$$

The aniline point (ASTM D 611) is the lower equilibrium solution temperature for equal volumes of aniline and solvent. *n*-Heptane is the standard hydrocarbon solvent with an aniline point of 69.3°C. The aniline point decreases from approx. 90°C for isoparaffins to 10°C for solvents with a medium aromatic content. Highly aromatic solvents result in values below 0°C so that the aniline can crystallise from the mixture. Mixed aniline points (ASTM D 1012) are used in these cases. This is the minimum equilibrium solution temperature of a mixture of two volumes of aniline, one volume of solvent and one volume of *n*-heptane. The relationship between the KB-number and aniline point for zero to medium aromatic content hydrocarbon solvents is shown in Figure 2.1. Figure 2.2 demonstrates the relationship between aniline and mixed aniline points. Toluene/*n*-heptane blends have been used for its determination. The relationship between KB-number and mixed aniline points for solvents with a medium to high aromatic contents is given in Figure 2.3.

The dilution ratio (ASTM D 1720) is a measure of solvent power of oxygenated solvents in relation to cellulose nitrate, although it can, in principle, also be obtained for other resins. The dilution ratio is a direct measure of the tolerance of a good or true solvent which contains a dissolved resin to a poor solvent or diluent added to it. ASTM D 1720 defines the maximum number of unit volumes of a diluent (e.g. toluene, xylene, SBP, *n*-butanol) that can be added to a unit volume of the solvent under investigation to cause the

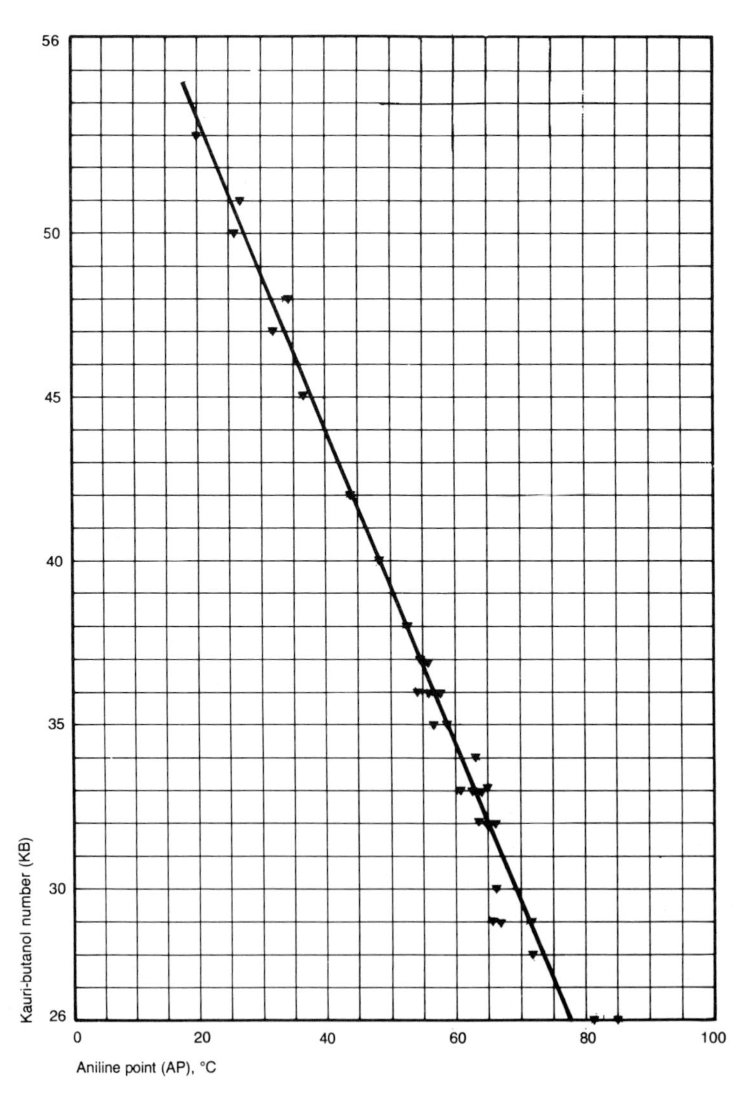

Figure 2.1 Solvent power: medium to zero aromatic content solvents.

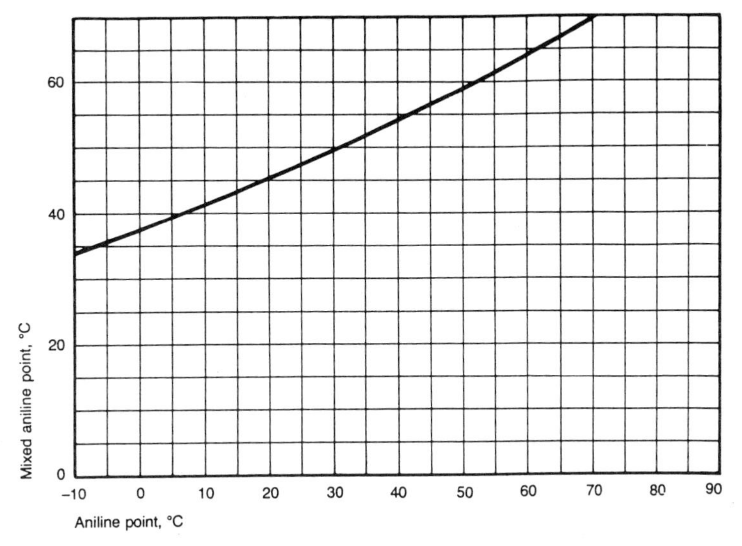

Figure 2.2 Relationship between the aniline and mixed aniline points.

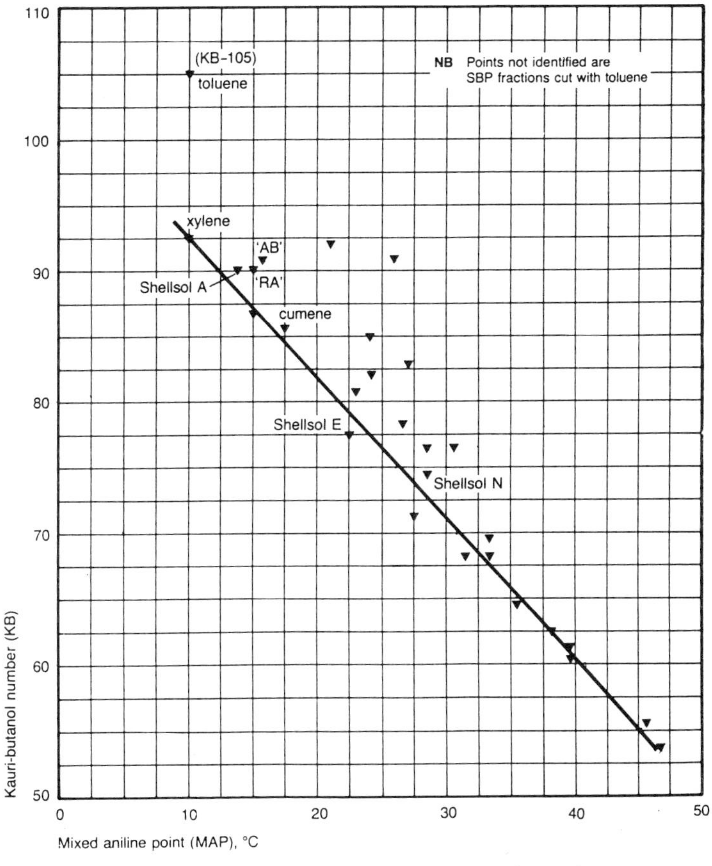

Figure 2.3 Solvent power: aromatic hydrocarbon solvents.

first persistent heterogeneity in the solution at a concentration of 8 g cellulose nitrate per 100 ml of combined solvent plus diluent and at a temperature of 25°C.

An illustrative example of dilution ratios using 1/4 second nitro-cellulose as the resin is given in Figure 2.4. It is obvious that considerably more aromatic hydrocarbon diluent can be tolerated than aliphatic. By the same token toluene is more of a solvent than xylene as would be expected on the basis of KB numbers or mixed aniline points.

The role of alcohols in cellulose nitrate formulations deserves special mention. Reference to Table 2.15 is sufficient to illustrate that very high quantities of alcohols can be added to solvents for cellulose nitrate before the resin finally precipitates from solution. Figure 2.5 shows that sometimes quite substantial amounts of true solvent can be replaced by the cheaper alcohols with, as often as not, an increase in solvent power.

This increase in the solvent power led to the term 'latent solvents' to describe the alcohols. As a general

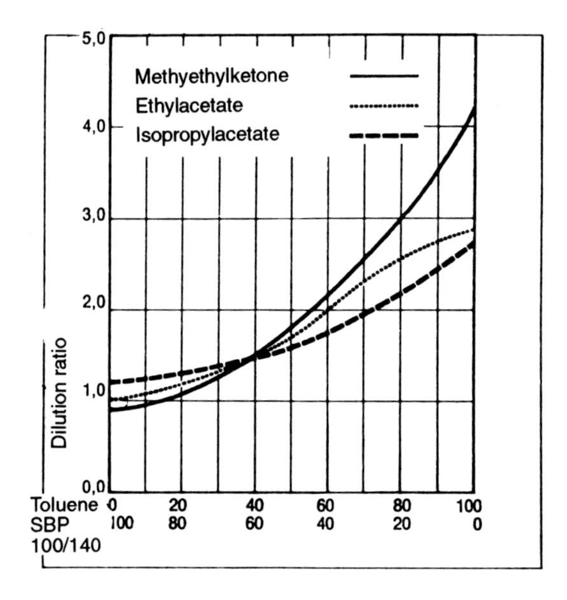

Figure 2.4 Influence of aromatic content on the dilution ratio of hydrocarbon diluents.

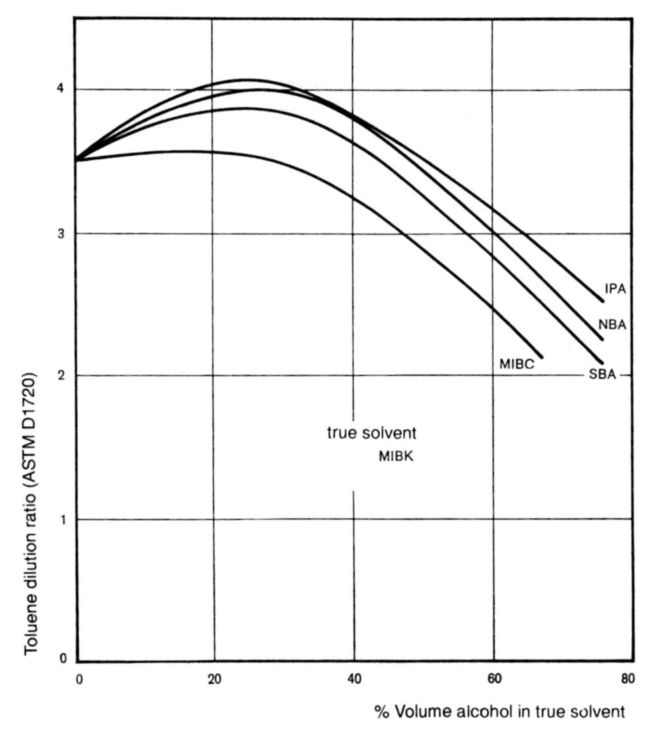

Figure 2.5 Latent solvent action of alcohols.

rule of thumb in formulating, using dilution ratios, one may assume that 20% of the true solvent may be replaced by an alcohol with no loss in solvent power. The attraction of a dilution ratio is that it is a 'hard' experimentally determined value. Its use in formulating is illustrated in the following short example of a very basic solvent blend for a cellulose nitrate spray application using methylisobutylketone as the true solvent. Its dilution ratio is 3.6 with respect to toluene. If we apply the latent solvent rule, using an alcohol which is more volatile than the ketone (such as isopropanol), we have:

$$3.6 = \frac{\text{volume of toluene}}{\text{volume of (80\% MiBK + 20\% IPA)}}$$

This gives a %volume formulation of 17.4/4.3/78.3 MiBK/IPA/toluene, as an initial trial blend for testing in the laboratory.

As is sadly all too often the case, the practical strength of the dilution ratio also contains the seeds of its limitations. Whilst it gives a good measure of the tolerance of a particular solvent for a diluent, it gives little indication of how mixtures of solvents are likely to behave. The rule of thumb for the latent solvency of alcohols for cellulose nitrates also fails when other polymer or resin systems are considered. For example, dilution ratios can never predict that mixtures of two non-solvents form a true solvent for a binder, as is the case when aromatic hydrocarbons/alcohols for epoxy resins. Although the dilution ratio concept can be extended to most other resins, it is almost exclusively associated with cellulose nitrate, as in times past this paint binder dominated the field of coating technology. Over the last decades, there has been a veritable explosion in the number of polymers available to many industries and consequently a need has arisen for a more general technique which provides information on solubility characteristics more quickly and more extensively than was possible before. The introduction of solubility parameters filled this need.

2.2.2 Modern solubility systems

The idea of solubility parameters was first introduced by Hildebrand using the concept of regular solutions and was further developed by Scatchard [6, 7]. Hildebrand's solubility parameter concept is based on Gibbs' free energy relationship. Thermodynamically two compounds (e.g. polymer in solvent) do mix when the change in their free energy is negative:

$$\Delta G = \Delta H - T\Delta S$$

in which: ΔG = free energy of mixing
ΔH = heat of mixing
ΔS = entropy of mixing, always positive.

In order to dissolve the polymer in the solvent, ΔH should thus be smaller than $T\Delta S$ and preferably as small as possible. ΔH is on its turn related to the cohesive energy density E/V_m of the compounds (E = potential energy per mole, V_m = molar volume):

$$\Delta H = V_t\left\{\sqrt{\frac{E_1}{V_{m1}}} - \sqrt{\frac{E_2}{V_{m2}}}\right\}^2 \varphi_1\varphi_2$$

where: V_t = total volume of the two compounds
φ_1 and φ_2 = their volume fractions

Mixing of the compounds would thus be facilitated when the difference in $\sqrt{E/V_m}$ is minimal.

As discussed before, Hildebrand has defined the solubility parameter of a liquid as the square root of the cohesive energy density where the cohesive energy density of a liquid is the energy required to separate the molecules present in 1 cm^3 and hence to vaporise it.

$$\text{solubility parameter } \delta = \sqrt{\frac{\Delta E}{V_m}}$$

(ΔE = energy of vaporisation per mole at a given temperature)

$$\Delta E = \Delta H_M - RT$$

in which: ΔH_M = latent heat of vaporisation per mole

A polymer will thus mix with a solvent when their solubility parameters are sufficiently close together, or in other words 'alike dissolves alike'. The solubility parameters of liquids thus can be calculated when their latent heat of vaporisation has been determined. Solubility parameters for polymers cannot be measured in this way and for this reason the solubility parameters of polymers has been measured by other methods such as mixing them with solvents covering a range of increasing solubility parameters and observing whether or not dissolution occurs. The solubility parameter of the polymer can then be taken as the centre of the range of solubility parameter values of solvents in which it is soluble.

Another method of determining polymer solubility parameters is to prepare a lightly crosslinked form of the polymer and to place this in a similar series of solvents. The amount of swelling should be greatest for the liquid having the same solubility parameter as the polymer. A further possibility involves the calculation of the polymer solubility parameter from group contributions as proposed by various authors.

As an example some solubility parameters of polymers are quoted from those given by Gardon [8]. Average values are given in case the various methods were leading to slightly different figures (Table 2.14).

Table 2.14 Solubility parameters of polymers

Polymers	Solubility parameters (cal/cm^3)
Methacrylate polymers	
Polymethylacrylate	10.0
Polyethylacrylate	9.3
Poly *n*-propylacrylate	9.0
Poly *n*-butylacrylate	8.75
Poly methylmethacrylate	9.3
Poly ethylmethacrylate	9.0
Poly *n*-propylmethacrylate	8.8
Poly *n*-butylmethacrylate	8.5
Hydrocarbon polymers	
Polyethylene	8.0
Polyisobutylene	7.8
Polybutadiene	8.45
Polyisoprene	8.1
Polystyrene	9.0
Halogen containing polymers	
Poly-tetrafluor ethylene	6.2
Polyvinyl chloride	9.55
Polychloroprene	9.05
Nitrile containing polymers	
Polyacrylonitrile	14.1
Polymethacrylonitrile	10.7
Cellulose derivatives	
Cellulose nitrate	10.05
Sec. cellulose acetate	11.1
Ethyl cellulose	9.6
Condensation polymers	
Polyethylene terephtholate	10.7
Nylon 66	13.6
Polyurethane	10.0
Medium oil length alkyd	10.4
Epoxy resin (Epikote 1004)	10.9

Many of the early solubility or swelling experiments were carried out by Burrell [9]. He already found it necessary to introduce a second parameter to explain the observed behaviour of polymer/solvent systems. Each solvent was placed into one of three categories: poorly, moderately or strongly hydrogen bonding. The first category included hydrocarbons, the second esters and ketones, and the third alcohols. The introduction of the second, hydrogen bonding parameter improved the separation of liquids into solvents and non-solvents for a particular polymer, but was still by no means entirely satisfactory. For instance, for ½ sec. nitrocellulose the solubility parameter range for the first category of solvents was 11.1–12.7, for the second category 7.8–14.7, and for the third one 12.7–14.5.

More recently a number of systems using three solubility parameters has therefore been suggested in order to predict the solubility characteristics of polymers more precisely. Hildebrand's original theory was intended only for application to mixtures of non-polar liquids and for such liquids the intermolecular forces are only one type, namely dispersion forces. For other liquids, as discussed, also polar and hydrogen bonding forces play a role, and it has been argued by several workers that a solvent should be described by a parameter representing each of these three types of intermolecular forces [10].

The first of these systems using three solubility parameters was proposed by Crowley, Teague and Lowe [11]. They chose for their parameters the Hildebrand solubility parameter for describing the dispersion forces, the dipole moment of the molecule to represent polar forces and a hydrogen bonding parameter based upon spectroscopic measurements of Gordy [12].

Since the work of Crowley, Teague and Lowe, a number of three-dimensional parameter systems have been suggested. Perhaps the best known of these systems are those of Hansen [13] and Nelson, Hemwall and Edwards [14]. These will be described in more detail and examples of their use will be given.

2.2.3 *Hansen's solubility parameter system*

Hansen argued that the cohesive energy density of a solvent as measured by the energy of vaporisation contains contributions from all intermolecular forces present in the liquid, i.e.:

$$\frac{\Delta E}{V_m} = \frac{\Delta E_d}{V_m} + \frac{\Delta E_p}{V_m} + \frac{\Delta E_h}{V_m}$$

Where ΔE_d, ΔE_p and ΔE_h are the energies/mole of solvent due to dispersion, polar and hydrogen bonding forces respectively. This is equivalent to writing the equation which defines the solubility parameter in the form

$$\delta^2 = {\delta_d}^2 + {\delta_p}^2 + {\delta_h}^2$$

Where δ_d, δ_p and δ_h are the solubility parameters representing dispersion, polar and hydrogen bonding forces respectively.

Hansen firstly determined δ_d for a solvent using the 'homomorph' concept. The energy of vaporisation of a hydrocarbon molecule of the same size and shape as the solvent molecule in question at the same reduced temperature (absolute temperature divided by the critical temperature) is assumed to be that due to dispersion forces existing in the solvent. The difference between the energy of vaporisation of the solvent, ΔE, and that calculated as the contribution due to dispersion forces, ΔE_d, is taken as that due to both polar and hydrogen bonding forces, i.e.:

$$\Delta E - \underset{\text{(homomorph)}}{\Delta E_d} = \Delta E_p + \Delta E_h$$

and

$$\delta^2 - \underset{\text{(homomorph)}}{{\delta_d}^2} = {\delta_p}^2 + {\delta_h}^2$$

The separate parameters δ_p and δ_h were in first instance determined in a rather arbitrary fashion by carrying out many solubility experiments and choosing values of δ_p and δ_h for each solvent as to provide the most coherent volumes of solubility for several polymers. At a later stage, Hansen calculated values of δ_p and δ_h independently. For instance, for the value of δ_h the parameter $5000\,N/V_m$ was introduced (N = number of H-bonds per molecule; 5000 is the approximate molar enthalpy of a hydrogen bridge formation in cal/mol.; V_m is the molar volume). The choice of δ_p is, of course, restricted by the above equation because the sum of the squares δ_d, δ_p and δ_h must equal the measured value of δ^2. Hansen generally found good agreement between his original empirical and calculated values, but, nevertheless, produced a revised set of parameters for solvents. Table 2.15 gives these values for a large range of solvents.

As with other three parameter systems, solvents are represented by points in a three dimensional model and polymer solubility by a volume. Solvents falling within this volume of solubility dissolve the polymer and those outside the volume do not. Hansen found that by doubling the scale of the δ_d axis relative to the other axes, the volumes of solubility of most polymers were approximately spherical. This means that each polymer may be described in terms of the centre of this sphere having coordinates δ_{do}, δ_{po} and δ_{ho}, and its radius (known as the radius of interaction), R_{Ao}. Values of the centre coordinates and radii of interaction for several polymers are given in Table 2.16.

These values are useful because the distance R_A between the solubility parameters of any given solvent from the centre of the 'sphere of solubility' for the polymer can be calculated using the following equation

$$R_A = \sqrt{4(\delta_{do} - \delta_d)^2 + (\delta_{po} - \delta_p)^2 + (\delta_{ho} - \delta_h)^2}$$

If R_A is greater than R_{Ao}, the radius of interaction for the polymer, then it should not dissolve the polymer but if it is less than R_{Ao} it should act as a solvent.

Figure 2.6 illustrates Hansen's 'sphere of solubility' and shows how three separate two-dimensional projections are required to obtain a full solubility picture by conventional means. This has been done for an acrylic polymer, 'Elvacite' 2013, in Figure 2.7. Only those points which fall within all three circles come within the 'sphere of solubility' and should therefore be true solvents for this particular polymer.

An alternative approach suggested by Teas [15] uses a triangular graph.

Since ${\delta_2}^2 + {\delta_p}^2 + {\delta_h}^2 = \delta^2$

Then $\dfrac{{\delta_d}^2}{\delta^2} + \dfrac{{\delta_p}^2}{\delta^2} + \dfrac{{\delta_h}^2}{\delta^2} = 1$

Using ${\delta_d}^2/\delta^2$, ${\delta_p}^2/\delta^2$ and ${\delta_h}^2/\delta^2$ as coordinates, we can use all three parameters on a single plot using triangular graph paper. However we have, of course, lost the 'sphere of solubility' and in its place have an

Table 2.15 Hansen's solubility parameters—solvents

Solvent type	Code No.	Solvent name	δ_d	δ_p	δ_h
Aliphatic hydrocarbons	1	*n*-Hexane	7.3	0	0
	2	*n*-Heptane	7.5	0	0
	3	Cyclohexane	8.2	0	0.1
Aromatic hydrocarbons	4	Benzene	9.0	0	1.0
	5	Toluene	8.8	0.7	1.0
	6	*o*-Xylene	8.7	0.5	1.5
	7	Ethyl benzene	8.7	0.3	0.7
Ketones	8	Acetone	7.6	5.1	3.4
	9	Methyl ethyl ketone	7.8	4.4	2.5
	10	Methyl iso-butyl ketone	7.5	3.0	2.0
	11	Cyclohexanone	8.7	3.1	2.5
	12	Isophorone	8.1	4.0	3.6
Esters	13	Methyl acetate	7.6	3.5	3.7
	14	Ethyl acetate	7.7	2.6	3.5
	15	*n*-Butyl acetate	7.7	1.8	3.1
	16	*iso*-Butyl acetate	7.4	1.8	3.1
	17	Oxitol acetate	7.8	2.3	5.2
Alcohols	18	Methanol	7.4	6.0	10.9
	19	Ethanol	7.7	4.3	9.5
	20	*n*-Propanol	7.8	3.3	8.5
	21	*iso*-Propanol	7.7	3.0	8.0
	22	*n*-Butanol	7.8	2.8	7.7
	23	*sec*-Butanol	7.7	2.8	7.1
	24	*iso*-Butanol	7.4	2.8	7.8
Glycol ethers	25	Methyl oxitol	7.9	4.5	8.0
	26	Oxitol	7.9	4.5	7.0
	27	Butyl oxitol	7.8	2.5	6.0
	28	Methyl dioxitol	7.9	3.8	6.2
	29	Dioxitol	7.9	3.4	6.0
	30	Butyl dioxitol	7.8	3.4	5.2
Ethers	31	Furan	8.7	0.9	2.6
	32	Tetrahydrofuran	8.2	2.8	3.9
	33	1,4-Dioxan	9.3	0.9	3.6
	34	Diethyl ether	7.1	1.4	2.5
Nitrogen compounds	35	Acetonitrile	7.5	8.8	3.0
	36	Nitromethane	7.7	9.2	2.5
	37	2-Nitropropane	7.9	5.9	2.0
	38	Nitrobenzene	9.8	4.2	2.0
	39	Ethanolamine	8.4	7.6	10.4
	40	Pyridine	9.3	4.3	2.9
	41	Aniline	9.5	2.5	5.0
	42	Formamide	8.4	12.8	9.3
	43	Dimethylformamide	8.5	6.7	5.5
Sulphur compounds	44	Carbon disulphide	10.0	0	0.3
	45	Dimethyl sulphoxide	9.0	8.0	5.0
Polyhydric alcohols	46	Ethylene glycol	8.3	5.4	12.7
	47	Propylene glycol	8.2	4.6	11.4
	48	Glycerol	8.5	5.9	14.3
	49	Diethylene glycol	7.9	7.2	10.0
Halogenated hydrocarbons	50	Dichloromethane	8.9	3.1	3.0
	51	Chloroform	8.7	1.5	2.8
	52	Carbon tetrachloride	8.7	0	0.3
Others	53	Diacetone alcohol	7.7	4.0	5.3
	54	Water	7.6	7.8	20.7
	55	Acetic acid	7.1	3.9	6.6

irregular area. Nevertheless, this is undoubtedly the most convenient graphical method for making use of Hansen's data, as can be seen from the solubility data illustrated by Figure 2.8.

2.2.4 The parameter system of Nelson, Hemwall and Edwards

Nelson, Hemwall and Edwards have proposed a three parameter system which shows some similarity with that of Crowley, Teague and Lowe. It also uses the Hildebrand solubility parameter and a fractional polarity as suggested by Gardon, together with a net hydrogen bond index. This latter parameter is an attempt to

Table 2.16 Hansen's solubility parameters—polymers

Polymer type	Trade name	Supplier	Computer parameters			
			δ_{do}	δ_{po}	δ_{ho}	R_{Ao}
Poly (ethyl methacrylate)	'Lucite' 2042	Du Pont	8.60	4.72	1.94	5.20
Poly (methyl methacrylate)		Rohm & Haas	9.11	5.14	3.67	4.20
Epoxy	Epikote 1001	Shell	9.95	5.88	5.61	6.20
Long oil alkyd (66% oil length)	'Plexal' P65	Polyplex	9.98	1.68	2.23	6.70
Alcohol soluble rosin resin	'Pentalyn' 830	Hercules	9.79	2.84	5.34	5.70
Poly (vinyl butyral)	'Butvar B76	Shawinigan	9.09	2.13	6.37	5.20
Polystyrene	Polystyrene LG	BASF	10.40	2.81	2.10	6.20
Poly (vinyl acetate)	'Mowilith' 50	Hoechst	10.23	5.51	4.72	6.70
Urea-formaldehyde	'Plastopal' H	BASF	10.17	4.05	7.31	6.20
Nitrocellulose	H23 (½ sec)	Hagedorn	7.53	7.20	4.32	5.60
Chlorinated polypropylene	'Parlon' P10	Hercules	9.90	3.09	2.64	5.20
Cellulose acetate	'Cellidora' A	Bayer	9.08	6.22	5.38	3.70
Pure phenolic	'Super Beckacite' 1001	Reichold	11.37	3.20	4.08	9.70
Phenolic resin (resole)	'Phenodur' 373U	Albert	9.65	5.68	7.13	6.20
Modified penta-ester of rosin	'Cellolyn' 102	Hercules	10.62	0.46	4.17	7.70
Alcohol soluble resin	'Pentalyn' 255	Hercules	8.58	4.58	7.00	5.20
Blocked isocyanate (phenol)	'Suprasec' F5100	ICI	9.87	6.43	6.39	5.70
Short oil alkyd (34% oil length)	'Plexal' C34	Polyplex	9.04	4.50	2.40	5.20
Saturated polyester	'Desmophen' 850	Bayer	10.53	7.30	6.00	8.20
Styrene-butadiene (SBR) elastomer	'Polysar' 5630	Polymer Corp	8.58	1.64	1.32	3.20
Acrylonitrile-butadiene elastomer	'Hycar' 1052	B.F. Goodrich	9.10	4.29	2.04	4.70
Isoprene elastomer	Cariflex IR305	Shell	8.10	0.69	−0.40	4.70
Poly (isobutylene)	'Lutonal' IC/123	BASF	7.10	1.23	2.28	6.20
cis-Polybutadiene elastomer	'Bunahuls' CB10	Huels	8.57	1.10	1.67	3.20
Thermoplastic polyamide	'Versamid' 930	General Mills	8.52	−0.94	7.28	4.70
Ester gum	Ester gum BL	Hercules	9.60	2.31	3.80	5.20
Hexamethoxymethyl malamine	'Cymel' 300	American Cynamid	9.95	4.17	5.20	7.20
Terpene resin	'Piccolyte' S-100	Pennsylvania Ind. Chem.	8.05	0.18	1.39	4.20
Furfuryl alcohol resin	'Durez' 14383	Hooker Chem.	10.34	6.63	6.26	6.70
Petroleum hydrocarbon resin	'Piccopale' 110	Pennsylvania Ind. Chem.	8.58	0.58	1.76	3.20
Poly (vinyl chloride)	'Vipla' KR (K = 50)	Montecatini	8.91	3.68	4.08	1.70
Coumarone-indene resin	'Piccoumarone' 450L	Pennsylvania Ind. Chem.	9.49	2.68	2.82	4.70

allow for various hydrogen bonding characteristics of different types of solvents. There are four main types of solvent as regards hydrogen bonding characteristics: non-hydrogen bonding (e.g. aliphatic hydrocarbons), hydrogen bond acceptors (e.g. esters and ketones), donors (e.g. chloroform) and donor/acceptors (e.g. alcohols). The hydrogen bond index is an attempt to allow for these categories. It is calculated from the

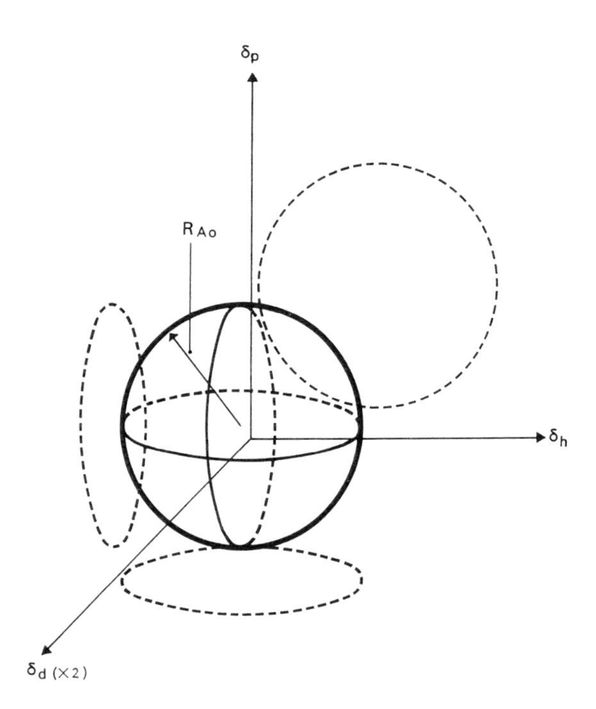

Figure 2.6 Diagrammatic representation of a polymer solubility plot based on Hansen's parameter system. The dotted circles represent projections of the 'sphere of solubility'. A specific example of these projections is shown in Figure 2.7.

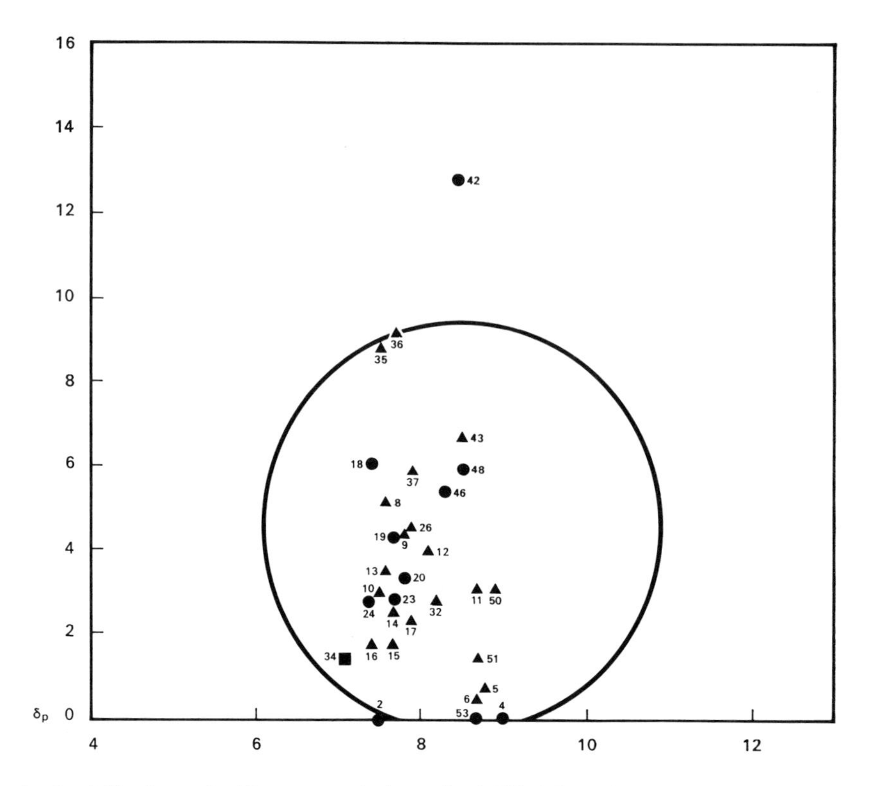

Figure 2.7 Projection in the δ_p/δ_d plane of a Hansen type 'sphere of solubility' for 'Elvacite' 2013 (solvent code numbers are given in Table 2.15). ▲, Soluble; ■, cloudy solution; ●, insoluble.

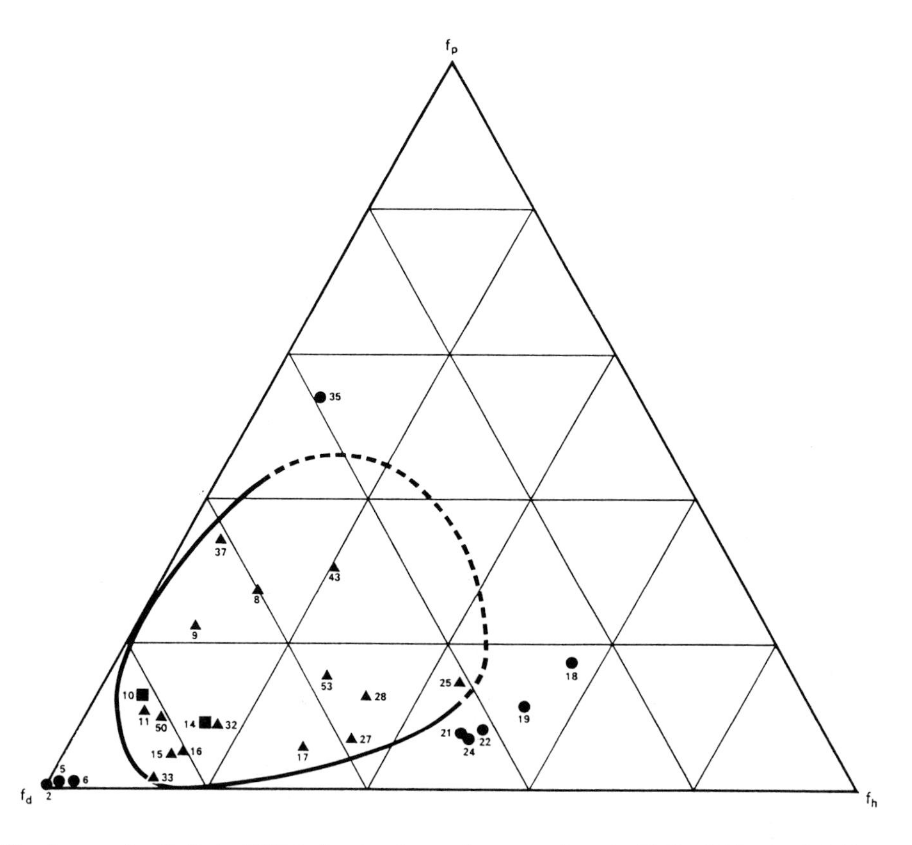

Figure 2.8 Solubility diagram for Epikote 1007 (solvent code numbers are given in Table 2.15). ▲, Soluble; ■, cloudy solution; ●, insoluble.

following equation for a simple solvent:

$$\text{hydrogen bond index} = k.\gamma$$

where γ is the infrared spectroscopic value for hydrogen bonding obtained by Gordy and K is a weighting factor taken as -1 for simple alcohols, zero for glycolethers and $+1$ for esters, ketones and all other solvents. Values for these parameters for a wide range of solvents are listed in Table 2.16. Although this type of

Table 2.17 Solubility parameters—Nelson, Hemwall and Edwards

	Solubilty parameter	Fractional polarity	Hydrogen bonding
Individual hydrocarbons			
n-Pentane	7.02	0	0
2-Methylbutane	6.75	0	0
n-Hexane	7.27	0	0
Methylpentanes	7.0–7.1	0	0
Dimethylbutanes	6.7–6.9	0	0
n-Heptane	7.5	0	0
Methylhexanes	7.25	0	0
3-Ethylpentane	7.37	0	0
Trimethylbutane	6.95	0	0
n-octane	7.54	0	0
Methylheptanes	7.35	0	0
Ethylhexanes	7.5	0	0
Dimethylhexanes	7.1–7.4	0	0
Ethyl methylpentanes	7.3	0	0
Nonane	7.64	0	0
Decane	7.74	0	0
Undecane	7.81	0	0
Dodecane	7.92	0	0
Cyclopentane	8.1	0	0
Ethyl cyclopentane	7.94	0	0
Methyl cyclopentane	7.85	0	0
Trimethyl cyclopentanes	7.3–7.55	0	0
Cyclohexane	8.2	0	0
Ethyl cyclohexane	7.96	0	0
Benzene	9.2	0.001	4.0
Toluene	8.9	0.001	4.2
Xylene (commercial)	8.85	0.001	4.5
Ethylbenzene	8.8	0.001	4.2
Trimethyl benzenes	ca 8.9	0.001	5.0
Commercial hydrocarbon solvents			
White spirit (LAWS)	7.85	0	0.5
Hexane	7.3	0	0
SBP 100/140	7.5	0	0
SBP 140/165	7.6	0	0
SBP 1	7.4	0	0
SBP 2	7.5	0	0
SBP 3	7.7	0	0.25
SBP 4	7.6	0	0.1
SBP 5	7.6	0	0
SBP 6	7.8	0	0.5
SBP 11	7.75	0	0.5
Rubber solvent	7.7	0	0.3
Shellsol A	8.8	0.001	5.0
Shellsol E	8.4	0.001	4.5
Shellsol AB	8.7	0.001	5.3
Shellsol R	8.4	0.001	5.0
Shellsol T	7.6	0	0
Shellsol TD	7.65	0	0
Shellsol K	7.7	0	0.1
Shellsol RA	8.65	0.001	5.5
Ketones			
Acetone	10.0	0.695	12.5
Methyl ethyl ketone	9.3	0.510	10.5
Methyl isobutyl ketone	8.4	0.315	10.5
Methyl isoamyl ketone	8.3	0.255	10.9
Cyclohexanone	9.9	0.380	13.7
Ethyl amyl ketone	8.2	0.223	8.5
PENT-o-XONE	8.5	0.190	12.5
Diisobutyl ketone	7.8	0.123	9.8
Diacetone alcohol	9.2	0.312	0
Isophorone	9.1	0.190	14.9
Mesityl oxide	9.0	0.332	12.0

Table 2.17 Continued

	Solubilty parameter	Fractional polarity	Hydrogen bonding
Esters			
Ethyl acetate (85–88%)	9.6	0.182	4.9
Ethyl acetate (99%)	9.1	0.167	8.4
Isopropyl acetate	8.6	0.100	8.5
n-Propyl acetate	8.75	0.129	8.5
sec-Butyl acetate	8.2	0.082	8.3
iso-Butyl acetate	8.3	0.097	8.7
n-Butyl acetate (90–92%)	9.0	0.118	5.7
n-Butyl acetate (99%)	8.6	0.120	8.0
Amyl acetate	8.45	0.067	8.2
Methyl amyl acetate	8.2	0.050	8.3
Isobutyl isobutyrate	7.7	0.042	8.0
Methyl Oxitol acetate	9.2	0.095	10.5
Ethyl Oxitol acetate	8.7	0.073	10.1
Ethers and glycol ethers			
Tetrahydrofuran	9.9	0.075	12.0
Dioxane	10.0	0.006	14.6
Methyl Oxitol	10.8	0.126	0
Ethyl Oxitol	9.9	0.086	0
Butyl Oxitol	8.9	0.048	0
Methyl Dioxitol	10.2	0.058	0
Ethyl Dioxitol	9.6	0.043	0
Butyl Dioxitol	8.9	0.028	0
Alcohols			
Methyl alcohol	14.5	0.388	−19.8
Isopropyl alcohol	11.5	0.178	−16.7
n-Propyl alcohol	11.9	0.152	−16.5
sec-Butyl alcohol	10.8	0.123	−17.5
iso-Butyl alcohol	10.7	0.111	−17.9
n-Butyl alcohol	11.4	0.096	−18.0
Methyl isobutyl carbinol	10.0	0.066	−18.7
n-Amyl alcohol	10.9	0.074	−18.2
Cyclohexanol	11.4	0.075	−16.5
2-Ethylhexanol	9.5	0.042	−18.7
Miscellaneous			
Dichoromethane	9.7	0.12	1.5
Chloroform	9.3	0.017	1.5
1,2-dichloroethane	9.8	0.043	1.5
1,2,2,2-tetrachloroethane	10.4	0.092	1.5
2-Nitropropane	9.9	0.72	4.0
Dimethylformamide	12.1	0.772	18.9
Chlorobenzene	9.50	0.058	2.7

treatment is theoretically more satisfactory than that of Hansen in that it distinguishes between solvents in which hydrogen bonding is present and those which are merely potential hydrogen bond donors or acceptors, it is perhaps a less convenient system as a means of predicting polymer solubility. This is because it results in a three-dimensional sphere of irregular form, similar to those formed in the system of Crowley, Teague and Lowe.

For this reason two-dimensional plots are generally made when using the Nelson, Hemwall and Edwards parameter system. In the first instance solubility maps were constructed of the fractional polarity against the Hildebrand solubility parameter, as shown in Figure 2.9.

In a later stage it appeared to be more convenient to plot the hydrogen bond index against the Hildebrand solubility parameter as demonstrated in Figure 2.10. In this way of presentation contours of the maximum permissible fractional polarity can be indicated, which gives a good, and in practice very satisfactory, presentation of the three-dimensional case.

Although Hansen's system, because it concerns regular, ball-shaped spheres, has the advantage of ease of mathematical treatment, the above concept of solubility mapping has the advantage that it gives qualitatively and even quantitatively good results and is therefore very useful in solving practical problems. This system will therefore be dealt with in some more detail.

2.2.5 Construction and use of solubility maps

This technique uses the system of solubility parameters proposed by Nelson, Hemwall and Edwards in conjunction with dilution ratios [16].

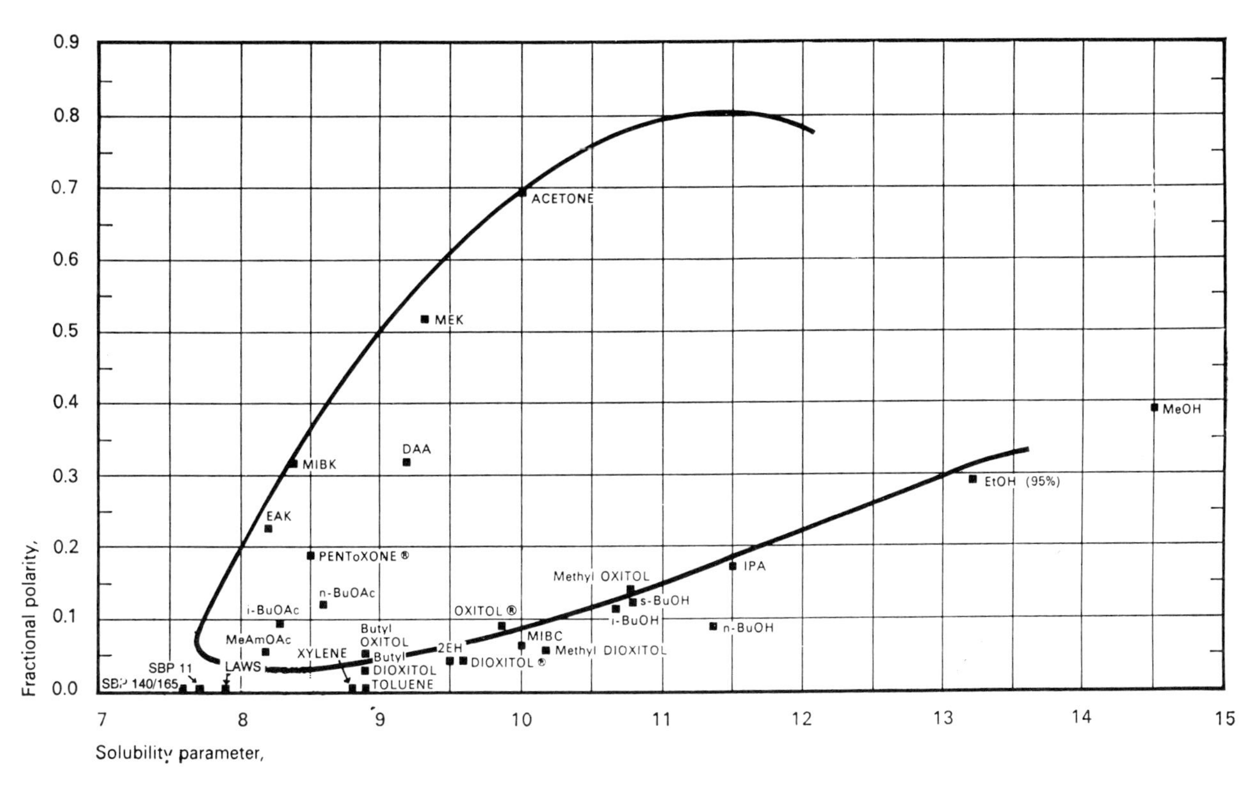

Figure 2.9 Solubility map for 25% w SS ¼, ½ and 5–6 sec nitrocellulose at 25°C.

Solubility parameters of solvent blends are calculated using the linear relationship:

$$P = \sum_{x=1}^{n} V_x . P_x$$

where P = parameter of an n-component solvent blend
P_x = value of the parameter of pure x (see Table 2.16)
V_x = volume of fraction of x in the solvent blend.

If a polymer(s) or resin(s) is dissolved in a true solvent and then titrated with a known diluent, there comes a point when the binder just starts to precipitate out of solution. In most cases, this point is sharply defined and the solution takes on a definite turbidity. The composition of the solvent at this point is assumed to be able to just dissolve the binder and its solubility parameters therefore indicate the absolute minimum acceptable solvent power at this point. By performing this titration and calculation with a variety of solvents and diluents, the limits of solubility of the binder are accurately defined. The construction of a map is then quite straightforward by using the following procedure:

1. Determine which are definite solvents (clear solutions) and non-solvents for the polymer at a given % weight concentration (say, 10–15%w) and note down examples of borderline solvent (turbid solutions). Very often this gives a clue to the area where solvency can be expected on a solubility map (Figure 2.10).
2. Pairs of solvents and diluents are then selected which make cross-sections through this area. Approximately 10 g of solution are weighed out accurately and titrated to a permanent turbidity with a diluent (or mixture of diluents).
3. A knowledge of the density of the solvent and the initial concentration of the solution allow a calculation of the volumes of solvent and diluent at the end point.
4. The solubility parameters of the end-points are then calculated using data given in Table 2.16.
5. These points are used to construct a map as in Figure 2.10.

Non-solvents have solubility parameters which are situated outside the boundaries of the solubility maps. Therefore, considering for example Figure 2.10, liquids such as the aliphatic hydrocarbons and alcohols have solubility parameters in this area and are therefore non-solvents for the Epikote resins. Aromatic hydrocarbons such as xylene, on the other hand, lie very close to the outer contour of the map but nonetheless outside it. Therefore, although they are not true solvents, they do have some partial solvent activity and are indeed compatible with the resins at high solids levels.

Solvents have solubility parameters which fall within the contours of the map. However, at a given Hildebrand solubility parameter and hydrogen bonding index, the higher the solvent fractional polarity, the lower

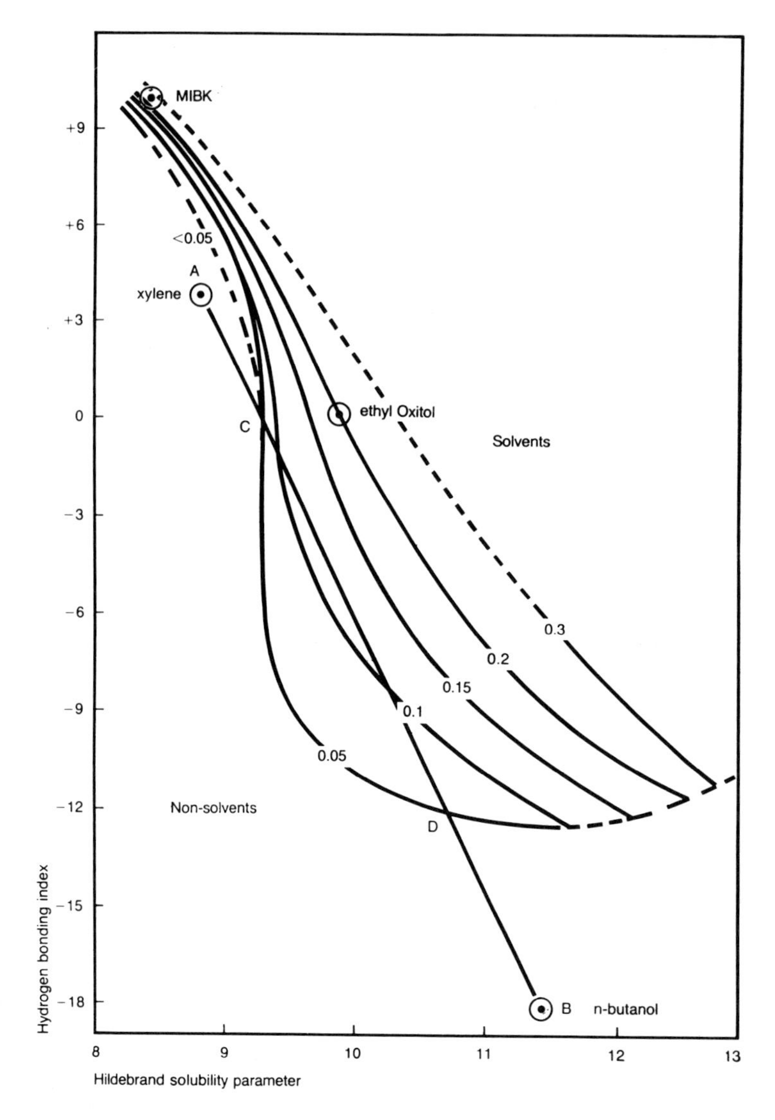

Figure 2.10 Solubility map for Epikote 1004 (solids > *c.* 30% w). Contours denote maximum permissible fractional polarity for true solvent.

is the power of the solvent for an Epikote resin. Ethyl oxitol with a fractional polarity of 0.086 is an example of a powerful solvent. When its solubility parameters are plotted onto, for example, the solubility map for Epikote 1004 (Figure 2.10) they fall on a map contour which denotes that true solvents for the resin at that point can have fractional polarities up to a maximum of 0.2. By contrast, on the same map, MIBK with a fractional polarity of 0.315 lies extremely close to a contour on the map of 0.3. In other words, the Hildebrand and hydrogen bonding index of MIBK place it within the solubility map but its fractional polarity is higher than the contour indicates is permissible for true solvency at that point. MIBK therefore exhibits only partial solubility for the solid Epikote resins.

The procedure employed for predicting the composition of solvent blends is explained below. As an example, consider the well-known phenomenon of xylene blended with *n*-butanol (i.e. two non-solvents) to form a true solvent for Epikote 1004 in Figure 2.10:

1. Mark out the points representing xylene (A) and *n*-butanol (B) on the solubility map.
2. Since solubility parameters combine linearly, the parameters of all blends of xylene with *n*-butanol fall on a straight line joining A with B in Figure 2.10. The intersects of this with the outer boundaries of the solubility map are C and D.
3. From the arguments presented in the previous paragraph, blends having solubility parameters falling between C and D are potentially true solvents for Epikote 1004.
4. The geometry of the map is such that the ratio of the lengths BC : AC represents the volume of xylene : *n*-butanol at C while BD : AD is that at point D. At point C, BC/AC = 85/15 and at D, BD/AD = 25/75. Therefore blends contain 85% to 25% volume xylene.
5. These blends now have to be checked with the fractional polarity contours of the map – neither of them

can have a fractional polarity higher than 0.05 if they are to be true solvents (see map contours at C and D).

6. The equation $P = \sum_{x=1}^{n} V_X.P_X$ is applied to calculate the fractional polarities of the blends at C and D. At C, polarity is 0.01 – therefore true solvency may be expected. At D, a fractional polarity of 0.07 indicates that only partial solvency may be expected from the blend. One therefore moves D further towards C (i.e. increase the xylene content) and recalculates the new blend fractional polarity until it complies with the limit set by the map.

Solvent maps also play an important role in precisely defining what is meant by 'solvent balance'. During evaporation of a blend, the more volatile component(s) tend to come off first, leaving behind the less volatile one(s). This may cause the solubility parameters, which in first instance fell well within the contours of a map, to gradually move outside of them. Due to this the solute will precipitate before all solvent has evaporated, which can for instance given undesired effects in paint formulations. With a well balanced blend, the solubility parameters will stay within the contours of the map during the entire evaporation period. This phenomenon will be discussed in more detail in section 2.3.

2.3 Volatility

Volatility, or more particularly evaporation rate under applicational conditions is also an important property of a solvent as it determines largely the behaviour of the formulation of which the solvent or blend forms a part.

In the section on basic solvent characteristics it was shown that molecular weight and intermolecular forces are the two major factors determining the volatility of a solvent. Several properties were shown to relate to volatility such as boiling point, vapour pressure and evaporation rate. These properties will be discussed in more detail and attention paid to the influence of solvent mixing on these properties and to the solvent evaporation from polymer films.

2.3.1 Boiling point and vapour pressure of single solvents

The boiling point of a pure liquid is the temperature at which its vapour pressure equals atmospheric pressure. Boiling points are usually quoted at a standard atmospheric pressure of 101.3 kPa. In practice however, a boiling range will often be quoted due to the fact that the solvent contains impurities, as is usually the case with industrial solvents, or because the product is a mixture of (isomeric) compounds distilled from a complex feedstock, which is the case for most commercial hydrocarbon solvents. For many solvents this range tends to be narrow, from a few tenths of a degree up to a few degrees C whereas for hydrocarbon solvents the boiling range may be considerably broader (from a few degrees up to over 100°C). Typical examples of solvents with a narrow, medium and broad boiling range respectively are acetone, industrial hexane and rubber solvent. The boiling point of pure acetone is 56.1°C. A typical boiling range for an industrial product with a minimum purity of 99.6%w is 55.8–56.6°C. Pure *n*-hexane has a boiling point of 68.7°C, whereas industrial hexane, which may consist of the components shown in Table 2.18, has a boiling range of 65–70°C. The boiling range of a rubber solvent is typically 110–156°C.

Attention should also be paid to the method by which the boiling range is determined. Different values can be obtained for the same product due to slight differences in the thermometers used, their position in the distillation flask and the use of different definitions for the beginning and the end of the boiling range.

IBP = *Initial Boiling Point*; the temperature at which the first drop of solvent is distilled over.
DP = *Dry Point*; the temperature at which the last liquid disappears from the bottom of the distillation flask. This value is in general used for lower boiling solvent types without a high boiling residue, thus with a sharply defined dry point.

Table 2.18

Component	Content
n-hexane	approx. 50%w
isohexane mixture	35–40
methyl-cyclopentane	10–12
cyclohexane	2–3
total	100%w

FBP = *Final Boiling Point*; the highest temperature registered during the distillation. When a small amount of high boiling residue is left in the distillation flask, the temperature at the top of the distillation column can decrease slightly towards the end of the distillation. In this case the FBP is higher than and is reached before the DP.

Two well known boiling range determination methods are ASTM D 1078 and D 86. The former is used for solvents with a low boiling range (up to approximately 200°C), whereas D 86 is used for higher boiling types. A typical example of the differences which result from distillation according to these methods is shown in Table 2.19.

Whereas the vapour pressure of a liquid at its boiling point equals that of atmospheric pressure, at room temperature the vapour pressure is in general well below that of atmospheric pressure, the difference being greater, the higher the boiling point of the solvent.

The Antoine equation can be used for an approximate calculation of the vapour pressure as a function of the temperature. Close examination of vapour pressure/temperature curves shows that in some instances the curves for different substances may actually cross. This phenomenon is observed, when the molecules of one of the liquids are associated at room temperature. This association causes a reduced vapour pressure of the substance, but the associated molecules break-up progressively as the temperature rises, causing the vapour pressure to rise faster. This phenomenon is observed particularly among substances having hydroxyl groups, as is the case with alcohols and acids. It is for this reason that low boiling alcohols have a lower vapour pressure than some of their higher boiling esters (Table 2.20).

Another instructive example is that of the closely similar substances cyclohexanone and cyclohexanol. They boil at nearly the same temperature (150–160°C), but cyclohexanone evaporates ten times as fast as cyclohexanol.

2.3.2 Boiling point and vapour pressure of solvent mixtures

The vapour pressure of a blend of two liquids, A and B, which are miscible in all proportions and do not react chemically or associate, is given by the Raoult expression [17]:

$$P_{A+B} = m_A P_A + (1 - m_A) P_B$$

in which: P_{A+B} = vapour pressure of the mixture
P_A, P_B = vapour pressure of A and B respectively
m_A = molar fraction of liquid A

If intermolecular attraction between the two liquids exists or if existing association forces are broken up, the mixtures have either lower or higher vapour pressures than given by the above equation. Thus, in the case of acetone and chloroform, the gradual addition of one component to the other causes a reduction of the vapour pressure of the blend until a minimum is reached. Upon continued addition, the vapour pressure increases again until the vapour pressure of the component that is present in excess is reached. The same phenomenon occurs with methylacetate and chloroform, but is not common with organic solvents. Gradual addition of one solvent to another usually causes the vapour pressure to increase progressively until a maximum

Table 2.19

Property	ASTM D 1078	ASTM D 86
IBP (°C)	195.6	192.7
50% (°C)	219.6	213.1
FBP (°C)	259.6	248.6
DP (°C)	255.6	—

Table 2.20

Product	Boiling point (°C)	Vapour pressure (kPa at 20°C)
Isopropanol	82.3	4.50
Isopropyl acetate	88.5	6.27
n-Butanol	117.7	0.57
n-Butyl acetate	126.0	1.26

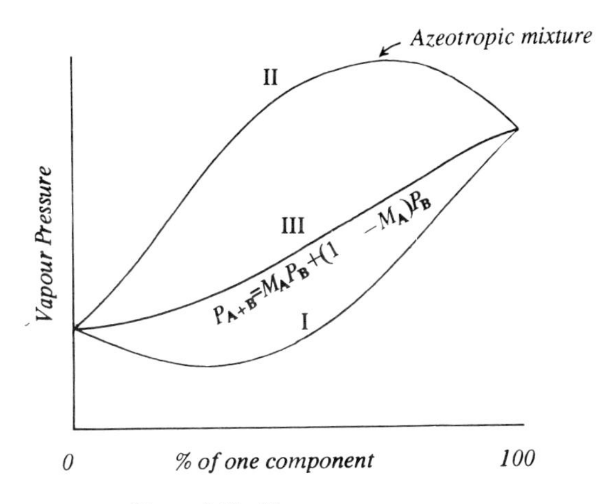

Figure 2.11 Vapour pressure curves.

is reached, followed by a reduction upon continued addition of that solvent. The mixture having the maximum vapour pressure is called an azeotropic mixture.

2.3.3 Azeotropy

Azeotropic mixtures are not uncommon and exist both between organic solvents (e.g. benzene/ethanol or ethyl acetate/ethanol) or with water (ethanol/water etc.). The vapour pressure curves of the three types of mixtures are shown in Figure 2.11.

Other types of curve are theoretically possible, but only one minimum or maximum can occur with substances that are miscible in all proportions.

Ternary mixtures are known where the mixture has a higher vapour pressure than any of its components, for example: ethyl alcohol, chloroform and hexane; isopropyl alcohol, ethyl acetate, and cyclohexane. In all probability the number of components may be still greater but no actual observations appear to have been recorded.

Azeotropic mixtures consisting of more than two components are rare since each component must be capable of forming a binary azeotrope with at least one of the others separately. The chance of the formation of an azeotropic mixture comprising all the components is therefore greatly reduced.

In practice azeotropic mixtures are important in several ways. The increase of vapour pressure which accompanies the formation of an azeotrope has a distinct influence on the flash point of a mixture. The increased vapour pressure causes a greater quantity of solvent to be evaporated into the air space of a container at any given temperature than if no azeotrope were formed, and since the flash point is largely governed by the proportion of inflammable vapour, the necessary proportion is reached at a lower temperature. Thus 25% of butanol (flash point 35°C), when mixed with 75% of xylene (flash point 23°C) gives a flash point of the mixture of 20°C. Similarly the proportion for maximum explosivity is reached at a lower temperature. The increase in vapour pressure will also cause the evaporation rate to be greater and may result in 'chilling' or other defects; furthermore, the deposition of water during chilling may give rise to ternary azeotropes with still higher rates of evaporation. The formation of an azeotrope may also affect the solvent power of a mixture.

The formation of an azeotrope affects the relative rates at which the constituents of a mixture evaporate, since the azeotrope tends to evaporate as though it were a single substance. It is advisable, therefore, when using a non-solvent diluent to choose the solvent so that either no azeotrope is formed or, if it is formed, that it is one having a higher proportion of non-solvent than solvent, so that the non-solvent will be removed more rapidly than the solvent; otherwise 'blushing' or 'whitening' may occur if the concentration of the solute gets too near the saturation point of the mixture.

When the temperature is increased above room temperature the vapour pressure composition of an azeotropic mixture tends to change slightly until the temperature is reached where the vapour pressure becomes equal to atmospheric pressure and the mixture starts to boil. This point is called the azeotropic boiling point and it is the minimum boiling temperature of a mixture of these components. It is lower than that of the single components. When a blend of two liquids, which form an azeotropic mixture, starts to boil, the composition of the vapour is the same as that of the azeotropic boiling point. The azeotropic mixture at its boiling point is therefore also called the constant boiling mixture. This composition remains constant until one of the components in the blend has become depleted.

An example of the change of the composition of the azeotropic mixture with increasing temperature and

Table 2.21

Temperature (°C)	Vapour pressure (mmHg)	Ethanol in azeotropic mixture with benzene (%v)
Room temp.	approx. 125	20
34.8	200	22
35.0	241	23.3
49.9	380	26
60.0	570	28
68.25	760	32.4

the composition at the azeotropic boiling point for benzene (b.pt. 80°C) and ethanol (b.pt. 78.3°C) is given in Table 2.21.

It is obvious that azeotropic mixtures will be formed in particular between miscible liquids, one of them having strong associate forces which are being broken by the other. In other words mixtures of miscible solvents with a high and a low heat of evaporation. Thus, for instance, water and alcohols and in particular an alcohol with hydrocarbon and halogenated solvent.

Where no azeotrope occurs, the vapour of a boiling blend always starts to be rich in the most volatile component and during evaporation the composition gradually shifts towards the less volatile one. The components of such blends can thus be separated by normal distillation.

2.3.4 Evaporation rate

The rate of evaporation of a single solvent is governed by several, sometimes related, factors:

- vapour pressure at the temperature under consideration
- rate at which heat is supplied
- heat conductivity of the liquid
- specific heat of the liquid
- latent heat of evaporation of the liquid
- surface tension
- molecular weight
- rate at which the vapour is removed
- vapour density of the solvent
- humidity of the atmosphere

When a solute is present, its influence on the evaporation characteristics needs to be taken into consideration, which is complex in view of the changing concentration of the solute during evaporation. The definition of evaporation rate is therefore obviously one of considerable complexity and needs an appropriate description for the classification of the evaporation rate.

The first attempt classified solvents as low, medium and high boilers, according to their boiling points. Although this classification was very rough, it permitted some degree of control to be obtained. It was customary to consider as low boilers those solvents with boiling points below 100°C, medium boilers those boiling from 100°C up to about 150°C, whilst those having still higher boiling points were termed high boilers, these last merging into the plasticiser class, which included substances like benzylalcohol (b.pt. 205°C).

The first methods of evaporation rate classification were based on a certain standardisation of the conditions under which the evaporation takes place and on comparison of the results with that of a reference solvent. An example is the evaporation of a small quantity of solvent from filter paper and comparing the drying time of the solvent under consideration to that of diethyl ether. The test procedure is given in the German DIN Method 53170 in more detail. It prescribes the use of equal volumes of solvent on a specified filter paper at 20°C, 65% relative humidity and no airflow. The relative evaporation time, RET equals the ratio of the evaporation times of test solvent and diethyl ether:

$$\text{RET} = \frac{\text{evaporation time of test solvent}}{\text{evaporation time of diethyl ether}}$$

The ASTM method D 3593 also uses equal amounts of test fluid and reference solvent on filter paper. In this case, *n*-butyl acetate is used as the reference. The evaporation, however, is determined in a special cabinet under well described conditions of airflow (21 l/m), temperature (25°C) and relative humidity (0–5%). The evaporation is not assessed visually but, more accurately, by monitoring the weight decrease and in order to avoid filter paper/solvent interaction effects only the first 90% of material evaporated is taken into consideration when measuring the relative evaporation rate:

$$\text{RER} = \frac{\text{time for evaporation of 90\% w. of } n\text{-butyl acetate}}{\text{time for evaporation of 90\% w. of test solvent}}$$

In Table 2.22 the evaporation data of some selected solvents are given as determined by both methods.

Evaporating a liquid at room temperature requires energy for its evaporation causing the temperature of the liquid surface to drop. For the evaporation to continue, heat has to be supplied and the temperature will drop until a balance is obtained between heat transport to the surface and heat required for evaporation. Due to the temperature reduction the evaporation of the solvent will be retarded and there will obviously be a difference in the rate whether evaporation is taking place under isothermal or free cooling conditions even when other factors like surface to volume ratio of the solvent, rate of airflow, relative humidity etc. are kept constant. The temperature drop is larger when the vapour pressure of the solvent is higher (lower boiling point) and when the heat required for the evaporation is lower. The specific heat of the solvent and its heat conductivity play a secondary role. The lowering of the temperature and the difference in RER under isothermal and free cooling conditions for a series of thin films of solvents evaporating at 21.5°C is given in Table 2.23.

It has been demonstrated by various authors [18] that the surface available for evaporation per volume of solvent applied in the test and the type of substrate can have a great effect on the evaporation rate even when other conditions are identical. Over the years there has therefore been some discussion on the question in how far evaporation measurement from filter paper relates to practical conditions e.g. from smooth paint films.

For instance, when measuring in the Shell Automatic Thin Film Evaporometer [19] which is used in the ASTM D 3529 method, during evaporation either from filter paper, from a smooth liquid thin film containing 10%w of tritolylphosphate or from smooth surface solvent film without plasticiser, quite different results are obtained as is demonstrated in Table 2.24.

In particular, the more volatile solvents evaporate slower from filter paper than from either of the smooth film surfaces. This may be due to greater cooling effects on the filter paper (closer to adiabatic conditions) or to the greater interaction between filter paper and solvent. The phenomenon is indeed most pronounced for

Table 2.22

Solvent	RET DIN 53107[a]	RER ASTM D3593[b]
Diethylether	1	11.0
Toluene	5.7	2.0
Isopropanol	11.0	1.5
n-Butyl acetate	11.8	1.0
n-Propanol	16.0	0.94
Xylene	15.1	0.76
n-Butanol	33	0.47
Shellsol A	46	0.20

[a] Diethylether = 1. [b] *n*-Butylacetate = 1.

Table 2.23

Product	Boiling point	Heat of evaporation at 20°C	RER		Temp. (°C) of solvent surface under free conditions
			Isothermal	Free cooling	
Acetone	56.2	560	19.4	7.8	3
Methanol	64.6	1170	7.9	3.2	5
n-Hexane	68.7	375	17.8	8.1	4
Ethylacetate	77.15	409	8.8	5.0	7
Ethanol	78.3	924	3.9	2.4	10
Methylethylketone	79.6	482	7.5	4.6	10
Isopropanol	82.3	754	2.6	2.2	13
Isopropylacetate	88.5	378	5.4	3.8	14
Isobutanol	107.9	692	0.8	0.85	19
Toluene	110.6	428	2.6	2.3	16
Methylisobutylketone	116.0	431	1.8	1.4	16
n-Butylacetate	126.0	375	1.0	1.0	20
p-Xylene	144.4	404	0.75	0.75	21

Table 2.24

Product	RER (n-butylacetate = 1)		
	ASTM D 3593 9 (filter paper)	Smooth liquid film with plasticiser	Smooth liquid film without plasticiser
Toluene	2.0	2.3	2.14
Isopropanol	1.5	2.2	2.36
n-Butylacetate	1.0	1.0	1.0
n-Propanol	0.94	0.94	1.22
Xylene	0.76	0.75	0.7 (est)
n-Butanol	0.47	0.46	0.48
SHELLSOL A	0.20	0.25	0.26

solvents which can interact with the substrate via hydrogen bonds. The ratio of the RER on filter paper to that of a film without plasticiser in relation to the vapour pressure of the solvent is given in Figure 2.12. It is clear that less volatile solvents with vapour pressures below that of *n*-butyl acetate are less affected. For the more volatile ones the ratio decreases to below 0.5, indicating that they evaporate more than twice as fast from a smooth film than from filter paper (Figure 2.12).

2.3.5 *Evaporation of blends*

It would be very convenient if solvents evaporated from blends in a way which was directly allied to their relative evaporation rates. Unfortunately this is not the case. For example if blends of toluene with *n*-butanol are considered, the evaporation rate data indicate that toluene is more volatile than *n*-butanol. It would be expected that during evaporation the residual solvent would always become richer in *n*-butanol. In practice, however, blends can grow richer in either of the two solvents or neither, depending on the composition of the mixture. Where the composition does not change during evaporation the mixture behaves like a pseudo-azeotrope. It should be noted that the pseudo-azeotropic composition for evaporation at room temperature will in general not be the same as the azeotropic composition under boiling conditions. For instance, using evaporation according to ASTM D 3593 as the criterion the pseudo-azeotropic composition is 15%v *n*-butanol and 85%v toluene, whereas at the azeotropic boiling point (105.5°C) the mixture contains 32%v alcohol.

Walsham and Edwards [20] have developed a model which satisfactorily describes the evaporation behaviour of solvent blends. They based their work on that of Sletmoe [21], who found that the solvent evaporation rate was limited by boundary layer diffusion into the vapour phase. On this basis the rate of evaporation should be proportional to the vapour pressure of the solvent. Deviations in evaporation behaviour from ideality were directly analogous to deviations from Raoult's law of vapour pressures and could be corrected using activity coefficients. Since for most solvent molecules activity coefficients are largely deter-

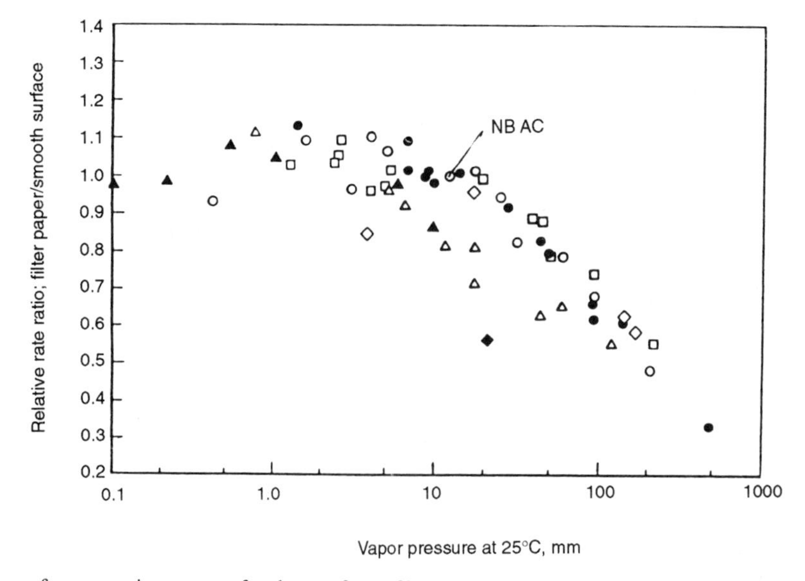

Figure 2.12 Comparison of evaporation rates of solvents from filter paper and from a smooth surface. △, Alcohols; ◆, water; ●, hydrocarbons; □, ketones; ○, esters; ▲, ether alcohols; ◇, miscellaneous.

mined by their chemical functional group(s) and only to a minor extent by the nature and size of the hydrocarbon part of the molecule, it is possible to prepare graphs of activity coefficients for classes of solvents. For practical purposes three classes are considered: hydrocarbons, alcohols and glycolethers (OH-group containing) and ketones and esters (C=O group containing). This classification in conjunction with data published by Deal and Derr [22] has been used to construct graphs of activity coefficients γ which may be used to calculate corrected solvent relative evaporation rates in a blend according to:

$$R_b = \gamma . R$$

An example for blends of hydrocarbons with OH or C=O group containing solvents are given in Figure 2.13. It is obvious that solvents belonging to all these three classes show positive deviations from Raoult's law when blended. This means that their evaporation in the blend is faster than when evaporated alone. This is in line with the observation that the vapour pressure of azeotropic mixtures is above that of the single components.

In a pseudo azeotropic blend the relative evaporation rates of both components are the same, or $\gamma_1 . R_1 = \gamma_2 . R_2$. When R_1 and R_2 are known, it can be determined with the help of the graphs at which composition γ_1 and γ_2 fulfil the above requirement. Such calculations for mixtures of the aromatic hydrocarbon solvents toluene, xylene and SHELLSOL A with *n*-propanol and *n*-butanol give the result in Table 2.25.

The evaporation rates as calculated by this technique and as actually measured for an ethyl glycol/white spirit blend are given in Figure 2.14, demonstrating the level of accuracy with which predictions of solvent blend evaporation rates can be made. In the latest approaches using computerised prediction techniques the rate of evaporation of single components in blends no activity coefficients for classes of solvents are used but the activity coefficients of the single solvents are calculated from contributions of molecular groups which compose the total solvent molecule. A method for doing so has, for instance, been developed using the UNIFAC system of group contributions [23]. When using this system and repeating the calculation of residual solvent composition after small steps of evaporation (e.g. after each 5% evaporated), the change in composition of the solvent blend during evaporation and thus how well it is balanced can be predicted. Such

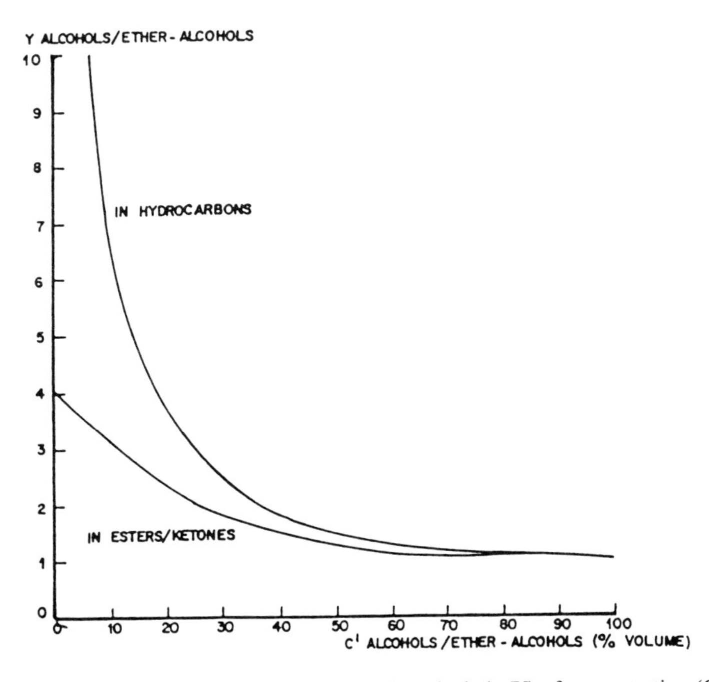

Figure 2.13 Activity coefficients of alcohols/ether-alcohols (Y) of concentration (C[1]).

Table 2.25

Product	Vol% alcohol in pseudo-azeotropic mixture with hydrocarbon	
	n-Propanol	*n*-Butanol
Toluene	27	15
Xylene	53.5	33
SHELLSOL A	⩾90	83.5

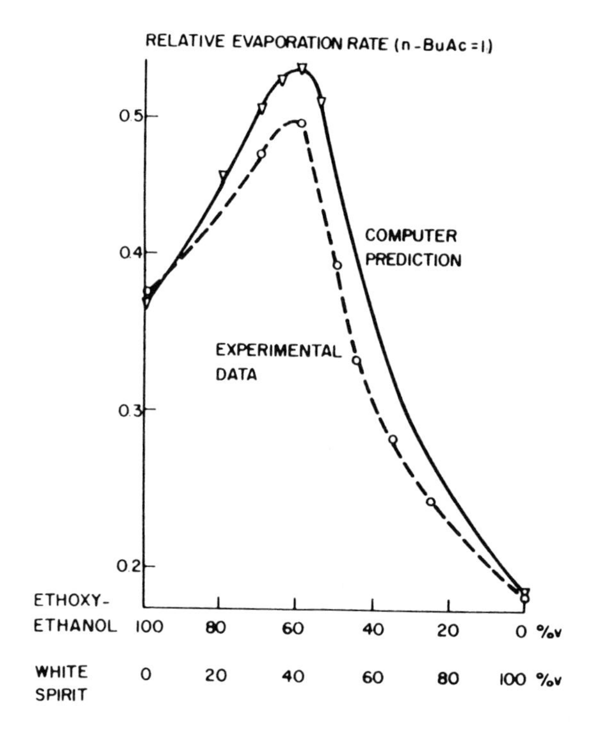

Figure 2.14 Predicted and measured evaporation rates for ethoxyethanol/white spirit system.

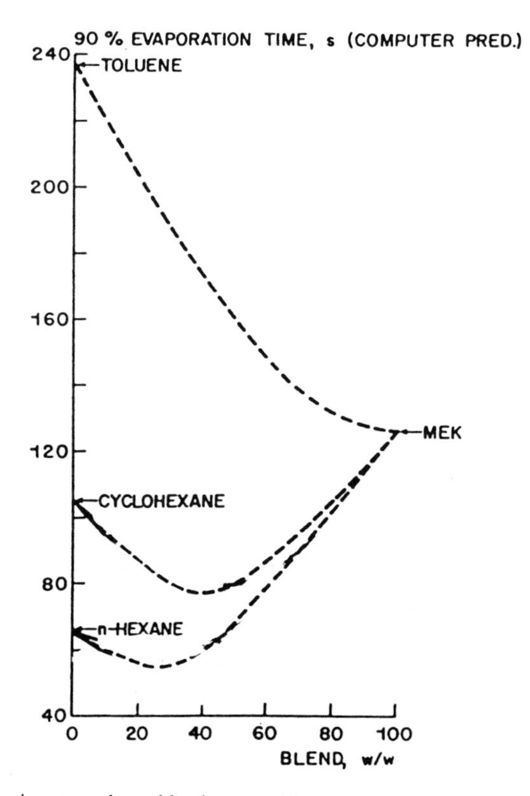

Figure 2.15 Evaporation time vs. solvent blend composition for various hydrocarbon solvent/MEK mixtures.

predictions are demonstrated for two hexane/MEK blends which have been selected just on the right and left hand side of the pseudo-azeotrope (see Figure 2.15 with evaporation curves for various hydrocarbon solvent/MEK mixtures).

From Figure 2.16 it is clear that at a hexane/MEK 61.5/38.5 ratio it is predicted that the blend becomes rich in MEK during evaporation. The 75/25 blend is so close to the pseudo-azeotrope that in the beginning hardly any change in composition is expected. When approaching the end of the evaporation a gradual enrichment of hexane has been calculated.

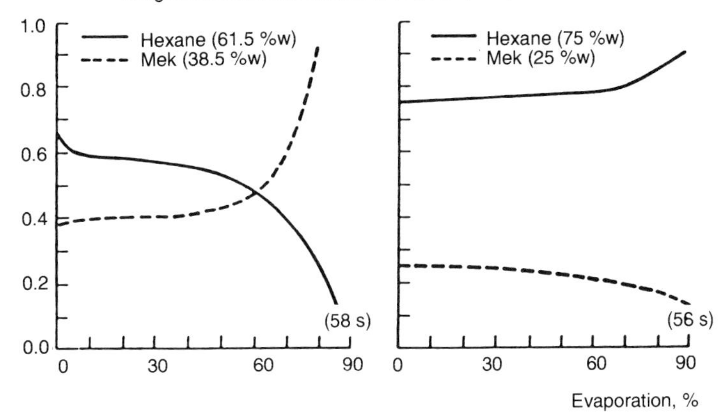

Figure 2.16 Change of solvent blend composition during evaporation on for two hexane/MEK blends.

2.3.6 Solvent loss from films of polymer solutions/solvent retention

Solvent loss from thin films of polymer or resin solutions is of importance in various applications, in particular in the paint and adhesive industry. When solvent evaporates from a film of a (diluted) polymer solution one can distinguish two phases. The first one ('wet' stage), where solvent evaporation is not noticeably hindered by the presence of other ingredients present. During this stage, in which usually over two thirds of the solvent are evaporated, evaporation takes place in a way similar to that of a pure solvent (blend). During the second or 'dry' stage the evaporation of residual solvent is controlled by diffusion processes through the polymeric film and is slowed down drastically. In this stage other factors like polymer type and sterical form of the solvent molecules are of more importance than the volatility of the pure solvent(s). This may be demonstrated with the examples below.

Hays [24] investigated the retention of toluene and 2-ethoxyethylacetate, in films of polymethylmethacrylate (PMMA). The compositions (%w) of the three acrylic lacquers used are as shown in Table 2.26.

Although the total amount of residual solvent was very similar for each of the lacquers, after 2 hours elapsed lacquer C contained less than 1% toluene, lacquer B approx. 4% and lacquer A approx. 15% toluene. The results show that the less volatile component in a binary solvent blend is the more retained, which is in line with the residual composition of a pure solvent blend when evaporated for approx. 80%. Baelz [25] investigated the retention of toluene and MEK in a series of natural and synthetic resins. He found that, in general, hard films retain significant amounts of solvent over a long period of time (at least several weeks), whereas soft resins retain little or no solvent after about 1 week. The work reported by a variety of investigators suggests that there are definite differences in retention behaviour between different polymers, but that for a series of individual solvents in any of these polymers, the order of retention is similar. This observation suggests that the general level of solvent retention is affected by the polymer and that specific solvent/polymer interactions are of secondary importance.

Hansen has shown that, if the logarithm of the concentration of residual solvent is plotted against the logarithm of drying time divided by the square of the film thickness, effects of differences in film thickness are eliminated [26]. When doing so for various polymers, an order of increasing retention can be drawn up for a large number of individual solvents, which seems to be independent of polymer type (see Table 2.27). If one

Table 2.26

	Relative evaporation rate[a]	A	B	C
PMMA	–	21.6	21.6	21.6
Toluene	2.3	78.4	66.9	32.4
2-Ethoxyethyl acetate	0.23	–	11.5	46.0

[a] *n*-Butylacetate = 1.

Table 2.27 Approximate order of increasing retention for individual solvents

Solvent (increasing retention ↓)	Evaporation rate (n-butyl acetate = 1.0)	Molar volume (cm^3/mole)
Methanol	4.1	40
Acetone	10.2	73
2-Methoxyethanol	0.51	79
MEK	4.5	90
Ethyl acetate	4.8	97
2-Ethoxyethanol	0.35	97
n-Heptane	3.3	146
2-Butoxyethanol	0.076	130
n-Butyl acetate	1.0	132
Benzene	5.4	88
2-Methoxyethyl acetate	0.35	117
2-Ethoxyethyl acetate	0.23	135
Dioxan	–	85
Toluene	2.3	106
Chlorobenzene	–	101
2-Nitropropane	1.5	90
m-Xylene	0.75	122
MIBK	1.4	124
Isobutyl acetate	1.7	133
2,4-Dimethyl pentane	5.6	148
Cyclohexane	5.9	108
Diacetone alcohol	0.095	123
Pent-Oxone	0.26	143
Methyl cyclohexane	3.5	126
Cyclohexanone	0.25	103
Methyl cyclohexanone	0.18	122
Cyclohexyl chloride	–	118

considers both molecular size and shape of these solvents, this gives an explanation for the order of retention if it is assumed that the rate-determining mechanism is the diffusion of solvent molecules through the polymer film [27].

It should be pointed out that the rate of diffusion of solvent molecules through a polymer film not only determines the rate of solvent release during the 'dry' stage, but is also a factor governing the concentration of solvent remaining at the beginning of this stage. The change over from boundary layer (wet stage) to diffusion-controlled (dry stage) release will occur when diffusion through the polymer film becomes slower than the rate of escape of solvent molecules from the film surface. For a more volatile but branched and thus slowly diffusing solvent molecule this changeover will occur at a higher concentration of residual solvent than for a slower evaporating, linear solvent molecule. When a blend of these solvents is being used, the blend will become richer in the linear solvent during the 'wet' stage but the reverse will occur during the 'dry' stage as is demonstrated in Figure 2.17.

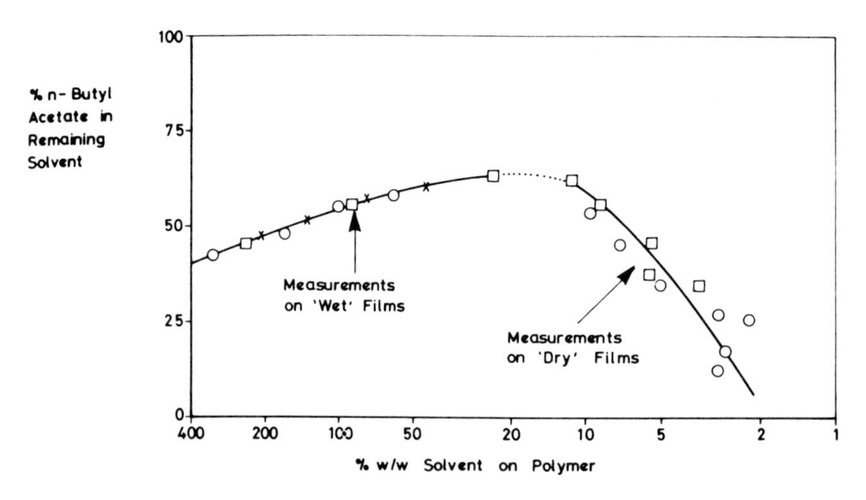

Figure 2.17 The 'wet' and 'dry' stages of solvent release for a 2:3 w/w blend of n-butyl acetate–isobutyl acetate in Elvacite 2013 and nitrocellulose FHX 30-50, at 23°C. ○, From 'Elvacite' 2013; □, from nitrocellulose FHX 30-50; ×, in the absence of resin.

2.4 Viscosity

2.4.1 Viscosity of solvent (blends)

The viscosity of a liquid is a description of its mobility or vice versa its resistance to flow. The viscosity of a solvent is, apart from its temperature dependence, influenced by both its inter-molecular interactions and to a lesser extent by its molecular weight. The kinematic viscosity at a given temperature is expressed in mm^2/s, formerly in Stokes or Centistokes,

$$1 \text{ Stoke (St)} = 100 \text{ Centistokes (cSt)} = 100\,mm^2/s$$

The dynamic viscosity is the kinematic viscosity multiplied with the solvent density at the same temperature. It is expressed in (milli) Pascalseconds (m)Pa.s, formerly in (centi) Poise.

$$1 \text{ Poise (P)} = 100 \text{ Centipoise (cP)} = 0.1 \text{ Pa.s} = 100 \text{ mPa.s}$$

$$1P = 1\,St \times d \quad \text{or} \quad 1\,mPa.s = 1\,mm^2/s \times d$$

In an ideal system the viscosity of a solvent blend can, according to Kendall, be calculated from the viscosity of the single components by means of the following equation [28]:

$$\log \eta = \sum fm_i.\log \eta_i$$

η = kinematic viscosity of the blend
η_i = kinematic viscosity of component i
fm_i = molar fraction of component i

In non-interacting solvent systems the use of volume fractions instead of molar fractions also gives satisfactory results (Figure 2.18). In interacting systems where the blend contains solvent with strong polar and/or hydrogenbonding interactions, appreciable deviations from ideal behaviour can be observed (Figure 2.19).

In such cases the viscosity of a blend which contains a limited (e.g. up to 30%v) amount of interacting solvent may be represented by using an effective instead of the real viscosity in the Kendall equation:

$$\log \eta = \sum \phi\upsilon_i.\log \eta_{i\,\text{effective}}$$

$\eta_{i\,\text{effective}}$ = effective viscosity of component i
$f\upsilon_i$ = volume fraction of component i

This pragmatic approach for predicting the viscosity of solvent blends has been developed by Rocklin and Edwards and is applicable both for hydrocarbon solvent based and aqueous blends with limited quantities of other components [29].

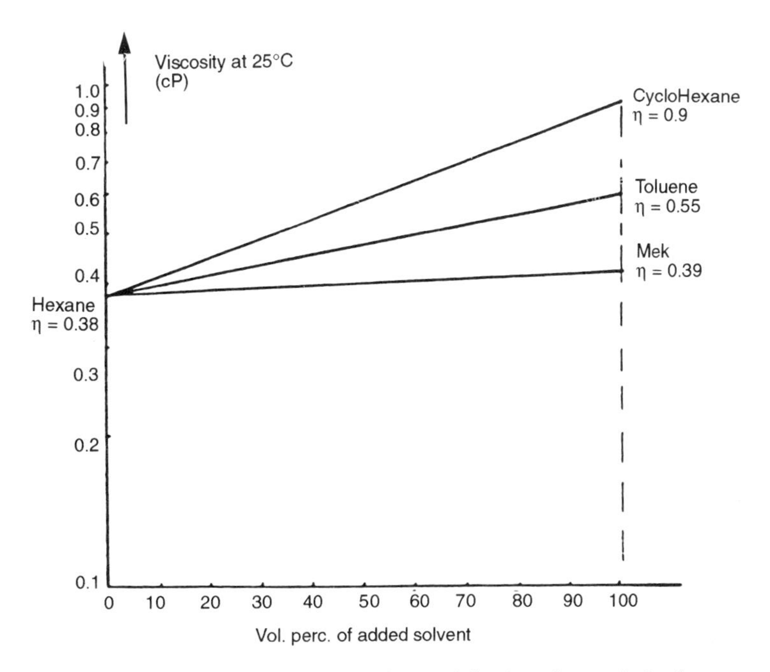

Figure 2.18 Viscosity of solvent blends containing 'non-interacting' solvents.

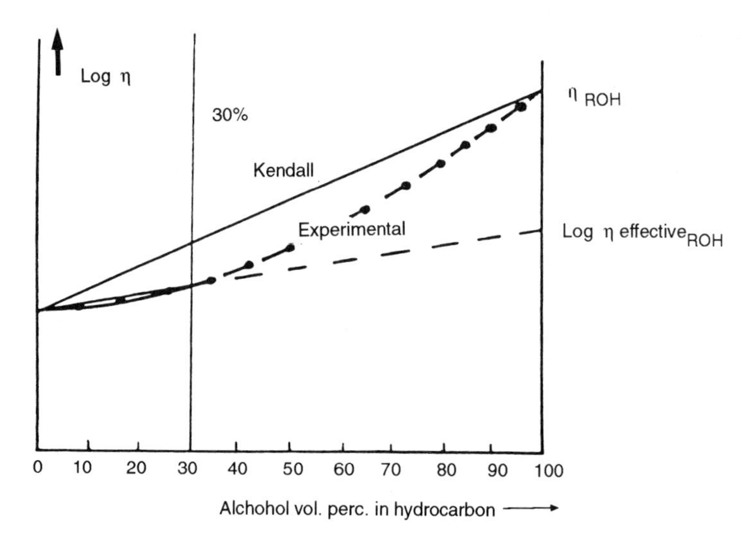

Figure 2.19 Viscosity of solvent blends containing strong 'interacting' solvents (e.g. alcohols, oxitols, etc.).

Table 2.28 Neat and effective viscosities, cps solvents at 25°C

Solvent	Neat viscosity	Effective viscosity in hydrocarbon solvents	Effective viscosity in aqueous blends
Ketones			
DMK	0.31	—	6.06
MEK	0.41	—	9.79
DEK	0.46	—	—
MIBK	0.55	—	—
EAK	0.79	—	—
DIBK	0.92	—	—
Me-6-K	1.52	—	15.6
MO	0.64	—	—
Isophorone	2.30	—	—
Cyclohexanone	2.00	—	—
Alcohols			
EtOH	1.30	1.05	35.0
n-POH	2.00	1.40	62.5
i-POH	2.40	1.10	59.3
n-BuOH	2.60	1.60	—
s-BuOH	2.90	1.40	110.0
t-BuOH	4.50	1.20	116.4
i-BuOH	3.40	1.80	80.0
DAA	2.90	2.00	22.2
MIBC	3.80	1.80	—
Hexane glycol	29.8	18.80	65.0
Esters			
Me OX Ac	1.14	—	10.1
OX Ac	1.20	—	33.9
BuDiOXAc	3.60	—	—
Glycerol ethers			
Methyl OXITOL	1.60	1.20	14.0
Ethyl OXITOL	1.90	1.35	24.4
i-Propyl OXITOL	2.50	1.54	29.8
Butyl OXITOL	2.90	2.00	32.1
Methyl DIOXITOL	3.80	1.50	17.3
Ethyl DIOXITOL	4.00	1.75	26.1
Butyl DIOXITOL	5.30	2.15	35.0
Ethyl TRIOXITOL	8.25	6.80	29.2
Other solvents			
Water	0.92	0.92	0.92
THF	0.55	0.50	13.8
DMF	0.86	0.82	14.4
Hydrocarbons			

When a solvent with strong intermolecular forces such as an alcohol is mixed into a non-interacting hydrocarbon solvent, the intermolecular forces are being broken and the alcohol is starting to behave like a non-interacting solvent where viscosity is mainly determined by molecular weight. In this case the apparent or effective viscosity of the alcohol is lower than its neat viscosity.

When the alcohol is blended with water, the intermolecular forces do increase due to further hydrogen bond formation with the result of a higher than expected effective viscosity. Examples are given in Table 2.28.

2.4.2 Viscosity of solutions

The viscosity of polymer solutions is much more complex and therefore is difficult to predict. It is only in ideal cases, for 'true' solvents, dependent on the viscosity of the solvent and polymer concentration. Apart from this polymer structure, polymer molecular weight and polymer/solvent interactions play an important if not overruling role. (For the theory behind the viscometry of 'ideal' diluted polymer solutions as developed

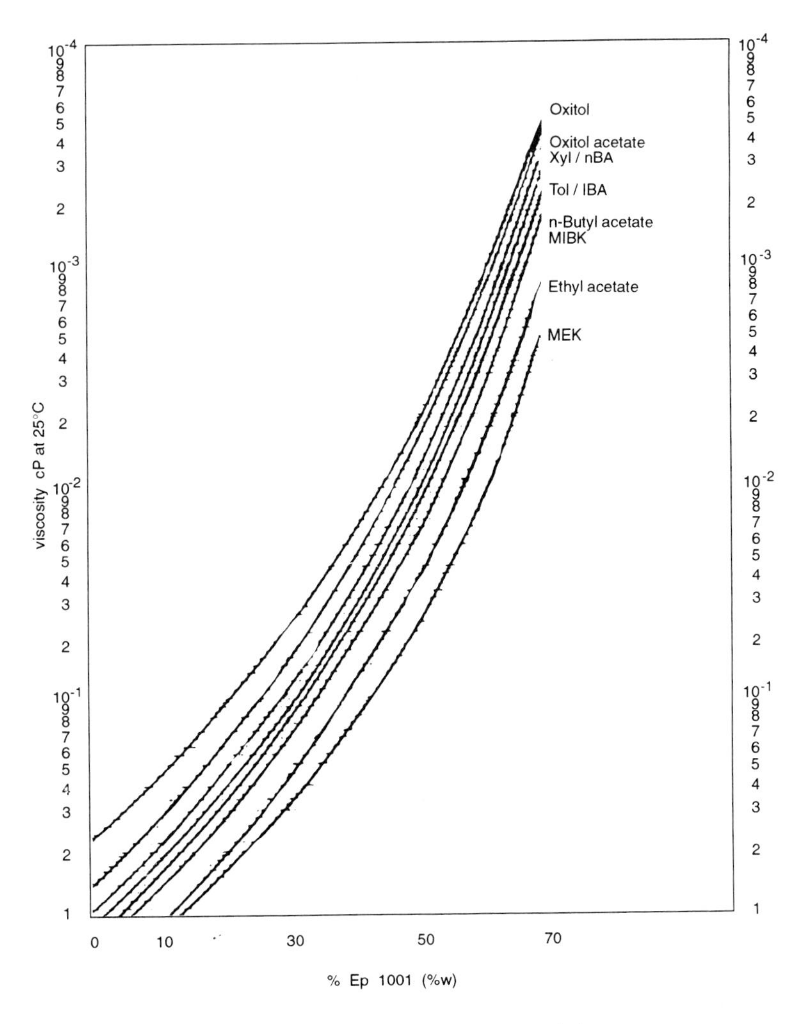

Figure 2.20 Viscosity of Epikote-1001 solutions.

Table 2.29 Viscosity (cps) of Epikote 1001 concentrations at 25°C

Epikote concn		MEK	*n*-Butyl acetate	Oxitol
0		0.46	0.84	2.27
10	exp.	0.8	1.5	4.2
	calc.	0.86	1.56	4.22
30	exp.	3.2	7.7	20.4
	calc.	4.06	7.41	20.04
50	exp.	21.1	73.1	177.9
	calc.	36.3	66.4	179.3

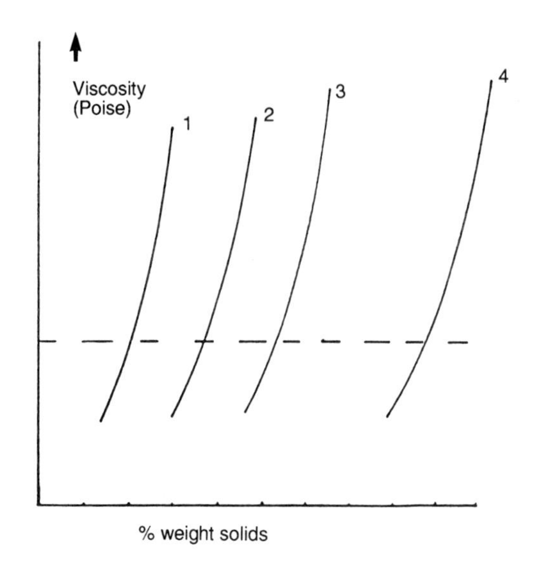

Figure 2.21 Influence of polymer molecular weight. 1, Acrylic polymer (molecular weight = 100.000); 2, acrylic polymer (25.000); 3, polyester polymer (6.000); 4, adipate plasticiser (1.500).

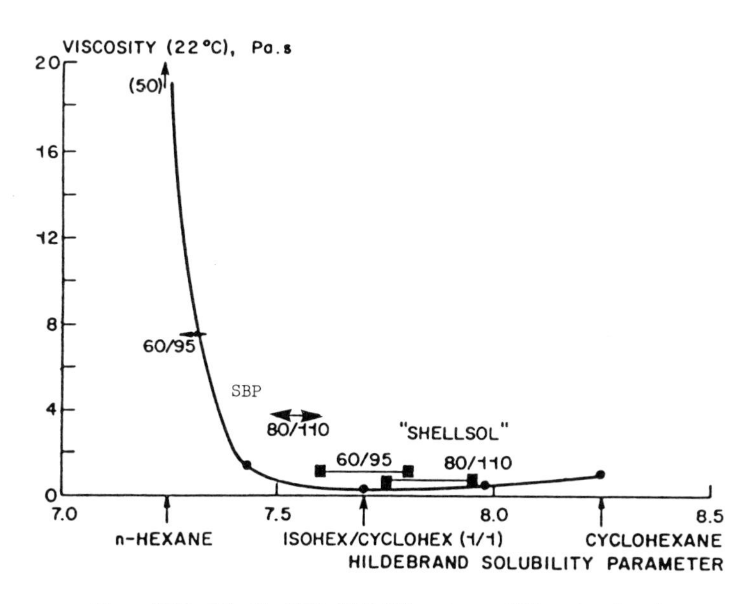

Figure 2.22 'Cariflex' TR-1107 (25 parts per 100 parts solvent).

by Flory–Huggins and Mark–Houwink reference is made to standard books on polymer physics.) Again there is a strong dependence on temperature as described for pure solvents and their blends. A rather ideal case is given in Figure 2.20, where the viscosity is given of a low viscosity epoxy resin dissolved at increasing concentrations in various solvents. At a given resin concentration the blend viscosity does increase with increasing viscosity of the neat solvent.

In such a case the solution viscosity at a given temperature is, according to Van Krevelen, related to the resin concentration:

$$\log \eta_{\text{rel}} = \frac{C}{K_{\text{a}} - K_{\text{b}}.c}$$

$\eta_{\text{rel}} = \eta/\eta_{\text{s}}$; η = kinematic solution viscosity;
η_{s} = kinematic solvent viscosity;
c = resin concentration, %w;
$K_{\text{a,b}}$ = constants.

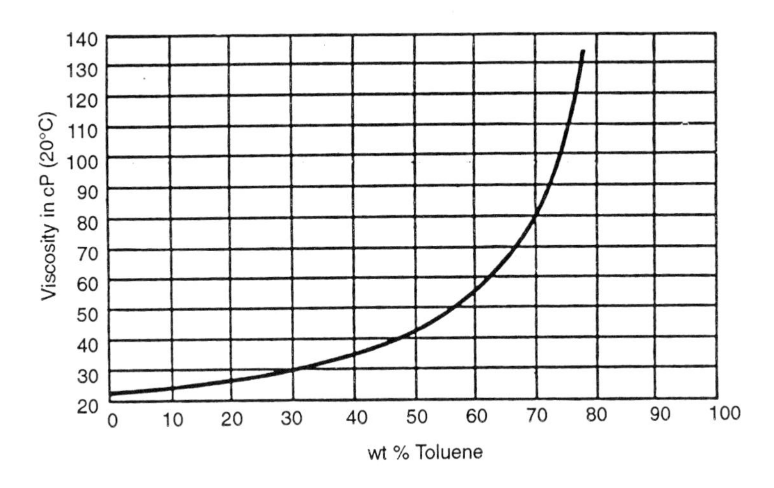

Figure 2.23 Influence of diluent (toluene) on viscosity of NC-solution in MIBK.

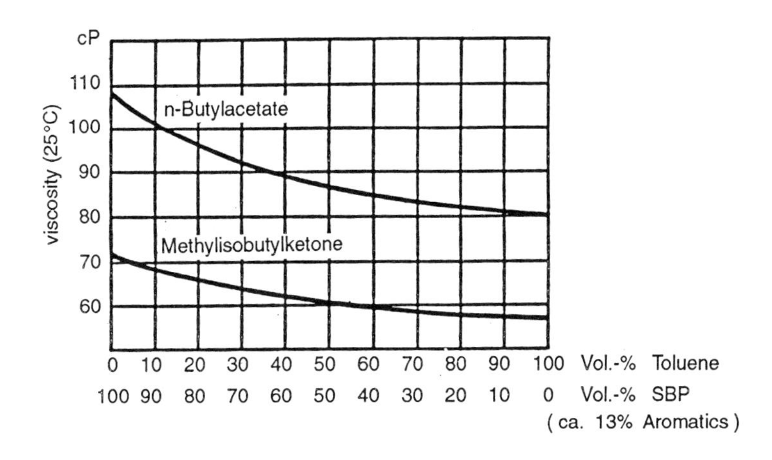

Figure 2.24 Influence of hydrocarbon solvent blend on viscosity of NC-solutions.

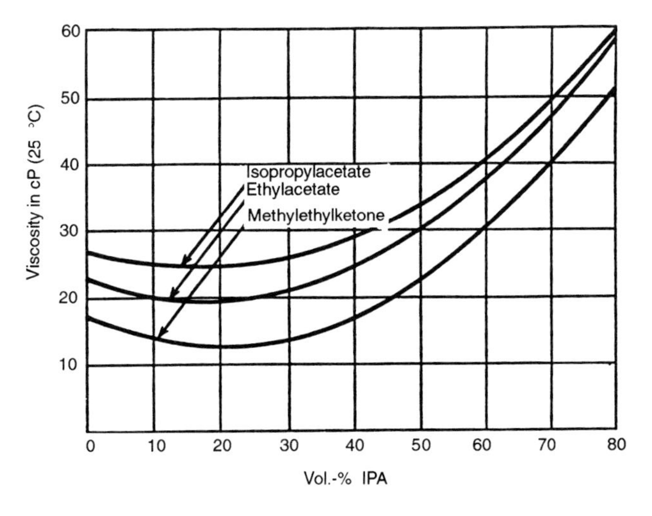

Figure 2.25 Influence of IPA on viscosity of NC-solutions.

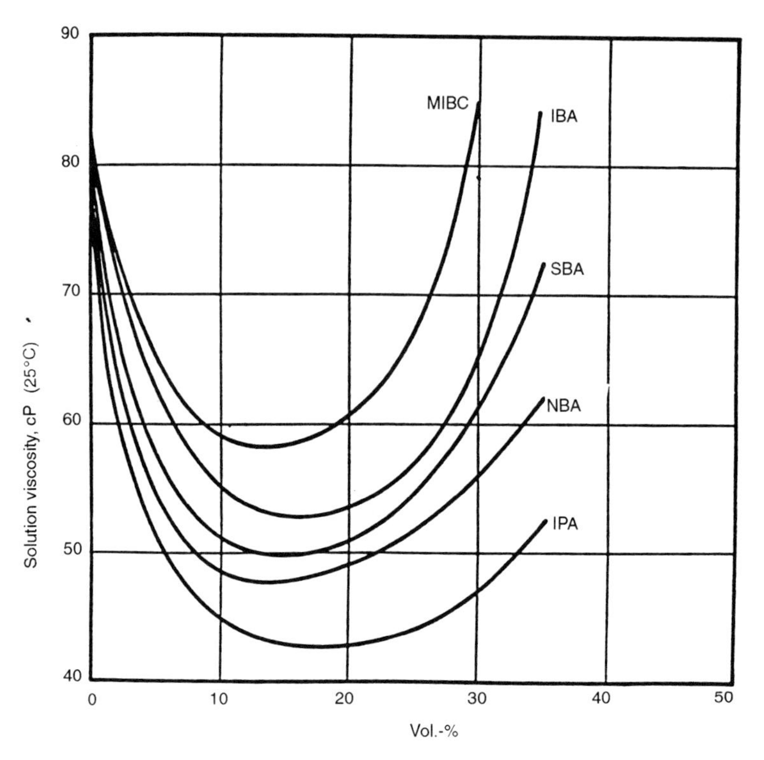

Figure 2.26 Influence of alcohols on viscosity of NC-solutions in a solvent/diluent blend. MIBK alcohol = 50 vol.–%, toluene = 50 vol.–%. (8 g RS ½ sec. NC/100 ml).

For instance, for Epitoke 1001, $K_a = 0.398$; $K_b = 0.269$. With the help of these constants, the viscosities in Table 2.29 were calculated.

The viscosity dependence on polymer molecular weight is demonstrated in Figure 2.21 whereas Figure 2.22 shows the effect of polymer/solvent interaction. In the latter, the viscosity of solutions of a thermoplastic rubber at constant concentration and temperature is given in various hydrocarbon(blend)s. In this example hydrogen bonding or polar effects play no or only a very limited role. It is clear that in this case the solution viscosity is very much influenced by the Hildebrand solubility parameter of the solvent in relation to that of the polymer (8.2–9.1).

At a Hildebrand solubility parameter of around 7.5 and below, where the polymer becomes less readily soluble, a sudden increase of viscosity can be observed, which is due to deviations from ideal behaviour. In this case the deviation results in an increase in viscosity, with other rubber types (e.g. natural rubber, SBR) in a similar situation a sudden decrease can be observed.

The change in viscosity of these solutions is due to changes in the dimensions of the coils of polymer molecules in solution. Depending on the type of solvent, the polymer molecules will either seek more contact with the solvent molecules (the coil will swell) or with itself (coils become more compact). Under certain, the so-called theta, conditions, the coil has undisturbed, 'ideal' dimensions. A solvent in which at room temperature a polymer forms such undisturbed coils is named a theta solvent for this polymer. In principle such a solvent has exactly the same solubility parameters as the polymer in question.

Another example of this type can be given in the case of nitrocellulose solutions (8 g RS ½ sec. NC/100 ml). The viscosity of these solutions in pure chemical solvents increases as in the case of the epoxy resin solutions with increasing resin concentration. Ketones, with lower neat viscosities than comparable esters result in lower solution viscosities (neat viscosity at 25°C in cS of MEK = 0.46, EtAc = 0.51, MiBK = 0.74, *n*-Bu Ac = 0.84). When ketone is used in combination with a non-solvent (e.g. an SPB) or diluent such as toluene, an increase in viscosity can be observed. This increase is rather sharp when coming close to the point where the resin becomes insoluble (Figure 2.23). When a fixed amount of an SBP/toluene blend is used as the diluent, the viscosity of similar solutions decreases with increasing toluene content of the diluent blend (Figure 2.24).

When an alcohol is added instead of a non-solvent or diluent to the ½ sec. NC-solutions, one can first observe a decrease in viscosity, hence the name latent solvent, before it starts to rise (Figure 2.25). This effect is even more pronounced when the alcohol is added to a solvent/diluent (MiBK/toluene) blend as demonstrated in Figure 2.26. The latent solvent action is very clearly demonstrated in this figure, which explains why in NC-formulations normally 10–20 vol % of (cheap) alcohol is added [30].

References

1. London, *Z. Physikal. Chemie*, **B II**, 222 (1930).
2. Debye, *Phys. Z.*, **13**, 97 (1912).
3. Trouton, *Phil. Mag.*, **18**, 54 (1884).
4. Hildebrand and Scott, *The Solubility of Nonelectrolytes*, 3rd edn, Reinhold, New York (1950).
5. Shell Industrial Chemicals, *Solvent Power*, ICS(X), 79/2 (1979).
6. Hildebrand and Scott, *Solubility of Non-Electrolytes*, 3rd edn, Reinhold, New York (1950).
7. Scatchard, *Chem. Rev.*, **8**, 321 (1931).
8. Gardon, *J. Paint Technol.*, **38**(492), 43 (1966).
9. Burrell, *Off. Digest*, **27**(369), 726 (1955).
10. Shell Chemicals, *Solubility Parameters*, IS 4.3.3, 3rd edn (1985).
11. Crowley, Teague and Lowe, *J. Paint Technol.*, **38**(496), 269 (1966); **39**(504), 19 (1967).
12. Gordy, *J. Chem. Phys.*, **7**, 93 (1939); **8**, 170 (1940); **9**, 204 (1941).
13. Hansen, *J. Paint Technol.*, **39**(505), 104 (1967); **39**(511), 505 (1967); and Hansen and Skaarmp, *J. Paint Technol.*, **39**(511), 511 (1967).
14. Nelson, Hemwall and Edwards, *J. Paint Technol.*, **42**(550), 636 (1970); and Nelson, Figurelli, Walsham and Edwards, *J. Paint Technol.*, **42**(550), 644 (1970).
15. Teas, *J. Paint Technol.*, **40**(516), 19 (1968).
16. Shell Chemicals, *A Guide to Solvent System Design*, IS 4.3.4, 3rd edn (1985).
17. Raoult, *Compt. rend.*, **103**, 1125 (1886); and *Ann. Chim. Phys.*, **15**, 375 (1888); **20**, 297 (1890).
18. Rocklin, *J. Coatings Technol.*, **48**(622), 45 (1976).
19. Shell Industrial Chemicals, *Evaporation of Organic Solvents from Surface Coatings*, IS 4.3.1, 4th edn (1990).
20. Walsham and Edwards, *J. Paint Technol.*, **43**(554), 64 (1971).
21. Sletmoe, *J. Paint Technol.*, **42**(543), 1970 (1970).
22. Deal and Derr, *Ind. and Eng. Chem.*, **60**, 28 (1968).
23. Friedenslung, Smekling and Rasmussen, *Vapor–Liquid Equilibria using UNiFAC*, Elsevier, Amsterdam (1977).
24. Hays, *Off. Dig.*, **36**, 605 (1964).
25. Baelz, *Solvent Retention by Lacquer Films*, Technical Paper No. 114, The Paint Research Station, Teddington, UK (1938).
26. Hansen, *Off. Dig.*, **37**, 57 (1965).
27. Newman and Num, *Progress in Org. Coatings*, **3**, 221 (1975).
28. Kendall, *Medd. K. Vatenskapsakad. Nobelinst.*, **2**(25), 1 (1913).
29. Rocklin and Edwards, *J. Coat. Technol.*, **48**, 68 (1976).
30. Shell Chemie, *Lösungsmittel in Nitrocelluloselacken*, SOLC-1.1 (1971).

3 Choosing a solvent – practical advice

W.C. ATEN

3.1 General advice

Solvents are used for various reasons and for a wide variety of industrial applications. Typical applications (and reasons for use) are:

- surface coatings, printing inks, adhesives, aerosols etc. (as carriers, for temporary viscosity reduction and to aid application)
- extraction or cleaning solvents (for the removal of valuable or undesired material)
- chemical industry, polymerisation industry (as reaction or heat transport medium)
- polyurethane foam (for temporary viscosity reduction and as blowing agent)
- transport industry, refrigeration industry (as defrosting agent, coolant)
- miscellaneous

The requirements for each application are different and are becoming more and more complex, due to technology shifts in industry and more stringent requirements regarding hazardous and environmental effects.

While it is not possible to give detailed practical advice on how to choose a solvent (blend), a general approach to the factors which can play a role can be presented. More detail will be given for surface coatings and related applications, as this is traditionally a major outlet for solvents.

In choosing a solvent, it is important to list the requirements for the application and to rank them in order of importance. In such a list the available data indicate how far the requirements can be met by a single solvent, or whether a more complex system is required.

Key properties for choosing a solvent are:

- price
- solvent power (solvency data [ASTM D 611/1133/1720], solubility parameters)
- volatility (boiling range [ASTM D 1078/86/850], vapour pressure, evaporation rate [ASTM D 3539/DIN 53170])
- viscosity (neat [ASTM D 455] and containing solute)
- hazardous properties (flashpoint [ASTM D 93/IP 170], autoignition temperature [ASTM E 659], explosion limits, corrosiveness, copper corrosion [ASTM D 130], doctor test [ASTM D 235], legal handling and transport classifications, R & S phrases)
- toxicologic properties (legal handling and transport classifications, R & S phrases, occupational exposure limits, TLV and MAC-values)
- environmental effects (legislation as regards water/air/soil pollution, waste regulations)
- water miscibility (partition coefficients for water/water-immiscible solvents)
- stability (chemical, physical under relevant conditions)
- composition (odour, colour [ASTM D 1209/156], purity [ASTM D 156], type/quantity of impurities, residue on evaporation [ASTM D 1353])
- density [ASTM D 1298]

Other properties which are often requested:

- electric properties (dielectric constant, electric conductivity [ASTM D 4308])
- refractive index [ASTM D 1218]
- surface tension [ASTM D 971]
- solidification temperature (pourpoint [ASTM D 97])
- caloric properties (specific heat [ASTM D 2766], thermal expansion coefficient, heat of evaporation, heat of combustion, thermal conductivity)
- registry number (CAS, EINECS)

Usually these properties have to be judged in relationship to a certain type of solute and/or container/packaging material or any other material with which the solvent comes in contact.

A typical example of the parameters which influence the choice of a solvent are, e.g. in the case of oilseed extraction:

- selective solvency for desired and rejection of undesired materials
- low boiler with narrow distillation range to minimise solvent loss and energy consumption
- low toxicity
- minimum of (toxic) impurities which can accumulate in the vegetable oil
- stable, non-corrosive product for repeated industrial use
- cost which allows economic vegetable oil production

The choice is high purity industrial hexane, a stable material with a boiling range 65–70°C, low benzene and lead content, low and neutral evaporation residue and low content of corrosive sulphur compounds. The same solvent is, for similar reasons, often used in polymerisation processes for the manufacture of certain polyethylene, polypropylene or rubber grades.

An example where the requirements recently have changed, because of the appearance of environmental legal constraints, is the use of trichloromonofluoromethane as blowing agent in polyurethane formulations. This is a non-flammable, very stable solvent with an attractive price, which fulfils three major technical requirements of, in particular the rigid, polyurethane industry:

- it is miscible with and lowers the viscosity of the polyol component of the two component formulation
- it is a low boiling liquid (boiling point 23.8°C) with a good compromise between handling and blowing/foaming properties
- its vapour has a low thermal conductivity and stays in the foam cells, which is very important to obtain good long term insulation properties with rigid foam

This solvent, because of its stability, has a negative influence on the stratospheric ozone layer, and in many countries legislation restricts its use. For this reason the polyurethane industry had to find other solutions, which lead to the use of other solvents such as e.g. 1,1-dichloro-1-fluoroethane or (cyclo)pentane (with boiling points of 32.1 and around 36°C respectively), which indeed fulfil the above requirements less optimally, but are more acceptable from an environmental point of view.

3.2 Solvent system design

Because a single solvent rarely meets all the requirements in certain applications, often blends have to be used and for this reason some of the approaches used in solvent system design are discussed. They were originally developed for the surface coating industry, but have a broader, more general applicability. To design a solvent system, whether this is a new formulation or the reformulation of an existing blend, there are sometimes conflicting requirements which need to be considered. The main application areas involve blends used in conjunction with oligomeric resin types or polymeric solutes. The main aspects to be considered for such applications are:

- *Solution properties* the solvent system should dissolve the solute to form a stable solution, usually with a solids content as high as possible and apart from seemingly trivial properties like colour and odour, viscosity is one of the most important characteristics.
- *Evaporation characteristics* in surface coatings the evaporation rate is of importance during various application stages. During the evaporation process the solute should not precipitate from the solvent blend, the composition of the system should remain 'balanced'. Air humidity should also be taken into consideration. After drying as little as possible solvent should remain in the solute and no negative effects should be present in the formed film.
- *Legislative constraints* the solution should fulfil the legislative requirements.
- *Price* the required properties should be obtained at an acceptable price.

3.2.1 Historic concepts

As discussed in the chapter on solvent power, most concepts are related historically to paint application, either as diluents or solvents for natural binders, synthetic resins and polymers. At first the requirements tended to be rather low, both in solvent power and sophistication. However, with the broadening of the spectrum of binders and application techniques the criteria for solvent selection became sharper and the formulations more complex.

In selecting a solvent for a certain binder, one of the oldest rules still holds: 'alike dissolves alike'. Thus a solvent with a similar basic structure as the solute and a volatility adapted to the application technique was chosen as the first approach. For instance, for a long oil alkyd a low aromatic hydrocarbon solvent and for a short oil alkyd, containing relatively more aromatic entities, a high aromatic spirit or a pure aromatic solvent were suitable starting points. For polyesters, ester type solvents came under consideration. Brush application required a medium fast evaporating type, whereas for spraying a fast evaporator was more suitable.

With, in the past, the increasing popularity of nitrocellulose based lacquers, the concept of true solvents (dissolving the binder at all concentrations), partial solvents (dissolving the binder within a limited concentration range), latent solvents (non-solvent when used alone but when added to a true solvent increasing its efficiency) and non-solvents or diluents (not dissolving the binder at any concentration but can be tolerated in true or partial solvents up to certain amounts) gained in importance.

Later the term synergistic solvent blend was added when two or more non-solvents together formed a (partial) solvent system for a binder. An example is an epoxy resin dissolved in an *n*-butanol/xylene blend. At the same time it was found that the applicational properties of a lacquer could be improved when the solvent blend consisted of more or less quickly evaporating solvents. For this purpose the solvents were grouped into fast evaporating types or low boilers (boiling range below 100°C), medium boilers (boiling range approximately 100–150°C) and high boilers or slow evaporators (boiling range above 150°C).

The ultimate blend had to be well balanced in its properties. For instance the true solvent should not evaporate before the diluent but vice versa. For a 'balanced' NC-lacquer the following guidelines were, for example, taken into consideration. The maximum possible amount of (cheap) diluent was used in order to minimise the blend price. The amount of hydrocarbon solvent (aromatic solvent, SBP 80/110 or 100/140) normally ranged from 40 up to 65% vol. of the total blend, the true solvents (ketones, esters, glycol esters) from 30–40% and latent solvents (alcohols, DAA, glycol ethers) from 10–20%. This gave the lowest viscosity to the formulation and the use of alcohol also had a price reducing effect.

The ratio low/medium/high boiler was in the order of 15/75/10, but this could change depending on the application. The high boiler should have a good solvent power in order to avoid film defects such as blushing (whitening) due to the binder becoming insoluble at the end of the solvent evaporation. Too much fast evaporating solvent should also be avoided. Due to this the surface of the lacquer film can cool down to cause water vapour condensation under humid conditions. This will also have a blushing effect on the lacquer film when the solvent blend does not contain sufficient slower evaporating polar solvent to accommodate this water. The presence of some (5% or less) water-compatibilising solvent like MIBC or preferably glycolether (e.g. isopropylglycol or MP) in the formulation is therefore in general beneficial. In the solvent system design of NC-lacquers also the advantages of using ketones instead of esters became important. The ketones resulted in lower viscosity solutions, which gave e.g. more latitude in the use of cheap diluents. Typical replacements were: MIBK/MIBC 83/17 v/v or MIBK/toluene 80/20 v/v instead of *n*-butylacetate and MEK/toluene 70/30 v/v i.s.o. ethyl acetate. Examples of the influence of solvent composition on NC-lacquer viscosity are given in section 2.4.

3.3 Systems for solvent-borne paints

Historic developments in the paint industry resulted in experience of which solvent(s) were the best to use in certain paint systems. A review of these solvent types is given in Table 3.1 in conjunction with convertible and in Table 3.2 in conjunction with non-convertible surface coating systems commonly used by present industry. The following examples provide details of the various binder types [1].

3.3.1 NC based lacquers

3.3.1.1 NC spray lacquer With the above guidelines in mind, a solvent blend can be formulated as follows:

low/medium boiling true solvent 1 : 2 e.g. MEK : MIBK
low/high boiling latent solvent 1 : 1 e.g. IPA : MIBC
true/latent solvent 3 or 4 : 1
solvent/diluent ratio 40 : 60 e.g. solvent : toluene

Resulting solvent blend:	% vol.
MEK	10
MIBK	20
IPA	5
MIBC	5
Toluene	60
Total	100

N.B. In this formulation a small quantity (5% or less) of MIBC is used both as a flow promotor and to suppress whitening.

Table 3.1 Convertible systems (drying by solvent evaporation and chemical reaction)

Type of system	Binder	Typical application	Solvent system	Comments
Air dry (all based on drying oils)	Long oil alkyd	'Oil' paints used for decoration and maintenance	LAWS SBP 100/140	Applied by brush Applied by spray
	Medium–short oil alkyds	Primers and finishes for large items of machinery. Car re-finish	HAWS Aromatic hydrocarbons, mainly xylene (the shorter the 'oil length', the higher the aromatic content required).	Applied by spray
	Urethane oil	'Oil' paints for decoration/ maintenance	LAWS	This may be used with long oil alkyds to improve hardness and chemical resisistance
	Long oil epoxy ester	Decorative/maintenance paints for use in harsh environments e.g. oil refineries, chemical works	LAWS	Brush application
Two-pack	Epoxy/polyamine or polyamide	Tank linings Marine hull paints	Ketones (glycol ethers) alcohols aromatic hydrocarbons	The appropriate combination of alcohol/aromatic can act as a true solvent. Brush applied systems are prone to retain high boiling solvents: This may affect performance especially water resistance.
	Polyurethane	Floor varnish Aircraft finish	Ketones esters EEAc Aromatic hydrocarbons	Users may insist on solvents with a very low water content e.g. 0.1% wt. max.
	Acid catalysed lacquers (Alkyd/UF or MF)	Furniture finish	Aromatic hydrocarbons Alcohol	The choice of alcohol has a marked effect on pot life. These sytems are often modified with nitro-cellulose
	Polyvinyl butyral (vinyl wash primers)	'Shop' primer for steel sheet	Lower boiling alcohols (IPA or industrial ethanol)	More of a metal pre-treatment than a paint
	Unsaturated polyester	Wood finish	—	Styrene acts as a solvent prior to taking part in the cross-linking reaction
Stoving paints	Short oil oleoresinous systems (drying type)	Cheap metal primer or finish for interior use	Aromatic hydrocarbon	These are really 'air-dry' types but may be stoved to improve both speed and degree of cure
	Short oil alkyd (drying type)	Metal primer	Aromatic hydrocarbon	
	Alkyd/MF or UF (short oil non-drying alkyd)	Car finish	Aromatic hydrocarbon Alcohol e.g. xylene/NBA 3/1	A widely used system, sometimes known as a 'stoving synthetic'
	Thermosetting acrylic			
	(a) Hydroxy acrylic/ MF	Car finish	Aromatic hydrocarbon Alcohol	Popularity largely linked with 'metallic' car finishes
	(b) Acrylamide type/ epoxy	Domestic appliance finish	Aromatic hydrocarbon Alcohol Ketone or glycol ether	
	Epoxy ester (short oil)	Car primer	Aromatic hydrocarbon Alcohol	
	Epoxy/phenolic	Car and drum linings	Ketone/glycol ether Aromatic hydrocarbon	

3.3.1.2 Slower evaporating general purpose NC/acrylic lacquer thinner

Solvent blend	% vol.
MIBK	27
EEAc	9
IPA	14
Toluene	30
Xylene	20
Total	100

N.B. In this formulation the EEAc may be replaced by MPAc, which has a higher TLV-value and a rather similar solvent power and evaporation rate.

Table 3.2 Non-convertible systems (drying solely by solvent evaporation)

Binder	Typical applications	Solvents	Diluents	Comments
Nitrocellulose (high nitrogen grades)	Car re-finishing Clear coating for wooden furniture	Ketones Esters Glycol ethers	Aromatic and aliphatic hydrocarbons	Mainly low and medium boiling solvents. Alcohols act as latent solvents.
Cellulose acetate butyrate	Clear coating for brass or aluminium As a modifying resin in e.g. acrylic lacquers	Ketones Esters	Aromatic hydrocarbons Alcohols	
Vinyl (solution types)	Aluminium foil coatings Can linings Marine paints	Ketones Esters	Aromatic hydrocarbons	Ketones show appreciable advantages over esters in respect of viscosity and maximum solids obtainable.
Vinyl (dispersion type – 'Organosols')	Finish for office equipment	Plasticiser	Aromatic and aliphatic hydrocarbons	Esentially a dispersion of PVC or a vinyl chloride copolymer in plasticiser/diluent. A high temperature is necessary to produce a continuous film. The solvent is required to reduce viscosity and assist dispersion. It must not swell the resin particles excessively or cause coagulation at room temp.
Acrylic (Thermoplastic)	Car finishing and re-finishing. Lacquers for PVC leathercloth	Ketones Esters Aromatic hydrocarbons	Aliphatic hydrocarbons Alcohols	Diluents are normally present at much lower concentrations than for example, in nitrocellulose lacquers. Thermoplastic acrylics may also be used in the form of non-aqueous dispersions, in which case the solvent blend will contain a high proportion of aliphatic hydrocarbons.
Chlorinated rubber	Marine paints	Aromatic hydrocarbons	Aliphatic hydrocarbons	Tends to 'string' on spray application unless suitably formulated. Current formulations are frequently heavily thixotroped and are applied in very thick films
Shellac	Wood sealer Modifying resins for e.g. nitrocellulose to improve adhesion	Lower alcohols – normally industrial ethanol	Not used	
Latex or emulsion paints	Decoration of plaster or masonry surfaces	Various coalescing solvents e.g. BEE BEEAc Hexylene glycol	Water	Dries by evaporation or migration of water followed by 'coalescence' of latex particles. This may be assisted by addition of small quantities of high boiling 'coalescing solvents'

3.3.1.3 Cheap blend for clear NC wood lacquer

Solvent blend	% vol.
DMK or MEK	6.5
MIBK	15.0
IPA	9.0
SBA	4.5
Xylene	65.0
Total	100.0

The resistance towards whitening or blushing due to water vapour condensation of various solvents for NC resins is expressed in the following table, which shows at which relative humidity at 26.7°C blushing occurs with a solution of 8 gr. ½ sec. NC in 100 ml solvent. The higher the humidity, the better the resistance.

Solvent blend	% relative humidity
DMK	below 20
MEK	36
MIBK	78
DAA	82
Isophorone	above 97
Ethylacetate (85%)	39
Ditto (99%)	44
NBA	82
EE	65
BE	96
EEAc	91

Solvent systems used for other paint systems are:

3.3.2 Alkyds

For the long-oil, airdrying types white spirit is the commonly used solvent type. The shorter oil types are dissolved in aromatic hydrocarbons like xylene. In stoving paint formulations, in combination with MF or UF resin, the aromatic solvents are combined with alcohols. A good thinner formulation is for instance: xylene/isobutanol 3/1 v/v.

3.3.3 Epoxy resin systems

3.3.3.1 Epoxy resin 2-component systems

For epoxy component:
True/partial solvents in order of decreasing volatility: DMK, MEK, MIBK, *n*-butylacetate, MP, EE, MPAc, EEAc, DAA, BE

Latent solvents: IPA, SBA, NBA MBC
Note: some alcohols may shorten the potlife of the 2-component system.

Diluents: toluene, xylene and higher boiling aromatic hydrocarbons such as for instance Shellsol A or AB*

For the amine component:
aromatic hydrocarbon/aliphatic alcohol blends.
Note: esters are less suitable as they may hydrolyse and react with the amine curing agent.

The solubility characteristics of two common epoxy resin types are given in Table 3.3.

Typical blends are:

- ketone/aromatic hydrocarbon 1/1 plus small quantities of glycolether or alcohol
- aromatic hydrocarbon/alcohol 2/1 to 3/1 (e.g. xylene/NBA)

Example for Epikote 1001* based spray paint:
MEK/IPA/toluene 17.3/18.8/63.9 %v.
Brush paint:
MEK/MIBK/xylene/EEAc 20.6/20.0/54.2/5.2 % v.

* Trade mark.

Table 3.3 Solubility characteristics of epoxy resins

Solvent	EPIKOTE 1001			EPIKOTE 1007		
	Approx. Viscosity (poises) 25°C		Limit of solubility[a]	Approx. Viscosity (poises) 25°C		Limit of solubility[a]
	50% wt/wt Resin Soln.	40% wt/wt Resin Soln.		50% wt/wt Resin Soln.	40% wt/wt Resin Soln.	
Ketones						
Acetone (DMK)	<0.5	<0.5	18	2.0	0.5	29
Methyl ethyl ketone (MEK)	<0.5	<0.5	CS	2.5	0.5	CS
Methyl isobutyl ketone (MIBK)	0.5	0.5	23	10.0	[b]	45
Methyl cyclohexanone	2.5	0.5	CS	60.0	15.0	CS
'Pent-Oxone'	—	—	CS	46.5[a]	6.4[c]	CS
Diacetone alcohol (DAA)	2.5	1.0	CS	98.5–148.0	25.0	CS
Esters						
Ethyl acetate	<0.5	<0.5	18	17.5	1.0	28
n-Butyl acetate	0.5	<0.5	20	13.0	2.0	37
MEAc	0.7	<0.5	CS	15.0	2.75	CS
EEAc	1.4	0.5	CS	25.0	5.0	CS
IPEAc	1.0	<0.5	CS	25.0	5.0	CS
Glycol ethers						
ME	1.0	<0.5	CS	20.0	4.0	CS
EE	1.4	0.5	CS	1–23	7.0	CS
IPE	1.25	0.5	CS	40.0	7.0	CS
BE	1.75	0.5	CS	46.0	10.0	CS

[a] Expressesd as % wt resin content. The figures represent the lowest solids content solution which is possible to prepare. Where complete solubility at all concentrations is obtained this is denoted by CS. [b] 45% is the limit of solubility. [c] At 23°C. *Note:* Although alcohols and aromatic hydrocarbons are not by themselves true solvents for 'EPIKOTE' resins, mixtures of them frequently are, especially for the lower molecular weight members. They may be used to improve the flow and film properties by providing a better solvents balance at lower raw material cost. In view of the occurrence in certain cases of limited tolerance, it is recommended that concentrated solution be first prepared and that this solution be subsequently diluted to the desired solids content.

3.3.3.2 Epoxy resin stoving systems. Higher boiling solvents are used than for 2-component, room temperature curing systems. The solvent choice also depends on the coresin used for crosslinking, e.g. PF, MF or UF resin. In general the blend contains 50% true solvent and 50% v. latent solvent/diluent. For instance EE/xylene 1/1 or EEAc/xylene 1/1.

3.3.3.3 Epoxy esters. Similar solvents are used as for alkyds, e.g. white spirit or xylene, depending on their type and amount of fatty acid, eventually with a small amount of alcohol.

3.3.4 Vinyl paints

Ketones are excellent solvents, used with up to 50% diluent (aromatic hydrocarbons); also esters like ethoxyethanol acetate are used. Vinyl resins cannot tolerate more than 10% alcohols. For stoving applications high-boilers like isophorone, DAA cyclohexanone and dibasic ester solvents are used. General examples are: MEK or EEAc/xylene 1/1.

A typical solvent blend for a vinyl rollercoating lacquer for tinplate is: isophorone/Shellsol E or N* 1/1. For paper: MEK/MIBK/toluene 40/25/35. For spray applied vinylorganosol coating: DIBK/Shellsol A*/SBP-3 10/45/45.

3.3.5 Acrylic resins

Acrylic resins are soluble in a broad range of solvents with the exception of aliphatic hydrocarbons. Usually a combination of a good and a cheap solvent is used for thermoplastic acrylics, e.g. ketone/aromatic hydrocarbon 1/1 or 2/1. For convertible stoving resins xylene/glycolether (ester) combinations are suitable. Small proportions of alcohol are used to improve solvent blend performance.

3.3.5.1 Thermoplastic acrylics

Typical examples for thermoplastic acrylate lacquers are:

DMK/EEAc/NBA/toluene/xylene/SBP-3 23/29.5/5/5/4.5/13 or DMK/toluene/xylene 20/55/25 %v.

Typical thinner formulations are:

Fast evaporating: DMK/EEAc/IPA/toluene 40/5/15/40
Slow evaporating: DMK/EEAc/IPA/xylene 25/20/15/40
or: DMK/MEK/toluene/xylene/SBP-3/NBA/EEAc 20/5/25/25/10/5/10.

Table 3.4 Solubility[a] of solid thermoplastic acrylic resins

Solvent	Methyl-methacrylate (low m.w.)	Ethyl-methacrylate (high m.w.)	n-Butyl-methacrylate
Ketones			
DMK	s	s	s
MEK	s	s	s
MIBK	s–sw	s	s
Cyclohexanone	i	s	s
Isophorone	i	s	s
DAA	s/js	s	s
Esters			
Methyl Acetate	s	s	s
Ethyl Acetate	s/i	s	s
n-Butyl Acetate	s/i	s	s
MEAc	s	s	s
EEAc	s/i	s	s
Alcohols			
MeOH	i	i	i
EtOH	i	i	i
IPA	i	i/sw	s
SBA	i	i/sw	s
NBA	i	js	s
Glycolethers			
ME	s/i	s/i	s
EE	s/js	s	s
BE	sw	s/sw	s
EEE	i	i	s
BEE	sw	s/sw	s
Hydrocarbons			
Toluene	s/sw	s/sw	s
Xylene	js/i	js/i	s
SSA	i	js/i	s
n-Hexane	i	i	s
Cyclohexane	i	i	s
SBPs	i	i	s
W.S.	i	i	s

[a] At R.T., 20% w solids. s = soluble, js = just soluble, sw = swellable, i = insoluble.

Typical solubility data of thermoplastic acrylic resins are given in Table 3.4.

3.3.5.2 *Thermosetting acrylics*

Good solvents are aromatic hydrocarbons, ketones, alcohols and glycolethers. Typical blends are:

White domestic appliance finish	
xylene	60
EEAc	20
NBA	20
Total	100

Acrylic car finish	
xylene	58
Shellsol A*	8
EEAc	30
NBA	4
Total	100

3.3.6 *Urethane lacquers (2-component)*

No hydroxy or acid group containing solvents should be used since they react with the isocyanate component, thus excluding alcohols or glycolethers. Other solvents should have a very low water and acid content.

Ketones, glycoletheresters and aromatic hydrocarbons are suitable. Normal esters (ethyl acetate, butyl acetate etc.) are less suitable (danger of hydrolysis to alcohol and acid).

Examples are: MIBK/EEAc/Shellsol A* 20/40/40 or EEAc/Shellsol A* 50/50 as brush thinners and MIBK/EEAc/xylene 35/30/35 as spray thinner.

3.3.7 *Chlorinated rubbers*

There is a broad range of suitable solvents with the exception of alcohols and aliphatic hydrocarbons. They dissolve in aromatic hydrocarbons, but to avoid 'cobwebbing' from sprayguns ketones or esters can be added. A low amount of alcohol or glycolether solvent may be used to lower the viscosity. Examples are:

xylene/NBA (95/5)	89 %v.
methoxy propanol acetate	11
Total	100

Brush application:		Spray application:	
Shellsol A*	45	Shellsol A*	40
White spirit	40	White spirit	20
EEAc	15	MEK/MIBK 1/3	40
Total	100	Total	100

3.4 Additional requirements

So far this chapter has mainly dealt with the basic solvent power, volatility and viscosity aspects of solvent systems. On many occasions, however, additional properties are required from a solvent blend. The dangers involved in the making, storage, handling and transport of solvents are for instance also of great importance, but the hazardous properties, toxicity and environmental aspects are nowadays so important that they will be the subject of separate chapters. The selection criteria according to some other aspects will be discussed below.

Surface tension is for instance a factor to be kept in mind when dealing with surface coatings. Good surface wetting is in general only obtained if the surface tension of the coating material is below that of the substrate. In former days when relatively high solvent contents were used and a high proportion, if not all, of the solvent consisted of hydrocarbon solvent with relatively low surface tension, little problem was encountered with the surface tension of the coating. Modern coating systems are often of more concern in this respect. For instance, aqueous coatings can have a high surface tension but the addition of glycoethers or alcohols can improve this situation as can be seen from Table 3.5.

Also the appropriate solvent choice in high solids paints is of importance as the (sometimes relatively high) surface tension of the binder plays an important role and as there is less solvent to compensate for this. The solvent power of the solvent must be rather high to accommodate for the high solids content at acceptable viscosity level. In such a case low surface tension aliphatic hydrocarbon solvents can in general not be used but in particular ketones give a good balance in viscosity reduction and low surface tension. A list of the surface tension and electrical conductivity of various solvent types is given in Table 3.5.

Another important property of solvents is their electric conductivity. This is for instance significant when industrial coatings are electrostatically applied. If the conductivity is too low, overspray losses increase markedly and when it is too high, bad surface coating properties due to the sticking together of the charged particles result. The optimum conductivity range, depending on the application, lies between 105 to 108 pS/m. This range can be reached either by adding chemical solvents to a hydrocarbon solvent based system or by adding a small amount of antistatic agent. From Figure 3.1 it can be seen that ketones, alcohols and glycol ethers are clearly more effective than esters in raising the conductivity to the required level. In Figure 3.2 the effect of antistatic agent ASA-3 on the conductivity of hydrocarbon solvents is shown. Also the addition of some water to water-miscible chemical solvents increases their conductivity, the most effective amount being approx. 0.5% v. (Figure 3.3).

In Figures 3.1–3.3 the conductivity is quoted as pS/m (picoSiemens/metre).

Resistance = 1/conductivity, thus ohm = 1/Siemens

Today also much attention is paid to the rheologic behaviour of wet paints and the flow during the wet stage of a drying paint just before its ultimate film formation. Typical examples are thixotropic, 'non-drip'

Table 3.5 Some properties of selected solvents

Solvent (cyclo)paraffins	Surface tension dyne/cm at 20°C	Electrical conductivity pS/m
n-Hexane	18.5	0.04
Cyclohexane	25.0	0.05
SBP 100/140	21.5	0.90
LAWS	25.0	0.20
SHELLSOL D40[a]	24.5	0.07
Aromatic hydrocarbons		
Toluene (technical grade)	28.5	10–50
Xylene	29.2	5–50
SHELLSOL A[a]	29.0	8
Alcohols		
Methanol	22.6	1.0×10^8
Isopropanol	21.4	6.0×10^6
n-Butanol	24.6	9.0×10^7
Methylisobutylcarbinol	22.8	3.0×10^5
Polyols		
Monoethyleneglycol	59	$>1.8 \times 10^7$
Monopropyleneglycol	36	—
Glycerol	71	2.8×10^6
Deionised water	71.4	1.14×10^8
Ketones		
Acetone	23.3	1.5×10^7
Methylethylketone	24.6	2.0×10^7
Methylisobutylketone	23.6	6.5×10^6
Diacetone alcohol	31	2.0×10^7
Esters		
Ethylacetate	23.9	2.0×10^5
n-Butylacetate	24.0	8.5×10^4
EEAc	28.2	4.5×10^5
BEAc	30.3	1.5×10^5
(Glycol)ethers		
Diisopropylether	18	180
Tetrahydrofuran	27	2.0×10^5
EE	28	7.5×10^7
BE	27.4	1.0×10^7
Chlorinated hydrocarbons		
Chloroform	27	1.0×10^5
Tetrachloromethane	27	—
Trichlorofluoromethane	24	—
1,1,1-Trichloroethane	25	1.6×10^5
Trichloroethylene	29.5	1.5×10^3
Tetrachloroethane	34.5	2.5×10^6
Nitrocompounds		
2-Nitropropane	30.0	—

[a] Trade marks.

paints, where the dripping of the wet paint from a brush and the sagging of the freshly applied coating are minimised, and paints for application by roller, with an improved 'spatter-resistance'. Such effects are usually achieved by formulating the rheology of the paint away from the, in general, Newtonian behaviour of the pure solvents, viz. away from the linear relationship between shear rate and shear stress towards a non-linear thixotropic behaviour.

This rheologic behaviour is, however, in most cases rather obtained by appropriate binder design in combination with the pigments, the total solids content or the addition of special agents such as thixotropic additives than by solvent choice.

The flow during the final part of the wet stage of film formation may, on the contrary, be influenced by the solvents. Very often such 'orange peel' defects or Benard-cell formation can be minimised by paying attention to solvent design. The phenomenon is caused by cooling due to rapid solvent evaporation and related rise of the surface tension of the film. The effect may be controlled by avoiding a solvent blend with entirely highly volatile solvents and not using excessively low paint viscosities or densities.

Although the flammability phenomena of solvents will be discussed separately, a few words will be said here about the (prediction of) flashpoints of solvent blends and solutions.

A simple, and in practice often used, method of estimating the flashpoint of blends is based on the concept that its flashpoint approaches that of the lower flashing component. This concept is, however, of limited

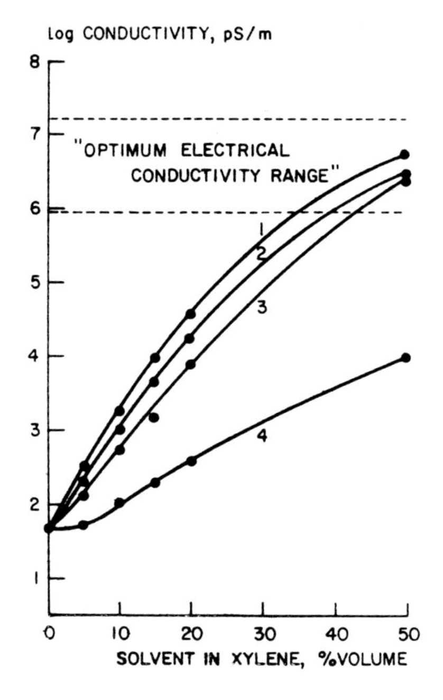

Figure 3.1 A comparison of some different chemical solvents. 1, EE; 2, methyl ethyl ketone; 3, *n*-butyl alcohol; 4, ethyl acetate.

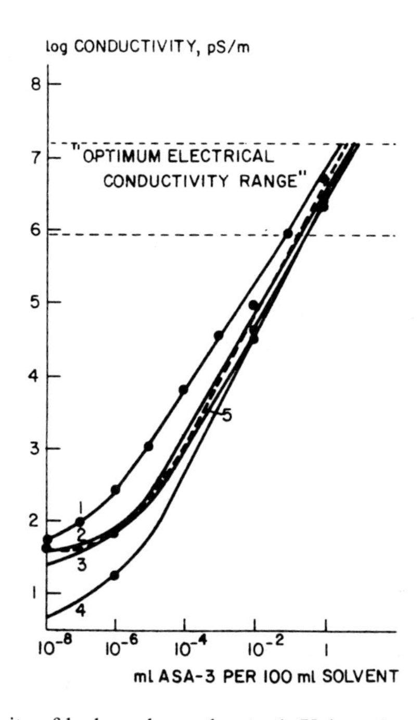

Figure 3.2 Effect of ASA-3 on the conductivity of hydrocarbon solvents. 1, Xylene; 2, octane; 3, SBP 140/165; 4, VLAWS; 5, toluene.

accuracy. A better approach is using the assumption that the flashpoint is related to the reciprocal of the vapour pressure of a blend. On the basis of Raoult's law for this vapour pressure (see the chapter on volatility), the following relationship can be developed:

$$\text{Flashpoint of a solvent blend} = \frac{F_1.V_1}{\gamma_1} + \frac{F_2.V_2}{\gamma_2} + \ldots$$

(F_x is flashpoint of pure component x, V_x is its volume fraction and γ_x is its activity coefficient)

Shell Research [2, 3] has developed a relatively easy applicable empirical method for obtaining the activity coefficients, calculated with help of graphs, an example of which is given in Figure 2.13. A more sophisticated computer-assisted technique has been described by Walsham [4] and later developments make use of the work by Gmelin and Rasmussen [5, 6] and utilise the UNIFAC group contribution method developed by Fredenslund [6, 7] for the calculation of the activity coefficients.

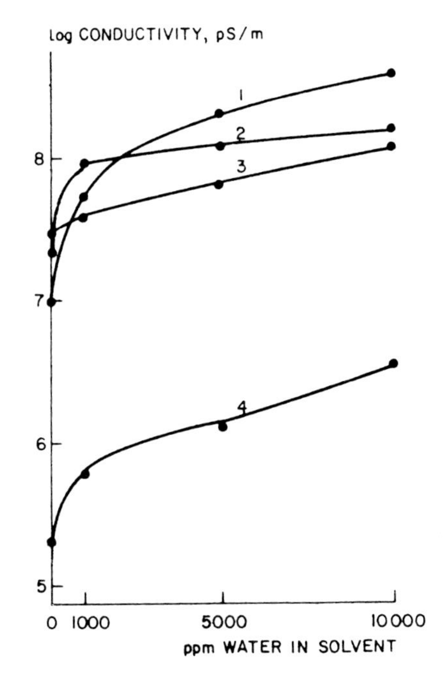

Figure 3.3 Influence of water content on conductivity of solvents. 1, Acetone; 2, ME; 3, isopropyl alcohol; 4, ethyl acetate.

The above expression predicts that the flash points of non-ideal combinations, such as hydrocarbon/alcohol mixtures, can fall below that of either component, as is indeed observed for some blends. The same effect can be observed especially with low concentrations of resin or polymer. The fact is that the solvent structure is broken and therefore the flashpoint of the blend falls. At higher concentrations they invariably rise again. The above expression also implies that the addition of solvents having an 'infinitely high' flash-point to a blend will eventually raise its flashpoint. Callaerts [8] has published data on ketones dosed with chlorinated hydrocarbons, from which it appears that no straightforward relationship can be drawn regarding the effectiveness of the individual chlorinated solvents. In this context a more general remark has to be made about these flashpoint prediction techniques. Although their accuracy has improved, practice has learned that they still have their limitations. Their outcomes are very useful as guidelines during the development of a blend, but it is important to check the actual flashpoint of the final system.

3.5 Aqueous paints

The coating industry is increasingly making use of aqueous systems alongside the more classical solvent borne paint types. Reasons for the changeover to aqueous paints are in general linked with safety (reduced flammability), health (less dangerous vapours, easier to handle) or environmental (less air/water pollution) aspects. Over the years the aqueous paints have also on price/performance become much more competitive with their solvent borne counterparts. Two major application areas for water based solvent systems can be distinguished: for emulsion (latex) paints and for so-called water soluble paints. The latter types in general do not contain truly water soluble binders but micro emulsions are formed. Their use is either by conventional brush, spray, rollercoat or dip application, or by electrodeposition in (large) paint baths.

3.5.1 Emulsion paints

In order to improve the film formation (down to approx. 5°C) of the polymer particles which are dispersed in the aqueous medium, low odour, high boiling so-called coalescence solvents are used in the paint formulation. These are solvents with only a limited water solubility, which partition mainly into the organic polymer phase and are there as a temporary plasticiser. Typical examples are BEEAc, which is very efficient in lowering the minimum temperature of film formation, but has the disadvantage of being an ester. It therefore will hydrolyse when used in a formulation of extreme pH (e.g. 8 or above). Other examples are Texanol, BEE, Dalpad A* and high boiling aromatic solvents such as Shellsol AB* etc. Before using a certain coalescence solvent in a paint formation, it should be checked whether its addition affects the dispersion/paint stability.

Apart from coalescence solvents to improve the freeze/thaw stability of emulsion paints are used. Typical examples are the glycols e.g. ethylene glycol. They partition mainly into the aqueous phase to act as an anti-freeze agent. Some types, like e.g. hexylene glycol contribute both to the freeze/thaw stability and film formation of the paint. For the open-time or wet-edge regulation (slowing down of the drying time) also glycols can be added. Propylene glycol is often used for this purpose.

The amounts of such solvents in the total emulsion paint formulation usually range from 1 to 4% wt.

3.5.2 *Conventional water-soluble paints*

For conventionally applied water-soluble paints, depending on whether they are air-drying or have to be stoved at elevated temperature, a wide range of (partly) water miscible solvents varying in volatility is used. These include low and high boiling glycolethers and alcohols. The presence of some relatively non-volatile glycol ether solvent which remains after the water has evaporated is of great help in ensuring that the binder forms a continuous uniform film.

3.5.2.1 Influence of solvent type on solids content/viscosity relationship. Usually the binder is made as a 60–75% wt solution in a water miscible solvent. A low viscosity in this stage facilitates preparation and handling of the binder. After neutralisation the binder is diluted with water to form a micellar solution which consequently is pigmented. Unlike solvent-borne binders, water soluble types peak or plateau in their viscosity upon dilution with water. Both solvent type and solids content of the original solution influence the viscosity curve obtained on solution of a certain binder (Figure 3.4). The results from various experiments indicate that solvents like NBA, BE and isopropoxy ethanol often result in the most desirable (flat) dilution curves.

3.5.2.2 Coupling efficiency of solvents. After dilution with water to a solids content which can range from 15 up to 40%, depending upon viscosity required for the specific application, the solvent present in the formulation will partition over the aqueous and the micellar organic phase. Roughly speaking, the more polar solvents will mainly be present in the aqueous, less polar solvents in the binder phase. The more homogeneous the aqueous binder solution, the less hazy it will be and the lower the possibilities for instability/phase separation. An impression about the ability of a solvent (blend) to homogenise an organic phase/water

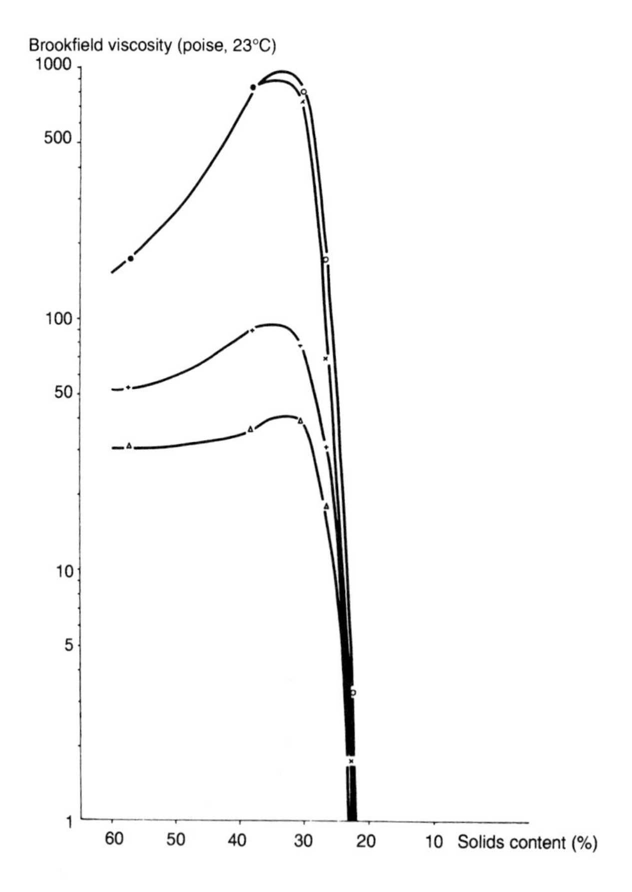

Figure 3.4 Viscosity/solids content relationship of 'Cardura' DX-28. Original Brookfield viscosity (25°C): 80% wt. in butanol = 325 poise; 80% wt. in BE = 460 poise. Cardura/DX-28 is neutralised with dimethylethanolamine. ○, Ethanol; △, butanol; ×, EE; +, BE.

mixture can be obtained by measuring its coupling efficiency for e.g. cyclohexane/water mixtures. Such coupling efficiency data are given for a series of solvent (blends) in Table 3.6. From this table it can be seen that BE, isopropoxy ethanol and EP are efficient single solvents but that also blends such as MP/NBA 1/1 m may be used.

Table 3.6 Coupling efficiency of various solvents and blends for cyclohexane/water mixtures

Solvent		Volume (ml) added at miscibility for cyclohexane/water[a]	
		50/50 m/m	10/90 m/m
Butoxyethanol	(BE)	50	34
Ethoxypropanol	(EP)	54	37
Isopropoxyethanol	(IPE)	58	38
Isopropoxypropanol	(IPP)	77	43
n-Propoxypropanol	(n-PP)	86	47
Methoxypropanol	(MP)	87	48
Butoxypropanol	(BP)	140	—
Hexoxyethanol	(HE)	105	140
n-Butyl alcohol	(NBA)	115	135
sec-Butyl alcohol	(SBA)	88	82
MP/NBA	1/1 m	50	34
MP/HE	1/1 m	58	47
EE/BE/HE	3/4/3 m	51	39
MP/BE/HE	1/1/1 m	54	39
MP/BE/NBA	1/1/1 m	50	37

[a] Combined mass: 30 grams.

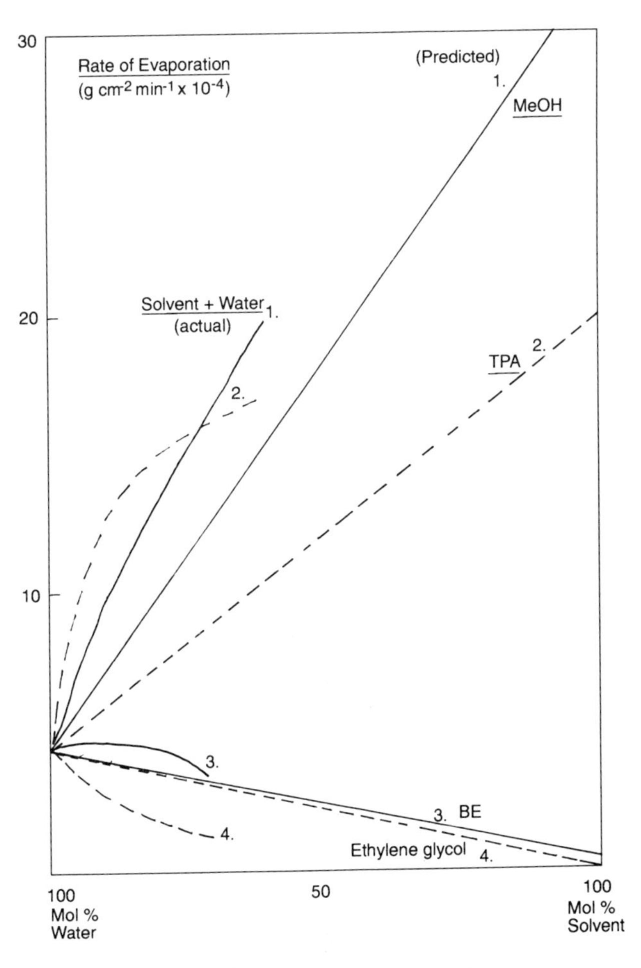

Figure 3.5 Evaporation rates of water/solvent blends.

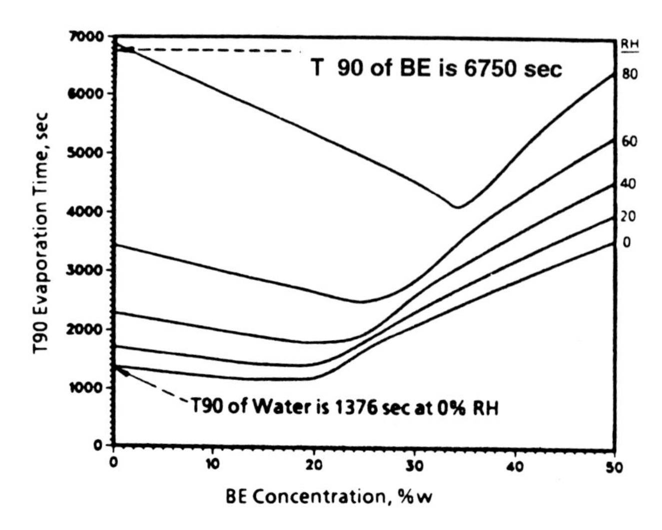

Figure 3.6 Calculated evaporation times of water/BE blends.

3.5.2.3 Influence of solvent (blend) on evaporation rate and solvent balance during evaporation. Alcohols and glycol ethers tend to evaporate faster in the presence of water than could be predicted from their single evaporation rates. An exception to this are solvents with a very strong hydrogen bonding tendency like ethylene glycol (see Figure 3.5). This is a matter which has to be accounted for in the final formulation. Therefore often combinations of a faster evaporating solvent like sec. butanol to obtain a fast initial viscosity increase of the paint (no sagging) and a slower evaporating type (e.g. BE) for the appropriate final film formation are used.

A complicating factor is the fact that the humidity of the air influences the evaporation rate of the water in the formulation (Figure 3.6) and hence the solvent balance during evaporation. A certain control of air humidity when applying this type of water-borne paints is therefore in general required [9].

In the final paint formulation, the amount of solvent ranges typically from 10–20%, the water/solvent ratio being in the order of 85/15 to 75/25. The most popular single solvent is BE but 1/1 to 1/3 blends of sec. or normal butanol or isopropanol with BE or EE may also be used. In order to improve the solvent balance during evaporation BE may be partly replaced by other glycolethers for instance EE/BE 1/1 instead of BE or SBA/EE/BEE or SBA/EE/EEE 2/1/1 instead of SBA/BE 1/1. In certain air drying applications strong solvents like glycol ethers have to be avoided as they can cause lifting of former coatings. In such cases they may (partially) be replaced by glycol/alcohol mixtures e.g. IPA/hexylene glycol/propylene glycol 1/2/1, or IPA/DAA 1/1.

3.5.3 Electrodeposition paints

One of the main differences of electrodeposition paints with conventional water soluble paints is their lower solids and thus solvent content. A typical binder content is around 10%w, the amount of solvent approximately 5%. The rest, apart from pigmentation, is water. The influence of solvent in the early stages of binder/paint formulation is very similar to the effects described for conventional aqueous paints which is also started from an approx. 70% solids binder solution in coupling solvent(s). The choice of the solvent (blend) is, however, less influenced by its evaporation characteristics as the deposited paint film does not contain much water and is stoved after application. Of more importance are paint stability and electrical properties (conductivity, rupture voltage).

Solvent selection for electrodeposition (ED) paints will therefore include coupling efficiency considerations as described above and also the partition coefficient of the solvent over the aqueous/micellar organic phase is of importance. The partition coefficient will for instance influence the amount of solvent which is deposited with the binder film and hence the final binder film formation/flow characteristics (Figure 3.7). Practice has shown that an equal distribution over water/organic phase as e.g. encountered by butoxyethanol results in a very satisfactory ED behaviour.

Also the conductivity of an aqueous binder solution is influenced by the type and amount of solvent present. This can be demonstrated in experiments with diluted aqueous potassium oleate emulsions. The oleate may be seen as a model for an anodic ED resin. Addition of solvent to pure water lowers its conductivity. In the presence of potassium oleate sometimes a decrease but also occasionally an increase in conductivity is

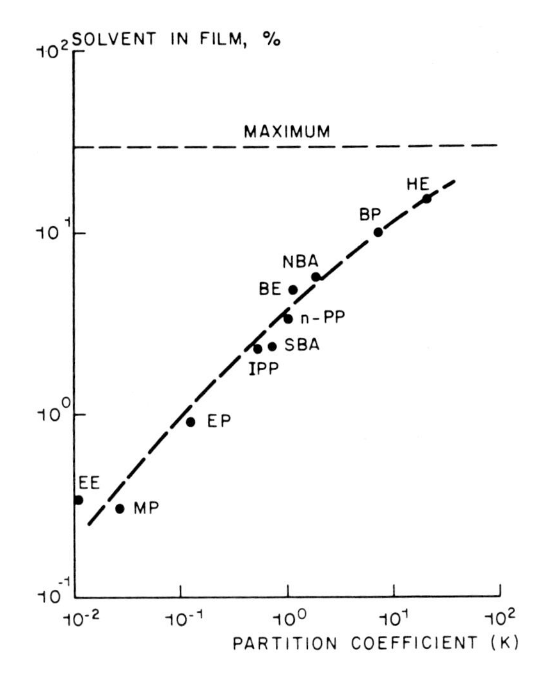

Figure 3.7 Relation between percentage solvent present in a deposited film and the partition coefficient in a cyclohexane/water blend (1:1) for glycol ethers and alcohols.

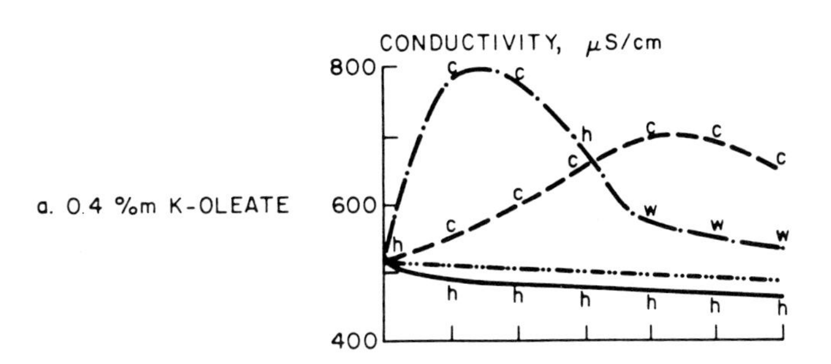

Figure 3.8 Conductivity variation on addition of various solvents to a potassium oleate emulsion in water. ———— EE; — — BE; –·–· HE; ··–··–·· Ref./(water dilution). Appearance: c = clear; h = hazy; w = white.

observed. The changes correlate well with changes in appearance and hence (micro) emulsion formation in the blends (Figure 3.8). The best performance in ED-paints is attributed to solvents which show a flat maximum in the test e.g. BE, NBA or SBA. Hexoxyethanol has a too sharp maximum and results in a hazy micellar potassium oleate solution when more than approx. 2%w is added. It obviously partitions too much into the organic phase and therefore is less suitable as a single coupling solvent but can be applied in a limited quantity as film formation prompting agent.

The effect which solvents can have on the electrical properties of an electrodeposition paint under practical application conditions is shown in Figure 3.9, where the throwing power of paints based on an anodic epoxy binder and their rupture voltage are plotted against application voltage. This figure shows again the excellent performance of butoxyethanol as both a high throwing power and rupture voltage can be obtained at a high application voltage.

From the above it will be clear that butoxyethanol is often the preferred single solvent, not only in normal aqueous but also in electrodeposition paints. However, for the latter application also other solvent systems are used, for instance blends of ethoxyethanol/butoxyethanol with hexoxyethanol and ethoxyethanolacetate as film coalescent in ratios of approx. 1/1/1. Sometimes propoxyethanol is used instead of ethoxyethanol and in some formulations 10–15% of the solvent blend is isopropanol, which is added because of its volatility and to optimise the partitioning of the solvent blend over the aqueous and organic microphase of the electrodeposition paint.

Table 3.7 Shell solvent property calculation program

Solvents	Volume percent
n-BuAc	44.00
SSA	20.00
Xylene	18.00
EEAc	14.00
BEAc	4.00
Total volume	100.00
Properties	
Viscosity, cps at 25°C	0.791
Solubility parameter	8.661
Fractional polarity	0.086
Evaporation time, sec	1655.399
Hydrogen bonding index	7.156
Cost, arbitrary units	184.600
Total aromatics, vol pct.	38.000

Table 3.8 Shell cost/performance optimisation program

Minimum cost blend solvents		Volume percent	
SSA		47.75	
MIBK		20.94	
Toluene		9.67	
MPAc		21.60	
BEEAc		0.04	
		Cost 182.69	
Property	**Value found**	**Limit set**	**Limits**
Viscosity	0.690	0.791	Less or equal
Solubility parameter	8.661	8.661	Greater or equal
Fractional polarity	0.086	0.086	Less or equal
Evaporation time	1655.001	1655.000	Equal
Hydrogen bonding	7.156	7.156	Greater or equal
Total aromatics	56.465		
Fast active	0.001	0.001	Less or equal
Slow active	21.635		

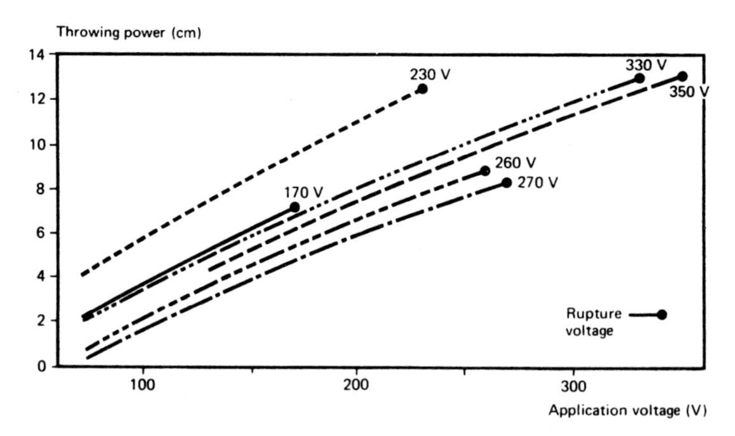

Figure 3.9 Effects of coupling solvents on throwing power. - - - - - - - - - - - - Shellsol A/(4:1); ———— diacetone alcohol; — ··· — — ··· – BE; – – – – – – – pentoxone; - - - - - - - - - - sec-butanol; — · — · — EE.

3.6 Computer assisted solvent system design

As has been discussed in the chapter dealing with solvent power, volatility and viscosity, there are various theoretical concepts which can either be used when designing a new solvent system or when reformulating or cost optimising an existing solvent blend. However, the majority of the solvent formulators has an interest in a combination of these techniques rather than in any single one. But this is time consuming and cumber-

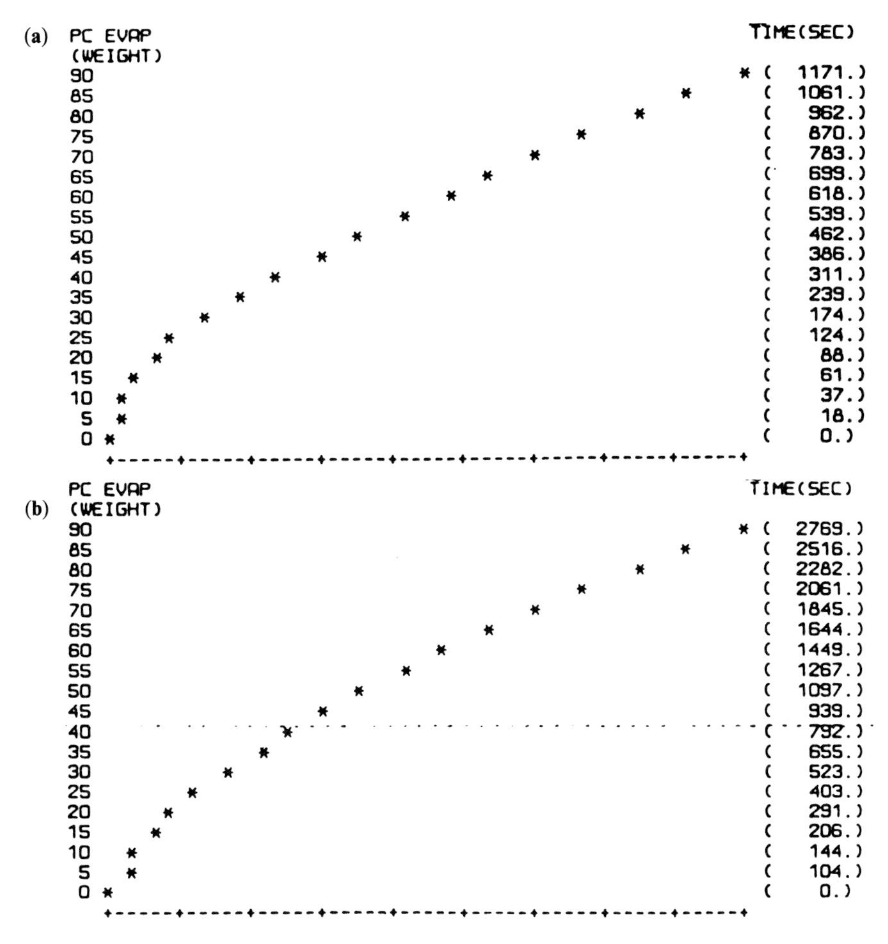

Figure 3.10 Shell solvent blend evaporation program.

Estimated properties for the following solvent blend:

Solvent blend	Vol. fraction	Wt. fraction	Mol. fraction
DAA	0.186	0.200	0.157
MEGK	0.192	0.200	0.140
EE	0.188	0.200	0.203
NBA	0.216	0.200	0.246
MEK	0.217	0.200	0.253
Total	1.000		

Blend properties: solubility parameter = 9.694; polarity = 0.242; hydrogen bonding = 2.009; viscosity (CPS) = 1.153; specific gravity = 0.875.
(a) 90% Evaporation time = 1171 sec at 25.0°C and 0.0% humidity.
(b) 90% Evaporation time = 2769 sec at 25.0°C and 67.5% humidity.
Note: This program does not take into account phase separation.

some, in particular when reiteration exercises have to be done in order to find the optimum balance between performance and price requirements. These calculations can nowadays easily be done with help of a computer as the theoretical concepts make use of, in principle simple, linear relationships. It is thus possible to arrive at a cost optimised solvent blend within the constraints one may wish to apply within the possibilities offered by a computer program.

Where such a program is available, in principle the following steps have to be taken. For the design of a completely new solvent system, the solubility 'sphere' within a three-dimensional solubility parameter system should either be known or has to be constructed for the solute in question. As a simplification, a solubility map described in the chapter on solvent power (Figure 2.10) may be used [10]. For many polymeric materials these data already exist for the Nelson, Hemwall and Edwards or the Hansen solubility parameter concepts. Alternatively as described in section 2.2 a 'sphere' or map can be constructed. Once the area of solubility is known, suitable solvent blends can be designed with solubility parameters falling within this area. When one has to choose one of the above concepts it should be noted that the idea of a 'sphere' of

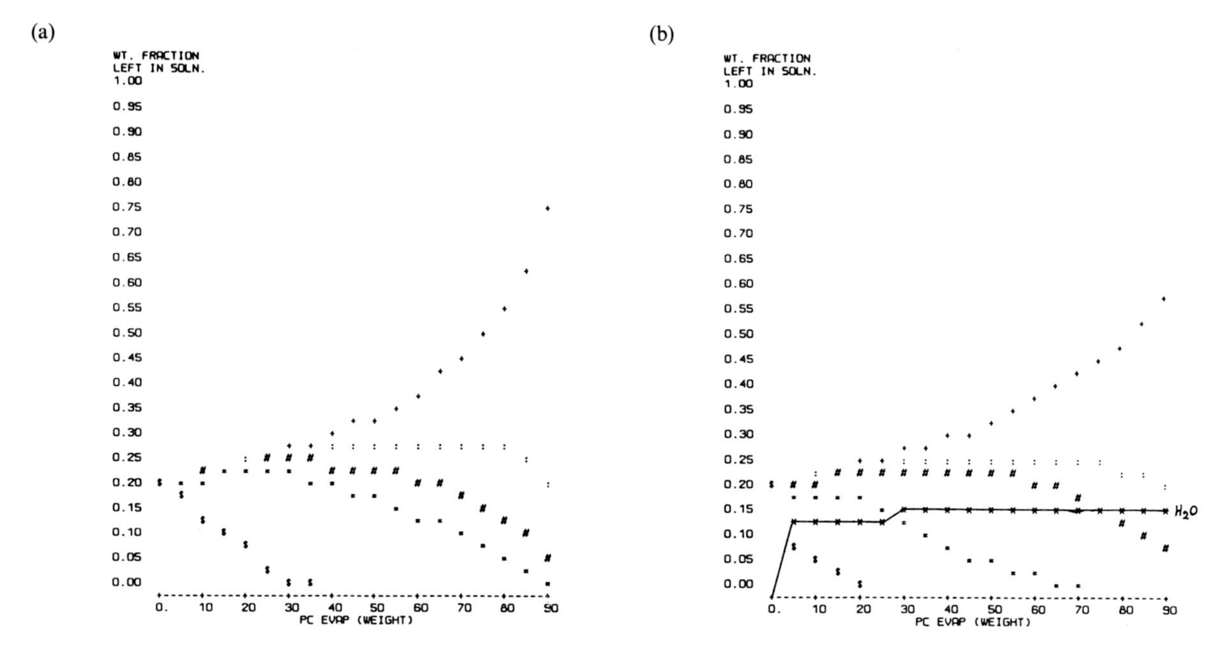

Figure 3.11 Calculation of solvent parameters. *, Water; †, DAA; :, MEGK; #, ETOX; = NBA; §, MEK.

(a) Solvent balance (wt%) of blend evaporated at 25.0°C and 0.0% humidity

Component	0	10	20	30	40	50	60	70	80	90
DAA	0.200	0.220	0.245	0.272	0.300	0.338	0.384	0.452	0.558	0.747
MEBK	0.200	0.218	0.238	0.256	0.264	0.270	0.275	0.276	0.265	0.200
EE	0.200	0.216	0.232	0.242	0.237	0.225	0.207	0.178	0.129	0.047
NBA	0.200	0.212	0.222	0.221	0.199	0.169	0.134	0.094	0.048	0.006
MEK	0.200	0.133	0.063	0.010	0.000	0.000	0.000	0.000	0.000	0.000

(b) Solvent balance (wt%) of blend evaporated at 25.0°C and 67.5% humidity.

Component	0	10	20	30	40	50	60	70	80	90
Water	0.000	0.132	0.135	0.138	0.143	0.146	0.149	0.149	0.150	0.160
DAA	0.200	0.218	0.240	0.265	0.293	0.327	0.367	0.417	0.483	0.578
MEBK	0.200	0.214	0.230	0.242	0.253	0.260	0.261	0.253	0.233	0.188
EE	0.200	0.210	0.223	0.229	0.230	0.225	0.208	0.178	0.134	0.073
NBA	0.200	0.187	0.168	0.126	0.082	0.042	0.015	0.003	0.000	0.000
MEK	0.200	0.039	0.004	0.000	0.000	0.000	0.000	0.000	0.000	0.000

solubility has a theoretical appeal, but that it has proved to be a little over-simplified for certain predictions. The more pragmatic approach of Nelson, Hemwall and Edwards tends to give quantitatively better acceptable results.

With these solubility parameters in hand, further optimisation can be achieved with help of the theories on viscosity, evaporation rate and flashpoint of solvent blends.

For viscosity, for instance, the approach of Rocklin and Edwards [11] with effective instead of neat viscosities in the Kendall equation is a very useful technique.

For the calculation of evaporation rates the model of Walsham and Edwards [12] who use activity coefficients to calculate corrected relative evaporation rates is a good starting point. Nowadays these activity coefficients are, however, calculated with the UNIFAC group contribution method [6, 7] and the evaporation rates take into account evaporative cooling of the blend [13].

In order to follow the changes in solvent composition during the evaporation process, the process is considered to take place stepwise. As a first step the quantity of the initial amounts of blend components evaporated is calculated. This leads to a different blend composition after a small amount (e.g. 5%) evaporation. The changed composition has to be recalculated and the whole process has to be repeated in further 5% steps until complete evaporation. Such a calculation process leads to a result as given in Figure 2.14.

The flashpoints can also be calculated using the UNIFAC group contribution method as described earlier in this chapter. Adequate computer programs for performing such calculations are already available and are rapidly gaining acceptance in the industry.

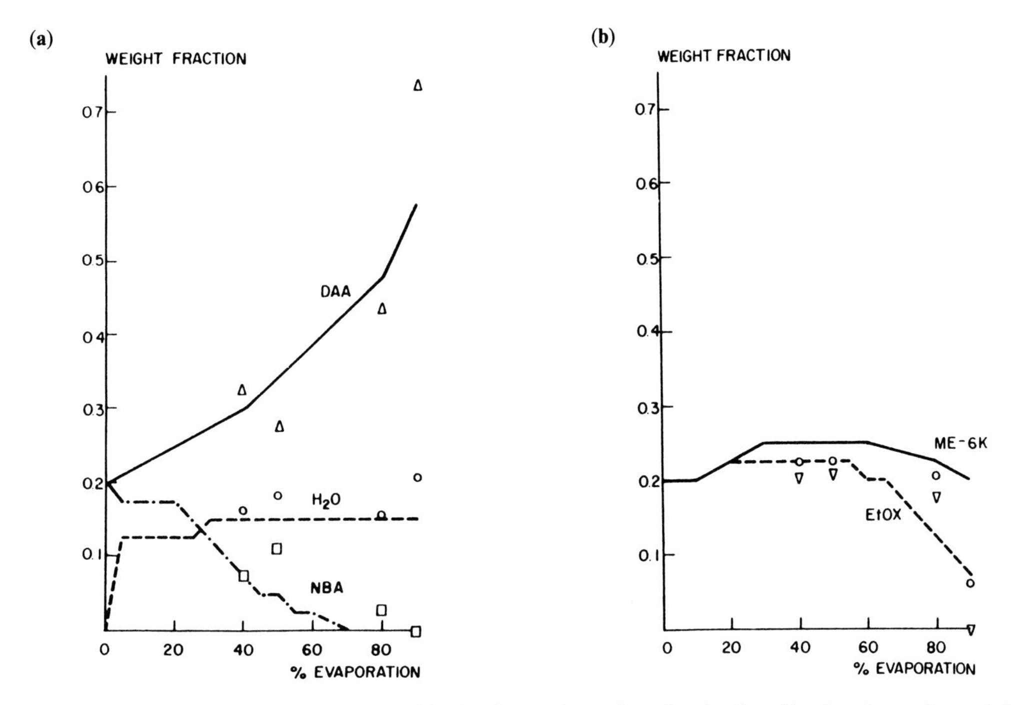

Figure 3.12 Solvent blend evaporation behaviour at 67.5% RH. Comparison of predicted values (lines) and experimental data (points). (a) △—DAA; ○- - -H_2O; □·-·-·NBA. (b) ○ — ME – 6K; ▽ - - - - EE.

A typical example of the possibilities of such calculations, using the Shell Solvents Computer Program [14] is as follows:

Cost optimisation: Within a number of constraints solvent blends can be cost optimised. The following constraints can be applied: solubility parameters, viscosity, evaporation time, aromatic hydrocarbon solvent content, low/high volatile true solvent content.

Evaporation characteristics: Actual evaporation time and solvent balance during evaporation of organic solvent and water based blends can be predicted under different humidity conditions. The most important parameters (see above) can be calculated and water pick-up by blends of oxygenated, polar solvents during evaporation predicted.

Aqueous viscosities: The viscosity of aqueous solvent blends with a minimum of 50%v water content can be predicted according to the guidelines as discussed in the chapter on viscosity.

Flashpoint predictions: The flashpoint of organic and also water-containing solvent blends can be calculated.

An example of the calculation of the properties of an existing blend using the Nelson–Hemwall–Edwards solubility parameter concept and the results of a consequent cost optimisation exercise with 14 different solvents and with constraints (set limits) as obtained from the starting point blend are given in Tables 3.7 and 3.8.

When using the evaporation program in relation to a solvent blend, one can for instance obtain information about the rate of evaporation under dry (e.g. 0.0% humidity) and humid (e.g. 67% humidity) conditions (Figure 3.10). Furthermore, the change in blend composition during evaporation under these humidity conditions is calculated (Figure 3.11). Note that under humid conditions a certain water pick-up is predicted during the evaporation. This is a phenomenon which – depending on the solvents used in the blend – actually takes place in practice both with pure solvents and in the presence of a binder (Figures 3.12 and 3.13). Such a water pick-up can for instance cause blushing effects in a drying nitrocellulose lacquer, as was discussed earlier in this chapter. Furthermore, the change in solvent power during the evaporation can be calculated. This is of importance to see whether the blend maintains the right balance during evaporation. If the solvent power of the blend for the solute decreases during the drying process, there would develop the possibility of preliminary precipitation of the binder. When this occurs film quality can also be negatively influenced (blushing, film integrity defects etc.). An example of the calculation of the flashpoint of a solvent blend is given in Figure 3.14.

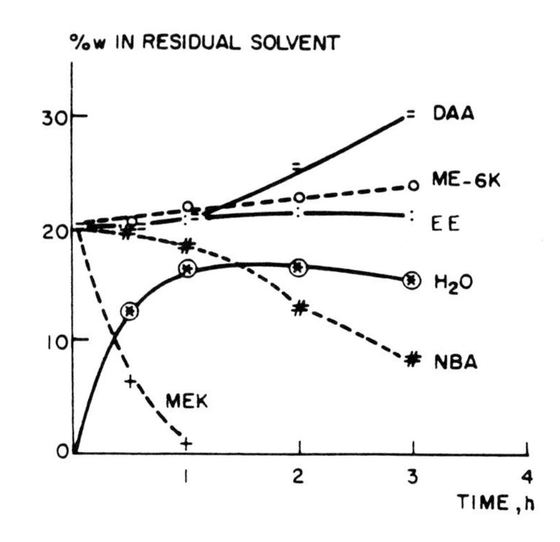

Figure 3.13 Evaporation behaviour in the presence of a binder. Solvent blend contains 20% epoxy resin. T = 23°C; RH = 67.5%.

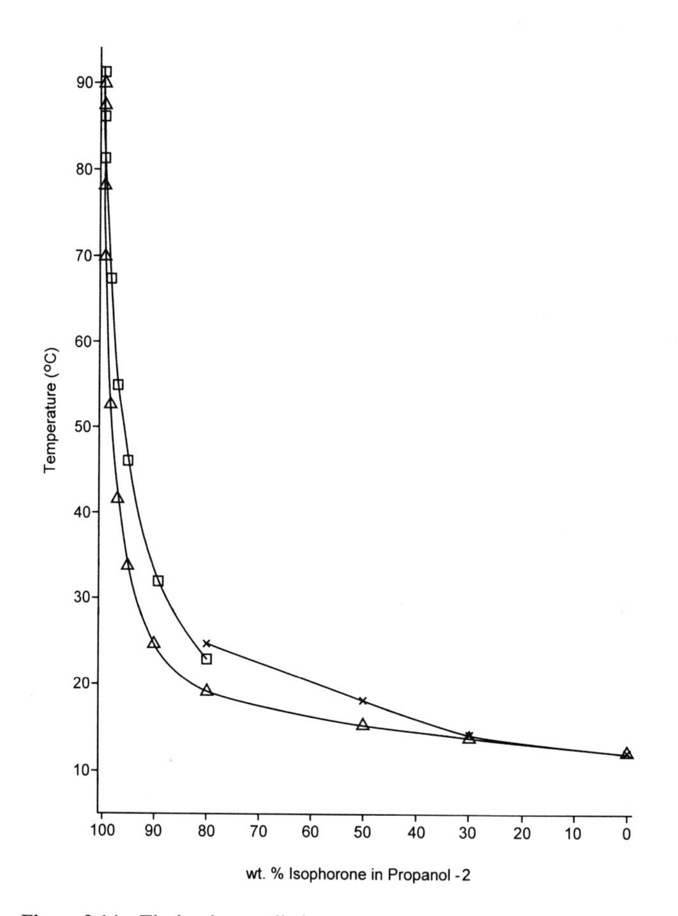

Figure 3.14 Flashpoint predictions. □, D93; ×, IPI70; △, calculated.

References

1. *Shell Chemicals Industrial Chemicals Technical Manual, Solvents*, Thinner formulations for the paint industry, IS 5.2.1.
2. *Shell Chemicals Industrial Chemicals Technical Manual, Solvents*, Predicting the flashpoint of solvent mixtures, IS 4.3.2.
3. Ramsbotham, *Progress in Organic Coatings*, **8**, 113 (1980).
4. Walsham, in *Solvents Theory and Practice* (R.W. Tess, ed.), ACS, Advances in Chemistry Series – 124, p. 56 (1973).
5. Gmelin and Rasmussen, *I. & E.C. Fundamentals*, **21**, 186 (1982).
6. Fredenslund, Gmelin and Rasmussen, *Vapor–Liquid Equilibria using UNIFAC*, Elsevier, Amsterdam (1977).
7. Fredenslund, Jones and Prausnitz, *AICHE J.*, **21**, 1086 (1975).
8. Callaerts, *J. Fire Flammability*, **9**, 229 (1978).
9. Rocklin, *J. Coatings Technol.*, **58**(732), 61 (1986).
10. Nelson, Figurelli, Walsham and Edwards, *J. Paint Technol.*, **42**(550), 644 (1970).
11. Rocklin and Edwards, *J. Coatings Technol.*, **48**, 68 (1976).
12. Walsham and Edwards, *J. Paint Technol.*, **43**, 64 (1971).
13. Rocklin and Bonner, *J. Coatings Technol.*, **52**(670), 27 (1980).
14. *Shell Chemicals Industrial Chemicals Technical Manual, Solvents*, The Shell Solvents Computer Program, IS 4.4.1.

Abbreviations

BE(Ac)	=	butoxy ethanol (or its acetate)
BEE(Ac)	=	butoxy ethoxy ethanol (or its acetate)
*n*BuAc	=	*n*-butylacetate
DAA	=	diacetonealcohol
DMK	=	dimethylketone (acetone)
EE(Ac)	=	ethoxy ethanol (or its acetate)
EEE(Ac)	=	ethoxy ethoxy ethanol (or its acetate)
EP(Ac)	=	ethoxy propanol (or its acetate)
EtAc	=	ethylacetate
EtOH	=	ethylalcohol (ethanol)
HAWS	=	high aromatic white spirit
HE	=	hexoxy ethanol
IPA	=	isopropylalcohol (isopropanol)
IPE(Ac)	=	isopropoxy ethanol (or its acetate)
LAWS	=	W.S. = low aromatic white spirit
Me(Ac)	=	methoxyethanol (or its acetate)
MEBK	=	pentoxone*
MEK	=	methylethylketone
MeOH	=	methylalcohol (methanol)
MIBC	=	methylisobutylcarbinol
MIBK	=	methylisobutylketone
MP(Ac)	=	methoxy propanol (or its acetate)
NBA	=	*n*-butylalcohol (*n*-butanol)
SBA	=	sec-butylalcohol (sec-butanol)
SBP	=	special boiling range product (aliphatic hydrocarbon)
SSA	=	Shellsol A*
VLAWS	=	very low aromatic white spirit

4 Solvent flammability and reactivity hazards

R.L. ROGERS

4.1 Introduction

The majority of solvents are combustible, that is, under certain conditions sufficient vapour is produced which, when mixed with air or another oxidant and ignited, will burn with a flame. An explosion with devastating consequences can also result particularly if this process occurs in a confined space. The hazards associated with the flammability of solvents are demonstrated by the many incidents that have occurred; these continue to cause death, injury and financial loss. Many incidents arise during relatively simple operations; for example, two operators sustained 2nd and 3rd degree burns of the face and body following a fire breaking out during the addition of resin powder to the solvent toluene [1]. The annual financial losses owing to fires and explosion are substantial. It was estimated in 1979 that, in the USA, property losses for explosions were over $150 million per year with a similar additional loss due to business interruption [2].

The hazards associated with all aspects of handling solvents including their transport, storage and use therefore must be carefully evaluated. In order to ensure the safe operation of the wide variety of plant and processes where solvents are used, due consideration must be given to the potential fire and explosion hazards that can arise and the procedures which need to be implemented to avoid these hazards.

This chapter describes the different flammability characteristics of solvents, considers the range of ignition sources, and explains the various options that are available for safe operation. In addition, the hazards that can result from chemical reactivity of solvents are also introduced. Finally, an overview is given of some of the main legislation and regulations that apply to the use of solvents.

4.2 Flammability

Flammability is the term used to describe the ability of a substance to burn or combust with a flame. Combustion is a usually rapid chemical process, described as an oxidation reaction between a fuel and an oxidiser which is accompanied by the evolution of heat and light. For this combustion reaction to take place, the presence of both fuel and oxidant are required together with an ignition source of sufficient energy to initiate the reaction. The well known ‘fire triangle’ (Figure 4.1) provides a useful means of illustrating this requirement that all three components must be present for combustion to occur. If any one component is removed and the triangle is broken, the combustion reaction will not take place.

Common examples of the three components of the fire triangle are wood, air and a match (as in a domestic fire), and petrol or gasoline, air and a spark (as in a car engine). Although in these examples the fuel element is either a solid or a liquid, the combustion reaction between the fuel and oxidant, with few exceptions, takes place in the vapour phase. Thus to start the reaction, i.e. for ignition to occur, enough energy must be supplied both to vaporise the fuel and initiate the reaction. For burning to continue, the combustion reaction must produce sufficient heat to ensure the propagation of the reaction and, in the case of solid or liquid fuels, to allow the continuation of the vaporisation process. The amount of heat produced depends on the nature of both the fuel and the oxidant and also on their respective concentrations in the mixture that burns. This mixture of fuel vapour and oxidant is called a flammable atmosphere and may be defined as one through which flame will be propagated, away from the influence of the ignition source. If the mixture contains too little fuel or too little oxidant then insufficient heat will be liberated by the combustion reaction and the burning will stop. There is therefore a concentration range over which fuel vapour is flammable, with a lower limit where there is too little fuel and an upper limit where there is too little oxidant.

The oxidant is often the oxygen in air; however, it should not be forgotten that other oxidisers such as the gases chlorine or nitrous oxide and even solids such as ammonium nitrate, as used in ammonium nitrate/fuel-oil explosives, can constitute the oxidant element of the fire triangle and thus allow the combustion reaction to occur.

In unconfined situations, combustion of the flammable mixture generally results in a fire. Under confined conditions the heat from the combustion reaction and also the production of gaseous reaction products can cause significant increases in pressure. This can occur in a matter of milliseconds and lead ultimately to an explosion. The severity of the explosion is influenced by a number of factors including the initial pressure, the mixture composition and the degree of confinement.

The flammability of a solvent is characterised therefore by: the conditions under which the solvent can be

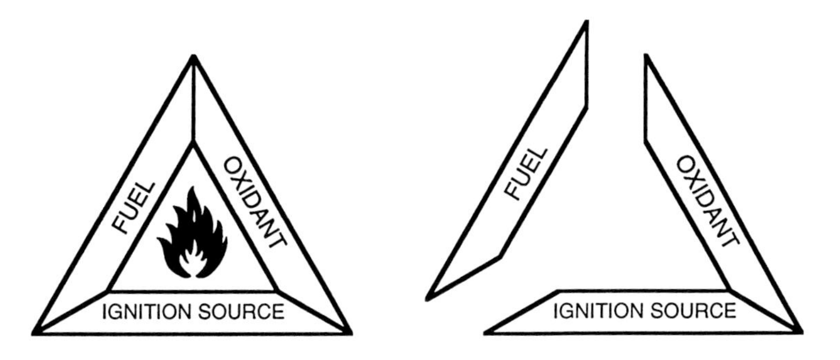

Figure 4.1 The fire triangle.

ignited and continue to burn, known respectively as the flash point and fire point; the concentration range over which the solvent vapour/oxidant mixture is flammable, i.e. the lower and upper flammability limits, which vary depending on the oxidant; and the minimum oxygen concentration required for the combustion reaction. In addition, the temperature above which a flammable mixture is capable of self igniting (the auto-ignition temperature) is also used as a flammability parameter.

4.2.1 Flash point

The flash point of a liquid is an important and practical flammability characteristic and is one of the main physical properties used to determine fire and explosion hazard. It is defined as the minimum temperature at which the liquid gives off sufficient vapour to form an ignitable mixture with air near the surface of the liquid or within a vessel. A fire may not necessarily develop at the flash point of a liquid. Instead, this will occur at the 'fire point', which is the lowest temperature at which the vapour above a liquid will continue to burn. The fire point is usually found a few degrees above the flash point.

A number of standard test methods exist to measure the flash point. All involve slowly heating the liquid in contact with air and, at intervals, applying an ignition source above its surface – the lowest temperature at which the vapour ignites is designated the flash point. There are essentially two techniques, namely the closed cup technique, for example the Pensky–Martin Closed Tester [3], illustrated in Figure 4.2 and the open cup technique, for example the Cleveland Open Cup Tester [4] shown in Figure 4.3. The fire point can be measured by continuing testing beyond the flash point in the Cleveland Open Cup Tester.

The closed cup flash point is the temperature at which the equilibrium concentration of vapour above the liquid is equal to the lower flammability limit. This is illustrated in Figure 4.4, which shows the relationship between the vapour pressure of a solvent, its temperature, the concentration range over which a mixture of its vapour in air is flammable and the flash point. In the open cup method, air convection above the liquid surface dilutes the fuel concentration with the net result that slightly higher temperatures are needed to reach the same fuel concentration in air as in the closed cup method. Consequently open cup flash points are usually a few degrees higher than those measured using closed cup techniques. This is shown in Table 4.1 which lists flash point values measured using the two techniques described above for some of the commonly encountered solvents. It can be seen that there is no simple relationship between the flash points determined using the different techniques.

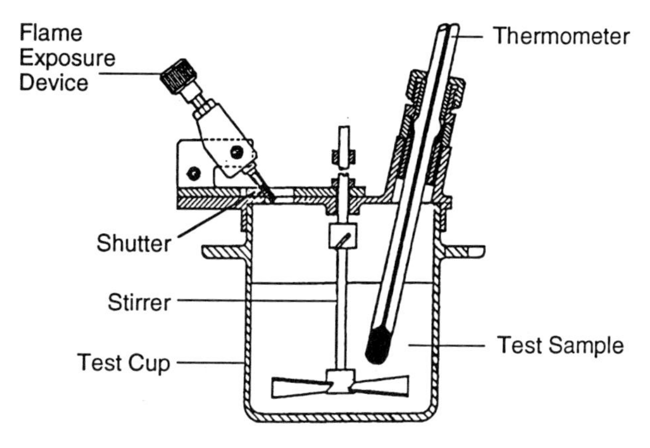

Figure 4.2 Pensky–Martin Closed Tester for measuring flash points.

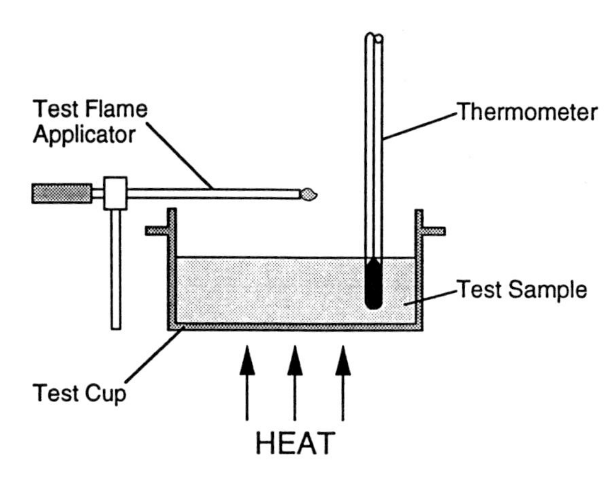

Figure 4.3 Cleveland Open Cup Tester for measuring flash points and fire points.

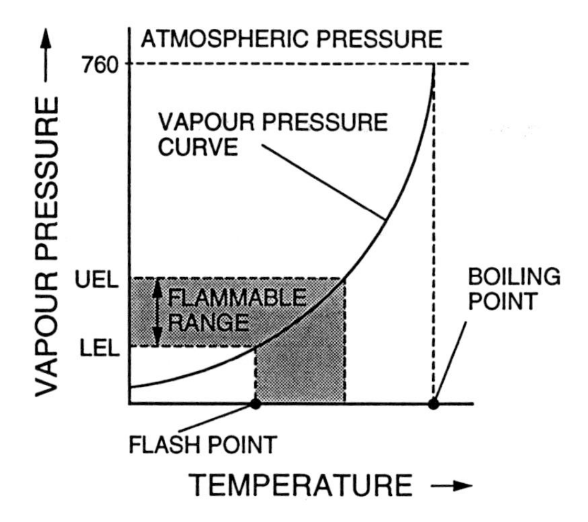

Figure 4.4 Relationship between pressure, temperature, flammability range and flash point.

Table 4.1 Closed cup and open cup flash point values for typical solvents

	Flash point (°C)	
	Closed cup	Open cup
Acetanilide	174	174
n-Amyl acetate	25	27
Butyl acetate	23	32
Ethyl benzene	15	24
2-Naphthol	152	171
n-Propanol	15	29
Toluene	4	7
o-Xylene	17	24

When using literature values it is important to identify the method used in obtaining the data. In general, closed cup values are quoted in the UK and Europe whereas both open cup and closed cup values are used in the USA.

The flash point is key for determining whether it is possible for a flammable vapour/air atmosphere to be present. It is the main parameter used in the classification of liquids for storage, transport, etc. If a liquid is below its flash point, a flammable vapour atmosphere will not be formed. However, if it is possible for a fine-droplet mist to be present then this may give rise to a flammable atmosphere. Fine-droplet mists are known to be flammable at temperatures lower than 100 K below the flash point. Such mists may be formed by mechanical means or during cooling. Similarly, foams may be flammable below the flash point temperature.

Factors affecting the flash point include atmospheric pressure and the composition of the liquid. As the pressure decreases, the flash point level declines and vice versa. This can be illustrated by considering the flash point of toluene as measured in the UK at sea level and in the mountains of Colorado.

UK (sea level)	Atmospheric pressure 760 mmHg	$FPt = 4.5°C$
Denver (5530 ft)	Atmospheric pressure 627 mmHg	$FPt = 1.0°C$

Flash point measurements can be corrected for atmospheric pressure using the following formula:

$$\Delta FPt = 0.033 \times (760 - P) \tag{4.1}$$

where ΔFPt is the difference in flash point between that measured at a pressure P (in mmHg) and that at a pressure of 760 mmHg.

Flash point values quoted in the literature, in general, are for pure solvents. The presence of contaminants can have a significant effect on the flash point, particularly if the contaminant is relatively more volatile. For example, pure ethylene glycol has a flash point of 111°C. However, the flash point is reduced to 29°C when acetaldehyde at a level of only 2% is present. Similarly, when handling solvent mixtures, a small change in the composition can have a significant effect on the flash point. Thus, whenever possible, flash points should be measured when it is known or suspected that more than one component will be present.

Adding water to a solvent can raise the flash point provided the solvent is miscible with water. If sufficient water is added the aqueous liquor can be rendered non-flammable (i.e. no flash point is detected below the boiling point of the mixture). This is illustrated in Figure 4.5 for ethanol. For solvents which are immiscible with water, the flash point will be largely unaffected by the addition of water.

It is possible to estimate the flash point of a multicomponent mixture provided only one of the components is flammable and its flash point and vapour pressure/temperature relationship is known [5]. The flash point temperature of the mixture is estimated by determining the temperature of the mixture at which the vapour pressure of the flammable component in the mixture, obtained using Raoult's Law, is equal to its vapour pressure at its flash point. Thus to determine the flash point of a mixture of 75% methanol and 25% water by weight, it is necessary to know the flash point of methanol (11°C) and that its vapour at this temperature is 53 mmHg. The mole fraction, a, of the flammable component in the mixture is needed (in this case the mole fraction of methanol is 0.63) in order to apply Raoult's Law. This is used to calculate the vapour pressure (P_{sat}) of pure methanol based on the partial pressure (p) required at the flash point:

$$p = a \times P_{sat}$$

$$P_{sat} = p/a = 53/0.63 = 84.1 \text{ mmHg} \tag{4.2}$$

The vapour pressure/temperature relationship for methanol is then used to determine the temperature that would result in this pressure. This temperature, in this case 17°C, is the flash point of the mixture.

Several attempts have been made to develop a general correlation for the estimation of closed cup flash points of organic compounds from their normal boiling points [6, 7, 8] using equations involving parabolic, hyperbolic and exponential functions. An equation containing an exponential function was used with the form:

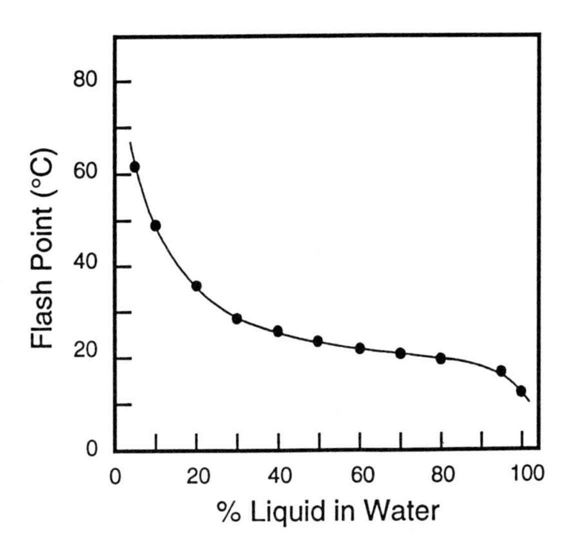

Figure 4.5 Variation of the flash point of ethanol/water mixtures with water content.

$$T_f = a + b(c/T)e^{-c/T}/(1 \times e^{-c/T})^2 \tag{4.3}$$

where T_f denotes the flash point temperature in K; T the normal boiling temperature in K; and a, b, c are constants to fit the flash points of 1200 organic compounds and 21 petroleum fractions as a function of their normal boiling temperature. An absolute average error of less than 1% was obtained between literature and predicted values of the flash point [8] when the compounds were divided into thirteen different chemical classes each with separate values for the constants a, b, and c.

4.2.2 *Flammable range*

Solvent vapour/air mixtures are flammable only over a limited concentration range, as illustrated in Figure 4.4. The boundaries are termed the Lower Flammable, or Explosible, Limit (LEL) and the Upper Flammable, or Explosible, Limit (UEL) with the words 'flammable' and 'explosible' being used interchangeably. These, in turn, are defined as the minimum and maximum fuel concentrations through which a flame will just propagate and are usually expressed as volume percent in air at atmospheric temperature and pressure. Typical values are given in Table 4.2.

Below the lower limit the mixture is too lean, or fuel deficient, to support combustion. Changes in the oxygen content of the air have little effect on the lower limit. At the upper limit the atmosphere becomes over rich in fuel and deficient in oxygen. The upper limit, consequently, is very dependent on the oxygen concentration in air, and is increased by an increase in oxygen concentration.

When handling a solvent mixture, data are often available for the individual components but not for the mixture itself. Application of Le Chatelier's Law enables estimation of the limits for the mixture provided the limits of the individual components are known. If P_n is the volume percentage of the combustible component n, with a lower flammable limit of L_n, then the lower limit, L_{mix}, of the mixture is:

$$L_{mix} = \frac{P_1 + P_2 + P_3 + \ldots + P_n}{P_1/L_1 + P_2/L_2 + P_3/L_3 + \ldots + P_n/L_n} \tag{4.4}$$

Similarly, the upper flammable limit, U_{mix}, for the mixture is given by:

$$U_{mix} = \frac{P_1 + P_2 + P_3 + \ldots + P_n}{P_1/U_1 + P_2/U_2 + P_3/U_3 + \ldots + P_n/U_n} \tag{4.5}$$

It is important to note that a mixture can be flammable even though the concentration of each constituent is below its lower limit.

Flammability limits are dependent on both temperature and pressure. Values quoted in the literature are generally for normal atmospheric temperature and pressure with little information available for elevated conditions.

An increase in temperature causes the lower limit to decrease and the upper limit to increase, the net result of which is an overall widening of the flammability range. For example, for ethanol, the lower limit was found to decrease from 3.4 vol% at 50°C to 3.0 vol% at 250°C, while the upper limit increased from 15.6 vol% to 22.5 vol% over the same temperature range [9]. Both the upper and the lower limits have been found to show a linear dependence on temperature. Based on this, a number of models have been developed to predict the effect of temperature on the limits, with the modified Burgess–Wheeler model giving closest agreement with measured values [9].

Using this model, the change in the lower limit is given by:

Table 4.2 Lower and upper flammability limits for typical solvents

	Vol% in air	
Compound	LEL	UEL
Acetone	2.15	13
Benzene	1.3	7.1
Cyclohexane	1.3	8
Ethanol	3.3	19
Ethyl ether	1.9	36
Heptane	1.05	6.7
Hexane	1.1	7.5
Methanol	6.7	36
Toluene	1.2	7.1

$$L_{t2} = L_{t1} - \frac{\mu(t_2 - t_1)}{\Delta H_c} \quad (4.6)$$

where L_{t1} = lower flammability limit at temperature $t1$ (% v/v)
L_{t2} = lower flammability limit at temperature $t2$ (% v/v)
t = temperature (°C)
μ = constant (5.35)
ΔHc = heat of combustion (kJ/mol)

The behaviour of the upper limit can be similarly described:

$$U_{t2} = U_{t1} + \frac{\mu(t_2 - t_1)}{\Delta H_c} \quad (4.7)$$

where U_{t1} = upper flammability limit at temperature $t1$ (% v/v)
U_{t2} = upper flammability limit at temperature $t2$ (% v/v)
t = temperature (°C)
μ = constant (39.5)
ΔHc = heat of combustion (kJ/mol)

Pressure apparently has an effect similar to that of temperature on flammability, with the effect being more marked on the upper limit than on the lower. Considerably less work has been carried out in studying the effect of pressure on the limits, with some inconsistencies apparent in the data being reported. Consequently, simple models have not been developed to enable the effect of pressure to be predicted with sufficient accuracy.

4.2.3 *Autoignition temperature*

The autoignition temperature (AIT) is the lowest temperature at which a gas or vapour spontaneously ignites owing to the heat from the environment or from contact with a heated surface without the need for any additional spark or flame. It is dependent on many factors including:

- ignition delay;
- vapour concentration (lean or rich mixtures give higher AITs);
- size of vessel (larger volume, lower AIT);
- catalytic material (surface coatings);
- pressure (increased pressure, decreased AIT);
- oxygen content (increased oxygen concentration, decreased AIT).

The standard test method for measuring AIT is illustrated in Figure 4.6. In the test, the temperature of the oven is varied and small quantities (μl) of the liquid under test are introduced into the heated flask with a syringe while observing whether ignition occurs. The autoignition temperature is the temperature at which ignition just fails to occur. The mixture and the surfaces of the vessel are all effectively at the same temperature. If a gas mixture was exposed to a hot surface, rather than being enclosed in a hot environment, only a portion of the mixture would come in contact with the surface and a higher AIT would result. AIT values of some of the more commonly encountered solvents are given in Table 4.3.

A special feature of the autoignition of many organic solvent vapours, including those from hydrocarbons, alcohols, ethers, aldehydes and acids is their ability to form 'cool flames' at temperatures well below their autoignition temperature as measured in the standard apparatus [10]. A cool flame is a form of incomplete combustion, usually involving the formation of an unstable peroxide and its decomposition to an aldehyde. Cool flames emit a pale blue light which is visible only in the dark and on their own produce a relatively modest (*ca.* 20–50 K) temperature rise, hence their name. The main hazard with cool flames lies in their potential for transition into true combustion, and that their products are often less stable and more reactive than the original compound.

4.2.4 *Minimum oxygen concentration and inerting*

The combustion reaction can be prevented by depletion of the oxidant, one of the three essential components making up the fire triangle. This forms the basis of inert gas blanketing in which the oxygen is replaced by an inert gas such as nitrogen, carbon dioxide or a halogenated hydrocarbon. By adequate depletion of oxygen, the atmosphere will be rendered non-flammable irrespective of the concentration of combustible material.

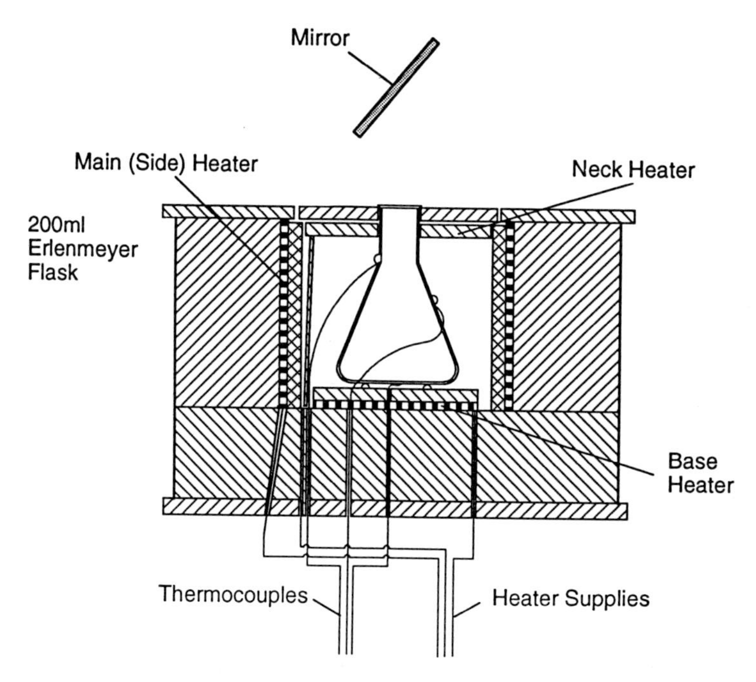

Figure 4.6 Autoignition temperature apparatus.

Table 4.3 Typical values for autoignition temperature

Compounds	AIT (°C)
Acetone	535
Acetic anhydride	334
Carbon disulphide	102
Diethylamine	312
Ethanol	363
Hexane	223
Pentane	258
Toluene	480

The minimum oxygen concentration (MOC) required for combustion for a given combustible varies depending on the inert gas used. For the inert gases mentioned above, the relative effectiveness increases in the order:

nitrogen < carbon dioxide < halogenated hydrocarbon

MOC values for a range of compounds are listed in Table 4.4. using nitrogen and carbon dioxide as the inerting medium. In general, organic solvent vapours will not propagate combustion if the oxygen content is < 10% v/v with nitrogen or < 13% v/v with carbon dioxide.

Nitrogen is the inerting medium used most frequently. When using nitrogen, the MOC can be estimated by multiplying the oxygen required for complete stoichiometric combustion by the lower explosive limit. An example of this is given for methanol:

$$CH_3OH + 1.5\,O_2 \rightarrow CO_2 + 2H_2O$$

$$\text{LEL for } CH_3OH \text{ in air} = 6.7\% \text{ v/v}$$

$$\text{MOC} = 6.7 \times 1.5 = 10\% \text{ v/v } O_2$$

$$\text{(Measured value} = 10\% \text{ v/v).}$$

4.2.5 *Effect of different oxidants on flammability*

In the preceding sections there has been an underlying assumption that the solvent would be handled in air with the oxidant being the oxygen which is present in normal air. However, most materials that are flammable in air are also flammable in other atmospheres such as oxygen, chlorine, nitrous oxide and nitric

Table 4.4 Typical values of minimum oxygen concentration with nitrogen and carbon dioxide as the inerts

Compounds	% v/v O_2	
	N_2–Air	CO_2–Air
Acetone	13.5	15.5
Benzene	11	14
Carbon disulphide	5	8
Ethyl alcohol	10.5	13
Methyl alcohol	10	13.5

oxide. Flammability limits in oxygen are very much wider than in air. In terms of flammability, the next most important oxidant after oxygen is chlorine. The amount of data on fuel/chlorine systems is limited and this is partly due to the difficulties involved in studying them. Whilst some fuels form unreactive systems with chlorine, others partially or totally react on contact. Increased pressure and temperature exacerbate the situation. Flammability limits in chlorine are generally wider than those in air and are close to those in oxygen for a given fuel.

4.2.6 *Sources of solvent flammability data*

Solvent flammability data can be obtained from many standard reference books [11, 12, 13]. However, the data quoted can often vary widely for the same solvent. This is mainly due to the different test methods that are used to measure what is nominally the same parameter. In addition, as explained above, small variations in composition can sometimes have a marked effect on the measured parameter.

Flammability data are also available from solvent manufacturers; however, where the flammability data are critical for the safety of an operation or process it may be necessary to carry out appropriate tests on samples taken from the specific process.

4.3 Ignition sources

The presence of a flammable solvent vapour atmosphere on its own does not constitute a hazard. An ignition source is also required that is capable of igniting the atmosphere. Sources of ignition range from the readily apparent, i.e. visible flames, to the electrostatic discharge which occurs when an insulating powder is poured from a plastic drum.

Once a flammable solvent vapour atmosphere has been formed, the energy required to ignite it is in the range 0.1–1 mJ. This is extremely small, particularly when it is recognised that the static discharge from a person initiated by walking across a synthetic carpet is of the order of tens of millijoules.

Ignition sources that need to be considered include flames and hot surfaces, electrical equipment, friction, static electricity, lightning discharges, pyrophoric materials, etc. In addition, it has to be recognised that a source of ignition can develop from the handling of the solvent itself; for example, the flow of insulating solvents such as toluene through pipes and pumps can generate high levels of electrostatic charge [14, 15].

4.4 Safe handling of flammable solvents

Whenever possible, the solvent chosen for a particular purpose should be non-flammable under the conditions of use, i.e. its flash point should be appreciably greater than the operating temperature. Where this is not possible it is generally preferable to operate processes outside the range of flammability. This can be achieved either by inert gas blanketing or by introducing well defined ventilation. To implement such precautions satisfactorily, the flammability characteristics of the materials must be well defined and clearly understood.

The most appropriate basis of safety for a given situation or plant/process combination is determined by many factors including the plant design, its operation and the materials handled. A careful and systematic assessment of the potential hazards is therefore required. There are essentially two basic approaches to process safety where flammable solvents are involved, namely, explosion prevention and explosion protection.

To prevent an explosion from occurring it is necessary either to avoid the formation of a flammable atmosphere or to ensure all potential sources of ignition are eliminated. The basic approach to explosion prevention can be illustrated by referring back to the fire triangle, which shows the three components necessary for combustion.

In many situations, it is not possible to ensure reliably that an explosion can be prevented – hence the need for explosion protection such that the effects of an explosion are limited to an acceptable level. The three methods of protection are containment, explosion relief venting and suppression. These are usually used in conjunction with appropriate preventive measures.

4.4.1 Avoidance of flammable atmospheres

To base safety on the avoidance of flammable atmospheres, it must be established that a flammable atmosphere will not be present. This may be achieved in a number of ways. The ideal approach is to adopt the inherently safe option of using only materials which are non-flammable. Unfortunately it is seldom practicable to limit the choice of process materials to those which are non-flammable.

Having established that flammable solvents will be present, formation of a flammable atmosphere can be prevented by appropriate temperature control and the use of well defined ventilation. If liquid transfers are involved, precautions will also be necessary to prevent the formation of a fine-droplet mist.

An alternative option is to render the atmosphere non-flammable by reducing the oxygen concentration below the minimum required for combustion by an inert gas blanket. Although this is often viewed as the simplest method of achieving safety in a plant, there are a number of points which need to be considered when using this technique. The integrity of the blanket must be maintained at all times. Due attention must be given, therefore, to operations which can introduce air to the process such as chargehole operations, additions and cooling cycles. Although the plant is blanketed, flammable atmospheres may still form in the vicinity of the plant outside process vessels. Lastly, operators need to understand the potential asphyxiation risk of the inert gases used.

A final option is to operate in the region above the upper flammability limit. Great care is needed in adopting this approach and it is rarely favoured. Any leakage of air into the process can bring the system into the flammable range. Similarly, at start up and shut down, the two stages in the process cycle in which incidents are more likely than usual to arise, the process will pass through the flammability range. Hence, alternative means of achieving safety are usually preferable.

4.4.2 Avoidance of ignition sources

The number of situations in which ignition can be reliably prevented by the avoidance of ignition sources are limited. Situations where prevention is viable include batch reaction processes and isolated liquid storage. However, it is good practice to ensure that all reasonable measures are implemented to eliminate the development of sources of ignition. This is particularly important when using an explosion protection technique in order to minimise down time and unnecessary refurbishment of the system which can be costly.

If basing safety on the avoidance of ignition sources, it is essential that all these sources are identified and controlled. Sources of ignition known to occur in chemical plant include electrostatic discharges, friction sparks, mechanical frictional heating, flames, hot surfaces, electrical equipment and pyrophoric materials (e.g. catalysts) to name but a few. Although these differ in incendivity, it should be assumed that most will be capable of igniting flammable vapour mixtures. If any doubt exists as to their effective exclusion, an alternative basis of safety should be sought.

4.4.3 Containment

To base safety on containment, the plant must be of adequate strength to withstand full explosion pressure and to contain the products of an explosion. For a typical vapour/air system, pressure ratios of up to 10 can develop, i.e. an explosion starting at atmospheric pressure can produce a pressure of up to 10 bar. If using oxygen, this can rise to 15. This requirement for strong plant results in significantly increased design pressure requirements with associated elevated costs.

Greater increases in pressure are encountered when two or more vessels are interconnected, owing to a phenomenon known as pressure piling. This phenomenon is characterised by the development of explosion pressure in the second and subsequent vessels, which is very much higher than when only one single volume is involved.

In practice containment is seldom specified as a suitable basis of safety. Its use tends to be limited to small items of processing equipment used in isolation.

4.4.4 Explosion relief venting

In a vented explosion, the pressure rise is limited to an acceptable level by the rapid opening of a vent to allow the escape of the explosion products and the relief of the developing pressure before damaging over-pressures are attained. Consequently, venting is not an acceptable technique if a toxic or polluting emission could result.

An explosion relief vent must be carefully sized to ensure that it will effectively relieve the potential explosion. Factors influencing the calculation include the material explosion pressure characteristics, the plant strength, vent position and the influence of ducting if required. Consideration must also be given to the plant configuration and the availability of an area into which the explosion can safely be discharged.

There is much guidance available in the literature regarding explosion relief venting with the nomograph method being, perhaps, the most widely used technique for deriving vent areas. This and other available techniques are considered in detail by Lunn [16] and the NFPA [17].

4.4.5 Explosion suppression

In explosion suppression, the incipient fire or explosion is detected and quenched by the injection of suppressant before significant damage or a destructive overpressure is reached. In the case of an explosion in a confined space, the technique relies on there being a finite time between detection of the incipient explosion and the development of destructive pressure. As the time scale is short, with the entire event being over in a matter of milliseconds, the suppressant has to discharge very rapidly. The rapid release of suppressant is achieved by using electrically fired detonators to initiate the discharge.

The design and installation of an explosion suppression system is generally carried out by a commercial supplier. Although the costs can be high, an advantage of explosion suppression is the ability to contain the explosion products fully without the need for robust plant.

4.5 Solvent reactivity

In addition to hazards caused by flammability, it is also necessary to consider the chemical reactivity of solvents and their interactions and compatibility with other materials in order to ensure the safe use of solvents. For example, the simple addition of waste chloroform solvent to a Winchester bottle containing a mixture of acetone, ether and petroleum ether chromatography solvent residues leads to a vigorous exothermic reaction which bursts the bottle.

Although many problems associated with solvent reactivity are listed in the literature [18], the absence of any literature information on a specific compatibility question cannot be taken to imply that a problem does not exist. This was illustrated by a major explosion and fire in 1990 which was caused by a previously unknown reaction whereby the dimethylacetamide solvent reacted with water, forming acetic acid which reduced the stability of the reaction mass [19].

It is often necessary to perform specific tests to identify potential reactivity hazards [20], particularly where elevated temperatures are used in a process, for example during solvent recovery.

4.6 Legislation and regulations

There is a vast range of legislation and regulations concerning flammable solvents; for example, in the UK alone a proposed new Health and Safety Executive guidance booklet on the use of flammable liquids lists 18 different legal Acts and Regulations which are concerned with different aspects of the handling of flammable liquids. Unfortunately legislation and the resulting Regulations continually change and, additionally, vary from country to country. The following describes some of the concepts that are used to cover flammability hazards. It is emphasised that the particular regulations in force in any specific place must be checked to obtain the most current information.

Legislation on flammability is mainly divided into the transport, storage and use (including labelling) aspects of the handling of solvents. Most legislation uses the characteristic of the flash point of a solvent as a basis for classifying solvents into different hazard categories, with solvents having the lowest flash point temperature and low boiling points being in the highest hazard category. Again, the actual flash point temperature used as the cut off point to specify whether a solvent is subject to regulation varies from legislation to legislation, even in the same country. Thus liquids with a flash point of less than 32°C and which support combustion when tested in the prescribed manner are defined as 'highly flammable' according to the Highly Flammable Liquids and Liquefied Gases Regulations 1972 [21], which include provisions that require precautions to be taken: during storage; against spills and leaks; controls for sources of ignition in areas where accumulations of vapours might occur; means to prevent the escape of vapours; and controls on smoking. In contrast, also in the UK, the Chemicals (Hazard Information and Packaging) Regulations 1993 [22] define highly flammable liquids as those with a flash point below 21°C. These Regulations, commonly referred to as the CHIP Regulations, contain requirements for the supply of chemicals. The supplier of chemicals is required to classify them, i.e. to identify their hazards and provide information in the form of labels and safety data sheets and to package the chemicals safely.

The CHIP Regulations contain the implementation in the UK of various European Directives concerned with safety. European Directives have to be enacted by legislation in each of the member states and therefore a degree of commonality is emerging in the classification of flammable solvents and their subsequent requirements, at least in Europe. Chemicals are grouped into three categories of danger according to their flash point as follows:

1. Extremely Flammable – liquids with a flash point lower than 0°C and a boiling point lower than or equal to 35°C.
2. Highly Flammable – liquids with a flash point of less than 21°C but which are not classified as extremely flammable.
3. Flammable – liquids with a flash point equal to or greater than 21°C and less than or equal to 55°C and which support combustion when tested in the prescribed manner at 55°C.

These definitions are embodied in many European Directives and are slowly becoming the norm across Europe.

In the USA flammable liquids are divided into six categories, again based on their flash point and boiling point as shown in Table 4.5. These are used in the NFPA hazard rating system [23] to provide a numerical rating of 0–4 for the three regular hazards: health; flammability; and reactivity. Thousands of materials including solvents are covered.

The necessity of international transport has ensured that at least in this area a reasonably common system of classification of flammability has arisen. The United Nations Transport of Dangerous Goods Working Group produces what is known as the Orange Book approximately every 2 years. The recommendations contained in the Orange Book are then debated by the regulators for the different modes of international transport, e.g. by IMO, the International Maritime Organisation (sea transport), and IATA (air transport) and, when deemed appropriate, are included in the respective transport mode regulations. Flammable liquids reside in Class 3 and the criteria used for classifying them into different packaging groups are similar to those which have been adopted for European Directives in this area.

Following the occurrence of several major accidents, many involving flammable materials, an inventory criterion has been adopted in some regulations to trigger certain requirements. The European Seveso Directive [24] and its subsequent amendments, enacted in the UK by the Control of Industrial Major Accident Hazards Regulations 1984, amended 1989/90 [25], apply to flammable solvents with a flash point below 21°C and a boiling point above 20°C. The first level requirements apply at premises where 5000 tonnes or more of these flammable liquids are involved in certain industrial activities, including processing operations and storage. The second level requirements apply where 50 000 tonnes or more are involved. The general requirements apply at both levels and require the person having control of an industrial process to demonstrate that the major accident hazards have been identified and that the process is being managed safely. The additional requirements that apply at the second level include the submission of a written safety report, preparation of an on-site emergency plan and the provision of certain information to the public. Similar legislation applies in all European states.

These regulations illustrate a trend that is occurring in safety legislation, which is now starting to move from a prescriptive approach with definitions of what has to be done in specific situations to a more risk-based approach where the precautions and actions taken are left to the individual user provided they meet certain risk-based criteria. A proposed European Directive concerning minimum requirements for improving the safety and health of workers potentially at risk from explosive atmospheres extends this concept to all work places where flammable atmospheres may be formed. When adopted, this will require all employers to carry out a hazard assessment to identify whether a flammable atmosphere is likely to exist and, where so, to demonstrate that appropriate technical, organisational and other measures appropriate to the nature of the operation have been taken to minimise the risk.

Table 4.5 NFPA classification of combustible liquids

Liquid type	Flash point (°C)	Boiling point (°C)
Flammable, Class IA	<22.8	<37.8
Flammable, Class IB	<22.8	≥ 37.8
Flammable, Class IC	≥ 22.8 and <37.8	
Combustible, Class II	≥ 37.8 and <60	
Combustible, Class IIIA	≥ 60 and <93.4	
Combustible, Class IIIB	≥ 93.4	

Legislation concerning flammable solvents is not limited to situations where such solvents are being handled. In addition, equipment manufacturers who supply equipment for use in such situations are also controlled by an extensive array of regulations. These requirements are illustrated by the recent European Directive concerning equipment and protective systems intended for use in potentially explosive atmospheres [26]. This sets out essential safety requirements which will have to be met by equipment manufacturers before a product can be placed on the market in the future.

It can be seen that legislation and regulations cover all aspects of flammable solvent handling. These need to be carefully studied to ensure compliance.

Acknowledgements

The author gratefully acknowledges the assistance provided by Dr Sheila Beattie and Elisabeth Houghton of Zeneca Specialties during the preparation of this manuscript.

References

1. Britton, L.G. (1993) Static hazards using flexible intermediate bulk containers for powder handling. *Process Safety Progress*, **12**(4), 240–50.
2. Bodurtha, F.T. (1980) *Industrial Explosion Prevention and Protection*, McGraw-Hill, New York.
3. ASTM D 93 (1979) *Standard Method of Test for Flash Point by Pensky–Martens Closed Tester*, American Society for Testing and Materials, Philadelphia, Pa.
4. ASTM D 92 (1979) *Standard Method of Test for Flash Point by Cleveland Open Cup*, American Society for Testing and Materials, Philadelphia, Pa.
5. Crowl, D.A. and Louvar, J.F. (1990) *Chemical Process Safety: Fundamentals with Applications*, Prentice Hall, New Jersey.
6. Riazi, M.R. and Daubert, T. E. (1987) Predicting flash and pour points. *Hydrocarbon Proces.*, **66**(9), 81–3.
7. Patil, G.S. (1988) Estimation of flash points. *Fire Mater.*, **12**, 127–31.
8. Satyanarayana, K. and Rao, P.G. (1992) Improved equation to estimate flash points of organic compounds. *Journal of Hazardous Materials*, **32**, 81–5.
9. Gibbon, H.J., Wainwright, J. and Rogers, R.L. (1994) Experimental determination of flammability limits of solvents at elevated temperatures and pressures. *IChemE Symposium*, Series No. 134, 1–12.
10. Coffee, R.D. (1982) Cool flames, in *Safety and Accident Prevention in Chemical Operations*, 2nd edn, (eds H.H. Fawcett and W.S. Wood), Wiley–Interscience, New York.
11. NFPA (1983) *Fire Protection Guide on Hazardous Materials*, National Fire Protection Association, Quincy, USA.
12. Sax, N.I. (1984) *Dangerous Properties of Industrial Materials*, 6th edn, Van Nostrand, New York.
13. Nabert, K. and Schön, G. (1991) *Sicherheitstechnische Kennzahlen brennbarer Gase und Dämpfe*, Deutscher Eichverlag, Braunschweig.
14. BS 5958: Part 1 (1991) *Control of Undesirable Static Electricity Part 1. General Considerations*, BSI, London.
15. BS 5958: Part 2 (1991) *Control of Undesirable Static Electricity Part 2. Recommendations for Particular Industrial Situations*, BSI, London.
16. Lunn, G. (1984) *Venting Gas and Dust Explosions*, IChemE, Rugby.
17. NFPA 68 (1988) *Venting of Deflagrations*, National Fire Protection Association, Quincy, USA.
18. Bretherick, L. (1990) *Handbook of Reactive Chemical Hazards*, 4th edn, Butterworths, London.
19. Mooney, D.G. (1991) An overview of the Shell fluoroaromatics plant explosion. *IChemE Symposium*, Series No. 124, 381–92.
20. Barton, J.A. and Rogers, R.L. (eds) (1993) *Chemical Reaction Hazards – A Guide*, IChemE, Rugby.
21. The Highly Flammable Liquids and Liquefied Gases Regulations (1972) (SI 1972/917) HMSO, UK.
22. The Chemicals (Hazard Information and Packaging) Regulations (1993) (SI 1993/1746) HMSO, UK.
23. NFPA Standard 704M (1977) *Identification of the Fire Hazards of Materials*, National Fire Protection Association, Quincy, USA.
24. EEC Council Directive 82/501/EEC (1982) The 'Seveso' Directive.
25. The Control of Industrial Major Accident Hazards Regulations (1984) (SI 1982/1357) HMSO, UK.
26. EC Directive 94/9/EC (1994) Equipment and protective systems intended for use in potentially explosive atmospheres.

5 Protection of health

A.M. MOSES

5.1 Principles of health protection

The fact that chemical substances have the potential to damage human health has been known for many years, yet it is only in the last fifty years or so that the concept of preventing adverse effects on health (as opposed to detecting them and compensating those suffering) has become firmly established.

Prior to the 1930s, the legislative framework covering occupational exposure to chemical substances in industrialised countries focused largely on compensation for the victims of such exposure. During the 1930s attitudes began to change, with increased emphasis on the principle of preventing exposure to chemicals rather than compensating those suffering damage to health. The same decade saw the beginnings of the profession of occupational hygiene, which initially had considerable difficulty in gaining acceptance for the principle that 'prevention is better than cure'. Resistance to this principle came both from management, who tended to regard such activities as expensive and counterproductive, and trade unions, who were still primarily concerned with 'danger money' and compensation. However, it became increasingly accepted that preventing ill-health in the first place was not only cheaper than paying compensation but was in fact in the best interests of everyone involved in industry. Legislative frameworks steadily developed to embody this principle and are today firmly based on the principle of preventing adverse effects on health.

The first stage in preventing adverse effects is to identify those situations in which such effects could occur. It is particularly important to recognise that adverse effects are only likely to arise (i.e. there is likely to be a *risk* to health) when *both* the following conditions are satisfied:

(i) the chemical substance involved has the potential to cause adverse effects (this is known as the *hazard*); *and*

(ii) there is sufficient *exposure* to the substance.

As a very simple example, aspirin tablets clearly have the potential to cause harm, that is, they represent a hazard. When kept in an appropriately labelled, closed container in a locked medicine cabinet there is very unlikely to be any unwanted exposure and thus the risk (the chance of harm occurring) will be very low. The same container, left unstoppered on a kitchen table in the presence of small children, would represent a very high risk, because of the high probability that exposure will occur.

Risks to health are thus clearly a function of both hazard and exposure,

$$\text{i.e. Risk} = f(\text{Hazard})(\text{Exposure})$$

Risks can be reduced either by reducing the hazard or by reducing the exposure. A highly toxic substance may be handled quite safely when exposure is reduced to a low level, while substances not normally regarded as particularly hazardous may cause adverse effects if handled in a way that leads to excessive exposure.

The process of identifying situations in which risk is likely to occur is known as 'risk assessment'. Risk assessment thus requires a knowledge of both hazards and exposure. Hazards are identified through a knowledge of which chemical substances are involved in a particular operation, together with their properties, while details of how the substances are handled are necessary to assess exposure. Simple risk assessment can then be carried out by comparing hazards with exposure, and decisions made on whether it is necessary and how best to reduce risks to health.

In its simplest form, risk assessment consists in a qualitative assessment of the hazards of a substance and the circumstances under which it is likely to cause harm. These circumstances can then be compared with actual handling practice and a judgement made as to whether such practice is likely to lead to harm. A straightforward example is the use of solvents in ways which may lead to splashes in the eyes. Since most solvents damage the eyes, such practices would not be regarded as acceptable and the risk would be reduced by the wearing of eye protection (which very significantly reduces the exposure).

Risk assessment may also be carried out in a quantitative manner. For example, a solvent may have an established Occupational Exposure Limit (OEL). An OEL represents the concentration of a substance in the workplace air which is believed to be sufficiently low to avoid adverse effects on health. It will have been derived from a knowledge of the hazards (i.e. the toxicity) of the solvent. The concentration of solvent in the workplace air to which people are actually exposed can then be measured and compared with this limit, to determine whether conditions are satisfactory or whether measures to reduce the risk (usually by reducing the exposure) are necessary.

5.2 Health hazards (toxicity)

5.2.1 Background

Information on the toxicity of solvents may be available in one or more of the following areas:

- human experience
- animal studies
- other laboratory studies
- structure–activity considerations

5.2.1.1 Human experience. Human experience is often regarded as the most useful and the most relevant source of information, but does vary considerably in both quality and quantity. For some solvents which have been in use in large quantities for many years, extensive studies of workplace health (known as 'epidemiology') may have been carried out. These studies may either have identified adverse effects or have shown the absence of such effects. A shortcoming is often the extent and quality of exposure estimation, absence of which considerably reduces the value of the study. Useful human data may also be available from studies in volunteers exposed to low levels. Individual case reports of adverse effects are of limited value, but may well indicate the need for further more detailed study.

5.2.1.2 Animal studies. In many cases, much of the information on the toxicity of a solvent will have been derived from experiments using laboratory animals (usually rats or mice). Such experiments may have involved exposure by inhalation or by the oral or dermal routes, with duration ranging from a single exposure to the lifespan of the animal. Experimental protocols are available to cover general toxic effects, as well as skin and eye irritation, skin sensitisation, reproductive effects and carcinogenicity.

Single-exposure studies. The simplest animal experiments are known as the LD_{50} and LC_{50} tests. In such tests, small groups of animals are exposed to different single oral or dermal doses (LD_{50}) or different concentrations in the air for a fixed period of time (LC_{50}). After exposure, the animals are observed for a fixed period of time and the number surviving at the end of this period is noted. The LD_{50} (Lethal Dose 50%) is the oral or dermal dose which kills 50% of a group of test animals. Similarly, the LC_{50} is the concentration of the substance in the air for a specified time period (usually 4 or 8 h) which results in a similar fatality rate. LD_{50} values are usually expressed in mg/kg bodyweight and LC_{50} values in mg/m^3 or parts per million (ppm) volume/volume in the air, together with a specified time (e.g. 4 h LC_{50}). Although more generally available than most other pieces of toxicological data (since the tests are relatively cheaply and quickly performed), LD_{50} values and LC_{50} values are not particularly useful in assessing the overall toxicity of a solvent. They do not give any information about the non-lethal effects of a single dose/exposure, neither do they give information on the effects of repeated exposure. A good example of how misleading an LD_{50} can be in isolation is afforded by carbon tetrachloride, which has an oral LD_{50} of 2800 mg/kg. This is quite a high value and implies a relatively low toxicity. To draw such conclusions would potentially be very dangerous, since it is well known from other animal studies and from human experience that carbon tetrachloride is a potent liver poison, producing such damage after single or repeated exposures at relatively low levels.

Because of the limited value of the information obtained from LD_{50} and LC_{50} studies and ethical concerns surrounding the use of animals, such tests are becoming much less common. Instead, increasing attention is being paid to single-dose/exposure studies performed to a protocol which has been modified to give an approximate indication of lethality (sometimes termed the Median Lethal Dose or MLD), while at the same time allowing more observations on the nature and severity of non-lethal toxic effects. Such studies give more useful information and use fewer animals than the traditional LD_{50} and LC_{50} tests.

Irritant effects on the skin and eyes can be readily studied in simple tests on animals. Rats have usually been used for skin irritation studies and rabbits for eye irritation. Because of ethical concerns, particularly surrounding the rabbit eye tests, it is becoming increasingly common to question the need for these studies, particularly if the required information can be obtained or inferred by means not involving the use of animals.

Studies can also be carried out, usually involving guinea pigs, to obtain information on the ability of a solvent to cause skin sensitisation. The guinea pig is used because its immune system appears to be more similar to that of humans than that of other common laboratory animals.

Repeated-exposure studies. Repeated-exposure studies in animals can provide much useful information. Such tests are commonly conducted over 28 days (subacute) or 90 days (subchronic). The route of exposure may again be oral, dermal or by inhalation. Animals receive a single oral or dermal exposure each day or are exposed daily to a test atmosphere for a fixed period (usually 6 h). There will normally be three groups of animals exposed to different dose/exposure levels, together with a control group receiving no exposure. Animals are weighed and observed for overt signs of toxicity daily. At the end of the study the animals are sacrificed and a range of body organs subjected to a postmortem examination (both macroscopic and microscopic) for signs of toxicity. A number of other observations (e.g. examination of the blood and urine for a range of biological parameters) are usually carried out. Observations made on the test (i.e. treated) animals are compared with those of control (i.e. untreated) animals and any treatment-related abnormalities listed. In a well-designed study, some adverse effects will be seen at the highest dose/exposure level, while the lowest level should show no adverse effect. Frequently, the adverse effects seen at the highest levels will be seen in a milder form at the middle level.

Subacute and subchronic studies give information which can be extremely useful in establishing the hazardous nature of a solvent. Information will be obtained on the nature of the toxic effects and the dose/exposure levels which lead to such effects in animals. A well-designed study will also give a 'no adverse effect level', which can be used as a basis for deriving acceptable human exposure levels for repeated exposure.

Long-term studies. Although most types of adverse effect likely to result from repeated exposure can be expected to become apparent during subacute or subchronic studies, there are some effects which will not be detected. One such effect is the ability of a chemical to cause cancer (carcinogenicity). Cancer, by its nature, is a condition with a long induction period, i.e. it usually takes a significant proportion of the lifespan of an animal to develop. This means that, in order to test a chemical adequately for carcinogenicity, it is necessary to expose animals for a significant proportion of their lifespan and to continue to observe them after exposure has ceased. It is usual in such studies (which may be by any route of exposure) to include three exposed groups and a control group of both male and female rats or mice. At the end of the study, the animals are sacrificed and their organs examined macroscopically and microscopically for signs of cancer. The incidence of cancer in exposed groups of animals is compared with the incidence in the control group in order to decide whether exposure has caused any carcinogenic effect. Such tests are known as 'chronic' or 'lifetime tests' and since the normal lifespan of rats and mice is two years or more, they are expensive and time-consuming. Such studies have nevertheless been carried out for many of the larger tonnage solvents and, apart from information on carcinogenicity, they can also give very useful information on the non-carcinogenic toxic effects of prolonged exposure, extending that obtained in subacute and subchronic studies. It is not surprising that a 'no adverse effect level' derived from a lifetime study is usually lower than that obtained from studies of shorter duration.

Reproductive toxicity studies. Although the study protocols so far described may give information on the effect (or lack of it) that a solvent may have on both male and female organs of reproduction, they do not give any information on the effect of exposure on the ability of the animals to reproduce normally. In order to obtain such information, special study designs are necessary, and protocols have been developed to permit the study of:

- male fertility
- female fertility
- effects on male sperm production and function
- effects on female oestrus cycle
- adverse effects on the foetus (foetotoxicity)
- abnormalities in the foetus (teratogenicity)
- adverse effects on offspring after delivery
- effects on lactation
- effects of exposing second or subsequent generations

The study of reproductive toxicity can thus become very complex, and many of the above protocols are expensive in terms of animals, time and money. It is relatively unusual for a solvent to have been exhaustively tested in this area, although legislation is increasingly requiring that information on reproductive effects be obtained, particularly for high tonnage chemicals.

Relevance of animal test results to humans. The results of tests on animals must be interpreted to establish their significance for humans. Although the vast majority of adverse effects seen in humans can also be seen in experimental animals, it does not follow automatically that an effect seen in animals will necessarily be seen in humans, or that, if it is, it will be seen at the same exposure levels. In general, effects which are caused by direct action of the parent molecule (e.g. anaesthetic effects or skin irritation) will be seen in both test animals and humans. On the other hand, effects which are caused by a metabolite of a parent molecule (i.e. a product of chemical change of the original molecule brought about by body systems) may appear in one species but not in another. This is because metabolic pathways and therefore the products of metabolism may be different between, say, rat and human. An example of this (mentioned later in this chapter, see p. 89) is the occurrence of liver tumours in mice (but not in rats) as a result of exposure to trichloroethylene. Detailed investigation of the metabolism of trichloroethylene in mouse, rat and human tissue has shown significant differences, which support the conclusion that these mouse tumours are unlikely to be relevant to humans. Such detailed studies are, however, the exception rather than the rule, and in the absence of scientifically sound evidence to the contrary, it must be assumed that humans will be at least as sensitive, both qualitatively and quantitatively, as the most sensitive experimental species for which results are available.

5.2.1.3 Other laboratory studies. Useful data may also be available from laboratory studies not involving animals (*in vitro* data). The number and quality of such tests has been increasing steadily in recent years, partly as a result of the ethical requirement to minimise the number of animals used in safety testing and partly because of a desire to reduce the time taken to obtain useful information and also the cost of obtaining that information. A good example of this is in testing substances for carcinogenic potential, where conventional lifetime animal studies not only use a large number of animals, but are also extremely expensive and take several years before useful results are obtained. A number of short-term tests have been developed, based on the principle that many carcinogens owe their carcinogenic activity to an ability to interact with the genetic material of a cell (genes or chromosomes) and thus cause mutations, which may lead to cancer. These tests detect mutagenic activity and can be used for screening purposes, since they can be carried out relatively quickly and cheaply. Some of these tests are true *in vitro* tests, using cell lines in culture, while others use small numbers of whole animals. An example of the former is the Ames test, which detects mutagenic activity at the gene level in bacteria, while the mouse micronucleus test is an example of the latter and detects activity at the chromosome level in mice. *In vitro* test methods have also been developed for the assessment of skin and eye irritation and progress is being made towards the availability of such tests for other toxicological endpoints.

5.2.1.4 Structure–activity considerations. Predictions of the adverse effects that a substance may have can sometimes be made on the basis of a knowledge of the chemical structure of the substance and an association of various features of that structure with particular types of toxicity. It must be emphasised that this is a field for the expert and uninformed structure–activity predictions can be dangerously misleading. There are some areas, however, in which experts may make such predictions with a fair degree of accuracy. Perhaps the simplest example would be the prediction of irritant or corrosive properties for a molecule likely to give rise to acidic conditions in contact with water. Predictions of mutagenic activity may also in some cases be made by experts on the basis of the potential of the molecule to interact with DNA.

5.2.2 *Sources of information*

In assessing hazards, it is important that the user embarks on a proactive search for sufficient information. The most immediately available source of information is the label on the solvent container, which should give brief summary information on the principle hazard(s).

A Safety Data Sheet should also be available from the supplier. This will identify the hazards of the substance, as well as giving advice on First Aid, handling procedures, safe disposal and other matters. Data sheets summarising the properties of solvents are also available from other sources (e.g. Royal Society of Chemistry [1]).

Some official bodies publish extensive reviews of hazard data, and many common solvents are covered by the USA NIOSH Criteria Documents [2], the USA ACGIH TLV Documentation [3] and the UK HSE Criteria Document Summaries [4]. Several textbooks contain useful data [5, 6], and the ultimate source of information is the open scientific literature, to which access may be gained by searching online databases, for example TOXLINE [7].

In summary, the label on the solvent container and the supplier's Safety Data Sheet should always be available to the user of a solvent. If the user considers that these do not provide sufficient or sufficiently clear information to enable the solvent to be handled safely, further information may be obtained from one or more of the following sources:

- the supplier or manufacturer of the solvent
- published data sheets (e.g. Royal Society of Chemistry)
- technical and scientific literature (books and journals)

- the relevant Regulatory Authority for the country in question
- professional institutions and learned societies
- specialist consultants

The sources above are not listed in any particular order of priority, and the user should decide which of these is likely to be most available and most useful.

5.2.3 *Adverse effects on health*

5.2.3.1 Types of effect. Adverse effects can be classified in many different ways, and textbooks on toxicology should be consulted by those with an interest in the subject. For practical purposes, it is important that those involved in hazard and risk assessment have a basic understanding of some of the more important distinctions between different types of toxic effect, which are summarised as follows.

Acute and chronic effects. An *acute effect* is one which shows itself during or after a single exposure or a few repeated exposures (e.g. eye, nose or throat irritation). A *chronic effect* is one which does not show until after exposure has been maintained over a considerable period of time, often months or years (e.g. cirrhosis of the liver). It should be noted that the terms 'acute' and 'chronic' are also used to describe exposure duration.

Local and systemic effects. A *local effect* is one which shows itself at the part of the body which has been directly exposed to the solvent (e.g. skin or respiratory tract irritation). A *systemic effect* is one which shows itself at some part of the body away from the site of entry (e.g. anaesthetic effects or effects on the liver).

Reversible and irreversible effects. The meaning of these terms is obvious. Examples of *reversible effects* are minor irritation or anaesthetic effects, while examples of *irreversible effects* are cancer or foetal malformations.

The above terms are used in combination, for example a corrosive skin burn is an acute, local, irreversible effect, liver cancer is a chronic, systemic, irreversible effect and anaesthesia is an acute, systemic effect, usually reversible unless exposure has been very severe.

Another important way of describing the toxic effects of a solvent is to define the organ or organs in the body most likely to be affected by exposure (the 'target' organ(s)). The commonest target organs are the skin, eyes and respiratory tract for local effects and the central nervous system (CNS or brain), the liver and the kidneys for systemic effects.

5.2.3.2 Factors influencing toxicity. There are a number of important factors which influence and to a degree determine the impact that exposure to a solvent will have on the human body. These factors are the extent to which the solvent is

- *absorbed into,*
- *distributed around,*
- *metabolised by,* and
- *excreted from*

the body. These factors clearly have a particular importance where systemic effects are concerned.

Because of their lipophilic properties (i.e. ability to dissolve in fats), most solvents are readily absorbed into the body by the inhalation or oral routes. This is because they readily dissolve in, and pass through, the lipid membranes of the cells lining these parts of the body. Many solvents will also penetrate the skin and can thus give rise to systemic effects by that route of exposure.

Once absorbed, most solvents are rapidly and widely distributed around the body by the bloodstream. Again, their lipophilicity facilitates their passage across cell membranes and their relatively small molecular size allows them to pass the so-called 'blood–brain barrier' and cause CNS effects.

Some solvents are metabolised or chemically changed by enzyme systems within the body. The human body is structured to facilitate the absorption of lipophilic substances, but its organs of excretion work most effectively for hydrophilic substances (i.e. those dissolving in water). Enzyme systems exist (principally in the liver) to convert unwanted lipophilic substances absorbed in food into more water-soluble substances for excretion, and many foreign chemical substances (including solvents) are also metabolised. The chemical changes brought about can be crucial in determining the overall toxicity of a molecule; metabolism may either increase or decrease the toxicity of a 'parent' molecule. The toxicity of a solvent may be due either to the parent molecule (e.g. CNS effects) or due to a metabolite (e.g. most liver effects). The fact that the liver is the principal organ of metabolism explains why it is so frequently a target for toxic effects and the role of the kidneys as organs of excretion (with some metabolic capability) also renders them liable to adverse effects.

Solvents are excreted from the body via urine or faeces if they have hydrophilic properties or have been metabolised, or they can be exhaled unchanged in expired air.

One further factor which can be of overriding importance is the relative rate at which the above processes take place (i.e. kinetics). This is of particular importance in determining the rate of recovery from CNS effects once exposure has ceased. Solvents with a low solubility in blood will be excreted rapidly in expired air and blood levels can fall dramatically within minutes, thus facilitating recovery from accidental overexposure (e.g. 1,1,1-trichloroethane). Other, more soluble solvents may persist much longer (e.g. alcohols). Generally speaking, solvents which do not undergo extensive metabolism will be exhaled fairly rapidly and blood levels will fall accordingly: it is the level in blood which determines the extent of systemic (e.g. CNS) effects. Solvents may well persist in fatty tissue for much longer, but the associated blood levels will be relatively low.

5.2.4 Hazards common to all solvents

The physicochemical properties which confer solvent properties on a molecule are also responsible for a number of generic adverse effects. These effects vary in importance from solvent to solvent, but all organic solvents have the potential to cause these effects to a greater or lesser extent. In some cases, these effects (which arise as a result of the properties of the parent molecule) may be the only manifestations of toxicity shown by the solvent, whereas in others (where metabolism leads to toxic metabolites) they may be secondary in importance to effects specific to that solvent.

The principal generic effects common to all solvents are:

- irritation effects
- anaesthetic (CNS) effects

In addition, many solvents have been shown to cause cardiac sensitisation at very high exposure levels and this may also be regarded as an effect related to the solvent properties of the molecule.

5.2.4.1 Irritation. All organic solvents have the potential to cause irritation. This arises because the lipophilic nature of the molecule leads to ready absorption into, or through, the fatty membrane surrounding cells, resulting (if the solvent concentration is sufficiently high) in cell disruption or death. This effect is most likely to occur at the site of contact between the solvent and the body. Liquid solvents in contact with the skin or eyes can lead to effects ranging from defatting and minor irritation, through inflammation to corrosive effects (if exposure is severe and prolonged). Solvent vapours can irritate the respiratory tract, with effects ranging (depending on the solvent and the concentration) from minor irritation to severe inflammation (which in extreme cases can be life-threatening due to fluid exudate in the lungs). This severe inflammation is sometimes known as pulmonary oedema or chemical pneumonitis. These effects vary considerably in practical significance from solvent to solvent. They are, not surprisingly, exacerbated by any acidity or alkalinity that the solvent may contain.

5.2.4.2 Anaesthetic effects. Overexposure to organic solvents may lead to anaesthetic effects. Solvent molecules which are small, non-polar and lipophilic can be carried into the brain by the bloodstream, where they dissolve in cell membranes. This may lead to effects on the CNS, which range in severity from the early warning signs of anaesthesia (slight dizziness, impaired coordination) to, in extreme cases, coma and death. Recovery from these acute effects is usually complete (in other than very severe cases) following cessation of exposure. High-exposure concentrations are needed to give rise to these effects, which may result either in stimulation or depression of the CNS. It is this property of solvents which gives rise to the phenomenon of solvent abuse, or ‘glue sniffing’.

5.2.4.3 Cardiac sensitisation. Many solvents, at very high exposure levels, are able to affect the heart such that it becomes particularly sensitive to the effects of adrenaline or similar agents, whether generated internally within the body or administered therapeutically. This is known as cardiac sensitisation. In extreme cases, adrenaline generated by exercise or sudden shock can cause the sensitised heart suddenly to cease functioning. This is one of the hazards specifically associated with solvent abuse.

5.2.4.4 Individual variability. There is considerable variability between solvents in respect of the practical importance of the above toxic effects. Irritation of the skin and eyes can occur with all solvents in the liquid phase and in some cases with the vapour phase. As vapour concentrations rise, respiratory tract irritation may occur and this is often the critical effect regulating exposure control. At higher concentrations, acute

CNS effects will occur. Cardiac sensitisation only occurs at very high concentrations (which for most solvents approach or exceed anaesthetic levels) and is thus unlikely to be of significance in normal industrial practice.

All these effects are acute, in that they can occur during or immediately after a single exposure. It has been claimed by some investigators that chronic exposure to low levels of certain solvents can lead to a chronic adverse effect known as 'Danish painters' syndrome'. This effect, which is claimed to involve subtle changes in behaviour and in some cognitive and motor functions, has not been confirmed as a problem associated with exposure to solvents and has not been a determining factor in the establishment of occupational exposure limits by most regulatory authorities.

5.2.5 Hazards specific to individual solvents

In addition to the generic hazards indicated above, individual solvents may display toxic properties specific to that solvent or group of solvents. These specific hazards may arise from the toxic properties of the parent molecule or, more frequently, from the product of metabolism. A wide range of such effects can occur, affecting a variety of target organs. Some of the most common are described below.

5.2.5.1 Liver toxicity. The liver has been described as the 'chemical factory' of the body, and holds a key place in normal metabolic functions. It is the organ principally responsible for protein synthesis, carbohydrate metabolism and fat metabolism. The products of digestion of food are absorbed from the gastrointestinal tract into the bloodstream and carried direct to the liver, where they are converted by one or other of the above processes into molecules which the body can use either for energy or for generating new tissue. Unwanted products of digestion are converted to more soluble forms for excretion via the kidneys.

In order to carry out these crucial body functions, the liver possesses a very considerable metabolic capability, and is the organ of the body in which most such activity takes place. Solvents are readily absorbed into the bloodstream and carried to the liver, where they become available to the latter's metabolic capability. It is thus not surprising that those solvents which are metabolised to more toxic components can have an adverse effect on this organ.

The liver is thus a relatively common target organ and liver effects can range in severity from minor, reversible increases in liver weight to serious liver damage and in some cases carcinogenicity. Alcohols and chlorinated solvents provide examples of solvents with specific toxic effects on the liver, although there is a wide range of potency in the latter group with some (e.g. 1,1,1-trichloroethane) having virtually no effect while others (e.g. carbon tetrachloride) can cause serious acute and chronic liver damage.

5.2.5.2 Kidney toxicity. The kidneys are the body's principal organ of excretion for water-soluble materials. They are structured in a very complex way to remove larger molecules from the blood by filtration and then to re-absorb most of the water and smaller molecules useful to the body, leaving an aqueous solution of unwanted molecules for excretion as urine.

The kidneys can therefore become a target organ for toxic metabolites produced by the liver and transported to them in the bloodstream. In addition, the kidney has some metabolic capability in its own right, and this can lead to the production of toxic metabolites within that organ. Kidney effects can range from mild to serious and are often associated with and secondary to liver effects. The more toxic halogenated hydrocarbon solvents (e.g. carbon tetrachloride and chloroform) and certain glycol ethers (e.g. ethylene glycol monomethyl and monoethyl ethers and their acetates) provide examples of kidney toxins.

5.2.5.3 Carcinogenicity. The ability to cause cancer (or carcinogenicity) is clearly a very serious adverse effect. Cancer is a malignant growth (or tumour) in the body in which the cells involved divide and the tumour grows in an uncontrolled manner. Cell division is a normal bodily process required to replace old or damaged cells, but in the case of a cancer the cells are abnormal in that:

- they have to some degree lost the characteristics of a normal cell from the organ involved;
- they have lost the normal property of 'contact inhibition of growth' such that the tumour invades surrounding tissues; and
- cells from a developing original tumour (a 'primary' tumour) have the ability to break away from their original site in the body and become attached to other organs, where they continue to divide and produce secondary tumours.

It is these last two properties which underlie the seriousness of cancer as an adverse health effect. Chemically induced cancer may arise as a result of mutagenic properties possessed by a substance or metabolite causing a mutation in the target cells, which leads to expression as cancer when these cells divide. A substance causing cancer by this mechanism is known as a 'genotoxic' carcinogen.

Although a number of solvents have been suspected of carcinogenic potential, there are very few for which arising in practice is likely to be acute CNS effects and irritation, although effects on the liver have been seen in animal experiments. Long-term studies in animals have demonstrated that trichloroethylene causes liver cancer in mice and kidney cancer in rats, although detailed metabolic studies have suggested that these tumours arise as a result of species-specific mechanisms and are thus of doubtful relevance for humans.

Tetrachloroethylene. Tetrachloroethylene (OES 50 ppm 8 h TWA) is broadly similar to trichloroethylene in its toxicological profile. Mild adverse liver and kidney effects have been reported in humans who have been overexposed and tumours have been seen in animals exposed over their lifetime. Tetrachloroethylene follows similar metabolic pathways to trichloroethylene, and similar arguments to those developed for trichloroethylene cast doubt on the relevance of these tumours for humans. Mild CNS disturbances and eye irritation are considered to be the critical health effects in humans. Cardiac sensitisation has not been reported in animal studies, unusual among the halogenated solvents.

1,2-Dichloroethane. 1,2-Dichloroethane (MEL 5 ppm 8 h TWA) is one of the more toxic chlorinated hydrocarbons. It is readily absorbed through the skin, by inhalation and by ingestion and has a similar toxicity to carbon tetrachloride, with marked effects on the CNS and liver. It also causes liver tumours in lifetime animal experiments and (unlike carbon tetrachloride and chloroform) displays mutagenic activity *in vitro*, suggesting a direct genotoxic mechanism of action. For this reason, it must be assumed that repeated over-exposure in humans could also lead to carcinogenic effects.

*1,1,1-Trichloroethane.** 1,1,1-Trichloroethane (OES 200 ppm 8 h TWA) is one of the least toxic members of this group, probably due to the small extent to which it is metabolised. Very little effect is seen on the liver in animal experiments and lifetime studies have shown no carcinogenic effect. Extremely high exposures may lead to cardiac sensitisation and there have been reports of sudden deaths, some of which have been associated with solvent abuse. This solvent is being phased out under the Montreal Protocol.

Dichloromethane. Dichloromethane or methylene chloride (MEL 100 ppm 8 h TWA) is the least toxic of the chlorinated methanes and acts principally on the CNS and as an irritant. It does not cause significant acute or chronic liver or kidney damage. Detailed studies of the comparative metabolism of dichloromethane in human and animal tissues have provided support for the view that tumours seen in lifetime animal experiments are not relevant for humans. An effect quite specific to dichloromethane arises from its metabolism to carbon monoxide. The carbon monoxide so produced combines with haemoglobin in the blood to produce carboxyhaemoglobin. While carboxyhaemoglobin poisoning has not been a marked feature of human exposure to dichloromethane, it should be noted as one of the effects that the solvent can produce, which could be of more significance in those individuals whose blood oxygen-carrying capacity is compromised in some way.

1,1,2-Trichloro-1,2,2-trifluoroethane. 1,1,2-Trichloro-1,2,2-trifluoroethane or fluorocarbon-113 (OES 1000 ppm 8 h TWA) is possibly the least toxic of the solvents that have been in common use within this group. Apart from CNS effects and the ability to cause cardiac sensitisation at very high exposure levels, this solvent has been shown to cause few adverse effects. This absence of systemic toxicity is attributable to a rapid rate of elimination from the body and a very low degree of metabolism. This solvent is being phased out of production under the Montreal Protocol.

5.2.6.4 Alcohols. Alcohol solvents, as well as possessing lipophilic properties, also possess some degree of hydrophilicity due to the hydroxyl group. This means that they are more soluble in body fluids and take longer to eliminate than molecules devoid of hydrophilicity. All alcohols possess the ability to cause effects on the CNS, which may take the form either of stimulation or depression. They are also in general readily metabolised by oxidative processes to give the corresponding aldehydes and acids. These metabolites are in many cases responsible for the specific toxic effects. The alcohols possess defatting properties and have an irritant action on the skin and respiratory system which increases with molecular weight, probably due to decreasing volatility.

Methanol. Methanol (OES 200 ppm 8 h TWA) has, in addition to its intoxicating properties, a specific effect on the optic nerve and retina, probably due to its metabolism to formaldehyde and formic acid. This can lead to vision impairment and slow clearance of the metabolites from the body means that repeated exposure can produce cumulative effects.

---

* *Note:* Care should be taken to distinguish 1,1,1-trichloroethane from 1,1,2-trichloroethane, which is significantly more toxic. There is confusion between the two isomers in some of the older literatur

Ethanol. Ethanol (OES 1000 ppm 8 h TWA) has well-known effects on the CNS and can cause irritation. Although metabolised to acetaldehyde and acetate, ethanol appears to have little ability to cause chronic toxic effects at exposure levels likely to be encountered in reasonable industrial and laboratory use. It should be noted that ethanol may compete with other substances present in the workplace atmosphere for the oxidative enzyme systems involved in this metabolism, thereby inhibiting their metabolism or excretion and increasing their toxic effects. This may occur with a number of solvents including trichloroethylene, xylene, benzene and dimethylformamide. The presence of denaturants (e.g. methanol, pyridines) greatly increases the toxicity of ethanol by ingestion and may also present a hazard by the inhalation route.

n-*Propanol and isopropanol.* *n*-Propanol (OES 200 ppm 8 h TWA) is rather more toxic than ethanol, while showing a broadly similar pattern of effects. Its lower volatility means that it will remain in contact with the skin for longer and thus may more readily give rise to skin irritation. Passage through the skin may contribute to the overall toxic effect. Isopropanol (OES 400 ppm 8 h TWA) is similar to but slightly less toxic than *n*-propanol.

n-*Butanol.* *n*-Butanol (OES 50 ppm 10 min TWA) is significantly more irritant to the respiratory tract than the propanols, and this is the reason for the lower OES and the shorter (10 min) reference period.

Amyl alcohol. Amyl alcohol (pentanol) produces anaemia and toxic effects on the liver and kidneys, reflecting a tendency to increased systemic toxicity as the molecular weight of the alcohol increases.

5.2.6.5 Ketones. The lower-molecular-weight ketones, e.g. acetone (OES 750 ppm 8 h TWA), methyl ethyl ketone (OES 200 ppm 8 h TWA) and methyl isobutyl ketone (OES 50 ppm 8 h TWA), show a similar pattern of toxicity to the alcohols, with CNS effects and irritation tending to increase in potency as the molecular weight increases. For those with higher molecular weight and lower OESs, passage through the skin may contribute significantly to systemic effects.

Methyl n-*butyl ketone.* Methyl *n*-butyl ketone (OES 25 ppm 8 h TWA), in contrast to other lower molecular-weight ketones, shows similar peripheral neurotoxic effects to *n*-hexane. This is due to its metabolism to 2,5-hexanediol, a metabolite which it has in common with *n*-hexane and which is responsible for the neurotoxicity.

5.2.6.6 Esters. Esters are, in general, fairly readily hydrolysed in the body to give the corresponding alcohol and acid anion. They thus show the generic properties of irritation and effects on the CNS, together with any systemic effects associated with the products of hydrolysis. Such effects are usually associated with the alcohol. For example, ethyl acetate (OES 400 ppm 8 h TWA) shows little in the way of specific toxicity whereas amyl acetate can give rise to anaemia and liver effects similar to those produced by amyl alcohol.

5.2.6.7 Glycol ethers. The lower-molecular-weight ethylene glycol ethers (2-methoxyethanol (MEL 5 ppm 8 h TWA) and 2-ethoxyethanol (MEL 10 ppm 8 h TWA) and their corresponding acetates display specific toxic effects on the CNS, blood, kidneys and on the reproductive systems of both males and females. Testicular damage has been seen in animals and maternal and fetal toxicity as well as major fetal abnormalities have been seen in pregnant animals. The acetates show similar effects, presumably due to hydrolysis to the parent glycol ether. If it is necessary to use one of these solvents, great care should be taken.

In contrast to the above, propylene glycol monomethyl ether or 1-methoxypropan-2-ol (OES 100 ppm 8 h TWA) has been shown in animal studies not to display these specific toxic effects, presumably due to differences in metabolic products. Eye and respiratory tract irritation are the critical effects with this solvent.

5.2.6.8 Other solvents. Other solvents display a wide variety of specific toxic effects.

Carbon disulphide. Carbon disulphide (MEL 10 ppm 8 h TWA) is an excellent solvent with a wide range of potential uses, but it is also highly toxic. It has been shown to produce peripheral neuropathy and peripheral and brain arterial damage in humans. Epidemiological studies have shown a relationship between exposure to carbon disulphide and an increased incidence of coronary heart disease. Animal studies have shown teratogenic and fetotoxic effects.

Nitrobenzene. Nitrobenzene (OES 1 ppm 8 h TWA) is also highly toxic, having its principal effect on the blood, causing methaemoglobinaemia, which leads to cyanosis. Rapid passage through the skin means that these effects may occur in situations in which there is no significant exposure by the inhalation route.

Dimethylformamide. Dimethylformamide (OES 10 ppm 8 h TWA) is moderately toxic to the liver, kidneys and CNS, whereas dimethylsulphoxide appears to be of low toxicity.

5.3 Exposure (recognition, evaluation and control)

5.3.1 Principles of occupational hygiene

For a work situation to carry a risk to health there must not only be hazardous substances present, there must also be the potential for exposure to those substances at levels which may give rise to harmful effects. Occupational hygiene may be viewed as the science and practice of:

(i) *recognising* situations which may give rise to risks to health;
(ii) *evaluating* those risks qualitatively and/or quantitatively;
(iii) *controlling* the risks to a level which is acceptably low.

This necessarily involves an understanding of how exposure to a solvent may occur, how any such exposure may be evaluated and also how exposures may be controlled.

5.3.2 Routes of exposure

In practice, important routes of exposure through which humans may come into contact with solvents are by inhalation (of vapour or mists), by skin contact or by ingestion. Other routes (e.g. subcutaneous injection) may be of importance in specific situations.

Inhalation is in many cases the most important route of exposure and such exposure may lead to both local and systemic effects. Local effects on the respiratory tract may range from nose and throat irritation to deep lung irritant effects, which could lead to fluid collecting in the lungs (pulmonary oedema). Mists may be significantly more irritating than the equivalent amount of vapour in the air, since droplets depositing in the respiratory tract can lead to a much higher concentration of solvent in contact with tissue. Most solvents are readily absorbed through the lungs and are rapidly distributed around the body by the bloodstream, thereby giving rise to systemic effects. The eyes may also be irritated by vapour or mist.

Skin contact can also obviously lead to local effects at the site of contact. Many solvents have the ability to penetrate the skin in the liquid phase, and, once absorbed, are widely distributed around the body by the bloodstream. In some cases, this can contribute significantly to the overall systemic toxicity. In practice, absorption through the skin is likely to be significant in situations where the solvent is readily absorbed and the occupational exposure limit (which of course applies only to the inhalation route) is set at a fairly low level to protect against systemic effects. In these situations, a relatively small amount of solvent would need to be absorbed through the skin to raise the total amount in the body to a level which would have adverse systemic effects. The contribution from this route of exposure is potentially easy to overlook (a good example is nitrobenzene). Those solvents and other substances which penetrate the skin and which may by so doing contribute significantly to overall exposure are identified by a 'skin' notation in official lists of occupational exposure limits. For a solvent with such a notation, the occupational exposure limit is only valid if suitable precautions are taken to prevent skin absorption.

Although most solvents, when ingested *orally*, will irritate the gastrointestinal tract and be readily absorbed into the bloodstream, this is a much less common route of exposure. The possibility of accidental ingestion (or even deliberate abuse) should however be considered.

The relative importance of each of these routes of exposure should be assessed for each solvent and work situation. In this assessment the following factors should be considered:

1. The physicochemical properties of the solvent. In particular the vapour pressure will give an indication of the potential for vapour and/or mist exposure.
2. The toxicity of the solvent. The occupational exposure limit will give a broad indication of atmospheric exposure levels which are likely to cause concern and the ability to penetrate skin ('skin' notation) may indicate concern for exposure by that route.
3. The nature of the operation or process in which the solvent is used. This factor may be of overriding importance in determining the potential for exposure. Uses can range from processes conducted entirely within sealed equipment to operations carried out in open pans or on an open bench.

5.3.3 Evaluation of exposure

The potential for exposure should be evaluated qualitatively for each work situation, on the basis of the above factors. For example, use of a highly volatile solvent with a low occupational exposure limit in open pans would give considerable cause for concern, whereas contained use would considerably restrict the potential for exposure. Even with contained use, exposure may arise during maintenance and cleaning operations or as a result of equipment failure.

Practical advice on monitoring strategies for the quantitative determination of levels of hazardous substances in workplace air has been published [8]. The best method of measuring such exposure involves personal monitoring, in which air from the breathing zone of the worker is sampled continuously during a work day or shift. This can be achieved either by a small pump (which draws air through a sample tube) worn by the worker or by means of a passive sampler or badge (which relies on diffusion). In both cases, the sample is subsequently analysed and the exposure calculated. Methodologies are available for the analytical determination in air of many solvents which have official occupational exposure limits [10, 11].

Personal exposure may also be estimated from fixed station monitors or 'snap' samples (e.g. Draeger tubes). Such estimates require a detailed knowledge of the activity pattern of workers and may be less accurate than personal monitoring.

Personal exposures estimated in one of the above ways are then compared with the occupational exposure limit for the solvent in question. If exposures exceed the limit there is clearly a need for action and if they are, say, less than one-tenth the situation may be regarded as satisfactory. Within this range, the need for action will be a matter for professional judgement, but it may be regarded as good practice to maintain exposures as low as reasonably practicable below such limits.

Exposure by skin absorption is much more difficult to assess quantitatively. It is always good practice to avoid exposure of the skin to solvents and any need for special precautions will have to be determined on the basis of qualitative considerations. The possible contribution that skin absorption can make to overall toxicity must not be forgotten.

For some solvents, overall exposure by all routes can be estimated using methods of biological monitoring. This technique involves measuring the level of the solvent or a metabolite in a body fluid (usually expired air, urine or blood). The level so determined would then be compared with a biological exposure limit. The method may have particular value in situations where skin absorption or non-occupational exposure are significant, but the number of solvents for which validated methods of biological monitoring are available is limited. Reference should be made to the Health and Safety Executive publication on this topic [12].

5.3.4 Control of exposure

Adequate control of exposure is a crucial element in the overall process of ensuring that solvents are used without presenting risks to health. The principles underlying control of exposure may be briefly summarised as follows and these principles should always be applied in this order of preference.

1. Elimination (or substitution) of the hazard
2. Containment of the hazard.
3. Ventilation (where exposure by inhalation is of concern).
4. Personal protective equipment.

The detail of their application will vary depending on the likely route of exposure.

5.3.4.1 Exposure by inhalation. The above principles are best illustrated in the context of inhalation exposure, which is a very common route of exposure for solvents.

Elimination (or substitution). In any situation in which an unacceptable risk to health has been identified the preferred course of action is to eliminate the toxic substance from the workplace or substitute it by a less hazardous substance. This is the most certain way of eliminating exposure and therefore removing the risk. However, elimination will in many cases prove impractical, since there may still remain a need for a solvent. Substitution, however, should be seriously considered and there are many situations in which highly toxic solvents have been successfully substituted by those of lower toxicity. Examples are:

- cyclohexane for benzene
- trichloroethylene or 1,1,1-trichloroethane for carbon tetrachloride or chloroform
- propylene glycol ethers for ethylene glycol ethers

In considering substitution, however, care should be taken that new hazards and risks are not introduced inadvertently. Where a non-flammable solvent could be replaced by a flammable one of lower toxicity, care should be taken that a fire risk is not introduced. Similarly, in the reverse situation, a fire risk may be inadvertently replaced by a toxic risk.

Careful thought should also be given to the substitution of a well-investigated solvent (with clearly identified and quantified toxic effects) by one which has not been so thoroughly investigated.

Containment. Once the possibilities for elimination and substitution have been exhausted, the preferred method for controlling exposure is to contain the solvent, both liquid and vapour, within a totally enclosed system such that there is a physical barrier at all times between the solvent and the personnel involved in the process. The use of such closed systems enables quite toxic solvents to be handled without risk to health during normal operations. There will, however, be potential exposure during maintenance and cleaning of such equipment and special measures may be necessary to ensure the safety of personnel involved.

Ventilation. Where it does not prove practicable to use a totally enclosed system, it may be necessary to provide ventilation. The purpose of ventilation is to remove solvent vapours and mists such that they do not enter the breathing zone of personnel. Ventilation engineering is a complex subject, but two broad approaches are available:

- general workplace ventilation
- local exhaust ventilation

General ventilation may be required to maintain reasonable working conditions in terms of temperature and humidity, and it can be effective in reducing exposure to solvents where such vapour in the workplace air arises from a multiplicity of small sources. It is less effective, however, in reducing exposure where the vapour or mist arises from a single source or a small number of discrete sources, such as open pans or tanks. In these cases properly designed and operated local exhaust ventilation is to be preferred. Such ventilation should be designed to draw clean air from the workers' breathing zone across the work area, exhausting contaminated air either to a solvent recovery unit or to a safe place. Care should be taken that contaminated air is not exhausted to places where it may create a risk to people not associated with the process in question (e.g. near another air intake). While effective, this type of ventilation can be expensive in terms of lost solvent or solvent recovery costs; in appropriate cases, the option of total containment should be considered. It is important that all ventilation equipment is regularly maintained and checked to ensure effective operation.

Personal protective equipment. Personal protective equipment designed to protect against exposure by inhalation is known as respiratory protective equipment (or RPE). Such equipment should not be used as a long-term method of ensuring that exposures are reduced to acceptable levels. This is because it is prone to mechanical failure and human error. RPE may be used appropriately for specific short-term operations (e.g. some maintenance tasks) where it is not practicable to reduce exposures to acceptable levels by other means. It may also be of value in emergency situations. Positive pressure breathing apparatus (line fed or self-contained) is preferable to canister-type respirators. For RPE to be effective when required, it must be adequately maintained and personnel fully trained in its use.

5.3.4.2 Exposure to, or through, the skin. Equally as important as the need to control exposure by inhalation is the need to ensure that exposure to, and through, the skin is properly controlled. Most of the principles outlined above are also applicable here, with the preferred option being elimination or substitution of particularly hazardous solvents. Where this is not practicable, equipment should be designed to contain the solvent and present physical barriers to skin contact. If there still remains a likelihood of skin contamination (as will be the case for many routine operations) it will be necessary to provide adequate protective clothing. The extent of such protection must be considered, as well as the material from which the clothing is made. The material must be adequately resistant to the solvent in question and the extent of protective clothing may range from gloves to full protective suits.

5.3.4.3 Exposure of the eyes. Almost all solvents are at least irritant to the eyes and some can cause serious damage. It is thus important to provide adequate eye protection in all situations where liquid splashes could occur.

5.3.4.4 Exposure by ingestion. Exposure by ingestion is less likely to be of practical significance than inhalation or dermal exposure, but the possibility should always be considered. Prevention of exposure by ingestion depends on good housekeeping practice and commonsense precautions. Eating, drinking and smoking should not be permitted in areas where solvents are used and steps should be taken to ensure that solvents required for process purposes are not used for general cleaning (this will also reduce the possibility of exposure by other routes).

5.3.4.5 Information, instruction and training. An important element in the safe handling of solvents is the information, instruction and training given to personnel involved in their use. These three components may be summarised as:

- *Information.* Factual information about the hazards, exposures and potential risks of a work activity, together with the precautions necessary to ensure safety.
- *Instruction.* Telling personnel involved how to apply the required safety precautions and how to use any protective equipment.
- *Training.* Ensuring that personnel are fully informed and familiar with the correct way to use equipment and that safety precautions are actually being observed in practice.

These elements are particularly important in the context of exposure, and satisfactory control of exposure is only likely to be achieved and maintained if the workforce are fully informed of the hazards and risks involved and the precautions necessary to reduce those risks to acceptably low levels, and these precautions are being fully observed in practice.

5.4 Criteria for hazard warning

Two basic methods are employed by suppliers to convey warnings about the hazards of a substance to the user. These are:

(i) the label on the container;
(ii) the safety data sheet accompanying the substance.

5.4.1 Classification and labelling

Schemes for classifying substances according to hazard and labelling and packaging them accordingly have become part of the regulatory framework within the European Union. The original scheme was set out in Directive 67/548/EEC ('Classification, Packaging and Labelling of Dangerous Substances' – known as the 'Substances Directive'). This Directive has been amended several times, the most recent amendment being the '7th Amendment to the Dangerous Substances Directive' (92/32/EEC). There have also been a considerable number of 'Adaptations to Technical Progress', the most recent being the '20th Adaptation to the Dangerous Substances Directive' (93/101/EC). This Directive (and its Amendments and Adaptations) apply to individual chemical substances.

Classification, packaging and labelling of mixtures of substances (or 'Preparations') are covered by Directive 88/379/EEC (known as the 'Dangerous Preparations Directive'). This Directive replaced the 'Solvents Directive' (73/173/EEC) and the 'Paints Directive' (77/728/EEC). It embodies essentially the same principles as those put forward in the Substances Directive.

These European Directives are binding on Member States as regards the results to be achieved and Member States must establish legislation or administrative provisions to achieve these results. Thus individual States may choose to introduce this legislation in a manner which fits into their pre-existing legislative framework, and detailed provisions will thus vary within the European Union.

As an example of this implementation by a Member State, the UK has developed a scheme which is consistent with these Directives and is described in the Chemicals (Hazard Information and Packaging for Supply) Regulations (CHIP 2). These Regulations (which are currently at the Consultative Document stage) implement the above Directives and their Amendments and Adaptations as well as the Safety Data Sheets Directive (93/112/EC).

Classification under these Directives is based on the hazardous properties of the substance or preparation in question. Criteria are laid down against which each substance or preparation is judged to determine whether it possesses any of the 'Dangerous' properties (or hazard categories) defined in the legislation. These cover toxicological, ecotoxicological and physicochemical hazards. Substances which satisfy one or more of the criteria and thereby qualify for one or more hazard category are defined by the legislation as 'Dangerous'. Substances not so qualifying are not legally regarded as 'Dangerous', but users must nevertheless be aware of the hazards that they do possess and control exposure accordingly.

Hazard categories are identified by a pictorial symbol (e.g. skull and crossbones, St Andrew's cross, etc.), a 'symbol letter' (e.g. T+, Xi, etc.) and an indication of danger (e.g. VERY TOXIC, HARMFUL, etc.). These must appear on the label of any container or solvent so qualifying. RISK ('R') Phrases are intended to warn users of the potential dangers of the use of a solvent. SAFETY ('S') Phrases are intended to provide brief summary advice to users on precautions which should be taken in handling the solvent. 'R' and 'S' Phrases are identified by number. They must appear in full on the label and in the Safety Data Sheet.

Classification for Substances is based on available experimental data and covers the following types of adverse effect on health:

- acute lethal effects
- non-lethal irreversible effects after a single exposure
- severe effects after repeated or prolonged exposure

- corrosive effects
- irritant effects
- sensitising effects
- carcinogenic, mutagenic and reproductive toxicity effects
- other toxicological properties

Depending on the adverse effect concerned and the dose/exposure levels producing a response, substances may attract the following indications of Danger:

VERY TOXIC	T+
TOXIC	T
HARMFUL	Xn
CORROSIVE	C
IRRITANT	Xi

Preparations are also classified on the basis of experimental data, but, in the absence of such data on a specific preparation, classification is based on the properties of the component substances and the levels at which they appear in the preparation.

An example of classification and labelling under this system is cyclohexane, for which the appropriate symbols and 'R' and 'S' Phrases are shown below:

- Hazard Category: HIGHLY FLAMMABLE
- Warning Symbol: Black and orange flame on orange background
- Symbol Letter: F
- Risk Phrase: R11 HIGHLY FLAMMABLE
- Safety Phrases:
  S9 KEEP CONTAINER IN A WELL-VENTILATED PLACE
  S16 KEEP AWAY FROM SOURCES OF IGNITION – NO SMOKING
  S33 TAKE PRECAUTIONARY MEASURES AGAINST STATIC DISCHARGES

It can be seen that cyclohexane satisfies the criteria for HIGHLY FLAMMABLE, but not those for toxic or environmental hazards. This solvent does, however, possess the generic ability to cause CNS effects and irritation and this makes the point that the label alone is insufficient to give the user full knowledge of the potential hazards of a substance. More extensive information should be available in the Safety Data Sheet, which is required to be supplied with all substances classified as DANGEROUS under this system.

The system described above applies within the European Union and its Member States. Suppliers and users of solvents in other countries should ascertain details of relevant local legislation in this field.

5.4.2 *Safety Data Sheets*

The purpose of Safety Data Sheets is to provide the user with information about the hazards of a substance and the precautions necessary to avoid risks to health. There is a legal requirement in many countries for such data sheets to be supplied with the product and the onus for such provision is on the supplier.

Within the EU, the Safety Data Sheets Directive (93/112/EC) requires that data sheets be produced for all substances classified as DANGEROUS and information must be supplied under the following headings:

5.4.2.1 Identification of the substance/preparation and the supplying company. The name of the substance or preparation should be identical with that used on the label. Name, address and telephone number of the supplying company are required.

5.4.2.2 Composition/information on ingredients. Sufficient information required to enable user readily to identify associated risks. Preparations need not have full composition specified, but DANGEROUS components and those subject to official occupational exposure limits must be mentioned if they are present in greater than certain specified concentrations.

5.4.2.3 Hazards identification. Clear and brief indication of the most important hazards, in particular the critical hazards to human health and the environment. Should be compatible with information on the label but need not repeat it.

5.4.2.4 First Aid measures. Information should summarise briefly the symptoms and effects and immediate treatment to be followed in the event of accidental exposure. It should be easy to understand by the victim, bystanders and first aiders. The likelihood of delayed effects and any need for further medical treatment should be covered.

5.4.2.5 Fire-fighting measures. This should cover suitable and unsuitable extinguishing media, special hazards that might arise as a result of fire and special protective equipment for fire fighters.

5.4.2.6 Accidental release measures. Personal and environmental precautions that should be taken in the event of spillage should be covered, as well as methods for cleaning up spillages.

5.4.2.7 Handling and storage. Advice should cover technical measures such as ventilation and any prohibited or recommended procedures or equipment. Recommended materials of construction for containers and any incompatible materials may be mentioned.

5.4.2.8 Exposure controls/personal protection. Engineering measures recommended to control exposure should be covered, together with any occupational exposure limits or biological standards. Where personal protection is needed, the type of equipment should be specified. This will cover:

- respiratory protection
- hand protection
- eye protection
- skin protection

5.4.2.9 Physical and chemical properties. Information should be included, where applicable, on:

- appearance (colour, physical state)
- odour
- pH
- boiling point/boiling range
- melting point/melting range
- flash point
- flammability
- autoflammability
- explosive properties
- oxidising properties
- vapour pressure
- relative density
- solubility (water, fat)
- partition coefficient (*n*-octanol/water)
- other relevant data.

5.4.2.10 Stability and reactivity. Conditions under which hazardous reactions could occur should be stated. Conditions and materials to avoid should be covered as should the possibility of hazardous substances being produced on decomposition.

5.4.2.11 Toxicological information. This should include a concise but comprehensive description of the adverse health effects that could occur on contact with the substance or preparation. The different routes of exposure should be covered and symptoms described. All types of adverse health effect associated with the material should be included.

5.4.2.12 Ecological information. This should include relevant information on:

- mobility
- degradability
- accumulation
- ecotoxicity (short- and long-term effects)
- other adverse effects (e.g. Ozone-Layer depletion).

5.4.2.13 Disposal considerations. Appropriate methods of disposal of surplus or waste should be indicated, together with information to ensure their safe handling.

5.4.2.14 Transport information. This should cover any special precautions necessary in connection with the transport of the substance or preparation.

5.4.2.15 Regulatory information. The information on the label should be repeated in this section, together with any statutory restrictions on, for example, marketing or use. The user should be reminded to refer to specific regulatory provisions applying in the country involved.

5.4.2.16 Other information. Further relevant information may be included here, for example:

- training advice
- recommended uses and restrictions

- references
- date of issue

Within the United States, the OSHA Hazard Communication Standard (HCS) requires that a Material Safety Data Sheet (MSDS) be prepared for chemicals that are hazardous according to the criteria described within the HCS. An American National Standards Institute (ANSI) Standard has been prepared by the Chemical Manufacturers Association (CMA) with the aim of assisting the compilation of these MSDSs. This Standard [13], in final draft form, recommends the same 16 data elements, outlined above, as those applying within the European Union, and gives detailed advice on the compilation of each section, together with an analysis of the regulatory requirements in the United States, Canada and Europe. Example MSDSs are also included.

Historically, data sheets have varied substantially in the quality and quantity of information that they contain, although this is changing with the advent of legislation such as that described above. Some data sheets contain very little useful information while others contain a mass of raw, uninterpreted data. Neither is satisfactory and the user should be satisfied that an adequate understanding of the hazards and precautions appropriate to a given solvent is achieved, if necessary by contacting the supplier or the relevant Regulatory Authority.

It is most important that data sheets are available at the place at which solvents are actually used and that the information that they contain has been communicated to, and understood by, the workforce involved.

5.5 Criteria for control of risk

Risks to health are a function of both hazard and exposure. They can be reduced and controlled by reducing either the hazard or the exposure. A risk assessment will identify likely routes of exposure and should indicate situations in which exposure needs to be controlled in order to ensure adequate control of risk. Criteria for control of exposure vary depending on the route of exposure.

5.5.1 Inhalation exposure

Quantitative criteria are available for control of exposure by the inhalation route. Systems have been developed for the establishment of 'occupational exposure limits' (OELs), which represent the amount of solvent (vapour or mist) in the workplace air which is regarded as being sufficiently low to prevent adverse effects on health. Such OELs may be either established by a regulatory authority or developed in-house by a supplier or user.

A number of independent approaches to the setting of OELs have developed over the last fifty years or so in industrial nations. These have been reviewed extensively by Alexiadis [14]. Among these, some of the best known are:

- The US American Conference of Governmental Industrial Hygienists (ACGIH) list of 'Threshold Limit Values' (TLVs) [15].
- The German system of 'Maximale Arbeitsplatzkonzentrationen' (MAK, Maximum Concentration Values in the Workplace) and 'Technische Richtkonzentrationen' (TRK, Technical Exposure Limits) [16].
- The Netherlands 'Nationale MAC-lijst' (Maximale Aanvaarde Concentratie) [17].
- The United Kingdom system of 'Occupational Exposure Standards' (OES) and 'Maximum Exposure Limits' (MEL) [9].

In addition, within the European Union, a system has been developed by Directorate-General V (DGV) for the establishment of Binding Limit Values (BLV) and Indicative Limit Values (ILV) [18]. The former are directly binding on Member States, while the latter must be taken into account by Member States when setting national occupational exposure limits.

Although the above systems have developed independently, and individual limits for specific solvents may vary, all of them rely on the judgement of groups of scientific experts to review the available data and establish appropriate OELs on an individual, case-by-case basis. All these systems incorporate a period for public comment before a limit is finally established and most of them include within their groups of experts representatives of industry and the trade unions as well as independent and government experts.

In the case of the German and UK systems a distinction needs to be drawn between solvents with TRKs or MELs on the one hand, and those with MAKs or OESs on the other. MAKs and OESs are established for those substances for which it is believed that there is a clear threshold of exposure, below which no adverse effect on health is likely to occur. For some substances (e.g. genotoxic carcinogens and respiratory sensitisers), there are theoretical grounds for considering that there is no such threshold.

In the case of respiratory sensitisers, extremely low concentrations may elicit a response in people already sensitised to the substance in question and there is no evidence to indicate a threshold below which such a response will not occur. With genotoxic carcinogens, the so-called 'one-hit theory' maintains that a single molecular interaction between such a molecule and cellular DNA may be sufficient to start the formation of a cancerous growth, although the risk of this actually occurring would be very low. Although the body has several defence systems capable of preventing such a growth, there is no clear evidence that these would lead, in practice, to a threshold of activity and it is thus regarded as prudent to reduce exposure to genotoxic carcinogens to levels low enough to reduce any risk to an acceptably low level. For such substances, TRKs and MELs have been established as pragmatic limits representing the lowest levels that can reasonably be achieved in practice, and at which the risk is regarded as acceptably low. Compliance requirements also differ for the two types of limit. Although Indicative Criteria now exist within the UK for determining whether a substance should have an MEL or an OES, it is worth noting that a number of substances were accorded MEL status before these Criteria were established.

The methodology used to establish an OEL for a substance is complex and time-consuming and is properly the province of experts in the fields of toxicology, occupational hygiene and occupational medicine [19]. The following steps are involved:

1. Full literature review to ensure that all relevant data on toxic hazards are available. This will include human experience (epidemiology, workplace studies, volunteer studies), animal data, *in vitro* data, and, in appropriate cases, structure–activity information and data on closely related substances.
2. Identification of the 'critical' effect (or effects) that will determine the level of the OEL. By this is meant the effect(s) most likely to occur in practice if exposure exceeds an OEL.
3. Establishment of a 'No Observed Adverse Effect Level' (NOAEL) from the key studies for the critical effect(s). In some cases it will not be possible to establish a NOAEL from the available database, and in these situations, a 'Lowest Observed Adverse Effect Level' (LOAEL) is determined.
4. Establishment of an OEL at an appropriate level. The OEL will almost always be set at a level lower than the NOAEL or LOAEL, and its relationship to these figures is a matter for expert judgement. Factors affecting the level at which an OEL is established will include:

   - degree of confidence in key studies
   - whether key studies are in humans or animals
   - nature and severity of the 'critical' effect
   - whether 'critical' effect is well understood or rare
   - extent of concordance in animal studies
   - whether the reference point is a NOAEL or a LOAEL
   - the slope of the dose–effect curve (i.e. extent to which severity of effect increases as dose is increased)
   - known species differences
   - whether the effect is local or systemic
   - whether the effect is due to the parent molecular or a metabolite
   - pharmacokinetic data (e.g. half-lives)
   - precedents established from other substances

While the above process can be followed in-house, and some of the larger companies have made recommendations for OELs, there are official limits available for most of the solvents in common use.

OELs are expressed either as parts per million (ppm) volume/volume or as mg/m^3, and are related to a time reference period (e.g. 8 h Time Weighted Average (TWA) or 15 min Short Term Exposure Limit (STEL)). They do not provide a rigid dividing line between 'safe' and 'unsafe' conditions and it is good practice to maintain exposure as low as reasonably practicable below the limit. Exposures measured as described earlier may be compared with these limits and a decision taken on whether exposure is satisfactorily low or whether further exposure reduction is necessary. Exposure should be monitored with a frequency depending on its likely variability and the extent to which it is below the relevant OEL. Continuous monitoring of exposure is rarely required. It is important to remember that for substances with a SKIN notation the OEL is only valid as a level likely to be sufficiently low to protect health when adequate precautions have been taken to prevent skin exposure.

5.5.2 *Skin and eye exposure*

Measurement of the amount of solvent coming into contact with the skin is much more difficult than for the inhalation route and in consequence quantitative criteria are not generally available. Criteria for control of risk by this route are thus based on recognition and evaluation of the risk (either of local effects on the skin/eyes or of systemic toxicity caused by skin absorption) and control of the risk by taking practical steps to ensure that exposure is adequately low. These measures will include:

- properly designed equipment
- appropriate protective equipment
- satisfactory operating practices with proper documentation
- adequate training
- adequate supervision

Personal protective equipment should never be relied upon to cover inadequacies in equipment or in operating practice, and steps should be taken to eliminate any 'unofficial' use of solvents (e.g. for general cleaning purposes).

5.5.3 Exposure by ingestion

Although this is the least likely route of exposure, adequate precautions should be taken to prevent eating, drinking and smoking in areas where solvents are used. It is also important to prevent the use of unapproved containers for solvents.

5.6 Regulatory issues

Much of the regulatory background to the safe use of solvents has been mentioned earlier in this chapter. The most important pieces of legislation in the European Union, the UK and the United States are summarised below. It should be stressed that the onus is on suppliers and users of solvents in these and other countries to establish in detail the legislative requirements specific to their areas of interest.

In particular, it should be noted that different Regulatory Authorities may place different interpretations on a given database, perhaps leading to different conclusions for the same solvent on issues such as classification.

5.6.1 European Union

The European Union has the objective of harmonising minimum standards for occupational safety and health within its Member States. While such harmonisation will raise standards where they are weakest, it will not achieve uniformity since in many cases Member States are allowed to establish or maintain pre-existing standards where they are more rigorous.

Central to the development of European Union legislation is the Commission of the European Communities (CEC), which operates through 23 Directorates General (DG). The DGs having most relevance for the occupational health field are:

- DG V (responsible for Employment, Industrial Relations and Social Affairs including Health and Safety)
- DG XI (responsible for Environment, Nuclear Safety and Civil Protection including Notification, Classification, Packaging and Labelling of Dangerous Substances and Preparations)

The Commission (through its DG) makes proposals for legislation which have to be submitted to other statutory institutions for an opinion before being approved by the Council of Ministers or a Technical Progress Committee, on which each Member State is represented.

A wide variety of relevant European Union legislation has been developed over recent years, covering such areas as:

- Classification, packaging and labelling of dangerous substances and preparations.
- Notification of new substances (requiring the submission to a Competent Authority within a Member State of a defined set of hazard information before the substance may be marketed. The amount of information required increases as the quantity of the substance marketed reaches certain tonnage triggers.
- Existing substances (requiring manufacturers to submit to a Competent Authority a dossier which includes all available relevant hazard information; this is currently being applied to substances at the 1000 tpa level, and will apply to lower tonnage substances in the future). This legislation is in the form of a Regulation, which is directly binding in Member States without the need to be implemented by national legislation.
- Establishment of Binding and Indicative Limit Values for occupational exposure.
- General workplace safety and health (the first and second 'Framework Directives'; these cover evaluation of risks, choice of equipment and substances, 'competence', information and training, first aid, record keeping and a number of other related issues).
- Risk assessment (covering both new and existing substances and requiring Competent Authorities to assess both hazard and exposure as part of an assessment of risk). This risk assessment is based on the data submitted with the notification of a new chemical or contained within an existing chemical dossier, and the outcome may be:

(i) no further action needed;
(ii) further information on hazard and/or exposure needed;
(iii) further information on hazard and/or exposure needed when a specific tonnage trigger is reached (for new substances);
(iv) immediate measures are required to reduce the risk.

Most of this legislation has appeared in the form of Directives, which are binding on Member States in respect of the results to be achieved. Member States must make provision within their legal systems to ensure that these results are achieved, but pre-existing national legislation has meant that the details of their implementation vary.

For example, Germany has implemented the majority of the relevant EEC/European Union Directives through the following national legislation:

- The Chemikaliengesetz (Chemicals Law). This dates from 1980 and covers protection against dangerous substances, requiring registration and toxicological testing of new chemicals and calling for toxicological and ecotoxicological data on existing substances.
- The Technische Regeln für Gefahrstoffe (Technical Regulations for Dangerous Substances), covering safety, industrial medicine, hygiene and ergonomic requirements for dangerous substances.
- The Verordnung über gefährliche Stoffe (Directive on Dangerous Substances), issued in 1986. This covers classification, packaging and labelling as well as protective measures, prohibited uses, official regulations, health monitoring and a number of other related areas.

5.6.2 United Kingdom

The following brief description of the position in the UK is given as an example of how the results required by the relevant EEC/European Union Directives are covered by modifications and additions, where necessary, to a pre-existing pattern of legislation in one particular Member State.

The Health and Safety at Work etc. Act 1974 (HASAWA) is an enabling Act from which most other relevant UK legislation stems. The Act lays down broad principles which are developed in detail in the form of Regulations or Approved Codes of Practice. The Act established the Health and Safety Commission (HSC) as an overseeing body representing employers, trade unions, local authorities and independents. The Act also created the Health and Safety Executive (HSE) as the executive arm of the Commission.

The Control of Substances Hazardous to Health (COSHH) Regulations 1988 is a comprehensive piece of legislation stemming from the above Act. It lays down a rational approach to the control of risks to health from potential exposure to chemicals in the workplace. The approach encapsulated by COSHH (and its supporting Codes of Practice) is based on the principles of good occupational hygiene practice and covers the following:

- collection and evaluation of hazard information
- assessing risks to health in all work activities
- deciding on appropriate measures for the prevention or control of risk
- ensuring proper maintenance of the control measures
- monitoring exposure levels, if necessary
- carrying out appropriate health surveillance
- providing to workers appropriate information, instruction and training
- keeping appropriate and adequate records

The COSHH General Approved Code of Practice describes appropriate means by which the objectives of the COSHH Regulations can be met. The HSE also publishes guidance documents in the area, of which Guidance Booklet EH40 (Occupational Exposure Limits) [9] and EH42 (Monitoring Strategies for Toxic Substances) [8] are particularly useful. EH40 lists current UK MELs and OESs and describes the procedure involved in setting them. EH42 provides practical advice on monitoring workplace atmospheres.

Other important Regulations relating to the control of chemicals at work and stemming from HASAWA cover: First Aid; Protection of the Eyes; Reporting of Injuries; Diseases and Dangerous Occurrences; Classification, Packaging and Labelling; and the Notification of New Substances.

The Classification, Packaging and Labelling Regulations have been replaced by the Chemicals (Hazard Information and Packaging for Supply) Regulations 1993 (CHIP 1), which cover the classification and labelling requirements described in section 5.4.1. CHIP 1 is shortly to be replaced by CHIP 2 [20], which will implement the additional requirements contained within the 7th Amendment to the Dangerous Substances Directive, as well as introducing into UK law the requirements for Safety Data Sheets contained within the European Community Safety Data Sheets Directive (91/155/EEC). In this manner, the requirements of the relevant EEC/European Union Directives have been incorporated into the established pattern of UK legislation.

5.6.3 *United States*

In the United States, the principal basis for the control of chemicals in an occupational context is the Occupational Safety and Health Act (OSHAct), which became law in 1970. This Act established the Occupational Safety and Health Administration (OSHA) within the Department of Labor. OSHA has established since that time a wide range of Standards covering most aspects of occupational health, the majority of which are published in Title 29 of the Code of Federal Regulations (CFR).

Part 1904 of the CFR deals with the recording and reporting of occupational injuries and illnesses. Part 1910, in addition to Standards related to a number of specific chemical agents, contains Standards covering the following subjects:

- access to employee exposure and medical records
- ventilation
- hazardous materials (flammable, explosive, etc.)
- hazardous waste
- respiratory protection
- medical services and first aid
- air contaminants
- hazard communication

Of particular importance is the OSHA Hazard Communication Standard (1910.1200), which came into force in 1986 and requires manufacturers and importers of chemicals to assess the hazards of their chemicals and pass this information on to users. This general requirement covers:

- labelling
- material safety data sheets (MSDSs)
- training
- access to written records

This Standard contains detailed criteria for health hazards and a carcinogenic hazard must be mentioned if the substance has been assessed as a carcinogen by the International Agency for Research on Cancer (IARC), the US National Toxicology Program (NTP) or OSHA.

Another important piece of US legislation is the Toxic Substances Control Act (TSCA), which has been in force since 1977. This Act (which is administered by the Environmental Protection Agency, EPA) provides the regulatory vehicle to control exposure to, and use of, industrial chemicals not covered by previous environmental laws covering food, drugs and pesticide chemicals. It thus has a wider field of application than the OSHAct, which is limited to occupational situations. TSCA has four major purposes:

- to screen new chemicals to see if they pose a risk (section 5)
- to require testing of chemicals identified as possible risks (section 4)
- to gather information on existing chemicals (section 8)
- to control chemicals proven to pose a risk (section 6)

New chemical screening involves a ‘Premanufacture Notification’ (PMN) programme, which requires EPA to assess the safety of new chemicals before manufacture or import. Chemicals not included in an inventory of existing chemicals are regarded as ‘new’.

A manufacturer may be required to conduct testing on a substance (new or existing) if:

- it poses an unreasonable risk to health or the environment and there are insufficient data

or

- if it is produced in large quantities, entailing substantial human or environmental exposure and there are insufficient data to evaluate its effects.

In such cases, the EPA specifies the chemical to be tested, tests to be carried out, standards to be met and timetable for submitting results. Several solvents have been subjected to Test Rules under this Section of TSCA.

EPA may require manufacturers and processors to collect, record and submit certain information needed to permit reasoned judgements on the safety of chemicals. Information may be required in the following areas:

- data on production, use, exposure and disposal
- records of allegations of significant adverse reactions (health or environmental)
- unpublished health and safety rules
- notification of previously unknown risks

TSCA also contains provisions for EPA to control or ban chemicals which it considers to pose an unreasonable risk to human health or to the environment. Such action may range from requiring a warning label to imposing a complete ban on production, import and use. As of mid-1992, action under this Section of TSCA had been limited to polychlorinated biphenyls, chlorofluorocarbon propellants, asbestos in schools and dioxin (TCDD) wastes.

References

1. Royal Society of Chemistry (RSC) (1992), *Chemical Safety Data Sheets, Volume 1 (Solvents)*, RSC, Cambridge.
2. National Institute for Occupational Safety and Health (NIOSH) (various), *Criteria for a Recommended Standard*, US Department of Health, Education, and Welfare.
3. American Conference of Governmental Industrial Hygienists (ACGIH) (1991), *Documentation of Threshold Limit Values for Substances in the Workroom Air* (6th Edition), ACGIH, Cincinnati.
4. Health and Safety Executive (HSE) (1993), *Occupational Exposure Limits: Criteria Document Summaries*, HMSO, London.
5. Ballantyne, B., Marrs, T. and Turner, P. (eds) (1993), *General and Applied Toxicology*, Stockton Press, New York.
6. Clayton, G.D. and Clayton, F.E. (eds) (1991), *Patty's Industrial Hygiene and Toxicology* (4th Edition), John Wiley and Sons, New York.
7. TOXLINE, Datastar – Dialog Europe, online subfile.
8. Health and Safety Executive (HSE) (1989), EH42, *Monitoring Strategies for Toxic Substances (rev.)*, HSE, Sheffield.
9. Health and Safety Executive (HSE) (1994), EH40/94, *Occupational Exposure Limits 1994*, HSE, Sheffield.
10. Health and Safety Executive (HSE) (various), *Methods for the Determination of Hazardous Substances* (MDHS) Series 1–74, HSE, Sheffield.
11. Health and Safety Executive (HSE) (1994), *Workplace Air and Biological Monitoring Database* (3.5- and 5.25-inch floppy disks), HSE, Sheffield.
12. Health and Safety Executive (HSE) (1992), *Biological Monitoring for Chemical Exposures in the Workplace*, HSE, Sheffield.
13. Chemical Manufacturers Association (CMA) (USA) (1993), draft ANSI Standard for the Preparation of Material Safety Data Sheets (ANSI Z400.1), CMA, Washington.
14. Alexiadis, A. (1990), Comparison of worldwide procedures for setting occupational exposure limits, *Occupational Health in Ontario*, **11**(4), 168–192.
15. American Conference of Governmental Industrial Hygienists (ACGIH) (Annual Publication), *Threshold Limit Values for Chemical Substances and Physical Agents and Biological Exposure Indices*, ACGIH, Cincinnati.
16. Deutsche Forschungsgemeinschaft (Annual Publication), *Maximale Arbeitsplatzkonzentrationen und biologische Arbeitsstofftoleranzwerte*, VCH Verlagsgesellschaft, Weinheim, Germany.
17. Directoraat van de Arbeit (Annual Publication), *De Nationale MAC-lijst, Arbeidsinspectie P 145*, Directoraat van de Arbeid, Voorburg, The Netherlands.
18. EEC (1988), Amendment (88/642/EEC) to 'Framework Directive' (80/1107/EEC).
19. Chemical Industries Association (CIA) (UK) (1990), *Guidance on Setting In-House Occupational Exposure Limits*, CIA, London.
20. Health and Safety Executive (HSE) (1994), Draft Proposals for the Chemicals (Hazard Information and Packaging for Supply) Regulations (CHIP 2) – Consultative Document, HSE, London.

6 Solvents in the environment

I.D. DOBSON, A.S. McCORMICK and D.A. KING

6.1 Introduction

Solvents are an essential component in the manufacture of many of the products in common use both in industry and domestic applications. From industrial car paints to cosmetics and pharmaceutical products, solvents are often required in either their production or application. They provide the performance and efficiency that we require from these products.

This chapter covers the most significant areas where solvents can have an environmental impact, stratospheric ozone depletion (restricted to certain halogenates), tropospheric ozone creation in areas of nitrogen oxide pollution and solvents in water and solid waste. It gives an overview of these issues, and looks at the options available to reduce overall environmental impact.

The environmental impact of solvents lies in their emission rather than their use; while contained within a process they will have no environmental impact. For the most part, solvent emissions are into air; the quantity emitted into water is very much smaller.

Analysis of solvent impact is more complex than analysis of a single pollutant issue (e.g. NO_x or lead). Therefore assessment of the methods to reduce emissions requires much more careful consideration. Issues such as alterations in the durability of products, different environmental impacts of different solvents, local concentration of other pollutants and energy use all play a part in the contribution that any given system makes to the local and global environment. In addition, there are many more varied sources of volatile organic compounds (VOCs) than for other categories of atmospheric pollutants. Nor must we forget that in some regions, man-made VOC emissions are swamped by the background of naturally emitted VOCs. In such regions, limiting man-made VOC emissions would not be expected to benefit the environment. Indeed, the reverse could well apply.

In assessing the impact of solvent emissions we should consider the broad picture, evaluating total environmental impact of existing operations using solvents and the various options to reduce emissions. In principle the ideal approach would involve analysis of all emissions from 'cradle to grave' using a life cycle analysis (LCA) since this brings out the various trade-offs in different aspects of pollution which different approaches involve. In practice, full LCA is too time consuming and complex an exercise but nevertheless we should try at least to avoid the pitfalls of analysing a process on the basis of a single issue; otherwise this can lead to transfer of pollution rather than a reduction in overall environmental impact.

6.2 Solvents in the atmosphere

The major environmental issues concerning solvents arise from their emission into the atmosphere. These concerns fall into three broad categories:

1. Stratospheric ozone depletion as a result of the reaction of certain chlorinated hydrocarbons (e.g. CFCs, HCFCs, CCl_4, trichloroethane) with ozone.
2. Low level (tropospheric) ozone formation as a result of photochemical oxidation reactions in levels of significant nitrogen oxide pollution.

These two effects are unrelated, though often confused. A third effect is:

3. The climate change issue – the question of possible global warming.

Only one commonly used solvent group has a significant global warming potential – namely HCFCs. For all other solvents, atmospheric degradation is so rapid that their emission broadly equates to a CO_2 emission. However, in almost all industrial processes CO_2 emission from energy use is very, very much greater than solvent emissions. As an example, CO_2 emissions for Europe in 1990 were estimated at 3483 million tonnes of which 12 million tonnes was estimated to come from solvent emissions (excluding halogenates). In addition, solvents very often play an important part in reducing process energy requirements (relative to alternative systems) and therefore in some cases can contribute a reduction to the CO_2 inventory which would result from avoiding the use of solvents (see section on LCA). Overall the impact of solvent emissions on the greenhouse gas inventory could only be theoretically assessed by LCA of existing systems and alternatives. Pragmatically, whilst we cannot say overall whether the net effect of solvent use gives a positive or negative

Table 6.1 Ambient air concentrations for toluene and trichloroethylene

Solvent	Typical urban air concentration (mg/m^3)	Threshold Limit Value (mg/m^3)
Toluene	0.0001–0.204	188
1,1,2-Trichloroethylene	ca 0.05	269

contribution to the problem, we can say that at around 12 parts in nearly three and a half thousand, and falling as emissions are reduced, it can only ever be a tiny part of the total picture.

Clearly, the focus of our attention in this chapter should be on stratospheric ozone depletion and tropospheric ozone formation.

It is important to note that all the above environmental impacts come from the secondary consequences of the atmospheric chemistry of solvents.

6.2.1 Direct effects of solvents in the atmosphere?

It is very rare in normal operation that solvent concentrations in ambient air would come anywhere near the levels determined to be safe for workplace use. Organic solvents generally have relatively low toxicity and are emitted in low concentrations, and are then further diluted by dispersion. The concentrations of toluene and trichloroethylene in the environment are given in Table 6.1, together with typical occupational exposure standards for these solvents. From this we can see that the concentration in urban ambient air is several orders of magnitude lower than levels which would be considered safe. In addition these materials do not persist in the atmosphere.

As an example from Table 6.1, trichloroethylene, one of the longer lived solvents with a lifetime of 7.7 days, is used on a significant scale. Other major solvents have even shorter lifetimes and also do not accumulate in the atmosphere. As can be seen, even for trichloroethylene, the ambient air concentration is more than 3 orders of magnitude lower than the workplace recommendation. Similar or greater safety margins for other solvents would be expected.

Environmental concerns are therefore not due to the presence of solvents in the atmosphere *per se*, but rather to the reactions they can participate in which can produce secondary environmental impacts.

6.2.2 Stratospheric ozone depletion

In a simplified picture, Earth's atmosphere consists of the troposphere and stratosphere (as shown in Figure 6.1). Many natural and man-made materials react together in the troposphere; the reaction products can then be brought to ground level by rain or air currents.

However, there are a small number of substances which are sufficiently unreactive to pass through the troposphere to the higher atmosphere by a process of slow mixing. These substances may then be able to react with stratospheric ozone, thus contributing to ozone depletion. Only very unreactive substances live long enough to reach the stratosphere.

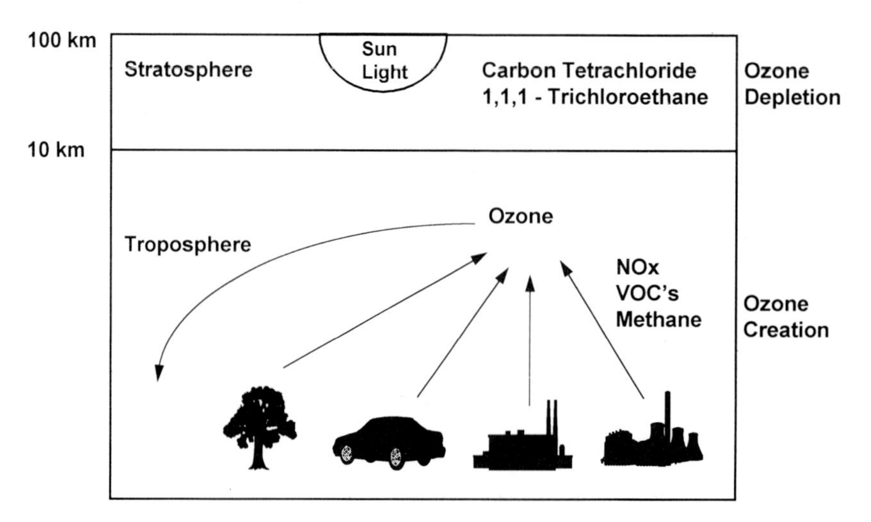

Figure 6.1 Solvents in the atmosphere.

Interestingly, this same natural process of slow mixing brings considerable quantities of stratospheric air down to ground level and provides part of the natural background of tropospheric ozone. It is important to be aware of the fact that there is an ever-present natural contribution to the photochemical pollution under current widespread scrutiny.

Low levels of the highly reactive form of oxygen, ozone, in the stratosphere are essential for the survival of life on Earth. Unfortunately, over the last two decades it has become clear that there is an ongoing and systematic depletion of part of the ozone-containing layer, most notably over the Antarctic. An increase in exposure of humans to UV radiation is expected to increase the incidence of some types of skin cancer and also cause cataracts. A decline in phytoplankton (microscopic plants) in the Southern Ocean has already been linked by some experts to the 'ozone hole'. In recognition of the need for action, at least 81 countries have signed the Montreal Protocol (originally formulated in 1987), which followed the Vienna Convention for the Protection of the Ozone Layer (1985). The Protocol, and its subsequent amendments (formulated as the problem was seen to be worse than originally thought), is perhaps the most visible example of international action in the face of a potential environmental crisis, and illustrates that concerted action can be taken when a problem is unambiguously identified and linked to specific substances.

Although the initial focus was on CFCs, used at the time for aerosol propellants (CFC-11 and -12), blowing agents (CFC-114), cleaners for electronics (CFC-113) and refrigerants (CFC-115), amendments to the Montreal Protocol have led to the control of such common solvents as carbon tetrachloride and 1,1,1-trichloroethane, which were used in very large quantities.

In response to the Protocol, except for certain medical applications, CFCs are now little used by the aerosol industry, alternatives such as hydrocarbons being used instead. Total phase-out of ozone depleters is expected by 2000, according to the Protocol, except in developing countries.

Why is the ozone layer so important and fragile? Ozone in the stratosphere protects life on Earth's surface through its ability to absorb much of the harmful ultraviolet (UV) radiation produced by the sun. As a molecule of ozone absorbs UV it splits into a molecule of conventional oxygen and a highly reactive oxygen radical (an electrically neutral single oxygen atom).

Normally, this radical reacts with the same, or occasionally another, normal oxygen molecule; this reforms ozone. In this way the harmful UV radiation is converted into harmless heat; for this reason the stratosphere is (perhaps) unexpectedly warm. A single ozone molecule can be responsible for mopping up several thousand particles (photons) of UV (Figure 6.2).

The current situation is an interesting example of the trade-off between environmental and other more easily predicted benefits. For example, on the one hand there is the well established (but perhaps exaggerated) risk of the use of flammable refrigerants, compared to the zero fire risk from highly stable halogen-containing compounds such as chlorofluorocarbons (CFCs). On the other hand there is the very serious environmental risk, which was almost completely unforeseen.

Non-halogenated solvents tend to have a very short atmospheric lifetime of only a few hours to a few weeks, depending on the climatic conditions. Even the presence of one hydrogen atom enormously shortens the lifetime of halogenated material as in, for example, methylene chloride. Therefore, both these kinds of material, in themselves, never build up in the atmosphere to levels that cause environmental or health problems. In contrast, some halogenated compounds which are not degraded in the troposphere, such as the

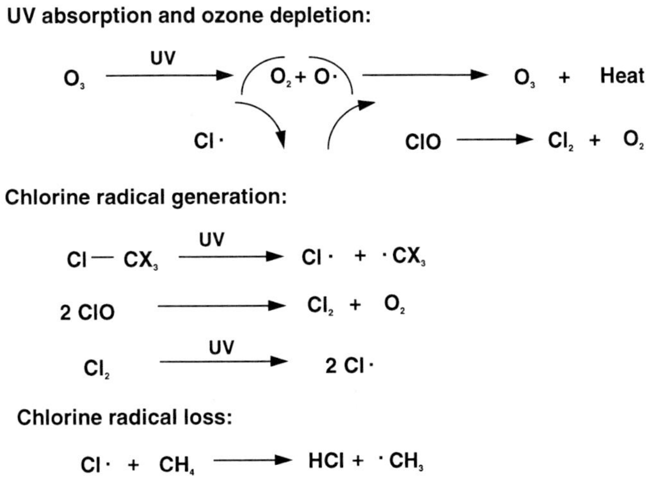

UV absorption and ozone depletion:

$$O_3 \xrightarrow{UV} O_2 + O\cdot \longrightarrow O_3 + \text{Heat}$$

$$Cl\cdot \qquad ClO \longrightarrow Cl_2 + O_2$$

Chlorine radical generation:

$$Cl{-}CX_3 \xrightarrow{UV} Cl\cdot + \cdot CX_3$$

$$2\,ClO \longrightarrow Cl_2 + O_2$$

$$Cl_2 \xrightarrow{UV} 2\,Cl\cdot$$

Chlorine radical loss:

$$Cl\cdot + CH_4 \longrightarrow HCl + \cdot CH_3$$

Figure 6.2 Some of the chemical reactions involved in ozone depletion.

Table 6.2 Ozone depletion potentials

Solvent formula	Ozone depletion potential
CCl_3F	1.0
CCl_2F_2	1.0
$C_2Cl_3F_3$	0.8
$C_2Cl_2F_4$	1.0
$CBrClF_2$	3.0
$CBrF_3$	10.0

CFCs, result in a potential build up which it is thought will be an environmental issue for about 100 years even if their emission stopped now.

The most problematic materials are CFCs and their bromine-containing relatives, because of their high stability in the lower atmosphere. However, they are not completely stable, and in the same UV-rich stratospheric environment that breaks down ozone (which then rapidly re-forms), the CFCs give rise to chlorine radicals, which under the right conditions live (and are re-formed) long enough to neutralise the effects of thousands of oxygen radicals, and hence molecules of ozone. Bromine radicals have a more serious effect, which is why halons (BCFCs etc.) used for fire-fighting have been limited to key essential uses.

Table 6.2 shows the ozone depletion potential of some of the more common ozone depleters.

How can a single chlorine atom react with a thousand or more other atoms? It does this through a radical chain reaction, similar to that used to make a number of industrial materials such as polymers and glues. In each stage of the reaction a radical (of one form or another) is preserved or is easily re-formed by UV; the chlorine radical is regenerated to knock out more ozone. A minority of reactions also, eventually, knock out the chlorine radical too (e.g. as the acidic gas HCl).

Why is there a hole only over the Antarctic? The reaction to destroy ozone in the stratosphere was initially predicted (via computer modelling) to have a much more widespread and noticeable effect than it has had. The initial model failed to take into account an inhibition of the destructive reaction in the gas phase by methane (which gives HCl). On certain surfaces (liquid or solid) normally inert forms of chlorine such as HCl can become re-activated, allowing halogen radicals to build up again.

Very high altitude ice clouds over the Antarctic provide such a surface, increasing the concentration of halogen radicals. This happens in the early spring as these clouds are lit by the sun; incidentally this illumination while darkness still pervades the Antarctic gives the clouds their name, noctilucent, literally shining at night. The spring build up of chlorine radicals prepares the way for large-scale ozone destruction when the UV radiation becomes intense enough to generate oxygen radicals in significant concentrations. As the stratospheric air warms up, the noctilucent clouds disappear and the ozone destruction stops, allowing the ozone layer to slowly rebuild, but not to the pre-spring levels.

Fortunately, such stratospheric ice clouds are rare because the UV absorption maintains the stratospheric temperature above zero. Indeed, the noctilucent clouds only occur in the upper, cooler parts of the Antarctic stratosphere. The weather patterns over the Southern Ocean also serve to confine stratospheric air over the Antarctic, perhaps exaggerating the scale of the problem. Even so, the Antarctic ozone layer has thinned from around 300 Dobson units to as low as 150 in recent years.

However, the problem is not confined to the Antarctic; lower levels of ozone are being observed over parts of South America. But this is nowhere near as marked in the northern hemisphere, and there is not yet clear evidence of an Arctic counterpart.

One way in which the global scale of the problem can be appreciated is the effect of powerful volcanic eruptions. These can thrust large quantities (many kilotons) of finely divided material deep into the stratosphere (including some chlorine also). Some of this material, especially sulphate aerosols, can provide catalytic surfaces for chlorine radical generation in the same way as the noctilucent clouds do. Strong stratospheric winds can spread this volcanic material around a hemisphere quite rapidly, and it eventually gets spread thinly round the globe producing a world-wide ozone drop, which has been measured from space. This effect is small, but it illustrates that at least the potential for effects is there.

A question that is often asked is whether the photochemical ozone pollution at ground level can have a desirable effect in counterbalancing the loss in the stratosphere. Although there is some benefit from ground level ozone in UV screening, policy is nevertheless targeted at limiting tropospheric ozone because of the concern to reduce the risk of possible irritant effects of peak levels of ozone at ground level.

6.2.3 Tropospheric ozone formation

6.2.3.1 Chemistry of ozone formation. Volatile organic compounds (VOCs), when present in the lower atmosphere, can affect reactions which produce photochemical oxidants. The most significant of these is the production of ozone. In the presence of nitrogen oxides and sunlight, VOCs can participate in the photochemical creation of ozone. The overall reaction scheme for this is shown in Figure 6.3.

The primary reaction to produce ozone (O_3) is therefore oxidation of oxygen by nitrogen dioxide (NO_2), which in turn is reduced to nitrogen monoxide (NO). This reaction requires sunlight. The equilibrium situation which exists in the absence of other pollutants is known as the photo-stationary state.

$$O_2 + NO_2 \longleftrightarrow O3 + NO$$

VOCs contribute (under certain circumstances) to increased peak ozone levels by assisting the conversion of nitrogen monoxide to nitrogen dioxide. This disturbs the kinetic equilibrium of the photo-stationary state by increasing the concentration of NO_2 available and simultaneously decreasing the NO concentration. This disturbs the photostationary state with the NO_2/NO equilibrium being re-established at a higher standing ozone concentration. It is wrong to say VOCs make ozone; they simply provide a reaction pathway to facilitate a higher standing concentration. Several other materials found in the atmosphere also contribute to this disturbance of the photo-stationary state (Figure 6.4) most notably methane, carbon monoxide and sulphur oxides. The presence of any of these materials in the atmosphere can increase NO_2 quantities relative to NO and therefore raise concentrations of ozone in the same way as VOCs can. However, because of a significant difference in the atmospheric reactivity of methane, carbon monoxide and SO_x compared to the majority of VOCs, their effect is manifested very much as a background ozone level upon which the effects of non-methane VOCs is superimposed. For this reason the effect of non-methane VOCs is classed as episodic, reflecting the fact that there is only a significant contribution during specific and short periods of particular meteorology. These events, which for the vast majority of locations are infrequent occurrences, are described as pollution episodes.

The atmospheric reaction chemistry is very complex. It should also be remembered that there is nothing special about solvent molecules in this respect. Natural VOC emissions play the same role and indeed in many regions actually dominate the chemistry. In those regions with a significant natural VOC inventory relative to NO_x pollution, modelling work shows that controls on man-made VOC emissions have essentially no effect.

Not only do natural emissions play a significant part in the episodic contribution to ozone, but addition-

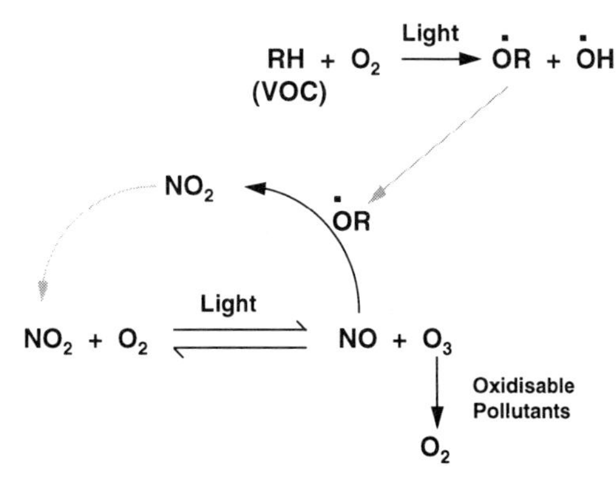

Figure 6.3 Atmospheric chemistry of VOC impact on ozone production.

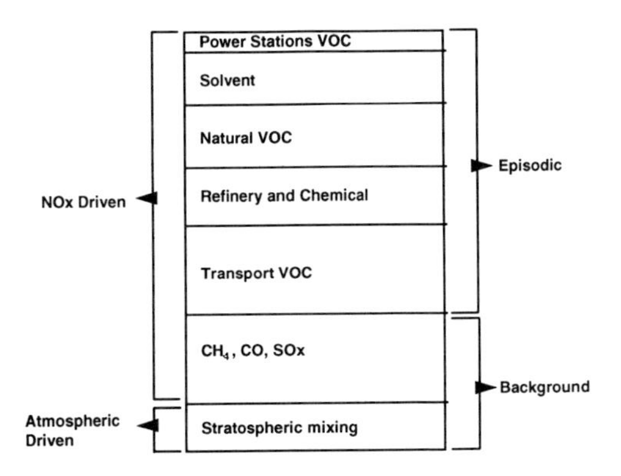

Figure 6.4 Illustration of comparative sources which contribute to an ozone episode in Northern Europe.

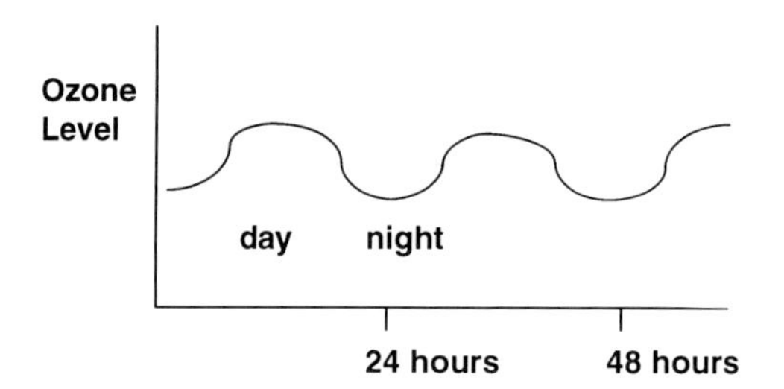

Figure 6.5 Variation of ozone concentrations with time.

ally, much of the background ozone level is from natural sources. As well as the methane CO and SO_x which are partly natural in origin, a significant part of the background ozone is from atmospheric mixing of the stratosphere with the troposphere. Jet stream mixing can in fact cause quite significant spikes in the usual ozone background.

Ozone is a highly reactive molecule. Therefore it can only persist in the atmosphere for a few days. Over time it will react with nitrogen oxide and other oxidisable pollutants and its concentration will be reduced. This includes ozone reacting with VOCs. Very high VOC levels act as a sink for ozone. Ozone formation has a diurnal cycle: levels rise during the day as the photochemical production occurs and potentially outpaces the rate of ozone disappearance, then decline at night as ozone production ceases and as the net reaction with other pollutants (Figure 6.5) removes ozone. Therefore tropospheric ozone formation is not a long term or persistent environmental problem. Since the reaction requires sunlight, ozone formation tends to be seasonal, being associated with summer.

VOCs, ozone and other pollutants can be transported by wind to regions remote from the site of emission. Tropospheric ozone formation is therefore a regional rather than a local issue.

The extent to which a VOC contributes to the formation of ozone depends on three main factors: (1) the nature of the particular VOC and its inventory; (2) the inventory of other materials in the atmosphere; and (3) weather conditions.

(a) Nature of the VOC – its potential to participate in the reactions to convert nitrogen monoxide to nitrogen dioxide. A measure of a VOC's reactivity in this is its Photochemical Ozone Creation Potential (POCP) or its Maximum Incremental Reactivity (MIR). In both these cases, values are determined for certain solvents by adding the solvent (experimentally or by modelling) to a system designed to mimic the atmosphere – containing the appropriate pollutants and subjected to relevant weather conditions. There is therefore some scope for different 'standard' conditions in this type of study. Therefore, while there is some variation in values determined by different groups of researchers, clear trends emerge in terms of those solvents which have a greater or lesser potential to contribute to ozone formation.

Table 6.3 shows the POCP and MIR values calculated for some of the most common solvents. POCP values are quoted relative to ethylene as 100; MIRs are determined as grams of ozone per gram of VOC.

Whereas aromatic hydrocarbons (especially the higher molecular weight materials) have high POCP and MIR values, many of the common esters and alcohols have lower values. Chlorinated solvents, because they are very unreactive, do not react significantly with nitrogen monoxide and therefore generally have very low POCP values (see section 6.2.1). Some of the natural VOCs (e.g. terpenes and limonene) have high POCP values.

(b) Other materials in the atmosphere – fundamentally, nitrogen oxides. As mentioned above, nitrogen dioxide is the actual source of ozone formation. In many cases the kinetics of ozone formation is zero order in VOC concentration (i.e. ozone levels are limited by the concentration of nitrogen dioxide). This situation exists where nitrogen oxide levels are significantly lower than the total VOC concentration. In such cases, limiting VOC emissions will have virtually no impact on ozone concentrations. Examples of such a situation are Scandinavia, Spain, Portugal and Greece (excluding Athens). Only where levels of both nitrogen oxides and VOCs are high, can reducing VOC emissions make a meaningful impact on ozone levels. Even then, some degree of caution is appropriate. Effectively all of the discussion over the effect of future policy on ozone levels is based on the prediction of atmospheric modelling. The most comprehensive of these models currently over-predicts actual UK background ozone levels by a factor of 100%. From the work of the UK Photochemical Oxidant Review Group, peak ozone levels also appear to be consistently over-predicted using the current models.

(c) Weather conditions. Since the reaction to form ozone is photochemical, ozone formation is naturally greater in more sunny climates and in summer. This is particularly true where nitrogen dioxide levels are

Table 6.3 POCP and MIR values for some common solvents

Solvent	POCP (rel. to ethylene = 100)	MIR ($gO_3/gVOC$)
Methane	1	0.01
n-Butane	40	0.62
n-Hexane	40	0.61
Branched C_{12} alkanes	40	0.79
Ethylene	100	5.3
α-Pinene	50	1.9
Benzene	20	0.28
Toluene	55	1.8
Xylene	65	4.6
1,3,5-Trimethylbenzene	115	7.4
Naphthalene	35	0.84
Methanol	10	0.39
Ethanol	25	0.77
n-Propanol	45	1.29
Isopropanol	15	0.38
n-Butanol	55	1.6
Ethylene glycol	40	1.1
Ethyl acetate	20	
n-Butyl acetate	45	
Isobutyl acetate	35	
Acetone	20	
MEK	40	
MIBK	65	
Trichloroethylene	7	
Methylene chloride	1	
Chloroform	1	

high and where low wind conditions prevent dispersion of pollutants. The most notorious examples of local geography playing a significant role are Los Angeles, Mexico City and Athens, where air quality is often poor. Indeed it is important that the local nature of these cities' problems is taken into account. It would be quite inappropriate and indeed impractical, to apply the same constraints on emissions generally as are required to regulate ozone in these locations. High ozone levels are relatively rare occurrences in cooler climates.

6.2.3.2 Sources of VOC emissions. There are many different sources of volatile organic compounds. The percentage split between different sources varies according to the country. Figure 6.6 shows values for European Union countries in 1985 (from Corinair).

23% of European VOC emissions come from natural sources – livestock farming and terpene emissions from trees are the major contributors to this. Of the man-made emissions, 35% is associated with transport – both the use and distribution of fuel, 16% from refinery and chemicals and 6% from power stations. Solvent use therefore constitutes only about 20% of the total VOC emissions in Western Europe. When these figures are weighted by POCP factors to reflect the ability of each source to contribute to ozone formation, the impact of solvents is reduced to 16%. In some countries, for example Spain and Canada, natural emissions far outweigh any other sources.

VOC emissions due to solvent use can be split by sector. Table 6.4 shows the figures for Western Europe in 1985 (from Corinair).

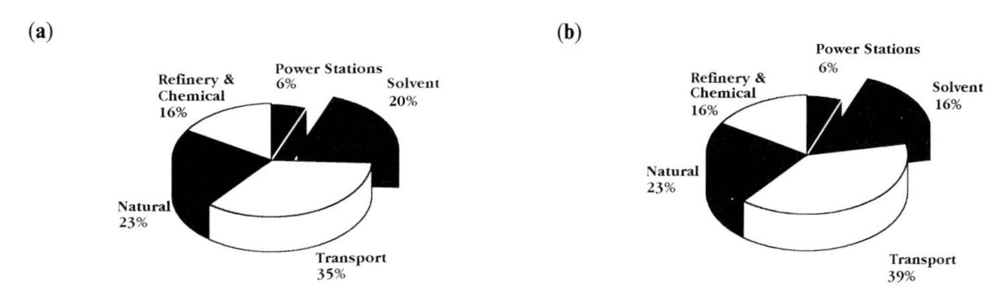

Figure 6.6 Contribution to non-methane VOC. (a) By mass. (b) POCP weighted. *NB*: These charts are for annual emissions. The natural emissions coincide with the summer ozone episode season. During the critical period, natural emissions may represent 60% or more of the total.

Table 6.4 Solvent emissions in Western Europe by sector

Sector	% Contribution
Surface cleaning	13.2
Painting and coating	
new vehicle painting	4.0
ship building	1.3
metal coating	10.2
wood coating	9.0
construction	15.3
vehicle refinish	2.3
domestic use	5.1
Printing	5.5
Adhesive and glue application	4.8
Dry cleaning	2.3
Fat and oil extraction	2.0
Private use (excluding paint)	11.4
Rubber manufacture and processing	6.6
Others	7.0

By application, about 84% of solvent emissions are from industrial solvent use, the remainder from domestic and trade activities.

6.2.3.3 Legislation to limit peak levels of tropospheric ozone.

(a) UNECE Protocols. With concern that guidelines on peak ozone levels can be exceeded and since ozone formation is a trans-boundary effect, the United Nations Economic Commission for Europe (UNECE) wish to achieve international commitment to reduce formation of low level ozone. To achieve this UNECE have opted to address emissions of both nitrogen dioxide and VOCs. It should be noted that NO_x is also being addressed as a pollutant in its own right as well as an ozone precursor.

Nitrogen dioxide emissions are the subject of a UNECE Protocol and of national and local regulations. The main sources of nitrogen dioxide are coal fired power stations and transport. Those countries which have signed up to this Protocol should ensure that their nitrogen dioxide emissions do not increase. Work is now beginning on a second stage NO_x Protocol which is expected to seek a genuine reduction in emissions beyond 2000.

There is also a UNECE Protocol on VOC emissions, the Geneva Protocol. This Protocol seeks a substantial reduction in VOC emissions. It sets targets of 30% reduction in anthropogenic (man-made) VOC emissions by the year 2000 referenced against various nationally selected baselines from 1984 to 1990 (for some signatories the target is a freeze in emissions rather than a reduction). This Protocol has been signed by the European Union (EU), its member states, the USA, Canada, Japan and some countries from Eastern Europe. Although the Protocol has not yet been ratified, this commitment underpins much of the national and local legislation on VOC emission control.

(b) Legal frameworks to control VOC emissions. Since the policy to control VOC emissions varies from one country to another it is beyond the scope of this chapter to discuss any legislation in detail. However, various mechanisms available to legislators to reach their emission reduction targets are explored.

Current trends in environmental legislation generally are starting to focus on cost effective reduction in overall environmental impact. Concentration on a single environmental issue is increasingly seen as an old fashioned approach which can result in regulations which increase pollution elsewhere; this can obviously have an adverse overall impact. By considering the total environmental impact of any emission reduction method, the total effect can be established. Such an approach is adopted in Integrated Pollution Control (IPC or IPPC) legislation. This requires users to look at all impacts of their current processes and at possible alternative techniques for improvement. Through this mechanism, any modifications must produce a genuine environmental benefit – they cannot simply move the environmental impact to another medium – e.g. water pollution. Life cycle analysis offers a tool to assess total environmental impacts of a system (see section 6.2.4).

Legislation which prescribes the use of one technology is unlikely to offer an optimum integrated approach in all cases and inevitably, by limiting the operator's flexibility, it increases costs. The most cost effective legislation sets end targets based on cost–benefit analysis, but it does not prescribe the mechanism for industry to use to achieve the target.

Figure 6.7 shows how the emission of solvent from a typical coating process can be reduced using three different technologies – conventional paint with end-of-pipe abatement, an alternative water-based paint and

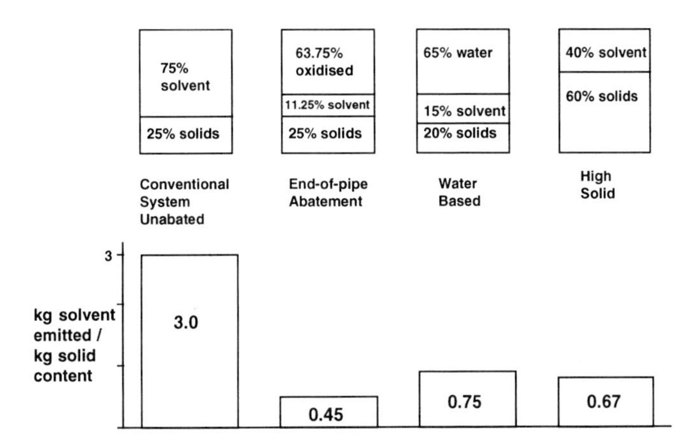

Figure 6.7 Options for reducing solvent emissions from a typical coating process.

thirdly a high solid solvent-based paint system. Each has its own technical, performance, and cost implications. All offer an effective means of limiting VOC emissions. Because of the different cost and performance profiles of each solution, the operator needs to be free to choose the option most appropriate for his needs if the cost of compliance is to be kept under control.

(c) Legislation in Europe. Some countries, such as the Netherlands, have opted for targets agreed with industry, although this is backed up with the mandatory Netherlands Emission Regulations (NER). However, most have chosen a fully legislative approach, controlling emissions from specified types and sizes of industrial processes. In most cases only certain industries are controlled. These are normally solvent end-using industries, particularly surface coating operations. In these industries solvent evaporation is an integral part of the process. This type of legislation will substantially reduce emissions from the industrial use of solvents.

Many European countries have opted for principles in line with the UK concept of Best Available Technique (Not Entailing Excessive Cost), the so-called BAT(NEEC) approach. There is an essentially equivalent American system, Reasonably Achievable Control Technique (RACT). As the names suggest, these require that sites covered by legislation use available and effective methods to reduce their solvent emissions, with the proviso that this should not involve 'excessive' costs. The costs (financial and also environmental) of reducing solvent emissions typically rises exponentially as more stringent targets are set. At the same time the environmental benefit of each extra tonne of emission limitation falls since, as total emission levels fall, peak ozone levels become more and more rare events. BAT(NEEC) or RACT based legislation accepts that excessive emission reduction targets can result in costs which would not be justified by any associated environmental gain.

In principle, BAT(NEEC) and RACT offer the user flexibility in choosing an emission reduction strategy – choice between a variety of end-of-pipe abatement techniques, low solvent technologies, waste minimisation or any combination of these. It recognises that the most appropriate solution will vary from one application to another with size, product quality required and the age and location of a plant.

In Europe there is already legislation controlling solvent emissions in many countries. The EU is also working on a draft legislative framework – the Directive on the Limitation of the Emission of Organic Compounds Due to the Use of Organic Solvents in Certain Processes and Industrial Installations. As currently drafted, this Directive is based on the principles of BAT; it will operate under Integrated Pollution Prevention and Control (IPPC) principles and require that any mechanism to reduce VOC emissions should be:

> without predetermining any specific technology or other techniques ... 'best' means effective in achieving a high level of protection for the environment as a whole, taking into account the benefits and costs which may result from action or lack of action. ... The techniques must be industrially feasible ... from a technical and economic point of view....

At the time of writing, the draft Directive sets out the options for BAT in industry-specific guidelines (annexes); these set out the various emission standards which constitute BAT within a given sector, together with the deadlines for compliance. It works on the principle of a reduction target for the industry sector. The end user is free to choose a mechanism to meet this target, using any technique available which does not simply move environmental impact elsewhere. US, French and German legislation is based on similar principles.

(d) Legislation in the USA. The USA has adopted a very different approach. At Federal level, emissions into air are controlled under the Clean Air Act. This creates many regulatory programmes to deal with

a variety of air pollution issues; most of these are not related to solvents. However, the Regulation to Control Ground Level Ozone and the Regulation of so called Hazardous Air Pollutants (HAPs) has a significant effect on many solvent-using industries. In addition, there is also the voluntary '33/50' programme.

The Environmental Protection Agency (EPA) has identified areas which do not meet national standards of 120 ppb for ground-level ozone concentrations; these have been designated as ozone non-attainment areas. In these areas there is a requirement to reduce VOC emissions (not including vehicle emissions) by 15% by 1999 and an additional 3% each year thereafter until the national standard is achieved. In ozone attainment areas these regulations do not apply. In regulated areas, there are different standards to control VOC emissions in various industry sectors. For example, in most coatings applications, targets are expressed as a notional average VOC content of the coatings; approaches which achieve equivalent emissions reductions are offered as alternatives (e.g. end-of-pipe control).

In addition to being regulated as VOCs, some solvents are also included in the HAPs list. Table 6.5 shows the regulatory status of most of the common solvents. The HAPs list contains a number of chemicals, including several common solvents, which are known to have very low toxicity. A recent EPA ranking of the relative hazards of all 189 HAPs found several solvents to be among the least hazardous materials on the list. Unlike the EU legislation, control of emissions classified as HAPs is based on Maximum Achievable Control Technology (MACT) rather than RACT.

The 33/50 Programme is a voluntary scheme designed to reduce emissions of 17 targeted chemicals. This list was created as much on the basis of the volumes of emissions as on environmental impact. For this reason, several solvents are included (Table 6.5). Companies (about 600 in total) that have joined this scheme were committed to achieving a 33% reduction in emissions of these particular materials by 1992 and a 50% reduction by 1995. The scheme appears to have made substantial progress towards these targets.

6.2.3.4 Options for complying with solvent emission legislation. To comply with solvent emission legislation, the options available will clearly depend on the exact wording of the legislation. MACT standards will often be more demanding and costly than BAT(NEEC) or RACT. However, since they only apply to HAPs-listed solvents, reformulation to other solvents is often the most effective method to meet such emission limits. Prescriptive legislation, unless carefully written to give maximum flexibility, can preclude certain environmentally acceptable and cost effective compliance options. Therefore not all solutions discussed below will necessarily be available to all solvent users.

Methods to reduce solvent emissions fall into four broad categories:

a. Waste reduction.
b. Reformulation to solvents which are not regulated.
c. End-of-pipe control technologies (abatement) – including recovery and destruction methods.
d. Process changes – low solvent systems or non-solvent technologies.

Choosing a mechanism for reducing VOC emissions will, for many companies, be a major strategic decision. Each of the above methods has its own advantages and disadvantages. The best compliance strategy will

Table 6.5 US regulatory status of some common solvents

Solvent	Regulated VOC	HAPs list	33/50 programme
Acetone			
Aromatic blends	•		
n-Butyl alcohol	•		
n-Butyl acetate	•		
Diacetone alcohol	•		
Dimethyl formamide	•	•	
Ethanol	•		
Ethyl acetate	•		
Ethylene glycol	•	•	
E-series glycol ethers	•	•	
n-Hexane	•	•	
MEK	•	•	•
Methanol	•	•	
Methylene chloride		•	•
MIBK	•	•	•
Perchloroethylene		•	•
n-Propyl acetate	•		
Isopropyl acetate	•		
Isopropanol	•		
Toluene	•	•	•
1,1,1,-Trichloroethane		•	•
Xylene	•	•	•

vary with size of operation from one operator to another, required productivity, product quality and many other factors all having a bearing. For many companies the optimal solution will involve a combination of the above options.

In weighing up the advantages and disadvantages of each of the above options, a solvent user will need to assess whether a proposed approach will:

- provide an acceptable product quality;
- maintain viable production rates;
- use existing equipment (or require acceptable modifications);
- be achievable with available resources (e.g. trained staff, R&D, buildings, capital);
- be cost effective;
- meet all legislative requirements (not just solvent emissions);
- be predictable and hold no surprises.

(a) Waste reduction. Reducing solvent waste will be an important part of any strategy to reduce solvent emission levels. Reduced losses from unnecessary solvent evaporation can often be achieved by better containment and mixing-vessel design. There are a variety of other site-specific options to reduce solvent losses. In many cases emission reductions through control of waste can be a very cost effective measure and in some cases can go a long way to meeting emission reduction targets.

(b) Reformulation. Substitution of one solvent by another can give environmental advantage if less of the new solvent is emitted or if the solvent has a lower potential to create ozone (i.e. a lower POCP or MIR value, see Table 6.3). If the mass of solvent emitted multiplied by the POCP or MIR of the solvent is lower for the new system, then a real environmental benefit can be achieved (all other impacts being equal). However, this will only offer a compliance option if this type of approach is incorporated into local legislation or, in the USA, if the solvents being replaced are on the HAPs list. Although not, strictly speaking, an environmental issue, reformulation can be used to replace materials which have toxicity concerns or to circumvent certain labelling requirements. For example, reducing the content of certain aromatic solvents in a paint can remove the need for a 'harmful' label.

Reformulation can clearly offer a low capital investment route to avoid HAPs regulations. R&D costs are often low and solvent/solvent reformulation does not normally require any of the process changes associated with low solvent systems. However, solvent choice is linked to other components of a formulation (e.g. resin in coatings). Therefore it may not always be possible to replace solvents without affecting product quality. Reformulation would typically result in a small overall increase in solvent costs.

(c) End-of-pipe control technologies (abatement). End-of-pipe abatement enables solvent users to continue to use established solvent-based systems. It prevents emission by some form of treatment after the process stage. It offers continued product quality with no major process modifications, therefore eliminating the complexity of many of the issues which surround changing to low solvent technologies (see below). In most cases an appropriate abatement technique will meet regulatory requirements. However, installation of end-of-pipe systems can represent a significant capital investment. Because of the capital element of the costs, it is usually the case that the larger the solvent-using process is, the more cost effective is an end-of-pipe solution. Process change options whilst having a capital element, are often more biased towards variable (process material) costs. For this reason there is a tendency for large operations to look towards end-of-pipe solutions whilst small operations may look more to process change.

End-of-pipe treatment has been, and will be, the compliance option chosen by very many solvent users. The merits of the most significant abatement options are described in detail in chapter 7 with particular reference to the cost implications.

There are a number of abatement methods available to most companies; these can be grouped into two categories – recovery and oxidation techniques.

(i) Recovery methods. Recovery can be split into two distinct operations:

- collection of the solvent;
- separation or purification to give a reusable product.

The cost of each will depend on the nature of the VOC stream, the mixture and type of solvents used and the purity required.

(a) Solvent purification. In evaluating the relative merits of recovery, one of the first considerations will be the need for and the potential cost and benefits of separation and/or purification. In some cases this consideration on its own will preclude the use of any recovery technique (particularly if a complex separation is required).

For on-site separation/purification of recovered solvent it is necessary to consider the number and complexity of distillations needed to obtain materials which are suitably pure for reuse. Where mixtures must be separated into individual solvents this can require several distillations, particularly where the solvents form azeotropes – this can significantly add to costs. The major costs associated with solvent purification are normally the capital required for distillation columns, energy and the additional staffing needs to oversee the operation. Where azeotropic distillations are required the cost of distillation columns can be greater than the capital cost of the recovery unit itself and staffing costs can be a significant variable cost (particularly if batch distillation is required).

(b) Recovery technologies. As described in chapter 7, there are three major types of recovery used to control VOC emissions – condensation, absorption and adsorption/desorption, with membrane technology also now being employed.

Condensation techniques (including sub-zero condensation) are normally only cost effective for very concentrated VOC streams, solvents with high boiling points or very high value solvents. Condensation is often used in the pharmaceutical and chemical production industries where very concentrated solvent streams at low flow rates are more common. However, it would often fail to meet emission limits as a stand alone technique and is therefore usually used in conjunction with an oxidation method.

Absorption can be a capital intensive technique and removal of solvent from the absorbent can be difficult and costly. It is normally only used if the solvents are considered particularly harmful, odorous or very valuable. It is also possible to use this as a non-recovery technique for small volumes of solvents (the absorbent plus solvent is normally oxidised); this is normally only done for odour control.

Adsorption/desorption involves trapping (adsorption) of the solvent on carbon (or another adsorbent), followed by removal of the solvent (desorption) using steam or nitrogen; after desorption, the stream is sufficiently concentrated for effective condensation to take place.

Adsorption with on-site recovery is generally best suited to single solvent systems where, ideally, the solvent is not soluble in water. The scale of solvent use must be large enough to justify the capital investment. The technique has found application in certain chemical and pharmaceutical manufacturing processes and in some of the largest printing installations.

Membrane technology involves the removal of VOCs by creating a pressure difference across the membrane which becomes more permeable to the solvent than to the carrier gas. It is best suited to high value solvents as the capital expenditure can be high.

(ii) Oxidation methods. There are three main disposal abatement methods used to reduce VOCs – biotreatment, adsorption with disposal, and thermal oxidation (considering catalytic oxidation as a variant of thermal oxidation). These are described in detail in chapter 7.

Biotreatment is gaining in popularity. It involves treatment of the VOC stream with bio-organisms and converts the solvent to carbon dioxide and water. It can have low investment and very low operating costs. However, it has a large space requirement and often has a low destruction efficiency (particularly when dealing with aromatic solvents); it can therefore fail to meet emission limits in some applications. The bio-organisms can be sensitive to sudden fluctuations in solvent concentration and to certain contaminants which may be present. Biotreatment is therefore best suited to low concentration solvent streams and low air flows. It is used effectively and extensively for odour control.

Adsorption with disposal of the spent adsorbent can offer a cost effective solution for abatement of relatively small quantities of solvent. However, for high total solvent emissions, costs for disposal of spent adsorbent can be high.

Thermal oxidation is proving to be one of the most widely used abatement methods, particularly within the surface-coatings industry; it is also used extensively in the pharmaceutical industry. It has the advantage of relatively low capital costs (compared to recovery and some low solvent technologies). With appropriate choice of equipment, thermal oxidation can also have very low operating costs. It is technically robust, being simpler to operate than some other options for VOC emission reduction. A particular advantage is that it can deal with virtually all solvents (though chlorinates need special consideration) and is well suited to solvent mixtures. As with all end-of-pipe solutions, it allows maintained product quality.

Thermal oxidation as a VOC abatement technique should not be confused with hazardous waste incineration – they have very different purposes and different feed streams. The issues surrounding hazardous waste incineration do not apply to VOC abatement. Thermal oxidation of hydrocarbon and oxygenated solvents is a clean process and converts a hydrocarbon fuel into carbon dioxide, water and recoverable heat.

Thermal oxidation involves heating up the waste gas stream to a temperature which is sufficient to sustain oxidation (Figure 6.8). The temperature required for oxidation will depend on the oxidation method

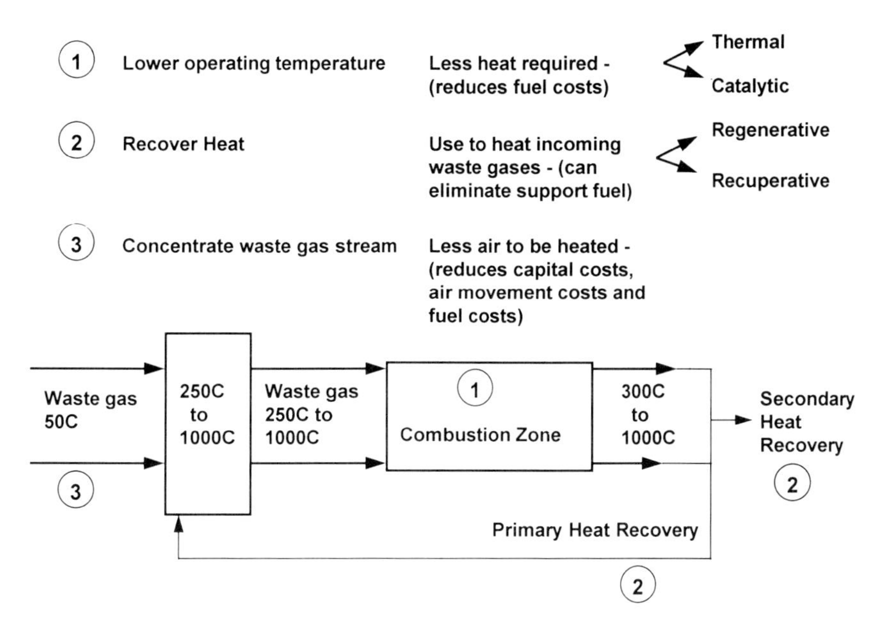

Figure 6.8 Reducing oxidation costs.

employed, destruction efficiency required and nature of the VOC stream. At very high concentrations, the temperature for thermal oxidation will be much lower. However, for most operations, the waste stream will be relatively dilute and the oxidation temperatures will be between 250 (using a catalyst) and 850°C.

To minimise costs, the heat generated in these systems is normally recovered. This can be primary heat recovered to heat incoming waste gas or additional secondary heat recovery to use elsewhere in the process or for space heating.

Important factors in the cost effective operation of a thermal oxidiser will normally be to reduce the volume of waste gas where possible and to choose the most cost effective mechanism to optimise oxidation temperature and value of recovered heat. Each of these is discussed below.

(a) Minimising air volumes. The volume of air to be treated is normally the main factor in the size (and therefore the capital investment needed) of a thermal oxidiser. Most operations have been designed to maximise drying, with no attempt to minimise air volumes. Therefore there is often potential to significantly reduce air volumes within the process without adverse impact on drying operations or retained solvent levels. Optimisation of fan and ventilation design and recirculation of air can offer effective mechanisms to reduce air flows in spray booths, printing processes and many coating operations. Air flows can be concentrated after the process by adsorbing solvent onto carbon (or other adsorbent) and desorbing in a lower volume of air, thus reducing the quantity of air required for oxidation.

(b) Maximising heat recovery. Oxidation of solvents in the waste gas stream generates heat. This, together with the heat used to increase the temperature of the incoming gas stream can be used either to heat the incoming stream (primary heat recovery) or to heat process air or space or water (secondary heat recovery). The value of secondary heat is very dependent on the nature and location of the heat generated and the end use required. For primary heat, there are two basic heat exchange mechanisms available – recuperative and regenerative. Recuperative heat exchange is normally less expensive to install, but it has lower efficiency (normally less than 75%).

Regenerative heat exchange can be more expensive to install but can readily offer 95% heat recovery efficiency. Regenerative systems are therefore best suited to more dilute streams or those with variable VOC concentration; they are widely used in coatings applications. Recuperative systems are more cost effective for higher-concentration VOC streams of constant composition; they find widespread use in chemical production. Both recuperative and regenerative heat recovery can be used with either thermal or catalytic oxidation systems and with secondary heat recovery.

(c) Reducing oxidation temperatures. Catalytic oxidation systems use a catalyst to reduce the oxidation temperature required. They therefore operate at the lower end of the range shown in Figure 6.8 – normally between 250 and 350°C. This can lower the cost of support fuel or heat recovery required. However, catalyst costs must be weighted against these potential benefits. Straightforward thermal oxidation systems are more robust, but they require a higher oxidation temperature and therefore need more effective heat recovery.

Catalytic systems are best suited to low flow, low solvent concentration streams which do not contain catalyst poisons. Thermal oxidation is generally better for high concentrations and high air flow rates.

With appropriate optimisation of air flow, heat recovery and oxidation method, it is normally possible to almost eliminate the need for support fuel. In some cases, the use of secondary heat exchange can generate a significant process credit (e.g. in applications where solvent concentrations in the waste gas are high and high temperatures are needed in the paint stoving process).

(d) Process changes. In certain applications, most notably coatings and surface cleaning, there may be potential to use technologies which require less organic solvent or are solvent free. These include water based (semi-aqueous) or high-solid formulations or non-solvent systems such as powder or radiation curable coatings or no-paint technologies such as coloured plastics.

The reduction in VOC emissions achievable may be sufficient to exempt the user from legislation. In other cases the process change is itself an accepted route to compliance with legislation.

Solvents are used in processes because they impart a certain performance to the system (see chapter 6). Removing them, or significantly reducing their concentration, almost always requires a major change in the rest of the formulation. For example, in coatings, solvents are needed to dissolve the resin; resins among other things give the coating adhesion, gloss and durability.

Therefore with less solvent, resin choice is more limited and as the systems technology is stretched, the performance characteristics of the coating are altered compared to a traditional system. Usually a low solvent product will be more specialised than the traditional solvent based version, often meaning more restricted flexibility in application or use.

For this reason, low solvent coatings do not always provide the product performance required. Low solvent systems generally require more expensive resins and may require the introduction of performance enhancing additives. Therefore these systems are generally more expensive than traditional solvent based systems. With domestic decorative paints for example, the cost increase on moving from solvent based to water based gloss has been estimated at 30–50%.

Although alternative technologies are often said to offer a low capital investment solution to lowering VOC emissions this is not always the case. Additionally, there can be significant costs associated with:

- increased costs of consumables;
- R&D;
- product evaluation;
- equipment modifications (for non-solvent systems new equipment will potentially be required);
- retraining staff in new application techniques;
- customer trials to check product performance;
- lost production during changeover;
- higher product rejection rates;
- for water based systems there may also be the need for a water treatment system;
- energy costs can increase, e.g. with water based cleaners or coatings use.

Nevertheless, for many coating processes encompassed by VOC legislation, low solvent systems are an important compliance option.

6.2.4 Life cycle assessment (LCA)

LCA is a developing technique to study the environmental effects of an activity, e.g. manufacturing a consumer item, by considering the whole product life cycle. The information gained by this process is used to identify options for reduction of the environmental effects, either by means of process improvements or substitution. Clearly, such a substitution can only be made by conducting two or more parallel studies. This is an area fraught with difficulty.

LCA data can also be used to investigate the validity of environmental claims made for a particular product (or process) over another similar one. This is a cornerstone of the various European or national ecolabelling schemes. Ecolabels are designed to protect the public from such unjustified or misleading claims, by setting criteria for defining 'greener' examples of a particular type of product on the basis of a thorough LCA.

The investigation described in the first paragraph must include: winning the most basic raw materials such as crude oil, mineral ores or trees; the steps to manufacture the article; the packaging and transport; in-use consumption (e.g. water and electricity); durability; and results of disposal, to be able to justify the term LCA. Less complete studies can only address limited issues and may well miss environmental trade-offs from other parts of the cycle.

By using this technique, environmental professionals have laid to rest a number of cherished myths; it turns out for example that plastic bags are often better than paper, that recycling plastics is not always the best option and, turning towards solvent use, that water based industrial paints are no better than the traditional alternatives.

There are four steps in the LCA process: working out exactly what is required; collecting the data; analysing the data; and finally, identifying the management options for improving environmental performance.

As in many walks of life, the biggest mistakes can be made at the beginning. There are many studies that have been flawed by either not looking at the whole process, or by failing to set up a fair comparison between systems. It is most important to establish a reference (Functional Unit) against which measurements are to be made. As its name suggests this is usually related to the function of the article in question, as it is employed. If one compares two items then they must have equivalent performance; so if 10% of consumers use doubled-up paper bags to avoid risk of breakage, then the functional unit must reflect this.

Process improvements can usually be made so it is important also to consider the degree of optimisation currently applied; studies should examine current best practice and not possibilities that have not yet been demonstrated to be economically feasible.

Although the methodology of data collection is well established, the trickiest part is analysing what the data mean; currently this is being vigorously debated. How do you judge between global warming and, say, a polluted local river? It is likely to prove an interesting discussion.

To date, the experience of LCA shows that there is generally no clear choice between the various alternative options. Most importantly, it shows on a rational basis how it is easy to be misled by unjustified perceptions.

6.2.5 Summary

Emission into air is the most significant pathway for solvents to enter the environment.

Two environmental issues are important, neither related to the solvents themselves but rather to the atmospheric chemical processes through which solvents are degraded.

The first of these issues, stratospheric ozone depletion, is caused by certain halogenated materials, some of which (CTC and 1,1,1-trichloroethane) are still used as solvents.

The second issue, tropospheric ozone formation, is a complex atmospheric process in which organic material (VOCs) in the atmosphere can react with nitrogen oxides and sunlight to give occasional ozone level peaks. Solvents are one such source of organic material.

A full assessment of the methods of limiting solvent emissions shows that care must be taken not to focus only on solvent emission, and in doing so, make other emissions worse. A variety of methods exist for limiting solvent emissions, but the cost is significant and because of the complexity of the issue, the benefits of limiting solvent emissions are difficult to predict. There are certainly many areas where limiting solvent emissions has no effect on the ozone concentration, and the issue must be addressed via NO_x controls.

In the EU and USA, legislation is in place or being developed not just for solvents, but for a range of VOC sources and also for NO_x. This will ensure a reduction in peak ozone levels through the next decade and beyond. Given the additional costs of further measures for solvents and the complexities of the environmental consequences which further emission reduction could bring, any further measures to limit solvent emissions into air would need to be very carefully analysed.

6.3 Aquatic toxicity of solvents

6.3.1 Introduction

Very few solvents pose a threat to the aqueous environment. Halogenated materials are the group of main concern with some lesser attention focused on aromatic and long chain aliphatic hydrocarbons. Often the main threat is not from a solvent, but instead from the dissolved material or else part of the additives package of a formulation (for example a biocide or a pigment). Physical effects can be just as important as chemical effects in the aquatic environment. For example, hot water discharge into rivers at power stations affects the local aquatic organisms. Similarly, adding fresh water to salt water causes changes at the point of discharge.

Aquatic toxicology and risk assessment are relatively young but rapidly expanding scientific fields. While a detailed coverage of aquatic toxicity is outside the scope of this book, the following sections develop the key concepts to the degree necessary to understand:

- potential aquatic harm;
- the data used to quantify that harm;

- the use of regulatory controls;
- solvent groupings where aqueous toxicity can be an issue.

6.3.2 *Toxicology and ecotoxicology*

Human toxicology is concerned with collecting data on the effects of exposure to a test substance in order to assess the potential of that substance to cause harm to man. Ecotoxicology evaluates the potential of a substance to accumulate, persist or cause harm in living organisms following release into the air, water and soil compartments of the environment.

Aquatic hazard testing gives an indication of the toxicity of a substance by utilising species representative of the three main levels of existence, i.e. primary producers (e.g. bacteria, algae), primary consumers (e.g. non-predatory invertebrates such as Daphnia) and secondary consumers (e.g. fish).

6.3.3 *Potential environmental effects*

(a) Aquatic toxicity. Chemical substances can have direct toxic activity against aquatic organisms. An example of substances which are very toxic to aquatic organisms is the group of organo-tin compounds, e.g. tributyl tin oxide, which have been used in marine anti-fouling paints to inhibit the growths on the hulls of ships and other submarine structures.

The normal measures of acute toxicity which are used for classification and initial risk assessment are:

1. The 72 hour IC_{50} (concentration which inhibits growth or growth rate of an alga by 50% during 72 hours' exposure).
2. The 48 hour EC_{50} (concentration which immobilises 50% of Daphnia during 48 hours' exposure).
3. The 96 hour LC_{50} (concentration which kills 50% of fish during 96 hours' exposure).

(b) Biodegradation, persistence and oxygen depletion. Micro-organisms in water courses, soil and waste water treatment plants act as 'decomposers', breaking down organic material to carbon dioxide and water. Substances for which there are no micro-organisms adapted to biodegrading them, or which are toxic or inhibiting to micro-organisms, are not broken down and are said to persist.

If a substance biodegrades in an aquatic environment, further information is needed on whether the substance has degraded to innocuous molecules or to relatively persistent and toxic derivatives.

There are two types of biodegradability. Primary biodegradability, the loss of the original substance, is indicated by a change of a specific property such as colour or surface activity. Ultimate biodegradability refers to complete breakdown to carbon dioxide and water.

In practical terms substances can be placed in one of three degrees of biodegradability:

1. Non-biodegradable substances are those which show no evidence of primary or ultimate biodegradability.
2. Inherently biodegradable substances are those which show biodegradation (primary or ultimate) in a recognised test.
3. Readily biodegradable substances achieve rapid degradation in ultimate biodegradability tests, which are so stringent that it can be assumed that the substances will rapidly and completely biodegrade in the natural, aerobic, aquatic environment.

A range of tests and criteria have been developed, which are listed in Table 6.6.

Biodegradability data need to be studied with care. Substances are sometimes claimed to be biodegradable, an imprecise term, while the actual data may show the material to be inherently not readily biodegradable.

In cases where only Chemical Oxygen Demand (COD) and biodegradation after 5 days (BOD5) data are available a substance is also deemed to be readily biodegradable if the ratio BOD5:COD is greater than or equal to 0.5. Substances may also be deemed to be readily biodegradable if there is other convincing evidence to show that the substance is degraded (biotically and/or abiotically) in the aquatic environment by $>70\%$ within 28 days.

Organic matter can sometimes be so rapidly biodegraded that the dissolved oxygen in the water is depleted. In waste water treatment plants, this is overcome by aeration but in the natural environment, oxygen depletion can result in the suffocation of the aquatic biota. A typical example of this is water course pollution by agricultural effluents such as silage liquor.

(c) Bioaccumulation, biomagnification and secondary poisoning. Bioaccumulation (bioconcentration) in fish arises by direct uptake from water through the skin, through the gills, or by ingestion of food in which the substance has already accumulated.

The bioconcentration factor (BCF) is a measure of the ability of a substance to become bioconcentrated. It is the ratio of concentration in the organism to concentration in the water. In the absence of measured data,

Table 6.6 Biodegradability testing methods and pass levels

OECD Test Method	Definition/Pass Level
Ready biodegradability	
301A: DOC Die-away	>/=70% DOC removal within 28 days and within a 10-day window after 10% DOC is reached
301B: CO_2 Evolution (modified Sturm Test)	>60% of theoretical CO_2 production within 28 days and within a 10-day window after 10% CO_2 is reached
301 C: MITI(I)	>60% of theoretical BOD removal within 28 days
301D: Closed bottle	>/=60% of theoretical BOD removal within 28 days and within a 10- or 14-day window after 10% BOD is reached
301E: Manometric respiratory	>/=60% of theoretical BOD removal within 28 days and within a 10-day window after 10% BOD is reached
Inherent biodegradability	
302A: Modified SCAS Test	20–70% daily DOC removal during a 12-week test period
302B: Zahn–Wellens	20–70% DOC removal within 28 days
302C: Modified MITI	20–70% BOD of parent compound removal within 28 days
304A: Test in soil	Not specified (Release of $^{14}CO_2$ from labelled substrate)
Simulation (confirmation test)	
303A: Aerobic Sewage Treatment/Coupled Units Test	Degradation rate is calculated from % DOC removal
Miscellaneous	
Biodegradation in seawater	Not specified. (% BOD removed)

OECD = Organisation for Economic Cooperation and Development; BOD = Biochemical oxygen demand; DOC = Dissolved organic carbon; SCAS = Semi continuous activated sludge; MITI = Ministry of International Trade and Industry (Japan).

a bioaccumulation potential can be predicted from the octanol–water partition coefficient (P_{ow}) of the substance. A $\log P_{ow}$ of 3 or more usually indicates bioconcentration potential. The estimation of BCF from the $\log P_{ow}$ value is unreliable for $\log P_{ow}$ values > 6.

When substances accumulate at increasingly higher tissue concentrations in each subsequent step of the food chain, the substance is said to be biomagnified. Humans consuming fish in Japan showed serious health effects as a result of organomercury compounds biomagnified in fish. Direct evidence of biomagnification may not be evident in such drastic terms but there may be subtle effects such as reduced fertility and, as in the case of the insecticide dieldrin, reduced eggshell thickness in predatory birds. These effects are manifestations of secondary poisoning.

Where a material is potentially bioaccumulative and there is evidence from animal toxicology of possible adverse health effects from exposure to such accumulation, secondary poisoning might result.

6.3.4 *Substances controlled by inclusion on Aquatic Pollution lists*

(a) Introduction. Several countries (e.g. Canada, the USA, the UK and the EU) have produced priority lists of substances to be considered for control. A recent article [1] compares the substances included in these lists. Readers outside Europe should refer to the relevant list of substances detailed in reference [1].

In Europe, a reference list has been developed by water pollution directives and by the impact of international pollution conventions. A detailed description of this development is given in reference [2].

(b) EU Reference List of Substances. The list is better known as the 3rd North Sea Conference Reference List of Substances. The conference agreed that it should be a common reference list for the further development of national lists of priority substances within the EU.

The list (see Appendix 6.1) contains 170 substances said to have the potential to cause significant water pollution. The majority are not solvents. The important solvents included on the list are: the major chlorinated hydrocarbons; toluene; *ortho*, *meta* and *para* xylenes; ethyl benzene; and some long chain aliphatics (C8 upwards). Pollution authorities in EU member states refer to this list in setting their own national lists for control.

Many of the materials will be subject to national controls in EU member states and elsewhere. If a nationally listed solvent or a preparation containing one may appear in a discharge, discussions should be held with the pollution authority to obtain approval or an assurance that none is required. It may be advisable to carry out a risk assessment as detailed in section 6.2.6 before detailed discussions take place with the pollution authority.

6.3.5 Hazard assessment by use of EU classification

Classification categorises substances by their inherent properties to cause harm to man or the environment, based on the results of standardised laboratory tests. Presently such classifications are used only in the EU, although their wider use via the Organisation for Economic Co-operation and Development (OECD) may be a future development. The classifications lend themselves as criteria for selection of solvents, which will be of value both within and outside Europe.

(a) EU environmental classifications. Detailed criteria for tests and subsequent classification for physical, health and environmental hazards are given in:

1. The Dangerous Substances Directive [3] as amended, for substances (defined as chemical elements and their compounds in the natural state, e.g. toluene or a substance obtained by any production process, e.g. white spirit, mixed xylenes).
2. The Dangerous Preparations Directive [4] as amended, for preparations (defined as mixtures or solutions composed of two or more substances, e.g. an intentional solvent blend such as methylated spirits).

The EU has laid down the classification for about 1500 substances based mainly on physical and health criteria. An exercise is underway to review these classifications and add environmental classifications where appropriate.

(b) EU self classification procedures for substances and preparations. Unless the EU lays down the environmental classification for a substance then it is for the supplier to carry out self classification. In the EU the classification and relevant data must be entered in the safety data sheet as required by the Safety Data Sheet Directive [5].

The aquatic toxicity classifications for substances are given in Table 6.7, which gives guidance on when a symbol (dead fish and tree), indication of danger to the environment (N) and individual risk phrases are to be used.

Preparations share the classifications used for substances, but the allocation is dependent either on the results of tests (except for those aimed at carcinogenic, mutagenic, and reprotoxic endpoints) or upon calculations using concentration limits. Unless specific concentration limits are given when the EU agrees the environmental classification for a substance, default concentration limits apply. Those which will apply when the Directive is amended are summarised in Table 6.8.

(c) Indication of potential environmental risk from health classification. In the absence of chronic ecotoxicity data there may be indications of other adverse effects and of long term environmental effects from toxi-

Table 6.7 Summary of EU Dangerous for the Environment classification

Criteria	R-Phrase					
	50	51	52	53	Symbol	Indication of danger
Acute toxicity </=1 mg/l	+				N	+
Acute toxicity </=1 mg/l and NRB or log P_{ow} >/=3*	+			+	N	+
Acute toxicity >1</=10 mg/l				Not classified		
Acute toxicity >1</=10 mg/l and NRB or log P_{ow} >/=3*		+		+	N	+
Acute toxicity >10</=100 mg/l				Not classified		
Acute toxicity >10</=100 mg/l and NRB (a)			+	+		
Water solubility <1 mg/l and NRB and log P_{ow} >/=3* (b)				+		
Criteria not specified			+			

Lowest Acute toxicity test performed in fish, Daphnia and algae is used; *Unless experimentally determined BCF = <100; NRB = Not readily biodegradable; (a) Unless substance is rapidly degradable or not chronically toxic at 1 mg/l; (b) Unless substance is rapidly degradable or not chronically toxic at its limit of water solubility.
R50 = Very toxic to aquatic organisms; R51 = Toxic to aquatic organisms; R52 = Harmful to aquatic organisms; R53 = May cause long-term adverse effects in the aquatic environment; N = symbol for 'dangerous for the environment.'
Indication of Danger is a pictogram showing a dead fish and tree.

Table 6.8 Summary of EC Dangerous for the Environment concentration limit criteria (%) for preparations

Substance classification	Preparation classification					
	N,R50	N,R50–53	N,R51–53	R52–53	R52	R53
N,R50	>/=25	–	–	–	–	–
N,R50–53	–	>/=25	>/=2.5 < 25	>/=0.25 < 2.5	–	–
N,R51–53	–	–	>/=25	>/=2.5 < 25	–	–
R52–53	–	–	–	>/=25	–	–
R52	–	–	–	–	>/=25	–
R53	–	–	–	–	–	>/=25

Calculation to determine the R-Phrase for a preparation with per cent constituents of different toxicities: R50 if sum of % R50/25 + % R50–R53/25 >/=1; R50–53 if sum of % R50–53/25 >/=1; R51–53 if sum of % R50–53/2.5 + % R51–53/25 >/=1;R52–53 if sum of % R50–53/0.25 + % R51–53/2.5 + % R52–53/25 >/=1.

city studies; e.g. classification as a mutagen, as toxic or very toxic or as harmful with risk-phrase R40 ('possible risk of irreversible effects') or R48 ('danger of serious damage to health by prolonged exposure'). These indicators, coupled with a potential to bioaccumulate, should raise concern and risk assessment is advised.

6.3.6 *Risk assessment*

By considering exposure as well as hazard, the process of risk assessment estimates the potential for the inherent hazards of a material actually to be realised during its life cycle. It should be carried out if a proposed solvent/preparation is subject to specific water pollution controls or has been classified as dangerous to the environment. The results will lend support to discussions with pollution authorities.

(a) Risk assessment guidance. Detailed guidance on the methodology of risk assessment for existing substances (i.e. those on the European Inventory of Existing Commercial Chemical Substances (EINECS) [6]) has been published [7]. The method makes allowance for there being very variable amounts of aquatic toxicity data for existing substances.

For new substances not in EINECS, a standardised set of toxicity and ecotoxicity data (known as the 'base set') will be available, on which assessment can be based. The detailed guidance [8] includes tables of values for specific release scenarios, organised by product type and use, which are an attempt to refine the risk assessment.

(b) The basic approach to risk assessment. Risk assessment involves iterative revision of the ratio of Predicted Environmental Concentration (PEC) to Predicted No Effect Concentration (PNEC). The guidance defines different PEC values (local, and regional) which may be calculated if enough is known about the actual environmental fate of the material.

The PNEC value is obtained by dividing the acute toxicity value for the most sensitive species (i.e. the lowest of IC_{50} for algae, EC_{50} for Daphnia and LC_{50} for fish) identified in ecotoxicity studies by a large assessment factor. If data from chronic studies with fish or Daphnia are available, the assessment factor is reduced.

If PEC is greater than PNEC there is environmental concern and measures would be recommended to minimise releases. If PEC is much less than PNEC, use of the material should present no environmental concern.

Thorough risk assessment requires a considerable amount of time and expertise for which experienced professional assistance may be required. Sources of such assistance may be trade and industry organisations, raw materials suppliers, consultants and hazard investigation laboratories.

6.3.7 *Summary*

Most solvents pose no threat to the aquatic environment because they are readily biodegraded by microorganisms and show low toxicity to representative aquatic species in tests. However, some chlorinated hydrocarbon, aromatic and long chain aliphatic solvents may pose a hazard. The selection of a solvent to show concern for harm to the aquatic environment should follow the following steps:

1. Is the solvent listed in Appendix 6.1 or in the relevant national list of possible aquatic pollutants?
2. Is the solvent classified as requiring any of the Environmental Risk Phrases R50–53?
3. If the answer to either question is yes, then carry out risk assessment hazard to estimate whether the environmental hazard is likely to be realised, prior to detailed discussion with the appropriate authorities.

6.4 Terrestrial toxicity of solvents

6.4.1 Introduction

The environmental impact of solvents upon soil is not an issue of major concern. Attention has been focused historically on spillages of chlorinated solvents, which because of their high density sink, persist, and eventually migrate into aquifers. Direct soil contamination from accidental spillages is therefore a possible but generally insignificant exposure. The main potential source of soil exposure is the spreading of sewage sludge and/or landfill. As with the aquatic compartment, it is more likely to be the solutes rather than the solvents which cause concern.

EU criteria for the classification of substances hazardous to the non-aquatic environment are developing from terrestrial toxicity tests in higher plants and earthworms. The R-phrases R56 (toxic to soil organisms), R54 (toxic to flora) and R55 (toxic to fauna) have been established and test methods are under development.

6.4.2 Risk assessment for contaminated sewage sludge

Elimination of substances in sewage treatment plants involves biodegradation and adsorption onto sludge. The adsorption constant for organic carbon (K_{oc}) can be measured or, at least for unipolar organic molecules, may be taken to be equal to P_{ow}. The product of K_{oc} and organic carbon per cent gives the adsorption coefficient K′. Adsorption onto sewage sludge and the resulting maximum concentration of a substance in the sludge may be then be estimated.

Given $\log K_{oc}$ (or $\log P_{ow}$), $\log H$ (Henry's constant in $Pa\, m^3 mol^{-1}$), and biodegradation data, it is possible to determine the rates of elimination for substances in waste water treatment plants to air, water and sludge (or in overall removal), by use of a computer model and/or by reference to published tables [9]. These tables indicate that for most solvents in general use, no more than a very small percentage of solvent is released in sewage sludge.

For the aquatic compartment, risk assessment involves PEC and PNEC values. The PECsoil value can be calculated as the product of the concentration in sewage sludge, annual application rates, and average bulk soil density. It may be revised to take into account anaerobic biodegradation and leaching of the substance from the soil into surface water.

If no toxicity data are available for soil organisms, the PNEC value for aquatic organisms may be used to derive a PECsoil pore water/PNECaquatic organisms ratio. This is based on the assumption that the bioavailability, and therefore the toxicity of chemicals for soil organisms, is determined by the pore water in the soil. If PECsoil pore water/PNECaquatic organisms is > 1, then tests with terrestrial organisms are recommended for refining the assessment. If toxicity data for plants and earthworms are available, an assessment factor (1000 has been suggested) is applied to the lowest LC_{50}, EC_{50} or IC_{50} value available in the calculation of a PNECsoil value.

6.4.3 Ecotoxicological aspects of solvent disposal

Choice of particular solvents is a major issue arising from health, safety and environmental legislation. Selection needs to be made carefully as there can be unexpected consequences. For example, whereas most organic solvent-based paints have an inherent biostatic activity preventing microbial spoilage, water-based paints often require the addition of biocides to prevent this. There is concern that the operation of waste water treatment plants may be adversely affected by effluents from such systems [10].

Spent solvents recovered from chemicals used in manufacturing plant or from those used in cleaning processes consist of a mixture of substances which can be difficult to process for re-use. Recovery also produces residues for which landfill is one option for disposal.

Thermal oxidation is a widely used route and the resultant energy is available to be utilised to save fuel and hence energy costs. A particularly efficient option is use of spent solvent as a fuel for cement manufacture, where it offers an attractive alternative to the practice of burning powdered coal. This combines a second use of the solvent molecules, with a net reduction in environmental emission compared to the coal fired case.

6.4.4 Summary

The methodology for terrestrial hazards assessment is developing but there are to date few data to allow classification. Methodologies for risk assessment are becoming available.

In conclusion, the general use of organic solvents results in insignificant exposure of the terrestrial compartment.

References

1. Hedgecott, S. (1994) Prioritisation and standards for hazardous chemicals, in *Handbook of Ecotoxicology*, Vol. 2 (ed. P.I. Calow), Blackwell Scientific Publications, pp. 368–93.
2. Edwards, P. (1992) *Dangerous Substances in Water: A Practical Guide*, Environmental Data Services, London.
3. EEC Council Directive 67/548/EC (1967) as amended, on the approximation of laws, regulations and administrative provisions relating to the classification, packaging and labelling of dangerous substances.
4. EEC Council Directive 88/379/EEC (1998) as amended, on the approximation of the laws, regulations and administrative provisions of the Member States relating to the classification, packaging and labelling of dangerous preparations.
5. EEC Council Directive 93/112/EEC (1993) 'Safety Data Sheets Directive' as amended.
6. European Inventory of Existing Commercial Substances (EINECS), Official Journal C146A, Vol. 33, 1990.
7. Technical Guidance on Risk Assessment of Existing Chemicals in Accordance with the Requirements of Council Regulation 93/793/EEC (July 1994 Draft), German UBA.
8. Health and Safety Executive, UK. Risk Assessment of New Substances – Technical Guidance Document (1994).
9. In references 7 and 8, based on Struijs, J., Stoltenkamp, J. and van de Meent, D. (1991). A Spreadsheet-based Box Model to predict the fate of xenobiotics in a municipal waste water treatment plant. *Water Res.*, **25**, 891–900.
10. Water Based Paints: A Treat to the Environment? I Kloefver (Swedish Environment Agency), Proceedings, Eurocoat Conference, Nice, 16 Sept 1993.

Appendix 6.1

3rd North Sea Conference Reference List of Substances

Acenaphthene
Aldicarb
2-Amino-4-chlorophenol
Amitrol
Anthracene
Bentazone
Benzene
Benzidine
Benzyl chloride
2-Benzyl-4-chlorophenol
Benzylidene chloride
Biphenyl
Bis(2-chloroisopropyl) ether
N(4-Bromophenyl)methyl-1,2-ethanediamine
Butylbenzyl phthalate
4-Tert-butyltoluene
Carbazole
Carbofuran
Chloroacetic acid
Chloroaminotoluene (all isomers)
2-Chloroaniline
3-Chloroaniline
4-Chloroaniline
2-Chloroantrachinone
Chlorobenzene
1-Chloro-2,4-dinitrobenzene
Chlorodinitrobenzene (mixed isomers)
2-Chloroethanol
1-Chlorohexane
4-Chloro-3-methylphenol
Chloronaphthalene (all isomers)
4-Chloro-2-nitroaniline
2-Chloronitrobenzene
3-Chloronitrobenzene
4-Chloronitrobenzene
Chloronitrotoluene (all isomers)
2-Chlorophenol
3-Chlorophenol
4-Chlorophenol
Chloroprene
3-Chloropropene
2-Chlorotoluene
3-Chlorotoluene
4-Chlorotoluene
Coumaphos
Cresyldiphenyl phosphate
Cyanoguanidine
Cyanuric chloride
Cyclohexane
Cyclohexylamine
2,4-D (including salts and esters)
Decanol
Demeton
Dibutylbis(oxylauroyl)tin
Dibutyl phthalate
Dibutyltin oxide
Dibutyltin salts (other than dibutyltin oxide)
Dichloroaniline (all isomers)
1,2-Dichlorobenzene
1,3-Dichlorobenzene
1,4-Dichlorobenzene
Dichlorobenzidine (all isomers)
2,6-Dichlorobenzonitrile
1,1-Dichloroethane
1,1-Dichloroethylene
1,2-Dichloroethylene
1,6-Dichlorohexane
Dichloromethane
Dichloronitrobenzene (all isomers)
2,4-Dichlorophenol
2,3-Dichlorophenol
2,4-Dichlorophenoxy-4-aniline
1,2-Dichloropropane
1,3-Dichloropropan-2-ol
1,3-Dichloropropene
2,3-Dichloropropene
2,2-Dichloropropionic acid
Dichlorprop
Dicofol
Diethylamine
Diethyl phthalate
Dihydrazine sulphate
Dimethoate
Dimethylamine
1,3-Dinitrobenzene
Dinitro-2-methylphenol
2,3-Dinitrotoluene
2,4-Dinitrotoluene
Dinoseb
Di-*n*-octyl phthalate
Diphenoxymethanal
N,N-Diphenylamine
Diphenylchloroarsine
Diphenyl ether
Diphenylmethane
Disulfoton
Dithiocarbamates
Dodecylphenol (mixed isomers)
Epichlorohydrin
Ethylbenzene
Ethyldichloroarsine
2-Ethyl-1-hexanol
Ethyltoluene (mixed isomers)
Fluoranthene
1-Fluoro-4-isocyanatobenzene
Foxim
Hexachloroethane
Hexachloronaphthalene
1,2,3,4,7,7-Hexachloronorbornadiene
Isodecanol
Isononanol
Isopropylbenzene
5-Iso-azolamine
Linuron
2-Methyl-4-chlorophenoxyacetic acid
2-Methyl-4-chlorophenoxypropanoic acid
Methamidophos
Methoxy 4-propenylphenol
Methylcyclohexane
4,4-Methylethylidenebisphenol
2-Methylphenol
Mevinphos
Mineral oil
Monolinuron
Naphthalene
Nitrobenzene
4-Nitro-1-isopropylbenzene
4-Nonylphenol
Octane
Octanol
Omethoate
Oxydemeton-methyl
Paraquat
Pentachlorobenzene
Pentachloroethane
Pentane
Phenanthrene
Phthalic acid
Propanil
Pyrazone
Tetrabromomethane
Tetrabutyltin
Tetracarbonylnickel
1,2,4,5-Tetrachlorobenzene
1,1,2,2-Tetrachloroethane
Tetraethyl lead
1,1,3,3-Tetramethyl-4-butylphenol
Toluene
Triazophos
Tributyl phosphate
Trichlorfon
Trichloroacetaldehyde
Trichloroacetic acid
1,1,2-Trichloroethane
Trichlorophenol (all isomers)
1,1,2-Trichlorotrifluoroethane
Tricrresyl phosphate
α-Trifluoro-3-nitrochlorotoluene
α-Trifluoro-2-nitrotoluene
α-Trifluoro-3-nitrotoluene
α-Trifluoro-4-nitrotoluene
3,5,5-Trimethyl-1-hexanol
Trioctyl phosphate
Triphenyl phosphate
Tris(2,3-bromo-1-propyl) phosphate
Trixylenyl phosphate
Vinyl chloride
1,2-Xylene
1,3-Xylene
1,4-Xylene

7 Solvent transport, storage, recovery and disposal

G.P. HOWE

Part 1 Transport and storage

7.1 Introduction

In general, the economic benefits of operating on a large scale have ensured that solvents are manufactured at a small number of sites on facilities capable of annually producing several thousand tonnes of high grade material. However, the use of these chemicals is almost entirely separate from the site of production. One solvent can often be used for a number of different applications in a variety of locations throughout the world. This diverse range of use and user, and the relationship between production and deployment, necessitates the widespread transportation and storage of these materials. This ranges from bulk facilities, as might be found next to the manufacturing unit or primary packing facility, to the storage and supply of less than one litre. It is essential for successful application of a solvent that the transportation and storage are performed to the highest possible standards with respect to safety, health and the environment. Indeed the identification of an ideal solvent becomes irrelevant if the methods used for transportation or storage are:

- unsafe
- uneconomic
- dangerous to the environment
- detrimental to the integrity or quality of the solvent
- poorly contained, resulting in fugitive losses
- unreliable
- geographically restricted

Worldwide demand and consumption of solvents increased rapidly in the early development of solvent applications and it was to the benefit of the producer, user and society as a whole that regulation and standards emerged for solvent transportation and storage. These have become of increasing importance as advances in chemical engineering, and the benefits in cost, have enlarged capacities for the transport and storage of solvents. Clearly with increased size the risks and possible consequences of error or mishap during this phase of use dramatically rise. Unfortunately, several recent examples in the petrochemicals industry can be cited to highlight the importance of safe handling and transport of all chemicals, including solvents:

The loss of crude oil from the Exxon Valdez which subsequently contaminated large areas of the coastline in Alaska [1].

In November 1984 a storage depot near Mexico City containing liquefied petroleum gas (LPG) ignited following a large gas leak killing 550 people with a further 2000 injured [2].

In July 1978 a road tanker carrying liquefied propylene burst into flames whilst passing a campsite near San Carlos in Spain killing 215 people [3].

However, considering the regularity and distance solvents are transported, coupled with the large number of storage facilities throughout the world, it is a credit to the industry and the current guidelines and regulations that such instances are so rare.

The recommendations and regulations governing solvent transportation are now discussed in origin and detail. Solvent storage regulations and guidelines are dealt with later.

7.2 Transport of solvents

7.2.1 United Nations Recommendations on the Transport of Dangerous Goods

In April 1957 a United Nations Committee was formed with the specific intention of providing universal guidelines and principles for the transport of dangerous goods. These recommendations would form the basis of a uniform scheme for national and international regulations governing all modes of transport of dangerous goods. The guidelines attempted to align a multitude of different standards that complicated the international movement of chemicals. All recommendations refer to products which are packed as this is the most common method for the transportation of dangerous goods. The movement of bulk product is normally covered by national legislation which is often based on the principles defined within the UN guidelines. Since the inaugural session, the Committee, comprising the current designated UN experts, has subsequently revised and improved the recommendations on seventy separate occasions with the latest session being held in November 1994. The recommendations are normally up-dated every two years. The widely revered manual is now in its ninth revised edition and is commonly known as the 'UN Orange Book' [4]. The recommendations are intended to be a universal reference for packed material and cover:

- the principles of classification of dangerous goods
- the definition of each Class
- general packing requirements for each Class
- the procedures for testing of packages for suitability to transport of each Class
- labelling, placarding and marking of packages relevant for each Class
- the need for, and format of, transport documents

All solvents fall into one or more of the classifications of dangerous goods as specified by the UN experts:

Class 1 Explosives
Class 2 Gases: compressed, liquefied or dissolved under pressure
Class 3 Flammable liquids
Class 4.1 Flammable solids
4.2 Substances liable to spontaneous combustion
4.3 Substances which, in contact with water, emit flammable gases
Class 5.1 Oxidising substances
5.2 Organic peroxides
Class 6.1 Toxic substances
6.2 Infectious or repugnant substances
Class 7 Radioactive material
Class 8 Corrosive substances
Class 9 Miscellaneous dangerous substances and articles

An example of each class is provided for illustration in Table 7.1.

The Classes of most relevance to solvents are 2, 3, 6.1 and 8. The definitions of each class within the UN Orange Book are comprehensive and materials can be easily assigned. If a material possess more than one risk the hierarchy of hazard is also specified. The Orange Book contains a list of all commonly transported dangerous goods and assigns a unique UN number to each specific substance or category of substance. Every substance or article is also assigned to the appropriate class. The packaging, labelling, and documentation requirements for each class are defined and therefore for each substance or article listed. The recommendations are suitable for all modes of transport and do not distinguish between road, rail, air or sea.

The intention of the UN Committee of Experts has broadly been met. The recommendations set out in the Orange Book are now being incorporated into, and form the basis of, the legally binding regulations governing the transport of goods in the developed world.

Table 7.1 Examples, by classification, of dangerous goods as defined in the UN Orange Book

UN Class	Example
1	TNT, fireworks
2	methane, butane
3	acetone, methanol
4.1	naphthalene, crude
4.2	activated carbon, white phosphorus
4.3	lithium aluminium hydride
5.1	hydrogen peroxide
5.2	cyclohexanone peroxide
6.1	toluene, dichloromethane
6.2	viruses, genetically modified organisms
7	uranium hexafluoride
8	glacial acetic acid
9	polyhalogenated biphenyls, solid

7.2.2 *Transport regulations summary*

Regulations now cover transport by road, rail, air and sea. One of the first to emerge covered transport by rail in Europe and dates back to 1893 as part of the Convention concerning International Carriage by Rail (COTIF) [5]. In the United States of America the Department of Transportation is the government agency that mandates over carriage by rail. Canada has its own system based on the TDG (Canadian Transportation of Dangerous Goods). Regulations in the Far East and Pacific rim are similar to those in COTIF.

The International Maritime Dangerous Goods Code began as part of the 1929 International Conference on Safety at Sea and was first published as general provisions in 1948. The coverage is worldwide and incorporates both national and international transportation over water.

In 1953 it was realised that a set of regulations covering transport by air was required and in 1956 the first edition of International Air Transport Association Dangerous Goods Regulations was published. Finally in 1959 the European Agreement concerning International Carriage of Dangerous Goods by Road was published. UK national compliance to European directives on the classification, packaging and labelling of dangerous goods is implemented by the new Chemicals (Hazard Information and Packaging for Supply) Regulations commonly known as the CHIP regulations. As for rail transport the US Department of Transportation covers carriage by road in the USA and the TDG for Canada. The regulations in the Far East and Pacific rim are similar to those operating in Europe.

In reviewing the text of each of these it is clear that the regulations for transport by road and air are the most complicated. Transport by road is also the most common method for carriage of solvents and hence is discussed first and in more detail providing an indication of the complexity of all transport regulations.

7.2.3 *European Agreement Concerning the International Carriage of Dangerous Goods by Road* [6]

This agreement is universally abbreviated to 'ADR' which is derived from its French title 'Accord Européen relatif au transport international des merchandises dangereuses par route'. The text of the agreement was first published as White papers in 1959 with the latest update printed in 1995. The purpose of ADR is to gain agreement to freely permit transport throughout the territory of a signatory if, on arrival at the frontier, dangerous goods in or on a road vehicle are packed and labelled and carried in containers or transport units which meet the guidelines set out in ADR. The regulations cover both bulk and packed product. The benefits of ADR mirror those of the UN Committee of experts such that the transport of dangerous goods is safe and can occur across national and international boundaries without unnecessary complication and delay.

In July 1992 the list of signatory countries comprised;

Austria, Belgium, Bosnia-Hercegovina, Belorussia, Czech Republic, Denmark, Finland, France, Germany, Greece, Hungary, Italy, Luxemburg, the Netherlands, Norway, Poland, Portugal, the Russian Federation, Spain, Sweden, Switzerland, the United Kingdom and Yugoslavia

In general, the competent authority for administration of ADR in each country is the government department associated with transport.

The ADR text is split into two Annexes, A and B. Annex A is a list of the goods which can be carried internationally under its provisions (defined as conditions or stipulations). It also refers to the responsibility of the consignor of the goods to correctly identify the goods as classified in ADR and to ensure the packing, labelling and transport documents are correct. Annex B provides the provisions for the vehicles and transport operations and clearly assigns the responsibilities of the carrier to provide and equip the vehicle and to ensure the activities of the driver are correct. An insight into the complexity of transport regulations is obtained from the text of Annexes A and B of ADR.

7.2.3.1 Annex A. Split into three parts in which Part I provides the list and definitions for each Class of dangerous goods and the general provisions. These classifications are identical to those set out in the UN Orange Book. However a special feature of ADR is that the Classes are either Restrictive or Non-Restrictive. For Restrictive Classes only goods specifically listed can be carried. Dangerous goods which are part of a Restrictive class but are not listed cannot be carried under the terms and conditions of ADR. Alternatively, a material which is within a Non-Restrictive Class can be carried under ADR unless otherwise specified within the regulations. The list of Restrictive Classes is provided in Table 7.2. In the majority of cases solvents are within a Non-Restrictive Class.

Part II consists of chapters for each Class and provides a list of all the goods covered by ADR and the special provisions for each Class. Goods are either specifically named, as in a Restrictive Class, or for a Non-Restrictive Class can be identified within a Section. The identification of a substance within a Class, and thereafter a Section, specifies the provisions which must be followed to allow transport under ADR. These include the net weights below which materials are exempt, packing, labelling and marking requirements, and

Table 7.2 List of Restrictive Classes in ADR

Class	Example
1	Explosives
2	Gases: compressed, liquified or dissolved under pressure
6.2	Repugnant or infectious
7	Radioactive material

Table 7.3 Classifications of dangerous goods within ADR of most relevance to solvents

Non-restrictive class	Example
Class 3	*Flammable liquids*
Section A	Substances having a flash point below 21°C, not toxic and not corrosive e.g. acetaldehyde, acetone, gasoline
Section B	Toxic substances having a flash point below 21°C, e.g. methanol, pyridine
Section C	Corrosive substances having a flash point below 21°C, e.g. pyrrolidone, acetyl chloride, triethylamine
Section D	Non-toxic and non-corrosive substances having a flash point between 21°C and 100°C, e.g. *n*-propanol, cumene, turpentine
Class 6.1	*Toxic substances*
Section B	Organic substances which have a flash point of 21°C and over or are non-flammable, e.g. aniline, toluidine, methylene chloride, trichloroethylene
Class 8	*Corrosive substances*
Section A	Acidic substances e.g. maleic anhydride, trifluoroecetic acid
Section B	Basic substances e.g. benzylamine, ethanolamine

the description of goods to be used on transport documents. Table 7.3 illustrates, with examples, the Classes and descriptions of dangerous goods of most relevance to solvents. In some cases the classification of a particular chemical is historic. For example, within ADR, benzene, a suspected human carcinogen, is Class 3 Section A (Flammable liquid – flash point below 21°C, not toxic and not corrosive) which seems somewhat unusual and perhaps reflects only the short term hazard of flammability!

Similar goods are further grouped within each Class by the use of an 'item' number which provides a more specific description than used for the Section head. All item numbers also possess a sub-group identification (a), (b) or (c) which denote high, medium or low danger, respectively. Therefore by assessing the properties and hazards associated with a material an ADR Class and Item Number can be assigned followed by an indication of the relative risk associated with transport. These are quoted on the Chemical Safety Datasheet and quickly identify the category of material and the provisions required for transport.

Part III of Annex A is a listing of appendices. The most relevant of these are A.5, general packing requirements, A.6, general conditions for the use of intermediate bulk containers (IBCs), both of which are based on the UN recommendations, and Appendix A.9, labelling requirements. The diamond shaped danger labels specified in Appendix A.9 are applied to both the package and the transport vehicle. These labels provide an instant indication of the classification of material carried and therefore an indication of the risk. All the labels used in ADR are displayed in Plate 1 (see *Colour Section*).

7.2.3.2 Annex B. Part I consists of a general section which specifies the gross consignment weights of materials from all classes for which many of the provisions do not apply. In addition, there are six listed sections

Section 1 refers to allowable forms of transport, e.g. bulk, tanks, and the goods that can be carried therein

Section 2 describes the types of vehicles, design features, equipment, approval of vehicles and the issue of certificates

Section 3 provides the general operating provisions, training of drivers, supervisors of vehicles, documents and emergency instructions

Section 4 covers loading, stowing and unloading

Section 5 covers vehicle marking, methods of parking and action to be taken in case of emergency, and

Section 6 describes how party states can agree trial movements between them in order to develop and extend the agreement.

Part II consists of specific chapters which provide additional provisions required by the Class and override those in Part I. The appendices to Annex B are technical in content and of most interest is Appendix B.5 which lists the identification numbers to be used on marking plates for transportation of bulk material. These plates are rectangular reflectorised orange coloured and displayed both the substance (upper part) and hazard (upper part) identification numbers which are listed in ADR for all hazardous materials. The substance identification number is the number specified in the UN Orange Book. The purpose is to enable a rapid identification of a substance carried and an assessment of the hazards associated with the material transported should an incident occur. The plate is constructed such that the numbers remain legible after 15 minutes engulfment in fire. An example of a transport plate is provided in Plate 2. (In certain countries national legislation allows the use of other systems or markings, such as Hazchem in the UK, but the principles are very similar.)

The hazard identification number is based upon a system derived from the Class numbers and is known as the Kemlar Code. Two or three figures are used and in general the figures used indicate the following short term hazards

2 emission of gas due to pressure or chemical reaction
3 flammability of liquids (vapours) and gases or self heating liquids
4 flammability of solids or self heating solids
5 oxidising (fire intensifying) effect
6 toxicity
7 radioactivity
8 corrosivity
9 risk of spontaneous violent reaction

Codes are constructed from one or more of these numbers such as 36, 38 or 423 with certain combinations possessing specific meanings. A doubling of a figure indicates an intensification of the identified hazard. Where the hazard can be adequately described by a single number this is followed by a zero. The inclusion of an X before the number indicates that the substance will react dangerously with water. Appendix B.5 of ADR provides a list of all materials which are given specific substance and hazard identification numbers. Table 7.4 provides examples of numbers quoted under ADR to be found on the chemical safety datasheet, vehicle marking plate and package label as described in the text above for a variety of solvents.

ADR was first based on the content of the provisions governing transport by rail and both regulations are now revised at joint meetings with the intention of harmonisation with the UN Orange Book whenever possible.

7.2.4 *Regulations Concerning International Carriage of Dangerous Goods by Rail* [7]

The purpose of these regulations, commonly known as RID is to set minimum standards for the packing and transportation by rail of dangerous goods such as solvents. If the conditions within RID are met then passage is allowed even if the local regulations are more stringent. The list of participants includes all those in agreement with ADR plus Albania, Algeria, Bulgaria, Iraq, Iran, Lebanon, Lichtenstein, Monaco, Morocco, Romania, Syria, Tunisia, and Turkey.

RID consists of three basic sections. Part I provides the general requirements and structure of the regulations, with definitions and lists of the units of measurement used. Part II comprises the special requirements for each Class of substance. The categories are identical to those specified in the Orange Book. Each Class has details of the relevant substances and conditions of carriage including packages, labelling, despatch methods and restrictions on forwarding, prohibitions on mixed packing, and the condition and marking of transport equipment. Materials are classified according to physical and chemical properties such as vapour pressure, critical temperature, flash point, flammability, toxicity, corrosivity, stability, reaction with water and

Table 7.4 Examples of class and item numbers, hazard and substance identification numbers and labels specified within ADR for a variety of solvents

Solvent	Class and item number	Hazard identification number	Substance identification number	Label type
Acetone	3, 3(b)	33	1090	3
Benzene	3, 3(b)	33	1114	3
Acetyl chloride	2, 25(b)	X338	1717	3+8
Methylene chloride	6.1, 15(c)	60	1593	6.1A
Trifluoroacetic acid	8, 32(a)	88	2699	8

viscosity. As in ADR the restrictive classes are 1, 2, 6.2 and 7. All labels are based on those used in the Orange Book and used in ADR (Plate 1). Part III consists of the technical appendices.

7.2.5 *International Maritime Dangerous Goods Code* [8]

The IMDG Code is different from both ADR and RID in that the purpose is not regulation but recommendation. The text amplifies the provisions for carriage of dangerous goods as defined in Regulation VII/1.3 of the 1974 International Convention for Safety of Life at Sea. The Convention detailed the mandatory provisions governing the carriage of dangerous goods in both packed form and bulk. The Code sets out in detail the requirements applicable for each individual substance, material or article which is classified as dangerous for transport by sea. Materials are either hazardous to health or to the environment. Classification follows the system used in the UN Orange Book. IMDG is set out in four volumes.

Volume I highlights the principles of the code, a general index of dangerous goods in alphabetical order, a numerical index of UN numbers with the corresponding IMDG Code page, and the general packing and labelling recommendations.
Volume II describes the specific requirements for Class 1 (explosives), 2 (gases), and 3 (flammable liquids) materials.
Volume III deals with Classes 4 (flammable solids) and 5 (oxidising and organic peroxides).
Volume IV covers Classes 6 (poisonous including toxic/infectious), 7 (radioactive), 8 (corrosive) and 9 (miscellaneous).

The labels used for each material are similar to those in ADR. In addition the package must be labelled with the UN Number and a Proper Shipping Name which provides either a specific or general description of the type of goods. The Master of the ship has ultimate authority with respect to the acceptability of labelling and containment and with controls implemented by the port authorities.

The International Maritime Organisation is the only authorised group able to change or adapt the recommendations in the IMDG Code. Materials can be moved across international waters from despatch point direct to warehouse or user under the IMDG classification regardless of national legislation which comes into force for all subsequent movements.

7.2.6 *International Air Transport Association (IATA) – Dangerous Goods Regulations* [9]

The purpose of these IATA regulations is to provide procedures by which articles and substances with hazardous properties can be safely transported by air on both cargo and commercial aircraft. The stipulations operate both for national and international travel. Certain national and airline regulations can also operate in addition to those specified by IATA. The text is available in four languages (English, French, German and Spanish), undergoes annual revision, and is in operation throughout the world. The IATA offices in Montreal cover North, Central and South America, Australasia and the Pacific regions. Europe, Africa and the Middle East is served by the IATA Centre in Geneva.

The philosophy behind IATA is similar to other transport regulations in that carriage is allowed providing the goods are packaged correctly. However, the quantity per package is also strictly limited and is dependent both upon the hazard associated with the goods and whether transport is by cargo or passenger aircraft. In addition, certain dangerous goods are forbidden under all circumstances from transport by air. Included in this group are materials which are unstable such as certain hydrazines, peroxides and azides. Other materials can only be transported by cargo and not passenger aircraft. The regulations are to be used as a step by step approach to ensuring the correct information and necessary provisions are followed such that a consignment of dangerous goods is correctly prepared for air transport.

The main body of the text is a table of dangerous goods which identifies:

- dangerous goods Classification, as per UN Orange Book
- hazard labels for packages, following UN Orange Book and ADR
- the package code (I, II, III), indicating the level of danger
- the packing instruction number, detailing the type and construction of the package
- the maximum quantity of material per package allowed for transport for both passenger and cargo aircraft, and
- special provisions, where applicable.

In addition, IATA also species ancillary information such as the documentation for shipping, package testing facilities, manufacturers' and suppliers' lists, training schools, competent authorities by country and a list of members of IATA. In general, the labelling requirements of IATA are more stringent than ADR, IMDG or RID.

The regulations described above clearly relate to the responsibility and actions required from both manufacturer and haulier for the supply of hazardous substances such as solvents. Transportation and delivery is performed worldwide and therefore the implementation and auditing of conformance to the regulations are undertaken on an international level. However, following successful delivery to the desired destination the responsibility for the materials now passes to the 'customer'. Whether the product is dispatched in bulk or in drums the user is required to have facilities suitable for receiving and subsequently storing the materials delivered. National, rather than international, legislation and regulations operate at this level and are designed to ensure that the storage and handling of hazardous products are performed in such a way as to minimise threat to human health and the environment. The basic principles, of particular relevance to solvents, for storage of both bulk and packed materials are now discussed.

7.3 Storage of solvents

7.3.1 Introduction

The hazard of most significance in the storage of solvents is flammability. Materials are often stored in substantial quantities, whether as bulk or packed, and therefore the consequences of inadequate practice or operation can be large and dramatic. The main hazards associated with the storage of flammable solvents are fire and explosion involving either bulk liquid or escaping liquid or vapour. In general, regulations only exist for materials which have a flash point of 55°C and below and cover storage of single drums through to tanks exceeding 10 000 m^3. The storage of materials which are not classified as flammable (flash point exceeds 55°C) or are non-flammable (e.g. chlorinated solvents) are not generally subject to specific regulations. In these circumstances guidelines for good working practice are often cited and will be described later.

The legislation and standards generally enforced are designed not only to minimise the potential for accidents or incidents, but also to contain the impact of failure and allow easy access for emergency services. In general, the Government Office associated with the maintenance of health and safety at work define and enforce the legislation. For example, in the United Kingdom the Health and Safety Executive administer the Health and Safety at Work Act of 1974 and to assist implementation publish a series of guidelines on the practical application of the regulations under the Act. Several of these refer to the storage of flammable liquids [10–12]. In addition, user trade associations and alliances often publish literature providing advice and guidelines on the safe storage of products used within an industry [13, 14].

Clearly it is advantageous to solvent manufacturers if their customers and end users are able to accept and store products within the bounds of legislation and good operating practice. Therefore producers, either individually or via the respective associations, often publish and supply information on products which assist end users in meeting both legislative requirements and minimum standards of safety [15, 16]. As the chemical industry further develops and extends the principles of 'Responsible Care' some suppliers of solvents are increasing the scope of their safety, health and environment policies. This can include the provision of pre-delivery inspections which assess the suitability of storage facilities for hazardous materials prior to supply to new users or customers. The Charters of Cooperation between the suppliers and distributors of chlorinated solvents in Europe include a commitment to meet minimum standards for the storage of materials such as trichloroethylene and perchloroethylene which are used extensively within the metal finishing and dry-cleaning industries.

The legislative requirements for correct storage of hazardous materials are dependent on the type of container, quantity and flash point of the material stored.

7.3.2 The storage of solvents with flash point less than 55°C

It is not possible within the confines of this text to describe the stipulations within each national regulation. However, the concepts described below provide a good illustration of typical requirements for the storage of flammable liquids in bulk, fixed tanks and in drums.

7.3.2.1 The storage of flammable liquids in fixed tanks. The basic principles operating for the storage of flammable liquids are all related to the risk of fire or explosion and can be summarised as:

- the provision of a tank compatible with the chemical and physical properties of the material stored and protected against corrosion and the environment in which it is located
- the provision of a tank capable of withstanding direct heat or a flame for a given period of time
- the segregation of the tank from other tanks and facilities or buildings to minimise the spread of fire or explosion
- location of the tank to allow early detection of leaks, faults, and ease of maintenance or repair

- the provision of secondary containment facilities in the event of tank rupture or failure
- the use of equipment and operating procedures in the vicinity of the tank which are compatible with the risks associated with flammable materials

Following consideration of the guidelines highlighted above it is easy to understand the constraints operating on the design and location of storage tanks used for flammable liquids.

In general, it is preferable for storage tanks to be located at ground level in open air in a well ventilated area away from sources of ignition. Although tanks located below ground provide better overall fire protection and clearly save space, problems often arise from long term corrosion and the difficulty and cost in detection and correction of leaks. It is extremely hazardous and unacceptable to locate tanks under a building or on a roof or in a position raised high above ground level. Tanks should not be stacked. Storage tanks can only be sited inside a building in exceptional circumstances. The location of the container should allow easy access for examinations, modifications, detection of leaks, and emergency services. The materials used in the construction of the tank, or tank lining, pipework and associated fittings, must be fully compatible with the solvents stored. In general tanks are manufactured from steel. The outer surfaces are suitably painted or coated to provide protection against corrosion. Tanks should be strength-tested to ensure the system is capable of withstanding the hydraulic loading of the material to be stored. In addition, the tank must be capable of withstanding direct heat or flame sufficient, in the event of fire, to allow an alarm to be raised, for people to escape, and for fire fighting to be implemented.

Strict guidelines exist for the siting of tanks in relation to other tanks, buildings and boundary walls. These are summarised in Table 7.5 which illustrates the relationship between the volume stored and the distance from surrounding buildings or tanks. In addition to the constraints listed in Table 7.5 tanks should be at least 2 m from any door, plain glazed window, ventilation opening, or any other building opening. The distance from a building can be reduced if the walls adjacent to the tanks have a fire resistance of at least 2 h.

Table 7.5 also highlights the necessity for bunding of all above-ground storage tanks. The bund must be capable of containing 110% of the capacity of the largest tank or group of interconnecting tanks within the bund and the height of the walls must not exceed 1.5 m. Ideally, the 110% containment should be 15 m away from the footprint of the storage vessels which minimises the escalation of an incident. Clearly in order to contain spillages due to tank or pipe failure the walls and floor of the bund must be impervious to all solvents stored within the area. In general bunds are constructed of brick or concrete. All pipes entering or leaving a bunded area should be above the height of the wall, or if through the wall, suitably sealed from possible leakage. The floor of the bund should also be impervious to the material stored and a gradient exist to ensure escaped liquid flows to the drainage point. A facility for removal of rainwater which is compatible with the hazards associated with the solvent stored is also required. Intermediate bund walls with a maximum height of 0.6 m are recommended for multi-tank facilities.

All tanks must be supported on non-combustible stands with a fire resistance of at least 4 h. All metal parts of the storage installation should be bonded together and earthed to avoid the build up of static charges which could cause sparks. The maximum allowable value of resistance to earth is 10 Ω. The integrity of the earth must be tested on a regular basis. Weather protection may be appropriate and this can be achieved by the use of a dutch-barn type structure of lightweight non-combustible construction with at least two open sides.

Fire fighting capability is required on all sites and should be discussed with the relevant local fire authority. All must have standard portable or trolley extinguishers, whether foam, water or dry powder of minimum 223B rating, easily accessible throughout the area. The type of extinguishant used must be compatible with the materials stored. For large installations it may be appropriate to fit integral fire fighting facilities. Typically fixed water sprays are used for cooling the tanks and foam sprays for damping down fires or

Table 7.5 Minimum separation distances for bulk storage of flammable liquids

Maximum single tank capacity	Maximum capacity of all tanks	Minimum distance between tanks	Minimum distance to bund wall	Minimum distance from building boundary, or ignition sources
Up to 1	3	0.5	1	1
over 1	15	1	1	4
over 5	100	1	1	6
over 33	300	1	1	8
over 100	750	2	2	10
over 250	1500	2	2	15

Note: All volumes are in cubic metres; all distances are in metres.

controlling evaporation from spillages which are not ignited. Other methods include the use of fire resisting thermal insulation, such as vermiculite cement, which provides protection for 2 h before overheating of the tank and contents occurs. Sites must have in place a procedure for raising the alarm and the local fire fighting service, dealing with small fires and controlling leaks and spills, and finally evacuation of the site. All sources of ignition should be removed from areas associated with the storage of solvents with a flash point below 55°C. Smoking is prohibited and 'No Smoking' signs should be posted. All electrical equipment required in the operation of the storage facility, such as instrumentation, lighting, switches, motors should be selected, installed and maintained in accordance with the zones and area classification operating in the site.

Hazardous areas within the tank and storage facility are classified according to the following:

Zone 0: an area in which an explosive vapour/air mixture is continuously present, or present for long periods

Zone 1: an area in which an explosive vapour/air mixture is likely to occur in normal operation

Zone 2: an area in which an explosive vapour/air mixture is not likely to occur in normal operation, and if it occurs will only exist for a short time

Areas outside the tank in which the conditions listed above do not exist are designated as non-hazardous. Table 7.6 provides an example of a typical hazardous area classification for storage tanks containing a highly flammable solvent (flash point less than 32°C). The procedure for identification of sources of release, determination of the type of hazardous zone and the process for assessment of the extent of the zone are detailed in British Standard 5345 Part 1 and Part 2 [17].

Tank fittings, connection lines and pipes must also be fitted with specific regard to the strict guidelines. Tubing used for pipework should be manufactured to the most current defined standard and the pipework must also meet a minimum standard (for example, BS 1387 and ANSI B31.3 plus supplement EEMU Supplement 153). All pipework should be protected against corrosion as per the tank. Only under exceptional circumstances should pipework be made of plastic.

The pipework should be designed to minimise the volume contained and potential for damage. It is preferable that all pipework, even for subterranean storage tanks, is above ground permitting easy access. It is advantageous for fill and empty lines to enter the tank from the top. In addition, the fill pipe of a tank should be extended within the tank to the lowest level of liquid likely to be present. This avoids the potential for build up of static electricity caused by splashing during filling. The fill pipe should have a siphon break facility fitted. The fill and empty connections and openings should be at least 4 m away from any source of ignition, building opening, trench, ditch or drain. For tanks above ground the openings should be outside the bund wall. Both the fill and discharge line should be provided with an isolation valve located close to the tank connection. The flexible hose from or to the tanker for filling and discharging, respectively, should be as short as possible. The use of an articulated arm is advisable.

Table 7.6 Typical area classification for storage tanks containing a solvent with flash point less than 32°C

Area	Extent of area zone	Classification
All tanks	within vapour space	0
Vent pipes	(a) within a 3 m radius in all directions of the open end	1
	(b) the area below the Zone 1 area of any vent pipe, for a radius of 3 m around the exit point and down to the ground	2
Above-ground tanks	(a) vertically from ground level to the height of the bund wall, and horizontally from the tank shell 1 m outside the bund wall	2
	(b) within 2 m of the tank shell	2
Underground tanks	within any manhole chamber containing fitting connections	0
All tank connections	within a horizontal radius of 4 m from tank filling connections, and vertically from ground level up to 1 m above the connections	2
Pumps and sample points	within a horizontal radius of 4 m and vertically from ground level to 2 m above the unit	2
Road and rail tankers (loading/unloading)	(a) within 300 mm in any direction of any opening on the tanker, and down to the ground	1
	(b) within 2 m of the shell of the tanker	2
	(c) within a horizontal radius of 4 m from tanker discharge connections and vertically from ground level up to 1 m above the connections	2
	(d) within a radius of 1.5 m of any opening on the tank top, and down to ground level	2
	(e) on the top of the tank within the valance	1

Vents are essential to eliminate the possibility of a pressure build up or the formation of a vacuum within the tank. The vents must be at least 7 mm in diameter and exceed the diameter of the fill or empty pipe. The top of a vent pipe must be at least 3 m and no more than 6 m above ground level, or if mounted on top of a tank at least 1 m above the height of the tank. Discharge must be away from window and door openings and into the open air. If the flash point of the solvent is less than 21°C then a flame arrester should be fitted to the exit of the vent pipe. The end of the vent pipe should face the ground and be fitted with an open metal cage. It is also advantageous that a mist eliminator is fitted to the vent pipe as the presence of a mist, which may arise during rapid filling, significantly increases the flammability risk even for materials with flash points above 21°C. The use of a vapour return between discharge vehicle and tank is also advisable.

All tanks should be fitted with a liquid level monitor which is accurate, reliable and robust. It is preferable that sight gauges are avoided. However, if they are used they must be magnetically coupled, or protected from impact damage and fitted with a top and bottom isolation valve. The use of a dip rod should also be avoided. However, when using a dip rod it is essential that the diameter of the rod is significantly smaller than the diameter of dip tube and the base of the tank should be protected by a wear pad. The rod should not be made of ferrous materials or aluminium. Automated gauges are preferable to the mechanical/manual measurement techniques described above. Gauges should be liquid and vapour tight when not in use. An independent high level trip system, with a different measuring technique to that of the contents measurement, should be fitted which automatically stops the filling pump when initiated. This arrangement has advantages even if the level device can be easily viewed during filling.

Pumps are essential to the successful operation of a solvent storage tank. All pumps should be located in open areas outside the bunds, on an impervious base, such as concrete, and at least 4 m from buildings, boundaries and sources of ignition. Clearly the characteristics of the motor must match the flash point of the solvent and the area classification operating at the point of location.

All metal parts within the entire storage installation should be bonded together and earthed. This process protects against the build up of static electricity capable of generating sparks. During the process of filling the storage tank a clamp, or other means of continuing the earth to the filling tanker, must be provided. It is advisable for this temporary earth to be fitted with an earth continuity monitor.

The storage tank should be clearly and boldly marked indicating the flash point of the solvent stored. This can either state the category of material, such as 'Highly Flammable' or 'Flammable', or more specifically 'Flash Point Below 32°C'. All pipes, valves and pumps should be clearly marked to indicate their purpose.

Regular inspection and maintenance of a correctly designed bulk storage facility is essential. All components should be listed with a schedule for the scope and regularity of inspection and overhaul. This is particularly appropriate for electrical equipment and the integrity of earthing points. Internal and external examinations should be performed by a competent person such as a specialist engineer. Time intervals should be between 3 and 10 years and may need to be less for older storage tanks. Any variations must be to the same standard as specified in the original construction and fully recorded.

Figure 7.1 provides an illustration of the principles described above applied to both horizontal and vertical storage tanks which are located above ground. These can be compared with the diagram of a typical underground storage tank provided in Figure 7.2.

The storage of solvents in bulk is a very common practice and is ideal for users requiring large volumes of material. However, the use of solvents from drums is also widespread and subject to strict legislation.

7.3.2.2 The storage of flammable liquids in drums. Fire and explosion are the principal hazards associated with the storage of non-toxic, non-corrosive, flammable solvents contained in drums. Regulations are designed to eliminate or minimise these two risks from either the liquid or vapour. The drums must meet the standards set out in the UN Orange Book as described earlier.

Ideally drums containing flammable solvents should be stored at ground level in an outdoor, well ventilated, designated open area. A notice should be displayed marked either 'Highly Flammable Liquids' or 'Flammable Liquids' as appropriate. Drums should not be stacked more than four high and should stand on pallets. If stacked on their side, drums must be firmly and securely chocked. Drums should be filled to allow an adequate ullage space for liquid expansion which is dependent upon the boiling point of the solvent. The base of the storage area should be impervious to the solvent stored and should be surrounded by a sill or low bund. The height of the wall is typically 150 mm. This sill or bund wall acts as a means of controlling spillages and must be large enough to contain 110% of the contents of the largest drum. Alternatively the base of the storage facility can slope into an open drain or to a local containment sump.

The total quantity of solvent stored determines the distance of the storage site away from nearby buildings and installations. Thus the distance from the edge of the storage area and an occupied building, boundary or fixed source of ignition should be 2 m for storage up to 1000 l, 4 m for up to 100 000 l, and 7.5 m for storage of any quantity above 100 000 l. If these distances cannot be satisfied or a boundary wall or a section of a

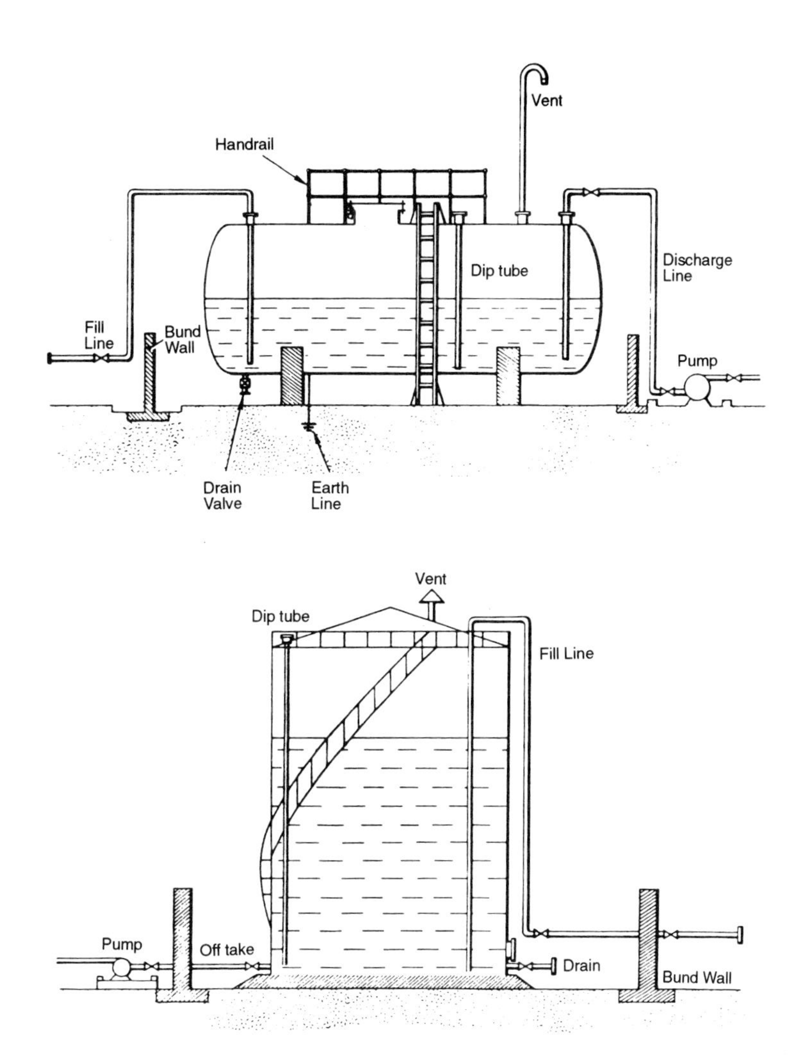

Figure 7.1 Diagram illustrating typical horizontal and vertical storage tanks suitable for bulk storage of flammable solvents.

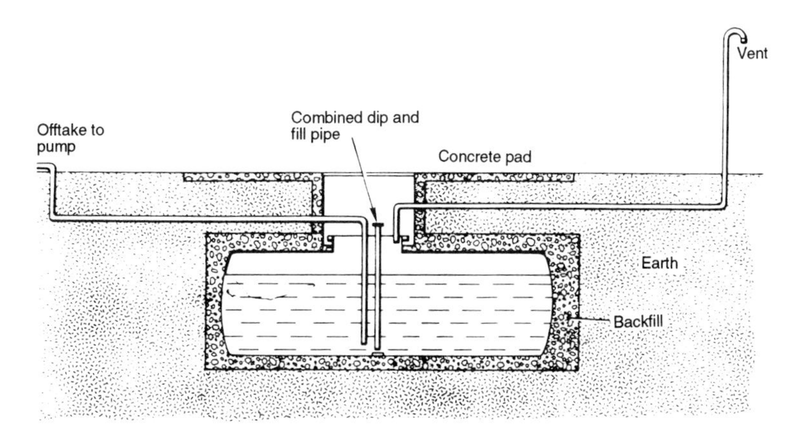

Figure 7.2 Diagram of a typical underground tank suitable for the bulk storage of flammable solvents.

building forms part of the storage site then it is necessary to use a fire wall with at least a 30 min fire resistance. The height of this wall must exceed 2 m and should extend above the height of the drum stack. Fire walls are usually made of concrete, masonry or brick. Ideally a wall should only form one part of the storage site such that the maximum amount of ventilation is achieved.

Shading from strong sunlight or weather protection is advisable and can be achieved in the form of a light non-combustible roof structure with open sides which do not restrict the flow of air and hence ventilation.

The storage site should be surrounded by either a wire mesh fence of height 1.8 m or brick or concrete wall of height 2.4 m. All facilities should possess a minimum of two separate exits fitted with non-self-locking doors which open outwards. These exits should be locked when not in use.

Flammable solvents can also be stored in drums in buildings. If the building is stand alone and only used for the storage of flammable materials then it need not be fire resistant, merely constructed of non-combustible materials, provided the distance from the nearest buildings complies with the minimum distances defined above. These are dependent upon the volume of solvent stored. If these distances (2 m, 4 m and 7.5 m) cannot be met, or the store forms part of another building, then the entire structure including windows, floors, doors and ceilings should be capable of withstanding accidental damage and a fire for at least 30 min. All doors must open outwards into open areas and remain locked when not in use. The windows should be non opening. The store must be located above ground. If the connecting building is used for residential accommodation then the fire resistance must exceed 60 minutes. The room must be clearly marked 'Highly Flammable' or 'Flammable' as appropriate.

All stores should be designed to provide at least five air changes per hour. This ventilation can only be achieved by air bricks on external walls which are not fire resisting. An alternative is the provision of mechanical ventilation. The roof of external buildings should be light weight to provide explosion relief. For indoor stores it may be more appropriate to have relief panels in one or more walls. A method for containing spills should be easily accessible. A 150 mm sill with ramps at door openings or a sloping floor is also necessary. The floor must be impermeable and solvent resistant.

If heating is required in the store then this can be achieved using hot water radiators or steam pipes. Electrical heaters can only be used if protected to Zone 2 standards. Indeed there should be no sources of ignition within a store containing flammable solvents. Electrical equipment must be capable of operation within the Zones identified. Typically all areas will be Zone 2. If vehicular access, such as forklift truck or other motorised handling device, is operating within a flammable solvent drum store room or site, then the vehicles should also be protected against igniting flammable vapour. Battery operated or diesel powered engines adapted for Zone 2 areas are preferred.

Adequate fire fighting equipment is also a necessary element of both indoor and outdoor facilities when storing flammable solvents in drums. Basic hand held or trolley fire extinguishers of foam, dry powder or water with a minimum 223B rating, should be accessible throughout the site and are generally located near exits and outdoors and therefore may need weather protection. The type of extinguishant should be compatible with the solvent stored. It is also advisable for extinguishers to be stored in pairs in case of failure. For larger installations it may be necessary to provide water in the event of using the local fire fighting service. A full emergency procedure should be created to deal with fires and leaks. This would include the process for raising the alarm, local fire fighting service, basic guidelines for tackling a small fire or leak, and finally the procedure for evacuation of the site.

The regulations described above have ensured that a facility for the storage of flammable solvents in drums is essentially remote from all other operations. This creates a certain degree of inconvenience for a user who frequently uses small quantities of solvent. Under these circumstances guidelines for temporary storage operate. Drums of solvent with a capacity of less than 250 l intended for use on the same day can be stored outside process buildings providing:

- the drum is closed and correctly labelled
- the drum is located 2 m away from any door, window, ventilation outlet or inlet, or fire escape
- the drum is not below any means of escape regardless of the vertical distance from the drum.

Storage within a workroom is acceptable provided that the minimum quantity consistent with the needs of the operation are kept and storage is in a drum at least 3 m and preferably 5 m away from any process activities. Storage within cupboards is also acceptable providing the maximum quantity stored is 50 l and the cupboard is constructed to a minimum standard, for example:

- all sides, including lids and floor must satisfy BS 476, Parts 20 and 22 with regard to freedom from collapse and resistance to passage of flame for at least 30 min
- all junctions and joints are bonded or fire stopped to prevent or retard the passage of flame and hot gases
- the structure should be robust enough to withstand forceable accidental damage
- all sides, top and floor should be supported and fastened to prevent failure of the structure in a fire for greater than 30 min

Provided the regulations and guidelines described above are followed then storage of both bulk and drummed solvents with flash points below 55°C is a safe process.

As indicated at the beginning of this section the storage of solvents not classified as flammable or which are non-flammable is not subject to regulation. Recommendations based on good operating practice should be followed.

7.3.3 The storage of chlorinated solvents

Only under extreme operating conditions do chlorinated solvents support combustion and hence the regulations and hazards identified above do not apply. However, it is necessary to follow strict guidance when storing these solvents as they are extremely volatile, potentially unstable, and are aggressive towards certain polymeric materials used as seals in pumps. Therefore the principal hazards during storage can be summarised as pressure, reactivity and solvency. All chlorinated solvents are stored and used fully stabilised thus reducing one of the potential areas of concern.

Advice on storage is normally provided by either a trade association or a producer of the materials. Different guidelines operate for the storage of bulk material in static tanks and for drummed or packed solvent.

7.3.3.1 Storage of chlorinated solvents in fixed tanks. Chlorinated solvents can be stored in mild steel tanks with welded joints. If necessary, perhaps for maintenance of very high quality material, stainless steel or a mild steel tank which has been internally galvanised or coated with a phenol formaldehyde resin may be more appropriate. Aluminium cannot be used for storage of chlorinated solvents due to the potential for reaction between the chemical and metal. Although there are no specific regulations for the siting of storage tanks the following guidelines should be followed. Installations above ground are preferred, allowing gravity flow discharge, and as with flammable solvents, ease of access, maintenance, inspection. The tank supports should be adequate to bear the weight of the tank when completely full of solvent. Tanks, where necessary, can be placed in a pit or under ground. Containment of leakages is essential. A bund or catchment area capable of holding the total volume of the largest storage tank in the facility is advisable. The use of concrete, which is not impervious to chlorinated solvents, for the bund floor and walls is possible provided it is laminated or coated with a solvent resistant material. Solvent proof laminates are normally based upon either phenolic or furane resins in combination with glass fibre which adds durability and strength. A recent development has been the availability of polyurethane concrete which is entirely impervious to chlorinated solvents. Bunds may also be constructed of metal provided they are protected against corrosion. Metal bunds are normally more appropriate for smaller installations. Another option to bunding is the use of a double-walled tank fitted with an alarm which detects pressure between the two walls of the tank should the inner wall fail.

All tanks should be fitted with liquid level indicators. Sight glasses are also not preferred due to the risk of accidental breakage. Methods based on dip pipes should be avoided. When a level transmitter is used the liquid level indicator signal can also serve as a level controller. When filling, either by gravity or pump, it is advisable that a high liquid level monitor is linked to an automatic shut off valve on the filling line and the stop circuit of the pump. Meters should be compatible with the solvent stored and aluminium, magnesium and zinc alloys should be avoided.

Chlorinated solvents are generally highly volatile and vents or relief systems are an essential element of the safety for storage of these chemicals. In particular a vapour return line should be fitted between the storage tank and the tank or vehicle under discharge. The storage tank should also be fitted with a vent or relief system. Chlorinated solvents can hydrolyse in the presence of water and therefore the vent should be fitted with an air dryer.

Drying agents such as anhydrous calcium sulphate or chloride are appropriate but sodium hydroxide flakes are not as they decompose in the presence of chlorinated solvents. The vent should also be fitted with a non return valve thus preventing saturation of the dryer with solvent vapour during filling. The use of positive pressure of dry nitrogen is often used as a method of 'padding' the vapour space of a tank and is useful for maintaining product quality. However, the costs can be high. A pressure–vacuum relief valve should be fitted to prevent possible damage to the tank should the vent become blocked.

Pipelines should be above ground and constructed of stainless, carbon or galvanised steel. Connections should be limited to a minimum and should be flanged or welded. Slip on or soldered connections are not acceptable. The sealants used for flanges must be resistant to the solvency of the chlorinated material stored. The external walls of all pipes must be protected against corrosion either by coating or wrapping. When pipelines are under ground they should be double walled and fitted with a leak alarm. If polymeric pipes are used then these are only suitable for above ground and when constructed of PTFE (polytetrafluoroethylene), PFEP (polyfluoroethylenepropylene), PCTFE/ECTFE (polychlorotrifluoroethylene) or PVDF (polyvinylidenefluoride). Gaskets must also be solvent resistant.

Cast iron or steel pumps are recommended and can either be centrifugal, hermetic or those operating with solvent resistant mechanical seals. All vessels and intake lines, marked at the delivery point, should be clearly labelled with the name of the solvent. Regular, recorded inspection should be undertaken to ensure the integrity of the storage tank, pipes and pumps.

The bulk storage of chlorinated solvents is practised widely throughout the world and providing the correct procedures are followed it is a safe process. As with flammable solvents large quantities of chlorinated solvents are stored in drums and again guidelines and not regulations should be followed.

7.3.3.2 The storage of chlorinated solvents in drums. In comparison to flammable solvents the storage of drums of chlorinated solvents is very straightforward. Clearly it is advisable, and necessary if involving transportation, that the United Nations specification of drum or container is used. Nevertheless, the drum must be capable of withstanding the weight and pressure exerted by the material contained. Care must be taken as chlorinated solvents typically have high density and vapour pressure at ambient temperatures.

Ideally the drum storage facilities for chlorinated solvents should be segregated from the storage of other materials. The site should be located in a cool, dry and well ventilated area out of direct sunlight. Protection from rain or ingress of water is advisable. Drums should be stored at ground level and if vertical stacking is necessary for larger drums, such as 200 l capacity, it should not exceed two high. Prevention of corrosion of the base of the drums is advisable. Sealed drums should not be stored in direct sunlight and should be remote from sources of heat to avoid the generation of pressure within the drum. If vertical stacking is not used then the use of permanent stillage is then preferred. In certain countries the area used for drum storage must be bunded by a solvent impervious material. This requirement, suitable both for the storage of solvent in bulk and in drums, relates to the elimination of soil or aquifer contamination which may occur from a leak or spillage of chlorinated solvent. The cost of remediation of such instances is extremely high and prevention, rather than cure, is clearly a more attractive option.

The description of the regulations and guidelines above provide only an insight into the requirements for the storage of commonly used solvents and definitely should not be used as the basis of designing or constructing a dedicated storage facility. Discussions and compliance with the requirements of the government authority for administration of health and safety are essential. For solvents with particularly hazardous characteristics, such as toxicity, then specialist assistance should be requested from the supplier and again agreement with the government office is necessary.

Part 2 Recovery and waste disposal of solvents

7.4 The significance of solvent recovery

An important phase in the life of solvents concerns the processes employed after they have performed their primary role and involves either the recovery or disposal of the solvent.

In virtually all of their applications, solvents are desirable because they facilitate another process while remaining inert within a particular system. For example, solvents are used in chemical transformations in which two or more materials react under the specified conditions yielding a desired product. A particular solvent is chosen for its ability to solubilise the reactants but also remain inert throughout the reaction, and process of separation, to yield the products. A formulator of an adhesive requires a solvent to solubilise the components which subsequently react when mixed, and harden, following solvent evaporation. The solvent is not consumed in the process and fundamentally remains unchanged. If a material remains unchanged during a particular operation then capture, or recovery, and reuse is clearly possible. This is extremely difficult for a completely emissive use, such as the application of an adhesive. However, for many applications of solvents recycling is possible.

In general solvent recovery is only economic when performed on a large scale. The contribution of a solvent to the total cost of a process is usually small in comparison to other chemicals used in a reaction or formulation. Therefore the use of an expensive recovery system only becomes economic when the volume of solvent used is large. For large installations the solvent recovery process or processes often form an integral part of the primary facility. This is particularly apparent for the manufacture of pharmaceuticals. Recovery systems are best incorporated into the design of the primary facility as retrofitting is very costly. In addition, the transportation of bulk volumes of waste solvent is expensive and time consuming and it is more efficient for a recovery process to be operating within the confines of the principal facility. Indeed the majority of solvent recovery systems are used in this way.

However, small scale users of solvents do not have the benefits of large volumes and the cost of operation of an 'on site' recovery system is unjustified. Nevertheless, the solvent remains recoverable and hence retains a value. Under these circumstances, provided the economics are viable, conglomeration of a solvent obtained from a number of small users is required. This role of collection and subsequent bulk recovery is performed by 'solvent recovery agents'. The reclaimed solvent obtained from such agents is generally of good quality and purity and can often be reused for less critical applications than those serviced by the virgin solvent.

This process can often be quite lucrative as the first user will invariably pay the recovery agent to remove the 'waste' and, following a recycling process, the material can be resold!

In addition to the economic benefits, the need for solvent recovery systems has been further compounded by recent legislation. Processes which result in significant losses to the environment are no longer socially or legally acceptable. Legislation such as the United Kingdom Environmental Protection Act, the USA Clean Air Act, and the forthcoming European Directive on VOC Emissions are examples in which very stringent limits are stipulated. The emphasis on the use of recovery is increased as a consequence of such legislation and significant investment implications may follow from the need to comply. Therefore the selection of a solvent not only has to focus on cost, efficacy and suitability for the primary role but now also has to consider the ease and cost of recovery (and consequently the reduction in the generation of waste). Indeed, in certain instances the elimination of an organic solvent in favour of an inorganic medium, such as water, is often undertaken to avoid the need to meet a stringent solvent emission regulation.

In many instances, the implications of regulation and the initial cost of instigating recycling and recovery has resulted in the elimination of solvents from processes and formulations. This is evident in trends such as the use of waterborne paints and coatings, aqueous based adhesives and metal pretreatments (although these changes may result in the need to comply to different legislation).

However, the unique properties of solvents make their continued use essential and guaranteed. The drive towards recycling and sustainable development led by environmentalists, and adopted by governments, legislators and major organisations, has enhanced the importance and significance of solvent recovery processes in the future use of these valuable chemicals. The potential cost savings from solvent recycling can be large and the effective use of a recovery process is very attractive particularly with increasing scale.

Recovery techniques exist for the separation of solvents from solids, residues, liquids, including water, and from gases. These are achieved by such technologies as decanting, distillation, liquid–liquid extraction, condensation, adsorption and absorption. In many instances several processes are used in series to achieve the most efficient recovery, or desired quality, from the recovered solvent. A summary of the principal technologies used in solvent recovery is discussed below.

7.5 Techniques of solvent recovery

As highlighted above a wide variety of technologies are available for the recovery of organic solvents. The choice of process can be confusing. Table 7.7 is provided as guidance on the appropriate techniques suitable for solvent recovery from gases, liquids and solids. Disposal methods are also listed for completeness. Detailed descriptions of each technology are provided in the text below.

7.5.1 Recovery from the gas phase

7.5.1.1 Adsorption

7.5.1.1.1 Activated carbon. This technique is perhaps the most common method for the capture and recovery of volatile organic materials, such as solvents, from gas streams and in particular from air [18]. The principle of adsorption is based on the passage of the mixture of vaporised solvent and carrier gas through a bed of material which will adsorb the organics onto the surface and inner pores of the solid adsorbant. The permeation of the gas through the solid results in the removal of the organics from the vapour phase. A second process is then used to liberate the organic from the adsorbant.

The most common adsorbant used is granular or powdered activated carbon. This material, which is available from almost all forms of organic carbon-containing matter, is a microcrystalline nongraphite form of carbon. The production of activated carbon can be achieved by use of rotary kilns, hearth furnaces, or furnaces of the vertical shaft or fluidised bed type, and each is suitable for the generation of different pore size and the source of carbon. The pore volume and size are influenced by both the carbon source and method of production. The adsorption properties are directly related to the pore volume, pore size distribution and the nature of the functional groups on the surface of the carbon. Activation is achieved chemically, by treatment by dehydration with zinc chloride or phosphoric acid, or by treatment with steam, hot carbon dioxide or a mixture of both. The activated carbon is available in three basic forms, powder, granules or as cylindrical or spherical pellets. For solvent recovery systems the carbon is usually obtained from either wood charcoal, petroleum residues or coconut shells and is often used in the form of pellets.

Activated carbon is capable of adsorption of a wide variety of chemicals. The source and pore size of the carbon directly influences the performance of the process and can be optimised for different solvents. A typical system operates with two horizontal or vertical containers or beds of activated carbon coupled to a

Table 7.7 A summary of solvent recovery and disposal techniques

Technique	Options	Comments
(a) Recovery		
Solvent in gas phase	Adsorption on activated carbon	Well established Wide applicability Efficient Continuous process Limited capacity Large spacial requirements Mixtures processed Low cost Generates solvent and water mixture
	Activated carbon by Rekusorb process	As above except: Compact in size Liberates solvent free from water
	Polymeric adsorbant	New technology High bed capacity Low flow rates Low volume Continuous process
	Membranes	New technology Multistage units for high efficiency Compact in size High capital costs Liberates pure solvent
	Absorption	Well established Highly selective High and low flows Liberates mixture requiring further separation
	Condensation	Well established Use cryogenics for high efficiency Contamination by water may result
Solvent in liquid phase[a]	Decanting	Well established Low cost Water immiscible solvents only Removal of large volumes of water Limit to performance Can be slow Useful as initial purification Continuous or batch
	Membranes	Well established Simple Low cost Further drying required Continuous or batch Recovery of low solvent content
	Stripping	Well established Batch only Recovery of low solvent content Further adsorption process required
	Adsorption	Highly efficient Suitable for water content of $<2\%$ Further filtration step required Low cost
	Absorption	Well established Very efficient Use in combination with distillation Batch only Removal of low water content High purity possible Further filtration required
	Distillation	Well established Yields high purity Batch or continuous Can be complex Entrainer may be required High or low water content in feed
	Liquid–liquid extraction	Well established General application Batch or continuous Removal of large water content Medium to large scale Requires secondary purification step
Solvent in liquid phase[b]	Distillation	Wide applications Small and large scale High or low contamination

Table 7.7 Continued

Technique	Options	Comments
		Yields high purity Cost effective Well established
	Liquid–liquid extraction	Second choice to distillation Relatively rare
Recovery from solids	Sedimentation	Batch or continuous Centrifugal process may be preferred Decanting may be required Low solids content
	Filtration	High or low solids content Cost effective Well used technology Small or large scale Highly efficient
	Expression	Very high solids content Batch or continuous Yields low purity
Recovery from residues	Distillation	Well established High efficiency Yields high purity Batch preferred Small and large scale Secondary process may be required
(b) Disposal		
Solvent in gas phase	Incineration	
	Thermal	All solvent groups High temperature High performance Well established Benign by-products High running costs Continuous process High or low solvent concentration High capital cost
	Catalytic	All solvent groups Low temperature High performance Benign by-products Low running costs Catalyst cost may be high Preheat required Filtration of feed High or low solvent concentration High capital cost
	Biotreatment	New technology Low flows Low concentration Low costs Benign by-products Wide range of solvent groups Highly efficient Continuous process

[a] Recovery from water.
[b] Recovery from organics.

series of valves which control the direction of the incoming and outgoing gas flows. The process operates by passage of the solvent-laden air through one bed for adsorption of the solvent until it has reached maximum capacity, or after a specific timescale. At this stage the air flow is switched by the automatic valves to the second bed. During adsorption on the second bed the solvent present on the first bed is stripped, as described below, and the carbon is regenerated. Regeneration of the carbon follows cooling and is achieved by the passage of air through the bed. The mixture of air and solvent is then switched back to the first bed, which is now reactivated, and the second bed undergoes stripping and regeneration in the same way. The process can be repeated over and over to provide a continuous method of solvent recovery.

The effective operation of a two-bed process typically achieves greater than 95% adsorption efficiency. Higher performances, in excess of 99%, can be achieved but the cost increases significantly particularly when the use of a third bed is involved. Beds are normally very large in comparison to the volume of solvent adsorbed. The typical capacity for solvent is 5% of the dry weight of the carbon. If the flow of air is very slow, in the region of 0.2–0.5 m/s, then up to 30% capacity can be achieved. The size and capacity of a bed is usually engineered to ensure a total 3 h cycle time divided equally between adsorption and desorption.

Factors other than the physical properties of the carbon can also affect the performance of the adsorption process. The molecular weight of the solvent must exceed that of air and the greater the difference, the more effective the adsorption. In the treatment of a gas stream containing a mixture of solvents the material with the lowest volatility is preferentially adsorbed. Clearly this is a critical issue if *all material* must be removed by the activated carbon recovery system.

The performance of the activated carbon bed can be severely affected by the presence of particulates in the incoming air stream. Filtration prior to the recovery system is often undertaken to ensure bed clogging does not occur. During the process of adsorption heat is generated which normally only results in a small increase in the temperature of the exit gases. However for adsorption of ketones, and in particular those of high molecular weight, the heat can be sufficient to ignite the carbon bed. Clearly care must be observed to ensure heat is adequately dissipated throughout the adsorption cycle. The concentration of the solvent in the incoming air will affect the total adsorption efficiency and time of operation. Flammable solvents are limited to approximately 30% of their lower flammable limit thus eliminating the potential for explosion within the system. There is no restriction on solvent concentration for non-flammable chlorinated solvents.

Solvent stripping is usually performed by passage of low pressure steam through the bed and the mixture is condensed by a water or air cooled heat exchanger. If the condensate of solvent, or solvents, and water is a two-phase system then gravity separation followed by decanting will liberate the organics. Further downstream processes will be required if the solvent and water are mutually soluble. In the presence of water certain solvents also hydrolyse and therefore rapid separation or treatment may be required. The stabilisers present in chlorinated solvents must also be replaced following water separation as these crucial inhibitors of chemical breakdown are removed during the processes operating within an activated carbon recovery system.

Table 7.8 provides an indication of the wide variety of applications in which an activated carbon recovery process, as described above has been used.

One disadvantage of the standard activated carbon system is the use of steam to strip the solvent from the carbon. The organics must subsequently be separated from water which can replace one recovery problem with another. This is eliminated in a related process which also uses activated carbon as the adsorbant and is called the Rekusorb Process [2].

The principal difference between the traditional activated carbon recovery unit (commonly abbreviated to ACRU) and the Rekusorb process concerns the desorption cycle. Although the adsorption is achieved in the same manner, desorption of the solvents is achieved by the passage of hot gas, not steam, through the carbon bed. Prior to desorption a purge of cold nitrogen is used to reduce the oxygen within the system to a level below the risk of explosion from flammable solvents. The nitrogen is then heated and this causes desorption of the solvent. Moisture in the carbon bed is also liberated and therefore the exit gases containing solvent vapours are dried by passage through molecular sieves. Condensation of the solvent is again achieved by a cooler and condenser. The heat removed during condensation is transferred to the gas heater and hot gas is returned to the bed. When all of the solvent has been desorbed, the bed is cooled by the circulating gas. The molecular sieves are also dehydrated by the hot gas, ready for the next desorption cycle. The Rekusorb process operates with very high energy efficiency and in general these units are more compact than the traditional ACRU. However, the most distinctive advantage is that the solvent, or solvents, do not require separation from water following desorption and in many cases are of sufficient purity for reuse without further treatment.

Other, less established, adsorption recovery technologies exist which do not use activated carbon as the adsorbant.

7.5.1.1.2 Synthetic polymeric adsorbants. D.H. Clarke details the use of polyurethane based forms which can absorb and subsequently desorb a range of volatile organic solvents such as chlorinated and aromatic hydrocarbons, or aliphatic ketones [20]. The precise composition of polyurethane foam has a profound effect on absorption performance. However providing the correct reaction and balance between specific isocyante

Table 7.8 Examples of the use of an activated carbon recovery system by industry group and solvesnt

Industry group	Solvent recovered
Printing	Toluene, *n*-hexane, petroleum spirits
Metal degreasing	Trichloroethylene, tetrachloroethylene
Rubber	Benzene, toluene, petroleum spirit
Synthetic leather and fibres	Alcohol, acetone, esters, hexane, dimethylformamide
Adhesives	Petroleum spirits, hexane, toluene

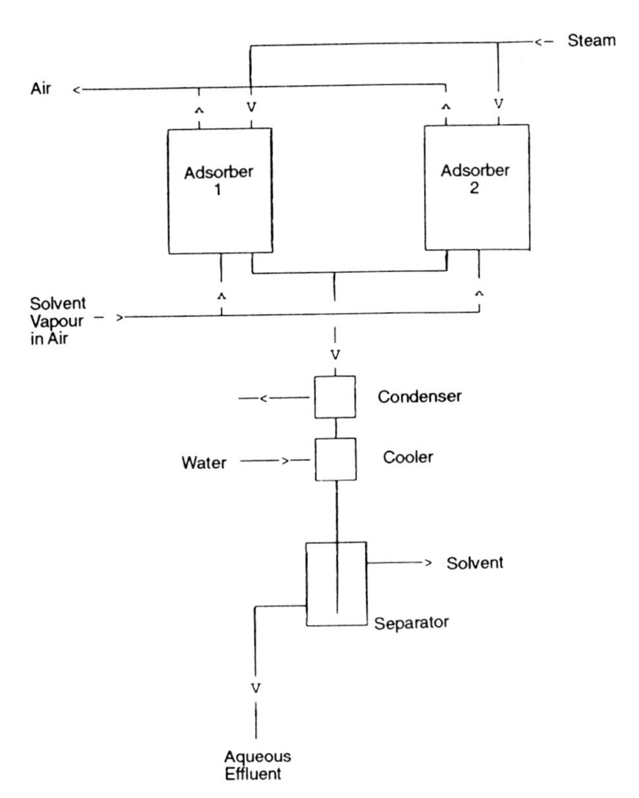

Figure 7.3 Diagram of a typical two bed activated carbon recovery system.

and polyol is achieved, good absorption characteristics can be obtained. Following contact between the polyurethane foam and the gas containing the volatile organic, adsorption quickly takes place. In general, 0.5–0.8 g of solvent can be absorbed per g of foam which is packed in a column as either granules or spheres. The majority of the adsorption occurs within the initial stages of contact. Desorption of the solvent is achieved by passage of solvent-free air through the foam and the material is then recovered by condensation. The principal advantage of using a polyurethane foam is a reduction in the size of absorption bed due to the higher solvent capacity of the foam.

A number of other systems are based on the use of polymers for adsorption of solvents, but perhaps of particular note is another process from the Dow Chemical Company [21]. 'Sorbathane' is the trade name for the resin which has been specifically developed for the recovery of chlorinated solvents such as perchloroethylene and trichloroethylene. Units which use this resin are usually two-tank systems which sequentially adsorb and desorb. Adsorption is achieved by passage of the solvent-laden air through the resin which is characterised by a high surface area, small pore size, a swellable polymer matrix and fast adsorption kinetics. Desorption of the solvent occurs when the resin is heated to 80–90°C and the application of a vacuum of less than 100 mbar. The novelty and advantage of using this system is that adsorption and desorption of the stabilisers, required for these solvents, also occurs and therefore the need for restabilisation, as necessary following activated carbon recovery, is eliminated.

One of the principal advantages of operating adsorption prior to condensation is the concentration of the solvent vapour which reduces the size and energy requirements of the condensation unit. Alternatively a number of fully loaded adsorbant beds can be collected for desorption and condensation through a dedicated larger scale unit. This conglomeration may improve the overall economics of the entire process and is often available from suppliers of activated carbon beds.

7.5.1.2 Membrane separations. Systems based on membranes have also been developed for the removal of volatile organics from air and rely on the use of a vacuum to achieve a pressure difference across a membrane which is 10–100 times more permeable to the solvent vapour than air [22]. Significant solvent enrichment occurs in the permeate and the solvent is subsequently recovered by condensation. Figure 7.4 represents a typical flow diagram of a membrane vapour separation process. For large-scale industrial applications a number of membrane modules are employed and often two-stage systems, in which the exit gas from the first membrane is fed to a second, are used to ensure up to 95% of the solvent is removed from the air. In general the modules are very compact but with high capital costs and therefore this technology is more appropriate for the recovery of high value solvents.

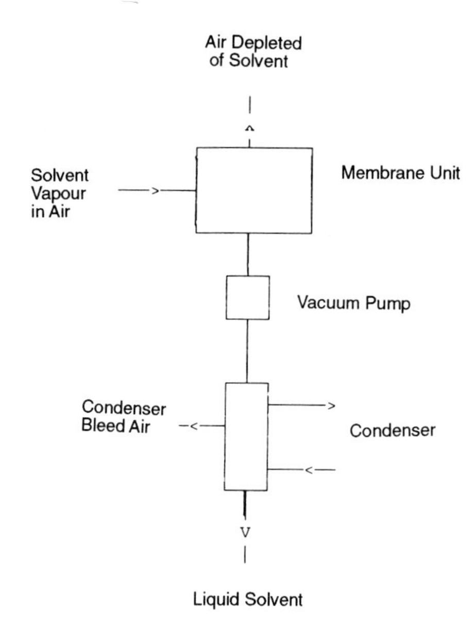

Figure 7.4 Diagram of flows within a typical membrane vapour separation process.

A related technique for the recovery of solvents from gaseous streams and flows involves washing or scrubbing the mixture of vaporised solvent and inert carrier gas with a suitable liquid which dissolves, and hence separates, one or more of the constituents from the gas.

7.5.1.3 Absorption. Absorption is the transfer of one or more components, termed absorbates, of a gas phase to a liquid phase in which it is soluble, called an absorbant [23]. For solvent recovery absorption is achieved by physical solution, a process in which the component being absorbed is more soluble in the liquid absorbant than the other gases with which it is mixed.

The equipment used for absorption processes is usually a vertical countercurrent column in which the gas stream, containing the absorbate, is fed vertically upwards against a downward flow of absorbant, usually another solvent. The absorbant dissolves the absorbate from the gas and forms a solution which is collected from the bottom of the column. The absorption process relies on the solubility of the gas to be absorbed in the absorbant and the rate of mass transfer between gas and liquid. Four basic designs are often used for the construction of the absorber all of which are used to maximise the contact between the two media and the mass transfer:

(i) Firstly a vertical cylindrical column which is packed with materials such as Rashig rings, Beryl saddles, Pall rings, or Intalox saddles, which are designed to disperse the liquid during downward flow and ensure close contact with the rising gas.
(ii) Alternatively, absorption towers in which the liquid flows in a cascade fashion over a series of plates or trays can also be used. The gas flows from a multitude of holes in each of the plates ensuring intimate contact between the gas and the solvent.
(iii) Absorption columns in which the absorbant is sprayed through the flow of gas is another effective design principal.
(iv) Bubble columns in which the gas is passed through a large volume of the absorbant which is contained within a vessel.

Each technique has advantages and disadvantages. Packed columns are flexible as the packing can be easily modified for optimisation of each recovery process, but they are best suited to small scale operations. Tray columns are used for large scale, low to medium flow rate applications involving non-corrosive materials. Spray systems are used almost entirely for applications where pressure drop is critical such as flue gas treatment. Types of equipment classified as spray columns include cyclone and venturi systems. Bubble column equipment is a low cost, low maintenance option; however back mixing can occur which reduces the countercurrent flows required for optimum absorption.

Following the scrubbing of the inert carrier gas the absorbent and absorbate are collected as a solution at the base of the column. A secondary process, desorption, is then required to achieve separation and liberate the absorbate. This separation is usually achieved by atmospheric pressure or vacuum distillation. Therefore the choice of absorbent is critical to ensure the materials have significant differences in boiling point and do not form an azeotrope.

7.5.1.4 Condensation. The use of simple water-cooled heat exchangers is perhaps one of the most commonly used condensation systems within both the laboratory and industrial environments [23]. However there is a limit to the capability of such systems and often the performance levels which might be required for end of pipe abatement technology or maximum recovery cannot be met. Under these circumstances lower temperatures are necessary from the condensation process and liquid nitrogen is often used as the cooling medium. Examples of such low temperature or cryogenic technology are 'Cryosolve' from Air Products and the 'Airco' process from the British Oxygen Company [19].

These liquid nitrogen-based systems are multistage and achieve highly efficient solvent condensation by a combination of water, or refrigerated cooling, followed by two or more stages of liquid nitrogen cooling. Nitrogen gas is also circulated throughout the system such that solvents can be handled well above, typically ten times, their lower explosive limit. The efficiency of the first stage is dependent on the boiling point of the solvent or solvents to be removed. The higher the boiling point then the greater the degree of condensation. It is possible to remove 90% of the solvent in the first stage, when operating at temperatures in the region of –30°C. The remaining solvent is removed by the two liquid nitrogen-based stages and the efficiency is such that the concentration of the solvent in the exit gases is extremely low. The presence of water or moisture in the solvent can seriously affect the efficiency of the heat exchangers throughout the systems by deposition of ice. Drying of the solvent is usually performed, using molecular sieves, prior to the first stage. Ice can also be removed by allowing the temperature of the heat exchanger to rise above 0°C. Clearly during this process it cannot be used for condensation and under these circumstances two units would operate in tandem and swing between condensation and de-icing. The presence of water in the recovered solvent via ingress of air from outside the system may also require subsequent drying or separation of the solvent if high purity is required.

Cryogenic systems that operate to the principles described above can be very compact and provide highly efficient solvent recovery. The costs from using large quantities of nitrogen as the carrier gas can be reduced by correctly balancing the process and the efficient use of any waste nitrogen generated from the system. As an alternative to cryogenic condensation, direct low-temperature condensation can be achieved by passage of solvent laden nitrogen through a bulk volume of the solvent which is refrigerated. Build up of ice can also occur in this process and therefore the water must subsequently be removed.

As can be seen, few technologies for the recovery of solvents in the vapour phase yield the desired product without the need for further treatment. This often involves the separation of two liquids, one of which is water.

7.5.2 Recovery from the liquid phase

The separation and subsequent recovery of one liquid from another is perhaps one of the most regular activities within the chemical industry and a number of methods have been developed to achieve this process. Two principal categories can be identified of relevance to solvents which are the recovery of a solvent from water, which includes drying, and the separation of one solvent from another. The techniques relevant to each case are discussed below.

7.5.2.1 Recovery from water

7.5.2.1.1 Decanting. Many organic solvents are only sparingly soluble in water and often two-phase systems exist. Decanting is a classical method for the separation of two liquids and relies on both poor mutual solubility and a difference in specific gravity [23]. Bulk separation of water and solvent is therefore possible providing a difference in density of 0.03, either above or below 1, exists between the two liquids. In general, the speed of separation is slow and therefore equipment is designed to minimise the vertical distance to be travelled and the residence time required to achieve partition. The most effective design of separator is a long horizontal cylinder with a narrow bore. The feed should enter the separator at a very low flow rate which reduces any turbulence within the system. The throughput and residence time required for separation of the two phases is dependent both on droplet size and viscosity of the organic solvent. Small globules settle more slowly than larger ones and the higher the viscosity of the solvent then the slower the speed of separation. The process can be improved by the addition of a plate or mesh which increases the rate of coalescence and size of droplet.

In the separation of large volumes at high flow rates then large, shallow rectangular basins, fitted with a plate to ensure rapid coalescence, should be used. Decanting is not a suitable process for separation of emulsions. However it is an ideal method for chlorinated solvents and hydrocarbon solvents with water solubilities of less than 0.2%. Clearly there is a limit to the performance of decanting and in instances of solubility between water and solvent it is not an appropriate method.

7.5.2.1.2 Membranes. Separation of solvents and water can also be achieved using membranes. This technique is called pervaporation as volatile organic compounds are removed from a liquid feed, such as a mixture of solvent and water, by passage through a semipermeable membrane into the gas phase followed by condensation to yield the solvent [24]. The separation is not only achieved by the difference in vapour pressure but also the permeation rate through the membrane. A schematic of a typical process is illustrated in Figure 7.5. The transport through the membrane is stimulated by maintaining a lower partial pressure of the solvent on the permeate side of the membrane. This is achieved by the application of a very small vacuum and by condensation of the permeate which establish a concentration gradient across the membrane. At the end of the pervaporation and condensation process the water content of the solvent is significantly reduced. However, although simple and often inexpensive the technique of permeation does not necessarily produce solvent of sufficient quality for reuse without further treatment. Permeation can also operate by the removal of water from the bulk liquid solvent. This is possible if the liquid feed is heated sufficiently to ensure vaporisation of the water on the vapour phase side of a membrane which is selective for permeation of water and not the solvent. Cooling and subsequent condensation of the water vapour creates the concentration gradient required to stimulate the separation. This is particularly appropriate if the technique is used as a pretreatment for aqueous effluent disposal.

Other methods suitable for the separation of solvents and water are the use of air or steam to 'strip out' the solvent from the aqueous phase [19].

7.5.2.1.3 Stripping. This technique is particularly useful for the recovery of low concentrations of solvents with low water solubilities and/or high volatilities with respect to water. Recapture of the solvent from the air stream can be accomplished by an adsorption process, such as activated carbon, as described above. This is a limitation of the system as the concentration of the solvent in the vapour phase may be very low resulting in excessive cost and complexity. The use of steam as the stripping medium results in easier methods for liberation of the solvent. Thus, for solvents which are immiscible with water then several methods, such as decanting, are now more appropriate as the proportion of solvent to water has increased.

A method for removing water from solvents which is appropriate for both the bench and industrial scale operations is adsorption.

7.5.2.1.4 Adsorption. Water and solvents can be separated using highly porous solids which preferentially adsorb water when in contact with the bulk liquid [19, 25]. Selectivity is achieved on the basis of the pore size of the solids which are small enough to exclude adsorption of solvents but not water. Molecular sieves, silica gel, activated alumina and sulphonic-type cationic exchange resins act as highly effective adsorbants. The process is best used for the removal of low levels of water as water capacities are usually small. Batch systems are preferred as the operation of a continuous system would necessitate the use of two beds which are sequentially adsorbed and regenerated. The process of reactivation or desorption of the water involves the passage of hot, inert gas through the bulk of the solid. Filtration is used to remove the adsorbant from the bulk solvent.

The advantages and properties of each adsorbant are different. Molecular sieves remove water with very

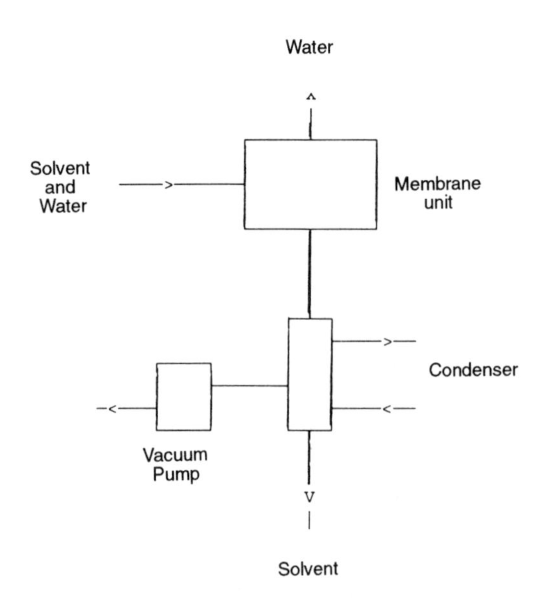

Figure 7.5 Schematic diagram of a typical pervaporation process.

little adsorption of the solvent and sieves with pore sizes in the range of 3, 4 and 5 Å can be used to dry all solvents with the exception of methanol. However regeneration requires temperatures in the region of 300°C which can result in high costs. The larger pore size of silica gel and alumina ensures greater water capacity. However, this also results in a reduction in selectivity and organics can be adsorbed leading to losses of valuable solvent. In addition, the presence of significant quantities of solvent in the regeneration process can be inconvenient and possibly hazardous. Nevertheless the lower temperature required for regeneration, 150°C for silica gel and 180–250°C for alumina, is attractive. Polymeric ion exchange resins have high capacity and selectivity for water, and can be regenerated at temperatures as low as 120°C using air rather than nitrogen. However, degradation in the presence of certain solvents is possible.

The process of adsorption releases heat. This is highest for molecular sieves at approximately 4000 J/g and lowest for silica gel which liberates in the region of 900 J/g. It is therefore necessary to ensure that control measures exist if large quantities of heat are liberated when drying flammable solvents. Although the process of adsorption is simple and relatively convenient it is only suitable for the removal of small percentages of water. The removal of water at greater than 1–2% would require very large quantities of adsorbant, long process times and very efficient methods for dissipation of any heat liberated during the process.

Several chemical, rather than physical approaches, to water and solvent separation are used in both small- and industrial-scale operations. Salting out relies on the use of a solid form, or highly saturated aqueous solution, of an electrolyte which on contact with a mixture of water and solvent will withdraw the water from the solvent and form a second phase. Separation is then achieved by decanting the two liquids. Chlorides and sulphates of Group 2 metals are commonly used as the electrolyte primarily due to their low cost, dehydrating power and the absence of reaction with common solvents.

7.5.2.1.5 Absorption. The use of a desiccant, a material with the ability to absorb water from a liquid, is a classical method for the chemical dehydration of a solvent and involves the addition of solid desiccant to a bulk volume of liquid followed by a secondary separation process, usually filtration. The selection of desiccant and solvent is crucial and depends upon the standard of dryness required and whether a reaction could occur between the solvent and the drying agent. A wide range of materials is available most of which are from Group 2 metals. Typical desiccants are the hydroxides of sodium and potassium, the hydride, oxide and sulphate of calcium, and the sulphates of sodium and magnesium.

A particularly powerful method for removing water from a solvent is the combination of a desiccant and distillation. In this batch process the solvent is distilled in the presence of the dehydrating agent. The desiccants used are distinctly different and significantly more hazardous than those listed above and again require careful selection with respect to the solvent to be dried. Indeed, lithium aluminium hydride, a commonly used desiccant, liberates hydrogen gas on contact with water! However, providing safe practices are adopted solvent containing extremely low levels of moisture, less than 50 ppm and in some cases below 10 ppm, can be obtained. Table 7.9 lists desiccants suitable for use with all the common groups of organic solvents in concert with filtration or distillation.

7.5.2.1.6 Distillation. Several methods under the collective name of distillation, even in the absence of a desiccant, are used for the separation of water and solvents [26]. The complexity of the process is dependent upon whether an azeotrope exists between the solvent, or solvents, and water to be separated. In the absence of an azeotrope, fractional distillation is possible providing the boiling point of the solvent is significantly higher or lower than 100°C. Thus, for solvents of very high boiling point the water can be removed by the application of heat and this can yield solvent with only traces of water present. The remaining moisture can be removed by dehydration. It should be noted that the water can contain traces of solvent and therefore may not be suitable for disposal direct to drain. Solvents which are very volatile in comparison to water must be separated by distillation at low temperature.

For situations in which an azeotrope exists then straightforward distillation may take place providing the solvent/water distillate splits into two phases on condensation. This is the case with chlorinated hydrocarbons and all hydrocarbons which are sparingly soluble in water. As above, the consequences of the likely presence of organics in the aqueous phase must be borne in mind prior to disposal. This issue is of increasing importance if the solvent is partly water miscible resulting in substantial quantities being present in the aqueous phase of the condensate.

Complications increase when there is an azeotrope between solvent and water which does not separate on condensation. Azeotropic mixtures of solvent and water can be separated by operating the distillation under high pressure. Under these circumstances the azeotropic composition changes and increases in water content. Separation can be achieved providing specialised equipment is used and measures are taken to ensure safe handling of the solvent at high pressure.

Alternatively, another solvent, or entrainer, can be added to a distillation process operating at atmospheric pressure. The entrainer, which has high solubility for the solvent to be recovered, also azeotropes with water

Table 7.9 A list of desiccants suitable for dehydration of solvents

Solvesnt	Desiccant (filtration)	Desiccant (distillation)
Hydrocarbons	Calcium chloride Calcium hydride Sodium (metal) Phosphorus pentoxide	Lithium or sodium aluminium hydride
Alcohols	Calcium oxide Calcium hydride Sodium (metal) Potassium carbonate	Magnesium iodide
Chlorocarbons	Calcium chloride Calcium hydride Calcium sulphate Sodium sulphate Potassium carbonate Magnesium sulphate	Phosphorus pentoxide
Ethers	Calcium oxide Sodium (metal) Sodium hydroxide Potassium hydroxide	Sodium or lithium aluminium hydride
Esters	Calcium hydride Calcium sulphate Sodium sulphate Potassium carbonate Magnesium sulphate	Phosphorus pentoxide
Ketones	Calcium sulphate Sodium sulphate Potassium carbonate Magnesium sulphate	
Glycol ethers	Calcium chloride Calcium hydride Sodium (metal)	Lithium aluminium hydride

but separates during the condensation. The solubility of the desired solvent is higher in the entrainer than in water and therefore separation from water occurs during condensation. Subsequent distillation of the organic phase, which comprises two organic materials of different boiling point, yields the solvent, which is now free of water.

7.5.2.1.7 Liquid–liquid extraction. The principles operating in the distillation process involving an entrainer have similarities to another method for separating water and organic solvents called liquid–liquid extraction [23]. This technique involves washing a mixture of a hydrophilic solvent and water with a hydrophobic organic, such as a chlorinated or non-chlorinated hydrocarbon. This process of washing facilitates a phase separation of the desired solvent which has a greater solubility for the organic phase. As in the case of the entrainer, distillation of the mixture of organics, which are of different boiling points, yields the solvent targeted for recovery, free from water. The material used for the extraction can also be recycled during this process. Equipment used to operate liquid–liquid extraction is designed to ensure maximum contact between the aqueous and organic phases. Large quantities of the extracting solvent will be present in the aqueous layer and therefore further treatment will be required prior to disposal.

7.5.2.2 The recovery of solvents from organic liquids. The principal method for the separation of one or more organic solvents is distillation and is both suitable for the bench and large scale [23]. The basic apparatus used for distillation is a vertical cylindrical column which contains either trays or packings which enhances the contact between vapour and liquid. A continuous process involves introduction of the feed at one or more points in the column. Batch processes consist of a vessel at the base of the column in which the material to be distilled is placed at the beginning of each process.

The difference in density of vapour and liquid results in the upward flow of the vapour and a downward flow of the liquid in the column. Material reaching the bottom of the column is reheated and partially vaporised back up the column. Vapour reaching the top of the column is partially removed, and becomes the distillate, and part is returned as the reflux. Therefore the overall flow pattern in the column is a countercurrent flow of vapour and liquid throughout all packings or trays. As a result the lighter boiling components tend to concentrate in the vapour phase and the higher boiling fractions tend toward the liquid phase. Therefore the separation of the components of the feed relies upon the difference in composition between the

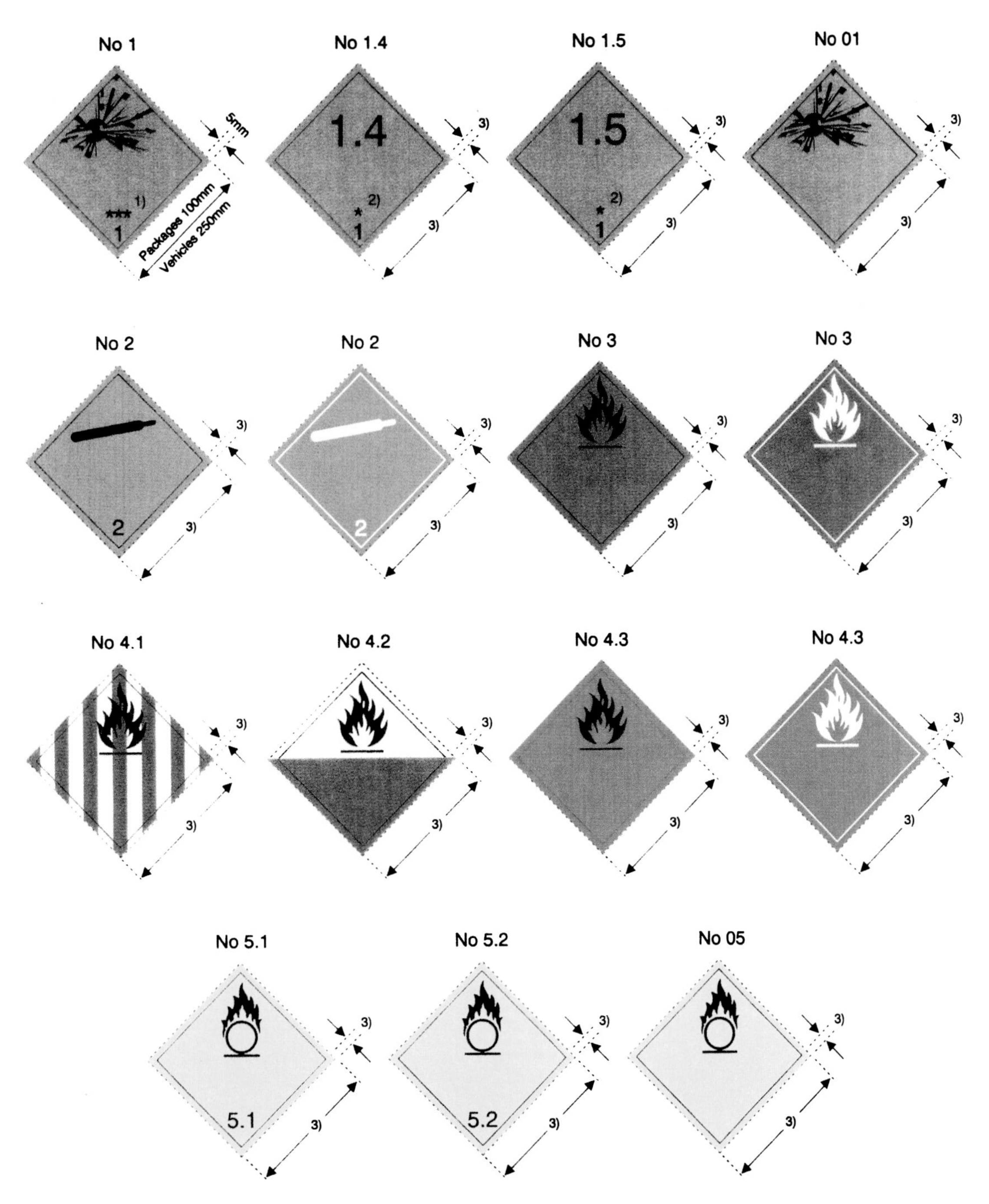

Plate 1a Danger labels for packages and transport vehicles specified in ADR.

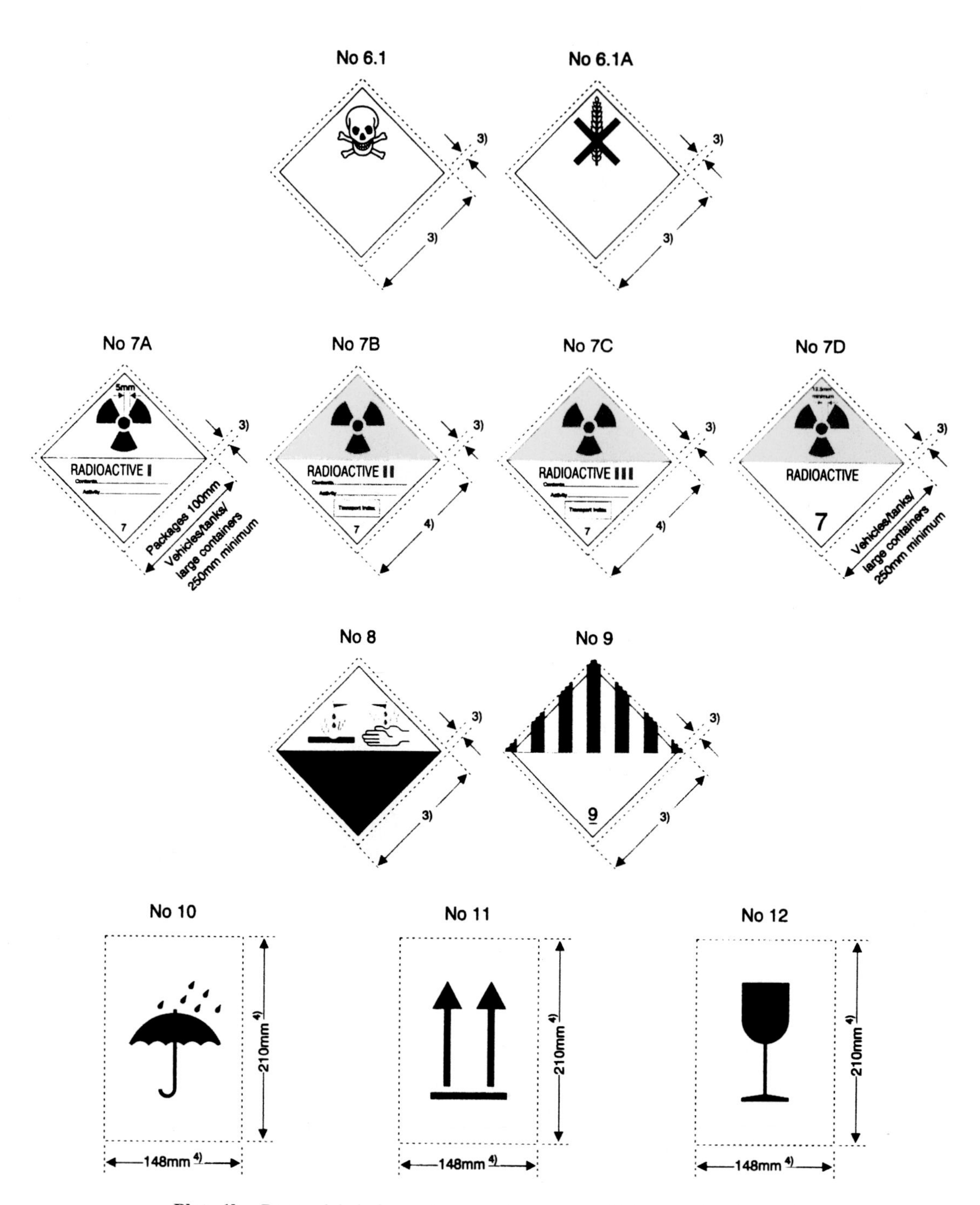

Plate 1b Danger labels for packages and transport vehicles specified in ADR.

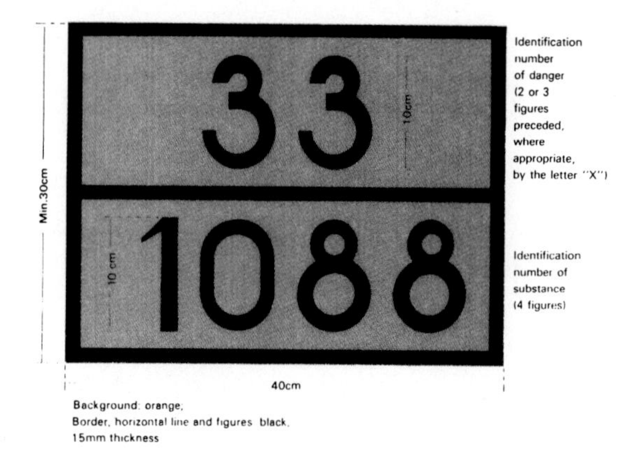

Plate 2 An example of a transport plate specified in ADR. The substance is acetal.

liquid mixture and the vapour formed from it. The relative volatilities of the components to be separated, the number of contacting trays, and the ratio of liquid phase flow and vapour phase flow all contribute to the efficiency of the process.

Distillation can be performed in a number of different ways:

- atmospheric
- vacuum
- steam
- azeotropic
- extractive

If the feed mixture is available as an isolated batch then separation can be achieved using all but extractive distillation. For continuous processing then separation can be achieved using all technologies. Most commercial distillations are operated on a continuous basis. Batch processes are generally more expensive per unit of product recovered and involve the use of complex monitoring equipment which has a limited throughput as a consequence of recharging the system at the end of each separation. In each type of distillation the intention of the system used is to maximise the difference in relative volatility under the conditions of operation.

7.5.2.2.1 Atmospheric distillation. In simple atmospheric distillation the liquid mixture is brought to boiling and the vapour formed is separated and condensed to form a product which comprises solely the more volatile component. This fractionation is applicable to a mixture of solvents in which the relative volatilities exceed 1.5 at standard pressure. Values for relative volatility are available for the majority of organic solvents [27]. The number of theoretical plates required to achieve effective dissection of the mixture of the solvents can be calculated using a number of equations [23]. If, under standard temperature and pressure, the difference in relative volatility between two materials is insufficient to enable fractionation then the distillation can be performed under conditions of reduced pressure.

7.5.2.2.2 Vacuum and steam distillation. The temperature at which distillation occurs directly affects relative volatility and differences are extenuated at lower temperature. Therefore by applying a vacuum the distillation temperature is reduced and two materials can be more readily separated. In addition, operating at lower temperature may reduce the cost of the process and can eliminate the potential for thermal decomposition of one or more of the components within the mixture. The same effect can be achieved by the injection of steam into an atmospheric distillation process. The steam reduces the overall boiling point of the process and therefore the differences in relative volatility between the components is increased. However, steam distillation is only a sensible approach for the recovery of water immiscible solvents, or for azeotropes of solvents which separate on condensation. The use of steam in a distillation process can also be used to lower the operating temperature of separations involving materials of very high boiling points, thus reducing costs.

7.5.2.2.3 Azeotropic distillation. A further development involves the addition of an entrainer, either another solvent or water, to the mixture of liquids to be separated. The purpose of this material is to form a selected azeotrope with one of the components. This results in a difference in relative volatility between the azeotrope and the non-azeotropic component allowing separation to be achieved. Typically the azeotrope will be of higher volatility and becomes the distillate, although the azeotrope can be such that it is removed as bottoms. An effective entrainer therefore must be selective for the solvent to be recovered, stable under the conditions of use, chemically compatible with all components, relatively inexpensive, readily available and must be easily separable from the desired product. Water is an ideal entrainer when used to form azeotropes with solvents which separate on condensation. Guidelines for entrainer selection have been provided by Berg and Gerster [28, 29]. Many examples of azeotropic distillation can be cited [23]. Examples include the separation of benzene from cyclohexane by the azeotrope of the latter with acetone followed by liquid–liquid extraction with water to yield the cyclic hydrocarbon. Similarly the use of methylene chloride as an entrainer for separation of an azeotropic mixture of methanol and acetone is achieved by addition of methylene chloride followed by the distillation of the selective azeotrope between the alcohol and chlorinated hydrocarbon.

7.5.2.2.4 Extractive distillation. Involves an entrainer which is used when the relative volatilities of the components of a mixture are similar or when an azeotrope exists. However rather than forming a new azeotropic mixture, the role of the entrainer is to influence the relative volatilities of the components of the mixture to be separated based on differences in chemical characteristics [23]. The more typical case is the addition of a high boiling solvent to a tray in the distillation column which reduces the vapour concentration of the material in the original mixture with the lower relative volatility. Alternatively low boiling point, highly volatile entrainers, can extract the more volatile components of a mixture providing a similar association exists. The process of entrainer selection is the key to the success of the process and is well described [28, 30]. Extractive distillation has been successfully used for the separation of toluene from naphthenes, benzene from paraffins, pentane from pentene and toluene from methyl cyclohexane.

7.5.2.2.5 Liquid–liquid extraction. Liquid–liquid extraction is primarily used when distillation is impractical or too costly. For example when the relative volatilities of the two components are less than 1.3. In much the same way as with separation of solvent and water, liquid–liquid extraction is a process of separating components based on their distribution between two immiscible liquids. A good example of this technology is the separation of aromatic and aliphatic hydrocarbons by washing with a mixture of diethylene glycol and water which selectively extracts the more polar aromatics [31]. However, the application of liquid–liquid extraction to the separation of two solvents is relatively rare. This is primarily driven by economics as further stages are required such as distillation to achieve complete recovery of the solvent.

7.5.3 Recovery from solids

In addition to separation from gases and other liquids, solvents are often recovered from solids or residues. This typically follows from the presence of a solid impurity, or the use of a solid drying agent, but may also follow a crystallisation process. Alternatively prior to disposal of the solid, wherein the presence of a solvent would be hazardous, wasteful or environmentally unacceptable. A number of techniques are applicable.

7.5.3.1 Sedimentation. Sedimentation is the partial separation or concentration of suspended solids from a liquid by gravity settling. The process of sedimentation depends upon the particle size, liquid viscosity, solid and solution densities and the characteristics of the particles in the slurry. At low concentrations particle settling will occur as the particles are sufficiently far apart to allow free settling. Recovery of the liquid is then achieved by decanting providing a clear demarcation exists at the interface between the liquid and solid phase. Alternatively the process of sedimentation may be achieved by the use of equipment such as thickeners or clarifiers which act by concentrating the suspending solids [23].

Rapid separation of solid and liquid can be achieved using centrifugal separators of which a wide variety of designs exist [23]. Centrifugal sedimentation requires a difference in density between the two phases. All components of a system comprising one or more particles in a continuous liquid phase that are enclosed in a rotating cylindrical vessel will experience centrifugal forces. This force causes the solids of density greater than the fluid to migrate radially toward the wall of the vessel causing sedimentation. Solids less dense than the fluid will migrate towards the axis of rotation until they reach the liquid–air interface. If the sides of the vessel are perforated then the liquid will flow through the sediment of solids. In laboratory scale systems the discharge of liquid may be intermittent but commercial systems usually operate on a continuous basis. Removal of the solids in continuous systems is necessary to maintain the efficiency of the process. If the solids are less dense than the liquid then a skimmer is employed at the liquid surface. Solids at the wall of the vessel are removed by a cutter knife or use of a screw conveyor.

7.5.3.2 Filtration. This involves the passage of the solids–liquid mixture through a porous medium which retains the particulates within the mixture. Filtration is facilitated by the flow of liquid through the medium by gravity, or can be assisted by the application of pressure upstream of the filter medium, or by a vacuum downstream. Centrifugal filtration is also possible and is closely related to centrifugal sedimentation. Separation is achieved by retention and build-up of the solids at the surface of the filter medium, called cake filtration, or by trapping in the pores of the medium, called depth filtration. A wide variety of materials have been used as filter media with different requirements for cake and depth processing.

Typical materials include woven fibres, metal screens or fabrics, pressed felt or cotton batting, sheets of synthetic polymers, paper, sand, coal, silica porcelain and many more. Filtration equipment is equally diverse for both batch and continuous processing [23].

Cross-flow filtration, in which the slurry of solid and liquid is applied tangentially to the filter bed, offers significant advantages over conventional filtration. These can be summarised as:

- rapid filtration rate as particle accumulation is minimised
- the absence of the need for flocculating agents to increase particulate size
- reduced energy requirements from the pump in moving the slurry across the filter bed or the provision of a pressure differential to drive filtration

The increase in filtration rate allows the use of microporous or ultrafiltration membranes for retention of micrometre sized particulates thus increasing the degree of separation of solid and liquid.

This highlights the emergence of the use of membranes in either reverse osmosis or ultrafiltration systems which are based on the selectivity of membranes. Reverse osmosis is the process of separating a solute from a solution by forcing the solvent to flow through a membrane by applying a pressure greater than the normal osmotic pressure. Ultrafiltration, as highlighted above, separates species by differences in molecular shape and size. Under pressure solvent passes through the membrane whereas large solute species remain [23, 26].

7.5.3.3 Expression. Both sedimentation and filtration are suitable separation techniques when the mixture of liquid and solids is sufficiently mobile to allow pumping, or similar method of motion, of the fluid to a barrier which retains the solid but not the liquid. If such movement is not possible then separation can be accomplished by compressing the mixture under conditions which permit the liquid to escape while retaining the solid between the compressing surfaces. This technique is called expression. Design of expression equipment is varied. Batch systems usually operate by the application of hydraulic pressure in units such as the box press, pot press, curb press and cage press. Continuous expression utilises equipment such as screw presses, roller mills, and belt presses [23].

The decision between the options of sedimentation, filtration and expression requires a thorough selection process to ensure the most appropriate technology is utilised for the particular application.

7.5.3.4 Distillation. Solids can be classified as materials with zero or negligible vapour pressure, the opposite of solvents, and therefore this difference offers the possibility of separation by distillation. There are a number of ways to achieve this process when considering a separation of solid, or more typically non-volatile viscous residues, from solvents.

Recovery of solvent is possible by batch distillation of the mixture of liquids and solids yielding solvent of very good purity. Batch systems are preferred to continuous processes as the former can limit the quantity of solids to be removed and are therefore easier to handle. The simplest form of distillation consists of the heated vessel, condenser and one or more collection tanks. Clearly there is only one theoretical plate in this system and this is only suitable for the recovery of a single solvent or for initial separation of a mixture of solvents from residues. Further treatment would be required to separate the individual components of a mixture.

The use of a rectifying column between the heated vessel and condenser permits a greater control on the distillation process and allows separation from the solids and fractionation of the solvent mixture in a single unit. The temperature of the process is controlled such that only a portion of the condensate is continuously withdrawn such that the column acts as an enriching section. As time proceeds the composition of the material under distillation increases in the less volatile components. Fractionation is achieved by switching to different receiving vessels during the distillation process.

However, the process of distilling mixtures of solvents and residues can be extremely hazardous due to the generation of an exothermic reaction in the bottoms. It is advisable to assess the potential for this exotherm by performing a laboratory scale trial prior to full scale operation. The process of adding new material on top of the bottoms from a previous process is also hazardous as the residence time and knowledge of the composition becomes unknown and the potential for an uncontrollable reaction increases. Therefore, each batch should be individually processed. One of the techniques used to avoid generation of an exotherm is to perform the distillation under vacuum and consequently at a temperature below the activation energy of the reaction.

Another problem associated with this method of separating solvents from non-volatile residues concerns the final stages of the distillation. As the solvent is depleted from the sump the solution becomes supersaturated and the residues begin to deposit on the heating surface reducing efficiency. Several methods can be used to eliminate or reduce this problem and include mechanically cleaning the heating surface or maintaining flow of the residues. If the boiling point of the solvent to be recovered is below 100°C then an alternative to dry distillation exists which reduces the risks associated with exotherms.

7.5.3.5 Steam distillation. The majority of residues will liberate the solvents contained therein following direct injection of steam. This process is ideal for materials which are water immiscible such as chlorinated or hydrocarbon solvents in which complete and rapid separation occurs after condensation. Materials which azeotrope on distillation with water but condense to form two phases are also appropriate candidates for this process. The use of steam distillation with water-miscible solvents can be effective. However, a secondary separation process is required, such as fractionation, which clearly adds to the overall cost and complexity. Indeed, the cost of steam and dry distillation is only comparable for water insoluble solvents. In all options the presence of organics in water may create an effluent disposal problem which may be reduced by recycling the water back into the steam generator. The solvent obtained from steam distillation may need a separate dehydration stage if dry product is required. The potential for exotherm from the residues is eliminated because steam distillation is a low temperature atmospheric method of distillation with insufficient activation energy for generation of an uncontrollable reaction. A different problem concerning the residues associated with steam distillation relates to the presence of water making them difficult to handle and are only suitable for disposal by incineration.

7.5.3.6 Superheated vapour distillation. An alternative to the use of steam distillation is to use a superheated vapour of the solvent to be recovered as the heat source and injecting it into the residues [19]. The superheated vapour is initially generated by extraction of solvent from the residues and passage through a heat exchanger. Injection of this superheated vapour into the residues liberates yet more vapour of the solvent and the excess is passed through a condenser and collected as a residue free liquid. The whole process is operated under a vacuum and therefore solvents with a boiling point below 180°C can be processed without decomposition. The use of vapour distillation eliminates the potential for fouling of the heating surface as achieved when using steam as the heat source. However, a further advantage is that the absence of water from the process eliminates the need for any further separation that might be required for water miscible solvents.

An interesting method of avoiding the potential for fouling of the heating surfaces is to eliminate the contact between the mixture of solvent and residues and the heating surface. Small-scale batch distillation units are available in which the feedstock is placed in an inert high temperature stable polymeric bag which is immersed in a bath of electrically heated oil [19]. The maximum distillation temperature must not exceed 200°C and therefore the application of a vacuum is often used for high boiling solvents. The process, which can be adapted for both flammable and non-flammable solvents, is also attractive because the entire bag can be removed from the equipment and sent for disposal thus avoiding any handling of the highly viscous or hazardous residues.

7.5.3.7 Thin film evaporator. The use of an agitated thin film evaporator is another very efficient method for applying heat to a maximum of residues and liquid without the potential for fouling of the heat transfer surfaces. The apparatus consists of a heating surface which is a large horizontal or vertical tube which is either tapered or straight. The liquid, which can be highly viscous, is spread on the tube wall by an assembly of rotating blades which either maintain a small clearance from the film or ride on the film of the liquid. The rotation is in the order of 12 m/sec and the process results in a residence time on the heating surface of a few seconds. Very rapid heat transfer is achieved and is ideal for temperature sensitive materials. The vapour of the solvent flows upwards through the separator and is thereafter condensed. If a mixture of solvents is liberated from the residues then a fractionation column can be fitted between the evaporator and the condenser [23].

The operation of an efficient recovery process should be the intention and primary goal of every user of solvents. However, this may not always be possible and a decision not to recycle solvent may be based on a cost/benefit analysis, or the technical difficulty. Indeed, even when operating a recovery process, waste will be generated which requires an environmentally acceptable disposal process.

7.6 Disposal of solvents

In the early applications of chemicals, the methods used for disposal were, in general, performed without consideration of the environmental impact. However, as the volumes of chemicals increased and the consequences of such practices became apparent it was necessary to develop non-polluting methods for the disposal of all chemicals, including solvents. New approaches were required which not only would benefit the environment but also the user, the chemical industry and society as a whole. Sensible legislation is the key driver to achieve a balance between use and impact minimisation. Sustainable development is achieved by establishing performance criteria, cooperation, commitment and by auditing compliance.

Disposal is the final stage in the life cycle of all solvents. It is clear that solvents are an essential element in the development and continued progress of today's society. However, responsible use of these valuable materials must be promoted by producers, suppliers, trade associations and regulatory bodies. An integral element of this responsibility is a commitment to disposal that is achieved in a manner which does not cause anthropological, natural or environment damage.

A number of technologies are now available and under development for the treatment of waste streams containing solvents. These cover waste in the vapour phase and as liquid or immobile residues.

7.6.1 Techniques for disposal of solvents in vapour phase

The use of this approach may be required for 'end of pipe' abatement when the quantity of solvent present in the vent stream is not sufficiently high to warrant a recovery process, as detailed above and is below what could be considered as best practice or exceeds an emission limit.

7.6.1.1 Thermal incineration. A process of chemical reaction which is activated by heat which occurs within a specifically designed reactor, and is often utilised for treatment of waste-gas streams that contain traces of organic matter. This can be achieved thermally, by a homogeneous gas-phase reaction, or catalytically, a heterogeneous gas–solid interfacial reaction. Both processes have been applied to the disposal of solvents in the vapour phase. Perhaps one of the simplest and most obvious forms of thermal gas-phase treatment are the flares typically observed from petrochemical installations. However, although adequate for their traditional use, flares are definitely not suitable for complete waste disposal.

More complete combustion is achieved using direct-flame incinerators in which the waste gases are heated to their autoignition temperatures in the presence of 25% excess of oxygen [32]. Factors affecting the performance of the process are residence time, temperature and turbulence within the gas stream. For hydrocarbons, a minimum temperature of 750°C and residence time of 0.5 sec is sufficient to achieve complete combustion without liberation of intermediates such as aldehydes, acids and carbon monoxide. Higher temperatures and longer residence times are required for chlorinated hydrocarbons to ensure that materials such as dioxins are not formed. Therefore combustion temperatures in excess of 1100°C for 2 sec are typical. In addition, rapid cooling is necessary to avoid the formation of dioxins when processing chlorinated organics. The incineration of halogenated materials, which have low calorific value, often requires the presence of a supplementary fuel such as methane. Fuels are also required during start up of all processes and may be used at various periods of operation.

The by-products obtained from incineration depend upon the solvents present. Saturated and unsaturated hydrocarbons will liberate carbon dioxide and water. Halogenated materials will also generate the corresponding hydrogen halides such as HCl or HF. These must be scrubbed from the vapour phase and neutralised prior to emission of the exit gases. Solvents containing sulphur or nitrogen will generate sulphur and nitrogen oxides respectively which also require treatment.

High turbulence is required within the incinerator. This is achieved by imparting shear within the gas streams which may be directed in opposing directions. Alternatively baffles can be placed in the incinerator. The design of the equipment is specific to the application and must take into account the composition of the waste gas, the required temperature and residence time and the degree of turbulence needed to achieve complete combustion.

7.6.1.2 Catalytic incineration. Catalytically induced incineration occurs at a lower temperature than the thermal process and hence running costs are significantly reduced. Catalytic species are generally precious metals such as platinum or palladium which are impregnated onto an inert support such as alumina or silica. A variety of support structures can be used such as ceramic honeycombs, metal honeycombs, spheres, pellets and fibres packed as either a single bed or in modules. Reactor design and operating conditions are again specific to each application. However critical issues can be identified. Firstly the need for filtration of the inlet gas to eliminate catalyst poisoning. Preheating of the waste stream to ensure complete combustion and operation at 25% of the lower explosive limits is necessary for flammables and combustibles.

Examples of thermal and catalytic incineration of gaseous waste streams are found in the literature [32]. These include the treatment of the solvent vapours emitting from beverage can coating processes. Destruction of the atmospheric emissions of hydrocarbons from the manufacture of acrylonitrile and of by-product organic acids from formation of phthalic/maleic anhydride. Catalytic oxidation of hydrocarbons emitted from primers and other coatings used in the manufacture of decorative or protected sheet metal items. High conversion of chlorinated hydrocarbons such as methylene chloride, perchloroethylene, to carbon dioxide and gaseous hydrogen chloride is achieved by treatment at 500°C over a palladium/aluminium trioxide catalyst in the presence of propane which is acting as a supplementary fuel [33].

7.6.1.3 Biological treatment. A rather different approach to the use of temperature and inorganic catalysts is the emergence of biological treatment. Advances in the understanding of the metabolic pathways existing within bacteria and other microorganisms have identified processes for the metabolism of alcohols, esters, ketones, aliphatic, aromatic and more recently halogenated hydrocarbons. This list of chemicals clearly overlaps with groups common in solvents. Processes have been developed for conditions of low flow rate and low concentration of materials relevant to solvents from gaseous vent streams [34]. In simple terms the volatile organics are fed into a bioreactor containing microorganisms capable of metabolism of the constituents of the stream. The majority of processes operate by absorption followed by biodegradation. The principal products are carbon dioxide, water and in the case of halogenated materials a dilute acidic aqueous solution of the halogen. Ester, alcohols and ketones are easily degraded and concentrations up to 1500 mg per cubic metre of inlet gas can be tolerated. Aromatics are more difficult and often specific inoculation, or the use of genetically engineered organisms, is required for maximum concentrations of 500 mg per cubic metre. Finally chlorinated solvents, which often require the addition of a co-substrate to stimulate microbial growth, are accommodated at levels of 20 mg per cubic metre. The typical operating temperature of the biomass is between 15°C and 35°C. Other process parameters include acclimatisation and the addition of nutrients such as nitrates, phosphates, metal ions and water to maintain and stimulate growth. The use of biosystems can require close supervision and maintenance but when operating correctly offers very efficient abatement.

The methods for treatment of liquid or solid waste are not so diverse and concentrate mainly upon thermal degradation. Open landfill, a typical practice for domestic and non-industrial waste, is no longer an acceptable process for chemical or solvent-containing waste. Discharge to any water course, sewer or sea, a practice previously considered appropriate, has long since been regarded as unsuitable for disposal of waste which, either whole or in part, consists of chemical solvents.

7.6.2 Technologies for the disposal of liquid or solid solvent-containing waste

Incineration is by far the most common method for the ultimate disposal of solvent-containing waste. There are many designs of incinerators for municipal waste. However the destruction of chemical waste requires specialist design of which the rotary kiln is one of the most popular [35]. These units are capable of handling any liquid waste which can be atomised and injected into the furnace which rotates on an axis slightly inclined from the horizontal. The units are very versatile and can also accommodate heavy tars, sludges, pallets, filter cakes and at the very height of the technology intact, but open, drums of solvent. Fluidised bed reactors are also extensively used for the incineration of industrial waste, including solvents.

The successful application of incineration involves balancing the temperature required for destruction, the residence time to ensure complete reaction and the turbulence within the incinerator to ensure good mixing and the absence of cold regions. The conditions will be specific to the nature of the waste. For example, combustible solvents will require lower temperatures than those solvents, such as chlorinated hydrocarbons, with a low calorific value. Indeed supplementary fuels may be required in treatment of poorly combustible materials. In the disposal of hydrocarbons minimum temperatures in the region of 1100°C are required, increased to 1400°C for chlorinated materials. The latter value is necessary to avoid the formation of dioxins resulting from incomplete combustion. Dioxin formation is also avoided by rapidly cooling the exit gases to temperatures below 250°C. Residence times in the kiln and any supplementary combustion chambers are dependent on the rate of gas flow, speed of rotation of the kiln, the composition of the waste and the design of the system. However, a usual standard is a minimum of 4 sec in the high temperature zones. Scrubbing of exit gases is to remove particulates, and possibly acids generated from the presence of halogenated species in the waste. Other effluents will include the slag, which is suitable for landfill, and the water from the scrubbing towers which following neutralisation can be sent to drain or watercourse. Carbon dioxide and water are emitted to atmosphere.

A further development in incinerator technology is the use of chemical waste as a supplementary fuel for use in cement of lime kilns. These processes operate at extremely high temperature sufficient for complete combustion of hydrocarbon waste. An extension to the disposal of halogenated materials may be possible providing the temperature and residence times required for elimination of dioxin formation can be consistently established within the kiln.

The goal of disposal technology is the conversion of waste which is hazardous and/or may cause environmental damage to products which are non-hazardous and essentially non-polluting. The text above illustrates the existence of processes for treatment of solvent and solvent-containing waste which satisfy these criteria. Technology is available for the manufacture, use, recycling and disposal of solvents which meet stringent standards with respect to health, safety and the environment. Clearly successful adoption of all of these is inherent in a commitment to Global Sustainable Development and the continued responsible use of valuable chemicals such as solvents.

Acknowledgements

Special thanks to my wife, Jacqui, for her patience and support, and to my colleagues Dr. Martin Smith, Keith Major, Ian Ball and David Stretch for their valuable assistance and comments, all so willingly provided during the preparation of this chapter.

References

1. Crude oil spillage from Exxon Valdez in Alaska in 1989.
2. Kharbanda, O.P. and Stallworthy, E.A. (eds) (1988) *Safety in the Chemical Industry: Lessons From Major Disasters*, Heinemann Professional Publishing, London. ISBN 0-434-910198.
3. Kletz, T.A. *What Went Wrong? Case Histories of Process Plant Disasters*, Gulf Publishing Co., Texas. ISBN 0-87201-339-1.
4. *United Nations Recommendations on the Transport of Dangerous Goods*, 9th edn, 1995. ISBN 92-1-139048-6.
5. The Convention concerning International Carriage by Rail has two appendices:
   CIV Uniform Rules governing carriage of passengers
   CIM Uniform Rules governing carriage of goods
   RID forms the Annex 1 of the CIM Uniform Rules.

6. *European Agreement Concerning the International Carriage of Goods by Road (ADR).* ISBN 0-11-55-1114-8.
7. *Regulations Concerning International Carriage of Dangerous Goods by Rail (RID).* ISBN 0-11-55-11-229.
8. *International Maritime Dangerous Goods Code (IMDG).* ISBN 92-801-1243-0.
9. International Air Transport Association, *Dangerous Goods Regulations (IATA).* ISBN 92-9035-543-3.
10. UK HSE, Guidance note HS(G)50, *The Storage of Flammable Liquids in Fixed Tanks (up to 10 000 cubic metres total capacity)*, 1990. ISBN 0-11-88532-8.
11. UK HSE, Guidance Note HS(G)51, *The Storage of Flammable Liquids in Containers*, 1990. ISBN 0-11-8855-336.
12. UK HSE, Guidance note HS(G)52, *The Storage of Flammable Liquids in Fixed Tanks (exceeding 10 000 cubic metres total capacity)*, in preparation.
13. The Loss Prevention Council, *Recommendations For Storage and Use of Flammable Liquids*, 1990.
14. The Institute of Petroleum, *Refining Safety Code*, 1981. ISBN 0-471-26196-3.
15. Chemical Industries Association, *Guidelines for the Safe Warehousing of Substances with Hazardous Characteristics*, 1983.
16. European Chlorinated Solvent Association, *Storage and Handling of Chlorinated Solvents*, 1989.
17. BS 5345 Part 1 (1976), Part 2, Code of practice for selection, installation and maintenance of electrical apparatus for use in potentially explosive atmospheres (other than mining applications or explosive processing or manufacture), 1983.
18. Soffel, R.W., Carbon (carbon and artificial graphite), in *Kirk–Othmer Encyclopedia of Chemical Technology*, 3rd edn, Vol. 3, Wiley Interscience, New York, 1978. pp. 556–631.
19. Smallwood, I.M. *Solvent Recovery Handbook*, Edward Arnold, London, 1993. ISBN 0-340-57467-4.
20. Clarke, D.H., Dow Chemical Company, US Patent 4 519 816, May 1985.
21. Hickman, J.C. and Glotz, H.R. *Proceedings* from the International CFC and Halons Alternatives Conference, Washington, DC, 1991. pp. 136–141.
22. Simmons, V.L. *et al.*, Membrane Technology and Research Inc., California, Membrane Vapour Separation Systems: Applications and Case Histories presented at AIChE 1992 Spring National Meeting, New Orleans.
23. Perry, R.H. and Green, D. (eds) *Perry's Chemical Engineer's Handbook*, 6th edn, McGraw-Hill, New York, 1984.
24. Bengey, P.M. *et al. Pervaporation in Synthetic Membranes: Science, Engineering and Applications*, D. Reidel Publishing Company, Boston, 1986.
    Wijmans, J.G. *et al. Environmental Progress*, **9**(4), 1990, 262–268.
25. Riddick, J.A., Bunger, W.B. and Sakano, T.K., Organic solvents physical properties and methods of purification, in *Techniques of Chemistry*, 4th edn, Vol. II (ed. A. Weissberger), John Wiley and Sons, 1986. pp. 795–799.
26. Rousseau, R.W., *Handbook of Separation Process Technology*, John Wiley and Sons, 1987. ISBN 0-471-89558-X. pp. 229–340.
27. Gmehling, J., Ohken, V. and Arlt, W., *Vapour Liquid Equilibrium Data Collection*, Dechema, Frankfurt, 1984.
28. Berg, L., *Chem. Eng. Progress*, **69**(9), 52, 1969.
29. Berster, J.A., *Chem. Eng. Progress*, **69**(9), 43, 1969.
30. Tassios, D.P., Rapid screening of extractive distillation solvents: Predictive and experimental techniques, in *Extractive and Azeotropic Distillation*, Advances in Chemistry Series No. 115, American Chemical Society, Washington, DC, 1972. Chapter 4.
31. Grote, *Chem. Eng. Progress*, **54**(8), 43, 1958.
32. Conklin, J.H., Sowards, D.M. and Kroehling, J.H. Exhaust control, industrial, in *Kirk–Othmer Encyclopedia of Chemical Technology*, 3rd edn, Vol. 9, Wiley Interscience, New York, 1989. pp. 511–541.
33. Bond, G.C. and Sadeghi, N., *J. Appl. Chem. Biotechnol.*, **25**, 1975, 241–248.
    Parkinson, G., *Chem. Eng.*, **98**, 1991, 37.
    Chaterjee, S. and Greene, M.L., *J. Catal.*, **130**(1), 1991, 76–85.
    Lester, G.R., *Ind. Eng. Chem. Res.*, **28**(10), 1989, 1449.
    Lester, G.R., *Environ. Sci. Tech.*, **23**(9), 1989, 1085.
34. Ottengraf, S.P.P. and Disks, R., *Chem. Oggi*, **8**(5), 1990, 41–45.
35. Crocker, B.B. and Bailie, R.C. Incinerators, in *Kirk–Othmer Encyclopedia of Chemical Technology*, 3rd edn, Vol. 13, Wiley Interscience, New York, 1978. pp. 182–206.

8 Major solvent applications overview

J.R. KELSEY

8.1 Introduction

Solvents have been a necessary part of life for centuries. Their use has grown increasingly more important in the past hundred years, some so much so that modern society could not function properly without them. If you look up from this book, there will be scarcely an item around you that has not required the use of solvents at some part of its manufacture, decoration or protection. Consider what would disappear from the typical house if it were not for solvents: paint on walls, most wooden furniture and soft furnishings, household cleaners, cosmetics and, of course, the car. The list is almost limitless. Solvents find use in a myriad of different applications. Excluding use as chemical intermediates for the manufacture of other materials, the greatest use of solvents is by far in the manufacture and application of paints and inks. These two, combined with the manufacture of resins which are key components of paints and inks, are often included, along with adhesives, as 'coatings' which account for about half of solvent use. Other major uses are in metal cleaning and degreasing and in pharmaceutical manufacture. Other important applications include use in oil dewaxing, coal washing, extraction processes, dry cleaning, agricultural products, rubber processing, detergents and cosmetics. There is a wealth of minor end uses; the more significant include explosives, flavours and fragrances. There are in addition some important end uses for individual solvents such as food and oil processing (mainly aliphatic hydrocarbons) and automotive de-icers (alcohols). Breakdowns of market size are shown in Figures 8.1 and 8.2.

Most applications are covered in more detail in this chapter but because of its importance and complexity, the paints sector is further broken down into its principal subcategories. The sectors covered are:

- Paints
- Inks
- Adhesives
- Metal cleaning
- Pharmaceutical and agrochemical manufacture
- Cosmetics and toiletries
- Detergents and household products
- Automotive de-icers
- Dry cleaning
- Miscellaneous

This review can only touch upon the major impacts of solvents in modern life. It will focus on key application areas to demonstrate our dependence upon them.

8.2 Paints

Owing to the size, complexity and importance of the paints market, it is treated in some detail but because of the varied nature of the end uses it is necessary to subdivide it. Coatings use can be classified into two categories according to whether the application is under contained or uncontained conditions (for example, painting in a factory production line or painting the outside of a building). Individual pieces of legislation tend to focus on one area or the other. Coatings can be viewed according to whether they are used professionally or in the 'Do It Yourself' market; this is a useful classification in some but not all areas of use. The most widely used approach, however, and the one used here is to organise by end use application; many of these can be identified but the main ones of importance are listed below:

- Decorative coatings
- Automotive OEM coatings
- Automotive refinish coatings
- Wood coatings and impregnation
- Metal packaging coatings
- Coil coatings
- Anticorrosion and marine coatings
- General industrial coatings

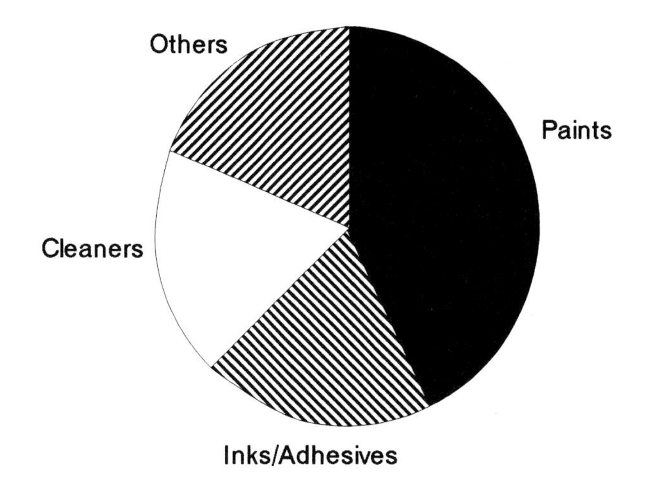

Figure 8.1 US solvent demand 1994 (excluding chemical intermediates) 4400 ktes.

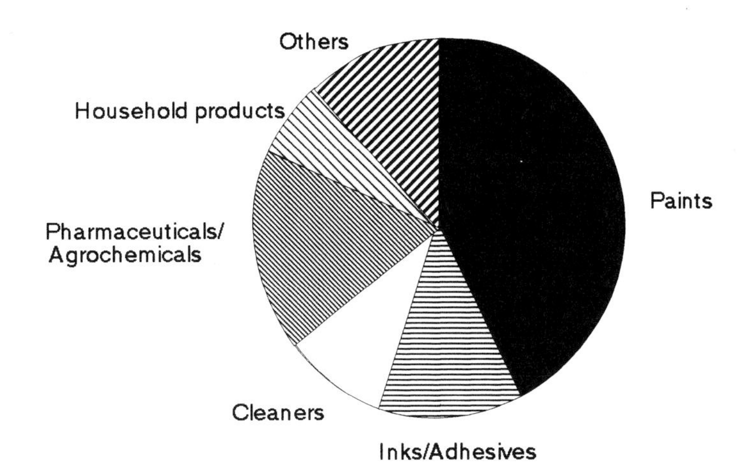

Figure 8.2 European solvent demand 1994 (excluding chemical intermediates) 4300 ktes.

8.2.1 The distinction between decorative and industrial paints

For the general public, 'paint' tends to mean decorative paint. Decorative paint is fundamentally different to the high-performance industrial coatings which comprise the other sections. This difference is all too often blurred in much literature on solvents in paint. The issues involved in industrial coatings are more complex; products are more application specific, of higher performance and often need thermal crosslinking (stoving) to cure them. When used in controlled environments, solvent emissions can be collected for reduction (abatement) by a variety of engineering control techniques. For industrial coatings, the solution to this environmental issue can often therefore lie just as successfully with the process as with the paint. This represents one major difference between decorative and industrial coatings.

Applications of solvents in these areas are covered in more detail in the following sections.

8.2.2 The solvents market in paints

Estimates of European and US solvent use (in thousands of tonnes per annum (ktpa)) in the paint and resins sectors are shown in Table 8.1. One key point worthy of note is the bias towards use of ketones in the USA and towards esters in Europe.

Chlorinated solvents are little used in the paints sector. The only application of note is in chemical paint strippers in products for both retail and domestic applications, but this is declining in some markets as chlorinates are replaced by other systems based on oxygenates such as glycol ethers and *N*-methyl pyrollidone. Total use in Europe is now thought to be less than 20 ktpa; use in the USA is substantially greater, at about 80 ktpa.

Table 8.1 Estimates of European and US solvent use in the paint and resins sectors

Solvent	Europe (ktpa)	USA (ktpa)
Hydrocarbons	900	950
Glycol ethers and esters	185	170
Ketones	150	225
Alcohols	165	200
Esters	250	150

8.2.3 Decorative coatings

Decorative paints, so called because they form a decorative coating inside and outside buildings, represent over 50% of coatings consumption in both the USA and Europe. This market can be further divided into trade and retail sectors. The former supplies the professional painter while the latter serves the retail outlets. One of the most noticeable developments in the retail sector in recent years has been the increase in paint sales from the large DIY stores at the expense of the smaller suppliers. Both retail and trade sectors can be further divided into waterborne 'emulsion paints', for walls, ceilings, etc., and solventborne gloss, primer, varnish, etc.

This market has not been characterised by complex technology; hence many paint producers exist. There are, however, several major companies who share approximately 40% of the decorative market. Resins for paint manufacture, both waterborne and solventborne, are produced by independent manufacturers who sell them to the paint producers, many of whom possess no (or limited) resin production facilities. The European decorative paint market is 2.7 million tonnes per annum (Mtpa), split 65% trade and 35% retail, while the US market is slightly smaller, at 2.5 Mtpa.

Emulsion (latex) paints are waterborne and use either vinyl or acrylic binders (resins) plus pigments and additives. They have existed for more than 50 years and now have a substantially larger market share, in volume terms, than solvent-based coatings in the decorative sector. This has occurred because of their virtual dominance of the wall-coatings market where their undoubted convenience scores highly, particularly for the home user. When coating large areas such as walls, the lower odour from water-based latex paints is also a marked advantage.

The level of binder polymer in a latex paint determines both the appearance and performance of the paint. High binder levels (18–23% as dry polymer) give a 'semi-gloss' finish, a sheen appearance is obtained with 14–16% polymer while 5–10% polymer will give a matt finish. Traditionally these paints have contained small quantities of solvents to act as coalescing agents, which lower the minimum film-forming temperature of the polymers. This enables the paint film to dry easily at ambient temperatures without the paint film becoming too soft. The most commonly used coalescing solvent is 2,2,4-trimethylpentane-1,3-diol monoisobutyrate (Texanol), although glycol ethers and esters can also be employed. There is a slight trend, however, towards coalescent-free emulsion paint systems but the consumer has, so far in most countries, shown little interest in such paints. Marketing of these products has concentrated on their odour-free nature.

Solvent-based decorative paints include gloss and sheen paints for wood, metal and walls plus wood varnish, undercoats and primers. The traditional binder is a long oil alkyd (approximately 60% oil length, i.e. the amount of oil, or fatty acid expressed as triglyceride oil, present as a percentage of the non-volatile content). The oil contains unsaturation which, in the presence of soluble metal catalysts, crosslinks in the presence of air to give a coherent paint film. Alkyds are branched polyesters and, because of the high levels of fatty acids present, they are generally soluble in simple hydrocarbons (white spirit, xylene, etc.). These paints traditionally contain about 65% solids (resin, pigments and additives) but the trend is to increase solids contents to around 85% and oil lengths to approximately 88%. This reduces VOC emissions from the paint.

Aqueous emulsions of alkyd resins are being developed for a new generation of decorative paints which contain only low levels of solvent, but these have yet to gain widespread approval in the market, especially in terms of ease of use and final appearance.

Estimates of European and US solvent use (ktpa) in decorative coatings are shown in Table 8.2.

The main application technique in the decorative area is still by hand (brush). Hence, future trends continue to reflect attention on worker exposure and environmental issues. This is already seen in the move to low aromatic content white spirits and isoparaffin solvents in conventional systems. High solids and waterborne technologies are being developed and both possess certain advantages and disadvantages, mainly relating to appearance and ease of use. Water-based systems bring, in principle, increased potential for water pollution, as consumers continue to rinse their brushes and paint rollers 'under the tap' and transfer the water-soluble components such as amines and biocides to the aqueous environment. The consequence of diffuse water emissions of this type is still under debate.

Table 8.2 Estimates of European and US solvent use in decorative coatings

Solvent	Europe (ktpa)	USA (ktpa)
Hydrocarbons	370	410
Esters	6	2
Glycol ethers and esters	24	24
Alcohols	13	11
Ketones	7	3
Others	30	30
Total	450	480

8.2.4 *Automotive OEM coatings*

The automotive OEM (Original Equipment Manufacture) sector is one of the easiest paint subsectors to define, and is arguably the one with the highest public profile; it includes all paints applied to the vehicle body shell in the paint shop of the manufacturer's production line. It does not include parts painted offline (e.g. plastic components, underbody or trim parts) or refinish items. The sector is dominated by car and light vans both in terms of unit volume production and paint tonnage used; other commercial vehicles only account for about 15% of European automotive coating use; the percentage is similar in the USA.

The auto OEM sector is characterised by having both a small number of end users (car companies) and an even smaller number of paint suppliers; 90% of paint is supplied from four companies in Europe and only three in the USA. Japan is the largest consumer of automotive coatings, accounting for one-third of worldwide production from only two major paint suppliers.

Car painting, or 'vehicle surface engineering' as it is officially called, involves a number of distinct steps. After cleaning and pretreatment, the steel body is electrocoated. The car body is connected to earth and passed automatically through a tank of positively charged water-based paint of low solids content. The paint is attracted to the body electrophoretically, being drawn into 'inaccessible' box sections as well as evenly coating more exposed areas. The 'E-coat' (as it is frequently referred to) provides the main anticorrosion protection for the vehicle. Although water based, it does contain up to 10% solvent, usually glycol ethers such as butyl glycol and butyl diglycol. These are essential components to achieve the necessary degree of resin solubilisation and surface wetting which are needed for the technical performance required.

The second coat applied is the primer surfacer. This is designed to hide minor surface imperfections from the pressing and assembly processes and to provide a good surface to show the colour coats to their best effect. These systems have traditionally been solvent based (mainly xylene and butanol with small amounts of glycol ether tail solvents) but water-based systems are also used, particularly in Europe. The primer surfacer is applied to the side and upper outer body panels. An anti-chip primer is alternatively applied to the lower sill areas.

The colour coat is either a single coat of solid colour material or a two-coat metallic system consisting of metallic basecoat plus a clear coat finish. The former systems are used predominantly on vans and commercial vehicles. Solvent systems and solids levels are similar to the primer surfacers. Metallic systems are the main finishes for cars. Figure 8.3 shows a schematic diagram of this part of the painting process for both solvent- and water-based base coats plus solvent-based clear coat. The base coat is a low solids coating with a solvent system which usually consists of a mix of very fast solvents such as acetone, ethyl acetate or MEK along with slow solvents such as the glycol ethers and glycol ether esters. Hydrocarbons (mainly xylene) are also used to keep formulation costs down. Metallic coatings such as the base coat must have a low solids content, because the film has to shrink significantly as it dries if the metallic flakes are to orientate to produce the desired visual appearance. Because of the low solids levels, most of the VOC emissions from the car-painting process come from the metallic base coat. The clear coat (which protects the metallic coat and gives the characteristic glossy finish) is applied 'wet on wet' to the base coat. Solids levels are in the region of 50–60% with xylene the main solvent used. Water-based metallic systems are being increasingly used. However, clear coats remain solvent based, with a few exceptions, because of the difficulties of achieving the desired performance from water-based systems. The relatively small change in VOC emission reduction that may be achieved in moving from a high solids clear coat to a water-based alternative is likely to be offset by trade-offs in other (typically energy related) emissions.

Estimates of European and US solvent use (ktpa) in automotive OEM coatings are shown in Table 8.3.

These figures include the use of paints, cleaning solvents and solvents used in anticorrosion waxes (predominantly hydrocarbons). The market size for automotive OEM paints alone (electrocoat, primers and all colour-coat systems) is approximately 260 ktpa in Europe and 210 ktpa in the USA.

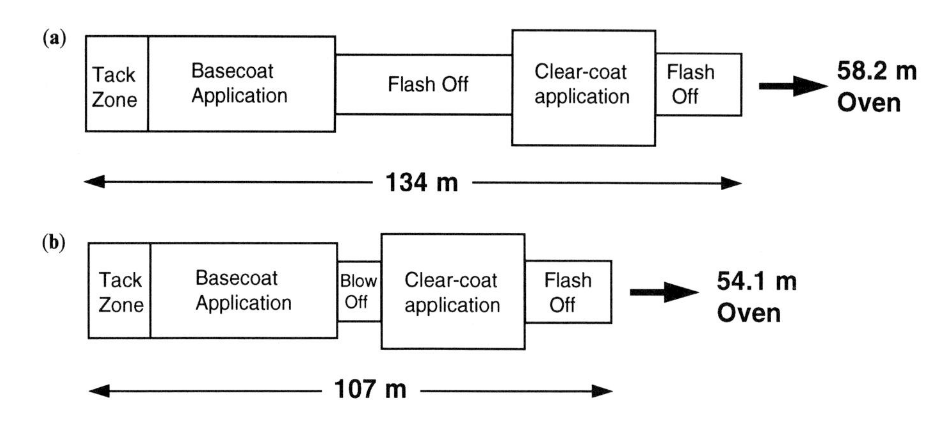

Figure 8.3 Schematic diagram of green-field car body (OEM) enamel booths showing space requirements in metres. (a) Water-based system. (b) Solvent-based system.

Table 8.3 Estimates of European and US solvent use in automotive OEM coatings

Solvent	Europe (ktpa)	USA (ktpa)
Hydrocarbons	70	70
Esters	20	45
Glycol ethers and esters	15	20
Alcohols	15	10
Ketones	10	30
Total	130	175

There are a number of significant differences between Europe, the USA and Japan. In Europe and the USA environmental regulations are affecting the technology while in Japan quality is clearly the dominant consideration. Japanese car manufacturers often use more coats than is implied in the previous section to ensure improved finish quality. In the USA, primer surfacers have not always been used and metallic coatings with a higher solids level are also used (> 30% compared to 20–25% in Europe).

Future technology trends are being driven simultaneously by environmental and cost considerations, which are often conflicting pressures. Base-coat VOC emissions are being increasingly reduced by incineration or by the use of water-based systems. The proven performance of water-based clear coats, however, does not match solvent-based products and the current focus is to increase the solids of the solvent-based systems. In the longer term, work is being done to develop powder coat alternatives for the clear coat. Odour emissions have caused problems with some water-based products. Since vehicle appearance is an important consideration of the car buyer, new technologies will not be adopted unless they can produce the same or better standards of finish as the existing products. Coloured primers are used to improve cosmetic appearance externally (reduced visual impact of stone chips) and eliminate the need for a base coat in some internal areas of the vehicle (underbonnet, boot, etc.). Pearlescent systems are increasingly used because of the high quality finish they produce, although this necessitates an extra paint application stage.

Because coating performance is so important to a car manufacturer there appears, at first glance, to be a conflict between adopting a technology marketed as being the 'green' alternative (water-based systems) and the maintenance of the existing quality delivered by solvent-based products. Faced with the objective of limiting VOC emissions from the manufacturing installation, the automotive paint engineer has two major options. One is to continue to use conventional paint technology to which a thermal oxidation unit is added to eliminate the VOC emissions; the other is to change the installation and use water-based coatings. VOCs are still emitted from such a process but the quantity present in the initial formulation is lower than in conventional solvent-based systems. Life cycle assessment (LCA, see chapter 6) can be used to examine the environmental impact of these two technology options. To do this, the steps to paint manufacture (as outlined in Figure 8.4) as well as paint application, the different mix of paint ingredients, application efficiencies, drying conditions and other factors need to be considered. On this basis, it proves impossible to distinguish, on

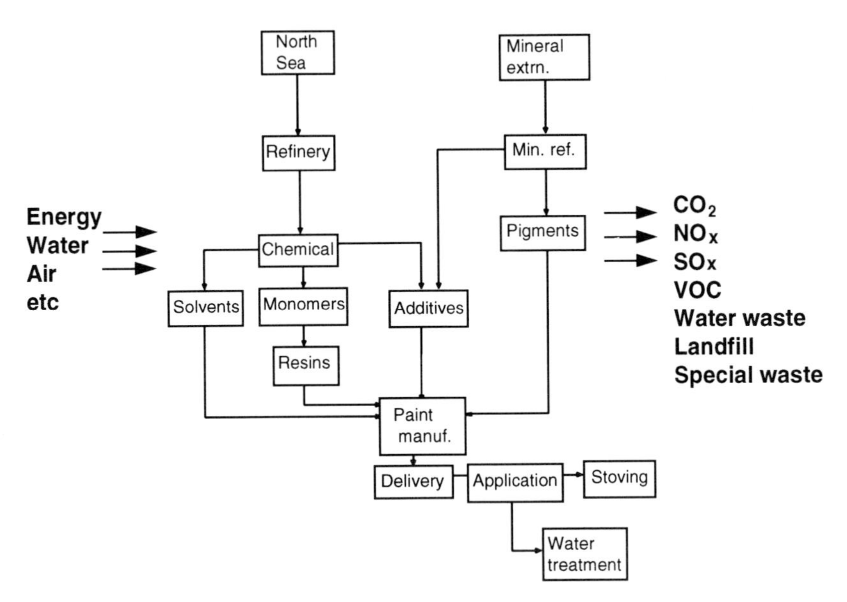

Figure 8.4 Schematic diagram for an automotive coating LCA.

environmental grounds, between the alternatives of conventional paint technology, using thermal destruction to eliminate VOC emissions, and water-based systems. Both meet the requirements to be defined as 'Best Available Techniques' (see chapter 6).

8.2.5 Automotive refinish coatings

Vehicle refinish (VR) coatings are used to repair damaged or corroded car paint systems. They must exhibit the same performance characteristics as the original paint system if they are to meet the car manufacturer's bodywork corrosion warranty. However, the most important feature of VR coatings is their colour. There are approximately 150 million cars on the roads of Europe, painted in over 20 000 different colours, and every year a further 600 new colours are added. Paints also 'age' and, depending on climatic conditions, a 10-year-old car may have changed colour considerably. A VR bodyshop must be able to match the colour of the car, irrespective of its age, colour and type of finish (which will usually be either a solid colour, metallic, or pearlescent). This is achieved by the skills of the VR bodyshop and the colour-matching system of the paint manufacturer.

Refinish painting of cars differs fundamentally from the original equipment process. There are fewer than 20 OEM customers in Europe (the major car producers) but there are many thousands of VR bodyshops in Europe (over 15 000 in the UK and around 100 000 in Europe). The USA has a much smaller number of body shops (15 000) but direct sales to the public for 'do-it-yourself' are much more important than in Europe. The size of the European VR paint market was approximately 190 000 tonnes in 1993, including thinners; the USA is about 50–60% larger. While there are some large VR operations, most use less than 2 tpa of paints and thinners; these are usually supplied by refinish distributors rather than the paint companies themselves. A high level of service is required from the suppliers with paint deliveries being made to the bodyshop several times a day.

The painting process for VR differs from that used for new cars. While new car bodies can be heated at 120–140°C to cure the OEM paint system, a car for repair, complete with a tank of petrol, electronic equipment, rubber tyres and other heat-sensitive components, must be treated with care. Hence, cure temperatures of 60°C maximum are used for VR coatings. The solvent systems used must be capable of providing the necessary solubility, application, viscosity and flow characteristics. It must also evaporate quickly enough to enable the bodyshop to complete the repair as soon as possible, often under widely varying temperature and humidity conditions.

During a repair, the damaged area is sanded down to bare metal and filled as necessary. An etch primer is often used to enable the VR coating to adhere to the surface of the metal. This is a very low solids coating which is applied at a low film thickness. Etch primers typically contain phosphoric acid and zinc chromate (although chromates are becoming unpopular for toxicity reasons), a binder (such as polyvinyl butyral) and a solvent system (usually based on ethanol, butanol and water).

The primer surface hides minor imperfections and provides a smooth surface for topcoat application. These coatings have traditionally been polyester or epoxy based and used a combination of aromatic solvents

Table 8.4 Estimates of European and US solvent use in automotive refinish coatings

Solvent	Europe (ktpa)	USA (ktpa)
Hydrocarbons	45	70
Esters	30	20
Glycol ethers and esters	10	3
Alcohols	10	15
Ketones	10	25
Total	105	133

(xylene, C9 aromatics) and oxygenates (esters, glycol ethers). However, water-based systems are becoming available that contain lower amounts of solvent; butyl glycol ether is the most important.

The colour coat is either a solid colour (red, blue, etc.) or a metallic or pearlescent finish, protected by a clear coating. The metallic or pearlescent base coats are low solids finishes and contain a combination of fast evaporating solvents (e.g. ethyl acetate, MEK), butyl acetate and slow 'tail' solvents such as the glycol ethers and glycol ether acetates to obtain the desired rheology and metallic appearance. Ketones are less favoured in Europe as their pungent odour is considered less acceptable to users and, with some resins, unacceptable yellowing of the resin occurs. Hydrocarbons are added to reduce cost (xylene, C9 aromatics) but the use of xylene is declining due to increasingly strict labelling requirements in Europe and concerns about atmospheric impact in the USA (e.g. listed as a hazardous air pollutant (HAP), reportable under SARA 313, covered by the 33/50 programme); aliphatic hydrocarbon/oxygenated solvent blends are being used instead. Similar types of solvents are used in the clear coat, which is predominantly based on acrylic–isocyanate technology, although glycol ether acetates are used instead of glycol ethers and fewer of the faster solvents are used. Metallic base coats are approximately 20% solids whilst the primers, colour coats and clear coat are between 40 and 60% solids. Solids in the USA tend to be higher, particularly for metallic coatings. Estimates of European and US solvent usage (ktpa) in automotive refinish coatings are shown in Table 8.4.

Waterborne variants have been introduced for primers and metallic coatings but still only have a very small market share because of reduced spray booth throughput and the costs of extra heaters and air movers to dry the coatings. All of these water-based systems still contain organic solvents. For clear coats, high solids solvent-based systems represent the only practical way of reducing emissions at present. The single biggest change to reduce emissions, already well underway, is the replacement of traditional air-atomised spray guns with HVLP (high volume low pressure) systems; improvements in transfer efficiencies of 100% are possible, with the reduction in paint wastage contributing greatly to reduced VOC emissions.

8.2.6 *Wood coating and impregnation*

8.2.6.1 Wood coating. The industrial wood-coating market includes protective and decorative coatings for furniture, both assembled and flat pack, and joinery (e.g. window frames, house cladding). The overall market is defined as being any coating applied to a wood substrate in a factory OEM environment and consumes approximately 400 000 tpa of coatings in Europe. The largest single market for wood coatings is furniture, which constitutes about 70% of the total.

While some other coatings sectors are moving towards a more global technology the wood industry is nationally based because of regional preferences in furniture choice, with different countries therefore favouring different technologies. Furniture finishes include nitrocellulose, acid-cured alkyd, water-based, polyurethane, radiation-cured and unsaturated polyester types. The coatings producers are typically smaller and more specialised than in some of the other sectors. Apart from a few major companies, most producers are national rather than multinational.

Nitrocellulose lacquers are fast drying, easy to apply and give an excellent gloss and grain appearance. However, their durability is not as great as coatings based on reactive resin systems, which means they are used mainly to protect fine furniture. Solids levels are typically about 20% at application. A mixture of ester, ketone, alcohol and hydrocarbon solvents is used, as these tend to evaporate more quickly (e.g. toluene, xylene, ethyl and butyl acetates, acetone, ethanol, butanol, methyl ethyl ketone, methyl isobutyl ketone). Nitrocellulose finishes are ideal for light duty applications on high-quality furniture.

Acid-curing systems are traditionally popular in the UK and Scandinavia although recent concerns have been expressed over formaldehyde emissions from the resin during cure, which may eventually limit their use. These coatings may be either one or two pack and nitrocellulose may be added, especially to the one-pack (precatalysed) systems to give faster drying. The alkyd resins, used as the main binders, are crosslinked with

Table 8.5 Estimates of European and US solvent use in the wood coating and impregnation sectors

Solvent	Europe (ktpa)	USA (ktpa)
Hydrocarbons	55	120
Esters	65	45
Glycol ethers and esters	20	40
Alcohols	25	40
Ketones	20	10
Total	185	255

either a urea or urea/melamine resin, using an acid catalyst. These amino resins require the presence of high levels of primary alcohols to inhibit self-condensation. If nitrocellulose is used as a modifier then the solvents mentioned above (oxygenates and hydrocarbons) are added to maintain compatibility, otherwise the main solvents used are alcohols (ethanol, butanol) and aromatic hydrocarbons such as xylene. Applications include coatings for general furniture (home and office) with the two-pack systems having improved durability.

Two-pack polyurethane systems (2K PU) give highly durable finishes with good gloss, flexibility, adhesion and build. They tend to be slower drying and have higher solids contents than traditional nitrocellulose systems. They are popular in Italy, Germany and Spain. The main binders (resins) used with them are hydroxy-functional alkyds/polyesters and acrylics which are crosslinked with isocyanates. Solvents include esters, ketones and hydrocarbons with smaller quantities of glycol ether esters as 'tail' solvents. As isocyanates are extremely reactive towards hydroxyl or other acidic protons, the solvents used must be dry and of suitable high quality. The chemistry of two-pack polyurethanes prevents the use of alcohols or glycol ethers.

Estimates of European and US solvent use (ktpa) are shown in Table 8.5. The US data includes wood impregnation and preservatives (predominantly white spirit and naphthas). A further 45 ktpa of such solvents are used in Europe for impregnation.

Ultraviolet (UV) radiation cured paint systems are the main technology for coating flat surfaces in large quantities. This method is not, however, suitable for smaller coating enterprises (costs are too high) or for coating objects after fabrication (difficulty of even irradiation). Use with pigmented coatings is also limited due to the problem of completely curing a non-transparent film since the UV rays only penetrate the outer film surface.

Waterborne coatings for wood have also been introduced but significant technical problems such as grain raising, appearance, clouding, lower build and consequent higher costs have still to be resolved.

High solids nitrocellulose systems (applied warm) are being introduced in the USA. In Europe, high solids polyurethane coatings are likely to be the main technology used in the future, especially where high gloss finishes are required.

8.2.6.2 Wood impregnation. Wood is impregnated with preservatives to protect it against weathering, fungal and insect attack. Approximately 140 000 tpa of preservatives are used in Europe and these fall into three main types. Creosote is derived from coal tar distillation (55% of the market) and is the oldest form of preservative. It is used for external applications, such as railway sleepers and telegraph poles, and is gradually being replaced by white spirit or waterborne systems, particularly in the DIY market, because of toxicity concerns. Creosote is used industrially in a pressure/vacuum impregnation process.

Waterborne systems (approximately 15% of the market) are applied in the same way as creosote but rely on the preservative effects of CCA salts (chrome copper arsenate: $CrO_3.CuO.As_2O_5$). Metal leaching is an issue with such systems.

Solventborne wood preservatives (about 30% of the market) consist of 10% active ingredients and 90% solvent, usually white spirit or similar petroleum-based hydrocarbons. The active ingredients include preservative/insecticides such as phenols, chlorinated and phosphated organics and copper/zinc naphthenates. Vacuum impregnation is used to apply this type of system. Timber treated by this method is not suitable for direct contact with soil; for exterior use, it should be painted to prevent extraction of the active ingredients. It is used where precise control of the dimensions of the timber is important and leaching can be avoided (e.g. in the construction industry for windows, doors, etc.).

Toxicity concerns surrounding creosote have promoted the newer solventborne and waterborne systems. The use of more controlled, enclosed processes plus emission abatement recovery techniques will restrict emissions from the new solventborne types. There has been little debate, so far, about the relative environmental impacts of the different active biocides and fungicides used in the newer solvent- and water-based systems.

8.2.7 *Metal-packaging coatings*

The metal-packaging (or 'Can Coating') segment can be split into a number of different packaging types, each of which requires different types of coating. There are possibilities for confusion in defining this segment; drum coatings are not always included and external coatings are also sometimes omitted. (In general, only white base coats ready for overprinting are counted as paints as opposed to inks.) A typical split would be:

- Beer and Beverage (spray internals sheet/coil coated ends)
- Food cans
- Closures
- Aerosols
- Other general line
- Drum coatings

However, for simplicity, it is sufficient to consider beer and beverage (B&B) as one category and the others collectively.

The metal-packaging sector shares a similarity with the automotive OEM coatings sector in having both a small number of coating suppliers and end users. In Europe 90% of paint is supplied by five companies, while only four account for nearly 95% in the USA. The USA accounts for over 40% of the world's metal packaging coatings, which reflects the large US demand for B&B containers.

B&B coatings are the largest volume products by a significant margin. Spray coatings are applied to two-piece cans after forming. For aluminium cans these are traditionally water-based epoxies for technical performance reasons. Glycol ethers (butyl glycol) and butanol are usually present at levels of 10–15%.

Can ends are stamped from sheets of precoated metal. These coatings need to be able to 'self heal' after stamping and for this reason are usually solvent-based thermoplastic coatings. Production speeds for a typical B&B line run into millions of units per day so application and cure times are measured in seconds with very hot (> 200°C) stoving temperatures. Most B&B products are now of two-piece, as opposed to three-piece, design. The market for this type of container is more mature in the USA than in Europe.

As a food contact sector, the food and closures market is necessarily conservative in nature, needing to use tried and tested products. The interaction of coatings with food materials is extremely complex. Changes to coating formulation require careful evaluation and testing to ensure that they do not lead to unexpected tainting or unpalatable odours. Likewise, coating integrity has to be guaranteed; defects such as pinholes can lead to disastrous consequences. There is also the obstacle for introducing new products of needing food contact approval, an expensive and time-consuming process.

General line products include such disparate items as giftware, boxes, etc. With these relatively low volume products decoration is a more important element than anticorrosion protection.

Aerosol coatings is a small but significant niche. Owing to the nature of the can contents, high-performance coatings are required, and these are usually applied at higher film thicknesses than other can coatings.

Estimates of European and US solvent use (ktpa) in metal-packaging coatings are shown in Table 8.6.

There are a number of conflicting pressures within the metal-packaging market, such as the competition between tinplate and aluminium and between metal and glass B&B containers. The battle between the former affects coating type but has little impact on overall coating volumes used, while the latter has potential for major impact on the coating use. Much work is being done to assess the life cycles of the two packaging types and it remains to be seen whether this will influence long-term trends. Plastic and composite containers, which require printing but no internal coating, are also starting to manage significant volume for the more sophisticated packaging such as convenience foods.

Environmental matters are again important in this sector but most attention is focused on the packaging

Table 8.6 Estimates of European and US solvent use in metal packaging coatings

Solvent	Europe (ktpa)	USA (ktpa)
Hydrocarbons	20	15
Glycol ethers and esters	12	20
Ketones	8	15
Alcohols	3	2
Esters	2	3
Total	45	55

issue and the issue of vinyl (chlorine containing) coatings rather than VOC emissions. Low VOC coatings, such as the water-based types, are expected to increase their market share slightly but this is essentially continuity of an established trend in B&B where specific technical performance factors are driving the change. Many can coaters need to retain process flexibility. VOC emission control here will be achieved by thermal oxidation, allowing the continued use of proven coatings and giving scope to recover calorific value from the solvents for use in the high-temperature stoving ovens.

Powder may find use in niche applications such as certain aerosol interior coatings and three-piece can side stripes.

8.2.8 Coil coatings

Coil coatings are used to create intermediate coated products which are then used in an end market sector. The largest single end use for coil-coated products (60% by area coated) is in the building industry. This market can be further subdivided according to substrate coated, mainly galvanised or zinc-coated steel and aluminium, but a more useful sub-split is by coating type:

- Primers
- Backing coats
- Plastisols
- Polyesters (including silicon modified)
- Polyvinylidene fluoride
- Epoxies
- Acrylics
- Alkyds
- Others

The coil sector is characterised by having both a small number of coatings suppliers and end users. In the USA there are five major coatings suppliers who hold more than 90% of market share. The market is a little more fragmented in Europe with up to a dozen significant suppliers. Both markets are of similar size, about 105 ktpa of coatings.

Polyester coatings, and their variants, account for more than 50% of coil coatings used in Europe. They require strong polar solvents, such as glycol ethers, to keep them in solution although high boiling aromatic hydrocarbons are also used in the formulations. The second largest volume is in plastisol coatings, which account for a further 25% (mainly in the UK and Scandinavia). These coatings are almost solvent free, using very small amounts of white spirit for viscosity adjustment only. Use of plastisols is declining as other systems are developed which can deliver similar performance at a fraction of the film thickness (and cost).

Production speeds for coil lines can be as high as 120 m/min for steel with peak temperatures during cure of up to 250°C for periods of up to 1 minute. Such curing schedules require that low volatility solvents are used. Most coil coaters currently use thermal oxidisers to recover the calorific value of the released solvents. This plays an important role in the economics of such a high-temperature curing process.

Estimates of European and US solvent use (ktpa) in coil coatings are shown in Table 8.7. Economically, the coil-coating market is one of the healthiest of the coating sectors. Traditionally it has grown at rates well above growth of gross domestic product since it offers the fabricator the ability to eliminate post-construction painting and hence the possibility of lower overall fabrication costs. The heavy reliance on one main end use does, however, mean that recession in the building industry has a heavy impact on demand for coil coatings, as has happened in the early 1990s, particularly in Europe.

Water-based products exist but are unlikely to be adopted in quantity in this sector. They eliminate the ability to recover energy from the solvents for use in the stoving ovens. Since the integrated incineration systems in present use can control solvent emissions very effectively, there is no VOC-related driving force to push their adoption. In the very long term, techniques such as radiation cure and even powder may be devel-

Table 8.7 Estimates of European and US solvent use in coil coatings

Solvent	Europe (ktpa)	USA (ktpa)
Hydrocarbons	28	25
Glycol ethers and esters	15	8
Alcohols	5	12
Ketones	1	3
Total	49	48

oped, but for the next 10 years at least, little change is expected in the pattern of solvent use beyond those changes resulting from the use of different coating types.

Long life and excellent weathering characteristics are required for the main application of coil coatings (architectural cladding). This means that new coil-coating systems have to be fully proved before they are widely adopted.

8.2.9 Anticorrosion and marine coatings

Anticorrosion and marine coatings overlap so much that they can be treated as one application area. Particular niches specific to each will be highlighted. The anticorrosion sector covers coatings for all structural steel and concrete constructions which are coated, generally but not always *in situ*, for new constructions or maintenance, to protect rather than decorate. The term 'marine' covers materials used for the purposes of coating the interiors and exteriors of ships and yachts. Many similar products can be found in both sectors. Exceptions include antifouling paints and some yacht products which are specific to marine applications. Powder coatings and some active primers are specific to the anticorrosion sector. Further classifications can be made within the sector by end use or by resin type used. The latter classification can be useful when considering solvent use. The list presented in Table 8.8 is not exhaustive and is designed merely to show examples.

These two sectors are served by major multinational coating suppliers and also by smaller national suppliers filling niche gaps in their respective markets. The latter is particularly true in the anticorrosion market. Marine coatings split into two main areas: ship and yacht paint. Worldwide distribution is the key to success in this sector, and for this reason the main market share is taken by a small number of major manufacturers with operations across Europe, Japan and the USA.

Estimates of European and US solvent use (ktpa) in anticorrosion and marine coatings are shown in Table 8.8. A comparison of data from different sources is difficult in this case because consistent definitions of what is and is not included are not used.

There are controls on the VOC content of certain anticorrosion products used in the USA, most notably California. The nature of the marine market has so far meant that these controls on VOC content have had little impact on the market. It is easy to move a vessel to a region where few controls exist for repainting and, in any case, much painting is also done while the vessel is underway; marine painting is now almost extinct in California. Water-based products (using glycol ether solvents) are growing very slowly and only in specific niche areas. (All products are applied manually by brush, spraygun or roller.) There will always be limitations on water-based materials in climates where outside application is required in a wide range of weather conditions; temperatures below 5°C or humidity extremes make application and drying problematic. High solids systems avoid these problems and, where technical performance criteria can be met, these systems are expected to show the highest growth rates. The main solvents used in high solids systems are esters, ketones and glycol ethers.

Table 8.8 Resins and solvents used in anticorrosion and marine coatings

Resin type	Main solvents typically used
Chlorinated rubber	Aromatic hydrocarbons (HCs)
Vinyl	Ketones, aromatic HCs
Epoxy	Ketones, esters, aromatic HCs, alcohols
Alkyd, epoxy ester	Aromatic and aliphatic HCs
Urethanes	Ketones, esters, aromatics HCs
Inorganics	Water, alcohols
Latex	Water, glycol ether

Table 8.9 Estimates of European and US solvent use in anticorrosion and marine coatings

Solvent	Europe (ktpa)	USA (ktpa)
Hydrocarbons	105	100
Glycol ethers and esters	12	10
Alcohols	15	12
Ketones	16	22
Esters	18	6
Total	166	156

8.2.10 General industrial coatings

Within the coatings sector, there are many other types of factory applied coatings that do not fit into any of the categories separately covered. These range from significantly sized operations to those that are quite minor. They are used to decorate and to protect primarily metal substrates but also include plastic coatings. This diverse sector is characterised by many paint manufacturers – usually national, sometimes regional, companies specialising in particular applications. The multinationals are active in this sector, but to a lesser extent than in any other. To describe fully the applications of solvents in this sector is beyond the scope of this review and for convenience they are grouped under this single heading. Table 8.10 lists many of the other manufactured products that fall into this category, although the list is by no means exhaustive.

There are many different types of paint in use, all relatively sophisticated materials that are usually formulated for a specific customer or application. Table 8.11 shows the types of binders used and the solvents associated with them. There are many methods of application, all of which can be used for the resins shown in the table. The choice is usually dictated by the type of object to be coated and the finish required. Solvent choice is dictated by the application technique.

Spraying is the most common application technique. It is simple and relatively cheap, quick, and with air-atomised systems in particular, produces good quality finishes. It can be used for virtually any size and shape of substrate. Apart from the air-atomised system, the other two main application types are airless and electrostatic, together with various derivative and combination techniques. Mixed solvent systems are usually required to obtain the rheology profile needed during application, particularly for air-atomised and electrostatic systems; blends often contain three or more solvents, varying from slow- to fast-evaporating types. Paints designed for airless application usually rely more on additives to achieve the desired rheology and are balanced more towards the slower evaporating solvents.

Dip coating (or tumble coating) is ideal for small, mass-produced articles requiring single coat paint application. Quality of finish is not as high as spray coating and control of film thickness is difficult. Solvents used tend to be relatively simple systems of either medium or slow-evaporating solvents, the exact choice

Table 8.10 End uses for coatings classed as 'General industrial OEM'

Application	Comments
Textile coatings	
Paper coatings	
Leather coatings	
White goods (appliances)	
Metal furniture	
Machinery/equipment	
Agricultural equipment	Sometimes classed separately
Commercial vehicles	Sometimes classed separately
Traffic paints	Can be included in anticorrosion
Electrical coatings	Wire, circuit board
Release coatings	
Plastic coatings	
Magnetic tape coatings	
Electrophoretic coatings	
Aircraft OEM	Sometimes separate class (small)
Railway coatings	Usually included with anticorrosion
Freight containers	Usually included with anticorrosion
Building components	
Racking	
Glass	
Small items	Anything from toys to gas meters, pencils to firearms

Table 8.11 Examples of binders and solvents used in industrial coatings

Resin type	Main solvents typically used
Amino	Aromatic hydrocarbons (HCs), alcohols, glycol ethers
Vinyl	Ketones, aromatic HCs
Epoxy, polyester, acrylics	Ketones, esters, aromatic HCs, alcohols
Alkyd, epoxy ester	Aromatic and aliphatic HCs
Urethane	Ketones, esters, aromatics HCs
Inorganic	Water, alcohols
Latex	Water, glycol ether/esters
WB alkyd, etc.	Water, glycol ether
Powders	None

being dependent on whether the article is to be stoved or just force dried. Barrel enamelling is a variant of this approach.

Flow or flood coating, where excess paint is poured onto an article with the surplus drained off for reuse, is often more economical than dip coating and avoids some of its other disadvantages. It is particularly useful for articles that are complex or too large to be dip coated automatically. Slow solvents are again typically used for such coatings.

Curtain coating (where paint falls through a slot as a 'curtain' of paint onto the substrate) is used to coat all types of substrate, including plastic and leather as well as metal. Material use is very efficient and careful process design allows close control of coating thickness. Slow solvents are again used.

Electrophoretic coatings are similar to those used and described in section 8.2 on automotive OEM coatings. Metallic (conductive) substrates are essential if this technique is to be used successfully. Water is obviously the main solvent to achieve the degree of conductivity needed, but significant quantities of solvents such as the glycol ethers are needed to solubilise the resins used and to produce the quality of film finish required.

Solvent choice is also influenced by the method used to cure the paint and can vary from simple air drying, through forced drying at moderately elevated temperatures, to high-temperature stoving. As a rule, the higher the temperature of cure, the higher the boiling point of the solvent used. It is essential that at no point during drying does any solvent remnant in the coating start to boil causing major disruption of the coating film.

This sector, despite being composed of many small producers and applicators, is faced with the same environmental challenges as the other sectors. The diverse nature of this sector means that all options for limiting VOC emissions will be used to some degree. Larger applicators will tend to use abatement (thermal or catalytic oxidation) to permit the flexibility in coating types often needed. Many coating operations are in fact a contract coating service working on different objects, substrates and specifications daily. Clearly such operations cannot constrain themselves to one line of coatings technology without restricting the work they can contract. For smaller users, the cost of abatement per unit of production is higher and hence there will be a greater tendency to use specific coating technologies such as high solids, water-based products and powder technology. This would imply 'jobbing coaters' having to specialise more on niche operations. For operators concerned with specific products where high-quality finish and multiple colour or overlays are not an issue, powder will continue its recent strong growth.

Technically, the bigger suppliers of general industrial coatings are trying to move away from small batch manufacture and seeking to adopt techniques currently seen only in the decorative sector; factory or depot mixing and tinting schemes are becoming increasingly common. This leaves the smaller coatings suppliers to focus on the varied demands for bespoke finishes.

8.3 Inks

The largest volume of solvents in the ink and printing industry is used in lacquers, varnishes and what are termed 'liquid inks' for gravure and flexographic printing. These printing techniques are characterised by high press speeds and the use of low-viscosity inks which must dry very quickly. Smaller quantities of higher boiling solvents are used in paste inks for screen, letterpress and lithographic printing.

About one-third of the total solvent used in liquid inks is used to produce concentrated (typically 50% solid) ink suitable for dilution by printers. The remaining solvent is used at printing sites, where concentrated ink is diluted, typically 1 : 1, with the required solvent system.

Liquid inks normally comprise pigments, resins, solvents and additives. Solvents are required to dissolve the resin system, control viscosity and allow the ink to flow. Once they have performed this function, they must evaporate at an appropriate rate determined by the press speed. Since flexography and gravure use very fast press speeds, rapidly evaporating solvents are needed. However, too rapid an evaporation can lead to trapped solvent bubbles and poor print quality; too slow an evaporation can require reduced press speeds or result in solvent retention. For this reason, mixed solvents are often used to provide the necessary evaporation profile.

Choice of solvent is clearly linked to the resin (or binder) used. Resin choice is determined by the substrate to be coated, print quality required and the conditions that the coating must withstand. Properties required could include water or oil resistance, high or low temperature resistance, flexibility or hardness and rub resistance. More than one resin is often used to produce the necessary combination of properties on the printed product; the use of solvent-based technology allows a wide choice of resin systems and therefore a broad range of properties can be achieved. The solvent system must be compatible with the substrate and printing rollers (some flexographic rollers can swell with certain solvents). It must also meet toxicity and residual odour requirements; in food packaging, for example, these are very stringent. Estimates of European and US solvent use (ktpa) in inks are shown in Table 8.12.

Table 8.12 Estimates of European and US solvent use in inks

Solvent	Europe (ktpa)	USA (ktpa)
Hydrocarbons	90	225
Glycol ethers and esters	15	
Esters	95	75
Alcohols	65	
Total	283	300

In Europe, toluene is the solvent of choice for publication gravure printing since it dissolves the resins used (often phenolics) and has a fast evaporation rate. There have been worker exposure concerns over the use of toluene, but the potential for process enclosure and the fact that most of this toluene is readily recovered for reuse means that an alternative to toluene is not imminent.

A large proportion of liquid inks is used in the printing of flexible packaging, normally on non-absorbent substrates such as film or foil and is often for food-packaging applications. In Europe the major solvents used are ethyl acetate and ethanol, with smaller volumes of ketones such as MEK and acetone, whereas in the USA, a greater proportion of ketones is used. Solvent mixtures are mainly used to dissolve the resins (most commonly nitrocellulose, with the addition of other resins to give the required properties) and they need fast evaporation rates, low toxicity and low residual odour. Slower evaporating solvents are used as 'retarders' to reduce the evaporation rate at the end of the drying process to improve print quality and prevent ink forming a skin or drying on the press. Retarding solvents are often glycol ethers or higher esters and alcohols, which have good solvency properties and slow evaporation rates.

Using mixed solvents (e.g. ethyl acetate, ethanol and retarder such as ethoxypropanol) enables the printer to adjust solvency or evaporation rates by altering the relative proportions; this is often necessitated by changes in ink composition, substrate or ambient air conditions. Where single solvents are used, printers are restricted by resin types and therefore the properties that can be achieved. In addition, adjustment to press conditions (e.g. web speed or air flow) is the only method available to control the evaporation rate and can be very time consuming leading to increased press downtime.

Flexographic printing on substrates such as paper and board is significantly different to printing on plastic film and because of the adsorbent nature of the substrate water-based inks are normally used. However, there are still a few applications where solvents are required for paper printing.

One of the most important issues in solvent use in printing is emission of VOCs at print works. Emission limits can normally be achieved by either reduced solvent use (e.g. switching to water-based or reactive diluent systems) or by continuing to use solvent-based inks and installing end of pipe abatement.

By the end of the decade, most printers using solvent-based inks will have installed engineering controls to abate their solvent emissions. Of the various abatement methods available, thermal oxidation and solvent recovery using carbon adsorption have found widespread use.

Oxidation is a widely used abatement method in printing. It has the advantage of lower capital cost, simplicity and is applicable for the mixed solvent systems used by many flexible packaging printers. It is likely to be the cheapest abatement option for most printers using liquid inks. Thermal oxidation is used widely in Germany where controls on VOC emissions from printing have been in place for many years.

For carbon adsorption and recovery there are three distinct operations – adsorption, desorption and purification – to give virgin quality solvent. Capital investment for such a system is substantial and is the reason why the technique is only adopted by the largest solvent users. Thermal desorption normally takes place using either steam or hot nitrogen as the heating medium. Steam can be used for solvents that are not water soluble (e.g. hydrocarbons). Where water-soluble solvents are used, the cost of using steam increases dramatically because of the increased costs for repurification. In addition, for hydrolysable systems such as esters, there is the possibility of hydrolysis leading to acid impurities in recovered material. A technically better alternative in this case is nitrogen, but capital and operating costs are even higher than for steam.

The mechanism for purification of recovered solvent depends on the nature of the material collected and the purity required. Where steam is used with water-immiscible solvents (e.g. toluene), purification can be done by simple decantation, but for water-soluble solvents, distillation is needed. Where water forms an azeotrope with the solvent (e.g. ethyl acetate) further complexity and cost is added to separate the mixture. With nitrogen desorption, purification can be simpler, but with mixed solvents which form azeotropes (e.g. ethyl acetate and ethanol) separation is still very difficult and costly. In this case, the cost of equipment needed to distil and purify the solvent can be higher than that for the recovery unit.

Recovery units are also more labour intensive and require more highly skilled operators than thermal destruction systems. Because of the capital investment in both the recovery unit and the purification system, and the technical limitations discussed above, recovery is normally only practical and economically sensible for single solvent inks at the largest printing sites.

Since many solvent-based liquid inks are used in food packaging, there are normally controls over retained solvent levels in the printed materials. Because the risk of taint and odour (and hence reproducible quality) are also important criteria in solvent choice, synthetic pure materials such as oxygenates are favoured. The use of rapidly evaporating solvents eradicates problems of solvent retention but care is needed to ensure that skinning problems do not exacerbate the problem through trapped solvent; retarder solvents help to prevent this while maintaining the technical advantages of fast, single solvent systems.

Screen printing uses slower drying and higher viscosity inks. It therefore uses smaller quantities of less volatile solvents. Glycol ethers (predominantly propylene glycol based) and esters are common, together with high boiling aliphatic and aromatic hydrocarbons.

Heatset lithographic printing uses relatively high drying temperatures and very high viscosity inks. The solvents used are predominantly high boiling hydrocarbons although small quantities of isopropanol are used in 'fountain solution' to improve wetting properties by reducing surface tension. Many heatset printers have installed thermal oxidation systems to control their solvent emissions.

8.4 Adhesives

Solvents are used in 'evaporative' adhesives, where the resin is dissolved in a solvent system and, once the adhesive is applied to the substrate, the solvent is evaporated through either forced or ambient drying. Evaporative adhesives are used in a broad range of applications, both for industrial and domestic use, and as the resin performance required varies enormously, a wide range of solvents are used.

Choice of resin is determined by the performance requirements of the adhesive, in terms of adhesive power, resistance and stability. The required evaporation rates depend on the application technique, which can vary from the rapid evaporation needed on a laminating machine to the slow drying necessary for manual brush applications. The major solvents employed are aromatic and aliphatic hydrocarbons (such as toluene, xylene and special boiling point aliphatics) and oxygenated solvents such as ethyl acetate, MEK and acetone. In many cases mixed solvent systems are used. Chlorinated solvents are found in some applications but their use is declining due to concerns over toxicity and environmental impact. Few alcohols are used in adhesives because they rarely have the solvency required to dissolve the resins.

Estimates of European and US solvent use (ktpa) in adhesives are shown in Table 8.13. Note that the US figures also include sealants.

As with other coating applications, VOC emissions from industrial adhesive use are, or soon will be, under mandatory control. For those opting to meet legal requirements by replacing solvents, there are a number of low/no solvent adhesive technologies available (e.g. hot melt, reactive systems and water based). In many applications, however, the cost and performance advantages of solvent-based adhesives will mean that end of pipe abatement will be the best method to reduce VOC emissions while still complying with statutory requirements. Many adhesive users find thermal oxidation to be the most cost-effective abatement technique to deal with their mixed solvent streams. Where single solvents or water-immiscible solvents are used, recovery by carbon adsorption may be economic at large sites using in excess of 1000 tpa solvent.

In cases where adhesives are applied manually (e.g. hand brushing in the shoe industry) there may be concerns over worker exposure to solvent vapour. In some cases there are moves to replace certain solvents with others that are considered less harmful.

Table 8.13 Estimates of European and US solvent use in adhesives

Solvent	Europe (ktpa)	USA (ktpa)
Hydrocarbons	65	200
Ketones	52	90
Chlorinates	5	140 (Chlorinates and Esters combined)
Esters	30	
Total	152	430

8.5 Metal cleaning

This is the removal of oil, grease or other contaminants from the surface of metals, and increasingly plastic components and assemblies. Substrates cleaned include mild steel, brass, copper, polypropylene and printed circuit boards, the level of cleanliness required depending on the application.

Cleaning usually occurs during production of a component as a precursor to subsequent processes, such as painting. For painting, a high level of cleanliness is required to ensure good paint adhesion. The cleaning of equipment during maintenance often requires the article to be free from particulate matter, but a submicron layer of residue can be tolerated, especially for moving parts where fresh lubricant will be applied before use. Apart from the cleanliness requirements, the principal difference between production and maintenance cleaning (the cleaning down of the equipment used to produce the components) is the type and level of contamination to be removed; maintenance cleaning usually involves removing much higher levels of contamination.

The main technique used is vapour phase cleaning, where the solvent (in an appropriately constructed tank) is heated to its boiling point and the resultant vapour is condensed further up the tank to produce a stable vapour zone. The article is cleaned by immersion in the hot vapour, which condenses on the cold article, dissolving the soil and washing it off. Contaminated solvent runs off the article back into the boiling reservoir. The vapour is effectively always fresh clean solvent. This process accounts for ~80% of the solvent used in metal cleaning. The lesser used processes include hand wiping (very common but collectively it still uses little solvent) and brushing, and cold cleaning techniques where the solvent is used below its boiling point (but generally above room temperature). Variations include ultrasonic cleaning (use of high-frequency sound to agitate the solvent), spray cleaning, and emulsion cleaning (where a non-water-soluble solvent is used in combination with water).

Chlorinated solvents are still the most important solvents in metal cleaning, with four principal products in use: perchloroethylene, trichloroethylene, methylene chloride (dichloromethane), and methylchloroform (1,1,1-trichloroethane) with a fifth halogenated material, CFC 113, almost exclusively used for printed circuit board cleaning. The reason for the popularity of chlorinated solvents has been their non-flammability. This, together with their high vapour density, means that they could be used for vapour degreasing without the need for expensive explosion-proofing measures. Hydrocarbons have been used in emulsion cleaners and for niche applications. While the use of oxygenated solvents was small, it is now growing markedly. The current approximate volumes for these products (ktpa) are shown in Table 8.14.

The types of solvent used are changing due to environmental legislation and agreements, principally the 'Montreal Protocol' restricting certain chlorinates and chlorofluorocarbons. As well as reduced use, recycling and improved hygiene and housekeeping are increasing, so shrinking the market. The Protocol is an international agreement to reduce the emissions of materials believed to be hazardous to the 'ozone layer' (a layer in the Earth's upper atmosphere which helps to filter out harmful UV radiation). Since chlorine is a key part of the mechanism for ozone depletion, some halogenated materials are being affected widely. All CFCs, some HCFCs and some chlorinated solvents (carbon tetrachloride and 1,1,1-trichloroethane) are or will be controlled. Methyl chloroform (1,1,1-trichloroethane) is included in this agreement and its production will consequently be phased out by the end of 1996. Many countries, including those of the European Union, will phase it out by the end of 1995. The introduction of legislation to limit VOC emission is additionally bringing changes in the design and operation of equipment.

These restrictions in the use of chlorinated solvents are increasing the demand for alternatives. Many are now available, either based on non-regulated chlorocarbons, non-chlorinated solvents or on non-solvent-based techniques. So far they have made little impact on the general market since chlorinates were still available in 1995. In the next few years this will no longer be the case. As the production of some of the common chlorinated solvents are phased out, users will be forced to adapt to new technologies using more environmentally benign materials. They have, however, made substantial inroads in printed circuit board cleaning where the pressure came earliest to eliminate CFC use; aqueous and semi-aqueous cleaning technologies are now in most prevalent use. Oxygenated solvent-based cleaning systems have been particularly successful in

Table 8.14 Approximate volumes of main solvents used in metal cleaning in Europe and the USA

Solvent	Europe (ktpa)	USA (ktpa)
Chlorinates	230	190
Hydrocarbons	10	25
Oxygenates	2	—
Total	242	215

this application in Europe where their excellent solvency properties and simple process requirements are well received. Some chlorinated solvents (trichloroethylene, perchloroethylene and methylene chloride) which are not being phased out, are being used as replacements in certain applications, providing the necessary operator safety standards can be met. Some alternatives, such as 'natural' citrus oil cleaners, have been promoted as alternatives. These terpene-containing materials, however, have a high photochemical ozone creation potential. In hand wiping, lower volatility solvents increasingly are being used to reduce emissions. Ultimately, much of the market may be expected to change to hydrocarbons and oxygenates. Although hydrocarbons have a longer track record, oxygenates are steadily increasing their market share owing to consistent and better drying characteristics and overall cleaning performance as a result of their higher solvency.

A continued demand for significant volumes of solvents is expected in this sector despite the uptake of aqueous detergent systems. This is due to recognition that, properly managed, solvents allow retention of the performance advantages to which the user has been accustomed without adverse environmental impact.

8.6 Pharmaceutical and agrochemical manufacture

The bulk active ingredients for pharmaceuticals and agrochemicals have much in common and many manufacturing companies are active in both businesses. In either case, the bulk active ingredients are of very high value and are ultimately supplied to internal and/or external businesses for conversion into final dosage forms.

The complex chemical nature of these active ingredients is reflected in the production routes; these are rarely straightforward and typically involve a high number of process steps. By definition, each of these process steps must be highly efficient to maximise both the reaction rate and yield of the overall process since minor inefficiencies, when combined throughout the process, can add up to a significant financial penalty in lost production. This drive for efficiency results in the selection or design of chemical processes that are dependent upon the use of organic solvents for their success.

The applications of solvents can be categorised into three main areas:

1. To solubilise reagents and/or minimise unwanted side reactions. In some cases the solvent itself may actually be a reagent or will allow the ideal, or safest, reaction conditions to be achieved (e.g. pH, low temperature, etc.).
2. To extract preferentially the reaction products or unwanted by-products once the reaction is complete.
3. To purify the final product or purify reaction intermediates prior to further processing.

In addition to these applications, solvents are also used in cleaning plant and process equipment (particularly important in the pharmaceutical industry) and in the extraction of natural products which cannot be synthesised economically. Their use in the manufacture of final formulations is significantly lower, with the exception of some agrochemical preparations.

The range of solvents used can be subdivided into four main categories: oxygenates (primarily ketones, alcohols and esters, but also ethers and glycol ethers); hydrocarbons (mainly aromatics such as toluene, but also some aliphatics); halogenates (primarily methylene chloride); and those containing a heteroatom (such as dimethyl sulphoxide, dimethylformamide and pyridine).

Estimates of European and US solvent use (ktpa) in the pharmaceutical/agrochemical sector are shown in Table 8.15. Note that the US figures are for agrochemical use only. However, the bulk (primary) pharmaceutical industry in the USA, for fiscal reasons, is very small compared to that in Europe and amounts to only a few tens of thousands of tonnes of solvent per year. Oxygenates and aromatic hydrocarbons are the most widely used general-purpose solvents. Solvents containing a heteroatom generally have more specialised, low volume applications.

Table 8.15 Estimates of European and US solvent use in in the pharmaceutical/agrochemical sector

Solvent	Europe (ktpa)	USA (ktpa)
Hydrocarbons	75	90
Oxygenates	360	130 mainly oxygenates
Chlorinates	30	
Total	465	220[a]

[a] Agrochemical use only.

The pharmaceutical sector is somewhat different to the others in that substantial internal recycling of solvents already occurs. The actual process throughput of solvents is 3–5 times the figures shown in the table, which represent make-up quantities. Most modern pharmaceutical plants are fully integrated for waste management, and solvent that cannot be recycled is destroyed by thermal oxidation. Some spent solvent is supplied to third party toll recoverers with the recovered material ending up in the thinners and cleaners market. The quantities are small, however, because of the reluctance of the manufacturers, for reasons of safety and confidentiality, to release waste process materials which could contain trace amounts of pharmaceutical products.

Environmental considerations are just some of the many issues currently faced by pharmaceutical and agrochemical producers. These encompass the direct effect of their products on the environment, the impact of their production processes on the local environment and the most appropriate disposal route for waste products. Other considerations affecting choice of raw materials used include issues such as security of supply for processes that are not easy to alter. At a time when some regulatory authorities now have the power to delay new product approvals if the process is deemed not to be environmentally acceptable, the stakes are high. As such, efficiency considerations are no longer the only factors in determining the optimum configuration for a new process.

Once a production process has been registered (e.g. through filing a Drug Master File), making changes to the process or materials used in the process is difficult. The producer must demonstrate that the change does not adversely affect the quality of the bulk pharmaceutical being manufactured. Where a significant change is considered this will involve lengthy and expensive testing procedures. As a consequence, many processes must remain unaltered, even where improvements are theoretically possible, because the costs of implementation outweigh the benefits. For example, for environmental reasons, a producer may wish to replace one solvent in the process with another. Testing may take years to complete, and a solution that may well be possible in another sector may not be economically justified here.

For some solvents, recent legislation in the USA, and proposed legislation in Europe, sets a maximum residual limit for processing chemicals in finished pharmaceutical products. The criteria are based on toxicity concerns; ten solvents are present in the *European Pharmacopoeia* issued in July 1995, including methylene chloride, acetonitrile and secondary butanol. While their use is not precluded, greater care is now required to ensure compliance with the rules. This may involve more intensive and expensive recrystallisation procedures to remove trace amounts of solvent when they are used in the latter stages of the manufacturing process.

The pharmaceutical industry is one which has developed a strong track record in the successful management of wastes, including solvents. Techniques exist which allow recovery efficiencies of well over 90% in terms of solvent usage. Pharmaceutical manufacturers have used thermal destruction as a responsible means for waste disposal when recovery is not economically credible. An uncontrolled drive to ever higher recovery rates would impose severe and unnecessary penalties upon the industry with negligible environmental benefits.

8.7 Cosmetics and toiletries

The major proportion of the world market for cosmetics and toiletries is in the hands of approximately 15 companies which account for 60% of sales. The demand for solvents in cosmetics and toiletries does not follow the sales value since solvents are used in only certain segments within this sector.

The major segments for solvents are:

Haircare	Hair spray, styling spray
Fragrance	Mass market – deodorants, colognes, fine fragrance
Oral care	Mouthwash

Solvents are used in nail care and some skin preparations but this is small in comparison to the three segments above. Each is covered in more detail in the following sections.

8.7.1 Haircare

Hair spray consists of a resin dissolved in a solvent (usually ethanol but can be isopropanol) and a propulsion system to place the product onto the hair. The propulsion system is mainly the condensed gas aerosol (hydrocarbon blends or dimethyl ether) but pump and compressed gas systems are available.

The traditional hair spray is a fixing spray which holds the style in place once created. These products are in decline, but new styling sprays, targeted at younger users, show good growth. Fashion changes impact strongly upon this segment.

8.7.2 *Fragrance*

Fragrances can be split into mass and fine market products. With mass market fragrances (deodorants, body sprays, aftershaves, certain colognes) the consumer is purchasing a function. Fine fragrances are more often purchased to bestow an image upon the user. Although in value terms the fine fragrance is a large proportion of the fragrance segment, mass market fragrances represent the bulk of the volume (with the solvent used almost entirely ethanol).

The latest major trend has been the rise of male-oriented products, such as in skin care.

8.7.3 *Oral care*

The growth of mouthwash has been dramatic over the last few years. The ethanol level in a mouthwash is much lower than in hair care or fragrance products and therefore the total volumes used are quite small in comparison.

8.7.4 *Solvents used*

All the major solvents used in cosmetics and toiletry formulations are from the most benign group and have low toxicity. Ethanol, the most common solvent in this sector, has a low environmental impact and, in particular, is among the least potent of VOCs in forming tropospheric ozone.

Ethanol is the dominant solvent in cosmetics and toiletries and in many countries is taxed for this application. To protect the excise duty (tax), the ethanol is denatured to prevent its being consumed as drink, and in each country the particular denaturing system adopted differs. Common denaturants include Bitrex (denatonium dibenzoate) or DEP (diethyl phthalate) together with a marker such as tertiary butanol. European countries are moving towards a position of mutual recognition where each country's denaturing system is allowed by their European partners in finished products. Each country, however, still has its own rules on the manufacture and original use of ethanol.

In mass market products such as hair sprays or deodorants, synthetic ethanol has distinct advantages for the manufacturer since the consistency and purity of the product ensures minimum plant and process changes from batch to batch. In the smaller fine fragrance market, however, the interaction of fragrance components and fermentation ethanol impurities is part of the artistry of the perfumer. This effect is much slower, or in certain cases cannot be repeated, with synthetic ethanol as the base. Hence fermentation ethanol has traditionally been used for these products and is likely to continue to be used in the future.

8.7.5 *Isopropanol*

Isopropanol is used in certain applications, but owing to its stronger intrinsic odour and slower evaporation characteristics is generally considered technically inferior to ethanol for most applications. The decision to use isopropanol is usually cost driven and related to the tax position for ethanol in that particular country.

8.7.6 *Aerosol propellants*

CFC propellants are no longer used in cosmetic products. The market has switched to a number of other delivery systems driven by the need to meet environmental concerns over CFCs. The aerosol industry has made enormous strides in reformulating its products to virtually eliminate CFCs from all but a few specialist (usually medical) applications. The aerosol is the major delivery system with condensed gas propellant systems based on hydrocarbons (propane/butane mixtures) or dimethyl ether (DME). Although compressed gas and pump delivery systems have been marketed by all the major companies, they have met with limited success largely on performance grounds and many have been withdrawn. In the deodorant segment, the aerosol is the preferred delivery system but roll-on, stick and pump systems have significant presence in the market.

8.7.7 *Other solvents*

Minor amounts of acetone, ethyl and butyl acetates are found in certain nail care products. Considering the overall demand for solvents and propellants in the cosmetics and toiletries market, the demand for these products is small (less than 2% of overall solvent/propellant demand).

Methylene chloride can still be used in aerosol hair sprays up to about 20% levels, but this is confined to a few manufacturers in southern Europe. Widespread usage of methylene chloride has virtually disappeared due to toxicity concerns in this application.

8.8 Detergents and household products

About 80% of detergents and household products supplied to the retail market come from not more than five multinational companies. Owing to the fact that water miscibility is an essential property for most of the applications in this sector, oxygenated solvents are virtually the only ones used.

The demand for solvents in detergents and household products is mainly in five areas:

- Laundry liquids
- Hard surface cleaning
- Dishwashing
- Fabric softeners
- Air fresheners

Other areas do exist, such as aliphatic hydrocarbon use in floor polishes, but these are relatively minor.

8.8.1 Laundry liquids

When considering the use of solvent in detergent and household products, we should be aware that only a small amount of solvent is released during each wash or cleaning operation. Also, under the particular conditions (washing machine waste, diluted floor cleaner) most of the solvent will be absorbed in the water, and as it will not be released into the atmosphere it will not contribute to atmospheric ozone formation. This is one reason why biodegradability is a very important issue in this sector and often should be considered more important than any VOC weighting.

8.8.2 Hard surface cleaning

The term 'hard surface' covers a wide range, including kitchen preparation surfaces, bathrooms, floors and glass. The formulations of the cleaning products for each of these subsegments differs and, in some cases, very different technologies are available for the same end use.

Surface cleaners for kitchen and bathroom can either be suspension (abrasive material held in suspension using no solvent) or isotropic (containing a minor amount of solvent). Many households purchase both, indicating that the two forms of cleaner are not truly competing products and are serving different functions.

Glass cleaners contain low amounts of solvents, usually a few percent of glycol ether and an alcohol.

8.8.3 Dishwashing

The hand dishwash liquid market has remained static for a number of years. This is due to an increase in the use of dishwasher machines, which will eventually cause a decline in the manual products. Ethanol is present at low levels in many hand dishwash products. Machine dishwasher products are mainly powder detergents. The machines also used a rinse aid which contains a short chain alcohol, usually isopropanol.

8.8.4 Fabric softeners

Laundry care is the largest segment, using ethanol/isopropanol and glycols in the liquid formulations. When laundry liquids first entered the market, ethanol and isopropanol were present in significant quantities, but due to reformulation and concentrated product development, this is no longer the case.

The active surfactant in a fabric softener is an ester quat (quaternary ammonium compound with ester links to improve biodegradability). These active ingredients are supplied to the product manufacturers as a mobile liquid which contains isopropanol. Other solvents, such as glycol ethers, can be used but, mainly for cost reasons, over 90% of the market uses isopropanol.

A small amount of isopropanol is therefore found in the final product, not as a directly added ingredient but as part of one of the raw materials used.

8.8.5 Air fresheners

The aerosol air freshener is a familiar sight in most homes but over the last couple of years changes have been made to the formulations which reversed the established technology trend.

During the late 1980s, many of these products were reformulated to 'dry' versions which were basically fragrance and condensed gas aerosol propellant; the propellant was almost exclusively mixtures of propane and butane to give the desired pressure.

In the early 1990s, concerns over the flammability of these products led manufacturers to reconsider their formulations. The latest products once again contain water and the change appears to have been accepted by the consumer since it is thought that the 'dry' property (allows the freshening of soft furnishings) of the product was not the primary function required by the consumer.

Other delivery mechanisms have also become available in the last few years, such as liquid or slow release sticks. In the future, these products will be developed further along with pumps and compressed gas aerosols but they tend to be more expensive to produce than the aerosol and are not as convenient for the consumer.

8.9 Automotive de-icers

The automotive windscreen de-icer and screenwash market is a significant user of methanol, ethanol and isopropanol. The alcohol present acts as a freezing point depressant. Western Europe uses 80–100 ktpa of these alcohols.

Freezing point depression is proportional to the molar rather than the mass concentration of the alcohol, and for a given mass of alcohol, efficiency decreases with increasing molecular mass. Methanol is therefore the most, and isopropanol the least, efficient freezing point depressant of the three alcohols. In practice this is only true in formulations containing greater than 30% mass. Below this concentration isopropanol efficiency is equivalent to, or marginally better than, ethanol. This anomalous behaviour of isopropanol can be explained by increased hydrate formation and hydrogen bonding.

Methanol is more volatile than ethanol or isopropanol and therefore evaporates more rapidly from the windscreen, making refreezing a problem with de-icers based on methanol. Ethylene glycol is usually added to retard the evaporation of the methanol and to improve further the de-icing efficiency. Ethanol is intermediate in evaporation rate between methanol and isopropanol, but this is also dependent upon concentration; in dilute aqueous solution (< 30% alcohol) the evaporation of ethanol is lower than that of isopropanol.

Because of its good freezing point depression efficiency, and its low cost, methanol has in the past been the favoured alcohol for this application. Typical de-icer formulations contained between 50 and 95% methanol, 9 and 30% ethylene glycol and the remainder water. Stricter labelling regulations and greater consumer awareness of toxicity issues have, however, led manufacturers to replace methanol in de-icer and screenwash formulations with much safer ethanol and isopropanol, while still retaining effective performance. In Europe, isopropanol is usually favoured for economic reasons, and methanol has all but disappeared, being replaced by 50–98% isopropanol in de-icers.

Screenwashes were typically 50 : 50 methanol and water but now usually contain typically 10–50% isopropanol although some mixed propanols are also used (*n*-propanol and isopropanol). In a small number of formulations, ethanol is used in similar proportions to isopropanol. Ethanol is the favoured alcohol in the Scandinavian countries (Norway, Sweden and Finland) where prices of isopropanol and ethanol are comparable. Methanol is used in the USA, where price is the dominant factor. Some changes are taking place as formulations based on isopropanol/ethanol are now comparable in price to the normal 20–30% methanol-based systems. The 'green' benefits of using ethanol from renewable fermentation sources is also causing a small but discernible shift.

8.10 Dry cleaning

Dry cleaning is the cleaning of fabrics in a mainly non-aqueous medium. When fabrics are cleaned in water, the hydrophilic fibres swell and distort when subjected to mechanical action. 'Dry' solvents avoid this change and, consequently, fabrics do not show wrinkles, shrinkage or other forms of distortion following cleaning. Dry cleaning is a major consumer of solvents world wide. The principal dry cleaning solvent is perchloroethylene, followed by CFC 113 (also known as R113, 1,1,2-trichloro-1,2,2-trifluoroethane) and hydrocarbons (such as 'Stoddard's solvent' or '140-F', which are inexpensive hydrocarbon mixtures similar to kerosene).

CFC 113 use is restricted to the UK, Scandinavia and the USA while hydrocarbons are only used significantly in the USA and Germany. However, in all regions perchloroethylene is the dominant solvent. Japan uses small quantities of 1,1,1-trichloroethane, where it represents approximately 5% of the total solvent demand.

Estimates of European and US solvent use (ktpa) in dry cleaning are shown in Table 8.16.

Of the common solvents, CFC 113 and 1,1,1-trichloroethane are under immediate threat as a result of legislation, based around the 'Montreal Protocol', to phase out their production. Furthermore, CFC producers in the European Union voluntarily agreed to cease production by the end of 1994. In practice, CFC 113 will no longer be used in dry cleaning after 1997. Since dry cleaning units are designed to be solvent specific, many CFC operators will need to find alternatives or exit the market.

Table 8.16 Estimates of European and US solvent use in the dry-cleaning sector

Solvent	Europe (ktpa)	USA (ktpa)
Perchlorethylene	85	90
CFC 113	5	3
Hydrocarbons	10	30
Total	100	133

Perchloroethylene also faces tightening usage controls as a result of toxicity concerns. The industry is responding to this challenge by engineering hardware solutions; modern closed circuit units are designed to minimise solvent losses (thus reducing potential worker exposure).

Within the European Union, driven by concerns over the toxicity of the solvents used, recommendations within the draft VOC Emissions Directive to reduce the amount of VOC emission from the existing 60 000 or more dry-cleaning outlets include use of 'new generation' closed circuit hardware (with refrigerated condensation and activated carbon adsorption vents) for new installations and retrofitting of existing installations with carbon filters within one year of the Directive. These recommendations would reduce dry-cleaning emissions by 75%.

Future developments may result in a move away from the traditional dry-cleaning solvents. In the USA, the EPA has established the 'Design for the Environment Program' to help find alternatives to current dry-cleaning technologies. Among the alternatives currently under study are the use of special water or steam treatments with ultrasonics or microwave processing and the use of supercritical carbon dioxide.

8.11 Miscellaneous

The main applications for solvents have already been described in the preceding sections. There are, however, many other minor applications, both in industry and the home, which demonstrate the wide use and importance of these materials. It would be impossible to produce an exhaustive list that could guarantee to cover all of them, but as an illustration some are described in this section.

Solvents are also used in food processing. The largest uses are of aliphatic hydrocarbons, such as *n*-hexane, which are important for oil seed extraction and purifying oils; good solvency and ease of separation from the extracted oils (low boiling point) are key properties. Extraction is important in many other parts of the food industry to produce, for example, flavour extracts; a variety of solvents from hydrocarbon to alcohols are used in such applications, but volumes are smaller. Isopropanol is used to extract fish protein from ground fish. Methylene chloride was used to extract caffeine from coffee, but has been largely superseded by liquid carbon dioxide. For all human food applications, food contact approval is vital if human consumption is involved.

Extraction is also important outside the food industry; for example, in the extraction of wood resins and rare earth metal extraction. Solvents, such as methyl ethyl ketone, are often used in wax production by the dewaxing of lubricating and other natural oils. Solvents are also used in the extraction of North American tar sands; and solvent extraction is important in many other parts of the petroleum industry but this often involves the use of relatively esoteric materials which are beyond the intended scope of this book. Solvents are also used in cleaning oil-contaminated aqueous effluents.

Solution casting of films such as cellulose acetate uses solvents such as acetone and methylene chloride (e.g. for photographic film). Most solvent is recovered for reuse in the process.

Solvents are widely used in the manufacture of resins both as reactants or, more usually, as processing solvents. Since these solvent-carried resins predominantly end up in surface coatings or inks, their use is usually classed in these sectors.

Aromatic (particularly toluene) and aliphatic hydrocarbons are used in large quantities in the rubber-processing industries; in synthetic rubber manufacture, aliphatic hydrocarbons act as catalyst carriers and can influence molecular weight through polymer precipitation.

Aliphatics such as kerosenes are used in coal washing to separate impurities. The principles are similar to froth flotation and involve preferential wetting of the coal by the hydrocarbon while the water-wetted impurities remain suspended in the water.

Acetone is an important solvent in textile manufacture where it is used in large quantities in the dry extrusion process of cellulose acetate manufacture. The filtered solution of cellulose acetate and acetone is passed through a spinneret into a heated chamber. Here the acetone evaporates, allowing the cellulose acetate to solidify as an extruded fibre which is passed on to the spinning machines for the production of yarn. Acetone is chosen as the solvent because of its rapid evaporation rate, and in large-scale applications the bulk of the solvent is recovered for reuse.

Acetone, isopropanol and ethanol are used in the manufacture of nitrocellulose explosives. Again solvency properties are important. The alcohols are also used as damping materials for nitrocellulose resins, making them safe to handle and transport.

Ethanol and isopropanol are used in antiseptics, although quantities are small.

8.12 Conclusions

The vast range of applications described in this chapter shows the vital part that solvents play in a civilised society. These important commodity and speciality chemicals have been used safely for many years and their use will undoubtedly continue in the future. There are, however, concerns and pressures, unfortunately not always well founded, which will lead to a change in patterns of use.

Industrial solvent use is predicted to decline, principally because of environmental pressures to reduce emissions of VOCs. This will occur through a mix of legislation and voluntary agreements. The solvents that will be most seriously affected are the chlorinates which, because of concerns over toxicity and in some cases ozone depletion, are expected to show a marked decline in use over the next decade. Likewise, hydrocarbon use will show a significant decline, which will be due mainly to a switch in the decorative coatings market from white spirit to water-based paints. Aromatic hydrocarbon use will also show a decline in industrial applications – particularly surface coatings – partly because of industrial coatings moving to higher solids content paints (which require more efficient solvents such as esters, ketones and glycol ethers) and partly because of additional toxicity labelling demands. Oxygenated solvent use is expected to show only a small decline; reduced use in sectors such as traditional solvent-based paints, household products and pharmaceuticals will be offset by increased use in high-performance solvent- and water-based coatings.

As can be seen, there has been an emotive reaction against solvents centred on the VOC issue alone. Rarely, until recently, has the wider issue of total environmental impact been considered. When implications for pollution to other media, such as soil and water, and energy use are taken into account (as in the science of life cycle assessment), solvent-based technologies look much more balanced. As impact assessment becomes more widely used and the issues better understood, the swing against solvent technologies will slow and may even reverse. For this reason, and because of the cost-effectiveness and reliable performance of solvent-based technologies, their wide use will continue. Solvents are an essential part in the life cycle of many consumer and industrial products and, through careful management and use, will remain so for many years to come.

Part Two—Solvent Data

Organisation of solvent data

1. Solvents are allocated a separate page and organised by the following groupings:
   ACIDS
   ALCOHOLS (KETO-ALCOHOLS)
   ALCOHOLS
   ALDEHYDES
   AMIDES
   ESTERS
   ETHERS
   FATTY ACID ESTERS
   GLYCOL ESTERS
   GLYCOL ETHER ESTERS
   GLYCOL ETHERS
   GLYCOLS
   HALOGENATED HYDROCARBONS
   HYDROCARBONS—ALIPHATIC
   HYDROCARBONS—ALIPHATIC/CYCLIC
   HYDROCARBONS—ALIPHATIC/CYCLIC/AROMATIC
   HYDROCARBONS—AROMATIC
   HYDROCARBONS—AROMATIC/ALIPHATIC
   HYDROCARBONS—CYCLIC
   HYDROCARBONS—UNSPECIFIED
   KETONES
   NITRO COMPOUNDS
   PHENOLS
   SULPHUR COMPOUNDS
   UNSPECIFIED

2. Within each functional grouping, the solvents are presented in a sequence with increasing boiling point (where the data are available).

3. Trade Names. The names quoted do not necessarily refer to Trade Marks, but may refer to the fact that the product is often traded under that name. No Trade Mark status is implied by any information presented in this volume.

4. Data Sets.
4.1 The units are in broad accordance with the SI convention.
4.2 A range of values is often quoted, since a given product may originate from several sources. The data should not be interpreted as specifications or typical values for any given supplier.
4.3 Flash points/flammability data. The flash point of a solvent is closely dependent on the test method used, and this is quoted where available. Due to discrepancies in data, individual values are quoted with the source of the information. Care must be exercised before using these data and the supplier of the solvent should always be consulted.
4.4 Solubility parameters. Nelson, Hemwall and Edwards values are often quoted with the units $(\text{cal/cm}^3)^{1/2}$. The conversion factor is:

$$2.05 \times (\text{cal/cm}^3)^{1/2} \equiv (\text{J/cm}^3)^{1/2}$$

4.5 UK Occupational Exposure Limits are taken from EH40/95, published by the UK Health & Safety Executive. Those marked with an * are internal company guidelines.
US Occupational Exposure Limits are taken from ACGIH 1994–1995, Threshold Limit Values for Chemical Substances and Physical Agents.
European Classification is taken from the Official Journal of the European Communities COM(93)638 1993.
IMO Classification is taken from International Maritime Dangerous Goods Code (IMDG Code), Amendment 26-91: 1991
ADR/RID Classification is taken from European Agreement concerning the International Carriage of Dangerous Goods by Road(ADR), ANNEX B 1995.
5. Key Parameter Tables.
The data are also presented in Key Parameter Tables, where the measured value, e.g. Boiling Point, is arranged with increasing value.

Solvent group		**ACIDS**
Chemical name		Acrylic acid
Chemical formula		$C_3H_4O_2$
Structural formula		$CH_2=CHCOOH$
CAS No.		79-10-7
Synonym/common name		Acrylic acid
Trade names		
Suppliers		UNION CARBIDE
Molecular mass		72.06
Boiling point/range	°C @ 760 mm Hg	141.2
Melting point	°C	13.5
Density	kg/m^3 @ 20 °C	1047
Evaporation time[a]		
Evaporation time[b]		
Vapour pressure	kPa @ 20 °C	1.03
Flash point	°C	
Auto ignition temperature	°C	
Upper flammability limit	%v/v in air	
Lower flammability limit	%v/v in air	
Viscosity	mPa.s (cP) @ 20 °C	
Refractive index	n 20 D	1.422
Solubility in water	%w/w @ 20 °C	Complete
Solubility of water	%w/w @ 20 °C	Complete
Solubility parameter[c]	(J/cm^3)$^{1/2}$	24.5
Hydrogen bond index		
Fractional polarity		
Nitro cellulose dil ratio[d]		
Nitro cellulose dil ratio[e]		
Surface tension	mN/m @ 20 °C	
Specific heat liquid	kJ/kg mole/°C @ 20 °C	
Latent heat	kJ/kg mole	
Dielectric constant	@ 20 °C	
Antoine constant A	kPa°C	6.7270
Antoine constant B	kPa°C	1954.6
Antoine constant C	kPa°C	286.7
Heat of combustion	kJ/kg mole	
UN number		2218
IMO classification		8/II
ADR/RID classification		8, 32 °C (b)2
UK exposure limits	mg/m^3 (ppm) 8hrTWA/MEL or OES	30(10)
USA exposure limits	mg/m^3 (ppm) TLV/TWA	5.9(2)
German exposure limits	mg/m^3 (ppm) MAK-TWA	
EU classification		R10
EU risk phrases		10–34
EU safety phrases		(1/2)-26-36-45

[a] Ether = 1. [b] n-Butyl acetate = 1. [c] Nelson, Hemwall, Edwards. [d] Toluene. [e] Hydrocarbon solvent.

Solvent group		**ACIDS**
Chemical name		Isopentanoic acid
Chemical formula		$C_5H_{10}O_2$
Structural formula		$(CH_3)_2CHCH_2COOH$
CAS No.		503-74-2
Synonym/common name		Isovaleric acid
Trade names		
Suppliers		UNION CARBIDE
Molecular mass		102.13
Boiling point/range	°C @ 760 mm Hg	175.0
Melting point	°C	
Density	kg/m^3 @ 20 °C	923
Evaporation time[a]		
Evaporation time[b]		
Vapour pressure	kPa @ 20 °C	
Flash point	°C	
Auto ignition temperature	°C	
Upper flammability limit	%v/v in air	
Lower flammability limit	%v/v in air	
Viscosity	mPa.s (cP) @ 20 °C	
Refractive index	n 20 D	
Solubility in water	%w/w @ 20 °C	
Solubility of water	%w/w @ 20 °C	
Solubility parameter[c]	(J/cm^3)$^{1/2}$	
Hydrogen bond index		
Fractional polarity		
Nitro cellulose dil ratio[d]		
Nitro cellulose dil ratio[e]		
Surface tension	mN/m @ 20 °C	
Specific heat liquid	kJ/kg mole/°C @ 20 °C	
Latent heat	kJ/kg mole	
Dielectric constant	@ 20 °C	
Antoine constant A	kPa°C	
Antoine constant B	kPa°C	
Antoine constant C	kPa°C	
Heat of combustion	kJ/kg mole	
UN number		
IMO classification		
ADR/RID classification		
UK exposure limits	mg/m^3 (ppm) 8hrTWA/MEL or OES	
USA exposure limits	mg/m^3 (ppm) TLV/TWA	
German exposure limits	mg/m^3 (ppm) MAK-TWA	
EU classification		
EU risk phrases		
EU safety phrases		

[a] Ether = 1. [b] n-Butyl acetate = 1. [c] Nelson, Hemwall, Edwards. [d] Toluene. [e] Hydrocarbon solvent.

Solvent group		**ACIDS**
Chemical name		2-Ethylhexoic Acid
Chemical formula		$C_8H_{16}O_2$
Structural formula		$C_4H_9CH(C_2H_5)COOH$
CAS No.		149-57-5
Synonym/common name		2-Ethylhexoic acid
Trade names		
Suppliers		UNION CARBIDE
Molecular mass		144.21
Boiling point/range	°C @ 760 mm Hg	227.0
Melting point	°C	
Density	kg/m^3 @ 20 °C	907
Evaporation time[a]		
Evaporation time[b]		
Vapour pressure	kPa @ 20 °C	
Flash point	°C	
Auto ignition temperature	°C	
Upper flammability limit	%v/v in air	
Lower flammability limit	%v/v in air	
Viscosity	mPa.s (cP) @ 20 °C	
Refractive index	n 20 D	
Solubility in water	%w/w @ 20 °C	0.1
Solubility of water	%w/w @ 20 °C	1.4
Solubility parameter[c]	(J/cm^3)$^{1/2}$	
Hydrogen bond index		
Fractional polarity		
Nitro cellulose dil ratio[d]		
Nitro cellulose dil ratio[e]		
Surface tension	mN/m @ 20 °C	
Specific heat liquid	kJ/kg mole/°C @ 20 °C	
Latent heat	kJ/kg mole	
Dielectric constant	@ 20 °C	
Antoine constant A	kPa°C	
Antoine constant B	kPa°C	
Antoine constant C	kPa°C	
Heat of combustion	kJ/kg mole	
UN number		
IMO classification		
ADR/RID classification		
UK exposure limits	mg/m^3 (ppm) 8hrTWA/MEL or OES	
USA exposure limits	mg/m^3 (ppm) TLV/TWA	
German exposure limits	mg/m^3 (ppm) MAK-TWA	
EU classification		
EU risk phrases		
EU safety phrases		

[a] Ether = 1. [b] n-Butyl acetate = 1. [c] Nelson, Hemwall, Edwards. [d] Toluene. [e] Hydrocarbon solvent.

Solvent group		**ALCOHOLS (Keto-alcohol)**
Chemical name		4-Hydroxy-4-methyl-2-pentanone
Chemical formula		$C_6H_{12}O_2$
Structural formula		$(CH_3)_2C(OH)CH_2COCH_3$
CAS No.		123-42-2
Synonym/common name		Diacetone alcohol
Trade names		DA,DAA
Suppliers		ELF ATOCHEM, SHELL
Molecular mass		116.16
Boiling point/range	°C @ 760 mm Hg	150–172
Melting point	°C	−43
Density	kg/m^3 @ 20 °C	937–940
Evaporation time[a]		150
Evaporation time[b]		0.15
Vapour pressure	kPa @ 20 °C	0.12–0.164
Flash point	°C	58(ASTM D93 & CLOSED CUP)
Auto ignition temperature	°C	603–620
Upper flammability limit	%v/v in air	6.9
Lower flammability limit	%v/v in air	1.8
Viscosity	mPa.s (cP) @ 20 °C	2.9–3.2
Refractive index	n 20 D	1.423
Solubility in water	%w/w @ 20 °C	Complete
Solubility of water	%w/w @ 20 °C	Complete
Solubility parameter[c]	(J/cm^3)$^{1/2}$	18.8
Hydrogen bond index		6.5
Fractional polarity		0.312
Nitro cellulose dil ratio[d]		3.2
Nitro cellulose dil ratio[e]		0.6
Surface tension	mN/m @ 20 °C	30.9
Specific heat liquid	kJ/kg mole/°C @ 20 °C	241.6
Latent heat	kJ/kg mole	43560
Dielectric constant	@ 20 °C	18.2
Antoine constant A	kPa°C	7.03103
Antoine constant B	kPa°C	2036.93
Antoine constant C	kPa°C	236.175
Heat of combustion	kJ/kg mole	−3355397
UN number		1148
IMO classification		3.3/III
ADR/RID classification		3,31°(c)
UK exposure limits	mg/m^3 (ppm) 8hrTWA/MEL or OES	240(50)
USA exposure limits	mg/m^3 (ppm) TLV/TWA	238(50)
German exposure limits	mg/m^3 (ppm) MAK-TWA	240(50)
EU classification		Xi
EU risk phrases		36
EU safety phrases		(2-)24/25

[a] Ether = 1. [b] n-Butyl acetate = 1. [c] Nelson, Hemwall, Edwards. [d] Toluene. [e] Hydrocarbon solvent.

Solvent group		**ALCOHOLS**
Chemical name		(2-Ethoxy-methylethoxy)-propanol
Chemical formula		
Structural formula		
CAS No.		30025-38-8
Synonym/common name		Ethoxy propoxypropanol
Trade names		ETHOXY PROPOXY PROPANOL
Suppliers		BP
Molecular mass		
Boiling point/range	°C @ 760 mm Hg	162–228
Melting point	°C	−51
Density	kg/m^3 @ 20 °C	942
Evaporation time[a]		
Evaporation time[b]		0.02
Vapour pressure	kPa @ 20 °C	0.14
Flash point	°C	86.5
Auto ignition temperature	°C	190
Upper flammability limit	%v/v in air	9.6
Lower flammability limit	%v/v in air	0.75
Viscosity	mPa.s (cP) @ 20 °C	
Refractive index	n 20 D	
Solubility in water	%w/w @ 20 °C	
Solubility of water	%w/w @ 20 °C	
Solubility parameter[c]	(J/cm^3)$^{1/2}$	
Hydrogen bond index		
Fractional polarity		
Nitro cellulose dil ratio[d]		
Nitro cellulose dil ratio[e]		
Surface tension	mN/m @ 20 °C	
Specific heat liquid	kJ/kg mole/°C @ 20 °C	
Latent heat	kJ/kg mole	
Dielectric constant	@ 20 °C	
Antoine constant A	kPa°C	
Antoine constant B	kPa°C	
Antoine constant C	kPa°C	
Heat of combustion	kJ/kg mole	
UN number		
IMO classification		none
ADR/RID classification		3,32°(c)
UK exposure limits	mg/m^3 (ppm) 8hrTWA/MEL or OES	
USA exposure limits	mg/m^3 (ppm) TLV/TWA	
German exposure limits	mg/m^3 (ppm) MAK-TWA	
EU classification		
EU risk phrases		
EU safety phrases		

[a] Ether = 1. [b] n-Butyl acetate = 1. [c] Nelson, Hemwall, Edwards. [d] Toluene. [e] Hydrocarbon solvent.

Solvent group		**ALCOHOLS**
Chemical name		(2-Methoxymethylethoxy)-propanol
Chemical formula		
Structural formula		
CAS No.		34590-94-8
Synonym/common name		
Trade names		METHOXY PROPOXY PROPANOL
Suppliers		BP
Molecular mass		
Boiling point/range	°C @ 760 mm Hg	184–197
Melting point	°C	−83
Density	kg/m^3 @ 20 °C	951
Evaporation time[a]		
Evaporation time[b]		0.04
Vapour pressure	kPa @ 20 °C	<1
Flash point	°C	79
Auto ignition temperature	°C	270
Upper flammability limit	%v/v in air	10.4
Lower flammability limit	%v/v in air	1.4
Viscosity	mPa.s (cP) @ 20 °C	
Refractive index	n 20 D	
Solubility in water	%w/w @ 20 °C	Complete
Solubility of water	%w/w @ 20 °C	
Solubility parameter[c]	(J/cm^3)$^{1/2}$	
Hydrogen bond index		
Fractional polarity		
Nitro cellulose dil ratio[d]		
Nitro cellulose dil ratio[e]		
Surface tension	mN/m @ 20 °C	
Specific heat liquid	kJ/kg mole/°C @ 20 °C	
Latent heat	kJ/kg mole	
Dielectric constant	@ 20 °C	
Antoine constant A	kPa°C	
Antoine constant B	kPa°C	
Antoine constant C	kPa°C	
Heat of combustion	kJ/kg mole	
UN number		
IMO classification		none
ADR/RID classification		3,32°(c)
UK exposure limits	mg/m^3 (ppm) 8hrTWA/MEL or OES	
USA exposure limits	mg/m^3 (ppm) TLV/TWA	
German exposure limits	mg/m^3 (ppm) MAK-TWA	
EU classification		
EU risk phrases		
EU safety phrases		

[a] Ether = 1. [b] n-Butyl acetate = 1. [c] Nelson, Hemwall, Edwards. [d] Toluene. [e] Hydrocarbon solvent.

Solvent group		**ALCOHOLS**
Chemical name		Methanol
Chemical formula		CH_4O
Structural formula		CH_3OH
CAS No.		67-56-1
Synonym/common name		Methyl alcohol/methanol
Trade names		
Suppliers		BP, ELF ATOCHEM, SHELL
Molecular mass		32.04
Boiling point/range	°C @ 760 mm Hg	64.7
Melting point	°C	−97.5 to −98
Density	kg/m^3 @ 20 °C	789–792
Evaporation time[a]		6.3
Evaporation time[b]		1.9–2.1
Vapour pressure	kPa @ 20 °C	13.1
Flash point	°C	10(IP 170)SHELL, 12 (PENSKY MARTEN) BP
Auto ignition temperature	°C	455–464
Upper flammability limit	%v/v in air	36.5
Lower flammability limit	%v/v in air	6.1
Viscosity	mPa.s (cP) @ 20 °C	0.59
Refractive index	n 20 D	1.329
Solubility in water	%w/w @ 20 °C	Complete
Solubility of water	%w/w @ 20 °C	Complete
Solubility parameter[c]	(J/cm^3)$^{1/2}$	29.7
Hydrogen bond index		−19.8
Fractional polarity		0.388
Nitro cellulose dil ratio[d]		2.5
Nitro cellulose dil ratio[e]		0.3
Surface tension	mN/m @ 20 °C	22.6
Specific heat liquid	kJ/kg mole/°C @ 20 °C	96.1
Latent heat	kJ/kg mole	37745
Dielectric constant	@ 20 °C	31.2
Antoine constant A	kPa°C	7.04389
Antoine constant B	kPa°C	1496.28
Antoine constant C	kPa°C	232.431
Heat of combustion	kJ/kg mole	−676374
UN number		1230
IMO classification		3.2/II
ADR/RID classification		3,17°(b)
UK exposure limits	mg/m^3 (ppm) 8hrTWA/MEL or OES	260(200)
USA exposure limits	mg/m^3 (ppm) TLV/TWA	262(200)
German exposure limits	mg/m^3 (ppm) MAK-TWA	260(200)
EU classification		F,T
EU risk phrases		11–23/25
EU safety phrases		(1/2)7-16-24-45

[a]Ether = 1. [b]n-Butyl acetate = 1. [c]Nelson, Hemwall, Edwards. [d]Toluene. [e]Hydrocarbon solvent.

Solvent group		**ALCOHOLS**
Chemical name		Ethanol
Chemical formula		C_2H_6O
Structural formula		CH_3CH_2OH
CAS No.		64-17-5
Synonym/common name		Ethyl alcohol, ethanol
Trade names		
Suppliers		BP, HAYMAN, HULS, UNION CARBIDE
Molecular mass		46.1
Boiling point/range	°C @ 760 mm Hg	78.2–80.0
Melting point	°C	−114
Density	kg/m^3 @ 20 °C	789–811
Evaporation time[a]		8.3
Evaporation time[b]		3.0–3.77
Vapour pressure	kPa @ 20 °C	5.8–5.9
Flash point	°C	12 to 13
Auto ignition temperature	°C	365–425
Upper flammability limit	%v/v in air	15–19
Lower flammability limit	%v/v in air	3.3–3.5
Viscosity	mPa.s (cP) @ 20 °C	1.2
Refractive index	n 20 D	
Solubility in water	%w/w @ 20 °C	Complete
Solubility of water	%w/w @ 20 °C	Complete
Solubility parameter[c]	(J/cm^3)$^{1/2}$	26.0
Hydrogen bond index		−17.8
Fractional polarity		0.296
Nitro cellulose dil ratio[d]		
Nitro cellulose dil ratio[e]		
Surface tension	mN/m @ 20 °C	22.4
Specific heat liquid	kJ/kg mole/°C @ 20 °C	
Latent heat	kJ/kg mole	
Dielectric constant	@ 20 °C	25.7
Antoine constant A	kPa°C	7.16879
Antoine constant B	kPa°C	1552.601
Antoine constant C	kPa°C	222.419
Heat of combustion	kJ/kg mole	
UN number		1170
IMO classification		3.2/II
ADR/RID classification		3,3°(b)
UK exposure limits	mg/m^3 (ppm) 8hrTWA/MEL or OES	1900(1000)
USA exposure limits	mg/m^3 (ppm) TLV/TWA	1880(1000)
German exposure limits	mg/m^3 (ppm) MAK-TWA	1900(1000)
EU classification		F
EU risk phrases		11
EU safety phrases		(2-)7-16

[a]Ether = 1. [b]n-Butyl acetate = 1. [c]Nelson, Hemwall, Edwards. [d]Toluene. [e]Hydrocarbon solvent.

Solvent group		**ALCOHOLS**
Chemical name		2-Propanol
Chemical formula		C_3H_8O
Structural formula		$CH_3CHOHCH_3$
CAS No.		67-63-0
Synonym/common name		isopropanol/isopropyl alcohol/IPA
Trade names		IPA, IPAC+
Suppliers		BP, EXXON, SHELL, UNION CARBIDE, HULS
Molecular mass		60.10
Boiling point/range	°C @ 760 mm Hg	82.0–83.0
Melting point	°C	−88
Density	kg/m^3 @ 20 °C	785–786
Evaporation time[a]		11
Evaporation time[b]		1.5–2.9
Vapour pressure	kPa @ 20 °C	4.1
Flash point	°C	12 (OPEN CUP) BP, EXXON, 12 (IP170) SHELL, 13 HULS
Auto ignition temperature	°C	399–425
Upper flammability limit	%v/v in air	12.0
Lower flammability limit	%v/v in air	2.0–2.5
Viscosity	mPa.s (cP) @ 20 °C	2.2–2.4
Refractive index	n 20 D	1.377
Solubility in water	%w/w @ 20 °C	Complete
Solubility of water	%w/w @ 20 °C	Complete
Solubility parameter[c]	(J/cm^3)$^{1/2}$	23.5
Hydrogen bond index		−16.7
Fractional polarity		0.178
Nitro cellulose dil ratio[d]		
Nitro cellulose dil ratio[e]		
Surface tension	mN/m @ 20 °C	22.7
Specific heat liquid	kJ/kg mole/°C @ 20 °C	180.3
Latent heat	kJ/kg mole	39966
Dielectric constant	@ 20 °C	18.6
Antoine constant A	kPa°C	6.86618
Antoine constant B	kPa°C	1360.131
Antoine constant C	kPa°C	197.592
Heat of combustion	kJ/kg mole	−1876742
UN number		1219
IMO classification		3.2/II
ADR/RID classification		3,3°(b)
UK exposure limits	mg/m^3 (ppm) 8hrTWA/MEL or OES	980(400)
USA exposure limits	mg/m^3 (ppm) TLV/TWA	983(400)
German exposure limits	mg/m^3 (ppm) MAK-TWA	980(400)
EU classification		F
EU risk phrases		11
EU safety phrases		(2-)7-16

[a] Ether = 1. [b] n-Butyl acetate = 1. [c] Nelson, Hemwall, Edwards. [d] Toluene. [e] Hydrocarbon solvent.

Solvent group		**ALCOHOLS**
Chemical name		1-Propanol
Chemical formula		C_3H_8O
Structural formula		$CH_3CH_2CH_2OH$
CAS No.		71-23-8
Synonym/common name		n-propyl alcohol/n-propanol
Trade names		
Suppliers		UNION CARBIDE, BASF
Molecular mass		60.10
Boiling point/range	°C @ 760 mm Hg	96.5–97.5
Melting point	°C	−126.2
Density	kg/m^3 @ 20 °C	805
Evaporation time[a]		16
Evaporation time[b]		
Vapour pressure	kPa @ 20 °C	
Flash point	°C	23 BASF
Auto ignition temperature	°C	360
Upper flammability limit	%v/v in air	13.5
Lower flammability limit	%v/v in air	2.1
Viscosity	mPa.s (cP) @ 20 °C	2.3
Refractive index	n 20 D	1.385
Solubility in water	%w/w @ 20 °C	Complete
Solubility of water	%w/w @ 20 °C	Complete
Solubility parameter[c]	(J/cm^3)$^{1/2}$	24.91
Hydrogen bond index		
Fractional polarity		
Nitro cellulose dil ratio[d]		
Nitro cellulose dil ratio[e]		
Surface tension	mN/m @ 20 °C	23.45
Specific heat liquid	kJ/kg mole/°C @ 20 °C	
Latent heat	kJ/kg mole	
Dielectric constant	@ 20 °C	
Antoine constant A	kPa°C	6.87613
Antoine constant B	kPa°C	1441.705
Antoine constant C	kPa°C	198.859
Heat of combustion	kJ/kg mole	
UN number		1274
IMO classification		3.2/II
ADR/RID classification		3,31°(c)
UK exposure limits	mg/m^3 (ppm) 8hrTWA/MEL or OES	500(200)
USA exposure limits	mg/m^3 (ppm) TLV/TWA	492(200)
German exposure limits	mg/m^3 (ppm) MAK-TWA	
EU classification		F
EU risk phrases		11
EU safety phrases		(2-)7-16

[a]Ether = 1. [b]n-Butyl acetate = 1. [c]Nelson, Hemwall, Edwards. [d]Toluene. [e]Hydrocarbon solvent.

Solvent group		**ALCOHOLS**
Chemical name		2-Methyl-2-propanol
Chemical formula		$C_4H_{10}O$
Structural formula		$(CH_3)_3COH$
CAS No.		75-65-0
Synonym/common name		tert-butyl alcohol
Trade names		
Suppliers		HULS
Molecular mass		74.1
Boiling point/range	°C @ 760 mm Hg	82–83
Melting point	°C	24.3
Density	kg/m^3 @ 20 °C	774
Evaporation time[a]		11
Evaporation time[b]		
Vapour pressure	kPa @ 20 °C	4.1
Flash point	°C	14
Auto ignition temperature	°C	490
Upper flammability limit	%v/v in air	8
Lower flammability limit	%v/v in air	2.3
Viscosity	mPa.s (cP) @ 20 °C	3.3
Refractive index	n 20 D	1.3877
Solubility in water	%w/w @ 20 °C	
Solubility of water	%w/w @ 20 °C	Complete
Solubility parameter[c]	(J/cm^3)$^{1/2}$	21.7
Hydrogen bond index		
Fractional polarity		
Nitro cellulose dil ratio[d]		
Nitro cellulose dil ratio[e]		
Surface tension	mN/m @ 20 °C	
Specific heat liquid	kJ/kg mole/°C @ 20 °C	
Latent heat	kJ/kg mole	39040
Dielectric constant	@ 20 °C	
Antoine constant A	kPa°C	6.35648
Antoine constant B	kPa°C	1107.060
Antoine constant C	kPa°C	172.102
Heat of combustion	kJ/kg mole	−2643950
UN number		1120
IMO classification		3.2/II
ADR/RID classification		3,3°(b)
UK exposure limits	mg/m^3 (ppm) 8hrTWA/MEL or OES	300(100)
USA exposure limits	mg/m^3 (ppm) TLV/TWA	303(100)
German exposure limits	mg/m^3 (ppm) MAK-TWA	300(100)
EU classification		F, Xn
EU risk phrases		11–20
EU safety phrases		(2-)9-16

[a] Ether = 1. [b] n-Butyl acetate = 1. [c] Nelson, Hemwall, Edwards. [d] Toluene. [e] Hydrocarbon solvent.

Solvent group		**ALCOHOLS**
Chemical name		1-Butanol
Chemical formula		$C_4H_{10}O$
Structural formula		$CH_3CH_2CH_2CH_2OH$
CAS No.		71-36-3
Synonym/common name		n-butanol/n-butyl alcohol
Trade names		NBA
Suppliers		BP, ELF ATOCHEM, HULS, SHELL, UNION CARBIDE, BASF
Molecular mass		74.12
Boiling point/range	°C @ 760 mm Hg	116.0–118.
Melting point	°C	−89, −90
Density	kg/m^3 @ 20 °C	803–811
Evaporation time[a]		33
Evaporation time[b]		0.44–0.5
Vapour pressure	kPa @ 20 °C	0.57–0.59
Flash point	°C	34 (DIN 51755 or DIN 51758) BASF, 34 (IP 170) SHELL, 37 BP
Auto ignition temperature	°C	340–372
Upper flammability limit	%v/v in air	11.2–11.3
Lower flammability limit	%v/v in air	1.4–1.5
Viscosity	mPa.s (cP) @ 20 °C	2.9
Refractive index	n 20 D	1.399
Solubility in water	%w/w @ 20 °C	7.5–8.5
Solubility of water	%w/w @ 20 °C	15.0–20.0
Solubility parameter[c]	$(J/cm^3)^{1/2}$	23.3
Hydrogen bond index		−18.0
Fractional polarity		0.096
Nitro cellulose dil ratio[d]		
Nitro cellulose dil ratio[e]		
Surface tension	mN/m @ 20 °C	24.4
Specific heat liquid	kJ/kg mole/°C @ 20 °C	193.5
Latent heat	kJ/kg mole	43730
Dielectric constant	@ 20 °C	17.8
Antoine constant A	kPa°C	6.54743
Antoine constant B	kPa°C	1338.769
Antoine constant C	kPa°C	177.042
Heat of combustion	kJ/kg mole	−2496658
UN number		1120
IMO classification		3.3/III
ADR/RID classification		3,31°(c)
UK exposure limits	mg/m^3 (ppm) 8hrTWA/MEL or OES	150(50)
USA exposure limits	mg/m^3 (ppm) TLV/TWA	152(50)
German exposure limits	mg/m^3 (ppm) MAK-TWA	300(100)
EU classification		Xn, R10
EU risk phrases		10–20
EU safety phrases		(2-)16

[a] Ether = 1. [b] n-Butyl acetate = 1. [c] Nelson, Hemwall, Edwards. [d] Toluene. [e] Hydrocarbon solvent.

Solvent group		**ALCOHOLS**
Chemical name		2-Butanol
Chemical formula		$C_4H_{10}O$
Structural formula		$CH_3CH_2CHOHCH_3$
CAS No.		78-92-2
Synonym/common name		see-butyl alcohol
Trade names		SBA, EXXON SBA
Suppliers		ELF ATOCHEM, EXXON, SHELL
Molecular mass		74.12
Boiling point/range	°C @ 760 mm Hg	99.4–101.5
Melting point	°C	−115
Density	kg/m^3 @ 20 °C	806–808
Evaporation time[a]		15
Evaporation time[b]		0.9–1.2
Vapour pressure	kPa @ 20 °C	1.53–1.6
Flash point	°C	23.9 (CLOSED CUP) ELF, 24 (IP 170) SHELL, 25 EXXON
Auto ignition temperature	°C	390–406
Upper flammability limit	%v/v in air	9.0–9.8
Lower flammability limit	%v/v in air	1.7
Viscosity	mPa.s (cP) @ 20 °C	3.9–4.2
Refractive index	n 20 D	1.397
Solubility in water	%w/w @ 20 °C	12.5
Solubility of water	%w/w @ 20 °C	37.5–45
Solubility parameter[c]	(J/cm^3)$^{1/2}$	22.1
Hydrogen bond index		−17.5
Fractional polarity		0.123
Nitro cellulose dil ratio[d]		
Nitro cellulose dil ratio[e]		
Surface tension	mN/m @ 20 °C	23.3
Specific heat liquid	kJ/kg mole/°C @ 20 °C	199.4
Latent heat	kJ/kg mole	41507
Dielectric constant	@ 20 °C	15.8
Antoine constant A	kPa°C	6.35457
Antoine constant B	kPa°C	1171.891
Antoine constant C	kPa°C	169.955
Heat of combustion	kJ/kg mole	−2492359
UN number		1120
IMO classification		3.3/III
ADR/RID classification		3,31°(c)
UK exposure limits	mg/m^3 (ppm) 8hrTWA/MEL or OES	300(100)
USA exposure limits	mg/m^3 (ppm) TLV/TWA	303(100)
German exposure limits	mg/m^3 (ppm) MAK-TWA	300(100)
EU classification		Xn, R10
EU risk phrases		10–20
EU safety phrases		(2-)16

[a] Ether = 1. [b] n-Butyl acetate = 1. [c] Nelson, Hemwall, Edwards. [d] Toluene. [e] Hydrocarbon solvent.

Property	Unit	Value
Solvent group		**ALCOHOLS**
Chemical name		2-Methyl-1-propanol
Chemical formula		$C_4H_{10}O$
Structural formula		$(CH_3)_2CHCH_2OH$
CAS No.		78-83-1
Synonym/common name		isobutanol/isobutyl alcohol
Trade names		
Suppliers		BP, ELF ATOCHEM, SHELL, BASF
Molecular mass		74.12
Boiling point/range	°C @ 760 mm Hg	106.0–109.
Melting point	°C	−103, −108
Density	kg/m^3 @ 20 °C	800–903
Evaporation time[a]		25
Evaporation time[b]		0.8
Vapour pressure	kPa @ 20 °C	1.07–1.2
Flash point	°C	27 (DIN 51755 or 51758) BASF, 28 BP/HULS, 28 (IP 170) SHELL
Auto ignition temperature	°C	390–427
Upper flammability limit	%v/v in air	10.9–12
Lower flammability limit	%v/v in air	1.2–1.7
Viscosity	mPa.s (cP) @ 20 °C	3.83–4.0
Refractive index	n 20 D	1.396
Solubility in water	%w/w @ 20 °C	8.5
Solubility of water	%w/w @ 20 °C	15–16.2
Solubility parameter[c]	(J/cm^3)$^{1/2}$	21.9
Hydrogen bond index		−17.9
Fractional polarity		0.111
Nitro cellulose dil ratio[d]		
Nitro cellulose dil ratio[e]		
Surface tension	mN/m @ 20 °C	23.0
Specific heat liquid	kJ/kg mole/°C @ 20 °C	170.5
Latent heat	kJ/kg mole	42871
Dielectric constant	@ 20 °C	18.8
Antoine constant A	kPa°C	7.22266
Antoine constant B	kPa°C	1669.95
Antoine constant C	kPa°C	212.219
Heat of combustion	kJ/kg mole	−2501401
UN number		1212
IMO classification		3.3/II
ADR/RID classification		3,31°(c)
UK exposure limits	mg/m^3 (ppm) 8hrTWA/MEL or OES	150(50)
USA exposure limits	mg/m^3 (ppm) TLV/TWA	152(50)
German exposure limits	mg/m^3 (ppm) MAK-TWA	300(100)
EU classification		Xn, R10
EU risk phrases		10–20
EU safety phrases		(2-)16

[a] Ether = 1. [b] n-Butyl acetate = 1. [c] Nelson, Hemwall, Edwards. [d] Toluene. [e] Hydrocarbon solvent.

Solvent group		**ALCOHOLS**
Chemical name		1-Pentanol
Chemical formula		$C_5H_{12}O$
Structural formula		$CH_3(CH_2)_3CH_2OH$
CAS No.		71-41-0
Synonym/common name		amyl alcohol/pentyl alcohol
Trade names		
Suppliers		UNION CARBIDE, BASF
Molecular mass		88.15
Boiling point/range	°C @ 760 mm Hg	137.0–139.
Melting point	°C	−78.2
Density	kg/m^3 @ 20 °C	816
Evaporation time[a]		96
Evaporation time[b]		0.18
Vapour pressure	kPa @ 20 °C	
Flash point	°C	49 (DIN 51755 or 51758) BASF
Auto ignition temperature	°C	300
Upper flammability limit	%v/v in air	10.5
Lower flammability limit	%v/v in air	1.3
Viscosity	mPa.s (cP) @ 20 °C	3.3–4.0
Refractive index	n 20 D	1.410
Solubility in water	%w/w @ 20 °C	2.6
Solubility of water	%w/w @ 20 °C	9.5
Solubility parameter[c]	(J/cm^3)$^{1/2}$	22.3
Hydrogen bond index		
Fractional polarity		
Nitro cellulose dil ratio[d]		
Nitro cellulose dil ratio[e]		
Surface tension	mN/m @ 20 °C	
Specific heat liquid	kJ/kg mole/°C @ 20 °C	
Latent heat	kJ/kg mole	44367
Dielectric constant	@ 20 °C	
Antoine constant A	kPa°C	6.30306
Antoine constant B	kPa°C	1286.333
Antoine constant C	kPa°C	161.307
Heat of combustion	kJ/kg mole	
UN number		1105
IMO classification		3.2/II
ADR/RID classification		3,31°(c)
UK exposure limits	mg/m^3 (ppm) 8hrTWA/MEL or OES	
USA exposure limits	mg/m^3 (ppm) TLV/TWA	
German exposure limits	mg/m^3 (ppm) MAK-TWA	
EU classification		Xn, R10
EU risk phrases		10–20
EU safety phrases		(2-)24/25

[a] Ether = 1. [b] n-Butyl acetate = 1. [c] Nelson, Hemwall, Edwards. [d] Toluene. [e] Hydrocarbon solvent.

Solvent group		**ALCOHOLS**
Chemical name		2-Methyl-1-butanol
Chemical formula		$C_5H_{12}O$
Structural formula		$CH_3CH_2CH(CH_3)CH_2OH$
CAS No.		137-32-6
Synonym/common name		2-Methyl-1-butanol
Trade names		
Suppliers		UNION CARBIDE
Molecular mass		88.15
Boiling point/range	°C @ 760 mm Hg	128.7
Melting point	°C	
Density	kg/m^3 @ 20 °C	816
Evaporation time[a]		
Evaporation time[b]		0.24
Vapour pressure	kPa @ 20 °C	
Flash point	°C	
Auto ignition temperature	°C	
Upper flammability limit	%v/v in air	
Lower flammability limit	%v/v in air	
Viscosity	mPa.s (cP) @ 20 °C	5.0
Refractive index	n 20 D	1.4107
Solubility in water	%w/w @ 20 °C	2.2
Solubility of water	%w/w @ 20 °C	8.3
Solubility parameter[c]	(J/cm^3)$^{1/2}$	
Hydrogen bond index		
Fractional polarity		
Nitro cellulose dil ratio[d]		
Nitro cellulose dil ratio[e]		
Surface tension	mN/m @ 20 °C	
Specific heat liquid	kJ/kg mole/°C @ 20 °C	
Latent heat	kJ/kg mole	
Dielectric constant	@ 20 °C	
Antoine constant A	kPa°C	
Antoine constant B	kPa°C	
Antoine constant C	kPa°C	
Heat of combustion	kJ/kg mole	−3325950
UN number		1105
IMO classification		3.2/II
ADR/RID classification		3,31°(c)
UK exposure limits	mg/m^3 (ppm) 8hrTWA/MEL or OES	
USA exposure limits	mg/m^3 (ppm) TLV/TWA	
German exposure limits	mg/m^3 (ppm) MAK-TWA	
EU classification		
EU risk phrases		
EU safety phrases		

[a] Ether = 1. [b] n-Butyl acetate = 1. [c] Nelson, Hemwall, Edwards. [d] Toluene. [e] Hydrocarbon solvent.

Property	Units	Value
Solvent group		**ALCOHOLS**
Chemical name		Cyclohexanol
Chemical formula		$C_6H_{12}O$
Structural formula	(cyclohexane ring) CHOH	
CAS No.		108-93-0
Synonym/common name		Cyclohexanol, cyclohexyl alcohol
Trade names		
Suppliers		ENICHEM
Molecular mass		100.16
Boiling point/range	°C @ 760 mm Hg	161.1
Melting point	°C	25.15
Density	kg/m^3 @ 20 °C	
Evaporation time[a]		
Evaporation time[b]		
Vapour pressure	kPa @ 20 °C	1.33
Flash point	°C	58 (OPEN CUP) 63 (CLOSED CUP) ENICHEM
Auto ignition temperature	°C	300
Upper flammability limit	%v/v in air	
Lower flammability limit	%v/v in air	2.4
Viscosity	mPa.s (cP) @ 20 °C	68
Refractive index	n 20 D	1.4648
Solubility in water	%w/w @ 20 °C	3.6
Solubility of water	%w/w @ 20 °C	
Solubility parameter[c]	$(J/cm^3)^{1/2}$	
Hydrogen bond index		
Fractional polarity		
Nitro cellulose dil ratio[d]		
Nitro cellulose dil ratio[e]		
Surface tension	mN/m @ 20 °C	
Specific heat liquid	kJ/kg mole/°C @ 20 °C	
Latent heat	kJ/kg mole	
Dielectric constant	@ 20 °C	
Antoine constant A	kPa°C	
Antoine constant B	kPa°C	
Antoine constant C	kPa°C	
Heat of combustion	kJ/kg mole	
UN number		1987
IMO classification		
ADR/RID classification		
UK exposure limits	mg/m^3 (ppm) 8hrTWA/MEL or OES	
USA exposure limits	mg/m^3 (ppm) TLV/TWA	
German exposure limits	mg/m^3 (ppm) MAK-TWA	
EU classification		
EU risk phrases		
EU safety phrases		

[a] Ether = 1. [b] n-Butyl acetate = 1. [c] Nelson, Hemwall, Edwards. [d] Toluene. [e] Hydrocarbon solvent.

Solvent group		**ALCOHOLS**
Chemical name		4-Methyl-2-pentanol
Chemical formula		$C_6H_{14}O$
Structural formula		$(CH_3)_2CHCH_2CHOHCH_3$
CAS No.		108-11-2
Synonym/common name		Methylisobutylcarbinol/ methylamylalcohol
Trade names		
Suppliers		ELF ATOCHEM, SHELL, UNION CARBIDE
Molecular mass	°C @ 760 mm Hg	102.18
Boiling point/range	°C	130.0–133.
Melting point	kg/m^3 @ 20 °C	−90
Density		808
Evaporation time[a]		66
Evaporation time[b]	kPa @ 20 °C	0.27–0.43
Vapour pressure	°C	0.42–0.49
Flash point	°C	41 (CLOSED CUP & IP170)
Auto ignition temperature	%v/v in air	305
Upper flammability limit	%v/v in air	5.5
Lower flammability limit	mPa.s (cP) @ 20 °C	1.0
Viscosity	n 20 D	5.1–5.2
Refractive index	%w/w @ 20 °C	1.411
Solubility in water	%w/w @ 20 °C	1.64–1.7
Solubility of water	$(J/cm^3)^{1/2}$	5.8–6.35
Solubility parameter[c]		20.5
Hydrogen bond index		−18.7
Fractional polarity		0.066
Nitro cellulose dil ratio[d]		
Nitro cellulose dil ratio[e]	mN/m @ 20 °C	
Surface tension	kJ/kg mole/°C @ 20 °C	22.7–23.0
Specific heat liquid	kJ/kg mole	236.0
Latent heat	@ 20 °C	43937
Dielectric constant	kPa°C	10.4
Antoine constant A	kPa°C	6.63977
Antoine constant B	kPa°C	1521.88
Antoine constant C	kJ/kg mole	196.767
Heat of combustion		−3720067
UN number		2053
IMO classification		3.3/III
ADR/RID classification	mg/m^3 (ppm) 8hrTWA/MEL or OES	3,31°(c)
UK exposure limits	mg/m^3 (ppm) TLV/TWA	100(25)
USA exposure limits	mg/m^3 (ppm) MAK-TWA	104(25)
German exposure limits		100(25)
EU classification		Xi, R10
EU risk phrases		10–37
EU safety phrases		(2-)24/25

[a] Ether = 1. [b] n-Butyl acetate = 1. [c] Nelson, Hemwall, Edwards. [d] Toluene. [e] Hydrocarbon solvent.

Solvent group		**ALCOHOLS**
Chemical name		Benzyl alcohol
Chemical formula		C_7H_8O
Structural formula		$C_6H_5CH_2OH$
CAS No.		100-51-6
Synonym/common name		Phenylmethyl alcohol
Trade names		
Suppliers		ELF ATOCHEM
Molecular mass		108.14
Boiling point/range	°C @ 760 mm Hg	205.4
Melting point	°C	−15
Density	kg/m^3 @ 20 °C	1050
Evaporation time[a]		
Evaporation time[b]		
Vapour pressure	kPa @ 20 °C	
Flash point	°C	
Auto ignition temperature	°C	436
Upper flammability limit	%v/v in air	13
Lower flammability limit	%v/v in air	1.3
Viscosity	mPa.s (cP) @ 20 °C	
Refractive index	n 20 D	
Solubility in water	%w/w @ 20 °C	
Solubility of water	%w/w @ 20 °C	
Solubility parameter[c]	(J/cm^3)$^{1/2}$	24.8
Hydrogen bond index		
Fractional polarity		
Nitro cellulose dil ratio[d]		
Nitro cellulose dil ratio[e]		
Surface tension	mN/m @ 20 °C	
Specific heat liquid	kJ/kg mole/°C @ 20 °C	
Latent heat	kJ/kg mole	
Dielectric constant	@ 20 °C	
Antoine constant A	kPa°C	8.963
Antoine constant B	kPa°C	3214
Antoine constant C	kPa°C	
Heat of combustion	kJ/kg mole	
UN number		
IMO classification		none
ADR/RID classification		none
UK exposure limits	mg/m^3 (ppm) 8hrTWA/MEL or OES	
USA exposure limits	mg/m^3 (ppm) TLV/TWA	
German exposure limits	mg/m^3 (ppm) MAK-TWA	
EU classification		Xn
EU risk phrases		20/22
EU safety phrases		(2-)26

[a] Ether = 1. [b] n-Butyl acetate = 1. [c] Nelson, Hemwall, Edwards. [d] Toluene. [e] Hydrocarbon solvent.

Solvent group		**ALCOHOLS**
Chemical name		n-Heptanol
Chemical formula		$C_7H_{16}O$
Structural formula		$CH_3(CH_2)_6OH$
CAS No.		111-70-6
Synonym/common name		n-Heptanol
Trade names		
Suppliers		ELF ATOCHEM
Molecular mass		116.2
Boiling point/range	°C @ 760 mm Hg	175
Melting point	°C	−35
Density	kg/m^3 @ 20 °C	820
Evaporation time[a]		
Evaporation time[b]		
Vapour pressure	kPa @ 20 °C	
Flash point	°C	
Auto ignition temperature	°C	350
Upper flammability limit	%v/v in air	
Lower flammability limit	%v/v in air	
Viscosity	mPa.s (cP) @ 20 °C	
Refractive index	n 20 D	
Solubility in water	%w/w @ 20 °C	
Solubility of water	%w/w @ 20 °C	
Solubility parameter[c]	(J/cm^3)$^{1/2}$	
Hydrogen bond index		
Fractional polarity		
Nitro cellulose dil ratio[d]		
Nitro cellulose dil ratio[e]		
Surface tension	mN/m @ 20 °C	
Specific heat liquid	kJ/kg mole/°C @ 20 °C	
Latent heat	kJ/kg mole	
Dielectric constant	@ 20 °C	
Antoine constant A	kPa°C	
Antoine constant B	kPa°C	
Antoine constant C	kPa°C	
Heat of combustion	kJ/kg mole	
UN number		
IMO classification		none
ADR/RID classification		3,32°(c)
UK exposure limits	mg/m^3 (ppm) 8hrTWA/MEL or OES	
USA exposure limits	mg/m^3 (ppm) TLV/TWA	
German exposure limits	mg/m^3 (ppm) MAK-TWA	
EU classification		
EU risk phrases		
EU safety phrases		

[a] Ether = 1. [b] n-Butyl acetate = 1. [c] Nelson, Hemwall, Edwards. [d] Toluene. [e] Hydrocarbon solvent.

Solvent group		**ALCOHOLS**
Chemical name		Methylbenzyl alcohol
Chemical formula		$C_8H_{10}O$
Structural formula		C_6H_5—$CHOHCH_3$
CAS No.		
Synonym/common name		Methyl benzyl alcohol
Trade names		
Suppliers		HULS
Molecular mass		122.2
Boiling point/range	°C @ 760 mm Hg	200–205
Melting point	°C	21.5
Density	kg/m^3 @ 20 °C	1008
Evaporation time[a]		920
Evaporation time[b]		
Vapour pressure	kPa @ 20 °C	0.03
Flash point	°C	90
Auto ignition temperature	°C	480
Upper flammability limit	%v/v in air	4.9
Lower flammability limit	%v/v in air	1.8
Viscosity	mPa.s (cP) @ 20 °C	11.2
Refractive index	n 20 D	1.525
Solubility in water	%w/w @ 20 °C	
Solubility of water	%w/w @ 20 °C	5.8
Solubility parameter[c]	(J/cm^3)$^{1/2}$	
Hydrogen bond index		
Fractional polarity		
Nitro cellulose dil ratio[d]		
Nitro cellulose dil ratio[e]		
Surface tension	mN/m @ 20 °C	
Specific heat liquid	kJ/kg mole/°C @ 20 °C	
Latent heat	kJ/kg mole	
Dielectric constant	@ 20 °C	
Antoine constant A	kPa°C	
Antoine constant B	kPa°C	
Antoine constant C	kPa°C	
Heat of combustion	kJ/kg mole	
UN number		2937
IMO classification		6.1/III
ADR/RID classification		6.1,14°(c)
UK exposure limits	mg/m^3 (ppm) 8hrTWA/MEL or OES	
USA exposure limits	mg/m^3 (ppm) TLV/TWA	
German exposure limits	mg/m^3 (ppm) MAK-TWA	
EU classification		
EU risk phrases		
EU safety phrases		

[a] Ether = 1. [b] n-Butyl acetate = 1. [c] Nelson, Hemwall, Edwards. [d] Toluene. [e] Hydrocarbon solvent.

Solvent group		**ALCOHOLS**
Chemical name		2-Ethylhexanol
Chemical formula		$C_8H_{18}O$
Structural formula		$CH_3(CH_2)_3CH(C_2H_5)CH_2OH$
CAS No.		104-76-7
Synonym/common name		2-Ethylhexyl alcohol
Trade names		
Suppliers		BP, ELF ATOCHEM, HULS, SHELL, UNION CARBIDE, BASF
Molecular mass		130.23
Boiling point/range	°C @ 760 mm Hg	183.0–185.0
Melting point	°C	−76
Density	kg/m^3 @ 20 °C	830–834
Evaporation time[a]		690
Evaporation time[b]		0.02
Vapour pressure	kPa @ 20 °C	0.05–0.09
Flash point	°C	76 (DIN 51755 or 51758) BASF, 77 BP, 82 HULS
Auto ignition temperature	°C	231–330
Upper flammability limit	%v/v in air	7.4–9.7
Lower flammability limit	%v/v in air	0.88–1.1
Viscosity	mPa.s (cP) @ 20 °C	8.8–11.9
Refractive index	n 20 D	1.432
Solubility in water	%w/w @ 20 °C	0.07
Solubility of water	%w/w @ 20 °C	2.6
Solubility parameter[c]	$(J/cm^3)^{1/2}$	19.4
Hydrogen bond index		
Fractional polarity		
Nitro cellulose dil ratio[d]		
Nitro cellulose dil ratio[e]		
Surface tension	mN/m @ 20 °C	
Specific heat liquid	kJ/kg mole/°C @ 20 °C	
Latent heat	kJ/kg mole	54200
Dielectric constant	@ 20 °C	
Antoine constant A	kPa°C	5.79628
Antoine constant B	kPa°C	1204.50
Antoine constant C	kPa°C	133.14
Heat of combustion	kJ/kg mole	−5287780
UN number		2282
IMO classification		3.3/III
ADR/RID classification		3,31°(c)
UK exposure limits	mg/m^3 (ppm) 8hrTWA/MEL or OES	
USA exposure limits	mg/m^3 (ppm) TLV/TWA	
German exposure limits	mg/m^3 (ppm) MAK-TWA	
EU classification		Xn
EU risk phrases		22
EU safety phrases		(2-)24/25

[a] Ether = 1. [b] n-Butyl acetate = 1. [c] Nelson, Hemwall, Edwards. [d] Toluene. [e] Hydrocarbon solvent.

Solvent group		ALCOHOLS
Chemical name		Trimethylcyclohexanol
Chemical formula		$C_9H_{18}O$
Structural formula		$(CH_3)_3C_6H_8$–OH
CAS No.		
Synonym/common name		Trimethylcyclohexanol
Trade names		
Suppliers		HULS
Molecular mass		142.2
Boiling point/range	°C @ 760 mm Hg	189–196
Melting point	°C	cis 37.3, trans 57.3
Density	kg/m^3 @ 20 °C	861 @ 60°C
Evaporation time[a]		109
Evaporation time[b]		
Vapour pressure	kPa @ 20 °C	
Flash point	°C	76
Auto ignition temperature	°C	375
Upper flammability limit	%v/v in air	
Lower flammability limit	%v/v in air	
Viscosity	mPa.s (cP) @ 20 °C	5.8(60°C)
Refractive index	n 20 D	1.439(60°C)
Solubility in water	%w/w @ 20 °C	
Solubility of water	%w/w @ 20 °C	
Solubility parameter[c]	(J/cm^3)$^{1/2}$	
Hydrogen bond index		
Fractional polarity		
Nitro cellulose dil ratio[d]		
Nitro cellulose dil ratio[e]		
Surface tension	mN/m @ 20 °C	
Specific heat liquid	kJ/kg mole/°C @ 20 °C	
Latent heat	kJ/kg mole	
Dielectric constant	@ 20 °C	
Antoine constant A	kPa°C	
Antoine constant B	kPa°C	
Antoine constant C	kPa°C	
Heat of combustion	kJ/kg mole	
UN number		
IMO classification		
ADR/RID classification		3,32°(c)
UK exposure limits	mg/m^3 (ppm) 8hrTWA/MEL or OES	
USA exposure limits	mg/m^3 (ppm) TLV/TWA	
German exposure limits	mg/m^3 (ppm) MAK-TWA	
EU classification		
EU risk phrases		
EU safety phrases		

[a] Ether = 1. [b] n-Butyl acetate = 1. [c] Nelson, Hemwall, Edwards. [d] Toluene. [e] Hydrocarbon solvent.

Solvent group		ALCOHOLS
Chemical name		2,6-Dimethyl-4-heptanol
Chemical formula		$C_9H_{20}O$
Structural formula		$(CH_3)_2CHCH_2CH(OH)CH_2CH(CH_3)_2$
CAS No.		108-82-7
Synonym/common name		Diisobutyl carbinol
Trade names		DIBC
Suppliers		SHELL
Molecular mass		144.26
Boiling point/range	°C @ 760 mm Hg	175–181
Melting point	°C	−65
Density	kg/m^3 @ 20 °C	810–814
Evaporation time[a]		300
Evaporation time[b]		0.04
Vapour pressure	kPa @ 20 °C	0.03
Flash point	°C	74(ASTM D93)
Auto ignition temperature	°C	290
Upper flammability limit	%v/v in air	6.1
Lower flammability limit	%v/v in air	0.8
Viscosity	mPa.s (cP) @ 20 °C	11.1
Refractive index	n 20 D	1.424
Solubility in water	%w/w @ 20 °C	0.05
Solubility of water	%w/w @ 20 °C	1.0
Solubility parameter[c]	$(J/cm^3)^{1/2}$	19.2
Hydrogen bond index		−17
Fractional polarity		0.040
Nitro cellulose dil ratio[d]		
Nitro cellulose dil ratio[e]		
Surface tension	mN/m @ 20 °C	30.4
Specific heat liquid	kJ/kg mole/°C @ 20 °C	295.7
Latent heat	kJ/kg mole	80785
Dielectric constant	@ 20 °C	
Antoine constant A	kPa°C	
Antoine constant B	kPa°C	
Antoine constant C	kPa°C	
Heat of combustion	kJ/kg mole	
UN number		
IMO classification		
ADR/RID classification		3,32°(c)
UK exposure limits	mg/m^3 (ppm) 8hrTWA/MEL or OES	
USA exposure limits	mg/m^3 (ppm) TLV/TWA	
German exposure limits	mg/m^3 (ppm) MAK-TWA	
EU classification		
EU risk phrases		52/53
EU safety phrases		61

[a] Ether = 1. [b] n-Butyl acetate = 1. [c] Nelson, Hemwall, Edwards. [d] Toluene. [e] Hydrocarbon solvent.

Solvent group		**ALCOHOLS**
Chemical name		Diisobutyl Carbinol
Chemical formula		$C_9H_{19}OH$ (mixed isomers)
Structural formula		
CAS No.		mixture
Synonym/common name		Diisobutyl carbinol
Trade names		
Suppliers		UNION CARBIDE
Molecular mass		144.26
Boiling point/range	°C @ 760 mm Hg	178.0
Melting point	°C	
Density	kg/m^3 @ 20 °C	812
Evaporation time[a]		
Evaporation time[b]		0.02
Vapour pressure	kPa @ 20 °C	
Flash point	°C	
Auto ignition temperature	°C	
Upper flammability limit	%v/v in air	
Lower flammability limit	%v/v in air	
Viscosity	mPa.s (cP) @ 20 °C	13.9
Refractive index	n 20 D	
Solubility in water	%w/w @ 20 °C	0.06
Solubility of water	%w/w @ 20 °C	1
Solubility parameter[c]	(J/cm^3)$^{1/2}$	19.3
Hydrogen bond index		−17.0
Fractional polarity		0.040
Nitro cellulose dil ratio[d]		
Nitro cellulose dil ratio[e]		
Surface tension	mN/m @ 20 °C	
Specific heat liquid	kJ/kg mole/°C @ 20 °C	
Latent heat	kJ/kg mole	
Dielectric constant	@ 20 °C	
Antoine constant A	kPa°C	
Antoine constant B	kPa°C	
Antoine constant C	kPa°C	
Heat of combustion	kJ/kg mole	
UN number		
IMO classification		
ADR/RID classification		
UK exposure limits	mg/m^3 (ppm) 8hrTWA/MEL or OES	
USA exposure limits	mg/m^3 (ppm) TLV/TWA	
German exposure limits	mg/m^3 (ppm) MAK-TWA	
EU classification		
EU risk phrases		
EU safety phrases		

[a] Ether = 1. [b] n-Butyl acetate = 1. [c] Nelson, Hemwall, Edwards. [d] Toluene. [e] Hydrocarbon solvent.

Solvent group		**ALCOHOLS**
Chemical name		Nonanol mixed isomers
Chemical formula		$C_9H_{20}O$
Structural formula		$C_9H_{19}OH$
CAS No.		27458-94-2
Synonym/common name		Isononanol
Trade names		Nonanol N
Suppliers		BASF
Molecular mass		144.3
Boiling point/range	°C @ 760 mm Hg	198.0–209.0
Melting point	°C	
Density	kg/m^3 @ 20 °C	836
Evaporation time[a]		2500
Evaporation time[b]		
Vapour pressure	kPa @ 20 °C	
Flash point	°C	96 (DIN 51755 or 51758) BASF
Auto ignition temperature	°C	270
Upper flammability limit	%v/v in air	6.0
Lower flammability limit	%v/v in air	0.9
Viscosity	mPa.s (cP) @ 20 °C	13.6
Refractive index	n 20 D	1.437
Solubility in water	%w/w @ 20 °C	
Solubility of water	%w/w @ 20 °C	
Solubility parameter[c]	$(J/cm^3)^{1/2}$	
Hydrogen bond index		
Fractional polarity		
Nitro cellulose dil ratio[d]		
Nitro cellulose dil ratio[e]		
Surface tension	mN/m @ 20 °C	
Specific heat liquid	kJ/kg mole/°C @ 20 °C	
Latent heat	kJ/kg mole	
Dielectric constant	@ 20 °C	
Antoine constant A	kPa°C	
Antoine constant B	kPa°C	
Antoine constant C	kPa°C	
Heat of combustion	kJ/kg mole	
UN number		
IMO classification		
ADR/RID classification		
UK exposure limits	mg/m^3 (ppm) 8hrTWA/MEL or OES	
USA exposure limits	mg/m^3 (ppm) TLV/TWA	
German exposure limits	mg/m^3 (ppm) MAK-TWA	
EU classification		
EU risk phrases		
EU safety phrases		

[a] Ether = 1. [b] n-Butyl acetate = 1. [c] Nelson, Hemwall, Edwards. [d] Toluene. [e] Hydrocarbon solvent.

Solvent group		**ALCOHOLS**
Chemical name		Isodecanol mixed isomers
Chemical formula		$C_{10}H_{22}O$
Structural formula		$C_{10}H_{21}OH$
CAS No.		25339-17-7
Synonym/common name		Isodecanol
Trade names		
Suppliers		BASF
Molecular mass		158.3
Boiling point/range	°C @ 760 mm Hg	214.0–222.0
Melting point	°C	
Density	kg/m^3 @ 20 °C	838
Evaporation time[a]		4500
Evaporation time[b]		
Vapour pressure	kPa @ 20 °C	
Flash point	°C	102 (DIN 51755 or 51758) BASF
Auto ignition temperature	°C	275
Upper flammability limit	%v/v in air	7.4
Lower flammability limit	%v/v in air	1.1
Viscosity	mPa.s (cP) @ 20 °C	18
Refractive index	n 20 D	1.440
Solubility in water	%w/w @ 20 °C	
Solubility of water	%w/w @ 20 °C	
Solubility parameter[c]	(J/cm^3)$^{1/2}$	
Hydrogen bond index		
Fractional polarity		
Nitro cellulose dil ratio[d]		
Nitro cellulose dil ratio[e]		
Surface tension	mN/m @ 20 °C	
Specific heat liquid	kJ/kg mole/°C @ 20 °C	
Latent heat	kJ/kg mole	
Dielectric constant	@ 20 °C	
Antoine constant A	kPa°C	
Antoine constant B	kPa°C	
Antoine constant C	kPa°C	
Heat of combustion	kJ/kg mole	
UN number		
IMO classification		
ADR/RID classification		
UK exposure limits	mg/m^3 (ppm) 8hrTWA/MEL or OES	
USA exposure limits	mg/m^3 (ppm) TLV/TWA	
German exposure limits	mg/m^3 (ppm) MAK-TWA	
EU classification		
EU risk phrases		
EU safety phrases		

[a] Ether = 1. [b] n-Butyl acetate = 1. [c] Nelson, Hemwall, Edwards. [d] Toluene. [e] Hydrocarbon solvent.

Solvent group		**ALCOHOLS**
Chemical name		Tridecanol mixed isomers
Chemical formula		$C_{13}H_{28}O$
Structural formula		$C_{13}H_{27}OH$
CAS No.		27458-92-0
Synonym/common name		Tridecanol
Trade names		Tridecanol A
Suppliers		BASF
Molecular mass		200.4
Boiling point/range	°C @ 760 mm Hg	248.0–264.0
Melting point	°C	
Density	kg/m³ @ 20 °C	844
Evaporation time[a]		>10000
Evaporation time[b]		
Vapour pressure	kPa @ 20 °C	
Flash point	°C	123 (DIN 51755 or 51758) BASF
Auto ignition temperature	°C	265
Upper flammability limit	%v/v in air	5.5
Lower flammability limit	%v/v in air	0.8
Viscosity	mPa.s (cP) @ 20 °C	40
Refractive index	n 20 D	1.448
Solubility in water	%w/w @ 20 °C	
Solubility of water	%w/w @ 20 °C	
Solubility parameter[c]	$(J/cm^3)^{1/2}$	
Hydrogen bond index		
Fractional polarity		
Nitro cellulose dil ratio[d]		
Nitro cellulose dil ratio[e]		
Surface tension	mN/m @ 20 °C	
Specific heat liquid	kJ/kg mole/°C @ 20 °C	
Latent heat	kJ/kg mole	
Dielectric constant	@ 20 °C	
Antoine constant A	kPa°C	
Antoine constant B	kPa°C	
Antoine constant C	kPa°C	
Heat of combustion	kJ/kg mole	
UN number		
IMO classification		
ADR/RID classification		
UK exposure limits	mg/m³ (ppm) 8hrTWA/MEL or OES	
USA exposure limits	mg/m³ (ppm) TLV/TWA	
German exposure limits	mg/m³ (ppm) MAK-TWA	
EU classification		
EU risk phrases		
EU safety phrases		

[a] Ether = 1. [b] n-Butyl acetate = 1. [c] Nelson, Hemwall, Edwards. [d] Toluene. [e] Hydrocarbon solvent.

Solvent group		**ALDEHYDES**
Chemical name		Propionaldehyde
Chemical formula		C_3H_6O
Structural formula		CH_3CH_2CHO
CAS No.		123-38-6
Synonym/common name		Propionaldehyde
Trade names		
Suppliers		UNION CARBIDE
Molecular mass		58.08
Boiling point/range	°C @ 760 mm Hg	48.0
Melting point	°C	−80
Density	kg/m^3 @ 20 °C	798
Evaporation time[a]		
Evaporation time[b]		
Vapour pressure	kPa @ 20 °C	
Flash point	°C	
Auto ignition temperature	°C	
Upper flammability limit	%v/v in air	
Lower flammability limit	%v/v in air	
Viscosity	mPa.s (cP) @ 20 °C	
Refractive index	n 20 D	1.3619
Solubility in water	%w/w @ 20 °C	22.0
Solubility of water	%w/w @ 20 °C	35.0
Solubility parameter[c]	(J/cm^3)$^{1/2}$	19.3
Hydrogen bond index		
Fractional polarity		
Nitro cellulose dil ratio[d]		
Nitro cellulose dil ratio[e]		
Surface tension	mN/m @ 20 °C	
Specific heat liquid	kJ/kg mole/°C @ 20 °C	
Latent heat	kJ/kg mole	28297
Dielectric constant	@ 20 °C	
Antoine constant A	kPa°C	6.2336
Antoine constant B	kPa°C	1180
Antoine constant C	kPa°C	231
Heat of combustion	kJ/kg mole	−1816000
UN number		1275
IMO classification		3.1/II
ADR/RID classification		3,3°(b)
UK exposure limits	mg/m^3 (ppm) 8hrTWA/MEL or OES	
USA exposure limits	mg/m^3 (ppm) TLV/TWA	
German exposure limits	mg/m^3 (ppm) MAK-TWA	
EU classification		F, Xi
EU risk phrases		11-36/37/38
EU safety phrases		(2-)9-16-29

[a] Ether = 1. [b] n-Butyl acetate = 1. [c] Nelson, Hemwall, Edwards. [d] Toluene. [e] Hydrocarbon solvent.

Solvent group		**ALDEHYDES**
Chemical name		Butyraldehyde
Chemical formula		C_4H_8O
Structural formula		$CH_3CH_2CH_2CHO$
CAS No.		123-72-8
Synonym/common name		Butanal
Trade names		
Suppliers		UNION CARBIDE
Molecular mass		72.11
Boiling point/range	°C @ 760 mm Hg	75.7
Melting point	°C	−96.4
Density	kg/m^3 @ 20 °C	803
Evaporation time[a]		
Evaporation time[b]		
Vapour pressure	kPa @ 20 °C	
Flash point	°C	
Auto ignition temperature	°C	
Upper flammability limit	%v/v in air	
Lower flammability limit	%v/v in air	
Viscosity	mPa.s (cP) @ 20 °C	
Refractive index	n 20 D	
Solubility in water	%w/w @ 20 °C	<0.2
Solubility of water	%w/w @ 20 °C	
Solubility parameter[c]	(J/cm^3)$^{1/2}$	18.4
Hydrogen bond index		
Fractional polarity		
Nitro cellulose dil ratio[d]		
Nitro cellulose dil ratio[e]		
Surface tension	mN/m @ 20 °C	
Specific heat liquid	kJ/kg mole/°C @ 20 °C	
Latent heat	kJ/kg mole	31497
Dielectric constant	@ 20 °C	
Antoine constant A	kPa°C	6.1461
Antoine constant B	kPa°C	1233.0
Antoine constant C	kPa°C	223.0
Heat of combustion	kJ/kg mole	−2479340
UN number		1129
IMO classification		3.2/II
ADR/RID classification		3,3°(b)
UK exposure limits	mg/m^3 (ppm) 8hrTWA/MEL or OES	
USA exposure limits	mg/m^3 (ppm) TLV/TWA	
German exposure limits	mg/m^3 (ppm) MAK-TWA	
EU classification		F
EU risk phrases		11
EU safety phrases		(2-)9-29-33

[a] Ether = 1. [b] n-Butyl acetate = 1. [c] Nelson, Hemwall, Edwards. [d] Toluene. [e] Hydrocarbon solvent.

Solvent group		**ALDEHYDES**
Chemical name		Isobutyraldehyde
Chemical formula		C_4H_8O
Structural formula		$(CH_3)_2CHCHO$
CAS No.		78-84-2
Synonym/common name		Isobutyraldehyde
Trade names		
Suppliers		UNION CARBIDE
Molecular mass		72.11
Boiling point/range	°C @ 760 mm Hg	64.1
Melting point	°C	−65
Density	kg/m^3 @ 20 °C	780
Evaporation time[a]		
Evaporation time[b]		
Vapour pressure	kPa @ 20 °C	
Flash point	°C	
Auto ignition temperature	°C	
Upper flammability limit	%v/v in air	
Lower flammability limit	%v/v in air	
Viscosity	mPa.s (cP) @ 20 °C	
Refractive index	n 20 D	1.3727
Solubility in water	%w/w @ 20 °C	11.0
Solubility of water	%w/w @ 20 °C	2.9
Solubility parameter[c]	$(J/cm^3)^{1/2}$	17.9
Hydrogen bond index		
Fractional polarity		
Nitro cellulose dil ratio[d]		
Nitro cellulose dil ratio[e]		
Surface tension	mN/m @ 20 °C	
Specific heat liquid	kJ/kg mole/°C @ 20 °C	
Latent heat	kJ/kg mole	
Dielectric constant	@ 20 °C	
Antoine constant A	kPa°C	5.14104
Antoine constant B	kPa°C	1946.999
Antoine constant C	kPa°C	0.005297
Heat of combustion	kJ/kg mole	−2287430
UN number		2045
IMO classification		3,3°(b)
ADR/RID classification		
UK exposure limits	mg/m^3 (ppm) 8hrTWA/MEL or OES	
USA exposure limits	mg/m^3 (ppm) TLV/TWA	
German exposure limits	mg/m^3 (ppm) MAK-TWA	
EU classification		
EU risk phrases		
EU safety phrases		

[a] Ether = 1. [b] n-Butyl acetate = 1. [c] Nelson, Hemwall, Edwards. [d] Toluene. [e] Hydrocarbon solvent.

Solvent group		**ALDEHYDES**
Chemical name		Valeraldehyde
Chemical formula		$C_5H_{10}O$
Structural formula		$CH_3CH_2CH_2CH_2CHO$
CAS No.		110-62-3
Synonym/common name		Valeraldehyde
Trade names		
Suppliers		UNION CARBIDE
Molecular mass		86.13
Boiling point/range	°C @ 760 mm Hg	103.7
Melting point	°C	
Density	kg/m^3 @ 20 °C	810.9
Evaporation time[a]		
Evaporation time[b]		
Vapour pressure	kPa @ 20 °C	
Flash point	°C	
Auto ignition temperature	°C	
Upper flammability limit	%v/v in air	
Lower flammability limit	%v/v in air	
Viscosity	mPa.s (cP) @ 20 °C	
Refractive index	n 20 D	
Solubility in water	%w/w @ 20 °C	1.4
Solubility of water	%w/w @ 20 °C	1.4
Solubility parameter[c]	$(J/cm^3)^{1/2}$	
Hydrogen bond index		
Fractional polarity		
Nitro cellulose dil ratio[d]		
Nitro cellulose dil ratio[e]		
Surface tension	mN/m @ 20 °C	
Specific heat liquid	kJ/kg mole/°C @ 20 °C	
Latent heat	kJ/kg mole	
Dielectric constant	@ 20 °C	
Antoine constant A	kPa°C	
Antoine constant B	kPa°C	
Antoine constant C	kPa°C	
Heat of combustion	kJ/kg mole	
UN number		2058
IMO classification		3.2/II
ADR/RID classification		3,3°(b)
UK exposure limits	mg/m^3 (ppm) 8hrTWA/MEL or OES	
USA exposure limits	mg/m^3 (ppm) TLV/TWA	176(50)
German exposure limits	mg/m^3 (ppm) MAK-TWA	
EU classification		
EU risk phrases		
EU safety phrases		

[a] Ether = 1. [b] n-Butyl acetate = 1. [c] Nelson, Hemwall, Edwards. [d] Toluene. [e] Hydrocarbon solvent.

Solvent group		**AMIDES**
Chemical name		N,N-Dimethylformamide
Chemical formula		C_3H_7ON
Structural formula		$HCON(CH_3)_2$
CAS No.		68-12-2
Synonym/common name		DMF, Dimethyl formamide
Trade names		
Suppliers		BASF
Molecular mass		73.1
Boiling point/range	°C @ 760 mm Hg	152.0-153.5
Melting point	°C	-60.4
Density	kg/m^3 @ 20 °C	950
Evaporation time[a]		100
Evaporation time[b]		
Vapour pressure	kPa @ 20 °C	
Flash point	°C	58 (DIN 51755 or 51758)
Auto ignition temperature	°C	410
Upper flammability limit	%v/v in air	16.0
Lower flammability limit	%v/v in air	2.2
Viscosity	mPa.s (cP) @ 20 °C	0.8
Refractive index	n 20 D	1.431
Solubility in water	%w/w @ 20 °C	
Solubility of water	%w/w @ 20 °C	
Solubility parameter[c]	(J/cm^3)$^{1/2}$	
Hydrogen bond index		
Fractional polarity		
Nitro cellulose dil ratio[d]		
Nitro cellulose dil ratio[e]		
Surface tension	mN/m @ 20 °C	36.76
Specific heat liquid	kJ/kg mole/°C @ 20 °C	
Latent heat	kJ/kg mole	38342
Dielectric constant	@ 20 °C	
Antoine constant A	kPa°C	6.2334
Antoine constant B	kPa°C	1537.78
Antoine constant C	kPa°C	210.39
Heat of combustion	kJ/kg mole	−1941630
UN number		
IMO classification		
ADR/RID classification		
UK exposure limits	mg/m^3 (ppm) 8hrTWA/MEL or OES	
USA exposure limits	mg/m^3 (ppm) TLV/TWA	
German exposure limits	mg/m^3 (ppm) MAK-TWA	
EU classification		
EU risk phrases		
EU safety phrases		

[a]Ether = 1. [b]n-Butyl acetate = 1. [c]Nelson, Hemwall, Edwards. [d]Toluene. [e]Hydrocarbon solvent.

Property	Unit	Value
Solvent group		**AMIDES**
Chemical name		N,N-Dimethylacetamide
Chemical formula		C_4H_9ON
Structural formula		$CH_3CON(CH_3)_2$
CAS No.		127-19-5
Synonym/common name		DMAC, Dimethyl acetamide
Trade names		
Suppliers		BASF
Molecular mass		87.1
Boiling point/range	°C @ 760 mm Hg	165.0–166.0
Melting point	°C	
Density	kg/m^3 @ 20 °C	940
Evaporation time[a]		172
Evaporation time[b]		
Vapour pressure	kPa @ 20 °C	
Flash point	°C	70 (DIN 51755 or 51758)
Auto ignition temperature	°C	400
Upper flammability limit	%v/v in air	11.5
Lower flammability limit	%v/v in air	1.7
Viscosity	mPa.s (cP) @ 20 °C	0.9
Refractive index	n 20 D	1.437
Solubility in water	%w/w @ 20 °C	
Solubility of water	%w/w @ 20 °C	
Solubility parameter[c]	$(J/cm^3)^{1/2}$	22.1
Hydrogen bond index		
Fractional polarity		
Nitro cellulose dil ratio[d]		
Nitro cellulose dil ratio[e]		
Surface tension	mN/m @ 20 °C	
Specific heat liquid	kJ/kg mole/°C @ 20 °C	
Latent heat	kJ/kg mole	43346
Dielectric constant	@ 20 °C	
Antoine constant A	kPa°C	6.88718
Antoine constant B	kPa°C	1889.10
Antoine constant C	kPa°C	221.0
Heat of combustion	kJ/kg mole	−2581990
UN number		
IMO classification		
ADR/RID classification		
UK exposure limits	mg/m^3 (ppm) 8hrTWA/MEL or OES	
USA exposure limits	mg/m^3 (ppm) TLV/TWA	
German exposure limits	mg/m^3 (ppm) MAK-TWA	
EU classification		
EU risk phrases		
EU safety phrases		

[a] Ether = 1. [b] n-Butyl acetate = 1. [c] Nelson, Hemwall, Edwards. [d] Toluene. [e] Hydrocarbon solvent.

Solvent group		**ESTERS**
Chemical name		Methyl acetate
Chemical formula		$C_3H_6O_2$
Structural formula		CH_3COOCH_3
CAS No.		79-20-9
Synonym/common name		Methyl ethanoate
Trade names		
Suppliers		CHEMOXY
Molecular mass		74.08
Boiling point/range	°C @ 760 mm Hg	57.5
Melting point	°C	−98
Density	kg/m^3 @ 20 °C	
Evaporation time[a]		
Evaporation time[b]		
Vapour pressure	kPa @ 20 °C	
Flash point	°C	−16
Auto ignition temperature	°C	
Upper flammability limit	%v/v in air	16
Lower flammability limit	%v/v in air	3.1
Viscosity	mPa.s (cP) @ 20 °C	
Refractive index	n 20 D	1.3614
Solubility in water	%w/w @ 20 °C	
Solubility of water	%w/w @ 20 °C	
Solubility parameter[c]	(J/cm^3)$^{1/2}$	19.6
Hydrogen bond index		
Fractional polarity		
Nitro cellulose dil ratio[d]		
Nitro cellulose dil ratio[e]		
Surface tension	mN/m @ 20 °C	24.8
Specific heat liquid	kJ/kg mole/°C @ 20 °C	
Latent heat	kJ/kg mole	30330
Dielectric constant	@ 20 °C	6.68
Antoine constant A	kPa°C	6.24410
Antoine constant B	kPa°C	1183.700
Antoine constant C	kPa°C	222.414
Heat of combustion	kJ/kg mole	−1592180
UN number		1231
IMO classification		3.2/II
ADR/RID classification		3,3°(b)
UK exposure limits	mg/m^3 (ppm) 8hrTWA/MEL or OES	610(200)
USA exposure limits	mg/m^3 (ppm) TLV/TWA	606(200)
German exposure limits	mg/m^3 (ppm) MAK-TWA	610(200)
EU classification		F
EU risk phrases		11
EU safety phrases		(2-)16-23-29-33

[a]Ether = 1. [b]n-Butyl acetate = 1. [c]Nelson, Hemwall, Edwards. [d]Toluene. [e]Hydrocarbon solvent.

Solvent group		**ESTERS**
Chemical name		Ethyl acetate
Chemical formula		$C_4H_8O_2$
Structural formula		$CH_3COOCH_2CH_3$
CAS No.		141-78-6
Synonym/common name		Ethyl acetate
Trade names		ETHYL ACETATE, EtAc
Suppliers		UNION CARBIDE, BP, HULS, SHELL
Molecular mass		88.11
Boiling point/range	°C @ 760 mm Hg	76.0–78.0
Melting point	°C	−82.4 to −84
Density	kg/m^3 @ 20 °C	900–902
Evaporation time[a]		2.9–3
Evaporation time[b]		4.2–8.02
Vapour pressure	kPa @ 20 °C	9.7–10.3
Flash point	°C	−4 (IP 170) SHELL
Auto ignition temperature	°C	425–530
Upper flammability limit	%v/v in air	10.2–11.5
Lower flammability limit	%v/v in air	2.1–2.5
Viscosity	mPa.s (cP) @ 20 °C	0.5
Refractive index	n 20 D	1.372
Solubility in water	%w/w @ 20 °C	7.7–8.7
Solubility of water	%w/w @ 20 °C	3.3
Solubility parameter[c]	(J/cm^3)$^{1/2}$	18.6
Hydrogen bond index		8.4
Fractional polarity		0.167
Nitro cellulose dil ratio[d]		2.9
Nitro cellulose dil ratio[e]		1.0
Surface tension	mN/m @ 20 °C	14.2
Specific heat liquid	kJ/kg mole/°C @ 20 °C	171.8
Latent heat	kJ/kg mole	35504
Dielectric constant	@ 20 °C	6.0
Antoine constant A	kPa°C	6.35132
Antoine constant B	kPa°C	1236.73
Antoine constant C	kPa°C	227.645
Heat of combustion	kJ/kg mole	−2099511
UN number		1173
IMO classification		
ADR/RID classification		3,3°(b)
UK exposure limits	mg/m^3 (ppm) 8hrTWA/MEL or OES	1400(400)
USA exposure limits	mg/m^3 (ppm) TLV/TWA	1440(400)
German exposure limits	mg/m^3 (ppm) MAK-TWA	1400(400)
EU classification		F
EU risk phrases		11
EU safety phrases		(2-)16-23-29-33

[a] Ether = 1. [b] n-Butyl acetate = 1. [c] Nelson, Hemwall, Edwards. [d] Toluene. [e] Hydrocarbon solvent.

Solvent group		**ESTERS**
Chemical name		2-Methoxy-2-methyl propane
Chemical formula		$C_5H_{12}O$
Structural formula		$CH_3OC(CH_3)_3$
CAS No.		1634-04-4
Synonym/common name		Methyl tert-butyl ether
Trade names		MTBE+
Suppliers		SHELL
Molecular mass		88.15
Boiling point/range	°C @ 760 mm Hg	
Melting point	°C	−109
Density	kg/m^3 @ 20 °C	740–745
Evaporation time[a]		1.6
Evaporation time[b]		8.4
Vapour pressure	kPa @ 20 °C	26.9
Flash point	°C	−34(IP 170)
Auto ignition temperature	°C	460
Upper flammability limit	%v/v in air	8
Lower flammability limit	%v/v in air	1
Viscosity	mPa.s (cP) @ 20 °C	0.35
Refractive index	n 20 D	1.369
Solubility in water	%w/w @ 20 °C	4.8
Solubility of water	%w/w @ 20 °C	1.5
Solubility parameter[c]	(J/cm^3)$^{1/2}$	14.3
Hydrogen bond index		10.5
Fractional polarity		0.016
Nitro cellulose dil ratio[d]		
Nitro cellulose dil ratio[e]		
Surface tension	mN/m @ 20 °C	18.5
Specific heat liquid	kJ/kg mole/°C @ 20 °C	192.2
Latent heat	kJ/kg mole	29530
Dielectric constant	@ 20 °C	4.5
Antoine constant A	kPa°C	6.26886
Antoine constant B	kPa°C	1250.15
Antoine constant C	kPa°C	238.394
Heat of combustion	kJ/kg mole	−3355694
UN number		2398
IMO classification		3.1/II
ADR/RID classification		3,3°(b)
UK exposure limits	mg/m^3 (ppm) 8hrTWA/MEL or OES	
USA exposure limits	mg/m^3 (ppm) TLV/TWA	
German exposure limits	mg/m^3 (ppm) MAK-TWA	
EU classification		F
EU risk phrases		
EU safety phrases		9-16-29

[a] Ether = 1. [b] n-Butyl acetate = 1. [c] Nelson, Hemwall, Edwards. [d] Toluene. [e] Hydrocarbon solvent.

Solvent group		ESTERS
Chemical name		Propylene carbonate
Chemical formula		$C_4H_6O_3$
Structural formula	CH_3–CH(–O–)–CH_2–O– ring with C=O	
CAS No.		108-32-7
Synonym/common name		Propylene carbonate
Trade names		
Suppliers		HULS
Molecular mass		102.1
Boiling point/range	°C @ 760 mm Hg	243.4
Melting point	°C	−54
Density	kg/m^3 @ 20 °C	1205
Evaporation time[a]		>1000
Evaporation time[b]		
Vapour pressure	kPa @ 20 °C	0.004
Flash point	°C	130
Auto ignition temperature	°C	510
Upper flammability limit	%v/v in air	2.8
Lower flammability limit	%v/v in air	1.5
Viscosity	mPa.s (cP) @ 20 °C	2.76(23°C)
Refractive index	n 20 D	1.421–1.42
Solubility in water	%w/w @ 20 °C	
Solubility of water	%w/w @ 20 °C	7.5
Solubility parameter[c]	$(J/cm^3)^{1/2}$	27.2
Hydrogen bond index		
Fractional polarity		
Nitro cellulose dil ratio[d]		
Nitro cellulose dil ratio[e]		
Surface tension	mN/m @ 20 °C	41.93
Specific heat liquid	kJ/kg mole/°C @ 20 °C	
Latent heat	kJ/kg mole	49800
Dielectric constant	@ 20 °C	
Antoine constant A	kPa°C	
Antoine constant B	kPa°C	
Antoine constant C	kPa°C	
Heat of combustion	kJ/kg mole	−1799710
UN number		
IMO classification		
ADR/RID classification		
UK exposure limits	mg/m^3 (ppm) 8hrTWA/MEL or OES	
USA exposure limits	mg/m^3 (ppm) TLV/TWA	
German exposure limits	mg/m^3 (ppm) MAK-TWA	
EU classification		Xi
EU risk phrases		36
EU safety phrases		(2)

[a] Ether = 1. [b] n-Butyl acetate = 1. [c] Nelson, Hemwall, Edwards. [d] Toluene. [e] Hydrocarbon solvent.

Solvent group		ESTERS
Chemical name		
Chemical formula		$C_4H_6O_3$
Structural formula		
CAS No.		108-32-7
Synonym/common name		
Trade names		Solvenon PC
Suppliers		BASF
Molecular mass		102.1
Boiling point/range	°C @ 760 mm Hg	240.0–243.0
Melting point	°C	
Density	kg/m^3 @ 20 °C	1204
Evaporation time[a]		>1500
Evaporation time[b]		
Vapour pressure	kPa @ 20 °C	
Flash point	°C	123 (DIN 51755 or 51758) BASF
Auto ignition temperature	°C	~465
Upper flammability limit	%v/v in air	14.3
Lower flammability limit	%v/v in air	1.8
Viscosity	mPa.s (cP) @ 20 °C	2.8
Refractive index	n 20 D	1.422
Solubility in water	%w/w @ 20 °C	
Solubility of water	%w/w @ 20 °C	
Solubility parameter[c]	(J/cm^3)$^{1/2}$	
Hydrogen bond index		
Fractional polarity		
Nitro cellulose dil ratio[d]		
Nitro cellulose dil ratio[e]		
Surface tension	mN/m @ 20 °C	
Specific heat liquid	kJ/kg mole/°C @ 20 °C	
Latent heat	kJ/kg mole	
Dielectric constant	@ 20 °C	
Antoine constant A	kPa°C	
Antoine constant B	kPa°C	
Antoine constant C	kPa°C	
Heat of combustion	kJ/kg mole	
UN number		
IMO classification		
ADR/RID classification		
UK exposure limits	mg/m^3 (ppm) 8hrTWA/MEL or OES	
USA exposure limits	mg/m^3 (ppm) TLV/TWA	
German exposure limits	mg/m^3 (ppm) MAK-TWA	
EU classification		
EU risk phrases		
EU safety phrases		

[a] Ether = 1. [b] n-Butyl acetate = 1. [c] Nelson, Hemwall, Edwards. [d] Toluene. [e] Hydrocarbon solvent.

Solvent group		ESTERS
Chemical name		1-Methylethyl acetate
Chemical formula		$C_5H_{10}O_2$
Structural formula		$CH_3COOCH(CH_3)_2$
CAS No.		108-21-4
Synonym/common name		Isopropyl acetate
Trade names		
Suppliers		RHÔNE POULENC, PARDIE ACETIQUES
Molecular mass		102.13
Boiling point/range	°C @ 760 mm Hg	88.4
Melting point	°C	−73.4
Density	kg/m^3 @ 20 °C	872
Evaporation time[a]		
Evaporation time[b]		
Vapour pressure	kPa @ 20 °C	
Flash point	°C	2 (CLOSED CUP)
Auto ignition temperature	°C	460
Upper flammability limit	%v/v in air	8
Lower flammability limit	%v/v in air	1.8
Viscosity	mPa.s (cP) @ 20 °C	
Refractive index	n 20 D	1.3772
Solubility in water	%w/w @ 20 °C	2.9–3.0
Solubility of water	%w/w @ 20 °C	1.8
Solubility parameter[c]	$(J/cm^3)^{1/2}$	
Hydrogen bond index		
Fractional polarity		
Nitro cellulose dil ratio[d]		3.0
Nitro cellulose dil ratio[e]		1.2
Surface tension	mN/m @ 20 °C	
Specific heat liquid	kJ/kg mole/°C @ 20 °C	
Latent heat	kJ/kg mole	
Dielectric constant	@ 20 °C	
Antoine constant A	kPa°C	
Antoine constant B	kPa°C	
Antoine constant C	kPa°C	
Heat of combustion	kJ/kg mole	
UN number		1220
IMO classification		3.2/II
ADR/RID classification		3.3°(b)
UK exposure limits	mg/m^3 (ppm) 8hrTWA/MEL or OES	840(200)
USA exposure limits	mg/m^3 (ppm) TLV/TWA	1040(250)
German exposure limits	mg/m^3 (ppm) MAK-TWA	840(200)
EU classification		F
EU risk phrases		11
EU safety phrases		(2-)16-23-29-33

[a] Ether = 1. [b] n-Butyl acetate = 1. [c] Nelson, Hemwall, Edwards. [d] Toluene. [e] Hydrocarbon solvent.

Solvent group		**ESTERS**
Chemical name		n-Propyl acetate
Chemical formula		$C_5H_{10}O_2$
Structural formula		$CH_3COOCH_2CH_2CH_3$
CAS No.		109-60-4
Synonym/common name		n-Propyl acetate
Trade names		
Suppliers		UNION CARBIDE, BASF
Molecular mass		102.13
Boiling point/range	°C @ 760 mm Hg	101.0–102.0
Melting point	°C	−95.0
Density	kg/m^3 @ 20 °C	888
Evaporation time[a]		5
Evaporation time[b]		2.80
Vapour pressure	kPa @ 20 °C	
Flash point	°C	10 (DIN 51755 or 51758) BASF
Auto ignition temperature	°C	430
Upper flammability limit	%v/v in air	8.0
Lower flammability limit	%v/v in air	1.7
Viscosity	mPa.s (cP) @ 20 °C	0.6
Refractive index	n 20 D	1.385
Solubility in water	%w/w @ 20 °C	2.0
Solubility of water	%w/w @ 20 °C	2.9
Solubility parameter[c]	(J/cm^3)$^{1/2}$	18.0
Hydrogen bond index		
Fractional polarity		
Nitro cellulose dil ratio[d]		3.2
Nitro cellulose dil ratio[e]		1.5
Surface tension	mN/m @ 20 °C	24.28
Specific heat liquid	kJ/kg mole/°C @ 20 °C	
Latent heat	kJ/kg mole	33860
Dielectric constant	@ 20 °C	6.002
Antoine constant A	kPa°C	6.14362
Antoine constant B	kPa°C	1284.080
Antoine constant C	kPa°C	208.786
Heat of combustion	kJ/kg mole	
UN number		1276
IMO classification		3.2/III
ADR/RID classification		3,3°(b)
UK exposure limits	mg/m^3 (ppm) 8hrTWA/MEL or OES	840(200)
USA exposure limits	mg/m^3 (ppm) TLV/TWA	835(200)
German exposure limits	mg/m^3 (ppm) MAK-TWA	840(200)
EU classification		F
EU risk phrases		11
EU safety phrases		(2-)16-23-29-33

[a] Ether = 1. [b] n-Butyl acetate = 1. [c] Nelson, Hemwall, Edwards. [d] Toluene. [e] Hydrocarbon solvent.

Solvent group		ESTERS
Chemical name		Isopropyl acetate
Chemical formula		$C_5H_{10}O_2$
Structural formula		$CH_3COOCH(CH_3)_2$
CAS No.		108-21-4
Synonym/common name		Isopropyl acetate
Trade names		
Suppliers		UNION CARBIDE
Molecular mass		102.13
Boiling point/range	°C @ 760 mm Hg	88.5
Melting point	°C	−73.4
Density	kg/m^3 @ 20 °C	873
Evaporation time[a]		
Evaporation time[b]		5.06
Vapour pressure	kPa @ 20 °C	
Flash point	°C	
Auto ignition temperature	°C	
Upper flammability limit	%v/v in air	
Lower flammability limit	%v/v in air	
Viscosity	mPa.s (cP) @ 20 °C	
Refractive index	n 20 D	0.6
Solubility in water	%w/w @ 20 °C	1.3773
Solubility of water	%w/w @ 20 °C	2.9
Solubility parameter[c]	$(J/cm^3)^{1/2}$	1.8
Hydrogen bond index		17.2
Fractional polarity		
Nitro cellulose dil ratio[d]		
Nitro cellulose dil ratio[e]		
Surface tension	mN/m @ 20 °C	21.2
Specific heat liquid	kJ/kg mole/°C @ 20 °C	
Latent heat	kJ/kg mole	33050
Dielectric constant	@ 20 °C	
Antoine constant A	kPa°C	6.12933
Antoine constant B	kPa°C	1237.232
Antoine constant C	kPa°C	211.436
Heat of combustion	kJ/kg mole	
UN number		1220
IMO classification		3.2/II
ADR/RID classification		3,3°(b)
UK exposure limits	mg/m^3 (ppm) 8hrTWA/MEL or OES	840(200)
USA exposure limits	mg/m^3 (ppm) TLV/TWA	1040(250)
German exposure limits	mg/m^3 (ppm) MAK-TWA	840(200)
EU classification		
EU risk phrases		
EU safety phrases		

[a] Ether = 1. [b] n-Butyl acetate = 1. [c] Nelson, Hemwall, Edwards. [d] Toluene. [e] Hydrocarbon solvent.

Solvent group		**ESTERS**
Chemical name		2,2-Oxybispropane
Chemical formula		$C_6H_{14}O$
Structural formula		$[(CH_3)_2CH]_2O$
CAS No.		108-20-3
Synonym/common name		Isopropyl ether
Trade names		IPE
Suppliers		SHELL
Molecular mass		102.18
Boiling point/range	°C @ 760 mm Hg	65–70
Melting point	°C	−85
Density	kg/m^3 @ 20 °C	722–726
Evaporation time[a]		1.4
Evaporation time[b]		8.2
Vapour pressure	kPa @ 20 °C	15.9
Flash point	°C	−24(IP 170)
Auto ignition temperature	°C	405
Upper flammability limit	%v/v in air	22
Lower flammability limit	%v/v in air	1.4
Viscosity	mPa.s (cP) @ 20 °C	0.32
Refractive index	n 20 D	1.368
Solubility in water	%w/w @ 20 °C	1.1
Solubility of water	%w/w @ 20 °C	0.5
Solubility parameter[c]	$(J/cm^3)^{1/2}$	14.3
Hydrogen bond index		14.0
Fractional polarity		0.013
Nitro cellulose dil ratio[d]		∞
Nitro cellulose dil ratio[e]		∞
Surface tension	mN/m @ 20 °C	17.5
Specific heat liquid	kJ/kg mole/°C @ 20 °C	219.7
Latent heat	kJ/kg mole	29121
Dielectric constant	@ 20 °C	3.9
Antoine constant A	kPa°C	5.97678
Antoine constant B	kPa°C	1143.97
Antoine constant C	kPa°C	219.34
Heat of combustion	kJ/kg mole	−3737540
UN number		1159
IMO classification		3.1/II
ADR/RID classification		3,3°(b)
UK exposure limits	mg/m^3 (ppm) 8hrTWA/MEL or OES	
USA exposure limits	mg/m^3 (ppm) TLV/TWA	1040(250)
German exposure limits	mg/m^3 (ppm) MAK-TWA	2100(500)
EU classification		F
EU risk phrases		19
EU safety phrases		9-16-33

[a] Ether = 1. [b] n-Butyl acetate = 1. [c] Nelson, Hemwall, Edwards. [d] Toluene. [e] Hydrocarbon solvent.

Solvent group		ESTERS
Chemical name		2-Methyl-2-methoxy-butane
Chemical formula		$C_6H_{14}O$
Structural formula		$CH_3OC(CH_3)_2CH_2CH_3$
CAS No.		994-05-8
Synonym/common name		tert-Amyl methyl ether
Trade names		TAME
Suppliers		EC Erdolchemie
Molecular mass		102.2
Boiling point/range	°C @ 760 mm Hg	86.3
Melting point	°C	−80
Density	kg/m^3 @ 20 °C	770
Evaporation time[a]		3
Evaporation time[b]		
Vapour pressure	kPa @ 20 °C	
Flash point	°C	−11
Auto ignition temperature	°C	415
Upper flammability limit	%v/v in air	7.1
Lower flammability limit	%v/v in air	1.0
Viscosity	mPa.s (cP) @ 20 °C	0.59
Refractive index	n 20 D	1.3885
Solubility in water	%w/w @ 20 °C	
Solubility of water	%w/w @ 20 °C	
Solubility parameter[c]	$(J/cm^3)^{1/2}$	
Hydrogen bond index		
Fractional polarity		
Nitro cellulose dil ratio[d]		
Nitro cellulose dil ratio[e]		
Surface tension	mN/m @ 20 °C	22.6
Specific heat liquid	kJ/kg mole/°C @ 20 °C	
Latent heat	kJ/kg mole	
Dielectric constant	@ 20 °C	
Antoine constant A	kPa°C	
Antoine constant B	kPa°C	
Antoine constant C	kPa°C	
Heat of combustion	kJ/kg mole	
UN number		1993
IMO classification		
ADR/RID classification		3,3°(b)
UK exposure limits	mg/m^3 (ppm) 8hrTWA/MEL or OES	
USA exposure limits	mg/m^3 (ppm) TLV/TWA	
German exposure limits	mg/m^3 (ppm) MAK-TWA	
EU classification		
EU risk phrases		
EU safety phrases		

[a] Ether = 1. [b] n-Butyl acetate = 1. [c] Nelson, Hemwall, Edwards. [d] Toluene. [e] Hydrocarbon solvent.

Solvent group		**ESTERS**
Chemical name		Methyl acetoacetate
Chemical formula		$C_5H_8O_3$
Structural formula		$CH_3COCH_2COOCH_3$
CAS No.		105-45-3
Synonym/common name		Methyl acetoacetate
Trade names		
Suppliers		BP
Molecular mass		116.1
Boiling point/range	°C @ 760 mm Hg	170–172
Melting point	°C	−80
Density	kg/m^3 @ 20 °C	1079
Evaporation time[a]		
Evaporation time[b]		
Vapour pressure	kPa @ 20 °C	0.09
Flash point	°C	77 (PENSKY MARTENS CLOSED CUP)
Auto ignition temperature	°C	280
Upper flammability limit	%v/v in air	16
Lower flammability limit	%v/v in air	3.1
Viscosity	mPa.s (cP) @ 20 °C	
Refractive index	n 20 D	
Solubility in water	%w/w @ 20 °C	
Solubility of water	%w/w @ 20 °C	
Solubility parameter[c]	(J/cm^3)$^{1/2}$	
Hydrogen bond index		
Fractional polarity		
Nitro cellulose dil ratio[d]		
Nitro cellulose dil ratio[e]		
Surface tension	mN/m @ 20 °C	
Specific heat liquid	kJ/kg mole/°C @ 20 °C	
Latent heat	kJ/kg mole	
Dielectric constant	@ 20 °C	
Antoine constant A	kPa°C	
Antoine constant B	kPa°C	
Antoine constant C	kPa°C	
Heat of combustion	kJ/kg mole	
UN number		
IMO classification		
ADR/RID classification		3,32°(c)
UK exposure limits	mg/m^3 (ppm) 8hrTWA/MEL or OES	
USA exposure limits	mg/m^3 (ppm) TLV/TWA	
German exposure limits	mg/m^3 (ppm) MAK-TWA	
EU classification		Xi
EU risk phrases		36
EU safety phrases		(2-)26

[a] Ether = 1. [b] n-Butyl acetate = 1. [c] Nelson, Hemwall, Edwards. [d] Toluene. [e] Hydrocarbon solvent.

Solvent group		**ESTERS**
Chemical name		n-Butyl acetate
Chemical formula		$C_6H_{12}O_2$
Structural formula		$CH_3COOCH_2CH_2CH_2CH_3$
CAS No.		123-86-4
Synonym/common name		n-Butyl acetate
Trade names		nBuAc
Suppliers		BP, SHELL, BASF, UNION CARBIDE
Molecular mass		116.16
Boiling point/range	°C @ 760 mm Hg	120-128
Melting point	°C	−76.8 to −78
Density	kg/m^3 @ 20 °C	879–881
Evaporation time[a]		11–12
Evaporation time[b]		1.0
Vapour pressure	kPa @ 20 °C	1.13–1.25
Flash point	°C	25 (IP 170) SHELL, 27 (DIN 51755 or 51758) BASF, 29 BP, 23 (CLOSED CUP), 38 (OPEN CUP) UNION CARBIDE
Auto ignition temperature	°C	370–421
Upper flammability limit	%v/v in air	7.5–10.4
Lower flammability limit	%v/v in air	1.0–3.0
Viscosity	mPa.s (cP) @ 20 °C	0.76–0.83
Refractive index	n 20 D	1.394
Solubility in water	%w/w @ 20 °C	0.70
Solubility of water	%w/w @ 20 °C	1.18–1.9
Solubility parameter[c]	(J/cm^3)$^{1/2}$	17.6
Hydrogen bond index		8.0
Fractional polarity		0.120
Nitro cellulose dil ratio[d]		2.7
Nitro cellulose dil ratio[e]		1.3
Surface tension	mN/m @ 20 °C	25.2
Specific heat liquid	kJ/kg mole/°C @ 20 °C	220.7
Latent heat	kJ/kg mole	43095
Dielectric constant	@ 20 °C	5.1
Antoine constant A	kPa°C	6.28085
Antoine constant B	kPa°C	1450.72
Antoine constant C	kPa°C	213.049
Heat of combustion	kJ/kg mole	−3327635
UN number		1123
IMO classification		3.3/III
ADR/RID classification		3,31°(c)
UK exposure limits	mg/m^3 (ppm) 8hrTWA/MEL or OES	710(150)
USA exposure limits	mg/m^3 (ppm) TLV/TWA	713(150)
German exposure limits	mg/m^3 (ppm) MAK-TWA	950(200)
EU classification		R10
EU risk phrases		10
EU safety phrases		(2)

[a]Ether = 1. [b]n-Butyl acetate = 1. [c]Nelson, Hemwall, Edwards. [d]Toluene. [e]Hydrocarbon solvent.

Solvent group		**ESTERS**
Chemical name		2-Methylpropyl acetate
Chemical formula		$C_6H_{12}O_2$
Structural formula		$CH_3COOCH_2CH(CH_3)_2$
CAS No.		110-19-0
Synonym/common name		Isobutyl acetate
Trade names		
Suppliers		UNION CARBIDE, BASF, BP
Molecular mass		116.2
Boiling point/range	°C @ 760 mm Hg	115.0–118.0
Melting point	°C	−98 to −99
Density	kg/m^3 @ 20 °C	871–873
Evaporation time[a]		8
Evaporation time[b]		1.70
Vapour pressure	kPa @ 20 °C	2.13
Flash point	°C	19 (DIN 51755 or 51758) BASF, 17 BP
Auto ignition temperature	°C	405–423
Upper flammability limit	%v/v in air	10.1–15.0
Lower flammability limit	%v/v in air	2.0–2.5
Viscosity	mPa.s (cP) @ 20 °C	0.7
Refractive index	n 20 D	1.390
Solubility in water	%w/w @ 20 °C	0.67
Solubility of water	%w/w @ 20 °C	1.6
Solubility parameter[c]	$(J/cm^3)^{1/2}$	17.0
Hydrogen bond index		
Fractional polarity		
Nitro cellulose dil ratio[d]		2.6
Nitro cellulose dil ratio[e]		1.3
Surface tension	mN/m @ 20 °C	23.7
Specific heat liquid	kJ/kg mole/°C @ 20 °C	
Latent heat	kJ/kg mole	35850
Dielectric constant	@ 20 °C	5.29
Antoine constant A	kPa°C	6.3546
Antoine constant B	kPa°C	1462.4
Antoine constant C	kPa°C	219.7
Heat of combustion	kJ/kg mole	−350900
UN number		1213
IMO classification		3.2/II
ADR/RID classification		3,3°(b)
UK exposure limits	mg/m^3 (ppm) 8hrTWA/MEL or OES	700(150)
USA exposure limits	mg/m^3 (ppm) TLV/TWA	713(150)
German exposure limits	mg/m^3 (ppm) MAK-TWA	950(200)
EU classification		F
EU risk phrases		11
EU safety phrases		(2-)16-23-29-33

[a] Ether = 1. [b] n-Butyl acetate = 1. [c] Nelson, Hemwall, Edwards. [d] Toluene. [e] Hydrocarbon solvent.

Solvent group		ESTERS
Chemical name		Ethyl acetoacetate
Chemical formula		$C_6H_{10}O_3$
Structural formula		$CH_3COCH_2COOCH_2CH_3$
CAS No.		141-97-9
Synonym/common name		Ethyl acetoacetate
Trade names		ETHYL ACETOACETATE
Suppliers		BP
Molecular mass		130.14
Boiling point/range	°C @ 760 mm Hg	180
Melting point	°C	−44
Density	kg/m^3 @ 20 °C	1030
Evaporation time[a]		
Evaporation time[b]		
Vapour pressure	kPa @ 20 °C	0.107
Flash point	°C	57
Auto ignition temperature	°C	295
Upper flammability limit	%v/v in air	9.5
Lower flammability limit	%v/v in air	1.4
Viscosity	mPa.s (cP) @ 20 °C	
Refractive index	n 20 D	
Solubility in water	%w/w @ 20 °C	
Solubility of water	%w/w @ 20 °C	
Solubility parameter[c]	$(J/cm^3)^{1/2}$	
Hydrogen bond index		
Fractional polarity		
Nitro cellulose dil ratio[d]		
Nitro cellulose dil ratio[e]		
Surface tension	mN/m @ 20 °C	
Specific heat liquid	kJ/kg mole/°C @ 20 °C	
Latent heat	kJ/kg mole	
Dielectric constant	@ 20 °C	
Antoine constant A	kPa°C	
Antoine constant B	kPa°C	
Antoine constant C	kPa°C	
Heat of combustion	kJ/kg mole	
UN number		
IMO classification		
ADR/RID classification		3,32°(c)
UK exposure limits	mg/m^3 (ppm) 8hrTWA/MEL or OES	
USA exposure limits	mg/m^3 (ppm) TLV/TWA	
German exposure limits	mg/m^3 (ppm) MAK-TWA	
EU classification		
EU risk phrases		
EU safety phrases		

[a] Ether = 1. [b] n-Butyl acetate = 1. [c] Nelson, Hemwall, Edwards. [d] Toluene. [e] Hydrocarbon solvent.

Solvent group		**ESTERS**
Chemical name		Amyl acetate
Chemical formula		$C_7H_{14}O_2$
Structural formula		$CH_3COOCH_2(CH_2)_3CH_3$ mixed isomers
CAS No.		mixture
Synonym/common name		Amyl acetate/pentyl acetate
Trade names		
Suppliers		UNION CARBIDE, BASF
Molecular mass		130.19
Boiling point/range	°C @ 760 mm Hg	146.0–149.0
Melting point	°C	
Density	kg/m^3 @ 20 °C	877
Evaporation time[a]		28
Evaporation time[b]		0.42
Vapour pressure	kPa @ 20 °C	
Flash point	°C	40 (DIN 51755 or 51758) BASF
Auto ignition temperature	°C	345
Upper flammability limit	%v/v in air	8.4
Lower flammability limit	%v/v in air	1.4
Viscosity	mPa.s (cP) @ 20 °C	0.9
Refractive index	n 20 D	1.403
Solubility in water	%w/w @ 20 °C	0.2
Solubility of water	%w/w @ 20 °C	0.9
Solubility parameter[c]	$(J/cm^3)^{1/2}$	17.4
Hydrogen bond index		
Fractional polarity		
Nitro cellulose dil ratio[d]		2.2
Nitro cellulose dil ratio[e]		1.2
Surface tension	mN/m @ 20 °C	
Specific heat liquid	kJ/kg mole/°C @ 20 °C	
Latent heat	kJ/kg mole	
Dielectric constant	@ 20 °C	
Antoine constant A	kPa°C	5.4315
Antoine constant B	kPa°C	1197
Antoine constant C	kPa°C	200
Heat of combustion	kJ/kg mole	−4428800
UN number		1104
IMO classification		3.3/III
ADR/RID classification		3,31°(c)
UK exposure limits	mg/m^3 (ppm) 8hrTWA/MEL or OES	
USA exposure limits	mg/m^3 (ppm) TLV/TWA	
German exposure limits	mg/m^3 (ppm) MAK-TWA	
EU classification		R10
EU risk phrases		10
EU safety phrases		(2-)23

[a] Ether = 1. [b] n-Butyl acetate = 1. [c] Nelson, Hemwall, Edwards. [d] Toluene. [e] Hydrocarbon solvent.

Solvent group		ESTERS
Chemical name		n-Butyl propionate
Chemical formula		$C_7H_{14}O_2$
Structural formula		$CH_3CH_2COOCH_2CH_2CH_2CH_3$
CAS No.		590-01-2
Synonym/common name		n-Butyl propionate
Trade names		UCAR n-Butyl Propionate
Suppliers		CHEMOXY, UNION CARBIDE
Molecular mass		130.19
Boiling point/range	°C @ 760 mm Hg	144.7–148
Melting point	°C	
Density	kg/m^3 @ 20 °C	875
Evaporation time[a]		
Evaporation time[b]		0.45
Vapour pressure	kPa @ 20 °C	
Flash point	°C	
Auto ignition temperature	°C	
Upper flammability limit	%v/v in air	
Lower flammability limit	%v/v in air	
Viscosity	mPa.s (cP) @ 20 °C	0.9
Refractive index	n 20 D	
Solubility in water	%w/w @ 20 °C	0.2
Solubility of water	%w/w @ 20 °C	<0.2
Solubility parameter[c]	(J/cm^3)$^{1/2}$	
Hydrogen bond index		
Fractional polarity		
Nitro cellulose dil ratio[d]		
Nitro cellulose dil ratio[e]		
Surface tension	mN/m @ 20 °C	
Specific heat liquid	kJ/kg mole/°C @ 20 °C	
Latent heat	kJ/kg mole	
Dielectric constant	@ 20 °C	
Antoine constant A	kPa°C	
Antoine constant B	kPa°C	
Antoine constant C	kPa°C	
Heat of combustion	kJ/kg mole	
UN number		1914
IMO classification		3.3/III
ADR/RID classification		3,31°(c)
UK exposure limits	mg/m^3 (ppm) 8hrTWA/MEL or OES	
USA exposure limits	mg/m^3 (ppm) TLV/TWA	
German exposure limits	mg/m^3 (ppm) MAK-TWA	
EU classification		R10
EU risk phrases		10
EU safety phrases		(2)

[a] Ether = 1. [b] n-Butyl acetate = 1. [c] Nelson, Hemwall, Edwards. [d] Toluene. [e] Hydrocarbon solvent.

Solvent group		**ESTERS**
Chemical name		2-Ethoxyethyl acetate
Chemical formula		$C_6H_{12}O_3$
Structural formula		$CH_3COOCH_2CH_2OCH_2CH_3$
CAS No.		111-15-9
Synonym/common name		
Trade names		CELLOSOLVE acetate
Suppliers		UNION CARBIDE, HULS
Molecular mass		132.16
Boiling point/range	°C @ 760 mm Hg	153–159
Melting point	°C	−62
Density	kg/m^3 @ 20 °C	973–975
Evaporation time[a]		57
Evaporation time[b]		0.20
Vapour pressure	kPa @ 20 °C	2.2
Flash point	°C	52
Auto ignition temperature	°C	400
Upper flammability limit	%v/v in air	7.2
Lower flammability limit	%v/v in air	3
Viscosity	mPa.s (cP) @ 20 °C	1.1
Refractive index	n 20 D	1.404
Solubility in water	%w/w @ 20 °C	22.9
Solubility of water	%w/w @ 20 °C	6.5
Solubility parameter[c]	$(J/cm^3)^{1/2}$	17.8
Hydrogen bond index		
Fractional polarity		
Nitro cellulose dil ratio[d]		
Nitro cellulose dil ratio[e]		
Surface tension	mN/m @ 20 °C	
Specific heat liquid	kJ/kg mole/°C @ 20 °C	
Latent heat	kJ/kg mole	
Dielectric constant	@ 20 °C	
Antoine constant A	kPa°C	
Antoine constant B	kPa°C	
Antoine constant C	kPa°C	
Heat of combustion	kJ/kg mole	
UN number		1172
IMO classification		3.3/III
ADR/RID classification		3,31°(c)
UK exposure limits	mg/m^3 (ppm) 8hrTWA/MEL or OES	54(10)
USA exposure limits	mg/m^3 (ppm) TLV/TWA	
German exposure limits	mg/m^3 (ppm) MAK-TWA	110(20)
EU classification		Repro Cat 2, Xn
EU risk phrases		60-61-20/21/22
EU safety phrases		53-45

[a] Ether = 1. [b] n-Butyl acetate = 1. [c] Nelson, Hemwall, Edwards. [d] Toluene. [e] Hydrocarbon solvent.

Solvent group		**ESTERS**
Chemical name		Methoxy propyl acetate
Chemical formula		
Structural formula		
CAS No.		108-65-6
Synonym/common name		Methoxypropyl acetate
Trade names		
Suppliers		BP
Molecular mass		132.2
Boiling point/range	°C @ 760 mm Hg	146
Melting point	°C	−88
Density	kg/m^3 @ 20 °C	968
Evaporation time[a]		
Evaporation time[b]		0.33
Vapour pressure	kPa @ 20 °C	0.49
Flash point	°C	50
Auto ignition temperature	°C	354
Upper flammability limit	%v/v in air	10.6
Lower flammability limit	%v/v in air	1.2
Viscosity	mPa.s (cP) @ 20 °C	1.23
Refractive index	n 20 D	1.403
Solubility in water	%w/w @ 20 °C	
Solubility of water	%w/w @ 20 °C	
Solubility parameter[c]	(J/cm^3)$^{1/2}$	
Hydrogen bond index		
Fractional polarity		
Nitro cellulose dil ratio[d]		
Nitro cellulose dil ratio[e]		
Surface tension	mN/m @ 20 °C	
Specific heat liquid	kJ/kg mole/°C @ 20 °C	
Latent heat	kJ/kg mole	
Dielectric constant	@ 20 °C	
Antoine constant A	kPa°C	
Antoine constant B	kPa°C	
Antoine constant C	kPa°C	
Heat of combustion	kJ/kg mole	
UN number		1993
IMO classification		3.3
ADR/RID classification		3,31°(c)
UK exposure limits	mg/m^3 (ppm) 8hrTWA/MEL or OES	
USA exposure limits	mg/m^3 (ppm) TLV/TWA	
German exposure limits	mg/m^3 (ppm) MAK-TWA	
EU classification		Xi, R10
EU risk phrases		10–36
EU safety phrases		(2-)25

[a] Ether = 1. [b] n-Butyl acetate = 1. [c] Nelson, Hemwall, Edwards. [d] Toluene. [e] Hydrocarbon solvent.

Property	Unit	Value
Solvent group		**ESTERS**
Chemical name		n-Pentyl propionate
Chemical formula		$C_8H_{16}O_2$
Structural formula		$CH_3CH_2COOCH_2CH_2CH_2CH_2CH_3$
CAS No.		624-52-4
Synonym/common name		n-Pentyl propionate
Trade names		UCAR n-Pentyl Propionate
Suppliers		UNION CARBIDE
Molecular mass		144.12
Boiling point/range	°C @ 760 mm Hg	164.9
Melting point	°C	
Density	kg/m^3 @ 20 °C	874
Evaporation time[a]		
Evaporation time[b]		0.18
Vapour pressure	kPa @ 20 °C	
Flash point	°C	
Auto ignition temperature	°C	
Upper flammability limit	%v/v in air	
Lower flammability limit	%v/v in air	
Viscosity	mPa.s (cP) @ 20 °C	1.0
Refractive index	n 20 D	
Solubility in water	%w/w @ 20 °C	<0.2
Solubility of water	%w/w @ 20 °C	<0.3
Solubility parameter[c]	(J/cm^3)$^{1/2}$	
Hydrogen bond index		
Fractional polarity		
Nitro cellulose dil ratio[d]		
Nitro cellulose dil ratio[e]		
Surface tension	mN/m @ 20 °C	
Specific heat liquid	kJ/kg mole/°C @ 20 °C	
Latent heat	kJ/kg mole	
Dielectric constant	@ 20 °C	
Antoine constant A	kPa°C	
Antoine constant B	kPa°C	
Antoine constant C	kPa°C	
Heat of combustion	kJ/kg mole	
UN number		
IMO classification		
ADR/RID classification		
UK exposure limits	mg/m^3 (ppm) 8hrTWA/MEL or OES	
USA exposure limits	mg/m^3 (ppm) TLV/TWA	
German exposure limits	mg/m^3 (ppm) MAK-TWA	
EU classification		R10
EU risk phrases		10
EU safety phrases		(2-)23

[a] Ether = 1. [b] n-Butyl acetate = 1. [c] Nelson, Hemwall, Edwards. [d] Toluene. [e] Hydrocarbon solvent.

Solvent group		ESTERS
Chemical name		Ethoxy propyl acetate
Chemical formula		
Structural formula		
CAS No.		54839-24-6
Synonym/common name		
Trade names		
Suppliers		BP
Molecular mass		146.2
Boiling point/range	°C @ 760 mm Hg	158
Melting point	°C	−89
Density	kg/m^3 @ 20 °C	941
Evaporation time[a]		
Evaporation time[b]		0.24
Vapour pressure	kPa @ 20 °C	0.227
Flash point	°C	54
Auto ignition temperature	°C	325
Upper flammability limit	%v/v in air	9.8
Lower flammability limit	%v/v in air	1.0
Viscosity	mPa.s (cP) @ 20 °C	1.4
Refractive index	n 20 D	1.405
Solubility in water	%w/w @ 20 °C	
Solubility of water	%w/w @ 20 °C	
Solubility parameter[c]	(J/cm^3)$^{1/2}$	
Hydrogen bond index		
Fractional polarity		
Nitro cellulose dil ratio[d]		
Nitro cellulose dil ratio[e]		
Surface tension	mN/m @ 20 °C	
Specific heat liquid	kJ/kg mole/°C @ 20 °C	
Latent heat	kJ/kg mole	
Dielectric constant	@ 20 °C	
Antoine constant A	kPa°C	
Antoine constant B	kPa°C	
Antoine constant C	kPa°C	
Heat of combustion	kJ/kg mole	
UN number		1993
IMO classification		3.3
ADR/RID classification		3,31°(c)
UK exposure limits	mg/m^3 (ppm) 8hrTWA/MEL or OES	
USA exposure limits	mg/m^3 (ppm) TLV/TWA	
German exposure limits	mg/m^3 (ppm) MAK-TWA	
EU classification		
EU risk phrases		10
EU safety phrases		16-24/25

[a]Ether = 1. [b]n-Butyl acetate = 1. [c]Nelson, Hemwall, Edwards. [d]Toluene. [e]Hydrocarbon solvent.

Solvent group		**ESTERS**
Chemical name		Dimethyl esters of adipic, glutaric and succinic acids
Chemical formula		
Structural formula		
CAS No.		106-65-0,627-93-0,1119-40-0
Synonym/common name		Estasol
Trade names		ESTASOL
Suppliers		CHEMOXY
Molecular mass		160
Boiling point/range	°C @ 760 mm Hg	200–230
Melting point	°C	−25
Density	kg/m^3 @ 20 °C	1090
Evaporation time[a]		
Evaporation time[b]		
Vapour pressure	kPa @ 20 °C	
Flash point	°C	108 (OPEN CUP)
Auto ignition temperature	°C	370
Upper flammability limit	%v/v in air	12.50
Lower flammability limit	%v/v in air	1.50
Viscosity	mPa.s (cP) @ 20 °C	
Refractive index	n 20 D	1.423–1.42
Solubility in water	%w/w @ 20 °C	5
Solubility of water	%w/w @ 20 °C	
Solubility parameter[c]	$(J/cm^3)^{1/2}$	
Hydrogen bond index		
Fractional polarity		
Nitro cellulose dil ratio[d]		
Nitro cellulose dil ratio[e]		
Surface tension	mN/m @ 20 °C	
Specific heat liquid	kJ/kg mole/°C @ 20 °C	
Latent heat	kJ/kg mole	
Dielectric constant	@ 20 °C	
Antoine constant A	kPa°C	
Antoine constant B	kPa°C	
Antoine constant C	kPa°C	
Heat of combustion	kJ/kg mole	
UN number		
IMO classification		
ADR/RID classification		
UK exposure limits	mg/m^3 (ppm) 8hrTWA/MEL or OES	
USA exposure limits	mg/m^3 (ppm) TLV/TWA	
German exposure limits	mg/m^3 (ppm) MAK-TWA	
EU classification		
EU risk phrases		
EU safety phrases		

[a] Ether = 1. [b] n-Butyl acetate = 1. [c] Nelson, Hemwall, Edwards. [d] Toluene. [e] Hydrocarbon solvent.

Solvent group		**ESTERS**
Chemical name		
Chemical formula		$C_8H_{16}O_3$
Structural formula		$CH_3COOCH_2CH_2OCH_2CH_2CH_2CH_3$
CAS No.		112-07-2
Synonym/common name		
Trade names		Butyl CELLOSOLVE acetate
Suppliers		UNION CARBIDE
Molecular mass		160.21
Boiling point/range	°C @ 760 mm Hg	192.3
Melting point	°C	
Density	kg/m^3 @ 20 °C	942
Evaporation time[a]		
Evaporation time[b]		0.04
Vapour pressure	kPa @ 20 °C	
Flash point	°C	
Auto ignition temperature	°C	
Upper flammability limit	%v/v in air	
Lower flammability limit	%v/v in air	
Viscosity	mPa.s (cP) @ 20 °C	1.8
Refractive index	n 20 D	
Solubility in water	%w/w @ 20 °C	1.5
Solubility of water	%w/w @ 20 °C	1.7
Solubility parameter[c]	$(J/cm^3)^{1/2}$	
Hydrogen bond index		
Fractional polarity		
Nitro cellulose dil ratio[d]		
Nitro cellulose dil ratio[e]		
Surface tension	mN/m @ 20 °C	
Specific heat liquid	kJ/kg mole/°C @ 20 °C	
Latent heat	kJ/kg mole	
Dielectric constant	@ 20 °C	
Antoine constant A	kPa°C	
Antoine constant B	kPa°C	
Antoine constant C	kPa°C	
Heat of combustion	kJ/kg mole	
UN number		
IMO classification		
ADR/RID classification		
UK exposure limits	mg/m^3 (ppm) 8hrTWA/MEL or OES	
USA exposure limits	mg/m^3 (ppm) TLV/TWA	
German exposure limits	mg/m^3 (ppm) MAK-TWA	
EU classification		
EU risk phrases		
EU safety phrases		

[a] Ether = 1. [b] n-Butyl acetate = 1. [c] Nelson, Hemwall, Edwards. [d] Toluene. [e] Hydrocarbon solvent.

Solvent group		**ESTERS**
Chemical name		2-Ethylhexyl acetate
Chemical formula		$C_{10}H_{20}O_2$
Structural formula		$CH_3COOCH_2CH(C_2H_5)(CH_2)_3CH_3$
CAS No.		103-09-3
Synonym/common name		2-Ethylhexyl acetate
Trade names		
Suppliers		HULS, BASF
Molecular mass		172.3
Boiling point/range	°C @ 760 mm Hg	192.0–205.0
Melting point	°C	−80
Density	kg/m^3 @ 20 °C	873
Evaporation time[a]		320–500
Evaporation time[b]		
Vapour pressure	kPa @ 20 °C	0.06
Flash point	°C	77 (DIN 51755 or 51758) BASF
Auto ignition temperature	°C	270–295
Upper flammability limit	%v/v in air	7.5
Lower flammability limit	%v/v in air	1.1
Viscosity	mPa.s (cP) @ 20 °C	1.5
Refractive index	n 20 D	1.420
Solubility in water	%w/w @ 20 °C	
Solubility of water	%w/w @ 20 °C	
Solubility parameter[c]	(J/cm^3)$^{1/2}$	
Hydrogen bond index		
Fractional polarity		
Nitro cellulose dil ratio[d]		
Nitro cellulose dil ratio[e]		
Surface tension	mN/m @ 20 °C	
Specific heat liquid	kJ/kg mole/°C @ 20 °C	
Latent heat	kJ/kg mole	
Dielectric constant	@ 20 °C	
Antoine constant A	kPa°C	
Antoine constant B	kPa°C	
Antoine constant C	kPa°C	
Heat of combustion	kJ/kg mole	
UN number		
IMO classification		
ADR/RID classification		3,32°(c)
UK exposure limits	mg/m^3 (ppm) 8hrTWA/MEL or OES	
USA exposure limits	mg/m^3 (ppm) TLV/TWA	
German exposure limits	mg/m^3 (ppm) MAK-TWA	
EU classification		
EU risk phrases		
EU safety phrases		

[a] Ether = 1. [b] n-Butyl acetate = 1. [c] Nelson, Hemwall, Edwards. [d] Toluene. [e] Hydrocarbon solvent.

Solvent group		**ESTERS**
Chemical name		2-(2-Ethoxyethoxy)ethyl acetate
Chemical formula		$C_8H_{16}O_3$
Structural formula		$C_2H_5OCH_2CH_2OCH_2CH_2OCOCH_3$
CAS No.		
Synonym/common name		
Trade names		
Suppliers		HULS
Molecular mass		176.2
Boiling point/range	°C @ 760 mm Hg	210–222
Melting point	°C	−25
Density	kg/m^3 @ 20 °C	1011
Evaporation time[a]		2400
Evaporation time[b]		
Vapour pressure	kPa @ 20 °C	0.013
Flash point	°C	107
Auto ignition temperature	°C	310
Upper flammability limit	%v/v in air	
Lower flammability limit	%v/v in air	1.5
Viscosity	mPa.s (cP) @ 20 °C	2.8
Refractive index	n 20 D	1.420–1.422
Solubility in water	%w/w @ 20 °C	
Solubility of water	%w/w @ 20 °C	Complete
Solubility parameter[c]	(J/cm^3)$^{1/2}$	19.0
Hydrogen bond index		
Fractional polarity		
Nitro cellulose dil ratio[d]		
Nitro cellulose dil ratio[e]		
Surface tension	mN/m @ 20 °C	
Specific heat liquid	kJ/kg mole/°C @ 20 °C	
Latent heat	kJ/kg mole	
Dielectric constant	@ 20 °C	
Antoine constant A	kPa°C	
Antoine constant B	kPa°C	
Antoine constant C	kPa°C	
Heat of combustion	kJ/kg mole	
UN number		
IMO classification		
ADR/RID classification		
UK exposure limits	mg/m^3 (ppm) 8hrTWA/MEL or OES	
USA exposure limits	mg/m^3 (ppm) TLV/TWA	
German exposure limits	mg/m^3 (ppm) MAK-TWA	
EU classification		
EU risk phrases		
EU safety phrases		

[a]Ether = 1. [b]n-Butyl acetate = 1. [c]Nelson, Hemwall, Edwards. [d]Toluene. [e]Hydrocarbon solvent.

Solvent group		ESTERS
Chemical name		
Chemical formula		$C_{10}H_{20}O_4$
Structural formula		$C_4H_9O[C_2H_4O]_2COCH_3$
CAS No.		124-17-4
Synonym/common name		
Trade names		Butyl CARBITOL acetate
Suppliers		UNION CARBIDE
Molecular mass		204.27
Boiling point/range	°C @ 760 mm Hg	246.7
Melting point	°C	
Density	kg/m^3 @ 20 °C	980
Evaporation time[a]		
Evaporation time[b]		<0.01
Vapour pressure	kPa @ 20 °C	
Flash point	°C	
Auto ignition temperature	°C	
Upper flammability limit	%v/v in air	
Lower flammability limit	%v/v in air	
Viscosity	mPa.s (cP) @ 20 °C	3.6
Refractive index	n 20 D	
Solubility in water	%w/w @ 20 °C	6.5
Solubility of water	%w/w @ 20 °C	3.7
Solubility parameter[c]	$(J/cm^3)^{1/2}$	
Hydrogen bond index		
Fractional polarity		
Nitro cellulose dil ratio[d]		
Nitro cellulose dil ratio[e]		
Surface tension	mN/m @ 20 °C	
Specific heat liquid	kJ/kg mole/°C @ 20 °C	
Latent heat	kJ/kg mole	
Dielectric constant	@ 20 °C	
Antoine constant A	kPa°C	
Antoine constant B	kPa°C	
Antoine constant C	kPa°C	
Heat of combustion	kJ/kg mole	
UN number		
IMO classification		
ADR/RID classification		
UK exposure limits	mg/m^3 (ppm) 8hrTWA/MEL or OES	
USA exposure limits	mg/m^3 (ppm) TLV/TWA	
German exposure limits	mg/m^3 (ppm) MAK-TWA	
EU classification		
EU risk phrases		
EU safety phrases		

[a] Ether = 1. [b] n-Butyl acetate = 1. [c] Nelson, Hemwall, Edwards. [d] Toluene. [e] Hydrocarbon solvent.

Solvent group		ESTERS
Chemical name		2-(Butoxyethoxy)ethyl acetate
Chemical formula		
Structural formula		
CAS No.		124-17-4
Synonym/common name		Butyl diglycol acetate
Trade names		BUTYL DIGLYCOL ACETATE
Suppliers		BP
Molecular mass		204.3
Boiling point/range	°C @ 760 mm Hg	247
Melting point	°C	−32
Density	kg/m^3 @ 20 °C	985
Evaporation time[a]		
Evaporation time[b]		
Vapour pressure	kPa @ 20 °C	0.0013
Flash point	°C	115 (OPEN CUP)
Auto ignition temperature	°C	299
Upper flammability limit	%v/v in air	
Lower flammability limit	%v/v in air	
Viscosity	mPa.s (cP) @ 20 °C	3.6
Refractive index	n 20 D	1.426
Solubility in water	%w/w @ 20 °C	
Solubility of water	%w/w @ 20 °C	
Solubility parameter[c]	$(J/cm^3)^{1/2}$	
Hydrogen bond index		
Fractional polarity		
Nitro cellulose dil ratio[d]		
Nitro cellulose dil ratio[e]		
Surface tension	mN/m @ 20 °C	
Specific heat liquid	kJ/kg mole/°C @ 20 °C	
Latent heat	kJ/kg mole	
Dielectric constant	@ 20 °C	
Antoine constant A	kPa°C	
Antoine constant B	kPa°C	
Antoine constant C	kPa°C	
Heat of combustion	kJ/kg mole	
UN number		
IMO classification		
ADR/RID classification		
UK exposure limits	mg/m^3 (ppm) 8hrTWA/MEL or OES	
USA exposure limits	mg/m^3 (ppm) TLV/TWA	
German exposure limits	mg/m^3 (ppm) MAK-TWA	
EU classification		
EU risk phrases		
EU safety phrases		

[a] Ether = 1. [b] n-Butyl acetate = 1. [c] Nelson, Hemwall, Edwards. [d] Toluene. [e] Hydrocarbon solvent.

Property	Unit	Value
Solvent group		**ESTERS**
Chemical name		[2-(2-Butoxyethoxy)ethyl acetate]
Chemical formula		$C_{10}H_{20}O_4$
Structural formula		$C_4H_9OCH_2CH_2OCH_2CH_2OCOCH_3$
CAS No.		
Synonym/common name		Butyl diglycol acetate
Trade names		
Suppliers		HULS
Molecular mass		204.3
Boiling point/range	°C @ 760 mm Hg	235–250
Melting point	°C	−32
Density	kg/m^3 @ 20 °C	978
Evaporation time[a]		3000
Evaporation time[b]		
Vapour pressure	kPa @ 20 °C	0.0013
Flash point	°C	108
Auto ignition temperature	°C	290
Upper flammability limit	%v/v in air	
Lower flammability limit	%v/v in air	0.6
Viscosity	mPa.s (cP) @ 20 °C	3.6
Refractive index	n 20 D	1.425–1.42
Solubility in water	%w/w @ 20 °C	
Solubility of water	%w/w @ 20 °C	
Solubility parameter[c]	(J/cm^3)$^{1/2}$	18.6
Hydrogen bond index		
Fractional polarity		
Nitro cellulose dil ratio[d]		
Nitro cellulose dil ratio[e]		
Surface tension	mN/m @ 20 °C	
Specific heat liquid	kJ/kg mole/°C @ 20 °C	
Latent heat	kJ/kg mole	
Dielectric constant	@ 20 °C	
Antoine constant A	kPa°C	
Antoine constant B	kPa°C	
Antoine constant C	kPa°C	
Heat of combustion	kJ/kg mole	
UN number		
IMO classification		
ADR/RID classification		
UK exposure limits	mg/m^3 (ppm) 8hrTWA/MEL or OES	
USA exposure limits	mg/m^3 (ppm) TLV/TWA	
German exposure limits	mg/m^3 (ppm) MAK-TWA	
EU classification		
EU risk phrases		
EU safety phrases		

[a] Ether = 1. [b] n-Butyl acetate = 1. [c] Nelson, Hemwall, Edwards. [d] Toluene. [e] Hydrocarbon solvent.

Property	Units	Value
Solvent group		**ESTERS**
Chemical name		Di-isobutyl esters of adipic acid, glutaric acid and succinic acid
Chemical formula		
Structural formula		
CAS No.		141-04-8, 71195-64-7, 925-06-4
Synonym/common name		
Trade names		COASOL
Suppliers		CHEMOXY
Molecular mass		244
Boiling point/range	°C @ 760 mm Hg	274–289
Melting point	°C	−60
Density	kg/m^3 @ 20 °C	
Evaporation time[a]		
Evaporation time[b]		
Vapour pressure	kPa @ 20 °C	
Flash point	°C	131
Auto ignition temperature	°C	400
Upper flammability limit	%v/v in air	
Lower flammability limit	%v/v in air	
Viscosity	mPa.s (cP) @ 20 °C	5.3
Refractive index	n 20 D	1.427–1.42
Solubility in water	%w/w @ 20 °C	0.1
Solubility of water	%w/w @ 20 °C	
Solubility parameter[c]	$(J/cm^3)^{1/2}$	
Hydrogen bond index		
Fractional polarity		
Nitro cellulose dil ratio[d]		
Nitro cellulose dil ratio[e]		
Surface tension	mN/m @ 20 °C	
Specific heat liquid	kJ/kg mole/°C @ 20 °C	
Latent heat	kJ/kg mole	
Dielectric constant	@ 20 °C	
Antoine constant A	kPa°C	
Antoine constant B	kPa°C	
Antoine constant C	kPa°C	
Heat of combustion	kJ/kg mole	
UN number		
IMO classification		
ADR/RID classification		
UK exposure limits	mg/m^3 (ppm) 8hrTWA/MEL or OES	
USA exposure limits	mg/m^3 (ppm) TLV/TWA	
German exposure limits	mg/m^3 (ppm) MAK-TWA	
EU classification		
EU risk phrases		
EU safety phrases		

[a] Ether = 1. [b] n-Butyl acetate = 1. [c] Nelson, Hemwall, Edwards. [d] Toluene. [e] Hydrocarbon solvent.

Solvent group		**ETHERS**
Chemical name		Tetrahydrofuran
Chemical formula		C_4H_8O
Structural formula		
CAS No.		109-99-9
Synonym/common name		THF
Trade names		
Suppliers		HULS, BASF
Molecular mass		72.1
Boiling point/range	°C @ 760 mm Hg	65.5–66.5
Melting point	°C	−108
Density	kg/m^3 @ 20 °C	888
Evaporation time[a]		2.0–2.3
Evaporation time[b]		
Vapour pressure	kPa @ 20 °C	17.3
Flash point	°C	−22 (DIN 51755 or 51758) BASF, −21.5 HULS
Auto ignition temperature	°C	212
Upper flammability limit	%v/v in air	12
Lower flammability limit	%v/v in air	1.5
Viscosity	mPa.s (cP) @ 20 °C	0.54
Refractive index	n 20 D	1.406–1.40
Solubility in water	%w/w @ 20 °C	
Solubility of water	%w/w @ 20 °C	Complete
Solubility parameter[c]	(J/cm^3)$^{1/2}$	19.2
Hydrogen bond index		
Fractional polarity		
Nitro cellulose dil ratio[d]		
Nitro cellulose dil ratio[e]		
Surface tension	mN/m @ 20 °C	27.31
Specific heat liquid	kJ/kg mole/°C @ 20 °C	
Latent heat	kJ/kg mole	29815
Dielectric constant	@ 20 °C	
Antoine constant A	kPa°C	6.79696
Antoine constant B	kPa°C	1557.06
Antoine constant C	kPa°C	260.05
Heat of combustion	kJ/kg mole	−3149700
UN number		2056
IMO classification		3.1/II
ADR/RID classification		3,3°(b)
UK exposure limits	mg/m^3 (ppm) 8hrTWA/MEL or OES	295(100)
USA exposure limits	mg/m^3 (ppm) TLV/TWA	590(200)
German exposure limits	mg/m^3 (ppm) MAK-TWA	590(200)
EU classification		F, Xi
EU risk phrases		11-19-36/37
EU safety phrases		(2-)16-29-33

[a] Ether = 1. [b] n-Butyl acetate = 1. [c] Nelson, Hemwall, Edwards. [d] Toluene. [e] Hydrocarbon solvent.

Solvent group		**ETHERS**
Chemical name		Diethyl ether
Chemical formula		$C_4H_{10}O$
Structural formula		$CH_3CH_2OCH_2CH_3$
CAS No.		60-29-7
Synonym/common name		Diethyl ether
Trade names		
Suppliers		HULS
Molecular mass		74.1
Boiling point/range	°C @ 760 mm Hg	34.5
Melting point	°C	−116
Density	kg/m^3 @ 20 °C	714
Evaporation time[a]		1
Evaporation time[b]		
Vapour pressure	kPa @ 20 °C	58.8
Flash point	°C	−40
Auto ignition temperature	°C	170
Upper flammability limit	%v/v in air	48
Lower flammability limit	%v/v in air	1.7
Viscosity	mPa.s (cP) @ 20 °C	0.24
Refractive index	n 20 D	1.353
Solubility in water	%w/w @ 20 °C	
Solubility of water	%w/w @ 20 °C	1.2
Solubility parameter[c]	$(J/cm^3)^{1/2}$	15.6
Hydrogen bond index		
Fractional polarity		
Nitro cellulose dil ratio[d]		
Nitro cellulose dil ratio[e]		
Surface tension	mN/m @ 20 °C	
Specific heat liquid	kJ/kg mole/°C @ 20 °C	
Latent heat	kJ/kg mole	26508
Dielectric constant	@ 20 °C	4.335
Antoine constant A	kPa°C	6.05115
Antoine constant B	kPa°C	1062.409
Antoine constant C	kPa°C	228.183
Heat of combustion	kJ/kg mole	−2741100
UN number		1155
IMO classification		3.1/I
ADR/RID classification		3,2°(a)
UK exposure limits	mg/m^3 (ppm) 8hrTWA/MEL or OES	1200(400)
USA exposure limits	mg/m^3 (ppm) TLV/TWA	1210(400)
German exposure limits	mg/m^3 (ppm) MAK-TWA	1200(400)
EU classification		F+
EU risk phrases		12-19
EU safety phrases		(2-)9-16-29-33

[a] Ether = 1. [b] n-Butyl acetate = 1. [c] Nelson, Hemwall, Edwards. [d] Toluene. [e] Hydrocarbon solvent.

Solvent group		ETHERS
Chemical name		Methyl tert-butyl ether
Chemical formula		$C_5H_{12}O$
Structural formula		$CH_3OC(CH_3)_3$
CAS No.		
Synonym/common name		
Trade names		DRIVERON S
Suppliers		HULS
Molecular mass		88.2
Boiling point/range	°C @ 760 mm Hg	55.3
Melting point	°C	−109
Density	kg/m^3 @ 20 °C	
Evaporation time[a]		1.6
Evaporation time[b]		
Vapour pressure	kPa @ 20 °C	27.1
Flash point	°C	−28
Auto ignition temperature	°C	460
Upper flammability limit	%v/v in air	8.4
Lower flammability limit	%v/v in air	1.6
Viscosity	mPa.s (cP) @ 20 °C	0.36
Refractive index	n 20 D	1.369
Solubility in water	%w/w @ 20 °C	
Solubility of water	%w/w @ 20 °C	1.3
Solubility parameter[c]	(J/cm^3)$^{1/2}$	
Hydrogen bond index		
Fractional polarity		
Nitro cellulose dil ratio[d]		
Nitro cellulose dil ratio[e]		
Surface tension	mN/m @ 20 °C	
Specific heat liquid	kJ/kg mole/°C @ 20 °C	
Latent heat	kJ/kg mole	
Dielectric constant	@ 20 °C	
Antoine constant A	kPa°C	
Antoine constant B	kPa°C	
Antoine constant C	kPa°C	
Heat of combustion	kJ/kg mole	
UN number		2398
IMO classification		3.1/II
ADR/RID classification		3,3°(b)
UK exposure limits	mg/m^3 (ppm) 8hrTWA/MEL or OES	
USA exposure limits	mg/m^3 (ppm) TLV/TWA	
German exposure limits	mg/m^3 (ppm) MAK-TWA	
EU classification		
EU risk phrases		
EU safety phrases		

[a] Ether = 1. [b] n-Butyl acetate = 1. [c] Nelson, Hemwall, Edwards. [d] Toluene. [e] Hydrocarbon solvent.

Solvent group		**FATTY ACID ESTER**
Chemical name		Ethoxylated fatty acid ester
Chemical formula		
Structural formula		
CAS No.		
Synonym/common name		
Trade names		PROTENE R
Suppliers		ELF ATOCHEM
Molecular mass		
Boiling point/range	°C @ 760 mm Hg	
Melting point	°C	
Density	kg/m^3 @ 20 °C	1000
Evaporation time[a]		
Evaporation time[b]		
Vapour pressure	kPa @ 20 °C	NEGLIGIBLE
Flash point	°C	>110 CLOSED CUP ASTM 3828 or IP 303 (SETA)
Auto ignition temperature	°C	
Upper flammability limit	%v/v in air	
Lower flammability limit	%v/v in air	
Viscosity	mPa.s (cP) @ 20 °C	60–70
Refractive index	n 20 D	
Solubility in water	%w/w @ 20 °C	Complete
Solubility of water	%w/w @ 20 °C	
Solubility parameter[c]	$(J/cm^3)^{1/2}$	
Hydrogen bond index		
Fractional polarity		
Nitro cellulose dil ratio[d]		
Nitro cellulose dil ratio[e]		
Surface tension	mN/m @ 20 °C	
Specific heat liquid	kJ/kg mole/°C @ 20 °C	
Latent heat	kJ/kg mole	
Dielectric constant	@ 20 °C	
Antoine constant A	kPa°C	
Antoine constant B	kPa°C	
Antoine constant C	kPa°C	
Heat of combustion	kJ/kg mole	
UN number		
IMO classification		
ADR/RID classification		
UK exposure limits	mg/m^3 (ppm) 8hrTWA/MEL or OES	
USA exposure limits	mg/m^3 (ppm) TLV/TWA	
German exposure limits	mg/m^3 (ppm) MAK-TWA	
EU classification		
EU risk phrases		
EU safety phrases		

[a] Ether = 1. [b] n-Butyl acetate = 1. [c] Nelson, Hemwall, Edwards. [d] Toluene. [e] Hydrocarbon solvent.

Property	Unit	Value
Solvent group		**FATTY ACID ESTER**
Chemical name		Formulated fatty acid ester
Chemical formula		
Structural formula		
CAS No.		
Synonym/common name		
Trade names		PROTENE F
Suppliers		ELF ATOCHEM
Molecular mass		
Boiling point/range	°C @ 760 mm Hg	230
Melting point	°C	
Density	kg/m^3 @ 20 °C	900
Evaporation time[a]		
Evaporation time[b]		
Vapour pressure	kPa @ 20 °C	NEGLIGIBLE
Flash point	°C	>110 CLOSED CUP ASTM 3828 or IP 303 (SETA)
Auto ignition temperature	°C	
Upper flammability limit	%v/v in air	
Lower flammability limit	%v/v in air	
Viscosity	mPa.s (cP) @ 20 °C	7.5
Refractive index	n 20 D	
Solubility in water	%w/w @ 20 °C	0.5
Solubility of water	%w/w @ 20 °C	
Solubility parameter[c]	$(J/cm^3)^{1/2}$	
Hydrogen bond index		
Fractional polarity		
Nitro cellulose dil ratio[d]		
Nitro cellulose dil ratio[e]		
Surface tension	mN/m @ 20 °C	
Specific heat liquid	kJ/kg mole/°C @ 20 °C	
Latent heat	kJ/kg mole	
Dielectric constant	@ 20 °C	
Antoine constant A	kPa°C	
Antoine constant B	kPa°C	
Antoine constant C	kPa°C	
Heat of combustion	kJ/kg mole	
UN number		
IMO classification		
ADR/RID classification		
UK exposure limits	mg/m^3 (ppm) 8hrTWA/MEL or OES	
USA exposure limits	mg/m^3 (ppm) TLV/TWA	
German exposure limits	mg/m^3 (ppm) MAK-TWA	
EU classification		
EU risk phrases		
EU safety phrases		

[a] Ether = 1. [b] n-Butyl acetate = 1. [c] Nelson, Hemwall, Edwards. [d] Toluene. [e] Hydrocarbon solvent.

Solvent group		GLYCOL ESTERS
Chemical name		
Chemical formula		
Structural formula		
CAS No.		
Synonym/common name		
Trade names		RHODIASOLV RPDE
Suppliers		RHÔNE POULENC
Molecular mass		160
Boiling point/range	°C @ 760 mm Hg	>200
Melting point	°C	20
Density	kg/m^3 @ 20 °C	1090
Evaporation time[a]		
Evaporation time[b]		
Vapour pressure	kPa @ 20 °C	0.08
Flash point	°C	>100
Auto ignition temperature	°C	
Upper flammability limit	%v/v in air	
Lower flammability limit	%v/v in air	
Viscosity	mPa.s (cP) @ 20 °C	
Refractive index	n 20 D	
Solubility in water	%w/w @ 20 °C	
Solubility of water	%w/w @ 20 °C	
Solubility parameter[c]	(J/cm^3)$^{1/2}$	
Hydrogen bond index		
Fractional polarity		
Nitro cellulose dil ratio[d]		
Nitro cellulose dil ratio[e]		
Surface tension	mN/m @ 20 °C	
Specific heat liquid	kJ/kg mole/°C @ 20 °C	
Latent heat	kJ/kg mole	
Dielectric constant	@ 20 °C	
Antoine constant A	kPa°C	
Antoine constant B	kPa°C	
Antoine constant C	kPa°C	
Heat of combustion	kJ/kg mole	
UN number		
IMO classification		
ADR/RID classification		
UK exposure limits	mg/m^3 (ppm) 8hrTWA/MEL or OES	
USA exposure limits	mg/m^3 (ppm) TLV/TWA	
German exposure limits	mg/m^3 (ppm) MAK-TWA	
EU classification		
EU risk phrases		
EU safety phrases		

[a] Ether = 1. [b] n-Butyl acetate = 1. [c] Nelson, Hemwall, Edwards. [d] Toluene. [e] Hydrocarbon solvent.

Solvent group		GLYCOL ESTERS
Chemical name		1,2-Diacetoxy-propane
Chemical formula		$C_7H_{12}O_4$
Structural formula		$CH_3COOCH_2CHCH_3OCOCH_3$
CAS No.		623-84-7
Synonym/common name		Propylene glycol diacetate
Trade names		DOWANOL PGDA
Suppliers		DOW
Molecular mass		160
Boiling point/range	°C @ 760 mm Hg	190
Melting point	°C	< −75
Density	kg/m^3 @ 20 °C	1056
Evaporation time[a]		250
Evaporation time[b]		0.04
Vapour pressure	kPa @ 20 °C	
Flash point	°C	86
Auto ignition temperature	°C	431
Upper flammability limit	%v/v in air	12.7
Lower flammability limit	%v/v in air	2.8
Viscosity	mPa.s (cP) @ 20 °C	2.97
Refractive index	n 20 D	1.415
Solubility in water	%w/w @ 20 °C	8
Solubility of water	%w/w @ 20 °C	4.3
Solubility parameter[c]	(J/cm^3)$^{1/2}$	
Hydrogen bond index		
Fractional polarity		
Nitro cellulose dil ratio[d]		
Nitro cellulose dil ratio[e]		
Surface tension	mN/m @ 20 °C	32.9
Specific heat liquid	kJ/kg mole/°C @ 20 °C	515.2 (25 °C)
Latent heat	kJ/kg mole	
Dielectric constant	@ 20 °C	
Antoine constant A	kPa°C	
Antoine constant B	kPa°C	
Antoine constant C	kPa°C	
Heat of combustion	kJ/kg mole	
UN number		
IMO classification		
ADR/RID classification		3,32°(c)
UK exposure limits	mg/m^3 (ppm) 8hrTWA/MEL or OES	
USA exposure limits	mg/m^3 (ppm) TLV/TWA	
German exposure limits	mg/m^3 (ppm) MAK-TWA	
EU classification		
EU risk phrases		
EU safety phrases		

[a] Ether = 1. [b] n-Butyl acetate = 1. [c] Nelson, Hemwall, Edwards. [d] Toluene. [e] Hydrocarbon solvent.

Solvent group		**GLYCOL ESTERS**
Chemical name		1n-Butoxy-propyl-acetate-2
Chemical formula		$C_9H_{18}O_3$
Structural formula		$C_4H_9OCH_2CH(CH_3)OCOCH_3$
CAS No.		85409-76-3
Synonym/common name		Propylene glycol n-butyl ether acetate
Trade names		DOWANOL PnBA
Suppliers		DOW
Molecular mass		174
Boiling point/range	°C @ 760 mm Hg	190
Melting point	°C	< -75
Density	kg/m^3 @ 20 °C	920
Evaporation time[a]		200
Evaporation time[b]		0.05
Vapour pressure	kPa @ 20 °C	
Flash point	°C	77 (PENSKY MARTENS, CLOSED CUP)
Auto ignition temperature	°C	279
Upper flammability limit	%v/v in air	
Lower flammability limit	%v/v in air	
Viscosity	mPa.s (cP) @ 20 °C	1.83
Refractive index	n 20 D	1.411
Solubility in water	%w/w @ 20 °C	0.3
Solubility of water	%w/w @ 20 °C	1.1
Solubility parameter[c]	$(J/cm^3)^{1/2}$	
Hydrogen bond index		
Fractional polarity		
Nitro cellulose dil ratio[d]		
Nitro cellulose dil ratio[e]		
Surface tension	mN/m @ 20 °C	24.7
Specific heat liquid	kJ/kg mole/°C @ 20 °C	285.4 (25 °C)
Latent heat	kJ/kg mole	
Dielectric constant	@ 20 °C	
Antoine constant A	kPa°C	
Antoine constant B	kPa°C	
Antoine constant C	kPa°C	
Heat of combustion	kJ/kg mole	
UN number		
IMO classification		
ADR/RID classification		
UK exposure limits	mg/m^3 (ppm) 8hrTWA/MEL or OES	
USA exposure limits	mg/m^3 (ppm) TLV/TWA	
German exposure limits	mg/m^3 (ppm) MAK-TWA	
EU classification		
EU risk phrases		36/38
EU safety phrases		41

[a] Ether = 1. [b] n-Butyl acetate = 1. [c] Nelson, Hemwall, Edwards. [d] Toluene. [e] Hydrocarbon solvent.

Property	Unit	GLYCOL ETHER ESTERS
Solvent group		**GLYCOL ETHER ESTERS**
Chemical name		Dipropylene glycol methyl ether acetate
Chemical formula		$C_9H_{18}O_4$
Structural formula		$CH_3O[CH_2(CH_3)O]_2OCH_3$
CAS No.		
Synonym/common name		
Trade names		DOWANOL DPMA
Suppliers		DOW
Molecular mass		
Boiling point/range	°C @ 760 mm Hg	209.3
Melting point	°C	
Density	kg/m^3 @ 20 °C	
Evaporation time[a]		
Evaporation time[b]		<0.01
Vapour pressure	kPa @ 20 °C	
Flash point	°C	
Auto ignition temperature	°C	
Upper flammability limit	%v/v in air	
Lower flammability limit	%v/v in air	
Viscosity	mPa.s (cP) @ 20 °C	
Refractive index	n 20 D	
Solubility in water	%w/w @ 20 °C	19.4
Solubility of water	%w/w @ 20 °C	3.5
Solubility parameter[c]	(J/cm^3)$^{1/2}$	
Hydrogen bond index		
Fractional polarity		
Nitro cellulose dil ratio[d]		
Nitro cellulose dil ratio[e]		
Surface tension	mN/m @ 20 °C	
Specific heat liquid	kJ/kg mole/°C @ 20 °C	
Latent heat	kJ/kg mole	
Dielectric constant	@ 20 °C	
Antoine constant A	kPa°C	
Antoine constant B	kPa°C	
Antoine constant C	kPa°C	
Heat of combustion	kJ/kg mole	
UN number		
IMO classification		
ADR/RID classification		
UK exposure limits	mg/m^3 (ppm) 8hrTWA/MEL or OES	
USA exposure limits	mg/m^3 (ppm) TLV/TWA	
German exposure limits	mg/m^3 (ppm) MAK-TWA	
EU classification		
EU risk phrases		
EU safety phrases		

[a] Ether = 1. [b] n-Butyl acetate = 1. [c] Nelson, Hemwall, Edwards. [d] Toluene. [e] Hydrocarbon solvent.

Solvent group		**GLYCOL ETHER ESTERS**
Chemical name		1-Methoxy-2-propanol acetate
Chemical formula		$C_6H_{12}O_3$
Structural formula		$CH_3OCH_2(CH_3)OCOCH_3$
CAS No.		108-65-6
Synonym/common name		Methoxy propyl acetate, methyl PROXITOL
Trade names		MePROXAc
Suppliers		BASF, SHELL
Molecular mass		132.2
Boiling point/range	°C @ 760 mm Hg	143.0–149.0
Melting point	°C	−65
Density	kg/m³ @ 20 °C	964–967
Evaporation time[a]		33
Evaporation time[b]		0.3
Vapour pressure	kPa @ 20 °C	0.42
Flash point	°C	45 (DIN 51755 or 51758) BASF, 45 (IP 170) SHELL
Auto ignition temperature	°C	315
Upper flammability limit	%v/v in air	7.0–10.8
Lower flammability limit	%v/v in air	1.5
Viscosity	mPa.s (cP) @ 20 °C	1.2–1.31
Refractive index	n 20 D	1.402
Solubility in water	%w/w @ 20 °C	23.0
Solubility of water	%w/w @ 20 °C	5.5
Solubility parameter[c]	$(J/cm^3)^{1/2}$	17.4
Hydrogen bond index		10
Fractional polarity		0.09
Nitro cellulose dil ratio[d]		2.5
Nitro cellulose dil ratio[e]		0.4
Surface tension	mN/m @ 20 °C	27.6
Specific heat liquid	kJ/kg mole/°C @ 20 °C	239.2
Latent heat	kJ/kg mole	40308
Dielectric constant	@ 20 °C	8.3
Antoine constant A	kPa°C	6.02968
Antoine constant B	kPa°C	1353.82
Antoine constant C	kPa°C	192.628
Heat of combustion	kJ/kg mole	−3198800
UN number		1993
IMO classification		3.2/III
ADR/RID classification		3,31°(c)
UK exposure limits	mg/m³ (ppm) 8hrTWA/MEL or OES	
USA exposure limits	mg/m³ (ppm) TLV/TWA	
German exposure limits	mg/m³ (ppm) MAK-TWA	
EU classification		Xi
EU risk phrases		10-36
EU safety phrases		25

[a] Ether = 1. [b] n-Butyl acetate = 1. [c] Nelson, Hemwall, Edwards. [d] Toluene. [e] Hydrocarbon solvent.

Solvent group		**GLYCOL ETHER ESTERS**
Chemical name		Butyl glycol acetate
Chemical formula		$C_8H_{16}O_3$
Structural formula		$CH_3(CH_2)_3OCH_2CH_2OCOCH_3$
CAS No.		112-07-2
Synonym/common name		Butyl glycol acetate
Trade names		
Suppliers		BASF
Molecular mass		160.2
Boiling point/range	°C @ 760 mm Hg	184.0–195.0
Melting point	°C	
Density	kg/m^3 @ 20 °C	942
Evaporation time[a]		190
Evaporation time[b]		
Vapour pressure	kPa @ 20 °C	
Flash point	°C	78 (DIN 51755 or 51758)
Auto ignition temperature	°C	280
Upper flammability limit	%v/v in air	8.4
Lower flammability limit	%v/v in air	1.7
Viscosity	mPa.s (cP) @ 20 °C	1.8
Refractive index	n 20 D	1.414
Solubility in water	%w/w @ 20 °C	
Solubility of water	%w/w @ 20 °C	
Solubility parameter[c]	$(J/cm^3)^{1/2}$	
Hydrogen bond index		
Fractional polarity		
Nitro cellulose dil ratio[d]		
Nitro cellulose dil ratio[e]		
Surface tension	mN/m @ 20 °C	
Specific heat liquid	kJ/kg mole/°C @ 20 °C	
Latent heat	kJ/kg mole	
Dielectric constant	@ 20 °C	
Antoine constant A	kPa°C	
Antoine constant B	kPa°C	
Antoine constant C	kPa°C	
Heat of combustion	kJ/kg mole	
UN number		
IMO classification		
ADR/RID classification		
UK exposure limits	mg/m^3 (ppm) 8hrTWA/MEL or OES	
USA exposure limits	mg/m^3 (ppm) TLV/TWA	
German exposure limits	mg/m^3 (ppm) MAK-TWA	
EU classification		
EU risk phrases		
EU safety phrases		

[a] Ether = 1. [b] n-Butyl acetate = 1. [c] Nelson, Hemwall, Edwards. [d] Toluene. [e] Hydrocarbon solvent.

Solvent group		**GLYCOL ETHER ESTERS**
Chemical name		2-(2-Butoxyethoxy)-ethanol acetate/ Diethylene glycol n-butyl ether
Chemical formula		$C_{10}H_{20}O_4$
Structural formula		$C_4H_9OCH_2CH_2OCH_2CH_2OCOCH_3$
CAS No.		124-17-4
Synonym/common name		Butyldiglycol acetate, butyl DIOXITOL
Trade names		nBuDIOXAc
Suppliers		BASF, SHELL
Molecular mass		204.3
Boiling point/range	°C @ 760 mm Hg	235.0–250.0
Melting point	°C	−32
Density	kg/m³ @ 20 °C	975–985
Evaporation time[a]		8200
Evaporation time[b]		0.001
Vapour pressure	kPa @ 20 °C	0.0003
Flash point	°C	116 (ASTM D92-COC) SHELL, 102 (DIN 51755 or 51758) BASF
Auto ignition temperature	°C	265–290
Upper flammability limit	%v/v in air	5.3
Lower flammability limit	%v/v in air	0.6–1.0
Viscosity	mPa.s (cP) @ 20 °C	3.0–3.5
Refractive index	n 20 D	1.426
Solubility in water	%w/w @ 20 °C	6.5
Solubility of water	%w/w @ 20 °C	3.7
Solubility parameter[c]	$(J/cm^3)^{1/2}$	17.4
Hydrogen bond index		9
Fractional polarity		0.02
Nitro cellulose dil ratio[d]		1.9
Nitro cellulose dil ratio[e]		0.7
Surface tension	mN/m @ 20 °C	32.2
Specific heat liquid	kJ/kg mole/°C @ 20 °C	394.4
Latent heat	kJ/kg mole	48745
Dielectric constant	@ 20 °C	7.0
Antoine constant A	kPa°C	6.64459
Antoine constant B	kPa°C	1910.67
Antoine constant C	kPa°C	167.032
Heat of combustion	kJ/kg mole	−5530809
UN number		
IMO classification		
ADR/RID classification		
UK exposure limits	mg/m³ (ppm) 8hrTWA/MEL or OES	
USA exposure limits	mg/m³ (ppm) TLV/TWA	
German exposure limits	mg/m³ (ppm) MAK-TWA	
EU classification		
EU risk phrases		
EU safety phrases		24

[a]Ether = 1. [b]n-Butyl acetate = 1. [c]Nelson, Hemwall, Edwards. [d]Toluene. [e]Hydrocarbon solvent.

Solvent group		**GLYCOL ETHERS**
Chemical name		
Chemical formula		$C_6H_{14}O_3$
Structural formula		$C_2H_5OCH_2CH_2OCH_2CH_2OH$
CAS No.		
Synonym/common name		
Trade names		SOLVENT APV SPECIAL
Suppliers		HULS
Molecular mass		
Boiling point/range	°C @ 760 mm Hg	198–203
Melting point	°C	
Density	kg/m^3 @ 20 °C	
Evaporation time[a]		1200
Evaporation time[b]		
Vapour pressure	kPa @ 20 °C	0.01
Flash point	°C	94
Auto ignition temperature	°C	220
Upper flammability limit	%v/v in air	11.6
Lower flammability limit	%v/v in air	1.2
Viscosity	mPa.s (cP) @ 20 °C	4.4
Refractive index	n 20 D	1.426–1.42
Solubility in water	%w/w @ 20 °C	
Solubility of water	%w/w @ 20 °C	Complete
Solubility parameter[c]	(J/cm^3)$^{1/2}$	
Hydrogen bond index		
Fractional polarity		
Nitro cellulose dil ratio[d]		
Nitro cellulose dil ratio[e]		
Surface tension	mN/m @ 20 °C	
Specific heat liquid	kJ/kg mole/°C @ 20 °C	
Latent heat	kJ/kg mole	
Dielectric constant	@ 20 °C	
Antoine constant A	kPa°C	
Antoine constant B	kPa°C	
Antoine constant C	kPa°C	
Heat of combustion	kJ/kg mole	
UN number		
IMO classification		
ADR/RID classification		3,32°(c)
UK exposure limits	mg/m^3 (ppm) 8hrTWA/MEL or OES	
USA exposure limits	mg/m^3 (ppm) TLV/TWA	
German exposure limits	mg/m^3 (ppm) MAK-TWA	
EU classification		
EU risk phrases		
EU safety phrases		

[a] Ether = 1. [b] n-Butyl acetate = 1. [c] Nelson, Hemwall, Edwards. [d] Toluene. [e] Hydrocarbon solvent.

Solvent group		GLYCOL ETHERS
Chemical name		Alkoxypropanol formulation
Chemical formula		
Structural formula		
CAS No.		
Synonym/common name		
Trade names		DOWANOL PX-16S
Suppliers		DOW
Molecular mass		
Boiling point/range	°C @ 760 mm Hg	186
Melting point	°C	
Density	kg/m^3 @ 20 °C	940
Evaporation time[a]		
Evaporation time[b]		
Vapour pressure	kPa @ 20 °C	0.07 (25 °C)
Flash point	°C	79
Auto ignition temperature	°C	
Upper flammability limit	%v/v in air	
Lower flammability limit	%v/v in air	
Viscosity	mPa.s (cP) @ 20 °C	
Refractive index	n 20 D	
Solubility in water	%w/w @ 20 °C	Complete
Solubility of water	%w/w @ 20 °C	Complete
Solubility parameter[c]	$(J/cm^3)^{1/2}$	
Hydrogen bond index		
Fractional polarity		
Nitro cellulose dil ratio[d]		
Nitro cellulose dil ratio[e]		
Surface tension	mN/m @ 20 °C	25.7 (25 °C)
Specific heat liquid	kJ/kg mole/°C @ 20 °C	
Latent heat	kJ/kg mole	
Dielectric constant	@ 20 °C	
Antoine constant A	kPa°C	
Antoine constant B	kPa°C	
Antoine constant C	kPa°C	
Heat of combustion	kJ/kg mole	
UN number		1993
IMO classification		
ADR/RID classification		3,32°(c)
UK exposure limits	mg/m^3 (ppm) 8hrTWA/MEL or OES	
USA exposure limits	mg/m^3 (ppm) TLV/TWA	(100)
German exposure limits	mg/m^3 (ppm) MAK-TWA	
EU classification		
EU risk phrases		
EU safety phrases		24

[a] Ether = 1. [b] n-Butyl acetate = 1. [c] Nelson, Hemwall, Edwards. [d] Toluene. [e] Hydrocarbon solvent.

Solvent group		**GLYCOL ETHERS**
Chemical name		Alkoxypropanol formulation
Chemical formula		
Structural formula		
CAS No.		
Synonym/common name		
Trade names		PRIMACLEAN 1601
Suppliers		DOW
Molecular mass		
Boiling point/range	°C @ 760 mm Hg	170–175
Melting point	°C	
Density	kg/m^3 @ 20 °C	880
Evaporation time[a]		150
Evaporation time[b]		0.15
Vapour pressure	kPa @ 20 °C	0.11
Flash point	°C	63
Auto ignition temperature	°C	242
Upper flammability limit	%v/v in air	8.4
Lower flammability limit	%v/v in air	1.0
Viscosity	mPa.s (cP) @ 20 °C	3.1
Refractive index	n 20 D	1.418
Solubility in water	%w/w @ 20 °C	6
Solubility of water	%w/w @ 20 °C	15
Solubility parameter[c]	(J/cm^3)$^{1/2}$	
Hydrogen bond index		
Fractional polarity		
Nitro cellulose dil ratio[d]		
Nitro cellulose dil ratio[e]		
Surface tension	mN/m @ 20 °C	
Specific heat liquid	kJ/kg mole/°C @ 20 °C	
Latent heat	kJ/kg mole	
Dielectric constant	@ 20 °C	
Antoine constant A	kPa°C	
Antoine constant B	kPa°C	
Antoine constant C	kPa°C	
Heat of combustion	kJ/kg mole	
UN number		1993
IMO classification		
ADR/RID classification		3,32°(c)
UK exposure limits	mg/m^3 (ppm) 8hrTWA/MEL or OES	
USA exposure limits	mg/m^3 (ppm) TLV/TWA	
German exposure limits	mg/m^3 (ppm) MAK-TWA	
EU classification		Xi
EU risk phrases		36/38
EU safety phrases		24/26

[a] Ether = 1. [b] n-Butyl acetate = 1. [c] Nelson, Hemwall, Edwards. [d] Toluene. [e] Hydrocarbon solvent.

Solvent group		**GLYCOL ETHERS**
Chemical name		Mixture
Chemical formula		$C_{10}H_{22}O_4/C_{12}H_{26}O_5$
Structural formula		
CAS No.		143-226/1559-34-8
Synonym/common name		Butyl triglycol
Trade names		
Suppliers		BASF
Molecular mass		
Boiling point/range	°C @ 760 mm Hg	265.0–350.0
Melting point	°C	
Density	kg/m^3 @ 20 °C	1440
Evaporation time[a]		~8000
Evaporation time[b]		
Vapour pressure	kPa @ 20 °C	
Flash point	°C	144 (DIN 15755 or 15758)
Auto ignition temperature	°C	203
Upper flammability limit	%v/v in air	3.8
Lower flammability limit	%v/v in air	0.8
Viscosity	mPa.s (cP) @ 20 °C	10.5
Refractive index	n 20 D	1.440
Solubility in water	%w/w @ 20 °C	
Solubility of water	%w/w @ 20 °C	
Solubility parameter[c]	(J/cm^3)$^{1/2}$	
Hydrogen bond index		
Fractional polarity		
Nitro cellulose dil ratio[d]		
Nitro cellulose dil ratio[e]		
Surface tension	mN/m @ 20 °C	
Specific heat liquid	kJ/kg mole/°C @ 20 °C	
Latent heat	kJ/kg mole	
Dielectric constant	@ 20 °C	
Antoine constant A	kPa°C	
Antoine constant B	kPa°C	
Antoine constant C	kPa°C	
Heat of combustion	kJ/kg mole	
UN number		
IMO classification		
ADR/RID classification		
UK exposure limits	mg/m^3 (ppm) 8hrTWA/MEL or OES	
USA exposure limits	mg/m^3 (ppm) TLV/TWA	
German exposure limits	mg/m^3 (ppm) MAK-TWA	
EU classification		
EU risk phrases		
EU safety phrases		

[a] Ether = 1. [b] n-Butyl acetate = 1. [c] Nelson, Hemwall, Edwards. [d] Toluene. [e] Hydrocarbon solvent.

Solvent group		**GLYCOL ETHERS**
Chemical name		2-Methoxyethanol
Chemical formula		$C_3H_8O_2$
Structural formula		$CH_3OCH_2CH_2OH$
CAS No.		109-86-4
Synonym/common name		Methyl glycol, methyl Cellosolve, ethylene glycol monomethyl ether
Trade names		
Suppliers		BASF
Molecular mass		76.1
Boiling point/range	°C @ 760 mm Hg	123.5–125.5
Melting point	°C	
Density	kg/m^3 @ 20 °C	966
Evaporation time[a]		34
Evaporation time[b]		
Vapour pressure	kPa @ 20 °C	
Flash point	°C	38 (DIN 51755 or 51758)
Auto ignition temperature	°C	310
Upper flammability limit	%v/v in air	20.6
Lower flammability limit	%v/v in air	2.4
Viscosity	mPa.s (cP) @ 20 °C	1.7
Refractive index	n 20 D	1.402
Solubility in water	%w/w @ 20 °C	Complete
Solubility of water	%w/w @ 20 °C	Complete
Solubility parameter[c]	(J/cm^3)$^{1/2}$	20.3
Hydrogen bond index		
Fractional polarity		
Nitro cellulose dil ratio[d]		4.7
Nitro cellulose dil ratio[e]		0.2
Surface tension	mN/m @ 20 °C	
Specific heat liquid	kJ/kg mole/°C @ 20 °C	
Latent heat	kJ/kg mole	40350
Dielectric constant	@ 20 °C	
Antoine constant A	kPa°C	6.9440
Antoine constant B	kPa°C	1801.9
Antoine constant C	kPa°C	230
Heat of combustion	kJ/kg mole	
UN number		
IMO classification		
ADR/RID classification		
UK exposure limits	mg/m^3 (ppm) 8hrTWA/MEL or OES	
USA exposure limits	mg/m^3 (ppm) TLV/TWA	
German exposure limits	mg/m^3 (ppm) MAK-TWA	
EU classification		R10, Repr, Cat 2, Xn
EU risk phrases		60-61-10-20/21/22
EU safety phrases		53-45

[a] Ether = 1. [b] n-Butyl acetate = 1. [c] Nelson, Hemwall, Edwards. [d] Toluene. [e] Hydrocarbon solvent.

Solvent group		GLYCOL ETHERS
Chemical name		
Chemical formula		$C_4H_{10}O_2$
Structural formula		$CH_3CH(OH)CH_2OCH_3$
CAS No.		107-98-2
Synonym/common name		
Trade names		Solvenon PM
Suppliers		BASF
Molecular mass		90.1
Boiling point/range	°C @ 760 mm Hg	119.0–122.0
Melting point	°C	
Density	kg/m^3 @ 20 °C	922
Evaporation time[a]		22
Evaporation time[b]		
Vapour pressure	kPa @ 20 °C	
Flash point	°C	32 (DIN 51755 or 51758)
Auto ignition temperature	°C	270
Upper flammability limit	%v/v in air	11.5
Lower flammability limit	%v/v in air	1.7
Viscosity	mPa.s (cP) @ 20 °C	2.0
Refractive index	n 20 D	1.403
Solubility in water	%w/w @ 20 °C	
Solubility of water	%w/w @ 20 °C	
Solubility parameter[c]	$(J/cm^3)^{1/2}$	
Hydrogen bond index		
Fractional polarity		
Nitro cellulose dil ratio[d]		
Nitro cellulose dil ratio[e]		
Surface tension	mN/m @ 20 °C	
Specific heat liquid	kJ/kg mole/°C @ 20 °C	
Latent heat	kJ/kg mole	
Dielectric constant	@ 20 °C	
Antoine constant A	kPa°C	
Antoine constant B	kPa°C	
Antoine constant C	kPa°C	
Heat of combustion	kJ/kg mole	
UN number		
IMO classification		
ADR/RID classification		
UK exposure limits	mg/m^3 (ppm) 8hrTWA/MEL or OES	
USA exposure limits	mg/m^3 (ppm) TLV/TWA	
German exposure limits	mg/m^3 (ppm) MAK-TWA	
EU classification		
EU risk phrases		
EU safety phrases		

[a]Ether = 1. [b]n-Butyl acetate = 1. [c]Nelson, Hemwall, Edwards. [d]Toluene. [e]Hydrocarbon solvent.

Solvent group		GLYCOL ETHERS
Chemical name		1-Methoxy-propanol-2
Chemical formula		$C_4H_{10}O_2$
Structural formula		$CH_3OCH_2CHOHCH_3$
CAS No.		107-98-2
Synonym/common name		Propylene glycol mono methyl ether, Methyl PROXITOL
Trade names		DOWANOL PM, METHOXY PROPANOL, MePROX
Suppliers		SHELL, BP, DOW
Molecular mass		90.1
Boiling point/range	°C @ 760 mm Hg	117–125
Melting point	°C	−96 to −100
Density	kg/m^3 @ 20 °C	920–926
Evaporation time[a]		22–25
Evaporation time[b]		0.70–0.85
Vapour pressure	kPa @ 20 °C	1.0–1.17
Flash point	°C	30 (IP 170) SHELL, 35 BP, 31 (SETA) DOW
Auto ignition temperature	°C	270–287
Upper flammability limit	%v/v in air	13.1–13.8
Lower flammability limit	%v/v in air	1.48–1.9
Viscosity	mPa.s (cP) @ 20 °C	1.9–1.93
Refractive index	n 20 D	1.404
Solubility in water	%w/w @ 20 °C	Complete
Solubility of water	%w/w @ 20 °C	Complete
Solubility parameter[c]	(J/cm^3)$^{1/2}$	19.4
Hydrogen bond index		0.0
Fractional polarity		0.110
Nitro cellulose dil ratio[d]		4.0
Nitro cellulose dil ratio[e]		0.8
Surface tension	mN/m @ 20 °C	28.3
Specific heat liquid	kJ/kg mole/°C @ 20 °C	198.3
Latent heat	kJ/kg mole	39202
Dielectric constant	@ 20 °C	12.3
Antoine constant A	kPa°C	7.01882
Antoine constant B	kPa°C	1780.17
Antoine constant C	kPa°C	236.322
Heat of combustion	kJ/kg mole	−2361504
UN number		3092
IMO classification		3.3/III
ADR/RID classification		3,31°(c)
UK exposure limits	mg/m^3 (ppm) 8hrTWA/MEL or OES	360(100)
USA exposure limits	mg/m^3 (ppm) TLV/TWA	369(100)
German exposure limits	mg/m^3 (ppm) MAK-TWA	375(100)
EU classification		R10
EU risk phrases		10
EU safety phrases		(2-)24

[a] Ether = 1. [b] n-Butyl acetate = 1. [c] Nelson, Hemwall, Edwards. [d] Toluene. [e] Hydrocarbon solvent.

Property	Unit	Value
Solvent group		**GLYCOL ETHERS**
Chemical name		Propylene glycol methyl ether
Chemical formula		$C_4H_{10}O_2$
Structural formula		$CH_3OCH_2CH(CH_3)OH$
CAS No.		
Synonym/common name		
Trade names		
Suppliers		UNION CARBIDE
Molecular mass		90.12
Boiling point/range	°C @ 760 mm Hg	121.0
Melting point	°C	
Density	kg/m^3 @ 20 °C	923
Evaporation time[a]		
Evaporation time[b]		
Vapour pressure	kPa @ 20 °C	
Flash point	°C	
Auto ignition temperature	°C	
Upper flammability limit	%v/v in air	
Lower flammability limit	%v/v in air	
Viscosity	mPa.s (cP) @ 20 °C	
Refractive index	n 20 D	
Solubility in water	%w/w @ 20 °C	Complete
Solubility of water	%w/w @ 20 °C	Complete
Solubility parameter[c]	$(J/cm^3)^{1/2}$	
Hydrogen bond index		
Fractional polarity		
Nitro cellulose dil ratio[d]		
Nitro cellulose dil ratio[e]		
Surface tension	mN/m @ 20 °C	
Specific heat liquid	kJ/kg mole/°C @ 20 °C	
Latent heat	kJ/kg mole	
Dielectric constant	@ 20 °C	
Antoine constant A	kPa°C	
Antoine constant B	kPa°C	
Antoine constant C	kPa°C	
Heat of combustion	kJ/kg mole	
UN number		
IMO classification		
ADR/RID classification		
UK exposure limits	mg/m^3 (ppm) 8hrTWA/MEL or OES	
USA exposure limits	mg/m^3 (ppm) TLV/TWA	
German exposure limits	mg/m^3 (ppm) MAK-TWA	
EU classification		
EU risk phrases		
EU safety phrases		

[a] Ether = 1. [b] n-Butyl acetate = 1. [c] Nelson, Hemwall, Edwards. [d] Toluene. [e] Hydrocarbon solvent.

Solvent group		GLYCOL ETHERS
Chemical name		2-Ethoxyethanol
Chemical formula		$C_4H_{10}O_2$
Structural formula		$CH_3CH_2OCH_2CH_2OH$
CAS No.		110-80-5
Synonym/common name		Ethylene glycol monoethyl ether
Trade names		CELLOSOLVE solvent
Suppliers		UNION CARBIDE, HULS
Molecular mass		90.12
Boiling point/range	°C @ 760 mm Hg	134–137
Melting point	°C	< −80
Density	kg/m^3 @ 20 °C	930–931
Evaporation time[a]		43
Evaporation time[b]		0.44
Vapour pressure	kPa @ 20 °C	0.5
Flash point	°C	43
Auto ignition temperature	°C	250
Upper flammability limit	%v/v in air	10.1
Lower flammability limit	%v/v in air	2.4
Viscosity	mPa.s (cP) @ 20 °C	2.08–2.3
Refractive index	n 20 D	1.406–1.40
Solubility in water	%w/w @ 20 °C	Complete
Solubility of water	%w/w @ 20 °C	Complete
Solubility parameter[c]	(J/cm^3)$^{1/2}$	20.3
Hydrogen bond index		
Fractional polarity		
Nitro cellulose dil ratio[d]		
Nitro cellulose dil ratio[e]		
Surface tension	mN/m @ 20 °C	
Specific heat liquid	kJ/kg mole/°C @ 20 °C	
Latent heat	kJ/kg mole	40350
Dielectric constant	@ 20 °C	
Antoine constant A	kPa°C	6.9440
Antoine constant B	kPa°C	1801.9
Antoine constant C	kPa°C	230
Heat of combustion	kJ/kg mole	
UN number		1171
IMO classification		3.3/III
ADR/RID classification		3,31°(c)
UK exposure limits	mg/m^3 (ppm) 8hrTWA/MEL or OES	37(10)
USA exposure limits	mg/m^3 (ppm) TLV/TWA	18(5)
German exposure limits	mg/m^3 (ppm) MAK-TWA	75(20)
EU classification		Repro Cat2, Xn, R10
EU risk phrases		60-61-10-20/21/22
EU safety phrases		53-45

[a] Ether = 1. [b] n-Butyl acetate = 1. [c] Nelson, Hemwall, Edwards. [d] Toluene. [e] Hydrocarbon solvent.

Property	Unit	Value
Solvent group		**GLYCOL ETHERS**
Chemical name		Ethoxy propanol
Chemical formula		$C_5H_{12}O_2$
Structural formula		$C_2H_5OCHCH(OH)CH_3$
CAS No.		
Synonym/common name		Ethoxypropanol, Propylene glycol monoethyl ether
Trade names		ETHOXY PROPANOL
Suppliers		BP
Molecular mass		104.1
Boiling point/range	°C @ 760 mm Hg	132
Melting point	°C	−90
Density	kg/m^3 @ 20 °C	896
Evaporation time[a]		
Evaporation time[b]		0.54
Vapour pressure	kPa @ 20 °C	<1
Flash point	°C	40
Auto ignition temperature	°C	255
Upper flammability limit	%v/v in air	12
Lower flammability limit	%v/v in air	1.3
Viscosity	mPa.s (cP) @ 20 °C	
Refractive index	n 20 D	
Solubility in water	%w/w @ 20 °C	
Solubility of water	%w/w @ 20 °C	
Solubility parameter[c]	(J/cm^3)$^{1/2}$	
Hydrogen bond index		
Fractional polarity		
Nitro cellulose dil ratio[d]		
Nitro cellulose dil ratio[e]		
Surface tension	mN/m @ 20 °C	
Specific heat liquid	kJ/kg mole/°C @ 20 °C	
Latent heat	kJ/kg mole	
Dielectric constant	@ 20 °C	
Antoine constant A	kPa°C	
Antoine constant B	kPa°C	
Antoine constant C	kPa°C	
Heat of combustion	kJ/kg mole	
UN number		1993
IMO classification		3.3
ADR/RID classification		3,31°(c)
UK exposure limits	mg/m^3 (ppm) 8hrTWA/MEL or OES	
USA exposure limits	mg/m^3 (ppm) TLV/TWA	
German exposure limits	mg/m^3 (ppm) MAK-TWA	
EU classification		
EU risk phrases		10
EU safety phrases		16-24-26

[a]Ether = 1. [b]n-Butyl acetate = 1. [c]Nelson, Hemwall, Edwards. [d]Toluene. [e]Hydrocarbon solvent.

Solvent group		GLYCOL ETHERS
Chemical name		
Chemical formula		$C_5H_{12}O_2$
Structural formula		$CH_3CH_2CH_2OCH_2CH_2OH$
CAS No.		2807-30-9
Synonym/common name		
Trade names		Propyl CELLOSOLVE solvent
Suppliers		UNION CARBIDE
Molecular mass		104.15
Boiling point/range	°C @ 760 mm Hg	150.1
Melting point	°C	
Density	kg/m³ @ 20 °C	916
Evaporation time[a]		
Evaporation time[b]		0.20
Vapour pressure	kPa @ 20 °C	
Flash point	°C	
Auto ignition temperature	°C	
Upper flammability limit	%v/v in air	
Lower flammability limit	%v/v in air	
Viscosity	mPa.s (cP) @ 20 °C	3.8
Refractive index	n 20 D	
Solubility in water	%w/w @ 20 °C	Complete
Solubility of water	%w/w @ 20 °C	Complete
Solubility parameter[c]	$(J/cm^3)^{1/2}$	
Hydrogen bond index		
Fractional polarity		
Nitro cellulose dil ratio[d]		
Nitro cellulose dil ratio[e]		
Surface tension	mN/m @ 20 °C	
Specific heat liquid	kJ/kg mole/°C @ 20 °C	
Latent heat	kJ/kg mole	
Dielectric constant	@ 20 °C	
Antoine constant A	kPa°C	
Antoine constant B	kPa°C	
Antoine constant C	kPa°C	
Heat of combustion	kJ/kg mole	
UN number		
IMO classification		
ADR/RID classification		
UK exposure limits	mg/m³ (ppm) 8hrTWA/MEL or OES	
USA exposure limits	mg/m³ (ppm) TLV/TWA	
German exposure limits	mg/m³ (ppm) MAK-TWA	
EU classification		
EU risk phrases		
EU safety phrases		

[a]Ether = 1. [b]n-Butyl acetate = 1. [c]Nelson, Hemwall, Edwards. [d]Toluene. [e]Hydrocarbon solvent.

Solvent group		**GLYCOL ETHERS**
Chemical name		2-(1-Methylethoxy)ethanol
Chemical formula		$C_5H_{12}O_2$
Structural formula		$(CH_3)_2CHOCH_2CH_2OH$
CAS No.		109-59-1
Synonym/common name		Isopropyl OXITOL
Trade names		iPrOX
Suppliers		SHELL
Molecular mass		104.15
Boiling point/range	°C @ 760 mm Hg	139.5–144.5
Melting point	°C	−60
Density	kg/m^3 @ 20 °C	902–905
Evaporation time[a]		52
Evaporation time[b]		0.3
Vapour pressure	kPa @ 20 °C	0.38
Flash point	°C	46 (IP 170)
Auto ignition temperature	°C	240
Upper flammability limit	%v/v in air	13.0
Lower flammability limit	%v/v in air	1.6
Viscosity	mPa.s (cP) @ 20 °C	2.55
Refractive index	n 20 D	1.410
Solubility in water	%w/w @ 20 °C	Complete
Solubility of water	%w/w @ 20 °C	Complete
Solubility parameter[c]	(J/cm^3)$^{1/2}$	18.8
Hydrogen bond index		0.0
Fractional polarity		0.065
Nitro cellulose dil ratio[d]		4.4
Nitro cellulose dil ratio[e]		1.7
Surface tension	mN/m @ 20 °C	37.2
Specific heat liquid	kJ/kg mole/°C @ 20 °C	231.2
Latent heat	kJ/kg mole	52075
Dielectric constant	@ 20 °C	9.9
Antoine constant A	kPa°C	7.09211
Antoine constant B	kPa°C	1946.24
Antoine constant C	kPa°C	241.229
Heat of combustion	kJ/kg mole	−2994520
UN number		1993
IMO classification		3.3/III
ADR/RID classification		3,31°(c)
UK exposure limits	mg/m^3 (ppm) 8hrTWA/MEL or OES	
USA exposure limits	mg/m^3 (ppm) TLV/TWA	105(25)
German exposure limits	mg/m^3 (ppm) MAK-TWA	
EU classification		Xn, Xi
EU risk phrases		20/21-36
EU safety phrases		(2-)24/25

[a] Ether = 1. [b] n-Butyl acetate = 1. [c] Nelson, Hemwall, Edwards. [d] Toluene. [e] Hydrocarbon solvent.

Solvent group		GLYCOL ETHERS
Chemical name		Propylene glycol n-propyl ether
Chemical formula		$C_6H_{14}O_2$
Structural formula		$C_3H_7OCH_2CH(CH_3)OH$
CAS No.		1569-01-3
Synonym/common name		
Trade names		DOWANOL PnP
Suppliers		DOW, UNION CARBIDE
Molecular mass		116.2–118.
Boiling point/range	°C @ 760 mm Hg	149.8
Melting point	°C	
Density	kg/m^3 @ 20 °C	886
Evaporation time[a]		
Evaporation time[b]		0.21
Vapour pressure	kPa @ 20 °C	
Flash point	°C	
Auto ignition temperature	°C	
Upper flammability limit	%v/v in air	
Lower flammability limit	%v/v in air	
Viscosity	mPa.s (cP) @ 20 °C	
Refractive index	n 20 D	
Solubility in water	%w/w @ 20 °C	Complete
Solubility of water	%w/w @ 20 °C	Complete
Solubility parameter[c]	$(J/cm^3)^{1/2}$	
Hydrogen bond index		
Fractional polarity		
Nitro cellulose dil ratio[d]		
Nitro cellulose dil ratio[e]		
Surface tension	mN/m @ 20 °C	
Specific heat liquid	kJ/kg mole/°C @ 20 °C	
Latent heat	kJ/kg mole	
Dielectric constant	@ 20 °C	
Antoine constant A	kPa°C	
Antoine constant B	kPa°C	
Antoine constant C	kPa°C	
Heat of combustion	kJ/kg mole	
UN number		
IMO classification		
ADR/RID classification		
UK exposure limits	mg/m^3 (ppm) 8hrTWA/MEL or OES	
USA exposure limits	mg/m^3 (ppm) TLV/TWA	
German exposure limits	mg/m^3 (ppm) MAK-TWA	
EU classification		
EU risk phrases		
EU safety phrases		

[a] Ether = 1. [b] n-Butyl acetate = 1. [c] Nelson, Hemwall, Edwards. [d] Toluene. [e] Hydrocarbon solvent.

Solvent group		**GLYCOL ETHERS**
Chemical name		2-Butoxyethanol
Chemical formula		$C_6H_{14}O_2$
Structural formula		$C_4H_9OCH_2CH_2OH$
CAS No.		111-76-2
Synonym/common name		Butyl glycol ether/ethylene glycol mono-butyl ether, butyl OXITOL
Trade names		nBuOX, Butyl CELLOSOLVE
Suppliers		BP, HULS, SHELL, UNION CARBIDE, BASF
Molecular mass		118.18
Boiling point/range	°C @ 760 mm Hg	168–173
Melting point	°C	−70
Density	kg/m^3 @ 20 °C	898–902
Evaporation time[a]		119–160
Evaporation time[b]		0.07–0.08
Vapour pressure	kPa @ 20 °C	0.08–0.13
Flash point	°C	67 (DIN 51755 or 51758) BASF, 67 (ASTM D93) SHELL, 69 (OPEN CUP) BP
Auto ignition temperature	°C	230–245
Upper flammability limit	%v/v in air	6.1–10.6
Lower flammability limit	%v/v in air	1.1–1.6
Viscosity	mPa.s (cP) @ 20 °C	3.3
Refractive index	n 20 D	1.419
Solubility in water	%w/w @ 20 °C	Complete
Solubility of water	%w/w @ 20 °C	Complete
Solubility parameter[c]	$(J/cm^3)^{1/2}$	18.2–18.6
Hydrogen bond index		0.0
Fractional polarity		0.048
Nitro cellulose dil ratio[d]		3.3
Nitro cellulose dil ratio[e]		1.9
Surface tension	mN/m @ 20 °C	28.9
Specific heat liquid	kJ/kg mole/°C @ 20 °C	235.2
Latent heat	kJ/kg mole	47272
Dielectric constant	@ 20 °C	9.4
Antoine constant A	kPa°C	6.95668
Antoine constant B	kPa°C	1920.77
Antoine constant C	kPa°C	317.774
Heat of combustion	kJ/kg mole	−3609571
UN number		2369
IMO classification		6.1/III
ADR/RID classification		6.1,14°(c)
UK exposure limits	mg/m^3 (ppm) 8hrTWA/MEL or OES	121 (25)
USA exposure limits	mg/m^3 (ppm) TLV/TWA	121 (25)
German exposure limits	mg/m^3 (ppm) MAK-TWA	100 (20)
EU classification		Xn, Xi
EU risk phrases		20/21/22-37
EU safety phrases		(2-)24/25

[a] Ether = 1. [b] n-Butyl acetate = 1. [c] Nelson, Hemwall, Edwards. [d] Toluene. [e] Hydrocarbon solvent.

Solvent group		GLYCOL ETHERS
Chemical name		Ethylene glycol n-butyl ether
Chemical formula		$C_6H_{14}O_2$
Structural formula		$C_4H_9OC_2H_4OH$
CAS No.		
Synonym/common name		
Trade names		DOWANOL EB
Suppliers		DOW
Molecular mass		118.2
Boiling point/range	°C @ 760 mm Hg	171.1
Melting point	°C	
Density	kg/m³ @ 20 °C	
Evaporation time[a]		
Evaporation time[b]		0.07
Vapour pressure	kPa @ 20 °C	
Flash point	°C	
Auto ignition temperature	°C	
Upper flammability limit	%v/v in air	
Lower flammability limit	%v/v in air	
Viscosity	mPa.s (cP) @ 20 °C	
Refractive index	n 20 D	
Solubility in water	%w/w @ 20 °C	Complete
Solubility of water	%w/w @ 20 °C	Complete
Solubility parameter[c]	$(J/cm^3)^{1/2}$	
Hydrogen bond index		
Fractional polarity		
Nitro cellulose dil ratio[d]		
Nitro cellulose dil ratio[e]		
Surface tension	mN/m @ 20 °C	
Specific heat liquid	kJ/kg mole/°C @ 20 °C	
Latent heat	kJ/kg mole	
Dielectric constant	@ 20 °C	
Antoine constant A	kPa°C	
Antoine constant B	kPa°C	
Antoine constant C	kPa°C	
Heat of combustion	kJ/kg mole	
UN number		
IMO classification		
ADR/RID classification		
UK exposure limits	mg/m³ (ppm) 8hrTWA/MEL or OES	
USA exposure limits	mg/m³ (ppm) TLV/TWA	
German exposure limits	mg/m³ (ppm) MAK-TWA	
EU classification		
EU risk phrases		
EU safety phrases		

[a] Ether = 1. [b] n-Butyl acetate = 1. [c] Nelson, Hemwall, Edwards. [d] Toluene. [e] Hydrocarbon solvent.

Solvent group		GLYCOL ETHERS
Chemical name		Diethylene glycol methyl ether
Chemical formula		$C_5H_{12}O_3$
Structural formula		$CH_3OC_2H_4OC_2H_4OH$
CAS No.		
Synonym/common name		
Trade names		DOWANOL DM
Suppliers		DOW
Molecular mass		120.1
Boiling point/range	°C @ 760 mm Hg	194.1
Melting point	°C	
Density	kg/m^3 @ 20 °C	
Evaporation time[a]		
Evaporation time[b]		0.02
Vapour pressure	kPa @ 20 °C	
Flash point	°C	
Auto ignition temperature	°C	
Upper flammability limit	%v/v in air	
Lower flammability limit	%v/v in air	
Viscosity	mPa.s (cP) @ 20 °C	
Refractive index	n 20 D	
Solubility in water	%w/w @ 20 °C	Complete
Solubility of water	%w/w @ 20 °C	Complete
Solubility parameter[c]	(J/cm^3)$^{1/2}$	
Hydrogen bond index		
Fractional polarity		
Nitro cellulose dil ratio[d]		
Nitro cellulose dil ratio[e]		
Surface tension	mN/m @ 20 °C	
Specific heat liquid	kJ/kg mole/°C @ 20 °C	
Latent heat	kJ/kg mole	
Dielectric constant	@ 20 °C	
Antoine constant A	kPa°C	
Antoine constant B	kPa°C	
Antoine constant C	kPa°C	
Heat of combustion	kJ/kg mole	
UN number		
IMO classification		
ADR/RID classification		
UK exposure limits	mg/m^3 (ppm) 8hrTWA/MEL or OES	
USA exposure limits	mg/m^3 (ppm) TLV/TWA	
German exposure limits	mg/m^3 (ppm) MAK-TWA	
EU classification		
EU risk phrases		
EU safety phrases		

[a]Ether = 1. [b]n-Butyl acetate = 1. [c]Nelson, Hemwall, Edwards. [d]Toluene. [e]Hydrocarbon solvent.

Solvent group		GLYCOL ETHERS
Chemical name		
Chemical formula		$C_5H_{12}O_3$
Structural formula		$CH_3OCH_2CH_2OCH_2CH_2OH$
CAS No.		111-77-3
Synonym/common name		Methyl carbitol/diethylene glycol monomethyl ether/methyl diglycol
Trade names		Methyl CARBITOL solvent
Suppliers		UNION CARBIDE, BASF
Molecular mass		120.16
Boiling point/range	°C @ 760 mm Hg	192.0–195.0
Melting point	°C	
Density	kg/m^3 @ 20 °C	1022–1026
Evaporation time[a]		~900
Evaporation time[b]		0.02
Vapour pressure	kPa @ 20 °C	
Flash point	°C	91 (DIN 51755 or 51758) BASF
Auto ignition temperature	°C	215
Upper flammability limit	%v/v in air	16.1
Lower flammability limit	%v/v in air	1.6
Viscosity	mPa.s (cP) @ 20 °C	4.1
Refractive index	n 20 D	1.427
Solubility in water	%w/w @ 20 °C	
Solubility of water	%w/w @ 20 °C	
Solubility parameter[c]	$(J/cm^3)^{1/2}$	
Hydrogen bond index		
Fractional polarity		
Nitro cellulose dil ratio[d]		
Nitro cellulose dil ratio[e]		
Surface tension	mN/m @ 20 °C	
Specific heat liquid	kJ/kg mole/°C @ 20 °C	
Latent heat	kJ/kg mole	
Dielectric constant	@ 20 °C	
Antoine constant A	kPa°C	
Antoine constant B	kPa°C	
Antoine constant C	kPa°C	
Heat of combustion	kJ/kg mole	
UN number		
IMO classification		
ADR/RID classification		
UK exposure limits	mg/m^3 (ppm) 8hrTWA/MEL or OES	
USA exposure limits	mg/m^3 (ppm) TLV/TWA	
German exposure limits	mg/m^3 (ppm) MAK-TWA	
EU classification		
EU risk phrases		
EU safety phrases		

[a] Ether = 1. [b] n-Butyl acetate = 1. [c] Nelson, Hemwall, Edwards. [d] Toluene. [e] Hydrocarbon solvent.

Solvent group		GLYCOL ETHERS
Chemical name		1-n-Butoxy-propanol-2
Chemical formula		$C_7H_{16}O_2$
Structural formula		$C_4H_9OCH_2CHOHCH_3$
CAS No.		29387-86-8
Synonym/common name		Propylene glycol n-butyl ether
Trade names		DOWANOL PnB
Suppliers		DOW
Molecular mass		132.2
Boiling point/range	°C @ 760 mm Hg	171
Melting point	°C	< −75
Density	kg/m^3 @ 20 °C	860
Evaporation time[a]		156
Evaporation time[b]		0.07
Vapour pressure	kPa @ 20 °C	
Flash point	°C	63 (PENSKY MARTENS)
Auto ignition temperature	°C	260
Upper flammability limit	%v/v in air	8.4
Lower flammability limit	%v/v in air	1.1
Viscosity	mPa.s (cP) @ 20 °C	3.63
Refractive index	n 20 D	1.418
Solubility in water	%w/w @ 20 °C	6
Solubility of water	%w/w @ 20 °C	~13
Solubility parameter[c]	(J/cm^3)$^{1/2}$	
Hydrogen bond index		
Fractional polarity		
Nitro cellulose dil ratio[d]		
Nitro cellulose dil ratio[e]		
Surface tension	mN/m @ 20 °C	27.9
Specific heat liquid	kJ/kg mole/°C @ 20 °C	349 (25 °C)
Latent heat	kJ/kg mole	
Dielectric constant	@ 20 °C	
Antoine constant A	kPa°C	
Antoine constant B	kPa°C	
Antoine constant C	kPa°C	
Heat of combustion	kJ/kg mole	
UN number		1993
IMO classification		
ADR/RID classification		
UK exposure limits	mg/m^3 (ppm) 8hrTWA/MEL or OES	
USA exposure limits	mg/m^3 (ppm) TLV/TWA	
German exposure limits	mg/m^3 (ppm) MAK-TWA	
EU classification		Xi
EU risk phrases		36/38
EU safety phrases		

[a] Ether = 1. [b] n-Butyl acetate = 1. [c] Nelson, Hemwall, Edwards. [d] Toluene. [e] Hydrocarbon solvent.

Solvent group		GLYCOL ETHERS
Chemical name		2-(2-Ethoxyethoxy)ethanol
Chemical formula		$C_6H_{14}O_3$
Structural formula		$C_2H_5OCH_2CH_2OCH_2CH_2OH$
CAS No.		111-90-0
Synonym/common name		Carbitol/diethylene glycol monoethyl ether/ethyl diglycol
Trade names		CARBITOL solvent low gravity
Suppliers		UNION CARBIDE, HULS
Molecular mass		134.18
Boiling point/range	°C @ 760 mm Hg	190–203
Melting point	°C	−90
Density	kg/m^3 @ 20 °C	990
Evaporation time[a]		1200
Evaporation time[b]		0.01
Vapour pressure	kPa @ 20 °C	0.02
Flash point	°C	94
Auto ignition temperature	°C	220
Upper flammability limit	%v/v in air	11.6
Lower flammability limit	%v/v in air	2
Viscosity	mPa.s (cP) @ 20 °C	4.4
Refractive index	n 20 D	1.426–1.42
Solubility in water	%w/w @ 20 °C	Complete
Solubility of water	%w/w @ 20 °C	Complete
Solubility parameter[c]	(J/cm^3)$^{1/2}$	19.6
Hydrogen bond index		
Fractional polarity		
Nitro cellulose dil ratio[d]		
Nitro cellulose dil ratio[e]		
Surface tension	mN/m @ 20 °C	
Specific heat liquid	kJ/kg mole/°C @ 20 °C	
Latent heat	kJ/kg mole	
Dielectric constant	@ 20 °C	
Antoine constant A	kPa°C	
Antoine constant B	kPa°C	
Antoine constant C	kPa°C	
Heat of combustion	kJ/kg mole	
UN number		
IMO classification		
ADR/RID classification		3,32°(c)
UK exposure limits	mg/m^3 (ppm) 8hrTWA/MEL or OES	
USA exposure limits	mg/m^3 (ppm) TLV/TWA	
German exposure limits	mg/m^3 (ppm) MAK-TWA	
EU classification		
EU risk phrases		
EU safety phrases		

[a] Ether = 1. [b] n-Butyl acetate = 1. [c] Nelson, Hemwall, Edwards. [d] Toluene. [e] Hydrocarbon solvent.

Solvent group		GLYCOL ETHERS
Chemical name		Ethylene glycol phenyl ether
Chemical formula		$C_8H_{10}O_2$
Structural formula		$C_6H_5OC_2H_4OH$
CAS No.		
Synonym/common name		
Trade names		DOWANOL EPh
Suppliers		DOW
Molecular mass		138.2
Boiling point/range	°C @ 760 mm Hg	245.2
Melting point	°C	
Density	kg/m^3 @ 20 °C	
Evaporation time[a]		
Evaporation time[b]		<0.01
Vapour pressure	kPa @ 20 °C	
Flash point	°C	
Auto ignition temperature	°C	
Upper flammability limit	%v/v in air	
Lower flammability limit	%v/v in air	
Viscosity	mPa.s (cP) @ 20 °C	
Refractive index	n 20 D	
Solubility in water	%w/w @ 20 °C	2.3
Solubility of water	%w/w @ 20 °C	10.8
Solubility parameter[c]	$(J/cm^3)^{1/2}$	
Hydrogen bond index		
Fractional polarity		
Nitro cellulose dil ratio[d]		
Nitro cellulose dil ratio[e]		
Surface tension	mN/m @ 20 °C	
Specific heat liquid	kJ/kg mole/°C @ 20 °C	
Latent heat	kJ/kg mole	
Dielectric constant	@ 20 °C	
Antoine constant A	kPa°C	
Antoine constant B	kPa°C	
Antoine constant C	kPa°C	
Heat of combustion	kJ/kg mole	
UN number		
IMO classification		
ADR/RID classification		
UK exposure limits	mg/m^3 (ppm) 8hrTWA/MEL or OES	
USA exposure limits	mg/m^3 (ppm) TLV/TWA	
German exposure limits	mg/m^3 (ppm) MAK-TWA	
EU classification		
EU risk phrases		
EU safety phrases		

[a] Ether = 1. [b] n-Butyl acetate = 1. [c] Nelson, Hemwall, Edwards. [d] Toluene. [e] Hydrocarbon solvent.

Solvent group		**GLYCOL ETHERS**
Chemical name		Aromatic based glycol ether
Chemical formula		$C_8H_{10}O_2$
Structural formula		$C_6H_5OC_2H_4OH$
CAS No.		
Synonym/common name		
Trade names		DOWANOL DALPAD A
Suppliers		DOW
Molecular mass		138.2
Boiling point/range	°C @ 760 mm Hg	245.6
Melting point	°C	
Density	kg/m^3 @ 20 °C	
Evaporation time[a]		
Evaporation time[b]		<0.01
Vapour pressure	kPa @ 20 °C	
Flash point	°C	
Auto ignition temperature	°C	
Upper flammability limit	%v/v in air	
Lower flammability limit	%v/v in air	
Viscosity	mPa.s (cP) @ 20 °C	
Refractive index	n 20 D	
Solubility in water	%w/w @ 20 °C	2.3
Solubility of water	%w/w @ 20 °C	10.8
Solubility parameter[c]	(J/cm^3)$^{1/2}$	
Hydrogen bond index		
Fractional polarity		
Nitro cellulose dil ratio[d]		
Nitro cellulose dil ratio[e]		
Surface tension	mN/m @ 20 °C	
Specific heat liquid	kJ/kg mole/°C @ 20 °C	
Latent heat	kJ/kg mole	
Dielectric constant	@ 20 °C	
Antoine constant A	kPa°C	
Antoine constant B	kPa°C	
Antoine constant C	kPa°C	
Heat of combustion	kJ/kg mole	
UN number		
IMO classification		
ADR/RID classification		
UK exposure limits	mg/m^3 (ppm) 8hrTWA/MEL or OES	
USA exposure limits	mg/m^3 (ppm) TLV/TWA	
German exposure limits	mg/m^3 (ppm) MAK-TWA	
EU classification		
EU risk phrases		
EU safety phrases		

[a]Ether = 1. [b]n-Butyl acetate = 1. [c]Nelson, Hemwall, Edwards. [d]Toluene. [e]Hydrocarbon solvent.

Solvent group		**GLYCOL ETHERS**
Chemical name		Phenyl glycol
Chemical formula		$C_8H_{10}O_2$
Structural formula		$C_6H_5OC_2H_4OH$
CAS No.		122-99-6
Synonym/common name		Phenyl glycol
Trade names		
Suppliers		BASF
Molecular mass		138.2
Boiling point/range	°C @ 760 mm Hg	244.0–250.0
Melting point	°C	
Density	kg/m^3 @ 20 °C	1110
Evaporation time[a]		> 10000
Evaporation time[b]		
Vapour pressure	kPa @ 20 °C	
Flash point	°C	121 (DIN 51755 or 51758)
Auto ignition temperature	°C	535
Upper flammability limit	%v/v in air	9.0
Lower flammability limit	%v/v in air	1.4
Viscosity	mPa.s (cP) @ 20 °C	29.0
Refractive index	n 20 D	1.538
Solubility in water	%w/w @ 20 °C	
Solubility of water	%w/w @ 20 °C	
Solubility parameter[c]	$(J/cm^3)^{1/2}$	
Hydrogen bond index		
Fractional polarity		
Nitro cellulose dil ratio[d]		
Nitro cellulose dil ratio[e]		
Surface tension	mN/m @ 20 °C	
Specific heat liquid	kJ/kg mole/°C @ 20 °C	
Latent heat	kJ/kg mole	
Dielectric constant	@ 20 °C	
Antoine constant A	kPa°C	
Antoine constant B	kPa°C	
Antoine constant C	kPa°C	
Heat of combustion	kJ/kg mole	
UN number		
IMO classification		
ADR/RID classification		
UK exposure limits	mg/m^3 (ppm) 8hrTWA/MEL or OES	
USA exposure limits	mg/m^3 (ppm) TLV/TWA	
German exposure limits	mg/m^3 (ppm) MAK-TWA	
EU classification		
EU risk phrases		
EU safety phrases		

[a] Ether = 1. [b] n-Butyl acetate = 1. [c] Nelson, Hemwall, Edwards. [d] Toluene. [e] Hydrocarbon solvent.

Solvent group		GLYCOL ETHERS
Chemical name		n-Hexyl glycol
Chemical formula		$C_8H_{18}O_2$
Structural formula		$C_6H_{13}OCH_2CH_2OH$
CAS No.		112-25-4
Synonym/common name		n-Hexyl glycol
Trade names		Hexyl CELLOSOLVE solvent
Suppliers		UNION CARBIDE, BASF
Molecular mass		146.2
Boiling point/range	°C @ 760 mm Hg	200.0–212.0
Melting point	°C	
Density	kg/m³ @ 20 °C	887–889
Evaporation time[a]		~1200
Evaporation time[b]		<0.01
Vapour pressure	kPa @ 20 °C	
Flash point	°C	94 (DIN 51755 or 51758) BASF
Auto ignition temperature	°C	220
Upper flammability limit	%v/v in air	8.4
Lower flammability limit	%v/v in air	1.2
Viscosity	mPa.s (cP) @ 20 °C	5.3
Refractive index	n 20 D	1.429
Solubility in water	%w/w @ 20 °C	1.0
Solubility of water	%w/w @ 20 °C	18.8
Solubility parameter[c]	$(J/cm^3)^{1/2}$	
Hydrogen bond index		
Fractional polarity		
Nitro cellulose dil ratio[d]		
Nitro cellulose dil ratio[e]		
Surface tension	mN/m @ 20 °C	
Specific heat liquid	kJ/kg mole/°C @ 20 °C	
Latent heat	kJ/kg mole	
Dielectric constant	@ 20 °C	
Antoine constant A	kPa°C	
Antoine constant B	kPa°C	
Antoine constant C	kPa°C	
Heat of combustion	kJ/kg mole	
UN number		
IMO classification		
ADR/RID classification		
UK exposure limits	mg/m³ (ppm) 8hrTWA/MEL or OES	
USA exposure limits	mg/m³ (ppm) TLV/TWA	
German exposure limits	mg/m³ (ppm) MAK-TWA	
EU classification		
EU risk phrases		
EU safety phrases		

[a] Ether = 1. [b] n-Butyl acetate = 1. [c] Nelson, Hemwall, Edwards. [d] Toluene. [e] Hydrocarbon solvent.

Solvent group		**GLYCOL ETHERS**
Chemical name		2(2-Methoxymethylethoxy)-propanol
Chemical formula		$C_7H_{16}O_3$
Structural formula		
CAS No.		34590-94-8
Synonym/common name		Methyl DIPROXITOL
Trade names		MeDIPROX
Suppliers		SHELL
Molecular mass		148.2
Boiling point/range	°C @ 760 mm Hg	184–196
Melting point	°C	−83
Density	kg/m^3 @ 20 °C	952–956
Evaporation time[a]		360
Evaporation time[b]		0.035
Vapour pressure	kPa @ 20 °C	0.052
Flash point	°C	79 (ASTM D93)
Auto ignition temperature	°C	205
Upper flammability limit	%v/v in air	8.7
Lower flammability limit	%v/v in air	1.3
Viscosity	mPa.s (cP) @ 20 °C	3.71
Refractive index	n 20 D	1.422
Solubility in water	%w/w @ 20 °C	Complete
Solubility of water	%w/w @ 20 °C	Complete
Solubility parameter[c]	$(J/cm^3)^{1/2}$	17.8
Hydrogen bond index		0.0
Fractional polarity		0.05
Nitro cellulose dil ratio[d]		4.2
Nitro cellulose dil ratio[e]		0.8
Surface tension	mN/m @ 20 °C	29.4
Specific heat liquid	kJ/kg mole/°C @ 20 °C	290.5
Latent heat	kJ/kg mole	48906
Dielectric constant	@ 20 °C	10.5
Antoine constant A	kPa°C	7.70707
Antoine constant B	kPa°C	1633.03
Antoine constant C	kPa°C	161.693
Heat of combustion	kJ/kg mole	−4080242
UN number		
IMO classification		
ADR/RID classification		3,32°(c)
UK exposure limits	mg/m^3 (ppm) 8hrTWA/MEL or OES	
USA exposure limits	mg/m^3 (ppm) TLV/TWA	606(100)
German exposure limits	mg/m^3 (ppm) MAK-TWA	
EU classification		
EU risk phrases		
EU safety phrases		24

[a] Ether = 1. [b] n-Butyl acetate = 1. [c] Nelson, Hemwall, Edwards. [d] Toluene. [e] Hydrocarbon solvent.

Solvent group		**GLYCOL ETHERS**
Chemical name		Dipropylene glycol methyl ether
Chemical formula		$C_5H_{14}O_3$
Structural formula		$CH_3O(CH_2CH_3O)_2H$
CAS No.		
Synonym/common name		
Trade names		
Suppliers		UNION CARBIDE
Molecular mass		148.20
Boiling point/range	°C @ 760 mm Hg	188.3
Melting point	°C	
Density	kg/m^3 @ 20 °C	951
Evaporation time[a]		
Evaporation time[b]		
Vapour pressure	kPa @ 20 °C	
Flash point	°C	
Auto ignition temperature	°C	
Upper flammability limit	%v/v in air	
Lower flammability limit	%v/v in air	
Viscosity	mPa.s (cP) @ 20 °C	
Refractive index	n 20 D	
Solubility in water	%w/w @ 20 °C	Complete
Solubility of water	%w/w @ 20 °C	Complete
Solubility parameter[c]	(J/cm^3)$^{1/2}$	
Hydrogen bond index		
Fractional polarity		
Nitro cellulose dil ratio[d]		
Nitro cellulose dil ratio[e]		
Surface tension	mN/m @ 20 °C	
Specific heat liquid	kJ/kg mole/°C @ 20 °C	
Latent heat	kJ/kg mole	
Dielectric constant	@ 20 °C	
Antoine constant A	kPa°C	
Antoine constant B	kPa°C	
Antoine constant C	kPa°C	
Heat of combustion	kJ/kg mole	
UN number		
IMO classification		
ADR/RID classification		
UK exposure limits	mg/m^3 (ppm) 8hrTWA/MEL or OES	
USA exposure limits	mg/m^3 (ppm) TLV/TWA	
German exposure limits	mg/m^3 (ppm) MAK-TWA	
EU classification		
EU risk phrases		
EU safety phrases		

[a] Ether = 1. [b] n-Butyl acetate = 1. [c] Nelson, Hemwall, Edwards. [d] Toluene. [e] Hydrocarbon solvent.

Solvent group		GLYCOL ETHERS
Chemical name		Mixture
Chemical formula		$C_7H_{16}O_3$ Mixed isomers
Structural formula		
CAS No.		34590-94-8
Synonym/common name		
Trade names		Solvenon DPM
Suppliers		BASF
Molecular mass		148.2
Boiling point/range	°C @ 760 mm Hg	185.0-195.0
Melting point	°C	
Density	kg/m^3 @ 20 °C	957
Evaporation time[a]		~380
Evaporation time[b]		
Vapour pressure	kPa @ 20 °C	
Flash point	°C	80 (DIN 51755 or 51758)
Auto ignition temperature	°C	205
Upper flammability limit	%v/v in air	10.4
Lower flammability limit	%v/v in air	1.4
Viscosity	mPa.s (cP) @ 20 °C	4.0
Refractive index	n 20 D	1.422
Solubility in water	%w/w @ 20 °C	
Solubility of water	%w/w @ 20 °C	
Solubility parameter[c]	(J/cm^3)$^{1/2}$	
Hydrogen bond index		
Fractional polarity		
Nitro cellulose dil ratio[d]		
Nitro cellulose dil ratio[e]		
Surface tension	mN/m @ 20 °C	
Specific heat liquid	kJ/kg mole/°C @ 20 °C	
Latent heat	kJ/kg mole	
Dielectric constant	@ 20 °C	
Antoine constant A	kPa°C	
Antoine constant B	kPa°C	
Antoine constant C	kPa°C	
Heat of combustion	kJ/kg mole	
UN number		
IMO classification		
ADR/RID classification		
UK exposure limits	mg/m^3 (ppm) 8hrTWA/MEL or OES	
USA exposure limits	mg/m^3 (ppm) TLV/TWA	
German exposure limits	mg/m^3 (ppm) MAK-TWA	
EU classification		
EU risk phrases		
EU safety phrases		

[a] Ether = 1. [b] n-Butyl acetate = 1. [c] Nelson, Hemwall, Edwards. [d] Toluene. [e] Hydrocarbon solvent.

Solvent group		**GLYCOL ETHERS**
Chemical name		1-(2-Methoxy-propoxy)-propanol 2/ 1-(2-Methoxy-methylethoxy)-propanol 2 mixture
Chemical formula		$C_7H_{16}O_3$
Structural formula		$CH_3OC_3H_6OCH_2CHOHCH_3$
CAS No.		34590-94-8
Synonym/common name		Dipropylene glycol methyl ether
Trade names		DOWANOL DPM
Suppliers		DOW
Molecular mass		148.2
Boiling point/range	°C @ 760 mm Hg	189
Melting point	°C	−83
Density	kg/m^3 @ 20 °C	950
Evaporation time[a]		~400
Evaporation time[b]		0.03
Vapour pressure	kPa @ 20 °C	
Flash point	°C	75 (SETA)
Auto ignition temperature	°C	270
Upper flammability limit	%v/v in air	14.0
Lower flammability limit	%v/v in air	1.1
Viscosity	mPa.s (cP) @ 20 °C	4.35
Refractive index	n 20 D	1.423
Solubility in water	%w/w @ 20 °C	Complete
Solubility of water	%w/w @ 20 °C	Complete
Solubility parameter[c]	$(J/cm^3)^{1/2}$	
Hydrogen bond index		
Fractional polarity		
Nitro cellulose dil ratio[d]		
Nitro cellulose dil ratio[e]		
Surface tension	mN/m @ 20 °C	29.1
Specific heat liquid	kJ/kg mole/°C @ 20 °C	336.4 (25 °C)
Latent heat	kJ/kg mole	
Dielectric constant	@ 20 °C	
Antoine constant A	kPa°C	
Antoine constant B	kPa°C	
Antoine constant C	kPa°C	
Heat of combustion	kJ/kg mole	
UN number		1993
IMO classification		
ADR/RID classification		3,32°(c)
UK exposure limits	mg/m^3 (ppm) 8hrTWA/MEL or OES	
USA exposure limits	mg/m^3 (ppm) TLV/TWA	(100)
German exposure limits	mg/m^3 (ppm) MAK-TWA	
EU classification		
EU risk phrases		
EU safety phrases		

[a] Ether = 1. [b] n-Butyl acetate = 1. [c] Nelson, Hemwall, Edwards. [d] Toluene. [e] Hydrocarbon solvent.

Solvent group		GLYCOL ETHERS
Chemical name		1-Phenoxy-propanol-2
Chemical formula		$C_9H_{12}O_2$
Structural formula		$C_6H_5OCH_2CHOHCH_3$
CAS No.		770-35-5 (3)
Synonym/common name		Propylene glycol phenyl ether
Trade names		DOWANOL PPh
Suppliers		DOW
Molecular mass		152.2
Boiling point/range	°C @ 760 mm Hg	242
Melting point	°C	13
Density	kg/m^3 @ 20 °C	1060
Evaporation time[a]		1200
Evaporation time[b]		0.01
Vapour pressure	kPa @ 20 °C	
Flash point	°C	127
Auto ignition temperature	°C	135
Upper flammability limit	%v/v in air	9.4
Lower flammability limit	%v/v in air	0.7
Viscosity	mPa.s (cP) @ 20 °C	35.39
Refractive index	n 20 D	1.523
Solubility in water	%w/w @ 20 °C	1.4
Solubility of water	%w/w @ 20 °C	6.5
Solubility parameter[c]	$(J/cm^3)^{1/2}$	
Hydrogen bond index		
Fractional polarity		
Nitro cellulose dil ratio[d]		
Nitro cellulose dil ratio[e]		
Surface tension	mN/m @ 20 °C	40.2
Specific heat liquid	kJ/kg mole/°C @ 20 °C	331.8 (25 °C)
Latent heat	kJ/kg mole	
Dielectric constant	@ 20 °C	
Antoine constant A	kPa°C	
Antoine constant B	kPa°C	
Antoine constant C	kPa°C	
Heat of combustion	kJ/kg mole	
UN number		
IMO classification		
ADR/RID classification		
UK exposure limits	mg/m^3 (ppm) 8hrTWA/MEL or OES	
USA exposure limits	mg/m^3 (ppm) TLV/TWA	
German exposure limits	mg/m^3 (ppm) MAK-TWA	
EU classification		
EU risk phrases		
EU safety phrases		

[a] Ether = 1. [b] n-Butyl acetate = 1. [c] Nelson, Hemwall, Edwards. [d] Toluene. [e] Hydrocarbon solvent.

Solvent group		GLYCOL ETHERS
Chemical name		Mixture
Chemical formula		$C_9H_{12}O_2$ Mixed isomers
Structural formula		
CAS No.		770-35-4/4169-04-4
Synonym/common name		
Trade names		Solvenon PP
Suppliers		BASF
Molecular mass		152.2
Boiling point/range	°C @ 760 mm Hg	241.0–246.0
Melting point	°C	
Density	kg/m^3 @ 20 °C	1064
Evaporation time[a]		~5000
Evaporation time[b]		
Vapour pressure	kPa @ 20 °C	
Flash point	°C	120 (DIN 51755 or 51758)
Auto ignition temperature	°C	480
Upper flammability limit	%v/v in air	6.4
Lower flammability limit	%v/v in air	1.0
Viscosity	mPa.s (cP) @ 20 °C	34
Refractive index	n 20 D	1.524
Solubility in water	%w/w @ 20 °C	
Solubility of water	%w/w @ 20 °C	
Solubility parameter[c]	(J/cm^3)$^{1/2}$	
Hydrogen bond index		
Fractional polarity		
Nitro cellulose dil ratio[d]		
Nitro cellulose dil ratio[e]		
Surface tension	mN/m @ 20 °C	
Specific heat liquid	kJ/kg mole/°C @ 20 °C	
Latent heat	kJ/kg mole	
Dielectric constant	@ 20 °C	
Antoine constant A	kPa°C	
Antoine constant B	kPa°C	
Antoine constant C	kPa°C	
Heat of combustion	kJ/kg mole	
UN number		
IMO classification		
ADR/RID classification		
UK exposure limits	mg/m^3 (ppm) 8hrTWA/MEL or OES	
USA exposure limits	mg/m^3 (ppm) TLV/TWA	
German exposure limits	mg/m^3 (ppm) MAK-TWA	
EU classification		
EU risk phrases		
EU safety phrases		

[a] Ether = 1. [b] n-Butyl acetate = 1. [c] Nelson, Hemwall, Edwards. [d] Toluene. [e] Hydrocarbon solvent.

Solvent group		GLYCOL ETHERS
Chemical name		Blend of isomers
Chemical formula		$C_8H_{18}O_3$
Structural formula		$CH_3OCH_2CH(CH_3)OCH_2CH$ $(CH_3)OCH_3$
CAS No.		111109-77-4
Synonym/common name		Dipropylene glycol dimethyl ether
Trade names		PROGLYDE DMM
Suppliers		DOW
Molecular mass		162
Boiling point/range	°C @ 760 mm Hg	175
Melting point	°C	< −71
Density	kg/m^3 @ 20 °C	900
Evaporation time[a]		131
Evaporation time[b]		0.16
Vapour pressure	kPa @ 20 °C	
Flash point	°C	65 (PENSKY MARTENS)
Auto ignition temperature	°C	165
Upper flammability limit	%v/v in air	
Lower flammability limit	%v/v in air	
Viscosity	mPa.s (cP) @ 20 °C	1.11
Refractive index	n 20 D	1.409
Solubility in water	%w/w @ 20 °C	52.6
Solubility of water	%w/w @ 20 °C	4.5
Solubility parameter[c]	(J/cm^3)$^{1/2}$	
Hydrogen bond index		
Fractional polarity		
Nitro cellulose dil ratio[d]		
Nitro cellulose dil ratio[e]		
Surface tension	mN/m @ 20 °C	26.8
Specific heat liquid	kJ/kg mole/°C @ 20 °C	294.8
Latent heat	kJ/kg mole	
Dielectric constant	@ 20 °C	
Antoine constant A	kPa°C	
Antoine constant B	kPa°C	
Antoine constant C	kPa°C	
Heat of combustion	kJ/kg mole	
UN number		
IMO classification		
ADR/RID classification		3,32°(c)
UK exposure limits	mg/m^3 (ppm) 8hrTWA/MEL or OES	
USA exposure limits	mg/m^3 (ppm) TLV/TWA	
German exposure limits	mg/m^3 (ppm) MAK-TWA	
EU classification		
EU risk phrases		
EU safety phrases		

[a] Ether = 1. [b] n-Butyl acetate = 1. [c] Nelson, Hemwall, Edwards. [d] Toluene. [e] Hydrocarbon solvent.

Solvent group		GLYCOL ETHERS
Chemical name		Diethylene glycol n-butyl ether
Chemical formula		$C_8H_{18}O_3$
Structural formula		$C_4H_9OC_2H_4OC_2H_4OH$
CAS No.		
Synonym/common name		
Trade names		DOWANOL DB
Suppliers		DOW
Molecular mass		162.2
Boiling point/range	°C @ 760 mm Hg	230.0
Melting point	°C	
Density	kg/m^3 @ 20 °C	
Evaporation time[a]		
Evaporation time[b]		0.003
Vapour pressure	kPa @ 20 °C	
Flash point	°C	
Auto ignition temperature	°C	
Upper flammability limit	%v/v in air	
Lower flammability limit	%v/v in air	
Viscosity	mPa.s (cP) @ 20 °C	
Refractive index	n 20 D	
Solubility in water	%w/w @ 20 °C	Complete
Solubility of water	%w/w @ 20 °C	Complete
Solubility parameter[c]	(J/cm^3)$^{1/2}$	
Hydrogen bond index		
Fractional polarity		
Nitro cellulose dil ratio[d]		
Nitro cellulose dil ratio[e]		
Surface tension	mN/m @ 20 °C	
Specific heat liquid	kJ/kg mole/°C @ 20 °C	
Latent heat	kJ/kg mole	
Dielectric constant	@ 20 °C	
Antoine constant A	kPa°C	
Antoine constant B	kPa°C	
Antoine constant C	kPa°C	
Heat of combustion	kJ/kg mole	
UN number		
IMO classification		
ADR/RID classification		
UK exposure limits	mg/m^3 (ppm) 8hrTWA/MEL or OES	
USA exposure limits	mg/m^3 (ppm) TLV/TWA	
German exposure limits	mg/m^3 (ppm) MAK-TWA	
EU classification		
EU risk phrases		
EU safety phrases		

[a] Ether = 1. [b] n-Butyl acetate = 1. [c] Nelson, Hemwall, Edwards. [d] Toluene. [e] Hydrocarbon solvent.

Solvent group		**GLYCOL ETHERS**
Chemical name		2-(2-Butoxyethoxy)ethanol
Chemical formula		$C_8H_{18}O_3$
Structural formula		$C_4H_9OCH_2CH_2OCH_2CH_2OH$
CAS No.		112-34-5
Synonym/common name		Butyl DIOXITOL, Butyl diglycol
Trade names		nBuDIOX, BUTYL DIGLYCOL
Suppliers		BP, HULS, SHELL, BASF
Molecular mass		162.22
Boiling point/range	°C @ 760 mm Hg	225–234
Melting point	°C	−68
Density	kg/m^3 @ 20 °C	951–956
Evaporation time[a]		3750
Evaporation time[b]		0.004
Vapour pressure	kPa @ 20 °C	0.002–0.01
Flash point	°C	105 (DIN 15755 or 15758) BASF, 105 (ASTM) SHELL, 105 HULS, 116 (OPEN CUP) BP
Auto ignition temperature	°C	210–230
Upper flammability limit	%v/v in air	5.3–5.9
Lower flammability limit	%v/v in air	0.7–0.9
Viscosity	mPa.s (cP) @ 20 °C	5.85–6.0
Refractive index	n 20 D	1.432
Solubility in water	%w/w @ 20 °C	Complete
Solubility of water	%w/w @ 20 °C	Complete
Solubility parameter[c]	$(J/cm^3)^{1/2}$	18.2
Hydrogen bond index		0.0
Fractional polarity		0.028
Nitro cellulose dil ratio[d]		3.9
Nitro cellulose dil ratio[e]		1.6
Surface tension	mN/m @ 20 °C	33.5
Specific heat liquid	kJ/kg mole/°C @ 20 °C	300.1
Latent heat	kJ/kg mole	52721
Dielectric constant	@ 20 °C	10.2
Antoine constant A	kPa°C	7.3791
Antoine constant B	kPa°C	2549.43
Antoine constant C	kPa°C	244.521
Heat of combustion	kJ/kg mole	−4711517
UN number		
IMO classification		
ADR/RID classification		
UK exposure limits	mg/m^3 (ppm) 8hrTWA/MEL or OES	
USA exposure limits	mg/m^3 (ppm) TLV/TWA	
German exposure limits	mg/m^3 (ppm) MAK-TWA	
EU classification		Xi
EU risk phrases		36
EU safety phrases		26

[a] Ether = 1. [b] n-Butyl acetate = 1. [c] Nelson, Hemwall, Edwards. [d] Toluene. [e] Hydrocarbon solvent.

Solvent group		**GLYCOL ETHERS**
Chemical name		
Chemical formula		$C_8H_{18}O_3$
Structural formula		$C_4H_9O(C_2H_4O)_2H$
CAS No.		112-34-5
Synonym/common name		
Trade names		Butyl CARBITOL solvent
Suppliers		UNION CARBIDE
Molecular mass		162.23
Boiling point/range	°C @ 760 mm Hg	230.6
Melting point	°C	
Density	kg/m^3 @ 20 °C	954
Evaporation time[a]		
Evaporation time[b]		<0.01
Vapour pressure	kPa @ 20 °C	
Flash point	°C	
Auto ignition temperature	°C	
Upper flammability limit	%v/v in air	
Lower flammability limit	%v/v in air	
Viscosity	mPa.s (cP) @ 20 °C	4.2
Refractive index	n 20 D	
Solubility in water	%w/w @ 20 °C	Complete
Solubility of water	%w/w @ 20 °C	Complete
Solubility parameter[c]	$(J/cm^3)^{1/2}$	
Hydrogen bond index		
Fractional polarity		
Nitro cellulose dil ratio[d]		
Nitro cellulose dil ratio[e]		
Surface tension	mN/m @ 20 °C	
Specific heat liquid	kJ/kg mole/°C @ 20 °C	
Latent heat	kJ/kg mole	
Dielectric constant	@ 20 °C	
Antoine constant A	kPa°C	
Antoine constant B	kPa°C	
Antoine constant C	kPa°C	
Heat of combustion	kJ/kg mole	
UN number		
IMO classification		
ADR/RID classification		
UK exposure limits	mg/m^3 (ppm) 8hrTWA/MEL or OES	
USA exposure limits	mg/m^3 (ppm) TLV/TWA	
German exposure limits	mg/m^3 (ppm) MAK-TWA	
EU classification		
EU risk phrases		
EU safety phrases		

[a] Ether = 1. [b] n-Butyl acetate = 1. [c] Nelson, Hemwall, Edwards. [d] Toluene. [e] Hydrocarbon solvent.

Solvent group		GLYCOL ETHERS
Chemical name		Methoxytriglycol
Chemical formula		$C_7H_{16}O_4$
Structural formula		$CH_3O(C_2H_4O)_3H$
CAS No.		112-35-6
Synonym/common name		
Trade names		
Suppliers		UNION CARBIDE
Molecular mass		164.20
Boiling point/range	°C @ 760 mm Hg	249.0
Melting point	°C	
Density	kg/m^3 @ 20 °C	1051
Evaporation time[a]		
Evaporation time[b]		<0.01
Vapour pressure	kPa @ 20 °C	
Flash point	°C	
Auto ignition temperature	°C	
Upper flammability limit	%v/v in air	
Lower flammability limit	%v/v in air	
Viscosity	mPa.s (cP) @ 20 °C	7.3
Refractive index	n 20 D	
Solubility in water	%w/w @ 20 °C	Complete
Solubility of water	%w/w @ 20 °C	Complete
Solubility parameter[c]	(J/cm^3)$^{1/2}$	
Hydrogen bond index		
Fractional polarity		
Nitro cellulose dil ratio[d]		
Nitro cellulose dil ratio[e]		
Surface tension	mN/m @ 20 °C	
Specific heat liquid	kJ/kg mole/°C @ 20 °C	
Latent heat	kJ/kg mole	
Dielectric constant	@ 20 °C	
Antoine constant A	kPa°C	
Antoine constant B	kPa°C	
Antoine constant C	kPa°C	
Heat of combustion	kJ/kg mole	
UN number		
IMO classification		
ADR/RID classification		
UK exposure limits	mg/m^3 (ppm) 8hrTWA/MEL or OES	
USA exposure limits	mg/m^3 (ppm) TLV/TWA	
German exposure limits	mg/m^3 (ppm) MAK-TWA	
EU classification		
EU risk phrases		
EU safety phrases		

[a] Ether = 1. [b] n-Butyl acetate = 1. [c] Nelson, Hemwall, Edwards. [d] Toluene. [e] Hydrocarbon solvent.

Property	Units	Value
Solvent group		**GLYCOL ETHERS**
Chemical name		Triethylene glycol methyl ether/Highers
Chemical formula		
Structural formula		$CH_3O(C_2H_4O)nH$ (n = 3, 4, 5)
CAS No.		
Synonym/common name		
Trade names		DOWANOL TMH
Suppliers		DOW
Molecular mass		173.0 Average
Boiling point/range	°C @ 760 mm Hg	232.0
Melting point	°C	
Density	kg/m^3 @ 20 °C	
Evaporation time[a]		
Evaporation time[b]		<0.01
Vapour pressure	kPa @ 20 °C	
Flash point	°C	
Auto ignition temperature	°C	
Upper flammability limit	%v/v in air	
Lower flammability limit	%v/v in air	
Viscosity	mPa.s (cP) @ 20 °C	
Refractive index	n 20 D	
Solubility in water	%w/w @ 20 °C	Complete
Solubility of water	%w/w @ 20 °C	
Solubility parameter[c]	(J/cm^3)$^{1/2}$	
Hydrogen bond index		
Fractional polarity		
Nitro cellulose dil ratio[d]		
Nitro cellulose dil ratio[e]		
Surface tension	mN/m @ 20 °C	
Specific heat liquid	kJ/kg mole/°C @ 20 °C	
Latent heat	kJ/kg mole	
Dielectric constant	@ 20 °C	
Antoine constant A	kPa°C	
Antoine constant B	kPa°C	
Antoine constant C	kPa°C	
Heat of combustion	kJ/kg mole	
UN number		
IMO classification		
ADR/RID classification		
UK exposure limits	mg/m^3 (ppm) 8hrTWA/MEL or OES	
USA exposure limits	mg/m^3 (ppm) TLV/TWA	
German exposure limits	mg/m^3 (ppm) MAK-TWA	
EU classification		
EU risk phrases		
EU safety phrases		

[a] Ether = 1. [b] n-Butyl acetate = 1. [c] Nelson, Hemwall, Edwards. [d] Toluene. [e] Hydrocarbon solvent.

Property	Units	Value
Solvent group		**GLYCOL ETHERS**
Chemical name		Dipropylene glycol n-propyl ether
Chemical formula		$C_9H_{20}O_3$
Structural formula		$C_3H_7O[CH_2(CH)CH_3O]_2H$
CAS No.		29911-27-1
Synonym/common name		
Trade names		DOWANOL DPnP
Suppliers		DOW
Molecular mass		176.2
Boiling point/range	°C @ 760 mm Hg	212.0
Melting point	°C	
Density	kg/m^3 @ 20 °C	
Evaporation time[a]		
Evaporation time[b]		0.015
Vapour pressure	kPa @ 20 °C	
Flash point	°C	
Auto ignition temperature	°C	
Upper flammability limit	%v/v in air	
Lower flammability limit	%v/v in air	
Viscosity	mPa.s (cP) @ 20 °C	
Refractive index	n 20 D	
Solubility in water	%w/w @ 20 °C	19.0
Solubility of water	%w/w @ 20 °C	20.5
Solubility parameter[c]	$(J/cm^3)^{1/2}$	
Hydrogen bond index		
Fractional polarity		
Nitro cellulose dil ratio[d]		
Nitro cellulose dil ratio[e]		
Surface tension	mN/m @ 20 °C	
Specific heat liquid	kJ/kg mole/°C @ 20 °C	
Latent heat	kJ/kg mole	
Dielectric constant	@ 20 °C	
Antoine constant A	kPa°C	
Antoine constant B	kPa°C	
Antoine constant C	kPa°C	
Heat of combustion	kJ/kg mole	
UN number		
IMO classification		
ADR/RID classification		
UK exposure limits	mg/m^3 (ppm) 8hrTWA/MEL or OES	
USA exposure limits	mg/m^3 (ppm) TLV/TWA	
German exposure limits	mg/m^3 (ppm) MAK-TWA	
EU classification		
EU risk phrases		
EU safety phrases		

[a] Ether = 1. [b] n-Butyl acetate = 1. [c] Nelson, Hemwall, Edwards. [d] Toluene. [e] Hydrocarbon solvent.

Solvent group		GLYCOL ETHERS
Chemical name		[2-[2-(Ethoxyethoxy)ethoxy]ethanol
Chemical formula		$C_8H_{18}O_4$
Structural formula		$C_2H_5(OCH_2CH_2)_3OH$
CAS No.		
Synonym/common name		Ethyl triglycol
Trade names		
Suppliers		HULS
Molecular mass		178.2
Boiling point/range	°C @ 760 mm Hg	235–280
Melting point	°C	−19
Density	kg/m^3 @ 20 °C	1040
Evaporation time[a]		>1200
Evaporation time[b]		
Vapour pressure	kPa @ 20 °C	<0.001
Flash point	°C	120
Auto ignition temperature	°C	220
Upper flammability limit	%v/v in air	
Lower flammability limit	%v/v in air	
Viscosity	mPa.s (cP) @ 20 °C	7.8
Refractive index	n 20 D	1.438–1.442
Solubility in water	%w/w @ 20 °C	
Solubility of water	%w/w @ 20 °C	Complete
Solubility parameter[c]	$(J/cm^3)^{1/2}$	
Hydrogen bond index		
Fractional polarity		
Nitro cellulose dil ratio[d]		
Nitro cellulose dil ratio[e]		
Surface tension	mN/m @ 20 °C	
Specific heat liquid	kJ/kg mole/°C @ 20 °C	
Latent heat	kJ/kg mole	
Dielectric constant	@ 20 °C	
Antoine constant A	kPa°C	
Antoine constant B	kPa°C	
Antoine constant C	kPa°C	
Heat of combustion	kJ/kg mole	
UN number		
IMO classification		
ADR/RID classification		
UK exposure limits	mg/m^3 (ppm) 8hrTWA/MEL or OES	
USA exposure limits	mg/m^3 (ppm) TLV/TWA	
German exposure limits	mg/m^3 (ppm) MAK-TWA	
EU classification		
EU risk phrases		
EU safety phrases		

[a] Ether = 1. [b] n-Butyl acetate = 1. [c] Nelson, Hemwall, Edwards. [d] Toluene. [e] Hydrocarbon solvent.

Solvent group		GLYCOL ETHERS
Chemical name		Ethoxytriglycol
Chemical formula		$C_8H_{18}O_4$
Structural formula		$CH_3CH_2O(C_2H_4O)_3H$
CAS No.		112-50-5
Synonym/common name		
Trade names		
Suppliers		UNION CARBIDE
Molecular mass		178.23
Boiling point/range	°C @ 760 mm Hg	256.0
Melting point	°C	
Density	kg/m^3 @ 20 °C	1025
Evaporation time[a]		
Evaporation time[b]		<0.1
Vapour pressure	kPa @ 20 °C	
Flash point	°C	
Auto ignition temperature	°C	
Upper flammability limit	%v/v in air	
Lower flammability limit	%v/v in air	
Viscosity	mPa.s (cP) @ 20 °C	7.8
Refractive index	n 20 D	
Solubility in water	%w/w @ 20 °C	Complete
Solubility of water	%w/w @ 20 °C	Complete
Solubility parameter[c]	$(J/cm^3)^{1/2}$	
Hydrogen bond index		
Fractional polarity		
Nitro cellulose dil ratio[d]		
Nitro cellulose dil ratio[e]		
Surface tension	mN/m @ 20 °C	
Specific heat liquid	kJ/kg mole/°C @ 20 °C	
Latent heat	kJ/kg mole	
Dielectric constant	@ 20 °C	
Antoine constant A	kPa°C	
Antoine constant B	kPa°C	
Antoine constant C	kPa°C	
Heat of combustion	kJ/kg mole	
UN number		
IMO classification		
ADR/RID classification		
UK exposure limits	mg/m^3 (ppm) 8hrTWA/MEL or OES	
USA exposure limits	mg/m^3 (ppm) TLV/TWA	
German exposure limits	mg/m^3 (ppm) MAK-TWA	
EU classification		
EU risk phrases		
EU safety phrases		

[a] Ether = 1. [b] n-Butyl acetate = 1. [c] Nelson, Hemwall, Edwards. [d] Toluene. [e] Hydrocarbon solvent.

Solvent group		GLYCOL ETHERS
Chemical name		
Chemical formula		$C_{10}H_{22}O_3$
Structural formula		$C_6H_{13}O(C_2H_4O)_2H$
CAS No.		112-59-4
Synonym/common name		
Trade names		Hexyl CARBITOL solvent
Suppliers		UNION CARBIDE
Molecular mass		190.28
Boiling point/range	°C @ 760 mm Hg	259.1
Melting point	°C	
Density	kg/m^3 @ 20 °C	935
Evaporation time[a]		
Evaporation time[b]		<0.01
Vapour pressure	kPa @ 20 °C	
Flash point	°C	
Auto ignition temperature	°C	
Upper flammability limit	%v/v in air	
Lower flammability limit	%v/v in air	
Viscosity	mPa.s (cP) @ 20 °C	8.5
Refractive index	n 20 D	
Solubility in water	%w/w @ 20 °C	3.0
Solubility of water	%w/w @ 20 °C	56.3
Solubility parameter[c]	(J/cm^3)$^{1/2}$	
Hydrogen bond index		
Fractional polarity		
Nitro cellulose dil ratio[d]		
Nitro cellulose dil ratio[e]		
Surface tension	mN/m @ 20 °C	
Specific heat liquid	kJ/kg mole/°C @ 20 °C	
Latent heat	kJ/kg mole	
Dielectric constant	@ 20 °C	
Antoine constant A	kPa°C	
Antoine constant B	kPa°C	
Antoine constant C	kPa°C	
Heat of combustion	kJ/kg mole	
UN number		
IMO classification		
ADR/RID classification		
UK exposure limits	mg/m^3 (ppm) 8hrTWA/MEL or OES	
USA exposure limits	mg/m^3 (ppm) TLV/TWA	
German exposure limits	mg/m^3 (ppm) MAK-TWA	
EU classification		
EU risk phrases		
EU safety phrases		

[a] Ether = 1. [b] n-Butyl acetate = 1. [c] Nelson, Hemwall, Edwards. [d] Toluene. [e] Hydrocarbon solvent.

Solvent group		**GLYCOL ETHERS**
Chemical name		Dipropylene glycol butylene ether
Chemical formula		$C_{12}H_{22}O_2$
Structural formula		$C_4H_9(CH_2CHCH_3CO)_2H$
CAS No.		
Synonym/common name		
Trade names		
Suppliers		UNION CARBIDE
Molecular mass		190.30
Boiling point/range	°C @ 760 mm Hg	227.9
Melting point	°C	
Density	kg/m^3 @ 20 °C	915
Evaporation time[a]		
Evaporation time[b]		
Vapour pressure	kPa @ 20 °C	
Flash point	°C	
Auto ignition temperature	°C	
Upper flammability limit	%v/v in air	
Lower flammability limit	%v/v in air	
Viscosity	mPa.s (cP) @ 20 °C	
Refractive index	n 20 D	
Solubility in water	%w/w @ 20 °C	3.0
Solubility of water	%w/w @ 20 °C	12.0
Solubility parameter[c]	(J/cm^3)$^{1/2}$	
Hydrogen bond index		
Fractional polarity		
Nitro cellulose dil ratio[d]		
Nitro cellulose dil ratio[e]		
Surface tension	mN/m @ 20 °C	
Specific heat liquid	kJ/kg mole/°C @ 20 °C	
Latent heat	kJ/kg mole	
Dielectric constant	@ 20 °C	
Antoine constant A	kPa°C	
Antoine constant B	kPa°C	
Antoine constant C	kPa°C	
Heat of combustion	kJ/kg mole	
UN number		
IMO classification		
ADR/RID classification		
UK exposure limits	mg/m^3 (ppm) 8hrTWA/MEL or OES	
USA exposure limits	mg/m^3 (ppm) TLV/TWA	
German exposure limits	mg/m^3 (ppm) MAK-TWA	
EU classification		
EU risk phrases		
EU safety phrases		

[a] Ether = 1. [b] n-Butyl acetate = 1. [c] Nelson, Hemwall, Edwards. [d] Toluene. [e] Hydrocarbon solvent.

Solvent group		**GLYCOL ETHERS**
Chemical name		1-(2-n-Butoxy-1-methyl-ethoxy)-propanol-2
Chemical formula		$C_{10}H_{22}O_3$
Structural formula		$C_4H_9OC_3H_6OCH_2CHOHCH_3$
CAS No.		35884-42-5
Synonym/common name		Dipropylene glycol n-butyl ether
Trade names		DOWANOL DPnB
Suppliers		DOW
Molecular mass		190.3
Boiling point/range	°C @ 760 mm Hg	229
Melting point	°C	−75
Density	kg/m^3 @ 20 °C	910
Evaporation time[a]		900
Evaporation time[b]		0.01
Vapour pressure	kPa @ 20 °C	
Flash point	°C	111 (PENSKY MARTENS)
Auto ignition temperature	°C	189
Upper flammability limit	%v/v in air	20.4
Lower flammability limit	%v/v in air	0.6
Viscosity	mPa.s (cP) @ 20 °C	5.72
Refractive index	n 20 D	1.427
Solubility in water	%w/w @ 20 °C	~5
Solubility of water	%w/w @ 20 °C	~10
Solubility parameter[c]	$(J/cm^3)^{1/2}$	
Hydrogen bond index		
Fractional polarity		
Nitro cellulose dil ratio[d]		
Nitro cellulose dil ratio[e]		
Surface tension	mN/m @ 20 °C	29.2
Specific heat liquid	kJ/kg mole/°C @ 20 °C	466.2 (25 °C)
Latent heat	kJ/kg mole	
Dielectric constant	@ 20 °C	
Antoine constant A	kPa°C	
Antoine constant B	kPa°C	
Antoine constant C	kPa°C	
Heat of combustion	kJ/kg mole	
UN number		
IMO classification		
ADR/RID classification		
UK exposure limits	mg/m^3 (ppm) 8hrTWA/MEL or OES	
USA exposure limits	mg/m^3 (ppm) TLV/TWA	
German exposure limits	mg/m^3 (ppm) MAK-TWA	
EU classification		
EU risk phrases		
EU safety phrases		

[a] Ether = 1. [b] n-Butyl acetate = 1. [c] Nelson, Hemwall, Edwards. [d] Toluene. [e] Hydrocarbon solvent.

Solvent group		**GLYCOL ETHERS**
Chemical name		Blend
Chemical formula		$C_{10}H_{22}O_4$
Structural formula		$CH_3OC_3H_6OC_3H_6OCH_2CHOHCH_3$
CAS No.		25498-49-1
Synonym/common name		Tripropylene glycol methyl ether
Trade names		DOWANOL TPM
Suppliers		DOW
Molecular mass		206.3
Boiling point/range	°C @ 760 mm Hg	243
Melting point	°C	−78
Density	kg/m^3 @ 20 °C	960
Evaporation time[a]		>1200
Evaporation time[b]		<0.01
Vapour pressure	kPa @ 20 °C	
Flash point	°C	116 (PENSKY MARTENS)
Auto ignition temperature	°C	270
Upper flammability limit	%v/v in air	7.0
Lower flammability limit	%v/v in air	1.1
Viscosity	mPa.s (cP) @ 20 °C	6.68
Refractive index	n 20 D	1.430
Solubility in water	%w/w @ 20 °C	Complete
Solubility of water	%w/w @ 20 °C	Complete
Solubility parameter[c]	(J/cm^3)$^{1/2}$	
Hydrogen bond index		
Fractional polarity		
Nitro cellulose dil ratio[d]		
Nitro cellulose dil ratio[e]		
Surface tension	mN/m @ 20 °C	30.3
Specific heat liquid	kJ/kg mole/°C @ 20 °C	441.5 (25 °C)
Latent heat	kJ/kg mole	
Dielectric constant	@ 20 °C	
Antoine constant A	kPa°C	
Antoine constant B	kPa°C	
Antoine constant C	kPa°C	
Heat of combustion	kJ/kg mole	
UN number		
IMO classification		
ADR/RID classification		
UK exposure limits	mg/m^3 (ppm) 8hrTWA/MEL or OES	
USA exposure limits	mg/m^3 (ppm) TLV/TWA	
German exposure limits	mg/m^3 (ppm) MAK-TWA	
EU classification		
EU risk phrases		
EU safety phrases		

[a] Ether = 1. [b] n-Butyl acetate = 1. [c] Nelson, Hemwall, Edwards. [d] Toluene. [e] Hydrocarbon solvent.

Solvent group		**GLYCOL ETHERS**
Chemical name		2-[2-(butoxyethoxy)ethoxy]ethanol]
Chemical formula		$C_{10}H_{22}O_4$
Structural formula		$C_4H_9(OCH_2CH_2)_3OH$
CAS No.		
Synonym/common name		Butyl triglycol
Trade names		
Suppliers		HULS
Molecular mass		206.3
Boiling point/range	°C @ 760 mm Hg	255-295
Melting point	°C	−35.1
Density	kg/m^3 @ 20 °C	985
Evaporation time[a]		>1200
Evaporation time[b]		
Vapour pressure	kPa @ 20 °C	<0.001
Flash point	°C	130
Auto ignition temperature	°C	230
Upper flammability limit	%v/v in air	
Lower flammability limit	%v/v in air	0.7
Viscosity	mPa.s (cP) @ 20 °C	9.42
Refractive index	n 20 D	1.439–1.441
Solubility in water	%w/w @ 20 °C	
Solubility of water	%w/w @ 20 °C	Complete
Solubility parameter[c]	$(J/cm^3)^{1/2}$	
Hydrogen bond index		
Fractional polarity		
Nitro cellulose dil ratio[d]		
Nitro cellulose dil ratio[e]		
Surface tension	mN/m @ 20 °C	
Specific heat liquid	kJ/kg mole/°C @ 20 °C	
Latent heat	kJ/kg mole	
Dielectric constant	@ 20 °C	
Antoine constant A	kPa°C	
Antoine constant B	kPa°C	
Antoine constant C	kPa°C	
Heat of combustion	kJ/kg mole	
UN number		
IMO classification		
ADR/RID classification		
UK exposure limits	mg/m^3 (ppm) 8hrTWA/MEL or OES	
USA exposure limits	mg/m^3 (ppm) TLV/TWA	
German exposure limits	mg/m^3 (ppm) MAK-TWA	
EU classification		
EU risk phrases		
EU safety phrases		

[a] Ether = 1. [b] n-Butyl acetate = 1. [c] Nelson, Hemwall, Edwards. [d] Toluene. [e] Hydrocarbon solvent.

Solvent group		GLYCOL ETHERS
Chemical name		Triethylene glycol butyl ether/Highers
Chemical formula		
Structural formula		$C_4H_9O(C_2H_4O)nH$ (n = 3, 4, 5)
CAS No.		
Synonym/common name		
Trade names		DOWANOL TBH
Suppliers		DOW
Molecular mass		231.2 Average
Boiling point/range	°C @ 760 mm Hg	283.0
Melting point	°C	
Density	kg/m^3 @ 20 °C	
Evaporation time[a]		
Evaporation time[b]		≪0.01
Vapour pressure	kPa @ 20 °C	
Flash point	°C	
Auto ignition temperature	°C	
Upper flammability limit	%v/v in air	
Lower flammability limit	%v/v in air	
Viscosity	mPa.s (cP) @ 20 °C	
Refractive index	n 20 D	
Solubility in water	%w/w @ 20 °C	Complete
Solubility of water	%w/w @ 20 °C	
Solubility parameter[c]	$(J/cm^3)^{1/2}$	
Hydrogen bond index		
Fractional polarity		
Nitro cellulose dil ratio[d]		
Nitro cellulose dil ratio[e]		
Surface tension	mN/m @ 20 °C	
Specific heat liquid	kJ/kg mole/°C @ 20 °C	
Latent heat	kJ/kg mole	
Dielectric constant	@ 20 °C	
Antoine constant A	kPa°C	
Antoine constant B	kPa°C	
Antoine constant C	kPa°C	
Heat of combustion	kJ/kg mole	
UN number		
IMO classification		
ADR/RID classification		
UK exposure limits	mg/m^3 (ppm) 8hrTWA/MEL or OES	
USA exposure limits	mg/m^3 (ppm) TLV/TWA	
German exposure limits	mg/m^3 (ppm) MAK-TWA	
EU classification		
EU risk phrases		
EU safety phrases		

[a] Ether = 1. [b] n-Butyl acetate = 1. [c] Nelson, Hemwall, Edwards. [d] Toluene. [e] Hydrocarbon solvent.

Solvent group		**GLYCOL ETHERS**
Chemical name		2-[2-(Butoxyethoxy)-ethoxy]ethanol
Chemical formula		$C_{12}H_{26}O_5$
Structural formula		$C_4H_9(OCH_2CH_2)_4OH$
CAS No.		
Synonym/common name		Butyl tetraglycol
Trade names		
Suppliers		HULS
Molecular mass		250.3
Boiling point/range	°C @ 760 mm Hg	300–340
Melting point	°C	
Density	kg/m^3 @ 20 °C	1015
Evaporation time[a]		>1200
Evaporation time[b]		
Vapour pressure	kPa @ 20 °C	<0.001
Flash point	°C	170
Auto ignition temperature	°C	250
Upper flammability limit	%v/v in air	
Lower flammability limit	%v/v in air	
Viscosity	mPa.s (cP) @ 20 °C	
Refractive index	n 20 D	1.446
Solubility in water	%w/w @ 20 °C	
Solubility of water	%w/w @ 20 °C	Complete
Solubility parameter[c]	$(J/cm^3)^{1/2}$	
Hydrogen bond index		
Fractional polarity		
Nitro cellulose dil ratio[d]		
Nitro cellulose dil ratio[e]		
Surface tension	mN/m @ 20 °C	
Specific heat liquid	kJ/kg mole/°C @ 20 °C	
Latent heat	kJ/kg mole	
Dielectric constant	@ 20 °C	
Antoine constant A	kPa°C	
Antoine constant B	kPa°C	
Antoine constant C	kPa°C	
Heat of combustion	kJ/kg mole	
UN number		
IMO classification		
ADR/RID classification		
UK exposure limits	mg/m^3 (ppm) 8hrTWA/MEL or OES	
USA exposure limits	mg/m^3 (ppm) TLV/TWA	
German exposure limits	mg/m^3 (ppm) MAK-TWA	
EU classification		
EU risk phrases		
EU safety phrases		

[a] Ether = 1. [b] n-Butyl acetate = 1. [c] Nelson, Hemwall, Edwards. [d] Toluene. [e] Hydrocarbon solvent.

Solvent group		**GLYCOL ETHERS**
Chemical name		Blend of isomers
Chemical formula		$C_{13}H_{28}O_4$
Structural formula		$C_4H_9OC_3H_6OC_3H_6OCH_2CHOHCH_3$
CAS No.		55924-93-5
Synonym/common name		Tripropylene glycol n-butyl ether
Trade names		DOWANOL TPnB
Suppliers		DOW
Molecular mass		284.4
Boiling point/range	°C @ 760 mm Hg	274
Melting point	°C	<-75
Density	kg/m^3 @ 20 °C	930
Evaporation time[a]		>1200
Evaporation time[b]		<0.01
Vapour pressure	kPa @ 20 °C	
Flash point	°C	124 (PENSKY MARTENS)
Auto ignition temperature	°C	186
Upper flammability limit	%v/v in air	
Lower flammability limit	%v/v in air	
Viscosity	mPa.s (cP) @ 20 °C	8.40
Refractive index	n 20 D	1.433
Solubility in water	%w/w @ 20 °C	3
Solubility of water	%w/w @ 20 °C	8
Solubility parameter[c]	$(J/cm^3)^{1/2}$	
Hydrogen bond index		
Fractional polarity		
Nitro cellulose dil ratio[d]		
Nitro cellulose dil ratio[e]		
Surface tension	mN/m @ 20 °C	29.9
Specific heat liquid	kJ/kg mole/°C @ 20 °C	
Latent heat	kJ/kg mole	
Dielectric constant	@ 20 °C	
Antoine constant A	kPa°C	
Antoine constant B	kPa°C	
Antoine constant C	kPa°C	
Heat of combustion	kJ/kg mole	
UN number		
IMO classification		
ADR/RID classification		
UK exposure limits	mg/m^3 (ppm) 8hrTWA/MEL or OES	
USA exposure limits	mg/m^3 (ppm) TLV/TWA	
German exposure limits	mg/m^3 (ppm) MAK-TWA	
EU classification		
EU risk phrases		
EU safety phrases		

[a] Ether = 1. [b] n-Butyl acetate = 1. [c] Nelson, Hemwall, Edwards. [d] Toluene. [e] Hydrocarbon solvent.

Solvent group		**GLYCOLS**
Chemical name		1,2-Ethanediol
Chemical formula		$C_2H_6O_2$
Structural formula		$(CH_2OH)_2$
CAS No.		107-21-1
Synonym/common name		Ethylene glycol
Trade names		
Suppliers		SHELL, UNION CARBIDE, BASF, ENICHEM
Molecular mass		62.07
Boiling point/range	°C @ 760 mm Hg	193–205
Melting point	°C	−11.5 to −12.7
Density	kg/m^3 @ 20 °C	1115.4
Evaporation time[a]		2500
Evaporation time[b]		<0.01
Vapour pressure	kPa @ 20 °C	0.012
Flash point	°C	111 CLOSED CUP/119 OPENCUP ENICHEM, 111 (DIN 51755 or 51758) BASF
Auto ignition temperature	°C	410
Upper flammability limit	%v/v in air	15.3
Lower flammability limit	%v/v in air	3.2
Viscosity	mPa.s (cP) @ 20 °C	13.7–19.9
Refractive index	n 20 D	1.4318
Solubility in water	%w/w @ 20 °C	Complete
Solubility of water	%w/w @ 20 °C	Complete
Solubility parameter[c]	$(J/cm^3)^{1/2}$	29.1
Hydrogen bond index		
Fractional polarity		
Nitro cellulose dil ratio[d]		
Nitro cellulose dil ratio[e]		
Surface tension	mN/m @ 20 °C	48.49
Specific heat liquid	kJ/kg mole/°C @ 20 °C	
Latent heat	kJ/kg mole	50460
Dielectric constant	@ 20 °C	37
Antoine constant A	kPa°C	6.83995
Antoine constant B	kPa°C	1818.591
Antoine constant C	kPa°C	178.651
Heat of combustion	kJ/kg mole	−1189720
UN number		
IMO classification		
ADR/RID classification		
UK exposure limits	mg/m^3 (ppm) 8hrTWA/MEL or OES	
USA exposure limits	mg/m^3 (ppm) TLV/TWA	
German exposure limits	mg/m^3 (ppm) MAK-TWA	26(10)
EU classification		Xn
EU risk phrases		22
EU safety phrases		(2)

[a] Ether = 1. [b] n-Butyl acetate = 1. [c] Nelson, Hemwall, Edwards. [d] Toluene. [e] Hydrocarbon solvent.

Solvent group		**GLYCOLS**
Chemical name		1,2-Propylene glycol
Chemical formula		$C_3H_8O_2$
Structural formula		$CH_3CHOHCH_2OH$
CAS No.		57-55-6
Synonym/common name		Propylene glycol
Trade names		
Suppliers		BASF, HULS
Molecular mass		76.1
Boiling point/range	°C @ 760 mm Hg	186.0–190.0
Melting point	°C	
Density	kg/m^3 @ 20 °C	1036
Evaporation time[a]		1000
Evaporation time[b]		
Vapour pressure	kPa @ 20 °C	
Flash point	°C	103 (DIN 51755 or 51758) BASF, 103 HULS
Auto ignition temperature	°C	410
Upper flammability limit	%v/v in air	12.6
Lower flammability limit	%v/v in air	2.6
Viscosity	mPa.s (cP) @ 20 °C	55–60
Refractive index	n 20 D	1.433
Solubility in water	%w/w @ 20 °C	Complete
Solubility of water	%w/w @ 20 °C	
Solubility parameter[c]	$(J/cm^3)^{1/2}$	
Hydrogen bond index		
Fractional polarity		
Nitro cellulose dil ratio[d]		
Nitro cellulose dil ratio[e]		
Surface tension	mN/m @ 20 °C	
Specific heat liquid	kJ/kg mole/°C @ 20 °C	
Latent heat	kJ/kg mole	
Dielectric constant	@ 20 °C	
Antoine constant A	kPa°C	
Antoine constant B	kPa°C	
Antoine constant C	kPa°C	
Heat of combustion	kJ/kg mole	
UN number		
IMO classification		
ADR/RID classification		
UK exposure limits	mg/m^3 (ppm) 8hrTWA/MEL or OES	
USA exposure limits	mg/m^3 (ppm) TLV/TWA	
German exposure limits	mg/m^3 (ppm) MAK-TWA	
EU classification		
EU risk phrases		
EU safety phrases		

[a] Ether = 1. [b] n-Butyl acetate = 1. [c] Nelson, Hemwall, Edwards. [d] Toluene. [e] Hydrocarbon solvent.

Solvent group		**GLYCOLS**
Chemical name		1,3-Butanediol
Chemical formula		$C_4H_{10}O_2$
Structural formula		$CH_3CHOHCH_2CH_2OH$
CAS No.		
Synonym/common name		1,3-Butanediol
Trade names		
Suppliers		HULS
Molecular mass		90.12
Boiling point/range	°C @ 760 mm Hg	207-210
Melting point	°C	
Density	kg/m^3 @ 20 °C	
Evaporation time[a]		
Evaporation time[b]		
Vapour pressure	kPa @ 20 °C	0.002
Flash point	°C	109
Auto ignition temperature	°C	440
Upper flammability limit	%v/v in air	9.4
Lower flammability limit	%v/v in air	1.8
Viscosity	mPa.s (cP) @ 20 °C	137
Refractive index	n 20 D	1.439–1.440
Solubility in water	%w/w @ 20 °C	
Solubility of water	%w/w @ 20 °C	Complete
Solubility parameter[c]	$(J/cm^3)^{1/2}$	23.7
Hydrogen bond index		
Fractional polarity		
Nitro cellulose dil ratio[d]		
Nitro cellulose dil ratio[e]		
Surface tension	mN/m @ 20 °C	
Specific heat liquid	kJ/kg mole/°C @ 20 °C	
Latent heat	kJ/kg mole	
Dielectric constant	@ 20 °C	
Antoine constant A	kPa°C	8.9284
Antoine constant B	kPa°C	3470.5
Antoine constant C	kPa°C	294.8
Heat of combustion	kJ/kg mole	
UN number		
IMO classification		
ADR/RID classification		
UK exposure limits	mg/m^3 (ppm) 8hrTWA/MEL or OES	
USA exposure limits	mg/m^3 (ppm) TLV/TWA	
German exposure limits	mg/m^3 (ppm) MAK-TWA	
EU classification		
EU risk phrases		
EU safety phrases		

[a] Ether = 1. [b] n-Butyl acetate = 1. [c] Nelson, Hemwall, Edwards. [d] Toluene. [e] Hydrocarbon solvent.

Solvent group		**GLYCOLS**
Chemical name		1,4-Butanediol
Chemical formula		$C_4H_{10}O_2$
Structural formula		$HOCH_2CH_2CH_2CH_2OH$
CAS No.		
Synonym/common name		1,4-Butanediol
Trade names		
Suppliers		HULS
Molecular mass		90.12
Boiling point/range	°C @ 760 mm Hg	229
Melting point	°C	19.7
Density	kg/m^3 @ 20 °C	
Evaporation time[a]		>1200
Evaporation time[b]		
Vapour pressure	kPa @ 20 °C	<0.01
Flash point	°C	135
Auto ignition temperature	°C	420
Upper flammability limit	%v/v in air	15.3
Lower flammability limit	%v/v in air	2.4
Viscosity	mPa.s (cP) @ 20 °C	71.5
Refractive index	n 20 D	1.445 (25 °C)
Solubility in water	%w/w @ 20 °C	
Solubility of water	%w/w @ 20 °C	Complete
Solubility parameter[c]	$(J/cm^3)^{1/2}$	24.7
Hydrogen bond index		
Fractional polarity		
Nitro cellulose dil ratio[d]		
Nitro cellulose dil ratio[e]		
Surface tension	mN/m @ 20 °C	44.6
Specific heat liquid	kJ/kg mole/°C @ 20 °C	
Latent heat	kJ/kg mole	
Dielectric constant	@ 20 °C	
Antoine constant A	kPa°C	8.51
Antoine constant B	kPa°C	3264
Antoine constant C	kPa°C	
Heat of combustion	kJ/kg mole	−2499940
UN number		
IMO classification		
ADR/RID classification		
UK exposure limits	mg/m^3 (ppm) 8hrTWA/MEL or OES	
USA exposure limits	mg/m^3 (ppm) TLV/TWA	
German exposure limits	mg/m^3 (ppm) MAK-TWA	
EU classification		
EU risk phrases		
EU safety phrases		

[a] Ether = 1. [b] n-Butyl acetate = 1. [c] Nelson, Hemwall, Edwards. [d] Toluene. [e] Hydrocarbon solvent.

Solvent group		**GLYCOLS**
Chemical name		Diethylene glycol
Chemical formula		$C_4H_{10}O_3$
Structural formula		$HOCH_2CH_2OCH_2CH_2OH$
CAS No.		111-46-6
Synonym/common name		Diethylene glycol
Trade names		
Suppliers		SHELL, UNION CARBIDE, BASF, ENICHEM
Molecular mass		106.12
Boiling point/range	°C @ 760 mm Hg	243.8–246.8
Melting point	°C	−7.8
Density	kg/m^3 @ 20 °C	1118.2
Evaporation time[a]		>1000
Evaporation time[b]		<0.001
Vapour pressure	kPa @ 20 °C	
Flash point	°C	140 (DIN 51755 or 51758) BASF
Auto ignition temperature	°C	345
Upper flammability limit	%v/v in air	12.2
Lower flammability limit	%v/v in air	1.8
Viscosity	mPa.s (cP) @ 20 °C	38.8
Refractive index	n 20 D	1.447
Solubility in water	%w/w @ 20 °C	Complete
Solubility of water	%w/w @ 20 °C	Complete
Solubility parameter[c]	(J/cm^3)$^{1/2}$	29.13
Hydrogen bond index		
Fractional polarity		
Nitro cellulose dil ratio[d]		
Nitro cellulose dil ratio[e]		
Surface tension	mN/m @ 20 °C	48.5
Specific heat liquid	kJ/kg mole/°C @ 20 °C	
Latent heat	kJ/kg mole	52262
Dielectric constant	@ 20 °C	31.7
Antoine constant A	kPa°C	6.67111
Antoine constant B	kPa°C	1897.637
Antoine constant C	kPa°C	161.067
Heat of combustion	kJ/kg mole	−2378600
UN number		
IMO classification		
ADR/RID classification		
UK exposure limits	mg/m^3 (ppm) 8hrTWA/MEL or OES	
USA exposure limits	mg/m^3 (ppm) TLV/TWA	
German exposure limits	mg/m^3 (ppm) MAK-TWA	
EU classification		
EU risk phrases		
EU safety phrases		

[a] Ether = 1. [b] n-Butyl acetate = 1. [c] Nelson, Hemwall, Edwards. [d] Toluene. [e] Hydrocarbon solvent.

Solvent group		**GLYCOLS**
Chemical name		2-Methyl-2,4-pentanediol
Chemical formula		$C_6H_{14}O_2$
Structural formula		$(CH_3)_2COHCH_2CHOHCH_3$
CAS No.		107-41-5
Synonym/common name		Hexylene glycol
Trade names		HG
Suppliers		ELF ATOCHEM, SHELL, UNION CARBIDE
Molecular mass		118.18
Boiling point/range	°C @ 760 mm Hg	195–200
Melting point	°C	−50
Density	kg/m^3 @ 20 °C	920–923
Evaporation time[a]		1680
Evaporation time[b]		0.007
Vapour pressure	kPa @ 20 °C	0.003–0.007
Flash point	°C	93 (ASTM D93) SHELL, 97.9 (CLOSED CUP) ELF ATOCHEM
Auto ignition temperature	°C	425
Upper flammability limit	%v/v in air	9.9
Lower flammability limit	%v/v in air	1.0
Viscosity	mPa.s (cP) @ 20 °C	38.9
Refractive index	n 20 D	1.428
Solubility in water	%w/w @ 20 °C	Complete
Solubility of water	%w/w @ 20 °C	Complete
Solubility parameter[c]	(J/cm^3)$^{1/2}$	23.1
Hydrogen bond index		−20.0
Fractional polarity		0.737
Nitro cellulose dil ratio[d]		
Nitro cellulose dil ratio[e]		
Surface tension	mN/m @ 20 °C	33.1
Specific heat liquid	kJ/kg mole/°C @ 20 °C	260.0
Latent heat	kJ/kg mole	51408
Dielectric constant	@ 20 °C	~7.7
Antoine constant A	kPa°C	6.55177
Antoine constant B	kPa°C	1606.29
Antoine constant C	kPa°C	156.045
Heat of combustion	kJ/kg mole	−3530627
UN number		
IMO classification		
ADR/RID classification		
UK exposure limits	mg/m^3 (ppm) 8hrTWA/MEL or OES	125(25)
USA exposure limits	mg/m^3 (ppm) TLV/TWA	121(25)
German exposure limits	mg/m^3 (ppm) MAK-TWA	
EU classification		
EU risk phrases		
EU safety phrases		

[a] Ether = 1. [b] n-Butyl acetate = 1. [c] Nelson, Hemwall, Edwards. [d] Toluene. [e] Hydrocarbon solvent.

Solvent group		**GLYCOLS**
Chemical name		Dipropylene glycol (mixed isomers)
Chemical formula		$C_6H_{14}O_3$
Structural formula		$CH_3CHOHCH_2OCH(CH_3)CH_2OH$
CAS No.		25265-71-8
Synonym/common name		Dipropylene glycol
Trade names		
Suppliers		BASF, HULS
Molecular mass		134.2
Boiling point/range	°C @ 760 mm Hg	228.0–236.0
Melting point	°C	
Density	kg/m^3 @ 20 °C	1023
Evaporation time[a]		> 10000
Evaporation time[b]		
Vapour pressure	kPa @ 20 °C	
Flash point	°C	138 (DIN 51755 or 51758) BASF, 120 HULS
Auto ignition temperature	°C	310–350
Upper flammability limit	%v/v in air	12.6
Lower flammability limit	%v/v in air	1.2–2.9
Viscosity	mPa.s (cP) @ 20 °C	107.0
Refractive index	n 20 D	1.411
Solubility in water	%w/w @ 20 °C	
Solubility of water	%w/w @ 20 °C	Complete
Solubility parameter[c]	(J/cm^3)$^{1/2}$	
Hydrogen bond index		
Fractional polarity		
Nitro cellulose dil ratio[d]		
Nitro cellulose dil ratio[e]		
Surface tension	mN/m @ 20 °C	
Specific heat liquid	kJ/kg mole/°C @ 20 °C	
Latent heat	kJ/kg mole	
Dielectric constant	@ 20 °C	
Antoine constant A	kPa°C	
Antoine constant B	kPa°C	
Antoine constant C	kPa°C	
Heat of combustion	kJ/kg mole	
UN number		
IMO classification		
ADR/RID classification		
UK exposure limits	mg/m^3 (ppm) 8hrTWA/MEL or OES	
USA exposure limits	mg/m^3 (ppm) TLV/TWA	
German exposure limits	mg/m^3 (ppm) MAK-TWA	
EU classification		
EU risk phrases		
EU safety phrases		

[a] Ether = 1. [b] n-Butyl acetate = 1. [c] Nelson, Hemwall, Edwards. [d] Toluene. [e] Hydrocarbon solvent.

Solvent group		**GLYCOLS**
Chemical name		2-Ethylhexane-1,3-diol
Chemical formula		$C_8H_{18}O_2$
Structural formula		$CH_3CH_2CH_2CHOHCH(C_2H_5)CH_2OH$
CAS No.		
Synonym/common name		
Trade names		HULS
Suppliers		146
Molecular mass		243–250
Boiling point/range		
Melting point	°C @ 760 mm Hg	
Density	°C	>1200
Evaporation time[a]	kg/m³ @ 20 °C	
Evaporation time[b]		<0.001
Vapour pressure		128
Flash point	kPa @ 20 °C	340
Auto ignition temperature	°C	6.4
Upper flammability limit	°C	0.9
Lower flammability limit	%v/v in air	294
Viscosity	%v/v in air	1.450–1.451
Refractive index	mPa.s (cP) @ 20 °C	
Solubility in water	n 20 D	13.0
Solubility of water	%w/w @ 20 °C	
Solubility parameter[c]	%w/w @ 20 °C	
Hydrogen bond index	$(J/cm^3)^{1/2}$	
Fractional polarity		
Nitro cellulose dil ratio[d]		
Nitro cellulose dil ratio[e]		
Surface tension		
Specific heat liquid	mN/m @ 20 °C	
Latent heat	kJ/kg mole/°C @ 20 °C	
Dielectric constant	kJ/kg mole	
Antoine constant A	@ 20 °C	
Antoine constant B	kPa°C	
Antoine constant C	kPa°C	
Heat of combustion	kPa°C	
UN number	kJ/kg mole	
IMO classification		
ADR/RID classification		
UK exposure limits		
USA exposure limits	mg/m³ (ppm) 8hrTWA/MEL or OES	
German exposure limits	mg/m³ (ppm) TLV/TWA	
EU classification	mg/m³ (ppm) MAK-TWA	
EU risk phrases		
EU safety phrases		

[a] Ether = 1. [b] n-Butyl acetate = 1. [c] Nelson, Hemwall, Edwards. [d] Toluene. [e] Hydrocarbon solvent.

Solvent group		**GLYCOLS**
Chemical name		Triethylene Glycol
Chemical formula		$C_6H_{14}O_4$
Structural formula		$HO[C_2H_4O]_3OH$
CAS No.		112-27-6
Synonym/common name		Triethylene glycol
Trade names		
Suppliers		UNION CARBIDE
Molecular mass		150.17
Boiling point/range	°C @ 760 mm Hg	288.0
Melting point	°C	−4.3
Density	kg/m^3 @ 20 °C	1125.2
Evaporation time[a]		
Evaporation time[b]		
Vapour pressure	kPa @ 20 °C	
Flash point	°C	
Auto ignition temperature	°C	
Upper flammability limit	%v/v in air	
Lower flammability limit	%v/v in air	
Viscosity	mPa.s (cP) @ 20 °C	
Refractive index	n 20 D	
Solubility in water	%w/w @ 20 °C	Complete
Solubility of water	%w/w @ 20 °C	Complete
Solubility parameter[c]	$(J/cm^3)^{1/2}$	21.9
Hydrogen bond index		
Fractional polarity		
Nitro cellulose dil ratio[d]		
Nitro cellulose dil ratio[e]		
Surface tension	mN/m @ 20 °C	45.2
Specific heat liquid	kJ/kg mole/°C @ 20 °C	
Latent heat	kJ/kg mole	71404
Dielectric constant	@ 20 °C	23.7
Antoine constant A	kPa°C	8.7645
Antoine constant B	kPa°C	3726.2
Antoine constant C	kPa°C	
Heat of combustion	kJ/kg mole	−3561000
UN number		
IMO classification		
ADR/RID classification		
UK exposure limits	mg/m^3 (ppm) 8hrTWA/MEL or OES	
USA exposure limits	mg/m^3 (ppm) TLV/TWA	
German exposure limits	mg/m^3 (ppm) MAK-TWA	
EU classification		
EU risk phrases		
EU safety phrases		

[a] Ether = 1. [b] n-Butyl acetate = 1. [c] Nelson, Hemwall, Edwards. [d] Toluene. [e] Hydrocarbon solvent.

Solvent group		**GLYCOLS**
Chemical name		2,2-[1,2-Ethanediylbis(oxy)]bisethanol
Chemical formula		$C_6H_{14}O_4$
Structural formula		$HOCH_2OCH_2CH_2OCH_2CH_2OH$
CAS No.		112-27-6
Synonym/common name		Triethylene glycol
Trade names		
Suppliers		SHELL
Molecular mass		150.17
Boiling point/range	°C @ 760 mm Hg	287.4
Melting point	°C	−7.2
Density	kg/m^3 @ 20 °C	
Evaporation time[a]		
Evaporation time[b]		<0.01
Vapour pressure	kPa @ 20 °C	
Flash point	°C	
Auto ignition temperature	°C	
Upper flammability limit	%v/v in air	
Lower flammability limit	%v/v in air	
Viscosity	mPa.s (cP) @ 20 °C	
Refractive index	n 20 D	1.4559
Solubility in water	%w/w @ 20 °C	Complete
Solubility of water	%w/w @ 20 °C	Complete
Solubility parameter[c]	(J/cm^3)$^{1/2}$	
Hydrogen bond index		
Fractional polarity		
Nitro cellulose dil ratio[d]		
Nitro cellulose dil ratio[e]		
Surface tension	mN/m @ 20 °C	
Specific heat liquid	kJ/kg mole/°C @ 20 °C	
Latent heat	kJ/kg mole	
Dielectric constant	@ 20 °C	
Antoine constant A	kPa°C	
Antoine constant B	kPa°C	
Antoine constant C	kPa°C	
Heat of combustion	kJ/kg mole	
UN number		
IMO classification		
ADR/RID classification		
UK exposure limits	mg/m^3 (ppm) 8hrTWA/MEL or OES	
USA exposure limits	mg/m^3 (ppm) TLV/TWA	
German exposure limits	mg/m^3 (ppm) MAK-TWA	
EU classification		
EU risk phrases		
EU safety phrases		

[a] Ether = 1. [b] n-Butyl acetate = 1. [c] Nelson, Hemwall, Edwards. [d] Toluene. [e] Hydrocarbon solvent.

Solvent group		**HALOGENATED HYDROCARBONS**
Chemical name		Mixture 1,1-Dichloro-1-fluoroethane/ Methyl alcohol/Nitromethane
Chemical formula		
Structural formula		
CAS No.		1717-00/67-56-1/75-52-5
Synonym/common name		HCFC-141b/Methanol azeotrope (stabilised)
Trade names		GENESOLV 2004
Suppliers		ALLIED SIGNAL INC
Molecular mass		
Boiling point/range	°C @ 760 mm Hg	29.4
Melting point	°C	
Density	kg/m^3 @ 20 °C	1216.88
Evaporation time[a]		
Evaporation time[b]		
Vapour pressure	kPa @ 20 °C	65.5
Flash point	°C	None
Auto ignition temperature	°C	550
Upper flammability limit	%v/v in air	20.3
Lower flammability limit	%v/v in air	6.0
Viscosity	mPa.s (cP) @ 20 °C	0.450
Refractive index	n 20 D	1.364 @ 25 °C
Solubility in water	%w/w @ 20 °C	0.17
Solubility of water	%w/w @ 20 °C	0.20
Solubility parameter[c]	(J/cm^3)$^{1/2}$	
Hydrogen bond index		
Fractional polarity		
Nitro cellulose dil ratio[d]		
Nitro cellulose dil ratio[e]		
Surface tension	mN/m @ 20 °C	18.5
Specific heat liquid	kJ/kg mole/°C @ 20 °C	
Latent heat	kJ/kg mole	29420 est
Dielectric constant	@ 20 °C	
Antoine constant A	kPa°C	
Antoine constant B	kPa°C	
Antoine constant C	kPa°C	
Heat of combustion	kJ/kg mole	
UN number		
IMO classification		
ADR/RID classification		
UK exposure limits	mg/m^3 (ppm) 8hrTWA/MEL or OES	
USA exposure limits	mg/m^3 (ppm) TLV/TWA	
German exposure limits	mg/m^3 (ppm) MAK-TWA	
EU classification		
EU risk phrases		59
EU safety phrases		7-16-24/25-59

[a] Ether = 1. [b] n-Butyl acetate = 1. [c] Nelson, Hemwall, Edwards. [d] Toluene. [e] Hydrocarbon solvent.

Property	Unit	Value
Solvent group		**HALOGENATED HYDROCARBONS**
Chemical name		1,1-Dichloro-1-fluoroethane/methanol mixture
Chemical formula		
Structural formula		
CAS No.		
Synonym/common name		141b, HCFC-141b
Trade names		FORANE 141b MGX
Suppliers		ELF ATOCHEM
Molecular mass		
Boiling point/range	°C @ 760 mm Hg	29.4
Melting point	°C	
Density	kg/m^3 @ 20 °C	
Evaporation time[a]		
Evaporation time[b]		
Vapour pressure	kPa @ 20 °C	
Flash point	°C	NONE
Auto ignition temperature	°C	532
Upper flammability limit	%v/v in air	17.7
Lower flammability limit	%v/v in air	5.6
Viscosity	mPa.s (cP) @ 20 °C	
Refractive index	n 20 D	
Solubility in water	%w/w @ 20 °C	
Solubility of water	%w/w @ 20 °C	
Solubility parameter[c]	$(J/cm^3)^{1/2}$	
Hydrogen bond index		
Fractional polarity		
Nitro cellulose dil ratio[d]		
Nitro cellulose dil ratio[e]		
Surface tension	mN/m @ 20 °C	
Specific heat liquid	kJ/kg mole/°C @ 20 °C	
Latent heat	kJ/kg mole	
Dielectric constant	@ 20 °C	
Antoine constant A	kPa°C	
Antoine constant B	kPa°C	
Antoine constant C	kPa°C	
Heat of combustion	kJ/kg mole	
UN number		
IMO classification		
ADR/RID classification		
UK exposure limits	mg/m^3 (ppm) 8hrTWA/MEL or OES	
USA exposure limits	mg/m^3 (ppm) TLV/TWA	
German exposure limits	mg/m^3 (ppm) MAK-TWA	
EU classification		
EU risk phrases		
EU safety phrases		

[a] Ether = 1. [b] n-Butyl acetate = 1. [c] Nelson, Hemwall, Edwards. [d] Toluene. [e] Hydrocarbon solvent.

Solvent group		**HALOGENATED HYDROCARBONS**
Chemical name		1-Chloro,4-(trifluoromethyl)benzene
Chemical formula		$C_7H_4ClF_3$
Structural formula		
CAS No.		98-56-6
Synonym/common name		p-Chlorobenzotrifluoride
Trade names		OXSOL 100, PCBTF
Suppliers		OCCIDENTAL
Molecular mass		
Boiling point/range	°C @ 760 mm Hg	139
Melting point	°C	−36
Density	kg/m^3 @ 20 °C	1336 (25 °C)
Evaporation time[a]		
Evaporation time[b]		0.9
Vapour pressure	kPa @ 20 °C	35.8
Flash point	°C	43 (TCC)
Auto ignition temperature	°C	>500
Upper flammability limit	%v/v in air	10.5
Lower flammability limit	%v/v in air	0.9
Viscosity	mPa.s (cP) @ 20 °C	0.88
Refractive index	n 20 D	
Solubility in water	%w/w @ 20 °C	0.0035
Solubility of water	%w/w @ 20 °C	0.0240
Solubility parameter[c]	(J/cm^3)$^{1/2}$	
Hydrogen bond index		
Fractional polarity		
Nitro cellulose dil ratio[d]		
Nitro cellulose dil ratio[e]		
Surface tension	mN/m @ 20 °C	25
Specific heat liquid	kJ/kg mole/°C @ 20 °C	
Latent heat	kJ/kg mole	
Dielectric constant	@ 20 °C	
Antoine constant A	kPa°C	
Antoine constant B	kPa°C	
Antoine constant C	kPa°C	
Heat of combustion	kJ/kg mole	
UN number		
IMO classification		
ADR/RID classification		
UK exposure limits	mg/m^3 (ppm) 8hrTWA/MEL or OES	
USA exposure limits	mg/m^3 (ppm) TLV/TWA	
German exposure limits	mg/m^3 (ppm) MAK-TWA	
EU classification		
EU risk phrases		
EU safety phrases		

[a] Ether = 1. [b] n-Butyl acetate = 1. [c] Nelson, Hemwall, Edwards. [d] Toluene. [e] Hydrocarbon solvent.

Solvent group		**HALOGENATED HYDROCARBONS**
Chemical name		Dichloromethane
Chemical formula		CH_2Cl_2
Structural formula		CH_2Cl_2
CAS No.		75-09-2
Synonym/common name		Methylene chloride
Trade names		METHOKLONE, UKLENE, AEROTHENE, SOLVACLENE
Suppliers		ICI, DOW, ELF ATOCHEM, AKZO, VULCAN, SOLVAY
Molecular mass		84.9
Boiling point/range	°C @ 760 mm Hg	39.8–40.2
Melting point	°C	−97
Density	kg/m^3 @ 20 °C	1325
Evaporation time[a]		1.5–1.8
Evaporation time[b]		7.0
Vapour pressure	kPa @ 20 °C	47
Flash point	°C	NONE
Auto ignition temperature	°C	624–662
Upper flammability limit	%v/v in air	22.0
Lower flammability limit	%v/v in air	13–15
Viscosity	mPa.s (cP) @ 20 °C	0.43
Refractive index	n 20 D	1.4244
Solubility in water	%w/w @ 20 °C	1.32–1.96
Solubility of water	%w/w @ 20 °C	0.14–0.15
Solubility parameter[c]	(J/cm^3)$^{1/2}$	20.21
Hydrogen bond index		2.2
Fractional polarity		
Nitro cellulose dil ratio[d]		
Nitro cellulose dil ratio[e]		
Surface tension	mN/m @ 20 °C	28
Specific heat liquid	kJ/kg mole/°C @ 20 °C	99.8
Latent heat	kJ/kg mole	27900
Dielectric constant	@ 20 °C	9.1
Antoine constant A	kPa°C	7.0803
Antoine constant B	kPa°C	1138.91
Antoine constant C	kPa°C	231.45
Heat of combustion	kJ/kg mole	−557890
UN number		1593
IMO classification		6.1/III
ADR/RID classification		6.1,15(c)
UK exposure limits	mg/m^3 (ppm) 8hrTWA/MEL or OES	350(100)
USA exposure limits	mg/m^3 (ppm) TLV/TWA	350(100)
German exposure limits	mg/m^3 (ppm) MAK-TWA	360(100)
EU classification		Xn, Carc 3
EU risk phrases		40
EU safety phrases		(2-)$_2$3-24/25-36/37

[a] Ether = 1. [b] n-Butyl acetate = 1. [c] Nelson, Hemwall, Edwards. [d] Toluene. [e] Hydrocarbon solvent.

Solvent group		**HALOGENATED HYDROCARBONS**
Chemical name		1,2-Dichloroethane
Chemical formula		$C_2H_4Cl_2$
Structural formula		CH_2ClCH_2Cl
CAS No.		107-06-2
Synonym/common name		EDC/ethylene dichloride
Trade names		
Suppliers		VULCAN, ENICHEM
Molecular mass		98.96
Boiling point/range	°C @ 760 mm Hg	83.5
Melting point	°C	−35.5
Density	kg/m^3 @ 20 °C	1250
Evaporation time[a]		
Evaporation time[b]		
Vapour pressure	kPa @ 20 °C	8.13
Flash point	°C	18.3 (OPEN CUP)/13 (CLOSED CUP) ENICHEM
Auto ignition temperature	°C	413
Upper flammability limit	%v/v in air	16
Lower flammability limit	%v/v in air	6.2
Viscosity	mPa.s (cP) @ 20 °C	
Refractive index	n 20 D	1.4454
Solubility in water	%w/w @ 20 °C	0.8
Solubility of water	%w/w @ 20 °C	
Solubility parameter[c]	(J/cm^3)$^{1/2}$	
Hydrogen bond index		
Fractional polarity		
Nitro cellulose dil ratio[d]		
Nitro cellulose dil ratio[e]		
Surface tension	mN/m @ 20 °C	
Specific heat liquid	kJ/kg mole/°C @ 20 °C	
Latent heat	kJ/kg mole	
Dielectric constant	@ 20 °C	10.65
Antoine constant A	kPa°C	
Antoine constant B	kPa°C	
Antoine constant C	kPa°C	
Heat of combustion	kJ/kg mole	
UN number		1184
IMO classification		
ADR/RID classification		
UK exposure limits	mg/m^3 (ppm) 8hrTWA/MEL or OES	
USA exposure limits	mg/m^3 (ppm) TLV/TWA	
German exposure limits	mg/m^3 (ppm) MAK-TWA	
EU classification		
EU risk phrases		
EU safety phrases		

[a] Ether = 1. [b] n-Butyl acetate = 1. [c] Nelson, Hemwall, Edwards. [d] Toluene. [e] Hydrocarbon solvent.

Solvent group		HALOGENATED HYDROCARBONS
Chemical name		Monochlorobenzene
Chemical formula		C_6H_5Cl
Structural formula		C_6H_5Cl
CAS No.		108-90-7
Synonym/common name		Monochlorobenzene, MCB
Trade names		
Suppliers		ELF ATOCHEM
Molecular mass		112.6
Boiling point/range	°C @ 760 mm Hg	132
Melting point	°C	−45.2
Density	kg/m^3 @ 20 °C	1110
Evaporation time[a]		
Evaporation time[b]		
Vapour pressure	kPa @ 20 °C	
Flash point	°C	
Auto ignition temperature	°C	637
Upper flammability limit	%v/v in air	7.1
Lower flammability limit	%v/v in air	1.3
Viscosity	mPa.s (cP) @ 20 °C	
Refractive index	n 20 D	
Solubility in water	%w/w @ 20 °C	
Solubility of water	%w/w @ 20 °C	
Solubility parameter[c]	$(J/cm^3)^{1/2}$	
Hydrogen bond index		
Fractional polarity		
Nitro cellulose dil ratio[d]		
Nitro cellulose dil ratio[e]		
Surface tension	mN/m @ 20 °C	
Specific heat liquid	kJ/kg mole/°C @ 20 °C	
Latent heat	kJ/kg mole	
Dielectric constant	@ 20 °C	
Antoine constant A	kPa°C	
Antoine constant B	kPa°C	
Antoine constant C	kPa°C	
Heat of combustion	kJ/kg mole	
UN number		1134
IMO classification		3.3/III
ADR/RID classification		3,31°(c)
UK exposure limits	mg/m^3 (ppm) 8hrTWA/MEL or OES	230(50)
USA exposure limits	mg/m^3 (ppm) TLV/TWA	46(10)
German exposure limits	mg/m^3 (ppm) MAK-TWA	230(50)
EU classification		Xn, R10
EU risk phrases		10-20
EU safety phrases		(2-)24/25

[a] Ether = 1. [b] n-Butyl acetate = 1. [c] Nelson, Hemwall, Edwards. [d] Toluene. [e] Hydrocarbon solvent.

Solvent group		**HALOGENATED HYDROCARBONS**
Chemical name		1,1-Dichloro-1-fluoroethane (Stabilised)
Chemical formula		$C_2H_3Cl_2F$
Structural formula		CCl_2FCH_3
CAS No.		1717-00-6
Synonym/common name		141b, HCFC-141b
Trade names		FORANE, GENESOLV
Suppliers		ELF ATOCHEM, ALLIED SIGNAL
Molecular mass		116.95
Boiling point/range	°C @ 760 mm Hg	32
Melting point	°C	−103.5
Density	kg/m^3 @ 20 °C	1240
Evaporation time[a]		1.2
Evaporation time[b]		
Vapour pressure	kPa @ 20 °C	65.855
Flash point	°C	NONE
Auto ignition temperature	°C	532–550
Upper flammability limit	%v/v in air	17.7
Lower flammability limit	%v/v in air	5.6–7.6
Viscosity	mPa.s (cP) @ 20 °C	
Refractive index	n 20 D	1.37
Solubility in water	%w/w @ 20 °C	0.5
Solubility of water	%w/w @ 20 °C	0.042
Solubility parameter[c]	$(J/cm^3)^{1/2}$	
Hydrogen bond index		
Fractional polarity		
Nitro cellulose dil ratio[d]		
Nitro cellulose dil ratio[e]		
Surface tension	mN/m @ 20 °C	
Specific heat liquid	kJ/kg mole/°C @ 20 °C	
Latent heat	kJ/kg mole	26290
Dielectric constant	@ 20 °C	
Antoine constant A	kPa°C	
Antoine constant B	kPa°C	
Antoine constant C	kPa°C	
Heat of combustion	kJ/kg mole	
UN number		
IMO classification		
ADR/RID classification		
UK exposure limits	mg/m^3 (ppm) 8hrTWA/MEL or OES	
USA exposure limits	mg/m^3 (ppm) TLV/TWA	
German exposure limits	mg/m^3 (ppm) MAK-TWA	
EU classification		
EU risk phrases		
EU safety phrases		

[a] Ether = 1. [b] n-Butyl acetate = 1. [c] Nelson, Hemwall, Edwards. [d] Toluene. [e] Hydrocarbon solvent.

Solvent group		**HALOGENATED HYDROCARBONS**
Chemical name		Trichloromethane
Chemical formula		$CHCl_3$
Structural formula		$CHCl_3$
CAS No.		67-66-3
Synonym/common name		Chloroform
Trade names		
Suppliers		ELF ATOCHEM, AKZO, VULCAN
Molecular mass		119.38
Boiling point/range	°C @ 760 mm Hg	59.4–61.2
Melting point	°C	−63.5
Density	kg/m^3 @ 20 °C	1448
Evaporation time[a]		2.5
Evaporation time[b]		10.5
Vapour pressure	kPa @ 20 °C	20.0
Flash point	°C	NONE
Auto ignition temperature	°C	624
Upper flammability limit	%v/v in air	
Lower flammability limit	%v/v in air	
Viscosity	mPa.s (cP) @ 20 °C	0.57
Refractive index	n 20 D	1.446
Solubility in water	%w/w @ 20 °C	0.77
Solubility of water	%w/w @ 20 °C	0.86
Solubility parameter[c]	$(J/cm^3)^{1/2}$	19.42
Hydrogen bond index		
Fractional polarity		
Nitro cellulose dil ratio[d]		
Nitro cellulose dil ratio[e]		
Surface tension	mN/m @ 20 °C	27.14
Specific heat liquid	kJ/kg mole/°C @ 20 °C	113.65
Latent heat	kJ/kg mole	29796
Dielectric constant	@ 20 °C	4.785
Antoine constant A	kPa°C	
Antoine constant B	kPa°C	
Antoine constant C	kPa°C	
Heat of combustion	kJ/kg mole	−401960
UN number		1888
IMO classification		6.1/II
ADR/RID classification		6.1,15°(c)
UK exposure limits	mg/m^3 (ppm) 8hrTWA/MEL or OES	9.8(2)
USA exposure limits	mg/m^3 (ppm) TLV/TWA	49(10)
German exposure limits	mg/m^3 (ppm) MAK-TWA	50(10)
EU classification		Xn, Xi, Carc 3
EU risk phrases		22-38-40-/20/22
EU safety phrases		(2-)36/37

[a] Ether = 1. [b] n-Butyl acetate = 1. [c] Nelson, Hemwall, Edwards. [d] Toluene. [e] Hydrocarbon solvent.

Solvent group		HALOGENATED HYDROCARBONS
Chemical name		1,2,2-Trichloroethylene
Chemical formula		C_2HCl_3
Structural formula		$CHCl=CCl_2$
CAS No.		79-01-6
Synonym/common name		TRI/TRIKE, Trichloroethylene
Trade names		TRIKLONE, NEU TRI, HI-TRI, TAVOXENE, ALTENE
Suppliers		ICI, ELF ATOCHEM, DOW, ENICHEM, SOLVAY
Molecular mass		131.39
Boiling point/range	°C @ 760 mm Hg	87
Melting point	°C	−88
Density	kg/m^3 @ 20 °C	1456–1464
Evaporation time[a]		3–3.8
Evaporation time[b]		6.39
Vapour pressure	kPa @ 20 °C	7.23–7.72
Flash point	°C	NONE
Auto ignition temperature	°C	410
Upper flammability limit	%v/v in air	10.5
Lower flammability limit	%v/v in air	8
Viscosity	mPa.s (cP) @ 20 °C	0.54–0.607
Refractive index	n 20 D	1.477
Solubility in water	%w/w @ 20 °C	0.11–1.1
Solubility of water	%w/w @ 20 °C	0.025
Solubility parameter[c]	$(J/cm^3)^{1/2}$	19.0
Hydrogen bond index		3.6
Fractional polarity		
Nitro cellulose dil ratio[d]		
Nitro cellulose dil ratio[e]		
Surface tension	mN/m @ 20 °C	26–29
Specific heat liquid	kJ/kg mole/°C @ 20 °C	122.6
Latent heat	kJ/kg mole	32532
Dielectric constant	@ 20 °C	3.39
Antoine constant A	kPa°C	6.15298
Antoine constant B	kPa°C	1315.04
Antoine constant C	kPa°C	230.0
Heat of combustion	kJ/kg mole	−401930
UN number		1710
IMO classification		6.1/III
ADR/RID classification		6.1,15°(c)
UK exposure limits	mg/m^3 (ppm) 8hrTWA/MEL or OES	535(100)
USA exposure limits	mg/m^3 (ppm) TLV/TWA	269(50)
German exposure limits	mg/m^3 (ppm) MAK-TWA	270(50)
EU classification		Xn, Carc 3
EU risk phrases		40-51/53
EU safety phrases		(2-)23-36/37-61

[a] Ether = 1. [b] n-Butyl acetate = 1. [c] Nelson, Hemwall, Edwards. [d] Toluene. [e] Hydrocarbon solvent.

Solvent group		HALOGENATED HYDROCARBONS
Chemical name		1,1,1-Trichloroethane
Chemical formula		$C_2H_3Cl_3$
Structural formula		CH_3CCl_3
CAS No.		71-55-6
Synonym/common name		Methyl chloroform
Trade names		BALTANE CF, CHLOROTHENE
Suppliers		ELF ATOCHEM, DOW, VULCAN
Molecular mass		133.41
Boiling point/range	°C @ 760 mm Hg	74
Melting point	°C	−37
Density	kg/m^3 @ 20 °C	1300
Evaporation time[a]		2.5
Evaporation time[b]		4.6
Vapour pressure	kPa @ 20 °C	13.3
Flash point	°C	NONE
Auto ignition temperature	°C	537
Upper flammability limit	%v/v in air	10.5
Lower flammability limit	%v/v in air	8
Viscosity	mPa.s (cP) @ 20 °C	1.1
Refractive index	n 20 D	1.435
Solubility in water	%w/w @ 20 °C	
Solubility of water	%w/w @ 20 °C	
Solubility parameter[c]	$(J/cm^3)^{1/2}$	17.4
Hydrogen bond index		2.6
Fractional polarity		
Nitro cellulose dil ratio[d]		
Nitro cellulose dil ratio[e]		
Surface tension	mN/m @ 20 °C	
Specific heat liquid	kJ/kg mole/°C @ 20 °C	
Latent heat	kJ/kg mole	29708
Dielectric constant	@ 20 °C	7.25
Antoine constant A	kPa°C	5.98755
Antoine constant B	kPa°C	1182.527
Antoine constant C	kPa°C	222.894
Heat of combustion	kJ/kg mole	−1108050
UN number		2831
IMO classification		6.1/III
ADR/RID classification		6.1,15°(c)
UK exposure limits	mg/m^3 (ppm) 8hrTWA/MEL or OES	1900(350)
USA exposure limits	mg/m^3 (ppm) TLV/TWA	1910(350)
German exposure limits	mg/m^3 (ppm) MAK-TWA	1080(200)
EU classification		Xn, N
EU risk phrases		20-59
EU safety phrases		(2-)24/25-59-61

[a] Ether = 1. [b] n-Butyl acetate = 1. [c] Nelson, Hemwall, Edwards. [d] Toluene. [e] Hydrocarbon solvent.

Solvent group		HALOGENATED HYDROCARBONS
Chemical name		Dichlorotoluene
Chemical formula		$C_7H_6Cl_2$
Structural formula		
CAS No.		
Synonym/common name		Dichlorotoluene
Trade names		
Suppliers		ELF ATOCHEM
Molecular mass		161.0
Boiling point/range	°C @ 760 mm Hg	202–208
Melting point	°C	
Density	kg/m^3 @ 20 °C	1250
Evaporation time[a]		
Evaporation time[b]		
Vapour pressure	kPa @ 20 °C	
Flash point	°C	
Auto ignition temperature	°C	>500
Upper flammability limit	%v/v in air	
Lower flammability limit	%v/v in air	
Viscosity	mPa.s (cP) @ 20 °C	
Refractive index	n 20 D	
Solubility in water	%w/w @ 20 °C	<1
Solubility of water	%w/w @ 20 °C	
Solubility parameter[c]	(J/cm^3)$^{1/2}$	
Hydrogen bond index		
Fractional polarity		
Nitro cellulose dil ratio[d]		
Nitro cellulose dil ratio[e]		
Surface tension	mN/m @ 20 °C	
Specific heat liquid	kJ/kg mole/°C @ 20 °C	
Latent heat	kJ/kg mole	
Dielectric constant	@ 20 °C	
Antoine constant A	kPa°C	
Antoine constant B	kPa°C	
Antoine constant C	kPa°C	
Heat of combustion	kJ/kg mole	
UN number		
IMO classification		
ADR/RID classification		3,32°(c)
UK exposure limits	mg/m^3 (ppm) 8hrTWA/MEL or OES	
USA exposure limits	mg/m^3 (ppm) TLV/TWA	
German exposure limits	mg/m^3 (ppm) MAK-TWA	
EU classification		Xi
EU risk phrases		38
EU safety phrases		

[a] Ether = 1. [b] n-Butyl acetate = 1. [c] Nelson, Hemwall, Edwards. [d] Toluene. [e] Hydrocarbon solvent.

Solvent group		**HALOGENATED HYDROCARBONS**
Chemical name		1,1,2,2-Tetrachloroethylene
Chemical formula		C_2Cl_4
Structural formula		$CCl_2{=}CCl_2$
CAS No.		127-18-4
Synonym/common name		Perchloroethylene/Per/Perk
Trade names		DOWPER, Pertene, PERKLONE, SOLTENE
Suppliers		DOW, ELF ATOCHEM, ICI, ENICHEM, VULCAN, SOLVAY
Molecular mass		165.83
Boiling point/range	°C @ 760 mm Hg	121
Melting point	°C	−22.3
Density	kg/m^3 @ 20 °C	1619
Evaporation time[a]		7.1–9.5
Evaporation time[b]		2.1
Vapour pressure	kPa @ 20 °C	1.71–1.87
Flash point	°C	NONE
Auto ignition temperature	°C	NONE
Upper flammability limit	%v/v in air	NONE
Lower flammability limit	%v/v in air	NONE
Viscosity	mPa.s (cP) @ 20 °C	0.8839
Refractive index	n 20 D	1.504
Solubility in water	%w/w @ 20 °C	0.015
Solubility of water	%w/w @ 20 °C	0.008–0.01
Solubility parameter[c]	$(J/cm^3)^{1/2}$	19.0
Hydrogen bond index		
Fractional polarity		
Nitro cellulose dil ratio[d]		
Nitro cellulose dil ratio[e]		
Surface tension	mN/m @ 20 °C	33.4
Specific heat liquid	kJ/kg mole/°C @ 20 °C	145.9
Latent heat	kJ/kg mole	35438
Dielectric constant	@ 20 °C	
Antoine constant A	kPa°C	6.10170
Antoine constant B	kPa°C	1386.90
Antoine constant C	kPa°C	217.52
Heat of combustion	kJ/kg mole	−510600
UN number		1897
IMO classification		6.1/III
ADR/RID classification		6.1,15°(c)
UK exposure limits	mg/m^3 (ppm) 8hrTWA/MEL or OES	335(50)
USA exposure limits	mg/m^3 (ppm) TLV/TWA	170(25)
German exposure limits	mg/m^3 (ppm) MAK-TWA	345(50)
EU classification		Xn, Carc 3
EU risk phrases		40
EU safety phrases		(2-)23-36/37

[a] Ether = 1. [b] n-Butyl acetate = 1. [c] Nelson, Hemwall, Edwards. [d] Toluene. [e] Hydrocarbon solvent.

Property	Unit	Value
Solvent group		**HALOGENATED HYDROCARBONS**
Chemical name		Trichlorotrifluoroethane
Chemical formula		$C_2H_2Cl_3F_3$
Structural formula		$CHCl_2FCHClF_2$
CAS No.		76-13-1
Synonym/common name		CFC113
Trade names		Forane
Suppliers		ELF ATOCHEM
Molecular mass		187.38
Boiling point/range	°C @ 760 mm Hg	47.6
Melting point	°C	−35
Density	kg/m^3 @ 20 °C	1570
Evaporation time[a]		
Evaporation time[b]		
Vapour pressure	kPa @ 20 °C	
Flash point	°C	
Auto ignition temperature	°C	
Upper flammability limit	%v/v in air	
Lower flammability limit	%v/v in air	
Viscosity	mPa.s (cP) @ 20 °C	
Refractive index	n 20 D	
Solubility in water	%w/w @ 20 °C	0.17
Solubility of water	%w/w @ 20 °C	
Solubility parameter[c]	$(J/cm^3)^{1/2}$	
Hydrogen bond index		
Fractional polarity		
Nitro cellulose dil ratio[d]		
Nitro cellulose dil ratio[e]		
Surface tension	mN/m @ 20 °C	
Specific heat liquid	kJ/kg mole/°C @ 20 °C	
Latent heat	kJ/kg mole	
Dielectric constant	@ 20 °C	
Antoine constant A	kPa°C	
Antoine constant B	kPa°C	
Antoine constant C	kPa°C	
Heat of combustion	kJ/kg mole	
UN number		
IMO classification		
ADR/RID classification		
UK exposure limits	mg/m^3 (ppm) 8hrTWA/MEL or OES	7600(1000)
USA exposure limits	mg/m^3 (ppm) TLV/TWA	7670(1000)
German exposure limits	mg/m^3 (ppm) MAK-TWA	3800(500)
EU classification		N
EU risk phrases		59
EU safety phrases		15-23-59-61

[a] Ether = 1. [b] n-Butyl acetate = 1. [c] Nelson, Hemwall, Edwards. [d] Toluene. [e] Hydrocarbon solvent.

Solvent group		HYDROCARBONS—ALIPHATIC
Chemical name		
Chemical formula		
Structural formula		
CAS No.		64741-73-7
Synonym/common name		
Trade names		SHELLSOL TM
Suppliers		SHELL
Molecular mass		
Boiling point/range	°C @ 760 mm Hg	215–260
Melting point	°C	
Density	kg/m^3 @ 20 °C	799
Evaporation time[a]		
Evaporation time[b]		0.01
Vapour pressure	kPa @ 20 °C	
Flash point	°C	86 (ASTM D93)
Auto ignition temperature	°C	
Upper flammability limit	%v/v in air	6.0
Lower flammability limit	%v/v in air	0.6
Viscosity	mPa.s (cP) @ 20 °C	2.64
Refractive index	n 20 D	1.442
Solubility in water	%w/w @ 20 °C	
Solubility of water	%w/w @ 20 °C	
Solubility parameter[c]	$(J/cm^3)^{1/2}$	
Hydrogen bond index		
Fractional polarity		
Nitro cellulose dil ratio[d]		
Nitro cellulose dil ratio[e]		
Surface tension	mN/m @ 20 °C	
Specific heat liquid	kJ/kg mole/°C @ 20 °C	
Latent heat	kJ/kg mole	
Dielectric constant	@ 20 °C	
Antoine constant A	kPa°C	
Antoine constant B	kPa°C	
Antoine constant C	kPa°C	
Heat of combustion	kJ/kg mole	
UN number		
IMO classification		
ADR/RID classification		3,32°(c)
UK exposure limits	mg/m^3 (ppm) 8hrTWA/MEL or OES	
USA exposure limits	mg/m^3 (ppm) TLV/TWA	
German exposure limits	mg/m^3 (ppm) MAK-TWA	
EU classification		
EU risk phrases		
EU safety phrases		

[a] Ether = 1. [b] n-Butyl acetate = 1. [c] Nelson, Hemwall, Edwards. [d] Toluene. [e] Hydrocarbon solvent.

Solvent group		HYDROCARBONS—ALIPHATIC
Chemical name		
Chemical formula		
Structural formula		
CAS No.		64742-49-0
Synonym/common name		
Trade names		HYDROSOL Ess.80/120
Suppliers		TOTAL
Molecular mass		
Boiling point/range	°C @ 760 mm Hg	83–108
Melting point	°C	
Density	kg/m^3 @ 20 °C	700
Evaporation time[a]		3.2
Evaporation time[b]		
Vapour pressure	kPa @ 20 °C	
Flash point	°C	<0
Auto ignition temperature	°C	
Upper flammability limit	%v/v in air	
Lower flammability limit	%v/v in air	
Viscosity	mPa.s (cP) @ 20 °C	
Refractive index	n 20 D	1.392
Solubility in water	%w/w @ 20 °C	
Solubility of water	%w/w @ 20 °C	
Solubility parameter[c]	$(J/cm^3)^{1/2}$	
Hydrogen bond index		
Fractional polarity		
Nitro cellulose dil ratio[d]		
Nitro cellulose dil ratio[e]		
Surface tension	mN/m @ 20 °C	
Specific heat liquid	kJ/kg mole/°C @ 20 °C	
Latent heat	kJ/kg mole	
Dielectric constant	@ 20 °C	
Antoine constant A	kPa°C	
Antoine constant B	kPa°C	
Antoine constant C	kPa°C	
Heat of combustion	kJ/kg mole	
UN number		1203
IMO classification		3.1
ADR/RID classification		3,2°(b)
UK exposure limits	mg/m^3 (ppm) 8hrTWA/MEL or OES	
USA exposure limits	mg/m^3 (ppm) TLV/TWA	
German exposure limits	mg/m^3 (ppm) MAK-TWA	
EU classification		F
EU risk phrases		11
EU safety phrases		9-16-23-29-33

[a] Ether = 1. [b] n-Butyl acetate = 1. [c] Nelson, Hemwall, Edwards. [d] Toluene. [e] Hydrocarbon solvent.

Solvent group		HYDROCARBONS—ALIPHATIC
Chemical name		
Chemical formula		
Structural formula		
CAS No.		90622-57-4
Synonym/common name		
Trade names		ISANE IP155
Suppliers		TOTAL
Molecular mass		
Boiling point/range	°C @ 760 mm Hg	155–173
Melting point	°C	
Density	kg/m^3 @ 20 °C	745
Evaporation time[a]		36
Evaporation time[b]		
Vapour pressure	kPa @ 20 °C	
Flash point	°C	41
Auto ignition temperature	°C	
Upper flammability limit	%v/v in air	
Lower flammability limit	%v/v in air	
Viscosity	mPa.s (cP) @ 20 °C	
Refractive index	n 20 D	1.416
Solubility in water	%w/w @ 20 °C	
Solubility of water	%w/w @ 20 °C	
Solubility parameter[c]	$(J/cm^3)^{1/2}$	
Hydrogen bond index		
Fractional polarity		
Nitro cellulose dil ratio[d]		
Nitro cellulose dil ratio[e]		
Surface tension	mN/m @ 20 °C	
Specific heat liquid	kJ/kg mole/°C @ 20 °C	
Latent heat	kJ/kg mole	
Dielectric constant	@ 20 °C	
Antoine constant A	kPa°C	
Antoine constant B	kPa°C	
Antoine constant C	kPa°C	
Heat of combustion	kJ/kg mole	
UN number		1300
IMO classification		3.3
ADR/RID classification		3,31°(c)
UK exposure limits	mg/m^3 (ppm) 8hrTWA/MEL or OES	
USA exposure limits	mg/m^3 (ppm) TLV/TWA	
German exposure limits	mg/m^3 (ppm) MAK-TWA	
EU classification		
EU risk phrases		10
EU safety phrases		

[a] Ether = 1. [b] n-Butyl acetate = 1. [c] Nelson, Hemwall, Edwards. [d] Toluene. [e] Hydrocarbon solvent.

Solvent group		HYDROCARBONS—ALIPHATIC
Chemical name		
Chemical formula		
Structural formula		
CAS No.		90622-58-5
Synonym/common name		
Trade names		ISANE IP175
Suppliers		TOTAL
Molecular mass		
Boiling point/range	°C @ 760 mm Hg	180–196
Melting point	°C	
Density	kg/m^3 @ 20 °C	765
Evaporation time[a]		65
Evaporation time[b]		
Vapour pressure	kPa @ 20 °C	
Flash point	°C	61
Auto ignition temperature	°C	
Upper flammability limit	%v/v in air	
Lower flammability limit	%v/v in air	
Viscosity	mPa.s (cP) @ 20 °C	
Refractive index	n 20 D	1.426
Solubility in water	%w/w @ 20 °C	
Solubility of water	%w/w @ 20 °C	
Solubility parameter[c]	(J/cm^3)$^{1/2}$	
Hydrogen bond index		
Fractional polarity		
Nitro cellulose dil ratio[d]		
Nitro cellulose dil ratio[e]		
Surface tension	mN/m @ 20 °C	
Specific heat liquid	kJ/kg mole/°C @ 20 °C	
Latent heat	kJ/kg mole	
Dielectric constant	@ 20 °C	
Antoine constant A	kPa°C	
Antoine constant B	kPa°C	
Antoine constant C	kPa°C	
Heat of combustion	kJ/kg mole	
UN number		1300
IMO classification		3.3
ADR/RID classification		3,32°(c)
UK exposure limits	mg/m^3 (ppm) 8hrTWA/MEL or OES	
USA exposure limits	mg/m^3 (ppm) TLV/TWA	
German exposure limits	mg/m^3 (ppm) MAK-TWA	
EU classification		
EU risk phrases		
EU safety phrases		

[a] Ether = 1. [b] n-Butyl acetate = 1. [c] Nelson, Hemwall, Edwards. [d] Toluene. [e] Hydrocarbon solvent.

Solvent group		HYDROCARBONS—ALIPHATIC
Chemical name		
Chemical formula		
Structural formula		
CAS No.		64742-48-9
Synonym/common name		
Trade names		SPIRDANE D60
Suppliers		TOTAL
Molecular mass		
Boiling point/range	°C @ 760 mm Hg	180–215
Melting point	°C	
Density	kg/m^3 @ 20 °C	782
Evaporation time[a]		200
Evaporation time[b]		
Vapour pressure	kPa @ 20 °C	
Flash point	°C	60
Auto ignition temperature	°C	
Upper flammability limit	%v/v in air	
Lower flammability limit	%v/v in air	
Viscosity	mPa.s (cP) @ 20 °C	
Refractive index	n 20 D	1.432
Solubility in water	%w/w @ 20 °C	
Solubility of water	%w/w @ 20 °C	
Solubility parameter[c]	$(J/cm^3)^{1/2}$	
Hydrogen bond index		
Fractional polarity		
Nitro cellulose dil ratio[d]		
Nitro cellulose dil ratio[e]		
Surface tension	mN/m @ 20 °C	
Specific heat liquid	kJ/kg mole/°C @ 20 °C	
Latent heat	kJ/kg mole	
Dielectric constant	@ 20 °C	
Antoine constant A	kPa°C	
Antoine constant B	kPa°C	
Antoine constant C	kPa°C	
Heat of combustion	kJ/kg mole	
UN number		1300
IMO classification		3.3
ADR/RID classification		3,32°(c)
UK exposure limits	mg/m^3 (ppm) 8hrTWA/MEL or OES	
USA exposure limits	mg/m^3 (ppm) TLV/TWA	
German exposure limits	mg/m^3 (ppm) MAK-TWA	
EU classification		
EU risk phrases		
EU safety phrases		

[a] Ether = 1. [b] n-Butyl acetate = 1. [c] Nelson, Hemwall, Edwards. [d] Toluene. [e] Hydrocarbon solvent.

Solvent group		**HYDROCARBONS—ALIPHATIC**
Chemical name		
Chemical formula		
Structural formula		
CAS No.		64742-49-0
Synonym/common name		
Trade names		HYDROSOL Ess F < 5
Suppliers		TOTAL
Molecular mass		
Boiling point/range	°C @ 760 mm Hg	105–155
Melting point	°C	
Density	kg/m^3 @ 20 °C	735
Evaporation time[a]		12
Evaporation time[b]		
Vapour pressure	kPa @ 20 °C	
Flash point	°C	<0
Auto ignition temperature	°C	
Upper flammability limit	%v/v in air	
Lower flammability limit	%v/v in air	
Viscosity	mPa.s (cP) @ 20 °C	
Refractive index	n 20 D	1.409
Solubility in water	%w/w @ 20 °C	
Solubility of water	%w/w @ 20 °C	
Solubility parameter[c]	$(J/cm^3)^{1/2}$	
Hydrogen bond index		
Fractional polarity		
Nitro cellulose dil ratio[d]		
Nitro cellulose dil ratio[e]		
Surface tension	mN/m @ 20 °C	
Specific heat liquid	kJ/kg mole/°C @ 20 °C	
Latent heat	kJ/kg mole	
Dielectric constant	@ 20 °C	
Antoine constant A	kPa°C	
Antoine constant B	kPa°C	
Antoine constant C	kPa°C	
Heat of combustion	kJ/kg mole	
UN number		1203
IMO classification		3.2
ADR/RID classification		3,3°(b)
UK exposure limits	mg/m^3 (ppm) 8hrTWA/MEL or OES	
USA exposure limits	mg/m^3 (ppm) TLV/TWA	
German exposure limits	mg/m^3 (ppm) MAK-TWA	
EU classification		F
EU risk phrases		11
EU safety phrases		9-16-29-33

[a] Ether = 1. [b] n-Butyl acetate = 1. [c] Nelson, Hemwall, Edwards. [d] Toluene. [e] Hydrocarbon solvent.

Solvent group		HYDROCARBONS—ALIPHATIC
Chemical name		
Chemical formula		
Structural formula		
CAS No.		64742-47-8
Synonym/common name		
Trade names		KERDANE
Suppliers		TOTAL
Molecular mass		
Boiling point/range	°C @ 760 mm Hg	185–240
Melting point	°C	
Density	kg/m^3 @ 20 °C	786
Evaporation time[a]		550
Evaporation time[b]		
Vapour pressure	kPa @ 20 °C	
Flash point	°C	65
Auto ignition temperature	°C	
Upper flammability limit	%v/v in air	
Lower flammability limit	%v/v in air	
Viscosity	mPa.s (cP) @ 20 °C	
Refractive index	n 20 D	1.436
Solubility in water	%w/w @ 20 °C	
Solubility of water	%w/w @ 20 °C	
Solubility parameter[c]	$(J/cm^3)^{1/2}$	
Hydrogen bond index		
Fractional polarity		
Nitro cellulose dil ratio[d]		
Nitro cellulose dil ratio[e]		
Surface tension	mN/m @ 20 °C	
Specific heat liquid	kJ/kg mole/°C @ 20 °C	
Latent heat	kJ/kg mole	
Dielectric constant	@ 20 °C	
Antoine constant A	kPa°C	
Antoine constant B	kPa°C	
Antoine constant C	kPa°C	
Heat of combustion	kJ/kg mole	
UN number		1223
IMO classification		
ADR/RID classification		3,32°(c)
UK exposure limits	mg/m^3 (ppm) 8hrTWA/MEL or OES	
USA exposure limits	mg/m^3 (ppm) TLV/TWA	
German exposure limits	mg/m^3 (ppm) MAK-TWA	
EU classification		
EU risk phrases		
EU safety phrases		

[a] Ether = 1. [b] n-Butyl acetate = 1. [c] Nelson, Hemwall, Edwards. [d] Toluene. [e] Hydrocarbon solvent.

Solvent group		HYDROCARBONS—ALIPHATIC
Chemical name		
Chemical formula		
Structural formula		
CAS No.		64742-82-1
Synonym/common name		
Trade names		SPIRDANE L2
Suppliers		TOTAL
Molecular mass		
Boiling point/range	°C @ 760 mm Hg	135–160
Melting point	°C	
Density	kg/m^3 @ 20 °C	760
Evaporation time[a]		22
Evaporation time[b]		
Vapour pressure	kPa @ 20 °C	
Flash point	°C	22
Auto ignition temperature	°C	
Upper flammability limit	%v/v in air	
Lower flammability limit	%v/v in air	
Viscosity	mPa.s (cP) @ 20 °C	
Refractive index	n 20 D	1.425
Solubility in water	%w/w @ 20 °C	
Solubility of water	%w/w @ 20 °C	
Solubility parameter[c]	$(J/cm^3)^{1/2}$	
Hydrogen bond index		
Fractional polarity		
Nitro cellulose dil ratio[d]		
Nitro cellulose dil ratio[e]		
Surface tension	mN/m @ 20 °C	
Specific heat liquid	kJ/kg mole/°C @ 20 °C	
Latent heat	kJ/kg mole	
Dielectric constant	@ 20 °C	
Antoine constant A	kPa°C	
Antoine constant B	kPa°C	
Antoine constant C	kPa°C	
Heat of combustion	kJ/kg mole	
UN number		1300
IMO classification		3.2
ADR/RID classification		3,31°(c)
UK exposure limits	mg/m^3 (ppm) 8hrTWA/MEL or OES	
USA exposure limits	mg/m^3 (ppm) TLV/TWA	
German exposure limits	mg/m^3 (ppm) MAK-TWA	
EU classification		
EU risk phrases		10
EU safety phrases		

[a] Ether = 1. [b] n-Butyl acetate = 1. [c] Nelson, Hemwall, Edwards. [d] Toluene. [e] Hydrocarbon solvent.

Solvent group		HYDROCARBONS—ALIPHATIC
Chemical name		
Chemical formula		
Structural formula		
CAS No.		64742-47-8
Synonym/common name		
Trade names		KETRUL BT
Suppliers		TOTAL
Molecular mass		
Boiling point/range	°C @ 760 mm Hg	185–240
Melting point	°C	
Density	kg/m^3 @ 20 °C	786
Evaporation time[a]		550
Evaporation time[b]		
Vapour pressure	kPa @ 20 °C	
Flash point	°C	65
Auto ignition temperature	°C	
Upper flammability limit	%v/v in air	
Lower flammability limit	%v/v in air	
Viscosity	mPa.s (cP) @ 20 °C	
Refractive index	n 20 D	1.436
Solubility in water	%w/w @ 20 °C	
Solubility of water	%w/w @ 20 °C	
Solubility parameter[c]	$(J/cm^3)^{1/2}$	
Hydrogen bond index		
Fractional polarity		
Nitro cellulose dil ratio[d]		
Nitro cellulose dil ratio[e]		
Surface tension	mN/m @ 20 °C	
Specific heat liquid	kJ/kg mole/°C @ 20 °C	
Latent heat	kJ/kg mole	
Dielectric constant	@ 20 °C	
Antoine constant A	kPa°C	
Antoine constant B	kPa°C	
Antoine constant C	kPa°C	
Heat of combustion	kJ/kg mole	
UN number		1223
IMO classification		
ADR/RID classification		3,32°(c)
UK exposure limits	mg/m^3 (ppm) 8hrTWA/MEL or OES	
USA exposure limits	mg/m^3 (ppm) TLV/TWA	
German exposure limits	mg/m^3 (ppm) MAK-TWA	
EU classification		
EU risk phrases		
EU safety phrases		

[a] Ether = 1. [b] n-Butyl acetate = 1. [c] Nelson, Hemwall, Edwards. [d] Toluene. [e] Hydrocarbon solvent.

Solvent group		HYDROCARBONS—ALIPHATIC
Chemical name		
Chemical formula		
Structural formula		
CAS No.		64741-65-7
Synonym/common name		
Trade names		SHELLSOL TK
Suppliers		SHELL
Molecular mass		
Boiling point/range	°C @ 760 mm Hg	180–199
Melting point	°C	
Density	kg/m^3 @ 20 °C	770
Evaporation time[a]		
Evaporation time[b]		0.09
Vapour pressure	kPa @ 20 °C	
Flash point	°C	59 (ASTM D93)
Auto ignition temperature	°C	
Upper flammability limit	%v/v in air	6.0
Lower flammability limit	%v/v in air	0.6
Viscosity	mPa.s (cP) @ 20 °C	1.29
Refractive index	n 20 D	1.428
Solubility in water	%w/w @ 20 °C	
Solubility of water	%w/w @ 20 °C	
Solubility parameter[c]	$(J/cm^3)^{1/2}$	
Hydrogen bond index		
Fractional polarity		
Nitro cellulose dil ratio[d]		
Nitro cellulose dil ratio[e]		
Surface tension	mN/m @ 20 °C	
Specific heat liquid	kJ/kg mole/°C @ 20 °C	
Latent heat	kJ/kg mole	
Dielectric constant	@ 20 °C	
Antoine constant A	kPa°C	
Antoine constant B	kPa°C	
Antoine constant C	kPa°C	
Heat of combustion	kJ/kg mole	
UN number		
IMO classification		
ADR/RID classification		3,32°(c)
UK exposure limits	mg/m^3 (ppm) 8hrTWA/MEL or OES	
USA exposure limits	mg/m^3 (ppm) TLV/TWA	
German exposure limits	mg/m^3 (ppm) MAK-TWA	
EU classification		
EU risk phrases		
EU safety phrases		

[a] Ether = 1. [b] n-Butyl acetate = 1. [c] Nelson, Hemwall, Edwards. [d] Toluene. [e] Hydrocarbon solvent.

Solvent group		HYDROCARBONS—ALIPHATIC
Chemical name		
Chemical formula		
Structural formula		
CAS No.		90622-58-5
Synonym/common name		
Trade names		ISANE IP235
Suppliers		TOTAL
Molecular mass		
Boiling point/range	°C @ 760 mm Hg	235–265
Melting point	°C	
Density	kg/m^3 @ 20 °C	795
Evaporation time[a]		>1000
Evaporation time[b]		
Vapour pressure	kPa @ 20 °C	
Flash point	°C	102
Auto ignition temperature	°C	
Upper flammability limit	%v/v in air	
Lower flammability limit	%v/v in air	
Viscosity	mPa.s (cP) @ 20 °C	
Refractive index	n 20 D	1.441
Solubility in water	%w/w @ 20 °C	
Solubility of water	%w/w @ 20 °C	
Solubility parameter[c]	(J/cm^3)$^{1/2}$	
Hydrogen bond index		
Fractional polarity		
Nitro cellulose dil ratio[d]		
Nitro cellulose dil ratio[e]		
Surface tension	mN/m @ 20 °C	
Specific heat liquid	kJ/kg mole/°C @ 20 °C	
Latent heat	kJ/kg mole	
Dielectric constant	@ 20 °C	
Antoine constant A	kPa°C	
Antoine constant B	kPa°C	
Antoine constant C	kPa°C	
Heat of combustion	kJ/kg mole	
UN number		
IMO classification		
ADR/RID classification		
UK exposure limits	mg/m^3 (ppm) 8hrTWA/MEL or OES	
USA exposure limits	mg/m^3 (ppm) TLV/TWA	
German exposure limits	mg/m^3 (ppm) MAK-TWA	
EU classification		
EU risk phrases		
EU safety phrases		

[a] Ether = 1. [b] n-Butyl acetate = 1. [c] Nelson, Hemwall, Edwards. [d] Toluene. [e] Hydrocarbon solvent.

Solvent group		**HYDROCARBONS—ALIPHATIC**
Chemical name		
Chemical formula		
Structural formula		
CAS No.		64742-47-8
Synonym/common name		
Trade names		KETRUL 210
Suppliers		TOTAL
Molecular mass		
Boiling point/range	°C @ 760 mm Hg	215–240
Melting point	°C	
Density	kg/m^3 @ 20 °C	795
Evaporation time[a]		1000
Evaporation time[b]		
Vapour pressure	kPa @ 20 °C	
Flash point	°C	85
Auto ignition temperature	°C	
Upper flammability limit	%v/v in air	
Lower flammability limit	%v/v in air	
Viscosity	mPa.s (cP) @ 20 °C	
Refractive index	n 20 D	1.440
Solubility in water	%w/w @ 20 °C	
Solubility of water	%w/w @ 20 °C	
Solubility parameter[c]	$(J/cm^3)^{1/2}$	
Hydrogen bond index		
Fractional polarity		
Nitro cellulose dil ratio[d]		
Nitro cellulose dil ratio[e]		
Surface tension	mN/m @ 20 °C	
Specific heat liquid	kJ/kg mole/°C @ 20 °C	
Latent heat	kJ/kg mole	
Dielectric constant	@ 20 °C	
Antoine constant A	kPa°C	
Antoine constant B	kPa°C	
Antoine constant C	kPa°C	
Heat of combustion	kJ/kg mole	
UN number		1223
IMO classification		
ADR/RID classification		3,32°(c)
UK exposure limits	mg/m^3 (ppm) 8hrTWA/MEL or OES	
USA exposure limits	mg/m^3 (ppm) TLV/TWA	
German exposure limits	mg/m^3 (ppm) MAK-TWA	
EU classification		
EU risk phrases		
EU safety phrases		

[a] Ether = 1. [b] n-Butyl acetate = 1. [c] Nelson, Hemwall, Edwards. [d] Toluene. [e] Hydrocarbon solvent.

Solvent group		**HYDROCARBONS—ALIPHATIC**
Chemical name		
Chemical formula		
Structural formula		
CAS No.		64742-48-9
Synonym/common name		
Trade names		SPIRDANE D40
Suppliers		TOTAL
Molecular mass		
Boiling point/range	°C @ 760 mm Hg	155–200
Melting point	°C	
Density	kg/m^3 @ 20 °C	768
Evaporation time[a]		65
Evaporation time[b]		
Vapour pressure	kPa @ 20 °C	
Flash point	°C	38
Auto ignition temperature	°C	
Upper flammability limit	%v/v in air	
Lower flammability limit	%v/v in air	
Viscosity	mPa.s (cP) @ 20 °C	
Refractive index	n 20 D	1.424
Solubility in water	%w/w @ 20 °C	
Solubility of water	%w/w @ 20 °C	
Solubility parameter[c]	(J/cm^3)$^{1/2}$	
Hydrogen bond index		
Fractional polarity		
Nitro cellulose dil ratio[d]		
Nitro cellulose dil ratio[e]		
Surface tension	mN/m @ 20 °C	
Specific heat liquid	kJ/kg mole/°C @ 20 °C	
Latent heat	kJ/kg mole	
Dielectric constant	@ 20 °C	
Antoine constant A	kPa°C	
Antoine constant B	kPa°C	
Antoine constant C	kPa°C	
Heat of combustion	kJ/kg mole	
UN number		1300
IMO classification		3.3
ADR/RID classification		3,31°(c)
UK exposure limits	mg/m^3 (ppm) 8hrTWA/MEL or OES	
USA exposure limits	mg/m^3 (ppm) TLV/TWA	
German exposure limits	mg/m^3 (ppm) MAK-TWA	
EU classification		
EU risk phrases		10
EU safety phrases		

[a] Ether = 1. [b] n-Butyl acetate = 1. [c] Nelson, Hemwall, Edwards. [d] Toluene. [e] Hydrocarbon solvent.

Solvent group		**HYDROCARBONS—ALIPHATIC**
Chemical name		
Chemical formula		
Structural formula		
CAS No.		64742-47-8
Synonym/common name		
Trade names		KETRUL D70
Suppliers		TOTAL
Molecular mass		
Boiling point/range	°C @ 760 mm Hg	185–240
Melting point	°C	
Density	kg/m^3 @ 20 °C	785
Evaporation time[a]		550
Evaporation time[b]		
Vapour pressure	kPa @ 20 °C	
Flash point	°C	65
Auto ignition temperature	°C	
Upper flammability limit	%v/v in air	
Lower flammability limit	%v/v in air	
Viscosity	mPa.s (cP) @ 20 °C	
Refractive index	n 20 D	1.435
Solubility in water	%w/w @ 20 °C	
Solubility of water	%w/w @ 20 °C	
Solubility parameter[c]	(J/cm^3)$^{1/2}$	
Hydrogen bond index		
Fractional polarity		
Nitro cellulose dil ratio[d]		
Nitro cellulose dil ratio[e]		
Surface tension	mN/m @ 20 °C	
Specific heat liquid	kJ/kg mole/°C @ 20 °C	
Latent heat	kJ/kg mole	
Dielectric constant	@ 20 °C	
Antoine constant A	kPa°C	
Antoine constant B	kPa°C	
Antoine constant C	kPa°C	
Heat of combustion	kJ/kg mole	
UN number		1223
IMO classification		
ADR/RID classification		3,32°(c)
UK exposure limits	mg/m^3 (ppm) 8hrTWA/MEL or OES	
USA exposure limits	mg/m^3 (ppm) TLV/TWA	
German exposure limits	mg/m^3 (ppm) MAK-TWA	
EU classification		
EU risk phrases		
EU safety phrases		

[a] Ether = 1. [b] n-Butyl acetate = 1. [c] Nelson, Hemwall, Edwards. [d] Toluene. [e] Hydrocarbon solvent.

Property	Unit	Value
Solvent group		**HYDROCARBONS—ALIPHATIC**
Chemical name		Mixture based on aliphatic hydrocarbons
Chemical formula		
Structural formula		
CAS No.		64742-47-8
Synonym/common name		Paraffinic, naphthenic solvent
Trade names		VISTA LPA SOLVENT
Suppliers		VISTA
Molecular mass		~166
Boiling point/range	°C @ 760 mm Hg	187-259
Melting point	°C	< −70
Density	kg/m^3 @ 20 °C	810
Evaporation time[a]		
Evaporation time[b]		0.02
Vapour pressure	kPa @ 20 °C	0.016
Flash point	°C	71 (PENSKY MARTENS)
Auto ignition temperature	°C	231
Upper flammability limit	%v/v in air	4.7
Lower flammability limit	%v/v in air	0.6
Viscosity	mPa.s (cP) @ 20 °C	1.9
Refractive index	n 20 D	
Solubility in water	%w/w @ 20 °C	Negligible
Solubility of water	%w/w @ 20 °C	Negligible
Solubility parameter[c]	(J/cm^3)$^{1/2}$	
Hydrogen bond index		
Fractional polarity		
Nitro cellulose dil ratio[d]		
Nitro cellulose dil ratio[e]		
Surface tension	mN/m @ 20 °C	
Specific heat liquid	kJ/kg mole/°C @ 20 °C	
Latent heat	kJ/kg mole	
Dielectric constant	@ 20 °C	
Antoine constant A	kPa°C	
Antoine constant B	kPa°C	
Antoine constant C	kPa°C	
Heat of combustion	kJ/kg mole	
UN number		1268
IMO classification		
ADR/RID classification		
UK exposure limits	mg/m^3 (ppm) 8hrTWA/MEL or OES	
USA exposure limits	mg/m^3 (ppm) TLV/TWA	525(100)*
German exposure limits	mg/m^3 (ppm) MAK-TWA	
EU classification		
EU risk phrases		
EU safety phrases		

[a] Ether = 1. [b] n-Butyl acetate = 1. [c] Nelson, Hemwall, Edwards. [d] Toluene. [e] Hydrocarbon solvent.

Solvent group		**HYDROCARBONS—ALIPHATIC**
Chemical name		Mixture based on naphtha solvent refined light
Chemical formula		
Structural formula		
CAS No.		64741-84-0
Synonym/common name		
Trade names		SBP, SOLSOL A1 1
Suppliers		PETROGAL, SA
Molecular mass		
Boiling point/range	°C @ 760 mm Hg	60–95
Melting point	°C	
Density	kg/m^3 @ 20 °C	675–688 (60 °F)
Evaporation time[a]		
Evaporation time[b]		
Vapour pressure	kPa @ 20 °C	
Flash point	°C	<-15
Auto ignition temperature	°C	220
Upper flammability limit	%v/v in air	7
Lower flammability limit	%v/v in air	0.8
Viscosity	mPa.s (cP) @ 20 °C	
Refractive index	n 20 D	1.38443
Solubility in water	%w/w @ 20 °C	
Solubility of water	%w/w @ 20 °C	
Solubility parameter[c]	$(J/cm^3)^{1/2}$	
Hydrogen bond index		
Fractional polarity		
Nitro cellulose dil ratio[d]		
Nitro cellulose dil ratio[e]		
Surface tension	mN/m @ 20 °C	
Specific heat liquid	kJ/kg mole/°C @ 20 °C	
Latent heat	kJ/kg mole	
Dielectric constant	@ 20 °C	
Antoine constant A	kPa°C	
Antoine constant B	kPa°C	
Antoine constant C	kPa°C	
Heat of combustion	kJ/kg mole	
UN number		1203
IMO classification		3.1
ADR/RID classification		3,3°(b)
UK exposure limits	mg/m^3 (ppm) 8hrTWA/MEL or OES	
USA exposure limits	mg/m^3 (ppm) TLV/TWA	1590
German exposure limits	mg/m^3 (ppm) MAK-TWA	
EU classification		F, Xn
EU risk phrases		11-48/20
EU safety phrases		9-16-24/25-29-51

[a] Ether = 1. [b] n-Butyl acetate = 1. [c] Nelson, Hemwall, Edwards. [d] Toluene. [e] Hydrocarbon solvent.

Solvent group		**HYDROCARBONS—ALIPHATIC**
Chemical name		Mixture of aliphatic hydrocarbons
Chemical formula		
Structural formula		
CAS No.		
Synonym/common name		
Trade names		EVOLVE CH12
Suppliers		ICI
Molecular mass		
Boiling point/range	°C @ 760 mm Hg	182-212
Melting point	°C	< -70
Density	kg/m^3 @ 20 °C	766
Evaporation time[a]		105
Evaporation time[b]		
Vapour pressure	kPa @ 20 °C	0.1
Flash point	°C	58 (SETA CLOSED CUP)
Auto ignition temperature	°C	375
Upper flammability limit	%v/v in air	5.20
Lower flammability limit	%v/v in air	0.55
Viscosity	mPa.s (cP) @ 20 °C	1.08
Refractive index	n 20 D	1.422
Solubility in water	%w/w @ 20 °C	<0.01
Solubility of water	%w/w @ 20 °C	Insoluble
Solubility parameter[c]	$(J/cm^3)^{1/2}$	
Hydrogen bond index		
Fractional polarity		
Nitro cellulose dil ratio[d]		
Nitro cellulose dil ratio[e]		
Surface tension	mN/m @ 20 °C	23.4
Specific heat liquid	kJ/kg mole/°C @ 20 °C	
Latent heat	kJ/kg mole	
Dielectric constant	@ 20 °C	
Antoine constant A	kPa°C	
Antoine constant B	kPa°C	
Antoine constant C	kPa°C	
Heat of combustion	kJ/kg mole	
UN number		1993
IMO classification		3
ADR/RID classification		3,31°(c)
UK exposure limits	mg/m^3 (ppm) 8hrTWA/MEL or OES	(300)*
USA exposure limits	mg/m^3 (ppm) TLV/TWA	
German exposure limits	mg/m^3 (ppm) MAK-TWA	
EU classification		
EU risk phrases		
EU safety phrases		

[a] Ether = 1. [b] n-Butyl acetate = 1. [c] Nelson, Hemwall, Edwards. [d] Toluene. [e] Hydrocarbon solvent.

Solvent group		**HYDROCARBONS—ALIPHATIC**
Chemical name		Mixture of aliphatic hydrocarbons
Chemical formula		
Structural formula		
CAS No.		
Synonym/common name		
Trade names		EVOLVE CH15
Suppliers		ICI
Molecular mass		
Boiling point/range	°C @ 760 mm Hg	150–190
Melting point	°C	< −50
Density	kg/m^3 @ 20 °C	750
Evaporation time[a]		
Evaporation time[b]		
Vapour pressure	kPa @ 20 °C	<0.4
Flash point	°C	42 (SETA CLOSED CUP)
Auto ignition temperature	°C	292
Upper flammability limit	%v/v in air	5.4
Lower flammability limit	%v/v in air	0.8
Viscosity	mPa.s (cP) @ 20 °C	1.05
Refractive index	n 20 D	
Solubility in water	%w/w @ 20 °C	Insoluble
Solubility of water	%w/w @ 20 °C	Insoluble
Solubility parameter[c]	$(J/cm^3)^{1/2}$	
Hydrogen bond index		
Fractional polarity		
Nitro cellulose dil ratio[d]		
Nitro cellulose dil ratio[e]		
Surface tension	mN/m @ 20 °C	17.6
Specific heat liquid	kJ/kg mole/°C @ 20 °C	
Latent heat	kJ/kg mole	
Dielectric constant	@ 20 °C	
Antoine constant A	kPa°C	
Antoine constant B	kPa°C	
Antoine constant C	kPa°C	
Heat of combustion	kJ/kg mole	
UN number		1993
IMO classification		3/III
ADR/RID classification		3,31°(c)
UK exposure limits	mg/m^3 (ppm) 8hrTWA/MEL or OES	(300)*
USA exposure limits	mg/m^3 (ppm) TLV/TWA	
German exposure limits	mg/m^3 (ppm) MAK-TWA	
EU classification		F
EU risk phrases		10
EU safety phrases		7-16-24/25

[a] Ether = 1. [b] n-Butyl acetate = 1. [c] Nelson, Hemwall, Edwards. [d] Toluene. [e] Hydrocarbon solvent.

Solvent group		HYDROCARBONS—ALIPHATIC
Chemical name		
Chemical formula		
Structural formula		
CAS No.		90622-58-5
Synonym/common name		
Trade names		ISANE IP260
Suppliers		TOTAL
Molecular mass		
Boiling point/range	°C @ 760 mm Hg	265–310
Melting point	°C	
Density	kg/m^3 @ 20 °C	810
Evaporation time[a]		>1000
Evaporation time[b]		
Vapour pressure	kPa @ 20 °C	
Flash point	°C	122
Auto ignition temperature	°C	
Upper flammability limit	%v/v in air	
Lower flammability limit	%v/v in air	
Viscosity	mPa.s (cP) @ 20 °C	
Refractive index	n 20 D	1.448
Solubility in water	%w/w @ 20 °C	
Solubility of water	%w/w @ 20 °C	
Solubility parameter[c]	$(J/cm^3)^{1/2}$	
Hydrogen bond index		
Fractional polarity		
Nitro cellulose dil ratio[d]		
Nitro cellulose dil ratio[e]		
Surface tension	mN/m @ 20 °C	
Specific heat liquid	kJ/kg mole/°C @ 20 °C	
Latent heat	kJ/kg mole	
Dielectric constant	@ 20 °C	
Antoine constant A	kPa°C	
Antoine constant B	kPa°C	
Antoine constant C	kPa°C	
Heat of combustion	kJ/kg mole	
UN number		
IMO classification		
ADR/RID classification		
UK exposure limits	mg/m^3 (ppm) 8hrTWA/MEL or OES	
USA exposure limits	mg/m^3 (ppm) TLV/TWA	
German exposure limits	mg/m^3 (ppm) MAK-TWA	
EU classification		
EU risk phrases		
EU safety phrases		

[a] Ether = 1. [b] n-Butyl acetate = 1. [c] Nelson, Hemwall, Edwards. [d] Toluene. [e] Hydrocarbon solvent.

Solvent group		HYDROCARBONS—ALIPHATIC
Chemical name		
Chemical formula		
Structural formula		
CAS No.		64742-49-0
Synonym/common name		
Trade names		HYDROSOL Ess A
Suppliers		TOTAL
Molecular mass		
Boiling point/range	°C @ 760 mm Hg	45–95
Melting point	°C	
Density	kg/m^3 @ 20 °C	665
Evaporation time[a]		1.2
Evaporation time[b]		
Vapour pressure	kPa @ 20 °C	
Flash point	°C	<0
Auto ignition temperature	°C	
Upper flammability limit	%v/v in air	
Lower flammability limit	%v/v in air	
Viscosity	mPa.s (cP) @ 20 °C	
Refractive index	n 20 D	1.374
Solubility in water	%w/w @ 20 °C	
Solubility of water	%w/w @ 20 °C	
Solubility parameter[c]	(J/cm^3)$^{1/2}$	
Hydrogen bond index		
Fractional polarity		
Nitro cellulose dil ratio[d]		
Nitro cellulose dil ratio[e]		
Surface tension	mN/m @ 20 °C	
Specific heat liquid	kJ/kg mole/°C @ 20 °C	
Latent heat	kJ/kg mole	
Dielectric constant	@ 20 °C	
Antoine constant A	kPa°C	
Antoine constant B	kPa°C	
Antoine constant C	kPa°C	
Heat of combustion	kJ/kg mole	
UN number		1203
IMO classification		3.1
ADR/RID classification		3,3°(b)
UK exposure limits	mg/m^3 (ppm) 8hrTWA/MEL or OES	
USA exposure limits	mg/m^3 (ppm) TLV/TWA	
German exposure limits	mg/m^3 (ppm) MAK-TWA	
EU classification		F
EU risk phrases		11
EU safety phrases		9-16-23-29-33

[a] Ether = 1. [b] n-Butyl acetate = 1. [c] Nelson, Hemwall, Edwards. [d] Toluene. [e] Hydrocarbon solvent.

Solvent group		HYDROCARBONS—ALIPHATIC
Chemical name		
Chemical formula		
Structural formula		
CAS No.		64742-49-0
Synonym/common name		
Trade names		HYDROSOL Ess95/160N
Suppliers		TOTAL
Molecular mass		
Boiling point/range	°C @ 760 mm Hg	98–153
Melting point	°C	
Density	kg/m^3 @ 20 °C	763
Evaporation time[a]		10
Evaporation time[b]		
Vapour pressure	kPa @ 20 °C	
Flash point	°C	−3
Auto ignition temperature	°C	
Upper flammability limit	%v/v in air	
Lower flammability limit	%v/v in air	
Viscosity	mPa.s (cP) @ 20 °C	
Refractive index	n 20 D	1.418
Solubility in water	%w/w @ 20 °C	
Solubility of water	%w/w @ 20 °C	
Solubility parameter[c]	$(J/cm^3)^{1/2}$	
Hydrogen bond index		
Fractional polarity		
Nitro cellulose dil ratio[d]		
Nitro cellulose dil ratio[e]		
Surface tension	mN/m @ 20 °C	
Specific heat liquid	kJ/kg mole/°C @ 20 °C	
Latent heat	kJ/kg mole	
Dielectric constant	@ 20 °C	
Antoine constant A	kPa°C	
Antoine constant B	kPa°C	
Antoine constant C	kPa°C	
Heat of combustion	kJ/kg mole	
UN number		1203
IMO classification		3.2
ADR/RID classification		3,3°(b)
UK exposure limits	mg/m^3 (ppm) 8hrTWA/MEL or OES	
USA exposure limits	mg/m^3 (ppm) TLV/TWA	
German exposure limits	mg/m^3 (ppm) MAK-TWA	
EU classification		F
EU risk phrases		11
EU safety phrases		9-16-29-33

[a] Ether = 1. [b] n-Butyl acetate = 1. [c] Nelson, Hemwall, Edwards. [d] Toluene. [e] Hydrocarbon solvent.

Solvent group		HYDROCARBONS—ALIPHATIC
Chemical name		
Chemical formula		
Structural formula		
CAS No.		64742-48-9
Synonym/common name		
Trade names		SPIRDANE L1
Suppliers		TOTAL
Molecular mass		
Boiling point/range	°C @ 760 mm Hg	135–160
Melting point	°C	
Density	kg/m^3 @ 20 °C	750
Evaporation time[a]		22
Evaporation time[b]		
Vapour pressure	kPa @ 20 °C	
Flash point	°C	22
Auto ignition temperature	°C	
Upper flammability limit	%v/v in air	
Lower flammability limit	%v/v in air	
Viscosity	mPa.s (cP) @ 20 °C	
Refractive index	n 20 D	1.416
Solubility in water	%w/w @ 20 °C	
Solubility of water	%w/w @ 20 °C	
Solubility parameter[c]	$(J/cm^3)^{1/2}$	
Hydrogen bond index		
Fractional polarity		
Nitro cellulose dil ratio[d]		
Nitro cellulose dil ratio[e]		
Surface tension	mN/m @ 20 °C	
Specific heat liquid	kJ/kg mole/°C @ 20 °C	
Latent heat	kJ/kg mole	
Dielectric constant	@ 20 °C	
Antoine constant A	kPa°C	
Antoine constant B	kPa°C	
Antoine constant C	kPa°C	
Heat of combustion	kJ/kg mole	
UN number		1300
IMO classification		3.2
ADR/RID classification		3,31°(c)
UK exposure limits	mg/m^3 (ppm) 8hrTWA/MEL or OES	
USA exposure limits	mg/m^3 (ppm) TLV/TWA	
German exposure limits	mg/m^3 (ppm) MAK-TWA	
EU classification		
EU risk phrases		10
EU safety phrases		

[a] Ether = 1. [b] n-Butyl acetate = 1. [c] Nelson, Hemwall, Edwards. [d] Toluene. [e] Hydrocarbon solvent.

Property	Unit	Value
Solvent group		**HYDROCARBONS—ALIPHATIC**
Chemical name		Mixture of aliphatic hydrocarbons/ester blend
Chemical formula		
Structural formula		
CAS No.		
Synonym/common name		
Trade names		EVOLVE CH11
Suppliers		ICI
Molecular mass		
Boiling point/range	°C @ 760 mm Hg	168–192
Melting point	°C	< −70
Density	kg/m^3 @ 20 °C	759
Evaporation time[a]		55
Evaporation time[b]		
Vapour pressure	kPa @ 20 °C	0.25
Flash point	°C	48 (SETA CLOSED CUP)
Auto ignition temperature	°C	370
Upper flammability limit	%v/v in air	5.40
Lower flammability limit	%v/v in air	0.57
Viscosity	mPa.s (cP) @ 20 °C	0.57
Refractive index	n 20 D	1.41
Solubility in water	%w/w @ 20 °C	<0.02
Solubility of water	%w/w @ 20 °C	Insoluble
Solubility parameter[c]	(J/cm^3)$^{1/2}$	
Hydrogen bond index		
Fractional polarity		
Nitro cellulose dil ratio[d]		
Nitro cellulose dil ratio[e]		
Surface tension	mN/m @ 20 °C	23
Specific heat liquid	kJ/kg mole/°C @ 20 °C	
Latent heat	kJ/kg mole	
Dielectric constant	@ 20 °C	
Antoine constant A	kPa°C	
Antoine constant B	kPa°C	
Antoine constant C	kPa°C	
Heat of combustion	kJ/kg mole	
UN number		1993
IMO classification		3/III
ADR/RID classification		3,31°(c)
UK exposure limits	mg/m^3 (ppm) 8hrTWA/MEL or OES	(300)*
USA exposure limits	mg/m^3 (ppm) TLV/TWA	
German exposure limits	mg/m^3 (ppm) MAK-TWA	
EU classification		F
EU risk phrases		10
EU safety phrases		7-16-24/25

[a] Ether = 1. [b] n-Butyl acetate = 1. [c] Nelson, Hemwall, Edwards. [d] Toluene. [e] Hydrocarbon solvent.

Solvent group		HYDROCARBONS—ALIPHATIC
Chemical name		
Chemical formula		
Structural formula		
CAS No.		64742-49-0
Synonym/common name		
Trade names		HYDROSOL Ess100/140
Suppliers		TOTAL
Molecular mass		
Boiling point/range	°C @ 760 mm Hg	106–140
Melting point	°C	
Density	kg/m^3 @ 20 °C	730
Evaporation time[a]		7.5
Evaporation time[b]		
Vapour pressure	kPa @ 20 °C	
Flash point	°C	<0
Auto ignition temperature	°C	
Upper flammability limit	%v/v in air	
Lower flammability limit	%v/v in air	
Viscosity	mPa.s (cP) @ 20 °C	
Refractive index	n 20 D	1.405
Solubility in water	%w/w @ 20 °C	
Solubility of water	%w/w @ 20 °C	
Solubility parameter[c]	(J/cm^3)$^{1/2}$	
Hydrogen bond index		
Fractional polarity		
Nitro cellulose dil ratio[d]		
Nitro cellulose dil ratio[e]		
Surface tension	mN/m @ 20 °C	
Specific heat liquid	kJ/kg mole/°C @ 20 °C	
Latent heat	kJ/kg mole	
Dielectric constant	@ 20 °C	
Antoine constant A	kPa°C	
Antoine constant B	kPa°C	
Antoine constant C	kPa°C	
Heat of combustion	kJ/kg mole	
UN number		1203
IMO classification		3.2
ADR/RID classification		3,3°(b)
UK exposure limits	mg/m^3 (ppm) 8hrTWA/MEL or OES	
USA exposure limits	mg/m^3 (ppm) TLV/TWA	
German exposure limits	mg/m^3 (ppm) MAK-TWA	
EU classification		F
EU risk phrases		11
EU safety phrases		9-16-29-33

[a] Ether = 1. [b] n-Butyl acetate = 1. [c] Nelson, Hemwall, Edwards. [d] Toluene. [e] Hydrocarbon solvent.

Solvent group		**HYDROCARBONS—ALIPHATIC**
Chemical name		
Chemical formula		
Structural formula		
CAS No.		64742-49-0
Synonym/common name		
Trade names		HYDROSOL Ess <5
Suppliers		TOTAL
Molecular mass		
Boiling point/range	°C @ 760 mm Hg	103–123
Melting point	°C	
Density	kg/m^3 @ 20 °C	725
Evaporation time[a]		6
Evaporation time[b]		
Vapour pressure	kPa @ 20 °C	
Flash point	°C	<0
Auto ignition temperature	°C	
Upper flammability limit	%v/v in air	
Lower flammability limit	%v/v in air	
Viscosity	mPa.s (cP) @ 20 °C	
Refractive index	n 20 D	1.404
Solubility in water	%w/w @ 20 °C	
Solubility of water	%w/w @ 20 °C	
Solubility parameter[c]	$(J/cm^3)^{1/2}$	
Hydrogen bond index		
Fractional polarity		
Nitro cellulose dil ratio[d]		
Nitro cellulose dil ratio[e]		
Surface tension	mN/m @ 20 °C	
Specific heat liquid	kJ/kg mole/°C @ 20 °C	
Latent heat	kJ/kg mole	
Dielectric constant	@ 20 °C	
Antoine constant A	kPa°C	
Antoine constant B	kPa°C	
Antoine constant C	kPa°C	
Heat of combustion	kJ/kg mole	
UN number		1203
IMO classification		3.2
ADR/RID classification		3,3°(b)
UK exposure limits	mg/m^3 (ppm) 8hrTWA/MEL or OES	
USA exposure limits	mg/m^3 (ppm) TLV/TWA	
German exposure limits	mg/m^3 (ppm) MAK-TWA	
EU classification		F
EU risk phrases		11
EU safety phrases		9-16-23-29-33

[a] Ether = 1. [b] n-Butyl acetate = 1. [c] Nelson, Hemwall, Edwards. [d] Toluene. [e] Hydrocarbon solvent.

Solvent group		HYDROCARBONS—ALIPHATIC
Chemical name		
Chemical formula		
Structural formula		
CAS No.		64742-82-1
Synonym/common name		
Trade names		SPIRDANE K2
Suppliers		TOTAL
Molecular mass		
Boiling point/range	°C @ 760 mm Hg	180–215
Melting point	°C	
Density	kg/m^3 @ 20 °C	795
Evaporation time[a]		200
Evaporation time[b]		
Vapour pressure	kPa @ 20 °C	
Flash point	°C	60
Auto ignition temperature	°C	
Upper flammability limit	%v/v in air	
Lower flammability limit	%v/v in air	
Viscosity	mPa.s (cP) @ 20 °C	
Refractive index	n 20 D	1.440
Solubility in water	%w/w @ 20 °C	
Solubility of water	%w/w @ 20 °C	
Solubility parameter[c]	$(J/cm^3)^{1/2}$	
Hydrogen bond index		
Fractional polarity		
Nitro cellulose dil ratio[d]		
Nitro cellulose dil ratio[e]		
Surface tension	mN/m @ 20 °C	
Specific heat liquid	kJ/kg mole/°C @ 20 °C	
Latent heat	kJ/kg mole	
Dielectric constant	@ 20 °C	
Antoine constant A	kPa°C	
Antoine constant B	kPa°C	
Antoine constant C	kPa°C	
Heat of combustion	kJ/kg mole	
UN number		1300
IMO classification		3.3
ADR/RID classification		3,32°(c)
UK exposure limits	mg/m^3 (ppm) 8hrTWA/MEL or OES	
USA exposure limits	mg/m^3 (ppm) TLV/TWA	
German exposure limits	mg/m^3 (ppm) MAK-TWA	
EU classification		
EU risk phrases		
EU safety phrases		

[a] Ether = 1. [b] n-Butyl acetate = 1. [c] Nelson, Hemwall, Edwards. [d] Toluene. [e] Hydrocarbon solvent.

Solvent group		HYDROCARBONS—ALIPHATIC
Chemical name		
Chemical formula		
Structural formula		
CAS No.		64742-48-9
Synonym/common name		
Trade names		SPIRDANE D30
Suppliers		TOTAL
Molecular mass		
Boiling point/range	°C @ 760 mm Hg	145–170
Melting point	°C	
Density	kg/m^3 @ 20 °C	760
Evaporation time[a]		27
Evaporation time[b]		
Vapour pressure	kPa @ 20 °C	
Flash point	°C	27
Auto ignition temperature	°C	
Upper flammability limit	%v/v in air	
Lower flammability limit	%v/v in air	
Viscosity	mPa.s (cP) @ 20 °C	
Refractive index	n 20 D	1.422
Solubility in water	%w/w @ 20 °C	
Solubility of water	%w/w @ 20 °C	
Solubility parameter[c]	(J/cm^3)$^{1/2}$	
Hydrogen bond index		
Fractional polarity		
Nitro cellulose dil ratio[d]		
Nitro cellulose dil ratio[e]		
Surface tension	mN/m @ 20 °C	
Specific heat liquid	kJ/kg mole/°C @ 20 °C	
Latent heat	kJ/kg mole	
Dielectric constant	@ 20 °C	
Antoine constant A	kPa°C	
Antoine constant B	kPa°C	
Antoine constant C	kPa°C	
Heat of combustion	kJ/kg mole	
UN number		1300
IMO classification		3.2
ADR/RID classification		3,31°(c)
UK exposure limits	mg/m^3 (ppm) 8hrTWA/MEL or OES	
USA exposure limits	mg/m^3 (ppm) TLV/TWA	
German exposure limits	mg/m^3 (ppm) MAK-TWA	
EU classification		
EU risk phrases		10
EU safety phrases		

[a] Ether = 1. [b] n-Butyl acetate = 1. [c] Nelson, Hemwall, Edwards. [d] Toluene. [e] Hydrocarbon solvent.

Solvent group		HYDROCARBONS—ALIPHATIC
Chemical name		
Chemical formula		
Structural formula		
CAS No.		64742-49-0
Synonym/common name		
Trade names		HYDROSOL Ess C
Suppliers		TOTAL
Molecular mass		
Boiling point/range	°C @ 760 mm Hg	73–98
Melting point	°C	
Density	kg/m^3 @ 20 °C	700
Evaporation time[a]		2
Evaporation time[b]		
Vapour pressure	kPa @ 20 °C	
Flash point	°C	<0
Auto ignition temperature	°C	
Upper flammability limit	%v/v in air	
Lower flammability limit	%v/v in air	
Viscosity	mPa.s (cP) @ 20 °C	
Refractive index	n 20 D	1.392
Solubility in water	%w/w @ 20 °C	
Solubility of water	%w/w @ 20 °C	
Solubility parameter[c]	(J/cm^3)$^{1/2}$	
Hydrogen bond index		
Fractional polarity		
Nitro cellulose dil ratio[d]		
Nitro cellulose dil ratio[e]		
Surface tension	mN/m @ 20 °C	
Specific heat liquid	kJ/kg mole/°C @ 20 °C	
Latent heat	kJ/kg mole	
Dielectric constant	@ 20 °C	
Antoine constant A	kPa°C	
Antoine constant B	kPa°C	
Antoine constant C	kPa°C	
Heat of combustion	kJ/kg mole	
UN number		1203
IMO classification		3.1
ADR/RID classification		3,3°(b)
UK exposure limits	mg/m^3 (ppm) 8hrTWA/MEL or OES	
USA exposure limits	mg/m^3 (ppm) TLV/TWA	
German exposure limits	mg/m^3 (ppm) MAK-TWA	
EU classification		F, Xn
EU risk phrases		11-48/20
EU safety phrases		9-16-24/25-29-51

[a] Ether = 1. [b] n-Butyl acetate = 1. [c] Nelson, Hemwall, Edwards. [d] Toluene. [e] Hydrocarbon solvent.

Property	Unit	HYDROCARBONS—ALIPHATIC
Solvent group		**HYDROCARBONS—ALIPHATIC**
Chemical name		
Chemical formula		
Structural formula		
CAS No.		64742-49-0
Synonym/common name		
Trade names		HYDROSOL Ess 60/95
Suppliers		TOTAL
Molecular mass		
Boiling point/range	°C @ 760 mm Hg	60–98
Melting point	°C	
Density	kg/m^3 @ 20 °C	678
Evaporation time[a]		2
Evaporation time[b]		
Vapour pressure	kPa @ 20 °C	
Flash point	°C	<0
Auto ignition temperature	°C	
Upper flammability limit	%v/v in air	
Lower flammability limit	%v/v in air	
Viscosity	mPa.s (cP) @ 20 °C	
Refractive index	n 20 D	1.382
Solubility in water	%w/w @ 20 °C	
Solubility of water	%w/w @ 20 °C	
Solubility parameter[c]	$(J/cm^3)^{1/2}$	
Hydrogen bond index		
Fractional polarity		
Nitro cellulose dil ratio[d]		
Nitro cellulose dil ratio[e]		
Surface tension	mN/m @ 20 °C	
Specific heat liquid	kJ/kg mole/°C @ 20 °C	
Latent heat	kJ/kg mole	
Dielectric constant	@ 20 °C	
Antoine constant A	kPa°C	
Antoine constant B	kPa°C	
Antoine constant C	kPa°C	
Heat of combustion	kJ/kg mole	
UN number		1203
IMO classification		3.1
ADR/RID classification		3,3°(b)
UK exposure limits	mg/m^3 (ppm) 8hrTWA/MEL or OES	
USA exposure limits	mg/m^3 (ppm) TLV/TWA	
German exposure limits	mg/m^3 (ppm) MAK-TWA	
EU classification		F
EU risk phrases		11
EU safety phrases		9-16-23-29-33

[a] Ether = 1. [b] n-Butyl acetate = 1. [c] Nelson, Hemwall, Edwards. [d] Toluene. [e] Hydrocarbon solvent.

Solvent group		HYDROCARBONS—ALIPHATIC
Chemical name		
Chemical formula		
Structural formula		
CAS No.		90622-58-5
Synonym/common name		
Trade names		ISANE IP185
Suppliers		TOTAL
Molecular mass		
Boiling point/range	°C @ 760 mm Hg	185–210
Melting point	°C	
Density	kg/m^3 @ 20 °C	768
Evaporation time[a]		150
Evaporation time[b]		
Vapour pressure	kPa @ 20 °C	
Flash point	°C	65
Auto ignition temperature	°C	
Upper flammability limit	%v/v in air	
Lower flammability limit	%v/v in air	
Viscosity	mPa.s (cP) @ 20 °C	
Refractive index	n 20 D	1.427
Solubility in water	%w/w @ 20 °C	
Solubility of water	%w/w @ 20 °C	
Solubility parameter[c]	(J/cm^3)$^{1/2}$	
Hydrogen bond index		
Fractional polarity		
Nitro cellulose dil ratio[d]		
Nitro cellulose dil ratio[e]		
Surface tension	mN/m @ 20 °C	
Specific heat liquid	kJ/kg mole/°C @ 20 °C	
Latent heat	kJ/kg mole	
Dielectric constant	@ 20 °C	
Antoine constant A	kPa°C	
Antoine constant B	kPa°C	
Antoine constant C	kPa°C	
Heat of combustion	kJ/kg mole	
UN number		1300
IMO classification		
ADR/RID classification		3,32°(c)
UK exposure limits	mg/m^3 (ppm) 8hrTWA/MEL or OES	
USA exposure limits	mg/m^3 (ppm) TLV/TWA	
German exposure limits	mg/m^3 (ppm) MAK-TWA	
EU classification		
EU risk phrases		
EU safety phrases		

[a] Ether = 1. [b] n-Butyl acetate = 1. [c] Nelson, Hemwall, Edwards. [d] Toluene. [e] Hydrocarbon solvent.

Solvent group		HYDROCARBONS—ALIPHATIC
Chemical name		
Chemical formula		
Structural formula		
CAS No.		64742-47-8
Synonym/common name		
Trade names		KETRUL D80
Suppliers		TOTAL
Molecular mass		
Boiling point/range	°C @ 760 mm Hg	198–240
Melting point	°C	
Density	kg/m^3 @ 20 °C	795
Evaporation time[a]		800
Evaporation time[b]		
Vapour pressure	kPa @ 20 °C	
Flash point	°C	75
Auto ignition temperature	°C	
Upper flammability limit	%v/v in air	
Lower flammability limit	%v/v in air	
Viscosity	mPa.s (cP) @ 20 °C	
Refractive index	n 20 D	1.438
Solubility in water	%w/w @ 20 °C	
Solubility of water	%w/w @ 20 °C	
Solubility parameter[c]	$(J/cm^3)^{1/2}$	
Hydrogen bond index		
Fractional polarity		
Nitro cellulose dil ratio[d]		
Nitro cellulose dil ratio[e]		
Surface tension	mN/m @ 20 °C	
Specific heat liquid	kJ/kg mole/°C @ 20 °C	
Latent heat	kJ/kg mole	
Dielectric constant	@ 20 °C	
Antoine constant A	kPa°C	
Antoine constant B	kPa°C	
Antoine constant C	kPa°C	
Heat of combustion	kJ/kg mole	
UN number		1223
IMO classification		
ADR/RID classification		3,32°(c)
UK exposure limits	mg/m^3 (ppm) 8hrTWA/MEL or OES	
USA exposure limits	mg/m^3 (ppm) TLV/TWA	
German exposure limits	mg/m^3 (ppm) MAK-TWA	
EU classification		
EU risk phrases		
EU safety phrases		

[a] Ether = 1. [b] n-Butyl acetate = 1. [c] Nelson, Hemwall, Edwards. [d] Toluene. [e] Hydrocarbon solvent.

Solvent group		**HYDROCARBONS—ALIPHATIC**
Chemical name		
Chemical formula		
Structural formula		
CAS No.		64742-48-9
Synonym/common name		
Trade names		SPIRDANE L30
Suppliers		TOTAL
Molecular mass		
Boiling point/range	°C @ 760 mm Hg	145–170
Melting point	°C	
Density	kg/m^3 @ 20 °C	765
Evaporation time[a]		27
Evaporation time[b]		
Vapour pressure	kPa @ 20 °C	
Flash point	°C	27
Auto ignition temperature	°C	
Upper flammability limit	%v/v in air	
Lower flammability limit	%v/v in air	
Viscosity	mPa.s (cP) @ 20 °C	
Refractive index	n 20 D	1.422
Solubility in water	%w/w @ 20 °C	
Solubility of water	%w/w @ 20 °C	
Solubility parameter[c]	(J/cm^3)$^{1/2}$	
Hydrogen bond index		
Fractional polarity		
Nitro cellulose dil ratio[d]		
Nitro cellulose dil ratio[e]		
Surface tension	mN/m @ 20 °C	
Specific heat liquid	kJ/kg mole/°C @ 20 °C	
Latent heat	kJ/kg mole	
Dielectric constant	@ 20 °C	
Antoine constant A	kPa°C	
Antoine constant B	kPa°C	
Antoine constant C	kPa°C	
Heat of combustion	kJ/kg mole	
UN number		1300
IMO classification		3.2
ADR/RID classification		3,31°(c)
UK exposure limits	mg/m^3 (ppm) 8hrTWA/MEL or OES	
USA exposure limits	mg/m^3 (ppm) TLV/TWA	
German exposure limits	mg/m^3 (ppm) MAK-TWA	
EU classification		
EU risk phrases		10
EU safety phrases		

[a] Ether = 1. [b] n-Butyl acetate = 1. [c] Nelson, Hemwall, Edwards. [d] Toluene. [e] Hydrocarbon solvent.

Solvent group		HYDROCARBONS—ALIPHATIC
Chemical name		
Chemical formula		
Structural formula		
CAS No.		64742-81-0
Synonym/common name		
Trade names		KETRUL HT
Suppliers		TOTAL
Molecular mass		
Boiling point/range	°C @ 760 mm Hg	185–240
Melting point	°C	
Density	kg/m^3 @ 20 °C	795
Evaporation time[a]		550
Evaporation time[b]		
Vapour pressure	kPa @ 20 °C	
Flash point	°C	65
Auto ignition temperature	°C	
Upper flammability limit	%v/v in air	
Lower flammability limit	%v/v in air	
Viscosity	mPa.s (cP) @ 20 °C	
Refractive index	n 20 D	1.446
Solubility in water	%w/w @ 20 °C	
Solubility of water	%w/w @ 20 °C	
Solubility parameter[c]	$(J/cm^3)^{1/2}$	
Hydrogen bond index		
Fractional polarity		
Nitro cellulose dil ratio[d]		
Nitro cellulose dil ratio[e]		
Surface tension	mN/m @ 20 °C	
Specific heat liquid	kJ/kg mole/°C @ 20 °C	
Latent heat	kJ/kg mole	
Dielectric constant	@ 20 °C	
Antoine constant A	kPa°C	
Antoine constant B	kPa°C	
Antoine constant C	kPa°C	
Heat of combustion	kJ/kg mole	
UN number		1223
IMO classification		
ADR/RID classification		3,32°(c)
UK exposure limits	mg/m^3 (ppm) 8hrTWA/MEL or OES	
USA exposure limits	mg/m^3 (ppm) TLV/TWA	
German exposure limits	mg/m^3 (ppm) MAK-TWA	
EU classification		
EU risk phrases		
EU safety phrases		

[a] Ether = 1. [b] n-Butyl acetate = 1. [c] Nelson, Hemwall, Edwards. [d] Toluene. [e] Hydrocarbon solvent.

Solvent group		HYDROCARBONS—ALIPHATIC
Chemical name		
Chemical formula		
Structural formula		
CAS No.		64742-48-9
Synonym/common name		
Trade names		SPIRDANE K1
Suppliers		TOTAL
Molecular mass		
Boiling point/range	°C @ 760 mm Hg	180–215
Melting point	°C	
Density	kg/m^3 @ 20 °C	783
Evaporation time[a]		200
Evaporation time[b]		
Vapour pressure	kPa @ 20 °C	
Flash point	°C	59
Auto ignition temperature	°C	
Upper flammability limit	%v/v in air	
Lower flammability limit	%v/v in air	
Viscosity	mPa.s (cP) @ 20 °C	
Refractive index	n 20 D	1.434
Solubility in water	%w/w @ 20 °C	
Solubility of water	%w/w @ 20 °C	
Solubility parameter[c]	$(J/cm^3)^{1/2}$	
Hydrogen bond index		
Fractional polarity		
Nitro cellulose dil ratio[d]		
Nitro cellulose dil ratio[e]		
Surface tension	mN/m @ 20 °C	
Specific heat liquid	kJ/kg mole/°C @ 20 °C	
Latent heat	kJ/kg mole	
Dielectric constant	@ 20 °C	
Antoine constant A	kPa°C	
Antoine constant B	kPa°C	
Antoine constant C	kPa°C	
Heat of combustion	kJ/kg mole	
UN number		1300
IMO classification		3.3
ADR/RID classification		3,32°(c)
UK exposure limits	mg/m^3 (ppm) 8hrTWA/MEL or OES	
USA exposure limits	mg/m^3 (ppm) TLV/TWA	
German exposure limits	mg/m^3 (ppm) MAK-TWA	
EU classification		
EU risk phrases		
EU safety phrases		

[a] Ether = 1. [b] n-Butyl acetate = 1. [c] Nelson, Hemwall, Edwards. [d] Toluene. [e] Hydrocarbon solvent.

Solvent group		HYDROCARBONS—ALIPHATIC
Chemical name		
Chemical formula		
Structural formula		
CAS No.		90622-57-4
Synonym/common name		
Trade names		ISANE IP165
Suppliers		TOTAL
Molecular mass		
Boiling point/range	°C @ 760 mm Hg	165–195
Melting point	°C	
Density	kg/m^3 @ 20 °C	755
Evaporation time[a]		50
Evaporation time[b]		
Vapour pressure	kPa @ 20 °C	
Flash point	°C	48
Auto ignition temperature	°C	
Upper flammability limit	%v/v in air	
Lower flammability limit	%v/v in air	
Viscosity	mPa.s (cP) @ 20 °C	
Refractive index	n 20 D	1.421
Solubility in water	%w/w @ 20 °C	
Solubility of water	%w/w @ 20 °C	
Solubility parameter[c]	$(J/cm^3)^{1/2}$	
Hydrogen bond index		
Fractional polarity		
Nitro cellulose dil ratio[d]		
Nitro cellulose dil ratio[e]		
Surface tension	mN/m @ 20 °C	
Specific heat liquid	kJ/kg mole/°C @ 20 °C	
Latent heat	kJ/kg mole	
Dielectric constant	@ 20 °C	
Antoine constant A	kPa°C	
Antoine constant B	kPa°C	
Antoine constant C	kPa°C	
Heat of combustion	kJ/kg mole	
UN number		1300
IMO classification		3.3
ADR/RID classification		3,31°(c)
UK exposure limits	mg/m^3 (ppm) 8hrTWA/MEL or OES	
USA exposure limits	mg/m^3 (ppm) TLV/TWA	
German exposure limits	mg/m^3 (ppm) MAK-TWA	
EU classification		
EU risk phrases		10
EU safety phrases		

[a] Ether = 1. [b] n-Butyl acetate = 1. [c] Nelson, Hemwall, Edwards. [d] Toluene. [e] Hydrocarbon solvent.

Solvent group		HYDROCARBONS—ALIPHATIC
Chemical name		
Chemical formula		
Structural formula		
CAS No.		64742-47-8
Synonym/common name		
Trade names		KETRUL 212
Suppliers		TOTAL
Molecular mass		
Boiling point/range	°C @ 760 mm Hg	215–245
Melting point	°C	
Density	kg/m^3 @ 20 °C	800
Evaporation time[a]		1000
Evaporation time[b]		
Vapour pressure	kPa @ 20 °C	
Flash point	°C	85
Auto ignition temperature	°C	
Upper flammability limit	%v/v in air	
Lower flammability limit	%v/v in air	
Viscosity	mPa.s (cP) @ 20 °C	
Refractive index	n 20 D	1.449
Solubility in water	%w/w @ 20 °C	
Solubility of water	%w/w @ 20 °C	
Solubility parameter[c]	$(J/cm^3)^{1/2}$	
Hydrogen bond index		
Fractional polarity		
Nitro cellulose dil ratio[d]		
Nitro cellulose dil ratio[e]		
Surface tension	mN/m @ 20 °C	
Specific heat liquid	kJ/kg mole/°C @ 20 °C	
Latent heat	kJ/kg mole	
Dielectric constant	@ 20 °C	
Antoine constant A	kPa°C	
Antoine constant B	kPa°C	
Antoine constant C	kPa°C	
Heat of combustion	kJ/kg mole	
UN number		1223
IMO classification		
ADR/RID classification		3,32°(c)
UK exposure limits	mg/m^3 (ppm) 8hrTWA/MEL or OES	
USA exposure limits	mg/m^3 (ppm) TLV/TWA	
German exposure limits	mg/m^3 (ppm) MAK-TWA	
EU classification		
EU risk phrases		
EU safety phrases		

[a] Ether = 1. [b] n-Butyl acetate = 1. [c] Nelson, Hemwall, Edwards. [d] Toluene. [e] Hydrocarbon solvent.

Solvent group		**HYDROCARBONS—ALIPHATIC**
Chemical name		
Chemical formula		
Structural formula		
CAS No.		64742-49-0
Synonym/common name		
Trade names		HYDROSOL Ess G
Suppliers		TOTAL
Molecular mass		
Boiling point/range	°C @ 760 mm Hg	45–65
Melting point	°C	
Density	kg/m^3 @ 20 °C	662
Evaporation time[a]		1
Evaporation time[b]		
Vapour pressure	kPa @ 20 °C	
Flash point	°C	<0
Auto ignition temperature	°C	
Upper flammability limit	%v/v in air	
Lower flammability limit	%v/v in air	
Viscosity	mPa.s (cP) @ 20 °C	
Refractive index	n 20 D	1.373
Solubility in water	%w/w @ 20 °C	
Solubility of water	%w/w @ 20 °C	
Solubility parameter[c]	(J/cm^3)$^{1/2}$	
Hydrogen bond index		
Fractional polarity		
Nitro cellulose dil ratio[d]		
Nitro cellulose dil ratio[e]		
Surface tension	mN/m @ 20 °C	
Specific heat liquid	kJ/kg mole/°C @ 20 °C	
Latent heat	kJ/kg mole	
Dielectric constant	@ 20 °C	
Antoine constant A	kPa°C	
Antoine constant B	kPa°C	
Antoine constant C	kPa°C	
Heat of combustion	kJ/kg mole	
UN number		1203
IMO classification		3.1
ADR/RID classification		3,3°(b)
UK exposure limits	mg/m^3 (ppm) 8hrTWA/MEL or OES	
USA exposure limits	mg/m^3 (ppm) TLV/TWA	
German exposure limits	mg/m^3 (ppm) MAK-TWA	
EU classification		
EU risk phrases		F
EU safety phrases		11
		9-16-23-29-33

[a] Ether = 1. [b] n-Butyl acetate = 1. [c] Nelson, Hemwall, Edwards. [d] Toluene. [e] Hydrocarbon solvent.

Solvent group		HYDROCARBONS—ALIPHATIC
Chemical name		
Chemical formula		
Structural formula		
CAS No.		64742-49-0
Synonym/common name		
Trade names		HYDROSOL Ess 70/80N
Suppliers		TOTAL
Molecular mass		
Boiling point/range	°C @ 760 mm Hg	73–79
Melting point	°C	
Density	kg/m^3 @ 20 °C	752
Evaporation time[a]		2.5
Evaporation time[b]		
Vapour pressure	kPa @ 20 °C	
Flash point	°C	<0
Auto ignition temperature	°C	
Upper flammability limit	%v/v in air	
Lower flammability limit	%v/v in air	
Viscosity	mPa.s (cP) @ 20 °C	
Refractive index	n 20 D	1.411
Solubility in water	%w/w @ 20 °C	
Solubility of water	%w/w @ 20 °C	
Solubility parameter[c]	$(J/cm^3)^{1/2}$	
Hydrogen bond index		
Fractional polarity		
Nitro cellulose dil ratio[d]		
Nitro cellulose dil ratio[e]		
Surface tension	mN/m @ 20 °C	
Specific heat liquid	kJ/kg mole/°C @ 20 °C	
Latent heat	kJ/kg mole	
Dielectric constant	@ 20 °C	
Antoine constant A	kPa°C	
Antoine constant B	kPa°C	
Antoine constant C	kPa°C	
Heat of combustion	kJ/kg mole	
UN number		1203
IMO classification		3.1
ADR/RID classification		3,3°(b)
UK exposure limits	mg/m^3 (ppm) 8hrTWA/MEL or OES	
USA exposure limits	mg/m^3 (ppm) TLV/TWA	
German exposure limits	mg/m^3 (ppm) MAK-TWA	
EU classification		F
EU risk phrases		11
EU safety phrases		9-16-23-29-33

[a] Ether = 1. [b] n-Butyl acetate = 1. [c] Nelson, Hemwall, Edwards. [d] Toluene. [e] Hydrocarbon solvent.

Property	Unit	Value
Solvent group		**HYDROCARBONS—ALIPHATIC**
Chemical name		
Chemical formula		
Structural formula		
CAS No.		64742-47-8
Synonym/common name		
Trade names		KETRUL 211
Suppliers		TOTAL
Molecular mass		
Boiling point/range	°C @ 760 mm Hg	215–240
Melting point	°C	
Density	kg/m^3 @ 20 °C	795
Evaporation time[a]		1000
Evaporation time[b]		
Vapour pressure	kPa @ 20 °C	
Flash point	°C	85
Auto ignition temperature	°C	
Upper flammability limit	%v/v in air	
Lower flammability limit	%v/v in air	
Viscosity	mPa.s (cP) @ 20 °C	
Refractive index	n 20 D	1.438
Solubility in water	%w/w @ 20 °C	
Solubility of water	%w/w @ 20 °C	
Solubility parameter[c]	(J/cm^3)$^{1/2}$	
Hydrogen bond index		
Fractional polarity		
Nitro cellulose dil ratio[d]		
Nitro cellulose dil ratio[e]		
Surface tension	mN/m @ 20 °C	
Specific heat liquid	kJ/kg mole/°C @ 20 °C	
Latent heat	kJ/kg mole	
Dielectric constant	@ 20 °C	
Antoine constant A	kPa°C	
Antoine constant B	kPa°C	
Antoine constant C	kPa°C	
Heat of combustion	kJ/kg mole	
UN number		1223
IMO classification		
ADR/RID classification		3,32°(c)
UK exposure limits	mg/m^3 (ppm) 8hrTWA/MEL or OES	
USA exposure limits	mg/m^3 (ppm) TLV/TWA	
German exposure limits	mg/m^3 (ppm) MAK-TWA	
EU classification		
EU risk phrases		
EU safety phrases		

[a] Ether = 1. [b] n-Butyl acetate = 1. [c] Nelson, Hemwall, Edwards. [d] Toluene. [e] Hydrocarbon solvent.

Solvent group		HYDROCARBONS—ALIPHATIC
Chemical name		
Chemical formula		
Structural formula		
CAS No.		64742-82-1
Synonym/common name		
Trade names		SPIRDANE HT
Suppliers		TOTAL
Molecular mass		
Boiling point/range	°C @ 760 mm Hg	155–200
Melting point	°C	
Density	kg/m^3 @ 20 °C	780
Evaporation time[a]		65
Evaporation time[b]		
Vapour pressure	kPa @ 20 °C	
Flash point	°C	38
Auto ignition temperature	°C	
Upper flammability limit	%v/v in air	
Lower flammability limit	%v/v in air	
Viscosity	mPa.s (cP) @ 20 °C	
Refractive index	n 20 D	1.435
Solubility in water	%w/w @ 20 °C	
Solubility of water	%w/w @ 20 °C	
Solubility parameter[c]	(J/cm^3)$^{1/2}$	
Hydrogen bond index		
Fractional polarity		
Nitro cellulose dil ratio[d]		
Nitro cellulose dil ratio[e]		
Surface tension	mN/m @ 20 °C	
Specific heat liquid	kJ/kg mole/°C @ 20 °C	
Latent heat	kJ/kg mole	
Dielectric constant	@ 20 °C	
Antoine constant A	kPa°C	
Antoine constant B	kPa°C	
Antoine constant C	kPa°C	
Heat of combustion	kJ/kg mole	
UN number		1300
IMO classification		3.2
ADR/RID classification		3,31°(c)
UK exposure limits	mg/m^3 (ppm) 8hrTWA/MEL or OES	
USA exposure limits	mg/m^3 (ppm) TLV/TWA	
German exposure limits	mg/m^3 (ppm) MAK-TWA	
EU classification		
EU risk phrases		10
EU safety phrases		

[a] Ether = 1. [b] n-Butyl acetate = 1. [c] Nelson, Hemwall, Edwards. [d] Toluene. [e] Hydrocarbon solvent.

Solvent group		HYDROCARBONS—ALIPHATIC
Chemical name		
Chemical formula		
Structural formula		
CAS No.		
Synonym/common name		
Trade names		Exxol DSP 60/95S
Suppliers		EXXON
Molecular mass		
Boiling point/range	°C @ 760 mm Hg	65–98
Melting point	°C	
Density	kg/m^3 @ 20 °C	
Evaporation time[a]		2.1
Evaporation time[b]		7.70
Vapour pressure	kPa @ 20 °C	
Flash point	°C	<0
Auto ignition temperature	°C	
Upper flammability limit	%v/v in air	
Lower flammability limit	%v/v in air	
Viscosity	mPa.s (cP) @ 20 °C	
Refractive index	n 20 D	1.393
Solubility in water	%w/w @ 20 °C	
Solubility of water	%w/w @ 20 °C	
Solubility parameter[c]	(J/cm^3)$^{1/2}$	
Hydrogen bond index		
Fractional polarity		
Nitro cellulose dil ratio[d]		
Nitro cellulose dil ratio[e]		
Surface tension	mN/m @ 20 °C	
Specific heat liquid	kJ/kg mole/°C @ 20 °C	
Latent heat	kJ/kg mole	
Dielectric constant	@ 20 °C	
Antoine constant A	kPa°C	
Antoine constant B	kPa°C	
Antoine constant C	kPa°C	
Heat of combustion	kJ/kg mole	
UN number		1271
IMO classification		3.1
ADR/RID classification		3,3°(b)
UK exposure limits	mg/m^3 (ppm) 8hrTWA/MEL or OES	
USA exposure limits	mg/m^3 (ppm) TLV/TWA	
German exposure limits	mg/m^3 (ppm) MAK-TWA	
EU classification		F
EU risk phrases		11
EU safety phrases		9-16-23-29-33-43a

[a] Ether = 1. [b] n-Butyl acetate = 1. [c] Nelson, Hemwall, Edwards. [d] Toluene. [e] Hydrocarbon solvent.

Solvent group		HYDROCARBONS—ALIPHATIC
Chemical name		
Chemical formula		
Structural formula		
CAS No.		
Synonym/common name		
Trade names		Norpar 15
Suppliers		EXXON
Molecular mass		
Boiling point/range	°C @ 760 mm Hg	256–279
Melting point	°C	
Density	kg/m^3 @ 20 °C	
Evaporation time[a]		>1000
Evaporation time[b]		<0.01
Vapour pressure	kPa @ 20 °C	
Flash point	°C	118
Auto ignition temperature	°C	
Upper flammability limit	%v/v in air	
Lower flammability limit	%v/v in air	
Viscosity	mPa.s (cP) @ 20 °C	
Refractive index	n 20 D	1.432
Solubility in water	%w/w @ 20 °C	
Solubility of water	%w/w @ 20 °C	
Solubility parameter[c]	$(J/cm^3)^{1/2}$	
Hydrogen bond index		
Fractional polarity		
Nitro cellulose dil ratio[d]		
Nitro cellulose dil ratio[e]		
Surface tension	mN/m @ 20 °C	
Specific heat liquid	kJ/kg mole/°C @ 20 °C	
Latent heat	kJ/kg mole	
Dielectric constant	@ 20 °C	
Antoine constant A	kPa°C	
Antoine constant B	kPa°C	
Antoine constant C	kPa°C	
Heat of combustion	kJ/kg mole	
UN number		
IMO classification		
ADR/RID classification		
UK exposure limits	mg/m^3 (ppm) 8hrTWA/MEL or OES	
USA exposure limits	mg/m^3 (ppm) TLV/TWA	
German exposure limits	mg/m^3 (ppm) MAK-TWA	
EU classification		
EU risk phrases		
EU safety phrases		

[a] Ether = 1. [b] n-Butyl acetate = 1. [c] Nelson, Hemwall, Edwards. [d] Toluene. [e] Hydrocarbon solvent.

Solvent group		HYDROCARBONS—ALIPHATIC
Chemical name		Solvent naphtha
Chemical formula		
Structural formula		
CAS No.		64742-88-7
Synonym/common name		White spirit
Trade names		
Suppliers		PETROGAL, SA
Molecular mass		
Boiling point/range	°C @ 760 mm Hg	145–210
Melting point	°C	
Density	kg/m^3 @ 20 °C	770–785 (15 °C)
Evaporation time[a]		
Evaporation time[b]		
Vapour pressure	kPa @ 20 °C	
Flash point	°C	>33
Auto ignition temperature	°C	230
Upper flammability limit	%v/v in air	8
Lower flammability limit	%v/v in air	0.6
Viscosity	mPa.s (cP) @ 20 °C	
Refractive index	n 20 D	1.4359
Solubility in water	%w/w @ 20 °C	
Solubility of water	%w/w @ 20 °C	
Solubility parameter[c]	$(J/cm^3)^{1/2}$	
Hydrogen bond index		
Fractional polarity		
Nitro cellulose dil ratio[d]		
Nitro cellulose dil ratio[e]		
Surface tension	mN/m @ 20 °C	
Specific heat liquid	kJ/kg mole/°C @ 20 °C	
Latent heat	kJ/kg mole	
Dielectric constant	@ 20 °C	
Antoine constant A	kPa°C	
Antoine constant B	kPa°C	
Antoine constant C	kPa°C	
Heat of combustion	kJ/kg mole	
UN number		1300
IMO classification		3.3
ADR/RID classification		3,31(c)
UK exposure limits	mg/m^3 (ppm) 8hrTWA/MEL or OES	
USA exposure limits	mg/m^3 (ppm) TLV/TWA	
German exposure limits	mg/m^3 (ppm) MAK-TWA	
EU classification		
EU risk phrases		10
EU safety phrases		24/25-16-29-33-37

[a] Ether = 1. [b] n-Butyl acetate = 1. [c] Nelson, Hemwall, Edwards. [d] Toluene. [e] Hydrocarbon solvent.

Solvent group		HYDROCARBONS—ALIPHATIC
Chemical name		
Chemical formula		
Structural formula		
CAS No.		
Synonym/common name		
Trade names		Isopar K
Suppliers		EXXON
Molecular mass		
Boiling point/range	°C @ 760 mm Hg	181–204
Melting point	°C	
Density	kg/m^3 @ 20 °C	
Evaporation time[a]		93
Evaporation time[b]		0.08
Vapour pressure	kPa @ 20 °C	
Flash point	°C	59
Auto ignition temperature	°C	
Upper flammability limit	%v/v in air	
Lower flammability limit	%v/v in air	
Viscosity	mPa.s (cP) @ 20 °C	
Refractive index	n 20 D	1.427
Solubility in water	%w/w @ 20 °C	
Solubility of water	%w/w @ 20 °C	
Solubility parameter[c]	(J/cm^3)$^{1/2}$	
Hydrogen bond index		
Fractional polarity		
Nitro cellulose dil ratio[d]		
Nitro cellulose dil ratio[e]		
Surface tension	mN/m @ 20 °C	
Specific heat liquid	kJ/kg mole/°C @ 20 °C	
Latent heat	kJ/kg mole	
Dielectric constant	@ 20 °C	
Antoine constant A	kPa°C	
Antoine constant B	kPa°C	
Antoine constant C	kPa°C	
Heat of combustion	kJ/kg mole	
UN number		1993
IMO classification		3.3
ADR/RID classification		3,32°(c)
UK exposure limits	mg/m^3 (ppm) 8hrTWA/MEL or OES	
USA exposure limits	mg/m^3 (ppm) TLV/TWA	
German exposure limits	mg/m^3 (ppm) MAK-TWA	
EU classification		
EU risk phrases		
EU safety phrases		

[a] Ether = 1. [b] n-Butyl acetate = 1. [c] Nelson, Hemwall, Edwards. [d] Toluene. [e] Hydrocarbon solvent.

Solvent group		HYDROCARBONS—ALIPHATIC
Chemical name		
Chemical formula		
Structural formula		
CAS No.		
Synonym/common name		
Trade names		Exxol D 60
Suppliers		EXXON
Molecular mass		
Boiling point/range	°C @ 760 mm Hg	182–215
Melting point	°C	
Density	kg/m^3 @ 20 °C	
Evaporation time[a]		150
Evaporation time[b]		0.04
Vapour pressure	kPa @ 20 °C	
Flash point	°C	62
Auto ignition temperature	°C	
Upper flammability limit	%v/v in air	
Lower flammability limit	%v/v in air	
Viscosity	mPa.s (cP) @ 20 °C	
Refractive index	n 20 D	1.434
Solubility in water	%w/w @ 20 °C	
Solubility of water	%w/w @ 20 °C	
Solubility parameter[c]	(J/cm^3)$^{1/2}$	
Hydrogen bond index		
Fractional polarity		
Nitro cellulose dil ratio[d]		
Nitro cellulose dil ratio[e]		
Surface tension	mN/m @ 20 °C	
Specific heat liquid	kJ/kg mole/°C @ 20 °C	
Latent heat	kJ/kg mole	
Dielectric constant	@ 20 °C	
Antoine constant A	kPa°C	
Antoine constant B	kPa°C	
Antoine constant C	kPa°C	
Heat of combustion	kJ/kg mole	
UN number		1270
IMO classification		3.3
ADR/RID classification		3,32°(c)
UK exposure limits	mg/m^3 (ppm) 8hrTWA/MEL or OES	
USA exposure limits	mg/m^3 (ppm) TLV/TWA	
German exposure limits	mg/m^3 (ppm) MAK-TWA	
EU classification		
EU risk phrases		
EU safety phrases		

[a] Ether = 1. [b] n-Butyl acetate = 1. [c] Nelson, Hemwall, Edwards. [d] Toluene. [e] Hydrocarbon solvent.

Solvent group		**HYDROCARBONS—ALIPHATIC**
Chemical name		
Chemical formula		
Structural formula		
CAS No.		
Synonym/common name		
Trade names		Exxol D 70
Suppliers		EXXON
Molecular mass		
Boiling point/range	°C @ 760 mm Hg	193–213
Melting point	°C	
Density	kg/m^3 @ 20 °C	
Evaporation time[a]		175
Evaporation time[b]		0.03
Vapour pressure	kPa @ 20 °C	
Flash point	°C	72
Auto ignition temperature	°C	
Upper flammability limit	%v/v in air	
Lower flammability limit	%v/v in air	
Viscosity	mPa.s (cP) @ 20 °C	
Refractive index	n 20 D	1.441
Solubility in water	%w/w @ 20 °C	
Solubility of water	%w/w @ 20 °C	
Solubility parameter[c]	$(J/cm^3)^{1/2}$	
Hydrogen bond index		
Fractional polarity		
Nitro cellulose dil ratio[d]		
Nitro cellulose dil ratio[e]		
Surface tension	mN/m @ 20 °C	
Specific heat liquid	kJ/kg mole/°C @ 20 °C	
Latent heat	kJ/kg mole	
Dielectric constant	@ 20 °C	
Antoine constant A	kPa°C	
Antoine constant B	kPa°C	
Antoine constant C	kPa°C	
Heat of combustion	kJ/kg mole	
UN number		
IMO classification		
ADR/RID classification		3,32°(c)
UK exposure limits	mg/m^3 (ppm) 8hrTWA/MEL or OES	
USA exposure limits	mg/m^3 (ppm) TLV/TWA	
German exposure limits	mg/m^3 (ppm) MAK-TWA	
EU classification		
EU risk phrases		
EU safety phrases		

[a] Ether = 1. [b] n-Butyl acetate = 1. [c] Nelson, Hemwall, Edwards. [d] Toluene. [e] Hydrocarbon solvent.

Solvent group		HYDROCARBONS—ALIPHATIC
Chemical name		
Chemical formula		
Structural formula		
CAS No.		
Synonym/common name		
Trade names		Exxol DSP 40/65
Suppliers		EXXON
Molecular mass		
Boiling point/range	°C @ 760 mm Hg	42–69
Melting point	°C	
Density	kg/m^3 @ 20 °C	
Evaporation time[a]		1.1
Evaporation time[b]		>40
Vapour pressure	kPa @ 20 °C	
Flash point	°C	<0
Auto ignition temperature	°C	
Upper flammability limit	%v/v in air	
Lower flammability limit	%v/v in air	
Viscosity	mPa.s (cP) @ 20 °C	
Refractive index	n 20 D	1.370
Solubility in water	%w/w @ 20 °C	
Solubility of water	%w/w @ 20 °C	
Solubility parameter[c]	(J/cm^3)$^{1/2}$	
Hydrogen bond index		
Fractional polarity		
Nitro cellulose dil ratio[d]		
Nitro cellulose dil ratio[e]		
Surface tension	mN/m @ 20 °C	
Specific heat liquid	kJ/kg mole/°C @ 20 °C	
Latent heat	kJ/kg mole	
Dielectric constant	@ 20 °C	
Antoine constant A	kPa°C	
Antoine constant B	kPa°C	
Antoine constant C	kPa°C	
Heat of combustion	kJ/kg mole	
UN number		1271
IMO classification		3.1
ADR/RID classification		3,3°(b)
UK exposure limits	mg/m^3 (ppm) 8hrTWA/MEL or OES	
USA exposure limits	mg/m^3 (ppm) TLV/TWA	
German exposure limits	mg/m^3 (ppm) MAK-TWA	
EU classification		F, Xn
EU risk phrases		11-20-48
EU safety phrases		9-16-24/25-29-33-43a-51

[a] Ether = 1. [b] n-Butyl acetate = 1. [c] Nelson, Hemwall, Edwards. [d] Toluene. [e] Hydrocarbon solvent.

Solvent group		HYDROCARBONS—ALIPHATIC
Chemical name		
Chemical formula		
Structural formula		
CAS No.		
Synonym/common name		
Trade names		Norpar 13
Suppliers		EXXON
Molecular mass		
Boiling point/range	°C @ 760 mm Hg	222–243
Melting point	°C	
Density	kg/m^3 @ 20 °C	
Evaporation time[a]		870
Evaporation time[b]		<0.01
Vapour pressure	kPa @ 20 °C	
Flash point	°C	95
Auto ignition temperature	°C	
Upper flammability limit	%v/v in air	
Lower flammability limit	%v/v in air	
Viscosity	mPa.s (cP) @ 20 °C	
Refractive index	n 20 D	1.428
Solubility in water	%w/w @ 20 °C	
Solubility of water	%w/w @ 20 °C	
Solubility parameter[c]	$(J/cm^3)^{1/2}$	
Hydrogen bond index		
Fractional polarity		
Nitro cellulose dil ratio[d]		
Nitro cellulose dil ratio[e]		
Surface tension	mN/m @ 20 °C	
Specific heat liquid	kJ/kg mole/°C @ 20 °C	
Latent heat	kJ/kg mole	
Dielectric constant	@ 20 °C	
Antoine constant A	kPa°C	
Antoine constant B	kPa°C	
Antoine constant C	kPa°C	
Heat of combustion	kJ/kg mole	
UN number		
IMO classification		
ADR/RID classification		3,32°(c)
UK exposure limits	mg/m^3 (ppm) 8hrTWA/MEL or OES	
USA exposure limits	mg/m^3 (ppm) TLV/TWA	
German exposure limits	mg/m^3 (ppm) MAK-TWA	
EU classification		
EU risk phrases		
EU safety phrases		

[a] Ether = 1. [b] n-Butyl acetate = 1. [c] Nelson, Hemwall, Edwards. [d] Toluene. [e] Hydrocarbon solvent.

Solvent group		HYDROCARBONS—ALIPHATIC
Chemical name		
Chemical formula		
Structural formula		
CAS No.		68557-20-2
Synonym/common name		
Trade names		Isopar P
Suppliers		EXXON
Molecular mass		
Boiling point/range	°C @ 760 mm Hg	239–275
Melting point	°C	
Density	kg/m^3 @ 20 °C	
Evaporation time[a]		
Evaporation time[b]		
Vapour pressure	kPa @ 20 °C	0.001
Flash point	°C	104
Auto ignition temperature	°C	
Upper flammability limit	%v/v in air	7.0
Lower flammability limit	%v/v in air	0.6
Viscosity	mPa.s (cP) @ 20 °C	
Refractive index	n 20 D	1.445
Solubility in water	%w/w @ 20 °C	
Solubility of water	%w/w @ 20 °C	
Solubility parameter[c]	$(J/cm^3)^{1/2}$	
Hydrogen bond index		
Fractional polarity		
Nitro cellulose dil ratio[d]		
Nitro cellulose dil ratio[e]		
Surface tension	mN/m @ 20 °C	
Specific heat liquid	kJ/kg mole/°C @ 20 °C	
Latent heat	kJ/kg mole	
Dielectric constant	@ 20 °C	
Antoine constant A	kPa°C	
Antoine constant B	kPa°C	
Antoine constant C	kPa°C	
Heat of combustion	kJ/kg mole	
UN number		
IMO classification		
ADR/RID classification		
UK exposure limits	mg/m^3 (ppm) 8hrTWA/MEL or OES	
USA exposure limits	mg/m^3 (ppm) TLV/TWA	
German exposure limits	mg/m^3 (ppm) MAK-TWA	
EU classification		
EU risk phrases		
EU safety phrases		

[a] Ether = 1. [b] n-Butyl acetate = 1. [c] Nelson, Hemwall, Edwards. [d] Toluene. [e] Hydrocarbon solvent.

Solvent group		**HYDROCARBONS—ALIPHATIC**
Chemical name		
Chemical formula		
Structural formula		
CAS No.		
Synonym/common name		
Trade names		Exxol D 80
Suppliers		EXXON
Molecular mass		
Boiling point/range	°C @ 760 mm Hg	199–237
Melting point	°C	
Density	kg/m^3 @ 20 °C	
Evaporation time[a]		600
Evaporation time[b]		<0.01
Vapour pressure	kPa @ 20 °C	
Flash point	°C	75
Auto ignition temperature	°C	
Upper flammability limit	%v/v in air	
Lower flammability limit	%v/v in air	
Viscosity	mPa.s (cP) @ 20 °C	
Refractive index	n 20 D	1.440
Solubility in water	%w/w @ 20 °C	
Solubility of water	%w/w @ 20 °C	
Solubility parameter[c]	$(J/cm^3)^{1/2}$	
Hydrogen bond index		
Fractional polarity		
Nitro cellulose dil ratio[d]		
Nitro cellulose dil ratio[e]		
Surface tension	mN/m @ 20 °C	
Specific heat liquid	kJ/kg mole/°C @ 20 °C	
Latent heat	kJ/kg mole	
Dielectric constant	@ 20 °C	
Antoine constant A	kPa°C	
Antoine constant B	kPa°C	
Antoine constant C	kPa°C	
Heat of combustion	kJ/kg mole	
UN number		
IMO classification		
ADR/RID classification		3,32°(c)
UK exposure limits	mg/m^3 (ppm) 8hrTWA/MEL or OES	
USA exposure limits	mg/m^3 (ppm) TLV/TWA	
German exposure limits	mg/m^3 (ppm) MAK-TWA	
EU classification		
EU risk phrases		
EU safety phrases		

[a] Ether = 1. [b] n-Butyl acetate = 1. [c] Nelson, Hemwall, Edwards. [d] Toluene. [e] Hydrocarbon solvent.

Solvent group		**HYDROCARBONS—ALIPHATIC**
Chemical name		
Chemical formula		
Structural formula		
CAS No.		
Synonym/common name		
Trade names		Exxol D 40
Suppliers		EXXON
Molecular mass		
Boiling point/range	°C @ 760 mm Hg	155–188
Melting point	°C	
Density	kg/m^3 @ 20 °C	
Evaporation time[a]		55
Evaporation time[b]		0.14
Vapour pressure	kPa @ 20 °C	
Flash point	°C	40
Auto ignition temperature	°C	
Upper flammability limit	%v/v in air	
Lower flammability limit	%v/v in air	
Viscosity	mPa.s (cP) @ 20 °C	
Refractive index	n 20 D	1.427
Solubility in water	%w/w @ 20 °C	
Solubility of water	%w/w @ 20 °C	
Solubility parameter[c]	$(J/cm^3)^{1/2}$	
Hydrogen bond index		
Fractional polarity		
Nitro cellulose dil ratio[d]		
Nitro cellulose dil ratio[e]		
Surface tension	mN/m @ 20 °C	
Specific heat liquid	kJ/kg mole/°C @ 20 °C	
Latent heat	kJ/kg mole	
Dielectric constant	@ 20 °C	
Antoine constant A	kPa°C	
Antoine constant B	kPa°C	
Antoine constant C	kPa°C	
Heat of combustion	kJ/kg mole	
UN number		1270
IMO classification		3.3
ADR/RID classification		3,31°(c)
UK exposure limits	mg/m^3 (ppm) 8hrTWA/MEL or OES	
USA exposure limits	mg/m^3 (ppm) TLV/TWA	
German exposure limits	mg/m^3 (ppm) MAK-TWA	
EU classification		
EU risk phrases		10
EU safety phrases		43a

[a] Ether = 1. [b] n-Butyl acetate = 1. [c] Nelson, Hemwall, Edwards. [d] Toluene. [e] Hydrocarbon solvent.

Solvent group		HYDROCARBONS—ALIPHATIC
Chemical name		
Chemical formula		
Structural formula		
CAS No.		
Synonym/common name		
Trade names		Isopar E
Suppliers		EXXON
Molecular mass		
Boiling point/range	°C @ 760 mm Hg	118–143
Melting point	°C	
Density	kg/m^3 @ 20 °C	
Evaporation time[a]		8
Evaporation time[b]		1.55
Vapour pressure	kPa @ 20 °C	
Flash point	°C	7
Auto ignition temperature	°C	
Upper flammability limit	%v/v in air	
Lower flammability limit	%v/v in air	
Viscosity	mPa.s (cP) @ 20 °C	
Refractive index	n 20 D	1.406
Solubility in water	%w/w @ 20 °C	
Solubility of water	%w/w @ 20 °C	
Solubility parameter[c]	(J/cm^3)$^{1/2}$	
Hydrogen bond index		
Fractional polarity		
Nitro cellulose dil ratio[d]		
Nitro cellulose dil ratio[e]		
Surface tension	mN/m @ 20 °C	
Specific heat liquid	kJ/kg mole/°C @ 20 °C	
Latent heat	kJ/kg mole	
Dielectric constant	@ 20 °C	
Antoine constant A	kPa°C	
Antoine constant B	kPa°C	
Antoine constant C	kPa°C	
Heat of combustion	kJ/kg mole	
UN number		1993
IMO classification		3.2
ADR/RID classification		3,3°(b)
UK exposure limits	mg/m^3 (ppm) 8hrTWA/MEL or OES	
USA exposure limits	mg/m^3 (ppm) TLV/TWA	
German exposure limits	mg/m^3 (ppm) MAK-TWA	
EU classification		F
EU risk phrases		11
EU safety phrases		9-16-29-33-43a

[a] Ether = 1. [b] n-Butyl acetate = 1. [c] Nelson, Hemwall, Edwards. [d] Toluene. [e] Hydrocarbon solvent.

Solvent group		HYDROCARBONS—ALIPHATIC
Chemical name		
Chemical formula		
Structural formula		
CAS No.		
Synonym/common name		
Trade names		Exxol DHN50
Suppliers		EXXON
Molecular mass		
Boiling point/range	°C @ 760 mm Hg	72–76
Melting point	°C	
Density	kg/m^3 @ 20 °C	
Evaporation time[a]		1.5
Evaporation time[b]		13
Vapour pressure	kPa @ 20 °C	
Flash point	°C	<0
Auto ignition temperature	°C	
Upper flammability limit	%v/v in air	
Lower flammability limit	%v/v in air	
Viscosity	mPa.s (cP) @ 20 °C	
Refractive index	n 20 D	1.393
Solubility in water	%w/w @ 20 °C	
Solubility of water	%w/w @ 20 °C	
Solubility parameter[c]	$(J/cm^3)^{1/2}$	
Hydrogen bond index		
Fractional polarity		
Nitro cellulose dil ratio[d]		
Nitro cellulose dil ratio[e]		
Surface tension	mN/m @ 20 °C	
Specific heat liquid	kJ/kg mole/°C @ 20 °C	
Latent heat	kJ/kg mole	
Dielectric constant	@ 20 °C	
Antoine constant A	kPa°C	
Antoine constant B	kPa°C	
Antoine constant C	kPa°C	
Heat of combustion	kJ/kg mole	
UN number		1271
IMO classification		3.2
ADR/RID classification		3,3°(b)
UK exposure limits	mg/m^3 (ppm) 8hrTWA/MEL or OES	
USA exposure limits	mg/m^3 (ppm) TLV/TWA	
German exposure limits	mg/m^3 (ppm) MAK-TWA	
EU classification		F, Xn
EU risk phrases		11
EU safety phrases		9-16-24/25-29-33-43a-51

[a] Ether = 1. [b] n-Butyl acetate = 1. [c] Nelson, Hemwall, Edwards. [d] Toluene. [e] Hydrocarbon solvent.

Solvent group		HYDROCARBONS—ALIPHATIC
Chemical name		
Chemical formula		
Structural formula		
CAS No.		
Synonym/common name		
Trade names		Norpar 12
Suppliers		EXXON
Molecular mass		
Boiling point/range	°C @ 760 mm Hg	193–220
Melting point	°C	
Density	kg/m^3 @ 20 °C	
Evaporation time[a]		185
Evaporation time[b]		0.03
Vapour pressure	kPa @ 20 °C	
Flash point	°C	74
Auto ignition temperature	°C	
Upper flammability limit	%v/v in air	
Lower flammability limit	%v/v in air	
Viscosity	mPa.s (cP) @ 20 °C	
Refractive index	n 20 D	1.421
Solubility in water	%w/w @ 20 °C	
Solubility of water	%w/w @ 20 °C	
Solubility parameter[c]	$(J/cm^3)^{1/2}$	
Hydrogen bond index		
Fractional polarity		
Nitro cellulose dil ratio[d]		
Nitro cellulose dil ratio[e]		
Surface tension	mN/m @ 20 °C	
Specific heat liquid	kJ/kg mole/°C @ 20 °C	
Latent heat	kJ/kg mole	
Dielectric constant	@ 20 °C	
Antoine constant A	kPa°C	
Antoine constant B	kPa°C	
Antoine constant C	kPa°C	
Heat of combustion	kJ/kg mole	
UN number		
IMO classification		
ADR/RID classification		3,32°(c)
UK exposure limits	mg/m^3 (ppm) 8hrTWA/MEL or OES	
USA exposure limits	mg/m^3 (ppm) TLV/TWA	
German exposure limits	mg/m^3 (ppm) MAK-TWA	
EU classification		
EU risk phrases		
EU safety phrases		

[a] Ether = 1. [b] n-Butyl acetate = 1. [c] Nelson, Hemwall, Edwards. [d] Toluene. [e] Hydrocarbon solvent.

Solvent group		HYDROCARBONS—ALIPHATIC
Chemical name		
Chemical formula		
Structural formula		
CAS No.		
Synonym/common name		
Trade names		Exxol D 140
Suppliers		EXXON
Molecular mass		
Boiling point/range	°C @ 760 mm Hg	288–310
Melting point	°C	
Density	kg/m^3 @ 20 °C	
Evaporation time[a]		>1000
Evaporation time[b]		<0.01
Vapour pressure	kPa @ 20 °C	
Flash point	°C	135
Auto ignition temperature	°C	
Upper flammability limit	%v/v in air	
Lower flammability limit	%v/v in air	
Viscosity	mPa.s (cP) @ 20 °C	
Refractive index	n 20 D	1.458
Solubility in water	%w/w @ 20 °C	
Solubility of water	%w/w @ 20 °C	
Solubility parameter[c]	(J/cm^3)$^{1/2}$	
Hydrogen bond index		
Fractional polarity		
Nitro cellulose dil ratio[d]		
Nitro cellulose dil ratio[e]		
Surface tension	mN/m @ 20 °C	
Specific heat liquid	kJ/kg mole/°C @ 20 °C	
Latent heat	kJ/kg mole	
Dielectric constant	@ 20 °C	
Antoine constant A	kPa°C	
Antoine constant B	kPa°C	
Antoine constant C	kPa°C	
Heat of combustion	kJ/kg mole	
UN number		
IMO classification		
ADR/RID classification		
UK exposure limits	mg/m^3 (ppm) 8hrTWA/MEL or OES	
USA exposure limits	mg/m^3 (ppm) TLV/TWA	
German exposure limits	mg/m^3 (ppm) MAK-TWA	
EU classification		
EU risk phrases		
EU safety phrases		

[a] Ether = 1. [b] n-Butyl acetate = 1. [c] Nelson, Hemwall, Edwards. [d] Toluene. [e] Hydrocarbon solvent.

Solvent group		HYDROCARBONS—ALIPHATIC
Chemical name		
Chemical formula		
Structural formula		
CAS No.		64742-82-1
Synonym/common name		
Trade names		HYDROSOL Ess F
Suppliers		TOTAL
Molecular mass		
Boiling point/range	°C @ 760 mm Hg	105–160
Melting point	°C	
Density	kg/m^3 @ 20 °C	745
Evaporation time[a]		12
Evaporation time[b]		
Vapour pressure	kPa @ 20 °C	
Flash point	°C	4
Auto ignition temperature	°C	
Upper flammability limit	%v/v in air	
Lower flammability limit	%v/v in air	
Viscosity	mPa.s (cP) @ 20 °C	
Refractive index	n 20 D	1.414
Solubility in water	%w/w @ 20 °C	
Solubility of water	%w/w @ 20 °C	
Solubility parameter[c]	(J/cm^3)$^{1/2}$	
Hydrogen bond index		
Fractional polarity		
Nitro cellulose dil ratio[d]		
Nitro cellulose dil ratio[e]		
Surface tension	mN/m @ 20 °C	
Specific heat liquid	kJ/kg mole/°C @ 20 °C	
Latent heat	kJ/kg mole	
Dielectric constant	@ 20 °C	
Antoine constant A	kPa°C	
Antoine constant B	kPa°C	
Antoine constant C	kPa°C	
Heat of combustion	kJ/kg mole	
UN number		1203
IMO classification		3.2
ADR/RID classification		3,3°(b)
UK exposure limits	mg/m^3 (ppm) 8hrTWA/MEL or OES	
USA exposure limits	mg/m^3 (ppm) TLV/TWA	
German exposure limits	mg/m^3 (ppm) MAK-TWA	
EU classification		F
EU risk phrases		11
EU safety phrases		9-16-29-33

[a] Ether = 1. [b] n-Butyl acetate = 1. [c] Nelson, Hemwall, Edwards. [d] Toluene. [e] Hydrocarbon solvent.

Solvent group		HYDROCARBONS—ALIPHATIC
Chemical name		
Chemical formula		
Structural formula		
CAS No.		
Synonym/common name		
Trade names		Exxol DSP100/120
Suppliers		EXXON
Molecular mass		
Boiling point/range	°C @ 760 mm Hg	100–119
Melting point	°C	
Density	kg/m^3 @ 20 °C	
Evaporation time[a]		5.3
Evaporation time[b]		3.15
Vapour pressure	kPa @ 20 °C	
Flash point	°C	<0
Auto ignition temperature	°C	
Upper flammability limit	%v/v in air	
Lower flammability limit	%v/v in air	
Viscosity	mPa.s (cP) @ 20 °C	
Refractive index	n 20 D	1.406
Solubility in water	%w/w @ 20 °C	
Solubility of water	%w/w @ 20 °C	
Solubility parameter[c]	$(J/cm^3)^{1/2}$	
Hydrogen bond index		
Fractional polarity		
Nitro cellulose dil ratio[d]		
Nitro cellulose dil ratio[e]		
Surface tension	mN/m @ 20 °C	
Specific heat liquid	kJ/kg mole/°C @ 20 °C	
Latent heat	kJ/kg mole	
Dielectric constant	@ 20 °C	
Antoine constant A	kPa°C	
Antoine constant B	kPa°C	
Antoine constant C	kPa°C	
Heat of combustion	kJ/kg mole	
UN number		1271
IMO classification		3.2
ADR/RID classification		3,3°(b)
UK exposure limits	mg/m^3 (ppm) 8hrTWA/MEL or OES	
USA exposure limits	mg/m^3 (ppm) TLV/TWA	
German exposure limits	mg/m^3 (ppm) MAK-TWA	
EU classification		F
EU risk phrases		11
EU safety phrases		9-16-29-33-43a

[a] Ether = 1. [b] n-Butyl acetate = 1. [c] Nelson, Hemwall, Edwards. [d] Toluene. [e] Hydrocarbon solvent.

Solvent group		**HYDROCARBONS—ALIPHATIC**
Chemical name		
Chemical formula		
Structural formula		
CAS No.		
Synonym/common name		
Trade names		Exxol DSP 145/160
Suppliers		EXXON
Molecular mass		
Boiling point/range	°C @ 760 mm Hg	146–159
Melting point	°C	
Density	kg/m^3 @ 20 °C	
Evaporation time[a]		22
Evaporation time[b]		
Vapour pressure	kPa @ 20 °C	
Flash point	°C	32
Auto ignition temperature	°C	
Upper flammability limit	%v/v in air	
Lower flammability limit	%v/v in air	
Viscosity	mPa.s (cP) @ 20 °C	
Refractive index	n 20 D	1.420
Solubility in water	%w/w @ 20 °C	
Solubility of water	%w/w @ 20 °C	
Solubility parameter[c]	$(J/cm^3)^{1/2}$	
Hydrogen bond index		
Fractional polarity		
Nitro cellulose dil ratio[d]		
Nitro cellulose dil ratio[e]		
Surface tension	mN/m @ 20 °C	
Specific heat liquid	kJ/kg mole/°C @ 20 °C	
Latent heat	kJ/kg mole	
Dielectric constant	@ 20 °C	
Antoine constant A	kPa°C	
Antoine constant B	kPa°C	
Antoine constant C	kPa°C	
Heat of combustion	kJ/kg mole	
UN number		1270
IMO classification		3.3
ADR/RID classification		3,31°(c)
UK exposure limits	mg/m^3 (ppm) 8hrTWA/MEL or OES	
USA exposure limits	mg/m^3 (ppm) TLV/TWA	
German exposure limits	mg/m^3 (ppm) MAK-TWA	
EU classification		
EU risk phrases		10
EU safety phrases		43a

[a] Ether = 1. [b] n-Butyl acetate = 1. [c] Nelson, Hemwall, Edwards. [d] Toluene. [e] Hydrocarbon solvent.

Solvent group		HYDROCARBONS—ALIPHATIC
Chemical name		
Chemical formula		
Structural formula		
CAS No.		
Synonym/common name		
Trade names		Norpar 5 S
Suppliers		EXXON
Molecular mass		
Boiling point/range	°C @ 760 mm Hg	35–38
Melting point	°C	
Density	kg/m^3 @ 20 °C	
Evaporation time[a]		<5
Evaporation time[b]		45
Vapour pressure	kPa @ 20 °C	
Flash point	°C	<0
Auto ignition temperature	°C	
Upper flammability limit	%v/v in air	
Lower flammability limit	%v/v in air	
Viscosity	mPa.s (cP) @ 20 °C	
Refractive index	n 20 D	1.359
Solubility in water	%w/w @ 20 °C	
Solubility of water	%w/w @ 20 °C	
Solubility parameter[c]	$(J/cm^3)^{1/2}$	
Hydrogen bond index		
Fractional polarity		
Nitro cellulose dil ratio[d]		
Nitro cellulose dil ratio[e]		
Surface tension	mN/m @ 20 °C	
Specific heat liquid	kJ/kg mole/°C @ 20 °C	
Latent heat	kJ/kg mole	
Dielectric constant	@ 20 °C	
Antoine constant A	kPa°C	
Antoine constant B	kPa°C	
Antoine constant C	kPa°C	
Heat of combustion	kJ/kg mole	
UN number		1265
IMO classification		3.1
ADR/RID classification		3,2°(b)
UK exposure limits	mg/m^3 (ppm) 8hrTWA/MEL or OES	
USA exposure limits	mg/m^3 (ppm) TLV/TWA	
German exposure limits	mg/m^3 (ppm) MAK-TWA	
EU classification		F+
EU risk phrases		12
EU safety phrases		3/7/9-16-29-33-43a

[a] Ether = 1. [b] n-Butyl acetate = 1. [c] Nelson, Hemwall, Edwards. [d] Toluene. [e] Hydrocarbon solvent.

Solvent group		HYDROCARBONS—ALIPHATIC
Chemical name		
Chemical formula		
Structural formula		
CAS No.		
Synonym/common name		
Trade names		Exxol D 110
Suppliers		EXXON
Molecular mass		
Boiling point/range	°C @ 760 mm Hg	239–275
Melting point	°C	
Density	kg/m^3 @ 20 °C	
Evaporation time[a]		>1000
Evaporation time[b]		<0.01
Vapour pressure	kPa @ 20 °C	
Flash point	°C	106
Auto ignition temperature	°C	
Upper flammability limit	%v/v in air	
Lower flammability limit	%v/v in air	
Viscosity	mPa.s (cP) @ 20 °C	
Refractive index	n 20 D	1.451
Solubility in water	%w/w @ 20 °C	
Solubility of water	%w/w @ 20 °C	
Solubility parameter[c]	(J/cm^3)$^{1/2}$	
Hydrogen bond index		
Fractional polarity		
Nitro cellulose dil ratio[d]		
Nitro cellulose dil ratio[e]		
Surface tension	mN/m @ 20 °C	
Specific heat liquid	kJ/kg mole/°C @ 20 °C	
Latent heat	kJ/kg mole	
Dielectric constant	@ 20 °C	
Antoine constant A	kPa°C	
Antoine constant B	kPa°C	
Antoine constant C	kPa°C	
Heat of combustion	kJ/kg mole	
UN number		
IMO classification		
ADR/RID classification		
UK exposure limits	mg/m^3 (ppm) 8hrTWA/MEL or OES	
USA exposure limits	mg/m^3 (ppm) TLV/TWA	
German exposure limits	mg/m^3 (ppm) MAK-TWA	
EU classification		
EU risk phrases		
EU safety phrases		

[a] Ether = 1. [b] n-Butyl acetate = 1. [c] Nelson, Hemwall, Edwards. [d] Toluene. [e] Hydrocarbon solvent.

Solvent group		HYDROCARBONS—ALIPHATIC
Chemical name		
Chemical formula		
Structural formula		
CAS No.		
Synonym/common name		
Trade names		Exxol D 240/260
Suppliers		EXXON
Molecular mass		
Boiling point/range	°C @ 760 mm Hg	239–266
Melting point	°C	
Density	kg/m^3 @ 20 °C	
Evaporation time[a]		>1000
Evaporation time[b]		<0.01
Vapour pressure	kPa @ 20 °C	
Flash point	°C	103
Auto ignition temperature	°C	
Upper flammability limit	%v/v in air	
Lower flammability limit	%v/v in air	
Viscosity	mPa.s (cP) @ 20 °C	
Refractive index	n 20 D	1.446
Solubility in water	%w/w @ 20 °C	
Solubility of water	%w/w @ 20 °C	
Solubility parameter[c]	$(J/cm^3)^{1/2}$	
Hydrogen bond index		
Fractional polarity		
Nitro cellulose dil ratio[d]		
Nitro cellulose dil ratio[e]		
Surface tension	mN/m @ 20 °C	
Specific heat liquid	kJ/kg mole/°C @ 20 °C	
Latent heat	kJ/kg mole	
Dielectric constant	@ 20 °C	
Antoine constant A	kPa°C	
Antoine constant B	kPa°C	
Antoine constant C	kPa°C	
Heat of combustion	kJ/kg mole	
UN number		
IMO classification		
ADR/RID classification		
UK exposure limits	mg/m^3 (ppm) 8hrTWA/MEL or OES	
USA exposure limits	mg/m^3 (ppm) TLV/TWA	
German exposure limits	mg/m^3 (ppm) MAK-TWA	
EU classification		
EU risk phrases		
EU safety phrases		

[a] Ether = 1. [b] n-Butyl acetate = 1. [c] Nelson, Hemwall, Edwards. [d] Toluene. [e] Hydrocarbon solvent.

Solvent group		HYDROCARBONS—ALIPHATIC
Chemical name		
Chemical formula		
Structural formula		
CAS No.		
Synonym/common name		
Trade names		Norpar 8
Suppliers		EXXON
Molecular mass		
Boiling point/range	°C @ 760 mm Hg	125–127
Melting point	°C	
Density	kg/m^3 @ 20 °C	
Evaporation time[a]		7
Evaporation time[b]		1.35
Vapour pressure	kPa @ 20 °C	
Flash point	°C	16
Auto ignition temperature	°C	
Upper flammability limit	%v/v in air	
Lower flammability limit	%v/v in air	
Viscosity	mPa.s (cP) @ 20 °C	
Refractive index	n 20 D	1.399
Solubility in water	%w/w @ 20 °C	
Solubility of water	%w/w @ 20 °C	
Solubility parameter[c]	$(J/cm^3)^{1/2}$	
Hydrogen bond index		
Fractional polarity		
Nitro cellulose dil ratio[d]		
Nitro cellulose dil ratio[e]		
Surface tension	mN/m @ 20 °C	
Specific heat liquid	kJ/kg mole/°C @ 20 °C	
Latent heat	kJ/kg mole	
Dielectric constant	@ 20 °C	
Antoine constant A	kPa°C	
Antoine constant B	kPa°C	
Antoine constant C	kPa°C	
Heat of combustion	kJ/kg mole	
UN number		1262
IMO classification		3.2
ADR/RID classification		3,3°(b)
UK exposure limits	mg/m^3 (ppm) 8hrTWA/MEL or OES	
USA exposure limits	mg/m^3 (ppm) TLV/TWA	
German exposure limits	mg/m^3 (ppm) MAK-TWA	
EU classification		F
EU risk phrases		11
EU safety phrases		9-16-29-33-43a

[a] Ether = 1. [b] n-Butyl acetate = 1. [c] Nelson, Hemwall, Edwards. [d] Toluene. [e] Hydrocarbon solvent.

Solvent group		HYDROCARBONS—ALIPHATIC
Chemical name		
Chemical formula		
Structural formula		
CAS No.		
Synonym/common name		
Trade names		Exxol D 120
Suppliers		EXXON
Molecular mass		
Boiling point/range	°C @ 760 mm Hg	260–294
Melting point	°C	
Density	kg/m^3 @ 20 °C	
Evaporation time[a]		>1000
Evaporation time[b]		<0.01
Vapour pressure	kPa @ 20 °C	
Flash point	°C	120
Auto ignition temperature	°C	
Upper flammability limit	%v/v in air	
Lower flammability limit	%v/v in air	
Viscosity	mPa.s (cP) @ 20 °C	
Refractive index	n 20 D	1.456
Solubility in water	%w/w @ 20 °C	
Solubility of water	%w/w @ 20 °C	
Solubility parameter[c]	(J/cm^3)$^{1/2}$	
Hydrogen bond index		
Fractional polarity		
Nitro cellulose dil ratio[d]		
Nitro cellulose dil ratio[e]		
Surface tension	mN/m @ 20 °C	
Specific heat liquid	kJ/kg mole/°C @ 20 °C	
Latent heat	kJ/kg mole	
Dielectric constant	@ 20 °C	
Antoine constant A	kPa°C	
Antoine constant B	kPa°C	
Antoine constant C	kPa°C	
Heat of combustion	kJ/kg mole	
UN number		
IMO classification		
ADR/RID classification		
UK exposure limits	mg/m^3 (ppm) 8hrTWA/MEL or OES	
USA exposure limits	mg/m^3 (ppm) TLV/TWA	
German exposure limits	mg/m^3 (ppm) MAK-TWA	
EU classification		
EU risk phrases		
EU safety phrases		

[a] Ether = 1. [b] n-Butyl acetate = 1. [c] Nelson, Hemwall, Edwards. [d] Toluene. [e] Hydrocarbon solvent.

Solvent group		HYDROCARBONS—ALIPHATIC
Chemical name		
Chemical formula		
Structural formula		
CAS No.		
Synonym/common name		
Trade names		Exxol D 100
Suppliers		EXXON
Molecular mass		
Boiling point/range	°C @ 760 mm Hg	239–266
Melting point	°C	
Density	kg/m^3 @ 20 °C	
Evaporation time[a]		>1000
Evaporation time[b]		<0.01
Vapour pressure	kPa @ 20 °C	
Flash point	°C	103
Auto ignition temperature	°C	
Upper flammability limit	%v/v in air	
Lower flammability limit	%v/v in air	
Viscosity	mPa.s (cP) @ 20 °C	
Refractive index	n 20 D	1.448
Solubility in water	%w/w @ 20 °C	
Solubility of water	%w/w @ 20 °C	
Solubility parameter[c]	(J/cm^3)$^{1/2}$	
Hydrogen bond index		
Fractional polarity		
Nitro cellulose dil ratio[d]		
Nitro cellulose dil ratio[e]		
Surface tension	mN/m @ 20 °C	
Specific heat liquid	kJ/kg mole/°C @ 20 °C	
Latent heat	kJ/kg mole	
Dielectric constant	@ 20 °C	
Antoine constant A	kPa°C	
Antoine constant B	kPa°C	
Antoine constant C	kPa°C	
Heat of combustion	kJ/kg mole	
UN number		
IMO classification		
ADR/RID classification		
UK exposure limits	mg/m^3 (ppm) 8hrTWA/MEL or OES	
USA exposure limits	mg/m^3 (ppm) TLV/TWA	
German exposure limits	mg/m^3 (ppm) MAK-TWA	
EU classification		
EU risk phrases		
EU safety phrases		

[a] Ether = 1. [b] n-Butyl acetate = 1. [c] Nelson, Hemwall, Edwards. [d] Toluene. [e] Hydrocarbon solvent.

Solvent group		**HYDROCARBONS—ALIPHATIC**
Chemical name		
Chemical formula		
Structural formula		
CAS No.		
Synonym/common name		
Trade names		Exxol D 220/230
Suppliers		EXXON
Molecular mass		
Boiling point/range	°C @ 760 mm Hg	225–233
Melting point	°C	
Density	kg/m^3 @ 20 °C	
Evaporation time[a]		500
Evaporation time[b]		<0.01
Vapour pressure	kPa @ 20 °C	
Flash point	°C	90
Auto ignition temperature	°C	
Upper flammability limit	%v/v in air	
Lower flammability limit	%v/v in air	
Viscosity	mPa.s (cP) @ 20 °C	
Refractive index	n 20 D	1.441
Solubility in water	%w/w @ 20 °C	
Solubility of water	%w/w @ 20 °C	
Solubility parameter[c]	$(J/cm^3)^{1/2}$	
Hydrogen bond index		
Fractional polarity		
Nitro cellulose dil ratio[d]		
Nitro cellulose dil ratio[e]		
Surface tension	mN/m @ 20 °C	
Specific heat liquid	kJ/kg mole/°C @ 20 °C	
Latent heat	kJ/kg mole	
Dielectric constant	@ 20 °C	
Antoine constant A	kPa°C	
Antoine constant B	kPa°C	
Antoine constant C	kPa°C	
Heat of combustion	kJ/kg mole	
UN number		
IMO classification		
ADR/RID classification		3,32°(c)
UK exposure limits	mg/m^3 (ppm) 8hrTWA/MEL or OES	
USA exposure limits	mg/m^3 (ppm) TLV/TWA	
German exposure limits	mg/m^3 (ppm) MAK-TWA	
EU classification		
EU risk phrases		
EU safety phrases		

[a] Ether = 1. [b] n-Butyl acetate = 1. [c] Nelson, Hemwall, Edwards. [d] Toluene. [e] Hydrocarbon solvent.

Solvent group		HYDROCARBONS—ALIPHATIC
Chemical name		
Chemical formula		
Structural formula		
CAS No.		
Synonym/common name		
Trade names		Isopar M
Suppliers		EXXON
Molecular mass		
Boiling point/range	°C @ 760 mm Hg	206–250
Melting point	°C	
Density	kg/m^3 @ 20 °C	
Evaporation time[a]		680
Evaporation time[b]		<0.01
Vapour pressure	kPa @ 20 °C	
Flash point	°C	78
Auto ignition temperature	°C	
Upper flammability limit	%v/v in air	
Lower flammability limit	%v/v in air	
Viscosity	mPa.s (cP) @ 20 °C	
Refractive index	n 20 D	1.437
Solubility in water	%w/w @ 20 °C	
Solubility of water	%w/w @ 20 °C	
Solubility parameter[c]	$(J/cm^3)^{1/2}$	
Hydrogen bond index		
Fractional polarity		
Nitro cellulose dil ratio[d]		
Nitro cellulose dil ratio[e]		
Surface tension	mN/m @ 20 °C	
Specific heat liquid	kJ/kg mole/°C @ 20 °C	
Latent heat	kJ/kg mole	
Dielectric constant	@ 20 °C	
Antoine constant A	kPa°C	
Antoine constant B	kPa°C	
Antoine constant C	kPa°C	
Heat of combustion	kJ/kg mole	
UN number		
IMO classification		
ADR/RID classification		3,32°(c)
UK exposure limits	mg/m^3 (ppm) 8hrTWA/MEL or OES	
USA exposure limits	mg/m^3 (ppm) TLV/TWA	
German exposure limits	mg/m^3 (ppm) MAK-TWA	
EU classification		
EU risk phrases		
EU safety phrases		

[a] Ether = 1. [b] n-Butyl acetate = 1. [c] Nelson, Hemwall, Edwards. [d] Toluene. [e] Hydrocarbon solvent.

Solvent group		HYDROCARBONS—ALIPHATIC
Chemical name		
Chemical formula		
Structural formula		
CAS No.		
Synonym/common name		
Trade names		Isopar L
Suppliers		EXXON
Molecular mass		
Boiling point/range	°C @ 760 mm Hg	189–210
Melting point	°C	
Density	kg/m^3 @ 20 °C	
Evaporation time[a]		125
Evaporation time[b]		0.04
Vapour pressure	kPa @ 20 °C	
Flash point	°C	65
Auto ignition temperature	°C	
Upper flammability limit	%v/v in air	
Lower flammability limit	%v/v in air	
Viscosity	mPa.s (cP) @ 20 °C	
Refractive index	n 20 D	1.429
Solubility in water	%w/w @ 20 °C	
Solubility of water	%w/w @ 20 °C	
Solubility parameter[c]	$(J/cm^3)^{1/2}$	
Hydrogen bond index		
Fractional polarity		
Nitro cellulose dil ratio[d]		
Nitro cellulose dil ratio[e]		
Surface tension	mN/m @ 20 °C	
Specific heat liquid	kJ/kg mole/°C @ 20 °C	
Latent heat	kJ/kg mole	
Dielectric constant	@ 20 °C	
Antoine constant A	kPa°C	
Antoine constant B	kPa°C	
Antoine constant C	kPa°C	
Heat of combustion	kJ/kg mole	
UN number		
IMO classification		
ADR/RID classification		3,32°(c)
UK exposure limits	mg/m^3 (ppm) 8hrTWA/MEL or OES	
USA exposure limits	mg/m^3 (ppm) TLV/TWA	
German exposure limits	mg/m^3 (ppm) MAK-TWA	
EU classification		
EU risk phrases		
EU safety phrases		

[a] Ether = 1. [b] n-Butyl acetate = 1. [c] Nelson, Hemwall, Edwards. [d] Toluene. [e] Hydrocarbon solvent.

Solvent group		HYDROCARBONS—ALIPHATIC
Chemical name		
Chemical formula		
Structural formula		
CAS No.		
Synonym/common name		
Trade names		Exxol Isohexane
Suppliers		EXXON
Molecular mass		
Boiling point/range	°C @ 760 mm Hg	55–62
Melting point	°C	
Density	kg/m^3 @ 20 °C	
Evaporation time[a]		1.3
Evaporation time[b]		19
Vapour pressure	kPa @ 20 °C	
Flash point	°C	<0
Auto ignition temperature	°C	
Upper flammability limit	%v/v in air	
Lower flammability limit	%v/v in air	
Viscosity	mPa.s (cP) @ 20 °C	
Refractive index	n 20 D	1.376
Solubility in water	%w/w @ 20 °C	
Solubility of water	%w/w @ 20 °C	
Solubility parameter[c]	(J/cm^3)$^{1/2}$	
Hydrogen bond index		
Fractional polarity		
Nitro cellulose dil ratio[d]		
Nitro cellulose dil ratio[e]		
Surface tension	mN/m @ 20 °C	
Specific heat liquid	kJ/kg mole/°C @ 20 °C	
Latent heat	kJ/kg mole	
Dielectric constant	@ 20 °C	
Antoine constant A	kPa°C	
Antoine constant B	kPa°C	
Antoine constant C	kPa°C	
Heat of combustion	kJ/kg mole	
UN number		1208
IMO classification		3.1
ADR/RID classification		3,3°(b)
UK exposure limits	mg/m^3 (ppm) 8hrTWA/MEL or OES	
USA exposure limits	mg/m^3 (ppm) TLV/TWA	
German exposure limits	mg/m^3 (ppm) MAK-TWA	
EU classification		F
EU risk phrases		11
EU safety phrases		9-16-23-29-33-43a

[a] Ether = 1. [b] n-Butyl acetate = 1. [c] Nelson, Hemwall, Edwards. [d] Toluene. [e] Hydrocarbon solvent.

Solvent group		HYDROCARBONS—ALIPHATIC
Chemical name		
Chemical formula		
Structural formula		
CAS No.		
Synonym/common name		
Trade names		Isopar C
Suppliers		EXXON
Molecular mass		
Boiling point/range	°C @ 760 mm Hg	100–105
Melting point	°C	
Density	kg/m^3 @ 20 °C	
Evaporation time[a]		2.8
Evaporation time[b]		5.45
Vapour pressure	kPa @ 20 °C	
Flash point	°C	<0
Auto ignition temperature	°C	
Upper flammability limit	%v/v in air	
Lower flammability limit	%v/v in air	
Viscosity	mPa.s (cP) @ 20 °C	
Refractive index	n 20 D	1.393
Solubility in water	%w/w @ 20 °C	
Solubility of water	%w/w @ 20 °C	
Solubility parameter[c]	$(J/cm^3)^{1/2}$	
Hydrogen bond index		
Fractional polarity		
Nitro cellulose dil ratio[d]		
Nitro cellulose dil ratio[e]		
Surface tension	mN/m @ 20 °C	
Specific heat liquid	kJ/kg mole/°C @ 20 °C	
Latent heat	kJ/kg mole	
Dielectric constant	@ 20 °C	
Antoine constant A	kPa°C	
Antoine constant B	kPa°C	
Antoine constant C	kPa°C	
Heat of combustion	kJ/kg mole	
UN number		1262
IMO classification		
ADR/RID classification		3,3°(b)
UK exposure limits	mg/m^3 (ppm) 8hrTWA/MEL or OES	
USA exposure limits	mg/m^3 (ppm) TLV/TWA	
German exposure limits	mg/m^3 (ppm) MAK-TWA	
EU classification		F
EU risk phrases		11
EU safety phrases		9-16-29-33-43a

[a] Ether = 1. [b] n-Butyl acetate = 1. [c] Nelson, Hemwall, Edwards. [d] Toluene. [e] Hydrocarbon solvent.

Solvent group		HYDROCARBONS—ALIPHATIC
Chemical name		
Chemical formula		
Structural formula		
CAS No.		
Synonym/common name		
Trade names		Exxol DSP 65/100
Suppliers		EXXON
Molecular mass		
Boiling point/range	°C @ 760 mm Hg	68–99
Melting point	°C	
Density	kg/m^3 @ 20 °C	
Evaporation time[a]		2.0
Evaporation time[b]		11
Vapour pressure	kPa @ 20 °C	
Flash point	°C	<0
Auto ignition temperature	°C	
Upper flammability limit	%v/v in air	
Lower flammability limit	%v/v in air	
Viscosity	mPa.s (cP) @ 20 °C	
Refractive index	n 20 D	1.386
Solubility in water	%w/w @ 20 °C	
Solubility of water	%w/w @ 20 °C	
Solubility parameter[c]	(J/cm^3)$^{1/2}$	
Hydrogen bond index		
Fractional polarity		
Nitro cellulose dil ratio[d]		
Nitro cellulose dil ratio[e]		
Surface tension	mN/m @ 20 °C	
Specific heat liquid	kJ/kg mole/°C @ 20 °C	
Latent heat	kJ/kg mole	
Dielectric constant	@ 20 °C	
Antoine constant A	kPa°C	
Antoine constant B	kPa°C	
Antoine constant C	kPa°C	
Heat of combustion	kJ/kg mole	
UN number		1271
IMO classification		3.1
ADR/RID classification		3,3°(b)
UK exposure limits	mg/m^3 (ppm) 8hrTWA/MEL or OES	
USA exposure limits	mg/m^3 (ppm) TLV/TWA	
German exposure limits	mg/m^3 (ppm) MAK-TWA	
EU classification		F, Xn
EU risk phrases		11
EU safety phrases		9-16-24/25-29-33-43a-51

[a] Ether = 1. [b] n-Butyl acetate = 1. [c] Nelson, Hemwall, Edwards. [d] Toluene. [e] Hydrocarbon solvent.

Solvent group		HYDROCARBONS—ALIPHATIC
Chemical name		
Chemical formula		
Structural formula		
CAS No.		
Synonym/common name		
Trade names		Exxol DSP 80/110
Suppliers		EXXON
Molecular mass		
Boiling point/range	°C @ 760 mm Hg	86–108
Melting point	°C	
Density	kg/m^3 @ 20 °C	
Evaporation time[a]		3.3
Evaporation time[b]		5.50
Vapour pressure	kPa @ 20 °C	
Flash point	°C	<0
Auto ignition temperature	°C	
Upper flammability limit	%v/v in air	
Lower flammability limit	%v/v in air	
Viscosity	mPa.s (cP) @ 20 °C	
Refractive index	n 20 D	1.398
Solubility in water	%w/w @ 20 °C	
Solubility of water	%w/w @ 20 °C	
Solubility parameter[c]	$(J/cm^3)^{1/2}$	
Hydrogen bond index		
Fractional polarity		
Nitro cellulose dil ratio[d]		
Nitro cellulose dil ratio[e]		
Surface tension	mN/m @ 20 °C	
Specific heat liquid	kJ/kg mole/°C @ 20 °C	
Latent heat	kJ/kg mole	
Dielectric constant	@ 20 °C	
Antoine constant A	kPa°C	
Antoine constant B	kPa°C	
Antoine constant C	kPa°C	
Heat of combustion	kJ/kg mole	
UN number		1271
IMO classification		3.2
ADR/RID classification		3,3°(b)
UK exposure limits	mg/m^3 (ppm) 8hrTWA/MEL or OES	
USA exposure limits	mg/m^3 (ppm) TLV/TWA	
German exposure limits	mg/m^3 (ppm) MAK-TWA	
EU classification		F
EU risk phrases		11
EU safety phrases		9-16-23-29-33-43a

[a] Ether = 1. [b] n-Butyl acetate = 1. [c] Nelson, Hemwall, Edwards. [d] Toluene. [e] Hydrocarbon solvent.

Solvent group		HYDROCARBONS—ALIPHATIC
Chemical name		
Chemical formula		
Structural formula		
CAS No.		
Synonym/common name		
Trade names		Exxol DSP 100/140
Suppliers		EXXON
Molecular mass		
Boiling point/range	°C @ 760 mm Hg	101–137
Melting point	°C	
Density	kg/m³ @ 20 °C	
Evaporation time[a]		9.0
Evaporation time[b]		2.10
Vapour pressure	kPa @ 20 °C	
Flash point	°C	<0
Auto ignition temperature	°C	
Upper flammability limit	%v/v in air	
Lower flammability limit	%v/v in air	
Viscosity	mPa.s (cP) @ 20 °C	
Refractive index	n 20 D	1.409
Solubility in water	%w/w @ 20 °C	
Solubility of water	%w/w @ 20 °C	
Solubility parameter[c]	$(J/cm^3)^{1/2}$	
Hydrogen bond index		
Fractional polarity		
Nitro cellulose dil ratio[d]		
Nitro cellulose dil ratio[e]		
Surface tension	mN/m @ 20 °C	
Specific heat liquid	kJ/kg mole/°C @ 20 °C	
Latent heat	kJ/kg mole	
Dielectric constant	@ 20 °C	
Antoine constant A	kPa°C	
Antoine constant B	kPa°C	
Antoine constant C	kPa°C	
Heat of combustion	kJ/kg mole	
UN number		1271
IMO classification		3.2
ADR/RID classification		3,3°(b)
UK exposure limits	mg/m³ (ppm) 8hrTWA/MEL or OES	
USA exposure limits	mg/m³ (ppm) TLV/TWA	
German exposure limits	mg/m³ (ppm) MAK-TWA	
EU classification		F
EU risk phrases		10
EU safety phrases		9-16-29-33-43a

[a] Ether = 1. [b] n-Butyl acetate = 1. [c] Nelson, Hemwall, Edwards. [d] Toluene. [e] Hydrocarbon solvent.

Solvent group		HYDROCARBONS—ALIPHATIC
Chemical name		
Chemical formula		
Structural formula		
CAS No.		
Synonym/common name		
Trade names		Isopar G
Suppliers		EXXON
Molecular mass		
Boiling point/range	°C @ 760 mm Hg	158–176
Melting point	°C	
Density	kg/m^3 @ 20 °C	
Evaporation time[a]		40
Evaporation time[b]		0.28
Vapour pressure	kPa @ 20 °C	
Flash point	°C	41
Auto ignition temperature	°C	
Upper flammability limit	%v/v in air	
Lower flammability limit	%v/v in air	
Viscosity	mPa.s (cP) @ 20 °C	
Refractive index	n 20 D	1.418
Solubility in water	%w/w @ 20 °C	
Solubility of water	%w/w @ 20 °C	
Solubility parameter[c]	(J/cm^3)$^{1/2}$	
Hydrogen bond index		
Fractional polarity		
Nitro cellulose dil ratio[d]		
Nitro cellulose dil ratio[e]		
Surface tension	mN/m @ 20 °C	
Specific heat liquid	kJ/kg mole/°C @ 20 °C	
Latent heat	kJ/kg mole	
Dielectric constant	@ 20 °C	
Antoine constant A	kPa°C	
Antoine constant B	kPa°C	
Antoine constant C	kPa°C	
Heat of combustion	kJ/kg mole	
UN number		1993
IMO classification		3.3
ADR/RID classification		3,31°(c)
UK exposure limits	mg/m^3 (ppm) 8hrTWA/MEL or OES	
USA exposure limits	mg/m^3 (ppm) TLV/TWA	
German exposure limits	mg/m^3 (ppm) MAK-TWA	
EU classification		
EU risk phrases		10
EU safety phrases		43a

[a] Ether = 1. [b] n-Butyl acetate = 1. [c] Nelson, Hemwall, Edwards. [d] Toluene. [e] Hydrocarbon solvent.

Solvent group		HYDROCARBONS—ALIPHATIC
Chemical name		
Chemical formula		
Structural formula		
CAS No.		
Synonym/common name		
Trade names		Norpar 7
Suppliers		EXXON
Molecular mass		
Boiling point/range	°C @ 760 mm Hg	98–99.6
Melting point	°C	
Density	kg/m^3 @ 20 °C	
Evaporation time[a]		3.8
Evaporation time[b]		4.30
Vapour pressure	kPa @ 20 °C	
Flash point	°C	<0
Auto ignition temperature	°C	
Upper flammability limit	%v/v in air	
Lower flammability limit	%v/v in air	
Viscosity	mPa.s (cP) @ 20 °C	
Refractive index	n 20 D	1.389
Solubility in water	%w/w @ 20 °C	
Solubility of water	%w/w @ 20 °C	
Solubility parameter[c]	$(J/cm^3)^{1/2}$	
Hydrogen bond index		
Fractional polarity		
Nitro cellulose dil ratio[d]		
Nitro cellulose dil ratio[e]		
Surface tension	mN/m @ 20 °C	
Specific heat liquid	kJ/kg mole/°C @ 20 °C	
Latent heat	kJ/kg mole	
Dielectric constant	@ 20 °C	
Antoine constant A	kPa°C	
Antoine constant B	kPa°C	
Antoine constant C	kPa°C	
Heat of combustion	kJ/kg mole	
UN number		1206
IMO classification		3.2
ADR/RID classification		3,3°(b)
UK exposure limits	mg/m^3 (ppm) 8hrTWA/MEL or OES	
USA exposure limits	mg/m^3 (ppm) TLV/TWA	
German exposure limits	mg/m^3 (ppm) MAK-TWA	
EU classification		F
EU risk phrases		11
EU safety phrases		9-16-23-29-33-43a

[a] Ether = 1. [b] n-Butyl acetate = 1. [c] Nelson, Hemwall, Edwards. [d] Toluene. [e] Hydrocarbon solvent.

Solvent group		HYDROCARBONS—ALIPHATIC
Chemical name		
Chemical formula		
Structural formula		
CAS No.		
Synonym/common name		
Trade names		Norpar 6
Suppliers		EXXON
Molecular mass		
Boiling point/range	°C @ 760 mm Hg	67–70
Melting point	°C	
Density	kg/m^3 @ 20 °C	
Evaporation time[a]		1.5
Evaporation time[b]		4.30
Vapour pressure	kPa @ 20 °C	
Flash point	°C	<0
Auto ignition temperature	°C	
Upper flammability limit	%v/v in air	
Lower flammability limit	%v/v in air	
Viscosity	mPa.s (cP) @ 20 °C	
Refractive index	n 20 D	1.357
Solubility in water	%w/w @ 20 °C	
Solubility of water	%w/w @ 20 °C	
Solubility parameter[c]	$(J/cm^3)^{1/2}$	
Hydrogen bond index		
Fractional polarity		
Nitro cellulose dil ratio[d]		
Nitro cellulose dil ratio[e]		
Surface tension	mN/m @ 20 °C	
Specific heat liquid	kJ/kg mole/°C @ 20 °C	
Latent heat	kJ/kg mole	
Dielectric constant	@ 20 °C	
Antoine constant A	kPa°C	
Antoine constant B	kPa°C	
Antoine constant C	kPa°C	
Heat of combustion	kJ/kg mole	
UN number		1208
IMO classification		3.1
ADR/RID classification		3,3°(b)
UK exposure limits	mg/m^3 (ppm) 8hrTWA/MEL or OES	
USA exposure limits	mg/m^3 (ppm) TLV/TWA	
German exposure limits	mg/m^3 (ppm) MAK-TWA	
EU classification		F, Xn
EU risk phrases		11-20-48
EU safety phrases		9-16-24/25-29-33-43a-51

[a] Ether = 1. [b] n-Butyl acetate = 1. [c] Nelson, Hemwall, Edwards. [d] Toluene. [e] Hydrocarbon solvent.

Solvent group		HYDROCARBONS—ALIPHATIC
Chemical name		
Chemical formula		
Structural formula		
CAS No.		
Synonym/common name		
Trade names		Exxol D 100S
Suppliers		EXXON
Molecular mass		
Boiling point/range	°C @ 760 mm Hg	235–265
Melting point	°C	
Density	kg/m^3 @ 20 °C	
Evaporation time[a]		>1000
Evaporation time[b]		<0.01
Vapour pressure	kPa @ 20 °C	
Flash point	°C	103
Auto ignition temperature	°C	
Upper flammability limit	%v/v in air	
Lower flammability limit	%v/v in air	
Viscosity	mPa.s (cP) @ 20 °C	
Refractive index	n 20 D	1.447
Solubility in water	%w/w @ 20 °C	
Solubility of water	%w/w @ 20 °C	
Solubility parameter[c]	(J/cm^3)$^{1/2}$	
Hydrogen bond index		
Fractional polarity		
Nitro cellulose dil ratio[d]		
Nitro cellulose dil ratio[e]		
Surface tension	mN/m @ 20 °C	
Specific heat liquid	kJ/kg mole/°C @ 20 °C	
Latent heat	kJ/kg mole	
Dielectric constant	@ 20 °C	
Antoine constant A	kPa°C	
Antoine constant B	kPa°C	
Antoine constant C	kPa°C	
Heat of combustion	kJ/kg mole	
UN number		
IMO classification		
ADR/RID classification		
UK exposure limits	mg/m^3 (ppm) 8hrTWA/MEL or OES	
USA exposure limits	mg/m^3 (ppm) TLV/TWA	
German exposure limits	mg/m^3 (ppm) MAK-TWA	
EU classification		
EU risk phrases		
EU safety phrases		

[a] Ether = 1. [b] n-Butyl acetate = 1. [c] Nelson, Hemwall, Edwards. [d] Toluene. [e] Hydrocarbon solvent.

Solvent group		HYDROCARBONS—ALIPHATIC
Chemical name		
Chemical formula		
Structural formula		
CAS No.		
Synonym/common name		
Trade names		Isopar H
Suppliers		EXXON
Molecular mass		
Boiling point/range	°C @ 760 mm Hg	179–192
Melting point	°C	
Density	kg/m^3 @ 20 °C	
Evaporation time[a]		75
Evaporation time[b]		0.09
Vapour pressure	kPa @ 20 °C	
Flash point	°C	58
Auto ignition temperature	°C	
Upper flammability limit	%v/v in air	
Lower flammability limit	%v/v in air	
Viscosity	mPa.s (cP) @ 20 °C	
Refractive index	n 20 D	1.425
Solubility in water	%w/w @ 20 °C	
Solubility of water	%w/w @ 20 °C	
Solubility parameter[c]	$(J/cm^3)^{1/2}$	
Hydrogen bond index		
Fractional polarity		
Nitro cellulose dil ratio[d]		
Nitro cellulose dil ratio[e]		
Surface tension	mN/m @ 20 °C	
Specific heat liquid	kJ/kg mole/°C @ 20 °C	
Latent heat	kJ/kg mole	
Dielectric constant	@ 20 °C	
Antoine constant A	kPa°C	
Antoine constant B	kPa°C	
Antoine constant C	kPa°C	
Heat of combustion	kJ/kg mole	
UN number		1993
IMO classification		3.3
ADR/RID classification		3,32°(c)
UK exposure limits	mg/m^3 (ppm) 8hrTWA/MEL or OES	
USA exposure limits	mg/m^3 (ppm) TLV/TWA	
German exposure limits	mg/m^3 (ppm) MAK-TWA	
EU classification		
EU risk phrases		
EU safety phrases		

[a] Ether = 1. [b] n-Butyl acetate = 1. [c] Nelson, Hemwall, Edwards. [d] Toluene. [e] Hydrocarbon solvent.

Solvent group		HYDROCARBONS—ALIPHATIC
Chemical name		
Chemical formula		
Structural formula		
CAS No.		
Synonym/common name		
Trade names		Isopar J
Suppliers		EXXON
Molecular mass		
Boiling point/range	°C @ 760 mm Hg	182–208
Melting point	°C	
Density	kg/m^3 @ 20 °C	
Evaporation time[a]		100
Evaporation time[b]		0.06
Vapour pressure	kPa @ 20 °C	
Flash point	°C	61
Auto ignition temperature	°C	
Upper flammability limit	%v/v in air	
Lower flammability limit	%v/v in air	
Viscosity	mPa.s (cP) @ 20 °C	
Refractive index	n 20 D	1.428
Solubility in water	%w/w @ 20 °C	
Solubility of water	%w/w @ 20 °C	
Solubility parameter[c]	(J/cm^3)$^{1/2}$	
Hydrogen bond index		
Fractional polarity		
Nitro cellulose dil ratio[d]		
Nitro cellulose dil ratio[e]		
Surface tension	mN/m @ 20 °C	
Specific heat liquid	kJ/kg mole/°C @ 20 °C	
Latent heat	kJ/kg mole	
Dielectric constant	@ 20 °C	
Antoine constant A	kPa°C	
Antoine constant B	kPa°C	
Antoine constant C	kPa°C	
Heat of combustion	kJ/kg mole	
UN number		1993
IMO classification		3.3
ADR/RID classification		3,32°(c)
UK exposure limits	mg/m^3 (ppm) 8hrTWA/MEL or OES	
USA exposure limits	mg/m^3 (ppm) TLV/TWA	
German exposure limits	mg/m^3 (ppm) MAK-TWA	
EU classification		
EU risk phrases		
EU safety phrases		

[a] Ether = 1. [b] n-Butyl acetate = 1. [c] Nelson, Hemwall, Edwards. [d] Toluene. [e] Hydrocarbon solvent.

Solvent group		HYDROCARBONS—ALIPHATIC
Chemical name		
Chemical formula		
Structural formula		
CAS No.		
Synonym/common name		
Trade names		Exxol DSP 100/160
Suppliers		EXXON
Molecular mass		
Boiling point/range	°C @ 760 mm Hg	103–160
Melting point	°C	
Density	kg/m^3 @ 20 °C	
Evaporation time[a]		9.0
Evaporation time[b]		1.30
Vapour pressure	kPa @ 20 °C	
Flash point	°C	<0
Auto ignition temperature	°C	
Upper flammability limit	%v/v in air	
Lower flammability limit	%v/v in air	
Viscosity	mPa.s (cP) @ 20 °C	
Refractive index	n 20 D	1.411
Solubility in water	%w/w @ 20 °C	
Solubility of water	%w/w @ 20 °C	
Solubility parameter[c]	$(J/cm^3)^{1/2}$	
Hydrogen bond index		
Fractional polarity		
Nitro cellulose dil ratio[d]		
Nitro cellulose dil ratio[e]		
Surface tension	mN/m @ 20 °C	
Specific heat liquid	kJ/kg mole/°C @ 20 °C	
Latent heat	kJ/kg mole	
Dielectric constant	@ 20 °C	
Antoine constant A	kPa°C	
Antoine constant B	kPa°C	
Antoine constant C	kPa°C	
Heat of combustion	kJ/kg mole	
UN number		1271
IMO classification		3.2
ADR/RID classification		3,3°(b)
UK exposure limits	mg/m^3 (ppm) 8hrTWA/MEL or OES	
USA exposure limits	mg/m^3 (ppm) TLV/TWA	
German exposure limits	mg/m^3 (ppm) MAK-TWA	
EU classification		F
EU risk phrases		10
EU safety phrases		9-16-29-33-43a

[a] Ether = 1. [b] n-Butyl acetate = 1. [c] Nelson, Hemwall, Edwards. [d] Toluene. [e] Hydrocarbon solvent.

Solvent group		HYDROCARBONS—ALIPHATIC
Chemical name		
Chemical formula		
Structural formula		
CAS No.		64742-48-9
Synonym/common name		
Trade names		SPIRDANE BT
Suppliers		TOTAL
Molecular mass		
Boiling point/range	°C @ 760 mm Hg	155–200
Melting point	°C	
Density	kg/m^3 @ 20 °C	770
Evaporation time[a]		65
Evaporation time[b]		
Vapour pressure	kPa @ 20 °C	
Flash point	°C	38
Auto ignition temperature	°C	
Upper flammability limit	%v/v in air	
Lower flammability limit	%v/v in air	
Viscosity	mPa.s (cP) @ 20 °C	
Refractive index	n 20 D	1.428
Solubility in water	%w/w @ 20 °C	
Solubility of water	%w/w @ 20 °C	
Solubility parameter[c]	$(J/cm^3)^{1/2}$	
Hydrogen bond index		
Fractional polarity		
Nitro cellulose dil ratio[d]		
Nitro cellulose dil ratio[e]		
Surface tension	mN/m @ 20 °C	
Specific heat liquid	kJ/kg mole/°C @ 20 °C	
Latent heat	kJ/kg mole	
Dielectric constant	@ 20 °C	
Antoine constant A	kPa°C	
Antoine constant B	kPa°C	
Antoine constant C	kPa°C	
Heat of combustion	kJ/kg mole	
UN number		1300
IMO classification		3.3
ADR/RID classification		3,31°(c)
UK exposure limits	mg/m^3 (ppm) 8hrTWA/MEL or OES	
USA exposure limits	mg/m^3 (ppm) TLV/TWA	
German exposure limits	mg/m^3 (ppm) MAK-TWA	
EU classification		
EU risk phrases		10
EU safety phrases		

[a] Ether = 1. [b] n-Butyl acetate = 1. [c] Nelson, Hemwall, Edwards. [d] Toluene. [e] Hydrocarbon solvent.

Solvent group		HYDROCARBONS—ALIPHATIC
Chemical name		
Chemical formula		
Structural formula		
CAS No.		
Synonym/common name		
Trade names		Exxol D 30
Suppliers		EXXON
Molecular mass		
Boiling point/range	°C @ 760 mm Hg	141–164
Melting point	°C	
Density	kg/m^3 @ 20 °C	
Evaporation time[a]		27
Evaporation time[b]		0.44
Vapour pressure	kPa @ 20 °C	
Flash point	°C	28
Auto ignition temperature	°C	
Upper flammability limit	%v/v in air	
Lower flammability limit	%v/v in air	
Viscosity	mPa.s (cP) @ 20 °C	
Refractive index	n 20 D	1.419
Solubility in water	%w/w @ 20 °C	
Solubility of water	%w/w @ 20 °C	
Solubility parameter[c]	$(J/cm^3)^{1/2}$	
Hydrogen bond index		
Fractional polarity		
Nitro cellulose dil ratio[d]		
Nitro cellulose dil ratio[e]		
Surface tension	mN/m @ 20 °C	
Specific heat liquid	kJ/kg mole/°C @ 20 °C	
Latent heat	kJ/kg mole	
Dielectric constant	@ 20 °C	
Antoine constant A	kPa°C	
Antoine constant B	kPa°C	
Antoine constant C	kPa°C	
Heat of combustion	kJ/kg mole	
UN number		1270
IMO classification		3.3
ADR/RID classification		3,31°(c)
UK exposure limits	mg/m^3 (ppm) 8hrTWA/MEL or OES	
USA exposure limits	mg/m^3 (ppm) TLV/TWA	
German exposure limits	mg/m^3 (ppm) MAK-TWA	
EU classification		
EU risk phrases		10
EU safety phrases		43a

[a] Ether = 1. [b] n-Butyl acetate = 1. [c] Nelson, Hemwall, Edwards. [d] Toluene. [e] Hydrocarbon solvent.

Solvent group		HYDROCARBONS—ALIPHATIC
Chemical name		Cyclopentane
Chemical formula		C_5H_{10}
Structural formula		
CAS No.		287-92-37
Synonym/common name		Cyclopentane
Trade names		EXXOL Cyclopentane
Suppliers		EXXON
Molecular mass		70.13
Boiling point/range	°C @ 760 mm Hg	49–56
Melting point	°C	−94
Density	kg/m^3 @ 20 °C	
Evaporation time[a]		
Evaporation time[b]		
Vapour pressure	kPa @ 20 °C	35.69
Flash point	°C	<0
Auto ignition temperature	°C	
Upper flammability limit	%v/v in air	
Lower flammability limit	%v/v in air	
Viscosity	mPa.s (cP) @ 20 °C	
Refractive index	n 20 D	1.40
Solubility in water	%w/w @ 20 °C	
Solubility of water	%w/w @ 20 °C	
Solubility parameter[c]	(J/cm^3)$^{1/2}$	16.57
Hydrogen bond index		
Fractional polarity		
Nitro cellulose dil ratio[d]		
Nitro cellulose dil ratio[e]		
Surface tension	mN/m @ 20 °C	
Specific heat liquid	kJ/kg mole/°C @ 20 °C	
Latent heat	kJ/kg mole	27296
Dielectric constant	@ 20 °C	1.968
Antoine constant A	kPa°C	6.04584
Antoine constant B	kPa°C	1142.3
Antoine constant C	kPa°C	233.463
Heat of combustion	kJ/kg mole	−3290930
UN number		1146
IMO classification		3.1/II
ADR/RID classification		3,3°(b)
UK exposure limits	mg/m^3 (ppm) 8hrTWA/MEL or OES	
USA exposure limits	mg/m^3 (ppm) TLV/TWA	1720(600)
German exposure limits	mg/m^3 (ppm) MAK-TWA	
EU classification		F
EU risk phrases		11
EU safety phrases		(2-)9-16-29-33

[a] Ether = 1. [b] n-Butyl acetate = 1. [c] Nelson, Hemwall, Edwards. [d] Toluene. [e] Hydrocarbon solvent.

Solvent group		**HYDROCARBONS—ALIPHATIC**
Chemical name		2-Methylbutane
Chemical formula		C_5H_{12}
Structural formula		$CH_3CH(CH_3)CH_2CH_3$
CAS No.		78-78-4
Synonym/common name		Isopentane
Trade names		Exxol Isopentane
Suppliers		SHELL, EXXON
Molecular mass		72
Boiling point/range	°C @ 760 mm Hg	24–35
Melting point	°C	
Density	kg/m³ @ 20 °C	625
Evaporation time[a]		<1
Evaporation time[b]		13
Vapour pressure	kPa @ 20 °C	80
Flash point	°C	−57 (IP 170) SHELL
Auto ignition temperature	°C	420
Upper flammability limit	%v/v in air	8.0
Lower flammability limit	%v/v in air	0.8
Viscosity	mPa.s (cP) @ 20 °C	0.25
Refractive index	n 20 D	1.354
Solubility in water	%w/w @ 20 °C	
Solubility of water	%w/w @ 20 °C	
Solubility parameter[c]	$(J/cm^3)^{1/2}$	13.9
Hydrogen bond index		0
Fractional polarity		0
Nitro cellulose dil ratio[d]		
Nitro cellulose dil ratio[e]		
Surface tension	mN/m @ 20 °C	
Specific heat liquid	kJ/kg mole/°C @ 20 °C	158.4
Latent heat	kJ/kg mole	24970
Dielectric constant	@ 20 °C	1.8
Antoine constant A	kPa°C	5.92023
Antoine constant B	kPa°C	1002.88
Antoine constant C	kPa°C	233.460
Heat of combustion	kJ/kg mole	−3497040
UN number		1265
IMO classification		3.1/I
ADR/RID classification		3,1°(a), 2°(b)
UK exposure limits	mg/m³ (ppm) 8hrTWA/MEL or OES	
USA exposure limits	mg/m³ (ppm) TLV/TWA	
German exposure limits	mg/m³ (ppm) MAK-TWA	2950(1000)
EU classification		F
EU risk phrases		11
EU safety phrases		(2-)9-16-29-33

[a] Ether = 1. [b] n-Butyl acetate = 1. [c] Nelson, Hemwall, Edwards. [d] Toluene. [e] Hydrocarbon solvent.

Solvent group		**HYDROCARBONS—ALIPHATIC**
Chemical name		n-Pentane
Chemical formula		C_5H_{12}
Structural formula		$CH_3CH_2CH_2CH_3$
CAS No.		109-66-0
Synonym/common name		n-pentane
Trade names		HYDROSOL Pentane
Suppliers		SHELL, TOTAL
Molecular mass		72
Boiling point/range	°C @ 760 mm Hg	35–37
Melting point	°C	−129.7
Density	kg/m^3 @ 20 °C	632
Evaporation time[a]		<1.0
Evaporation time[b]		12
Vapour pressure	kPa @ 20 °C	65
Flash point	°C	−50 (IP 170) SHELL
Auto ignition temperature	°C	285
Upper flammability limit	%v/v in air	8.0
Lower flammability limit	%v/v in air	0.8
Viscosity	mPa.s (cP) @ 20 °C	0.28
Refractive index	n 20 D	1.358
Solubility in water	%w/w @ 20 °C	
Solubility of water	%w/w @ 20 °C	
Solubility parameter[c]	$(J/cm^3)^{1/2}$	14.3
Hydrogen bond index		0
Fractional polarity		0
Nitro cellulose dil ratio[d]		
Nitro cellulose dil ratio[e]		
Surface tension	mN/m @ 20 °C	
Specific heat liquid	kJ/kg mole/°C @ 20 °C	172.8
Latent heat	kJ/kg mole	25560
Dielectric constant	@ 20 °C	1.8
Antoine constant A	kPa°C	6.04243
Antoine constant B	kPa°C	1094.56
Antoine constant C	kPa°C	235.108
Heat of combustion	kJ/kg mole	−3501720
UN number		1265
IMO classification		3.1/II
ADR/RID classification		3,1°(a)
UK exposure limits	mg/m^3 (ppm) 8hrTWA/MEL or OES	
USA exposure limits	mg/m^3 (ppm) TLV/TWA	
German exposure limits	mg/m^3 (ppm) MAK-TWA	
EU classification		F
EU risk phrases		11
EU safety phrases		(2-)9-16-29-33

[a] Ether = 1. [b] n-Butyl acetate = 1. [c] Nelson, Hemwall, Edwards. [d] Toluene. [e] Hydrocarbon solvent.

Solvent group		HYDROCARBONS—ALIPHATIC
Chemical name		Isohexane
Chemical formula		C_6H_{14}
Structural formula		
CAS No.		64742-49-0
Synonym/common name		Isohexane
Trade names		HYDROSOL Hexane
Suppliers		TOTAL
Molecular mass		86
Boiling point/range	°C @ 760 mm Hg	50–62
Melting point	°C	
Density	kg/m^3 @ 20 °C	660
Evaporation time[a]		1.3
Evaporation time[b]		
Vapour pressure	kPa @ 20 °C	
Flash point	°C	<0
Auto ignition temperature	°C	
Upper flammability limit	%v/v in air	
Lower flammability limit	%v/v in air	
Viscosity	mPa.s (cP) @ 20 °C	
Refractive index	n 20 D	1.375
Solubility in water	%w/w @ 20 °C	
Solubility of water	%w/w @ 20 °C	
Solubility parameter[c]	(J/cm^3)$^{1/2}$	
Hydrogen bond index		
Fractional polarity		
Nitro cellulose dil ratio[d]		
Nitro cellulose dil ratio[e]		
Surface tension	mN/m @ 20 °C	
Specific heat liquid	kJ/kg mole/°C @ 20 °C	
Latent heat	kJ/kg mole	
Dielectric constant	@ 20 °C	
Antoine constant A	kPa°C	
Antoine constant B	kPa°C	
Antoine constant C	kPa°C	
Heat of combustion	kJ/kg mole	
UN number		1208
IMO classification		3.1
ADR/RID classification		3,3°(b)
UK exposure limits	mg/m^3 (ppm) 8hrTWA/MEL or OES	
USA exposure limits	mg/m^3 (ppm) TLV/TWA	
German exposure limits	mg/m^3 (ppm) MAK-TWA	
EU classification		F
EU risk phrases		11
EU safety phrases		9-16-23-29-33

[a] Ether = 1. [b] n-Butyl acetate = 1. [c] Nelson, Hemwall, Edwards. [d] Toluene. [e] Hydrocarbon solvent.

Solvent group		**HYDROCARBONS—ALIPHATIC**
Chemical name		Hexane (branched and linear)
Chemical formula		C_6H_{14}
Structural formula		
CAS No.		92112-69-1
Synonym/common name		Hexane
Trade names		
Suppliers		PETROGAL, SA
Molecular mass		86
Boiling point/range	°C @ 760 mm Hg	64–70
Melting point	°C	< -95
Density	kg/m^3 @ 20 °C	665–680
Evaporation time[a]		
Evaporation time[b]		
Vapour pressure	kPa @ 20 °C	
Flash point	°C	< -20 (IP 170) PETROGAL
Auto ignition temperature	°C	> 240
Upper flammability limit	%v/v in air	7.5
Lower flammability limit	%v/v in air	0.8
Viscosity	mPa.s (cP) @ 20 °C	
Refractive index	n 20 D	
Solubility in water	%w/w @ 20 °C	
Solubility of water	%w/w @ 20 °C	
Solubility parameter[c]	(J/cm^3)$^{1/2}$	
Hydrogen bond index		
Fractional polarity		
Nitro cellulose dil ratio[d]		
Nitro cellulose dil ratio[e]		
Surface tension	mN/m @ 20 °C	
Specific heat liquid	kJ/kg mole/°C @ 20 °C	
Latent heat	kJ/kg mole	
Dielectric constant	@ 20 °C	
Antoine constant A	kPa°C	
Antoine constant B	kPa°C	
Antoine constant C	kPa°C	
Heat of combustion	kJ/kg mole	
UN number		
IMO classification		
ADR/RID classification		
UK exposure limits	mg/m^3 (ppm) 8hrTWA/MEL or OES	
USA exposure limits	mg/m^3 (ppm) TLV/TWA	
German exposure limits	mg/m^3 (ppm) MAK-TWA	
EU classification		
EU risk phrases		
EU safety phrases		

[a] Ether = 1. [b] n-Butyl acetate = 1. [c] Nelson, Hemwall, Edwards. [d] Toluene. [e] Hydrocarbon solvent.

Solvent group		HYDROCARBONS—ALIPHATIC
Chemical name		Hexane
Chemical formula		C_6H_{14}
Structural formula		$CH_3(CH_2)_4CH_3$
CAS No.		110-54-3
Synonym/common name		Hexane/n-hexane
Trade names		Exxol Hexane, HYDROSOL Hexane
Suppliers		EXXON, TOTAL
Molecular mass		86.18
Boiling point/range	°C @ 760 mm Hg	64–70
Melting point	°C	−95.3
Density	kg/m^3 @ 20 °C	670
Evaporation time[a]		1.5
Evaporation time[b]		14
Vapour pressure	kPa @ 20 °C	
Flash point	°C	<0
Auto ignition temperature	°C	
Upper flammability limit	%v/v in air	
Lower flammability limit	%v/v in air	
Viscosity	mPa.s (cP) @ 20 °C	
Refractive index	n 20 D	1.377
Solubility in water	%w/w @ 20 °C	
Solubility of water	%w/w @ 20 °C	
Solubility parameter[c]	$(J/cm^3)^{1/2}$	14.87
Hydrogen bond index		
Fractional polarity		
Nitro cellulose dil ratio[d]		
Nitro cellulose dil ratio[e]		
Surface tension	mN/m @ 20 °C	
Specific heat liquid	kJ/kg mole/°C @ 20 °C	
Latent heat	kJ/kg mole	28853
Dielectric constant	@ 20 °C	1.886
Antoine constant A	kPa°C	6.0091
Antoine constant B	kPa°C	1171.17
Antoine constant C	kPa°C	224.408
Heat of combustion	kJ/kg mole	−4163330
UN number		1208
IMO classification		3.1/II
ADR/RID classification		3,3°(b)
UK exposure limits	mg/m^3 (ppm) 8hrTWA/MEL or OES	70(20)
USA exposure limits	mg/m^3 (ppm) TLV/TWA	176(50)
German exposure limits	mg/m^3 (ppm) MAK-TWA	180(50)
EU classification		F, Xn
EU risk phrases		11-48/20
EU safety phrases		(2-)9-16-24/25-29-51

[a] Ether = 1. [b] n-Butyl acetate = 1. [c] Nelson, Hemwall, Edwards. [d] Toluene. [e] Hydrocarbon solvent.

Solvent group		HYDROCARBONS—ALIPHATIC
Chemical name		
Chemical formula		
Structural formula		
CAS No.		
Synonym/common name		
Trade names		SHELL SOL B HT
Suppliers		SHELL
Molecular mass		88
Boiling point/range	°C @ 760 mm Hg	61–77
Melting point	°C	
Density	kg/m^3 @ 20 °C	
Evaporation time[a]		
Evaporation time[b]		9.4
Vapour pressure	kPa @ 20 °C	
Flash point	°C	
Auto ignition temperature	°C	
Upper flammability limit	%v/v in air	
Lower flammability limit	%v/v in air	
Viscosity	mPa.s (cP) @ 20 °C	
Refractive index	n 20 D	
Solubility in water	%w/w @ 20 °C	
Solubility of water	%w/w @ 20 °C	
Solubility parameter[c]	$(J/cm^3)^{1/2}$	
Hydrogen bond index		
Fractional polarity		
Nitro cellulose dil ratio[d]		
Nitro cellulose dil ratio[e]		
Surface tension	mN/m @ 20 °C	
Specific heat liquid	kJ/kg mole/°C @ 20 °C	
Latent heat	kJ/kg mole	
Dielectric constant	@ 20 °C	
Antoine constant A	kPa°C	
Antoine constant B	kPa°C	
Antoine constant C	kPa°C	
Heat of combustion	kJ/kg mole	
UN number		
IMO classification		
ADR/RID classification		
UK exposure limits	mg/m^3 (ppm) 8hrTWA/MEL or OES	
USA exposure limits	mg/m^3 (ppm) TLV/TWA	
German exposure limits	mg/m^3 (ppm) MAK-TWA	
EU classification		
EU risk phrases		
EU safety phrases		

[a] Ether = 1. [b] n-Butyl acetate = 1. [c] Nelson, Hemwall, Edwards. [d] Toluene. [e] Hydrocarbon solvent.

Solvent group		**HYDROCARBONS—ALIPHATIC**
Chemical name		
Chemical formula		
Structural formula		
CAS No.		
Synonym/common name		
Trade names		SHELL RUBBER SOLVENT
Suppliers		SHELL
Molecular mass		90
Boiling point/range	°C @ 760 mm Hg	64–114
Melting point	°C	
Density	kg/m^3 @ 20 °C	
Evaporation time[a]		
Evaporation time[b]		5.3
Vapour pressure	kPa @ 20 °C	
Flash point	°C	
Auto ignition temperature	°C	
Upper flammability limit	%v/v in air	
Lower flammability limit	%v/v in air	
Viscosity	mPa.s (cP) @ 20 °C	
Refractive index	n 20 D	
Solubility in water	%w/w @ 20 °C	
Solubility of water	%w/w @ 20 °C	
Solubility parameter[c]	$(J/cm^3)^{1/2}$	
Hydrogen bond index		
Fractional polarity		
Nitro cellulose dil ratio[d]		
Nitro cellulose dil ratio[e]		
Surface tension	mN/m @ 20 °C	
Specific heat liquid	kJ/kg mole/°C @ 20 °C	
Latent heat	kJ/kg mole	
Dielectric constant	@ 20 °C	
Antoine constant A	kPa°C	
Antoine constant B	kPa°C	
Antoine constant C	kPa°C	
Heat of combustion	kJ/kg mole	
UN number		
IMO classification		
ADR/RID classification		
UK exposure limits	mg/m^3 (ppm) 8hrTWA/MEL or OES	
USA exposure limits	mg/m^3 (ppm) TLV/TWA	
German exposure limits	mg/m^3 (ppm) MAK-TWA	
EU classification		
EU risk phrases		
EU safety phrases		

[a] Ether = 1. [b] n-Butyl acetate = 1. [c] Nelson, Hemwall, Edwards. [d] Toluene. [e] Hydrocarbon solvent.

Solvent group		HYDROCARBONS—ALIPHATIC
Chemical name		
Chemical formula		
Structural formula		
CAS No.		
Synonym/common name		
Trade names		SHELL TOLU-SOL 19 EC SOLVENT
Suppliers		SHELL
Molecular mass		98
Boiling point/range	°C @ 760 mm Hg	91–104
Melting point	°C	
Density	kg/m^3 @ 20 °C	
Evaporation time[a]		
Evaporation time[b]		4.3
Vapour pressure	kPa @ 20 °C	
Flash point	°C	
Auto ignition temperature	°C	
Upper flammability limit	%v/v in air	
Lower flammability limit	%v/v in air	
Viscosity	mPa.s (cP) @ 20 °C	
Refractive index	n 20 D	
Solubility in water	%w/w @ 20 °C	
Solubility of water	%w/w @ 20 °C	
Solubility parameter[c]	(J/cm^3)$^{1/2}$	
Hydrogen bond index		
Fractional polarity		
Nitro cellulose dil ratio[d]		
Nitro cellulose dil ratio[e]		
Surface tension	mN/m @ 20 °C	
Specific heat liquid	kJ/kg mole/°C @ 20 °C	
Latent heat	kJ/kg mole	
Dielectric constant	@ 20 °C	
Antoine constant A	kPa°C	
Antoine constant B	kPa°C	
Antoine constant C	kPa°C	
Heat of combustion	kJ/kg mole	
UN number		
IMO classification		
ADR/RID classification		
UK exposure limits	mg/m^3 (ppm) 8hrTWA/MEL or OES	
USA exposure limits	mg/m^3 (ppm) TLV/TWA	
German exposure limits	mg/m^3 (ppm) MAK-TWA	
EU classification		
EU risk phrases		
EU safety phrases		

[a] Ether = 1. [b] n-Butyl acetate = 1. [c] Nelson, Hemwall, Edwards. [d] Toluene. [e] Hydrocarbon solvent.

Solvent group		HYDROCARBONS—ALIPHATIC
Chemical name		
Chemical formula		
Structural formula		
CAS No.		
Synonym/common name		
Trade names		SHELL TOLU-SOL 25 SOLVENT
Suppliers		SHELL
Molecular mass		98
Boiling point/range	°C @ 760 mm Hg	92–107
Melting point	°C	
Density	kg/m^3 @ 20 °C	
Evaporation time[a]		
Evaporation time[b]		4.1
Vapour pressure	kPa @ 20 °C	
Flash point	°C	
Auto ignition temperature	°C	
Upper flammability limit	%v/v in air	
Lower flammability limit	%v/v in air	
Viscosity	mPa.s (cP) @ 20 °C	
Refractive index	n 20 D	
Solubility in water	%w/w @ 20 °C	
Solubility of water	%w/w @ 20 °C	
Solubility parameter[c]	(J/cm^3)$^{1/2}$	
Hydrogen bond index		
Fractional polarity		
Nitro cellulose dil ratio[d]		
Nitro cellulose dil ratio[e]		
Surface tension	mN/m @ 20 °C	
Specific heat liquid	kJ/kg mole/°C @ 20 °C	
Latent heat	kJ/kg mole	
Dielectric constant	@ 20 °C	
Antoine constant A	kPa°C	
Antoine constant B	kPa°C	
Antoine constant C	kPa°C	
Heat of combustion	kJ/kg mole	
UN number		
IMO classification		
ADR/RID classification		
UK exposure limits	mg/m^3 (ppm) 8hrTWA/MEL or OES	
USA exposure limits	mg/m^3 (ppm) TLV/TWA	
German exposure limits	mg/m^3 (ppm) MAK-TWA	
EU classification		
EU risk phrases		
EU safety phrases		

[a] Ether = 1. [b] n-Butyl acetate = 1. [c] Nelson, Hemwall, Edwards. [d] Toluene. [e] Hydrocarbon solvent.

Solvent group		**HYDROCARBONS—ALIPHATIC**
Chemical name		
Chemical formula		
Structural formula		
CAS No.		
Synonym/common name		
Trade names		SHELL TOLU-SOL 10 SOLVENT
Suppliers		SHELL
Molecular mass		99
Boiling point/range	°C @ 760 mm Hg	91–105
Melting point	°C	
Density	kg/m^3 @ 20 °C	
Evaporation time[a]		
Evaporation time[b]		4.5
Vapour pressure	kPa @ 20 °C	
Flash point	°C	
Auto ignition temperature	°C	
Upper flammability limit	%v/v in air	
Lower flammability limit	%v/v in air	
Viscosity	mPa.s (cP) @ 20 °C	
Refractive index	n 20 D	
Solubility in water	%w/w @ 20 °C	
Solubility of water	%w/w @ 20 °C	
Solubility parameter[c]	(J/cm^3)$^{1/2}$	
Hydrogen bond index		
Fractional polarity		
Nitro cellulose dil ratio[d]		
Nitro cellulose dil ratio[e]		
Surface tension	mN/m @ 20 °C	
Specific heat liquid	kJ/kg mole/°C @ 20 °C	
Latent heat	kJ/kg mole	
Dielectric constant	@ 20 °C	
Antoine constant A	kPa°C	
Antoine constant B	kPa°C	
Antoine constant C	kPa°C	
Heat of combustion	kJ/kg mole	
UN number		
IMO classification		
ADR/RID classification		
UK exposure limits	mg/m^3 (ppm) 8hrTWA/MEL or OES	
USA exposure limits	mg/m^3 (ppm) TLV/TWA	
German exposure limits	mg/m^3 (ppm) MAK-TWA	
EU classification		
EU risk phrases		
EU safety phrases		

[a] Ether = 1. [b] n-Butyl acetate = 1. [c] Nelson, Hemwall, Edwards. [d] Toluene. [e] Hydrocarbon solvent.

Solvent group		HYDROCARBONS—ALIPHATIC
Chemical name		
Chemical formula		
Structural formula		
CAS No.		
Synonym/common name		
Trade names		SHELL TOLU-SOL W HT SOLVENT
Suppliers		SHELL
Molecular mass		100
Boiling point/range	°C @ 760 mm Hg	98–110
Melting point	°C	
Density	kg/m^3 @ 20 °C	
Evaporation time[a]		
Evaporation time[b]		3.8
Vapour pressure	kPa @ 20 °C	
Flash point	°C	
Auto ignition temperature	°C	
Upper flammability limit	%v/v in air	
Lower flammability limit	%v/v in air	
Viscosity	mPa.s (cP) @ 20 °C	
Refractive index	n 20 D	
Solubility in water	%w/w @ 20 °C	
Solubility of water	%w/w @ 20 °C	
Solubility parameter[c]	(J/cm^3)$^{1/2}$	
Hydrogen bond index		
Fractional polarity		
Nitro cellulose dil ratio[d]		
Nitro cellulose dil ratio[e]		
Surface tension	mN/m @ 20 °C	
Specific heat liquid	kJ/kg mole/°C @ 20 °C	
Latent heat	kJ/kg mole	
Dielectric constant	@ 20 °C	
Antoine constant A	kPa°C	
Antoine constant B	kPa°C	
Antoine constant C	kPa°C	
Heat of combustion	kJ/kg mole	
UN number		
IMO classification		
ADR/RID classification		
UK exposure limits	mg/m^3 (ppm) 8hrTWA/MEL or OES	
USA exposure limits	mg/m^3 (ppm) TLV/TWA	
German exposure limits	mg/m^3 (ppm) MAK-TWA	
EU classification		
EU risk phrases		
EU safety phrases		

[a] Ether = 1. [b] n-Butyl acetate = 1. [c] Nelson, Hemwall, Edwards. [d] Toluene. [e] Hydrocarbon solvent.

Solvent group		HYDROCARBONS—ALIPHATIC
Chemical name		
Chemical formula		
Structural formula		
CAS No.		
Synonym/common name		
Trade names		SHELL TOLU-SOL 6 SOLVENT
Suppliers		SHELL
Molecular mass		100
Boiling point/range	°C @ 760 mm Hg	99–108
Melting point	°C	
Density	kg/m^3 @ 20 °C	
Evaporation time[a]		
Evaporation time[b]		3.7
Vapour pressure	kPa @ 20 °C	
Flash point	°C	
Auto ignition temperature	°C	
Upper flammability limit	%v/v in air	
Lower flammability limit	%v/v in air	
Viscosity	mPa.s (cP) @ 20 °C	
Refractive index	n 20 D	
Solubility in water	%w/w @ 20 °C	
Solubility of water	%w/w @ 20 °C	
Solubility parameter[c]	$(J/cm^3)^{1/2}$	
Hydrogen bond index		
Fractional polarity		
Nitro cellulose dil ratio[d]		
Nitro cellulose dil ratio[e]		
Surface tension	mN/m @ 20 °C	
Specific heat liquid	kJ/kg mole/°C @ 20 °C	
Latent heat	kJ/kg mole	
Dielectric constant	@ 20 °C	
Antoine constant A	kPa°C	
Antoine constant B	kPa°C	
Antoine constant C	kPa°C	
Heat of combustion	kJ/kg mole	
UN number		
IMO classification		
ADR/RID classification		
UK exposure limits	mg/m^3 (ppm) 8hrTWA/MEL or OES	
USA exposure limits	mg/m^3 (ppm) TLV/TWA	
German exposure limits	mg/m^3 (ppm) MAK-TWA	
EU classification		
EU risk phrases		
EU safety phrases		

[a] Ether = 1. [b] n-Butyl acetate = 1. [c] Nelson, Hemwall, Edwards. [d] Toluene. [e] Hydrocarbon solvent.

Solvent group		HYDROCARBONS—ALIPHATIC
Chemical name		
Chemical formula		
Structural formula		
CAS No.		
Synonym/common name		
Trade names		SHELL TOLU-SOL 3 SOLVENT
Suppliers		SHELL
Molecular mass		100
Boiling point/range	°C @ 760 mm Hg	91–97
Melting point	°C	
Density	kg/m^3 @ 20 °C	
Evaporation time[a]		
Evaporation time[b]		4.7
Vapour pressure	kPa @ 20 °C	
Flash point	°C	
Auto ignition temperature	°C	
Upper flammability limit	%v/v in air	
Lower flammability limit	%v/v in air	
Viscosity	mPa.s (cP) @ 20 °C	
Refractive index	n 20 D	
Solubility in water	%w/w @ 20 °C	
Solubility of water	%w/w @ 20 °C	
Solubility parameter[c]	$(J/cm^3)^{1/2}$	
Hydrogen bond index		
Fractional polarity		
Nitro cellulose dil ratio[d]		
Nitro cellulose dil ratio[e]		
Surface tension	mN/m @ 20 °C	
Specific heat liquid	kJ/kg mole/°C @ 20 °C	
Latent heat	kJ/kg mole	
Dielectric constant	@ 20 °C	
Antoine constant A	kPa°C	
Antoine constant B	kPa°C	
Antoine constant C	kPa°C	
Heat of combustion	kJ/kg mole	
UN number		
IMO classification		
ADR/RID classification		
UK exposure limits	mg/m^3 (ppm) 8hrTWA/MEL or OES	
USA exposure limits	mg/m^3 (ppm) TLV/TWA	
German exposure limits	mg/m^3 (ppm) MAK-TWA	
EU classification		
EU risk phrases		
EU safety phrases		

[a] Ether = 1. [b] n-Butyl acetate = 1. [c] Nelson, Hemwall, Edwards. [d] Toluene. [e] Hydrocarbon solvent.

Solvent group		HYDROCARBONS—ALIPHATIC
Chemical name		
Chemical formula		
Structural formula		
CAS No.		
Synonym/common name		
Trade names		SHELL TOLU-SOL A HT SOLVENT
Suppliers		SHELL
Molecular mass		100
Boiling point/range	°C @ 760 mm Hg	90–97
Melting point	°C	
Density	kg/m^3 @ 20 °C	
Evaporation time[a]		
Evaporation time[b]		4.8
Vapour pressure	kPa @ 20 °C	
Flash point	°C	
Auto ignition temperature	°C	
Upper flammability limit	%v/v in air	
Lower flammability limit	%v/v in air	
Viscosity	mPa.s (cP) @ 20 °C	
Refractive index	n 20 D	
Solubility in water	%w/w @ 20 °C	
Solubility of water	%w/w @ 20 °C	
Solubility parameter[c]	$(J/cm^3)^{1/2}$	
Hydrogen bond index		
Fractional polarity		
Nitro cellulose dil ratio[d]		
Nitro cellulose dil ratio[e]		
Surface tension	mN/m @ 20 °C	
Specific heat liquid	kJ/kg mole/°C @ 20 °C	
Latent heat	kJ/kg mole	
Dielectric constant	@ 20 °C	
Antoine constant A	kPa°C	
Antoine constant B	kPa°C	
Antoine constant C	kPa°C	
Heat of combustion	kJ/kg mole	
UN number		
IMO classification		
ADR/RID classification		
UK exposure limits	mg/m^3 (ppm) 8hrTWA/MEL or OES	
USA exposure limits	mg/m^3 (ppm) TLV/TWA	
German exposure limits	mg/m^3 (ppm) MAK-TWA	
EU classification		
EU risk phrases		
EU safety phrases		

[a] Ether = 1. [b] n-Butyl acetate = 1. [c] Nelson, Hemwall, Edwards. [d] Toluene. [e] Hydrocarbon solvent.

Solvent group		**HYDROCARBONS—ALIPHATIC**
Chemical name		
Chemical formula		
Structural formula		
CAS No.		
Synonym/common name		
Trade names		SHELL TOLU-SOL 5 SOLVENT
Suppliers		SHELL
Molecular mass		100
Boiling point/range	°C @ 760 mm Hg	91–98
Melting point	°C	
Density	kg/m^3 @ 20 °C	
Evaporation time[a]		
Evaporation time[b]		4.7
Vapour pressure	kPa @ 20 °C	
Flash point	°C	
Auto ignition temperature	°C	
Upper flammability limit	%v/v in air	
Lower flammability limit	%v/v in air	
Viscosity	mPa.s (cP) @ 20 °C	
Refractive index	n 20 D	
Solubility in water	%w/w @ 20 °C	
Solubility of water	%w/w @ 20 °C	
Solubility parameter[c]	$(J/cm^3)^{1/2}$	
Hydrogen bond index		
Fractional polarity		
Nitro cellulose dil ratio[d]		
Nitro cellulose dil ratio[e]		
Surface tension	mN/m @ 20 °C	
Specific heat liquid	kJ/kg mole/°C @ 20 °C	
Latent heat	kJ/kg mole	
Dielectric constant	@ 20 °C	
Antoine constant A	kPa°C	
Antoine constant B	kPa°C	
Antoine constant C	kPa°C	
Heat of combustion	kJ/kg mole	
UN number		
IMO classification		
ADR/RID classification		
UK exposure limits	mg/m^3 (ppm) 8hrTWA/MEL or OES	
USA exposure limits	mg/m^3 (ppm) TLV/TWA	
German exposure limits	mg/m^3 (ppm) MAK-TWA	
EU classification		
EU risk phrases		
EU safety phrases		

[a] Ether = 1. [b] n-Butyl acetate = 1. [c] Nelson, Hemwall, Edwards. [d] Toluene. [e] Hydrocarbon solvent.

Solvent group		**HYDROCARBONS—ALIPHATIC**
Chemical name		Heptane
Chemical formula		C_7H_{16}
Structural formula		$CH_3(CH_2)_5CH_3$
CAS No.		142-82-5
Synonym/common name		Heptane/n-heptane
Trade names		Exxol Heptane, HYDROSOL Heptane
Suppliers		EXXON, TOTAL
Molecular mass		100.2
Boiling point/range	°C @ 760 mm Hg	90–99
Melting point	°C	−90.58
Density	kg/m^3 @ 20 °C	695
Evaporation time[a]		2.8–3.2
Evaporation time[b]		5.5
Vapour pressure	kPa @ 20 °C	
Flash point	°C	<0
Auto ignition temperature	°C	
Upper flammability limit	%v/v in air	
Lower flammability limit	%v/v in air	
Viscosity	mPa.s (cP) @ 20 °C	
Refractive index	n 20 D	1.390–1.397
Solubility in water	%w/w @ 20 °C	
Solubility of water	%w/w @ 20 °C	
Solubility parameter[c]	(J/cm^3)$^{1/2}$	15.20
Hydrogen bond index		
Fractional polarity		
Nitro cellulose dil ratio[d]		
Nitro cellulose dil ratio[e]		
Surface tension	mN/m @ 20 °C	
Specific heat liquid	kJ/kg mole/°C @ 20 °C	
Latent heat	kJ/kg mole	31700
Dielectric constant	@ 20 °C	1.925
Antoine constant A	kPa°C	6.02167
Antoine constant B	kPa°C	1254.90
Antoine constant C	kPa°C	216.544
Heat of combustion	kJ/kg mole	−4817000
UN number		1206
IMO classification		3.2/II
ADR/RID classification		3,3°(b)
UK exposure limits	mg/m^3 (ppm) 8hrTWA/MEL or OES	1600(400)
USA exposure limits	mg/m^3 (ppm) TLV/TWA	1640(400)
German exposure limits	mg/m^3 (ppm) MAK-TWA	2000(500)
EU classification		F
EU risk phrases		11
EU safety phrases		(2-)9-16-23-29-33

[a] Ether = 1. [b] n-Butyl acetate = 1. [c] Nelson, Hemwall, Edwards. [d] Toluene. [e] Hydrocarbon solvent.

Solvent group		HYDROCARBONS—ALIPHATIC
Chemical name		2,4,4,-Trimethylpentene-1 and 2
Chemical formula		C_8H_{16}
Structural formula		
CAS No.		91052-99-2
Synonym/common name		Codibutylene
Trade names		
Suppliers		EC Erdolchemie
Molecular mass		112.2
Boiling point/range	°C @ 760 mm Hg	107–120
Melting point	°C	−93
Density	kg/m^3 @ 20 °C	
Evaporation time[a]		5
Evaporation time[b]		
Vapour pressure	kPa @ 20 °C	
Flash point	°C	2
Auto ignition temperature	°C	300
Upper flammability limit	%v/v in air	5.3
Lower flammability limit	%v/v in air	0.6
Viscosity	mPa.s (cP) @ 20 °C	0.55
Refractive index	n 20 D	1.421
Solubility in water	%w/w @ 20 °C	
Solubility of water	%w/w @ 20 °C	
Solubility parameter[c]	(J/cm^3)$^{1/2}$	
Hydrogen bond index		
Fractional polarity		
Nitro cellulose dil ratio[d]		
Nitro cellulose dil ratio[e]		
Surface tension	mN/m @ 20 °C	23.7
Specific heat liquid	kJ/kg mole/°C @ 20 °C	
Latent heat	kJ/kg mole	
Dielectric constant	@ 20 °C	
Antoine constant A	kPa°C	
Antoine constant B	kPa°C	
Antoine constant C	kPa°C	
Heat of combustion	kJ/kg mole	
UN number		1216
IMO classification		
ADR/RID classification		3,3°(b)
UK exposure limits	mg/m^3 (ppm) 8hrTWA/MEL or OES	
USA exposure limits	mg/m^3 (ppm) TLV/TWA	
German exposure limits	mg/m^3 (ppm) MAK-TWA	
EU classification		
EU risk phrases		
EU safety phrases		

[a] Ether = 1. [b] n-Butyl acetate = 1. [c] Nelson, Hemwall, Edwards. [d] Toluene. [e] Hydrocarbon solvent.

Solvent group		**HYDROCARBONS—ALIPHATIC**
Chemical name		2,2,4-Trimethyl pentane
Chemical formula		C_8H_{18}
Structural formula		
CAS No.		540-84-1
Synonym/common name		Isooctane
Trade names		Isooctane 100
Suppliers		EC Erdolchemie
Molecular mass		114.2
Boiling point/range	°C @ 760 mm Hg	99–104
Melting point	°C	−103
Density	kg/m^3 @ 20 °C	
Evaporation time[a]		
Evaporation time[b]		
Vapour pressure	kPa @ 20 °C	
Flash point	°C	−14
Auto ignition temperature	°C	420
Upper flammability limit	%v/v in air	5.5
Lower flammability limit	%v/v in air	0.7
Viscosity	mPa.s (cP) @ 20 °C	0.50
Refractive index	n 20 D	
Solubility in water	%w/w @ 20 °C	
Solubility of water	%w/w @ 20 °C	
Solubility parameter[c]	(J/cm^3)$^{1/2}$	14.01
Hydrogen bond index		
Fractional polarity		
Nitro cellulose dil ratio[d]		
Nitro cellulose dil ratio[e]		
Surface tension	mN/m @ 20 °C	20.1
Specific heat liquid	kJ/kg mole/°C @ 20 °C	
Latent heat	kJ/kg mole	
Dielectric constant	@ 20 °C	
Antoine constant A	kPa°C	5.92885
Antoine constant B	kPa°C	1253.36
Antoine constant C	kPa°C	220.241
Heat of combustion	kJ/kg mole	
UN number		1262
IMO classification		3.2/II
ADR/RID classification		3,3°(b)
UK exposure limits	mg/m^3 (ppm) 8hrTWA/MEL or OES	
USA exposure limits	mg/m^3 (ppm) TLV/TWA	
German exposure limits	mg/m^3 (ppm) MAK-TWA	
EU classification		
EU risk phrases		
EU safety phrases		

[a] Ether = 1. [b] n-Butyl acetate = 1. [c] Nelson, Hemwall, Edwards. [d] Toluene. [e] Hydrocarbon solvent.

Solvent group		HYDROCARBONS—ALIPHATIC
Chemical name		
Chemical formula		
Structural formula		
CAS No.		
Synonym/common name		
Trade names		SHELL VM&P NAPHTHA EC
Suppliers		SHELL
Molecular mass		117
Boiling point/range	°C @ 760 mm Hg	121–134
Melting point	°C	
Density	kg/m^3 @ 20 °C	
Evaporation time[a]		
Evaporation time[b]		1.4
Vapour pressure	kPa @ 20 °C	
Flash point	°C	
Auto ignition temperature	°C	
Upper flammability limit	%v/v in air	
Lower flammability limit	%v/v in air	
Viscosity	mPa.s (cP) @ 20 °C	
Refractive index	n 20 D	
Solubility in water	%w/w @ 20 °C	
Solubility of water	%w/w @ 20 °C	
Solubility parameter[c]	$(J/cm^3)^{1/2}$	
Hydrogen bond index		
Fractional polarity		
Nitro cellulose dil ratio[d]		
Nitro cellulose dil ratio[e]		
Surface tension	mN/m @ 20 °C	
Specific heat liquid	kJ/kg mole/°C @ 20 °C	
Latent heat	kJ/kg mole	
Dielectric constant	@ 20 °C	
Antoine constant A	kPa°C	
Antoine constant B	kPa°C	
Antoine constant C	kPa°C	
Heat of combustion	kJ/kg mole	
UN number		
IMO classification		
ADR/RID classification		
UK exposure limits	mg/m^3 (ppm) 8hrTWA/MEL or OES	
USA exposure limits	mg/m^3 (ppm) TLV/TWA	
German exposure limits	mg/m^3 (ppm) MAK-TWA	
EU classification		
EU risk phrases		
EU safety phrases		

[a] Ether = 1. [b] n-Butyl acetate = 1. [c] Nelson, Hemwall, Edwards. [d] Toluene. [e] Hydrocarbon solvent.

Solvent group		HYDROCARBONS—ALIPHATIC
Chemical name		
Chemical formula		
Structural formula		
CAS No.		
Synonym/common name		
Trade names		SHELL VM&P NAPHTHA HT
Suppliers		SHELL
Molecular mass		118
Boiling point/range	°C @ 760 mm Hg	119–139
Melting point	°C	
Density	kg/m^3 @ 20 °C	
Evaporation time[a]		
Evaporation time[b]		1.5
Vapour pressure	kPa @ 20 °C	
Flash point	°C	
Auto ignition temperature	°C	
Upper flammability limit	%v/v in air	
Lower flammability limit	%v/v in air	
Viscosity	mPa.s (cP) @ 20 °C	
Refractive index	n 20 D	
Solubility in water	%w/w @ 20 °C	
Solubility of water	%w/w @ 20 °C	
Solubility parameter[c]	(J/cm^3)$^{1/2}$	
Hydrogen bond index		
Fractional polarity		
Nitro cellulose dil ratio[d]		
Nitro cellulose dil ratio[e]		
Surface tension	mN/m @ 20 °C	
Specific heat liquid	kJ/kg mole/°C @ 20 °C	
Latent heat	kJ/kg mole	
Dielectric constant	@ 20 °C	
Antoine constant A	kPa°C	
Antoine constant B	kPa°C	
Antoine constant C	kPa°C	
Heat of combustion	kJ/kg mole	
UN number		
IMO classification		
ADR/RID classification		
UK exposure limits	mg/m^3 (ppm) 8hrTWA/MEL or OES	
USA exposure limits	mg/m^3 (ppm) TLV/TWA	
German exposure limits	mg/m^3 (ppm) MAK-TWA	
EU classification		
EU risk phrases		
EU safety phrases		

[a] Ether = 1. [b] n-Butyl acetate = 1. [c] Nelson, Hemwall, Edwards. [d] Toluene. [e] Hydrocarbon solvent.

Solvent group		HYDROCARBONS—ALIPHATIC
Chemical name		
Chemical formula		
Structural formula		
CAS No.		
Synonym/common name		
Trade names		SHELL MINERAL SPIRITS 145 EC
Suppliers		SHELL
Molecular mass		131
Boiling point/range	°C @ 760 mm Hg	162–201
Melting point	°C	
Density	kg/m^3 @ 20 °C	
Evaporation time[a]		
Evaporation time[b]		0.14
Vapour pressure	kPa @ 20 °C	
Flash point	°C	
Auto ignition temperature	°C	
Upper flammability limit	%v/v in air	
Lower flammability limit	%v/v in air	
Viscosity	mPa.s (cP) @ 20 °C	
Refractive index	n 20 D	
Solubility in water	%w/w @ 20 °C	
Solubility of water	%w/w @ 20 °C	
Solubility parameter[c]	$(J/cm^3)^{1/2}$	
Hydrogen bond index		
Fractional polarity		
Nitro cellulose dil ratio[d]		
Nitro cellulose dil ratio[e]		
Surface tension	mN/m @ 20 °C	
Specific heat liquid	kJ/kg mole/°C @ 20 °C	
Latent heat	kJ/kg mole	
Dielectric constant	@ 20 °C	
Antoine constant A	kPa°C	
Antoine constant B	kPa°C	
Antoine constant C	kPa°C	
Heat of combustion	kJ/kg mole	
UN number		
IMO classification		
ADR/RID classification		
UK exposure limits	mg/m^3 (ppm) 8hrTWA/MEL or OES	
USA exposure limits	mg/m^3 (ppm) TLV/TWA	
German exposure limits	mg/m^3 (ppm) MAK-TWA	
EU classification		
EU risk phrases		
EU safety phrases		

[a] Ether = 1. [b] n-Butyl acetate = 1. [c] Nelson, Hemwall, Edwards. [d] Toluene. [e] Hydrocarbon solvent.

Solvent group		HYDROCARBONS—ALIPHATIC
Chemical name		
Chemical formula		
Structural formula		
CAS No.		
Synonym/common name		
Trade names		SHELL MINERAL SPIRITS 200 HT
Suppliers		SHELL
Molecular mass		132
Boiling point/range	°C @ 760 mm Hg	162–206
Melting point	°C	
Density	kg/m^3 @ 20 °C	
Evaporation time[a]		
Evaporation time[b]		0.13
Vapour pressure	kPa @ 20 °C	
Flash point	°C	
Auto ignition temperature	°C	
Upper flammability limit	%v/v in air	
Lower flammability limit	%v/v in air	
Viscosity	mPa.s (cP) @ 20 °C	
Refractive index	n 20 D	
Solubility in water	%w/w @ 20 °C	
Solubility of water	%w/w @ 20 °C	
Solubility parameter[c]	(J/cm^3)$^{1/2}$	
Hydrogen bond index		
Fractional polarity		
Nitro cellulose dil ratio[d]		
Nitro cellulose dil ratio[e]		
Surface tension	mN/m @ 20 °C	
Specific heat liquid	kJ/kg mole/°C @ 20 °C	
Latent heat	kJ/kg mole	
Dielectric constant	@ 20 °C	
Antoine constant A	kPa°C	
Antoine constant B	kPa°C	
Antoine constant C	kPa°C	
Heat of combustion	kJ/kg mole	
UN number		
IMO classification		
ADR/RID classification		
UK exposure limits	mg/m^3 (ppm) 8hrTWA/MEL or OES	
USA exposure limits	mg/m^3 (ppm) TLV/TWA	
German exposure limits	mg/m^3 (ppm) MAK-TWA	
EU classification		
EU risk phrases		
EU safety phrases		

[a] Ether = 1. [b] n-Butyl acetate = 1. [c] Nelson, Hemwall, Edwards. [d] Toluene. [e] Hydrocarbon solvent.

Solvent group		HYDROCARBONS—ALIPHATIC
Chemical name		
Chemical formula		
Structural formula		
CAS No.		
Synonym/common name		
Trade names		SHELL MINERAL SPIRITS 150 EC
Suppliers		SHELL
Molecular mass		132
Boiling point/range	°C @ 760 mm Hg	162–200
Melting point	°C	
Density	kg/m^3 @ 20 °C	
Evaporation time[a]		
Evaporation time[b]		0.13
Vapour pressure	kPa @ 20 °C	
Flash point	°C	
Auto ignition temperature	°C	
Upper flammability limit	%v/v in air	
Lower flammability limit	%v/v in air	
Viscosity	mPa.s (cP) @ 20 °C	
Refractive index	n 20 D	
Solubility in water	%w/w @ 20 °C	
Solubility of water	%w/w @ 20 °C	
Solubility parameter[c]	(J/cm^3)$^{1/2}$	
Hydrogen bond index		
Fractional polarity		
Nitro cellulose dil ratio[d]		
Nitro cellulose dil ratio[e]		
Surface tension	mN/m @ 20 °C	
Specific heat liquid	kJ/kg mole/°C @ 20 °C	
Latent heat	kJ/kg mole	
Dielectric constant	@ 20 °C	
Antoine constant A	kPa°C	
Antoine constant B	kPa°C	
Antoine constant C	kPa°C	
Heat of combustion	kJ/kg mole	
UN number		
IMO classification		
ADR/RID classification		
UK exposure limits	mg/m^3 (ppm) 8hrTWA/MEL or OES	
USA exposure limits	mg/m^3 (ppm) TLV/TWA	
German exposure limits	mg/m^3 (ppm) MAK-TWA	
EU classification		
EU risk phrases		
EU safety phrases		

[a] Ether = 1. [b] n-Butyl acetate = 1. [c] Nelson, Hemwall, Edwards. [d] Toluene. [e] Hydrocarbon solvent.

Solvent group		**HYDROCARBONS—ALIPHATIC**
Chemical name		Para-menthadienes
Chemical formula		$C_{10}H_{16}$
Structural formula		
CAS No.		
Synonym/common name		
Trade names		TABS D5
Suppliers		BUSH BOAKE ALLEN LTD
Molecular mass		136
Boiling point/range	°C @ 760 mm Hg	173–185
Melting point	°C	<20
Density	kg/m^3 @ 20 °C	850
Evaporation time[a]		
Evaporation time[b]		0.1
Vapour pressure	kPa @ 20 °C	
Flash point	°C	51
Auto ignition temperature	°C	
Upper flammability limit	%v/v in air	5.9
Lower flammability limit	%v/v in air	0.8
Viscosity	mPa.s (cP) @ 20 °C	<1.5
Refractive index	n 20 D	1.485
Solubility in water	%w/w @ 20 °C	<0.1
Solubility of water	%w/w @ 20 °C	
Solubility parameter[c]	(J/cm^3)$^{1/2}$	
Hydrogen bond index		
Fractional polarity		
Nitro cellulose dil ratio[d]		
Nitro cellulose dil ratio[e]		
Surface tension	mN/m @ 20 °C	
Specific heat liquid	kJ/kg mole/°C @ 20 °C	
Latent heat	kJ/kg mole	
Dielectric constant	@ 20 °C	
Antoine constant A	kPa°C	
Antoine constant B	kPa°C	
Antoine constant C	kPa°C	
Heat of combustion	kJ/kg mole	
UN number		2319
IMO classification		3.3
ADR/RID classification		3,31°(c)
UK exposure limits	mg/m^3 (ppm) 8hrTWA/MEL or OES	(100)
USA exposure limits	mg/m^3 (ppm) TLV/TWA	
German exposure limits	mg/m^3 (ppm) MAK-TWA	
EU classification		Xi
EU risk phrases		10-36/38
EU safety phrases		26

[a] Ether = 1. [b] n-Butyl acetate = 1. [c] Nelson, Hemwall, Edwards. [d] Toluene. [e] Hydrocarbon solvent.

Solvent group		HYDROCARBONS—ALIPHATIC
Chemical name		
Chemical formula		
Structural formula		
CAS No.		
Synonym/common name		
Trade names		SHELL MINERAL SPIRITS 135
Suppliers		SHELL
Molecular mass		138
Boiling point/range	°C @ 760 mm Hg	164–202
Melting point	°C	
Density	kg/m^3 @ 20 °C	
Evaporation time[a]		
Evaporation time[b]		0.10
Vapour pressure	kPa @ 20 °C	
Flash point	°C	
Auto ignition temperature	°C	
Upper flammability limit	%v/v in air	
Lower flammability limit	%v/v in air	
Viscosity	mPa.s (cP) @ 20 °C	
Refractive index	n 20 D	
Solubility in water	%w/w @ 20 °C	
Solubility of water	%w/w @ 20 °C	
Solubility parameter[c]	(J/cm^3)$^{1/2}$	
Hydrogen bond index		
Fractional polarity		
Nitro cellulose dil ratio[d]		
Nitro cellulose dil ratio[e]		
Surface tension	mN/m @ 20 °C	
Specific heat liquid	kJ/kg mole/°C @ 20 °C	
Latent heat	kJ/kg mole	
Dielectric constant	@ 20 °C	
Antoine constant A	kPa°C	
Antoine constant B	kPa°C	
Antoine constant C	kPa°C	
Heat of combustion	kJ/kg mole	
UN number		
IMO classification		
ADR/RID classification		
UK exposure limits	mg/m^3 (ppm) 8hrTWA/MEL or OES	
USA exposure limits	mg/m^3 (ppm) TLV/TWA	
German exposure limits	mg/m^3 (ppm) MAK-TWA	
EU classification		
EU risk phrases		
EU safety phrases		

[a] Ether = 1. [b] n-Butyl acetate = 1. [c] Nelson, Hemwall, Edwards. [d] Toluene. [e] Hydrocarbon solvent.

Solvent group		HYDROCARBONS—ALIPHATIC
Chemical name		
Chemical formula		
Structural formula		
CAS No.		
Synonym/common name		
Trade names		SHELL SOL 340 HT
Suppliers		SHELL
Molecular mass		143
Boiling point/range	°C @ 760 mm Hg	159–176
Melting point	°C	
Density	kg/m^3 @ 20 °C	
Evaporation time[a]		
Evaporation time[b]		0.27
Vapour pressure	kPa @ 20 °C	
Flash point	°C	
Auto ignition temperature	°C	
Upper flammability limit	%v/v in air	
Lower flammability limit	%v/v in air	
Viscosity	mPa.s (cP) @ 20 °C	
Refractive index	n 20 D	
Solubility in water	%w/w @ 20 °C	
Solubility of water	%w/w @ 20 °C	
Solubility parameter[c]	$(J/cm^3)^{1/2}$	
Hydrogen bond index		
Fractional polarity		
Nitro cellulose dil ratio[d]		
Nitro cellulose dil ratio[e]		
Surface tension	mN/m @ 20 °C	
Specific heat liquid	kJ/kg mole/°C @ 20 °C	
Latent heat	kJ/kg mole	
Dielectric constant	@ 20 °C	
Antoine constant A	kPa°C	
Antoine constant B	kPa°C	
Antoine constant C	kPa°C	
Heat of combustion	kJ/kg mole	
UN number		
IMO classification		
ADR/RID classification		
UK exposure limits	mg/m^3 (ppm) 8hrTWA/MEL or OES	
USA exposure limits	mg/m^3 (ppm) TLV/TWA	
German exposure limits	mg/m^3 (ppm) MAK-TWA	
EU classification		
EU risk phrases		
EU safety phrases		

[a] Ether = 1. [b] n-Butyl acetate = 1. [c] Nelson, Hemwall, Edwards. [d] Toluene. [e] Hydrocarbon solvent.

Solvent group		HYDROCARBONS—ALIPHATIC
Chemical name		
Chemical formula		
Structural formula		
CAS No.		
Synonym/common name		
Trade names		SHELL SOL 71
Suppliers		SHELL
Molecular mass		149
Boiling point/range	°C @ 760 mm Hg	179–204
Melting point	°C	
Density	kg/m^3 @ 20 °C	
Evaporation time[a]		
Evaporation time[b]		<0.1
Vapour pressure	kPa @ 20 °C	
Flash point	°C	
Auto ignition temperature	°C	
Upper flammability limit	%v/v in air	
Lower flammability limit	%v/v in air	
Viscosity	mPa.s (cP) @ 20 °C	
Refractive index	n 20 D	
Solubility in water	%w/w @ 20 °C	
Solubility of water	%w/w @ 20 °C	
Solubility parameter[c]	$(J/cm^3)^{1/2}$	
Hydrogen bond index		
Fractional polarity		
Nitro cellulose dil ratio[d]		
Nitro cellulose dil ratio[e]		
Surface tension	mN/m @ 20 °C	
Specific heat liquid	kJ/kg mole/°C @ 20 °C	
Latent heat	kJ/kg mole	
Dielectric constant	@ 20 °C	
Antoine constant A	kPa°C	
Antoine constant B	kPa°C	
Antoine constant C	kPa°C	
Heat of combustion	kJ/kg mole	
UN number		
IMO classification		
ADR/RID classification		
UK exposure limits	mg/m^3 (ppm) 8hrTWA/MEL or OES	
USA exposure limits	mg/m^3 (ppm) TLV/TWA	
German exposure limits	mg/m^3 (ppm) MAK-TWA	
EU classification		
EU risk phrases		
EU safety phrases		

[a] Ether = 1. [b] n-Butyl acetate = 1. [c] Nelson, Hemwall, Edwards. [d] Toluene. [e] Hydrocarbon solvent.

Solvent group		**HYDROCARBONS—ALIPHATIC**
Chemical name		Mixture of aliphatic hydrocarbons
Chemical formula		
Structural formula		
CAS No.		
Synonym/common name		
Trade names		EVOLVE CH_{10}
Suppliers		ICI
Molecular mass		156
Boiling point/range	°C @ 760 mm Hg	173–192
Melting point	°C	< -50
Density	kg/m^3 @ 20 °C	770
Evaporation time[a]		75
Evaporation time[b]		0.09
Vapour pressure	kPa @ 20 °C	0.066
Flash point	°C	58 (SETA CLOSED CUP)
Auto ignition temperature	°C	205
Upper flammability limit	%v/v in air	6.5
Lower flammability limit	%v/v in air	0.6
Viscosity	mPa.s (cP) @ 20 °C	1.13
Refractive index	n 20 D	1.425
Solubility in water	%w/w @ 20 °C	Insoluble
Solubility of water	%w/w @ 20 °C	Insoluble
Solubility parameter[c]	(J/cm^3)$^{1/2}$	
Hydrogen bond index		
Fractional polarity		
Nitro cellulose dil ratio[d]		
Nitro cellulose dil ratio[e]		
Surface tension	mN/m @ 20 °C	23.4
Specific heat liquid	kJ/kg mole/°C @ 20 °C	
Latent heat	kJ/kg mole	
Dielectric constant	@ 20 °C	
Antoine constant A	kPa°C	
Antoine constant B	kPa°C	
Antoine constant C	kPa°C	
Heat of combustion	kJ/kg mole	
UN number		1993
IMO classification		3/III
ADR/RID classification		3,31°(c)
UK exposure limits	mg/m^3 (ppm) 8hrTWA/MEL or OES	(300)*
USA exposure limits	mg/m^3 (ppm) TLV/TWA	
German exposure limits	mg/m^3 (ppm) MAK-TWA	
EU classification		
EU risk phrases		
EU safety phrases		

[a] Ether = 1. [b] n-Butyl acetate = 1. [c] Nelson, Hemwall, Edwards. [d] Toluene. [e] Hydrocarbon solvent.

Solvent group		HYDROCARBONS—ALIPHATIC
Chemical name		
Chemical formula		
Structural formula		
CAS No.		
Synonym/common name		
Trade names		SHELL SOL 142 HT
Suppliers		SHELL
Molecular mass		161
Boiling point/range	°C @ 760 mm Hg	190–207
Melting point	°C	
Density	kg/m^3 @ 20 °C	
Evaporation time[a]		
Evaporation time[b]		<0.1
Vapour pressure	kPa @ 20 °C	
Flash point	°C	
Auto ignition temperature	°C	
Upper flammability limit	%v/v in air	
Lower flammability limit	%v/v in air	
Viscosity	mPa.s (cP) @ 20 °C	
Refractive index	n 20 D	
Solubility in water	%w/w @ 20 °C	
Solubility of water	%w/w @ 20 °C	
Solubility parameter[c]	(J/cm^3)$^{1/2}$	
Hydrogen bond index		
Fractional polarity		
Nitro cellulose dil ratio[d]		
Nitro cellulose dil ratio[e]		
Surface tension	mN/m @ 20 °C	
Specific heat liquid	kJ/kg mole/°C @ 20 °C	
Latent heat	kJ/kg mole	
Dielectric constant	@ 20 °C	
Antoine constant A	kPa°C	
Antoine constant B	kPa°C	
Antoine constant C	kPa°C	
Heat of combustion	kJ/kg mole	
UN number		
IMO classification		
ADR/RID classification		
UK exposure limits	mg/m^3 (ppm) 8hrTWA/MEL or OES	
USA exposure limits	mg/m^3 (ppm) TLV/TWA	
German exposure limits	mg/m^3 (ppm) MAK-TWA	
EU classification		
EU risk phrases		
EU safety phrases		

[a] Ether = 1. [b] n-Butyl acetate = 1. [c] Nelson, Hemwall, Edwards. [d] Toluene. [e] Hydrocarbon solvent.

Solvent group		HYDROCARBONS—ALIPHATIC
Chemical name		Mixture based on aliphatic hydrocarbons
Chemical formula		
Structural formula		
CAS No.		64742-47-8
Synonym/common name		Paraffinic, naphthenic solvent
Trade names		VISTA LPA-150 SOLVENT
Suppliers		VISTA
Molecular mass		164
Boiling point/range	°C @ 760 mm Hg	192–230
Melting point	°C	< -70
Density	kg/m^3 @ 20 °C	811
Evaporation time[a]		
Evaporation time[b]		
Vapour pressure	kPa @ 20 °C	0.029
Flash point	°C	66 (TAG CLOSED CUP)
Auto ignition temperature	°C	228
Upper flammability limit	%v/v in air	4.7
Lower flammability limit	%v/v in air	0.6
Viscosity	mPa.s (cP) @ 20 °C	1.9
Refractive index	n 20 D	
Solubility in water	%w/w @ 20 °C	Negligible
Solubility of water	%w/w @ 20 °C	Negligible
Solubility parameter[c]	$(J/cm^3)^{1/2}$	
Hydrogen bond index		
Fractional polarity		
Nitro cellulose dil ratio[d]		
Nitro cellulose dil ratio[e]		
Surface tension	mN/m @ 20 °C	
Specific heat liquid	kJ/kg mole/°C @ 20 °C	
Latent heat	kJ/kg mole	
Dielectric constant	@ 20 °C	
Antoine constant A	kPa°C	
Antoine constant B	kPa°C	
Antoine constant C	kPa°C	
Heat of combustion	kJ/kg mole	
UN number		1268
IMO classification		
ADR/RID classification		
UK exposure limits	mg/m^3 (ppm) 8hrTWA/MEL or OES	
USA exposure limits	mg/m^3 (ppm) TLV/TWA	525(100)*
German exposure limits	mg/m^3 (ppm) MAK-TWA	
EU classification		
EU risk phrases		
EU safety phrases		

[a] Ether = 1. [b] n-Butyl acetate = 1. [c] Nelson, Hemwall, Edwards. [d] Toluene. [e] Hydrocarbon solvent.

Solvent group		HYDROCARBONS—ALIPHATIC
Chemical name		
Chemical formula		
Structural formula		
CAS No.		64741-65-7
Synonym/common name		
Trade names		SHELLSOL TD
Suppliers		SHELL
Molecular mass		164
Boiling point/range	°C @ 760 mm Hg	172–190
Melting point	°C	
Density	kg/m^3 @ 20 °C	751
Evaporation time[a]		70
Evaporation time[b]		0.18
Vapour pressure	kPa @ 20 °C	0.2
Flash point	°C	44 (IP 170)
Auto ignition temperature	°C	405
Upper flammability limit	%v/v in air	6.0
Lower flammability limit	%v/v in air	0.6
Viscosity	mPa.s (cP) @ 20 °C	1.20
Refractive index	n 20 D	1.420
Solubility in water	%w/w @ 20 °C	
Solubility of water	%w/w @ 20 °C	
Solubility parameter[c]	$(J/cm^3)^{1/2}$	14.9
Hydrogen bond index		0
Fractional polarity		0
Nitro cellulose dil ratio[d]		
Nitro cellulose dil ratio[e]		
Surface tension	mN/m @ 20 °C	23
Specific heat liquid	kJ/kg mole/°C @ 20 °C	328.0
Latent heat	kJ/kg mole	
Dielectric constant	@ 20 °C	2.0
Antoine constant A	kPa°C	6.4188
Antoine constant B	kPa°C	1745.6
Antoine constant C	kPa°C	222.16
Heat of combustion	kJ/kg mole	−7216000
UN number		1300
IMO classification		3.3/III
ADR/RID classification		3,31°(c)
UK exposure limits	mg/m^3 (ppm) 8hrTWA/MEL or OES	
USA exposure limits	mg/m^3 (ppm) TLV/TWA	
German exposure limits	mg/m^3 (ppm) MAK-TWA	
EU classification		
EU risk phrases		10
EU safety phrases		

[a] Ether = 1. [b] n-Butyl acetate = 1. [c] Nelson, Hemwall, Edwards. [d] Toluene. [e] Hydrocarbon solvent.

Solvent group		HYDROCARBONS—ALIPHATIC
Chemical name		Triisobutylene
Chemical formula		$C_{12}H_{24}$
Structural formula		
CAS No.		91053-01-9
Synonym/common name		Isododecene
Trade names		
Suppliers		EC Erdolchemie
Molecular mass		168.3
Boiling point/range	°C @ 760 mm Hg	170–183
Melting point	°C	−91
Density	kg/m^3 @ 20 °C	763
Evaporation time[a]		
Evaporation time[b]		
Vapour pressure	kPa @ 20 °C	
Flash point	°C	42
Auto ignition temperature	°C	355
Upper flammability limit	%v/v in air	6.5
Lower flammability limit	%v/v in air	0.5
Viscosity	mPa.s (cP) @ 20 °C	1.49
Refractive index	n 20 D	1.432
Solubility in water	%w/w @ 20 °C	
Solubility of water	%w/w @ 20 °C	
Solubility parameter[c]	(J/cm^3)$^{1/2}$	
Hydrogen bond index		
Fractional polarity		
Nitro cellulose dil ratio[d]		
Nitro cellulose dil ratio[e]		
Surface tension	mN/m @ 20 °C	24
Specific heat liquid	kJ/kg mole/°C @ 20 °C	
Latent heat	kJ/kg mole	
Dielectric constant	@ 20 °C	
Antoine constant A	kPa°C	
Antoine constant B	kPa°C	
Antoine constant C	kPa°C	
Heat of combustion	kJ/kg mole	
UN number		2324
IMO classification		3.3/III
ADR/RID classification		3,31°(c)
UK exposure limits	mg/m^3 (ppm) 8hrTWA/MEL or OES	
USA exposure limits	mg/m^3 (ppm) TLV/TWA	
German exposure limits	mg/m^3 (ppm) MAK-TWA	
EU classification		
EU risk phrases		
EU safety phrases		

[a] Ether = 1. [b] n-Butyl acetate = 1. [c] Nelson, Hemwall, Edwards. [d] Toluene. [e] Hydrocarbon solvent.

Solvent group		HYDROCARBONS—ALIPHATIC
Chemical name		
Chemical formula		
Structural formula		
CAS No.		64741-65-7
Synonym/common name		
Trade names		SHELLSOL T
Suppliers		SHELL
Molecular mass		169
Boiling point/range	°C @ 760 mm Hg	187–215
Melting point	°C	
Density	kg/m^3 @ 20 °C	761
Evaporation time[a]		110
Evaporation time[b]		0.09
Vapour pressure	kPa @ 20 °C	0.1
Flash point	°C	59 (ASTM D93)
Auto ignition temperature	°C	405
Upper flammability limit	%v/v in air	6.0
Lower flammability limit	%v/v in air	0.6
Viscosity	mPa.s (cP) @ 20 °C	1.41
Refractive index	n 20 D	1.424
Solubility in water	%w/w @ 20 °C	
Solubility of water	%w/w @ 20 °C	
Solubility parameter[c]	$(J/cm^3)^{1/2}$	15.0
Hydrogen bond index		0
Fractional polarity		0
Nitro cellulose dil ratio[d]		
Nitro cellulose dil ratio[e]		
Surface tension	mN/m @ 20 °C	23.5
Specific heat liquid	kJ/kg mole/°C @ 20 °C	338.0
Latent heat	kJ/kg mole	
Dielectric constant	@ 20 °C	2.0
Antoine constant A	kPa°C	6.71506
Antoine constant B	kPa°C	2009.16
Antoine constant C	kPa°C	241.891
Heat of combustion	kJ/kg mole	−7408115
UN number		1300
IMO classification		3.3/III
ADR/RID classification		3,32°(c)
UK exposure limits	mg/m^3 (ppm) 8hrTWA/MEL or OES	
USA exposure limits	mg/m^3 (ppm) TLV/TWA	
German exposure limits	mg/m^3 (ppm) MAK-TWA	
EU classification		
EU risk phrases		
EU safety phrases		

[a] Ether = 1. [b] n-Butyl acetate = 1. [c] Nelson, Hemwall, Edwards. [d] Toluene. [e] Hydrocarbon solvent.

Solvent group		**HYDROCARBONS—ALIPHATIC**
Chemical name		2,2,4,6,6-Pentamethyl heptane
Chemical formula		$C_{12}H_{26}$
Structural formula		
CAS No.		93685-81-5/13475-82-6
Synonym/common name		Isododecane
Trade names		
Suppliers		BP, Erdolchemie
Molecular mass		170.3
Boiling point/range	°C @ 760 mm Hg	176–192
Melting point	°C	−81
Density	kg/m^3 @ 20 °C	747
Evaporation time[a]		43–46
Evaporation time[b]		0.2
Vapour pressure	kPa @ 20 °C	0.1
Flash point	°C	46
Auto ignition temperature	°C	410
Upper flammability limit	%v/v in air	4.0
Lower flammability limit	%v/v in air	0.5
Viscosity	mPa.s (cP) @ 20 °C	1.38
Refractive index	n 20 D	1.421
Solubility in water	%w/w @ 20 °C	
Solubility of water	%w/w @ 20 °C	
Solubility parameter[c]	(J/cm^3)$^{1/2}$	
Hydrogen bond index		
Fractional polarity		
Nitro cellulose dil ratio[d]		
Nitro cellulose dil ratio[e]		
Surface tension	mN/m @ 20 °C	24.3
Specific heat liquid	kJ/kg mole/°C @ 20 °C	
Latent heat	kJ/kg mole	
Dielectric constant	@ 20 °C	2.12
Antoine constant A	kPa°C	
Antoine constant B	kPa°C	
Antoine constant C	kPa°C	
Heat of combustion	kJ/kg mole	
UN number		2286
IMO classification		3.3
ADR/RID classification		3,31°(c)
UK exposure limits	mg/m^3 (ppm) 8hrTWA/MEL or OES	
USA exposure limits	mg/m^3 (ppm) TLV/TWA	
German exposure limits	mg/m^3 (ppm) MAK-TWA	
EU classification		
EU risk phrases		10
EU safety phrases		9-16-29-33

[a] Ether = 1. [b] n-Butyl acetate = 1. [c] Nelson, Hemwall, Edwards. [d] Toluene. [e] Hydrocarbon solvent.

Solvent group		HYDROCARBONS—ALIPHATIC
Chemical name		Mixture based on aliphatic hydrocarbons
Chemical formula		
Structural formula		
CAS No.		64742-47-8
Synonym/common name		Paraffinic, naphthenic solvent
Trade names		
Suppliers		VISTA
Molecular mass		171
Boiling point/range	°C @ 760 mm Hg	213–237
Melting point	°C	< −70
Density	kg/m^3 @ 20 °C	815
Evaporation time[a]		
Evaporation time[b]		0.03
Vapour pressure	kPa @ 20 °C	0.012
Flash point	°C	82 (TAG CLOSED CUP)
Auto ignition temperature	°C	228
Upper flammability limit	%v/v in air	4.7
Lower flammability limit	%v/v in air	0.6
Viscosity	mPa.s (cP) @ 20 °C	2.1
Refractive index	n 20 D	1.445
Solubility in water	%w/w @ 20 °C	Negligible
Solubility of water	%w/w @ 20 °C	Negligible
Solubility parameter[c]	$(J/cm^3)^{1/2}$	
Hydrogen bond index		
Fractional polarity		
Nitro cellulose dil ratio[d]		
Nitro cellulose dil ratio[e]		
Surface tension	mN/m @ 20 °C	
Specific heat liquid	kJ/kg mole/°C @ 20 °C	
Latent heat	kJ/kg mole	
Dielectric constant	@ 20 °C	
Antoine constant A	kPa°C	
Antoine constant B	kPa°C	
Antoine constant C	kPa°C	
Heat of combustion	kJ/kg mole	
UN number		1268
IMO classification		
ADR/RID classification		
UK exposure limits	mg/m^3 (ppm) 8hrTWA/MEL or OES	
USA exposure limits	mg/m^3 (ppm) TLV/TWA	525(100)*
German exposure limits	mg/m^3 (ppm) MAK-TWA	
EU classification		
EU risk phrases		
EU safety phrases		

[a] Ether = 1. [b] n-Butyl acetate = 1. [c] Nelson, Hemwall, Edwards. [d] Toluene. [e] Hydrocarbon solvent.

Solvent group		**HYDROCARBONS—ALIPHATIC**
Chemical name		Mixture based on aliphatic hydrocarbons
Chemical formula		
Structural formula		
CAS No.		64742-47-8
Synonym/common name		Paraffinic, naphthenic solvent
Trade names		VISTA LPA-210 SOLVENT
Suppliers		VISTA
Molecular mass		194
Boiling point/range	°C @ 760 mm Hg	240–282
Melting point	°C	< −51
Density	kg/m^3 @ 20 °C	826
Evaporation time[a]		
Evaporation time[b]		0.004
Vapour pressure	kPa @ 20 °C	0.008
Flash point	°C	105 (PENSKY MARTENS)
Auto ignition temperature	°C	218
Upper flammability limit	%v/v in air	4.7
Lower flammability limit	%v/v in air	0.5
Viscosity	mPa.s (cP) @ 20 °C	3.3
Refractive index	n 20 D	1.451
Solubility in water	%w/w @ 20 °C	Negligible
Solubility of water	%w/w @ 20 °C	Negligible
Solubility parameter[c]	$(J/cm^3)^{1/2}$	
Hydrogen bond index		
Fractional polarity		
Nitro cellulose dil ratio[d]		
Nitro cellulose dil ratio[e]		
Surface tension	mN/m @ 20 °C	
Specific heat liquid	kJ/kg mole/°C @ 20 °C	
Latent heat	kJ/kg mole	
Dielectric constant	@ 20 °C	
Antoine constant A	kPa°C	
Antoine constant B	kPa°C	
Antoine constant C	kPa°C	
Heat of combustion	kJ/kg mole	
UN number		
IMO classification		
ADR/RID classification		
UK exposure limits	mg/m^3 (ppm) 8hrTWA/MEL or OES	
USA exposure limits	mg/m^3 (ppm) TLV/TWA	525(100)*
German exposure limits	mg/m^3 (ppm) MAK-TWA	
EU classification		
EU risk phrases		
EU safety phrases		

[a] Ether = 1. [b] n-Butyl acetate = 1. [c] Nelson, Hemwall, Edwards. [d] Toluene. [e] Hydrocarbon solvent.

Property	Unit	Value
Solvent group		**HYDROCARBONS—ALIPHATIC**
Chemical name		Mixture based on saturated aliphatic hydrocarbons
Chemical formula		
Structural formula		
CAS No.		64742-47-8/629-59-4/629-62-9/544-76-3
Synonym/common name		Blend of paraffinic, naphthenic: normal paraffins
Trade names		VISTA 47 SOLVENT
Suppliers		VISTA
Molecular mass		197
Boiling point/range	°C @ 760 mm Hg	240–277
Melting point	°C	−18
Density	kg/m^3 @ 20 °C	815
Evaporation time[a]		
Evaporation time[b]		<0.004
Vapour pressure	kPa @ 20 °C	0.013
Flash point	°C	107 (PENSKY MARTENS)
Auto ignition temperature	°C	
Upper flammability limit	%v/v in air	4.7
Lower flammability limit	%v/v in air	0.5
Viscosity	mPa.s (cP) @ 20 °C	3.2
Refractive index	n 20 D	
Solubility in water	%w/w @ 20 °C	Negligible
Solubility of water	%w/w @ 20 °C	Negligible
Solubility parameter[c]	$(J/cm^3)^{1/2}$	
Hydrogen bond index		
Fractional polarity		
Nitro cellulose dil ratio[d]		
Nitro cellulose dil ratio[e]		
Surface tension	mN/m @ 20 °C	
Specific heat liquid	kJ/kg mole/°C @ 20 °C	
Latent heat	kJ/kg mole	
Dielectric constant	@ 20 °C	
Antoine constant A	kPa°C	
Antoine constant B	kPa°C	
Antoine constant C	kPa°C	
Heat of combustion	kJ/kg mole	
UN number		
IMO classification		
ADR/RID classification		
UK exposure limits	mg/m^3 (ppm) 8hrTWA/MEL or OES	
USA exposure limits	mg/m^3 (ppm) TLV/TWA	525(100)*
German exposure limits	mg/m^3 (ppm) MAK-TWA	
EU classification		
EU risk phrases		
EU safety phrases		

[a] Ether = 1. [b] n-Butyl acetate = 1. [c] Nelson, Hemwall, Edwards. [d] Toluene. [e] Hydrocarbon solvent.

Solvent group		HYDROCARBONS—ALIPHATIC
Chemical name		Tetradecane
Chemical formula		$C_{14}H_{3}0$
Structural formula		$CH_3(CH_2)_{12}CH_3$
CAS No.		629-59-4
Synonym/common name		Tetradecane
Trade names		VISTA C14 PARAFFIN
Suppliers		VISTA
Molecular mass		201
Boiling point/range	°C @ 760 mm Hg	243–249
Melting point	°C	2
Density	kg/m^3 @ 20 °C	768
Evaporation time[a]		
Evaporation time[b]		0.002
Vapour pressure	kPa @ 20 °C	0.009
Flash point	°C	112 (PENSKY MARTENS)
Auto ignition temperature	°C	
Upper flammability limit	%v/v in air	4.7
Lower flammability limit	%v/v in air	0.5
Viscosity	mPa.s (cP) @ 20 °C	2.4
Refractive index	n 20 D	
Solubility in water	%w/w @ 20 °C	Negligible
Solubility of water	%w/w @ 20 °C	Negligible
Solubility parameter[c]	(J/cm^3)$^{1/2}$	
Hydrogen bond index		
Fractional polarity		
Nitro cellulose dil ratio[d]		
Nitro cellulose dil ratio[e]		
Surface tension	mN/m @ 20 °C	
Specific heat liquid	kJ/kg mole/°C @ 20 °C	
Latent heat	kJ/kg mole	
Dielectric constant	@ 20 °C	
Antoine constant A	kPa°C	
Antoine constant B	kPa°C	
Antoine constant C	kPa°C	
Heat of combustion	kJ/kg mole	
UN number		
IMO classification		
ADR/RID classification		
UK exposure limits	mg/m^3 (ppm) 8hrTWA/MEL or OES	
USA exposure limits	mg/m^3 (ppm) TLV/TWA	525(100)*
German exposure limits	mg/m^3 (ppm) MAK-TWA	
EU classification		
EU risk phrases		
EU safety phrases		

[a] Ether = 1. [b] n-Butyl acetate = 1. [c] Nelson, Hemwall, Edwards. [d] Toluene. [e] Hydrocarbon solvent.

Solvent group		**HYDROCARBONS—ALIPHATIC**
Chemical name		Tetradecane, Pentadecane, Hexadecane
Chemical formula		$C_{14}H_{30}$, $C_{15}H_{32}$, $C_{16}H_{34}$
Structural formula		$CH_3(CH_2)_{12}CH_3/CH_3(CH_2)_{13}CH_3/$ $CH_3(CH_2)_{14}$
CAS No.		629-59-4, 629-62-9, 544-76-3
Synonym/common name		Tetradecane, Pentadecane, Hexadecane
Trade names		VISTA C14–C16 PARAFFINS
Suppliers		VISTA
Molecular mass		211
Boiling point/range	°C @ 760 mm Hg	253–266
Melting point	°C	8
Density	kg/m^3 @ 20 °C	776
Evaporation time[a]		
Evaporation time[b]		0.003
Vapour pressure	kPa @ 20 °C	0.004
Flash point	°C	121 (PENSKY MARTENS)
Auto ignition temperature	°C	
Upper flammability limit	%v/v in air	4.7
Lower flammability limit	%v/v in air	0.5
Viscosity	mPa.s (cP) @ 20 °C	2.9
Refractive index	n 20 D	
Solubility in water	%w/w @ 20 °C	Negligible
Solubility of water	%w/w @ 20 °C	Negligible
Solubility parameter[c]	$(J/cm^3)^{1/2}$	
Hydrogen bond index		
Fractional polarity		
Nitro cellulose dil ratio[d]		
Nitro cellulose dil ratio[e]		
Surface tension	mN/m @ 20 °C	
Specific heat liquid	kJ/kg mole/°C @ 20 °C	
Latent heat	kJ/kg mole	
Dielectric constant	@ 20 °C	
Antoine constant A	kPa°C	
Antoine constant B	kPa°C	
Antoine constant C	kPa°C	
Heat of combustion	kJ/kg mole	
UN number		
IMO classification		
ADR/RID classification		
UK exposure limits	mg/m^3 (ppm) 8hrTWA/MEL or OES	
USA exposure limits	mg/m^3 (ppm) TLV/TWA	525(100)*
German exposure limits	mg/m^3 (ppm) MAK-TWA	
EU classification		
EU risk phrases		
EU safety phrases		

[a] Ether = 1. [b] n-Butyl acetate = 1. [c] Nelson, Hemwall, Edwards. [d] Toluene. [e] Hydrocarbon solvent.

Solvent group		**HYDROCARBONS—ALIPHATIC**
Chemical name		2,2,4,4,6,8,8,-Heptamethyl nonane
Chemical formula		$C_{16}H_{34}$
Structural formula		
CAS No.		93685-80-4
Synonym/common name		Isohexadecane
Trade names		
Suppliers		EC Erdolchemie
Molecular mass		226.5
Boiling point/range	°C @ 760 mm Hg	210–250
Melting point	°C	−70
Density	kg/m^3 @ 20 °C	
Evaporation time[a]		800
Evaporation time[b]		
Vapour pressure	kPa @ 20 °C	
Flash point	°C	102
Auto ignition temperature	°C	420
Upper flammability limit	%v/v in air	
Lower flammability limit	%v/v in air	
Viscosity	mPa.s (cP) @ 20 °C	
Refractive index	n 20 D	1.44
Solubility in water	%w/w @ 20 °C	
Solubility of water	%w/w @ 20 °C	
Solubility parameter[c]	(J/cm^3)$^{1/2}$	
Hydrogen bond index		
Fractional polarity		
Nitro cellulose dil ratio[d]		
Nitro cellulose dil ratio[e]		
Surface tension	mN/m @ 20 °C	
Specific heat liquid	kJ/kg mole/°C @ 20 °C	
Latent heat	kJ/kg mole	
Dielectric constant	@ 20 °C	
Antoine constant A	kPa°C	
Antoine constant B	kPa°C	
Antoine constant C	kPa°C	
Heat of combustion	kJ/kg mole	
UN number		
IMO classification		
ADR/RID classification		
UK exposure limits	mg/m^3 (ppm) 8hrTWA/MEL or OES	
USA exposure limits	mg/m^3 (ppm) TLV/TWA	
German exposure limits	mg/m^3 (ppm) MAK-TWA	
EU classification		
EU risk phrases		
EU safety phrases		

[a] Ether = 1. [b] n-Butyl acetate = 1. [c] Nelson, Hemwall, Edwards. [d] Toluene. [e] Hydrocarbon solvent.

Solvent group		**HYDROCARBONS—ALIPHATIC/CYCLIC**
Chemical name		
Chemical formula		
Structural formula		
CAS No.		64742-47-8
Synonym/common name		
Trade names		SHELLSOL D100
Suppliers		SHELL
Molecular mass		
Boiling point/range	°C @ 760 mm Hg	240–262
Melting point	°C	
Density	kg/m^3 @ 20 °C	801
Evaporation time[a]		>3900
Evaporation time[b]		<0.01
Vapour pressure	kPa @ 20 °C	0.003
Flash point	°C	105 (ASTM D93)
Auto ignition temperature	°C	220
Upper flammability limit	%v/v in air	5.5
Lower flammability limit	%v/v in air	0.5
Viscosity	mPa.s (cP) @ 20 °C	2.56
Refractive index	n 20 D	1.443
Solubility in water	%w/w @ 20 °C	
Solubility of water	%w/w @ 20 °C	
Solubility parameter[c]	(J/cm^3)$^{1/2}$	15.3
Hydrogen bond index		0
Fractional polarity		0
Nitro cellulose dil ratio[d]		
Nitro cellulose dil ratio[e]		
Surface tension	mN/m @ 20 °C	26
Specific heat liquid	kJ/kg mole/°C @ 20 °C	
Latent heat	kJ/kg mole	
Dielectric constant	@ 20 °C	2.1
Antoine constant A	kPa°C	7.4189
Antoine constant B	kPa°C	2603.5
Antoine constant C	kPa°C	241.46
Heat of combustion	kJ/kg mole	
UN number		
IMO classification		
ADR/RID classification		
UK exposure limits	mg/m^3 (ppm) 8hrTWA/MEL or OES	
USA exposure limits	mg/m^3 (ppm) TLV/TWA	
German exposure limits	mg/m^3 (ppm) MAK-TWA	
EU classification		
EU risk phrases		
EU safety phrases		

[a] Ether = 1. [b] n-Butyl acetate = 1. [c] Nelson, Hemwall, Edwards. [d] Toluene. [e] Hydrocarbon solvent.

Solvent group		**HYDROCARBONS—ALIPHATIC/CYCLIC**
Chemical name		
Chemical formula		
Structural formula		
CAS No.		64742-47-8
Synonym/common name		
Trade names		SHELLSOL D90
Suppliers		SHELL
Molecular mass		
Boiling point/range	°C @ 760 mm Hg	220-270
Melting point	°C	
Density	kg/m^3 @ 20 °C	799
Evaporation time[a]		>3900
Evaporation time[b]		<0.01
Vapour pressure	kPa @ 20 °C	0.02
Flash point	°C	95 (ASTM D93)
Auto ignition temperature	°C	220
Upper flammability limit	%v/v in air	5.5
Lower flammability limit	%v/v in air	0.6
Viscosity	mPa.s (cP) @ 20 °C	2.16
Refractive index	n 20 D	1.440
Solubility in water	%w/w @ 20 °C	
Solubility of water	%w/w @ 20 °C	
Solubility parameter[c]	(J/cm^3)$^{1/2}$	
Hydrogen bond index		
Fractional polarity		
Nitro cellulose dil ratio[d]		
Nitro cellulose dil ratio[e]		
Surface tension	mN/m @ 20 °C	
Specific heat liquid	kJ/kg mole/°C @ 20 °C	
Latent heat	kJ/kg mole	
Dielectric constant	@ 20 °C	
Antoine constant A	kPa°C	
Antoine constant B	kPa°C	
Antoine constant C	kPa°C	
Heat of combustion	kJ/kg mole	
UN number		
IMO classification		
ADR/RID classification		3,32°(c)
UK exposure limits	mg/m^3 (ppm) 8hrTWA/MEL or OES	
USA exposure limits	mg/m^3 (ppm) TLV/TWA	
German exposure limits	mg/m^3 (ppm) MAK-TWA	
EU classification		
EU risk phrases		
EU safety phrases		

[a] Ether = 1. [b] n-Butyl acetate = 1. [c] Nelson, Hemwall, Edwards. [d] Toluene. [e] Hydrocarbon solvent.

Solvent group		HYDROCARBONS—ALIPHATIC/CYCLIC
Chemical name		
Chemical formula		
Structural formula		
CAS No.		64742-49-0
Synonym/common name		
Trade names		SBP 80/95 LNH
Suppliers		SHELL
Molecular mass		
Boiling point/range	°C @ 760 mm Hg	84–94
Melting point	°C	
Density	kg/m^3 @ 20 °C	720
Evaporation time[a]		2.9
Evaporation time[b]		4.6
Vapour pressure	kPa @ 20 °C	8.5
Flash point	°C	<0 (IP 170)
Auto ignition temperature	°C	275
Upper flammability limit	%v/v in air	8.0
Lower flammability limit	%v/v in air	0.8
Viscosity	mPa.s (cP) @ 20 °C	0.49
Refractive index	n 20 D	1.398
Solubility in water	%w/w @ 20 °C	
Solubility of water	%w/w @ 20 °C	
Solubility parameter[c]	$(J/cm^3)^{1/2}$	15.1
Hydrogen bond index		0
Fractional polarity		0
Nitro cellulose dil ratio[d]		
Nitro cellulose dil ratio[e]		
Surface tension	mN/m @ 20 °C	
Specific heat liquid	kJ/kg mole/°C @ 20 °C	
Latent heat	kJ/kg mole	
Dielectric constant	@ 20 °C	
Antoine constant A	kPa°C	
Antoine constant B	kPa°C	
Antoine constant C	kPa°C	
Heat of combustion	kJ/kg mole	
UN number		1271
IMO classification		3.1/II
ADR/RID classification		3,3°(b)
UK exposure limits	mg/m^3 (ppm) 8hrTWA/MEL or OES	
USA exposure limits	mg/m^3 (ppm) TLV/TWA	
German exposure limits	mg/m^3 (ppm) MAK-TWA	
EU classification		F
EU risk phrases		
EU safety phrases		9-16-29-33

[a] Ether = 1. [b] n-Butyl acetate = 1. [c] Nelson, Hemwall, Edwards. [d] Toluene. [e] Hydrocarbon solvent.

Solvent group		**HYDROCARBONS—ALIPHATIC/CYCLIC**
Chemical name		
Chemical formula		
Structural formula		
CAS No.		64742-47-8
Synonym/common name		
Trade names		SHELLSOL DMA
Suppliers		SHELL
Molecular mass		
Boiling point/range	°C @ 760 mm Hg	220–270
Melting point	°C	
Density	kg/m^3 @ 20 °C	799
Evaporation time[a]		>3900
Evaporation time[b]		<0.01
Vapour pressure	kPa @ 20 °C	0.02
Flash point	°C	95 (ASTM D93)
Auto ignition temperature	°C	220
Upper flammability limit	%v/v in air	5.5
Lower flammability limit	%v/v in air	0.6
Viscosity	mPa.s (cP) @ 20 °C	2.16
Refractive index	n 20 D	1.440
Solubility in water	%w/w @ 20 °C	
Solubility of water	%w/w @ 20 °C	
Solubility parameter[c]	(J/cm^3)$^{1/2}$	
Hydrogen bond index		
Fractional polarity		
Nitro cellulose dil ratio[d]		
Nitro cellulose dil ratio[e]		
Surface tension	mN/m @ 20 °C	
Specific heat liquid	kJ/kg mole/°C @ 20 °C	
Latent heat	kJ/kg mole	
Dielectric constant	@ 20 °C	
Antoine constant A	kPa°C	
Antoine constant B	kPa°C	
Antoine constant C	kPa°C	
Heat of combustion	kJ/kg mole	
UN number		
IMO classification		
ADR/RID classification		3,32°(c)
UK exposure limits	mg/m^3 (ppm) 8hrTWA/MEL or OES	
USA exposure limits	mg/m^3 (ppm) TLV/TWA	
German exposure limits	mg/m^3 (ppm) MAK-TWA	
EU classification		
EU risk phrases		
EU safety phrases		

[a] Ether = 1. [b] n-Butyl acetate = 1. [c] Nelson, Hemwall, Edwards. [d] Toluene. [e] Hydrocarbon solvent.

Solvent group		HYDROCARBONS—ALIPHATIC/CYCLIC
Chemical name		
Chemical formula		
Structural formula		
CAS No.		64742-49-0
Synonym/common name		
Trade names		SBP 40/65 LNH
Suppliers		SHELL
Molecular mass		82
Boiling point/range	°C @ 760 mm Hg	44–62
Melting point	°C	
Density	kg/m^3 @ 20 °C	658
Evaporation time[a]		1.0
Evaporation time[b]		10
Vapour pressure	kPa @ 20 °C	33
Flash point	°C	−43 (IP 170)
Auto ignition temperature	°C	245
Upper flammability limit	%v/v in air	8.0
Lower flammability limit	%v/v in air	0.8
Viscosity	mPa.s (cP) @ 20 °C	0.27
Refractive index	n 20 D	1.370
Solubility in water	%w/w @ 20 °C	
Solubility of water	%w/w @ 20 °C	
Solubility parameter[c]	(J/cm^3)$^{1/2}$	14.1
Hydrogen bond index		0
Fractional polarity		0
Nitro cellulose dil ratio[d]		
Nitro cellulose dil ratio[e]		
Surface tension	mN/m @ 20 °C	17.5
Specific heat liquid	kJ/kg mole/°C @ 20 °C	184.5
Latent heat	kJ/kg mole	27060
Dielectric constant	@ 20 °C	
Antoine constant A	kPa°C	5.98987
Antoine constant B	kPa°C	1112.97
Antoine constant C	kPa°C	233.242
Heat of combustion	kJ/kg mole	−3639160
UN number		1271
IMO classification		3.1/II
ADR/RID classification		3,2°(b)
UK exposure limits	mg/m^3 (ppm) 8hrTWA/MEL or OES	
USA exposure limits	mg/m^3 (ppm) TLV/TWA	
German exposure limits	mg/m^3 (ppm) MAK-TWA	
EU classification		F+
EU risk phrases		
EU safety phrases		9-16-29-33

[a] Ether = 1. [b] n-Butyl acetate = 1. [c] Nelson, Hemwall, Edwards. [d] Toluene. [e] Hydrocarbon solvent.

Solvent group		**HYDROCARBONS—ALIPHATIC/CYCLIC**
Chemical name		
Chemical formula		
Structural formula		
CAS No.		64742-49-0
Synonym/common name		SBP 58/62
Trade names		Isohexane LNH
Suppliers		SHELL
Molecular mass		86
Boiling point/range	°C @ 760 mm Hg	57–63
Melting point	°C	
Density	kg/m^3 @ 20 °C	665
Evaporation time[a]		1.2
Evaporation time[b]		9.2
Vapour pressure	kPa @ 20 °C	24
Flash point	°C	−33 (IP 170)
Auto ignition temperature	°C	260
Upper flammability limit	%v/v in air	8.0
Lower flammability limit	%v/v in air	0.8
Viscosity	mPa.s (cP) @ 20 °C	0.30
Refractive index	n 20 D	1.375
Solubility in water	%w/w @ 20 °C	
Solubility of water	%w/w @ 20 °C	
Solubility parameter[c]	$(J/cm^3)^{1/2}$	14.7
Hydrogen bond index		0
Fractional polarity		0
Nitro cellulose dil ratio[d]		
Nitro cellulose dil ratio[e]		
Surface tension	mN/m @ 20 °C	18
Specific heat liquid	kJ/kg mole/°C @ 20 °C	189.2
Latent heat	kJ/kg mole	28380
Dielectric constant	@ 20 °C	
Antoine constant A	kPa°C	6.25922
Antoine constant B	kPa°C	1303.04
Antoine constant C	kPa°C	252.917
Heat of combustion	kJ/kg mole	−4149930
UN number		1208
IMO classification		3.1/II
ADR/RID classification		3,3°(b)
UK exposure limits	mg/m^3 (ppm) 8hrTWA/MEL or OES	
USA exposure limits	mg/m^3 (ppm) TLV/TWA	
German exposure limits	mg/m^3 (ppm) MAK-TWA	
EU classification		F
EU risk phrases		
EU safety phrases		9-16-23-29-33

[a] Ether = 1. [b] n-Butyl acetate = 1. [c] Nelson, Hemwall, Edwards. [d] Toluene. [e] Hydrocarbon solvent.

Solvent group		**HYDROCARBONS—ALIPHATIC/CYCLIC**
Chemical name		
Chemical formula		
Structural formula		
CAS No.		64742-49-0
Synonym/common name		
Trade names		Hexane Extraction
Suppliers		SHELL
Molecular mass		86
Boiling point/range	°C @ 760 mm Hg	66–69
Melting point	°C	
Density	kg/m^3 @ 20 °C	675
Evaporation time[a]		1.4
Evaporation time[b]		8.4
Vapour pressure	kPa @ 20 °C	19
Flash point	°C	−27 (IP 170)
Auto ignition temperature	°C	240
Upper flammability limit	%v/v in air	8.0
Lower flammability limit	%v/v in air	0.8
Viscosity	mPa.s (cP) @ 20 °C	0.31
Refractive index	n 20 D	1.379
Solubility in water	%w/w @ 20 °C	
Solubility of water	%w/w @ 20 °C	
Solubility parameter[c]	$(J/cm^3)^{1/2}$	14.9
Hydrogen bond index		0
Fractional polarity		0
Nitro cellulose dil ratio[d]		
Nitro cellulose dil ratio[e]		
Surface tension	mN/m @ 20 °C	18.5
Specific heat liquid	kJ/kg mole/°C @ 20 °C	189.2
Latent heat	kJ/kg mole	28810
Dielectric constant	@ 20 °C	1.9
Antoine constant A	kPa°C	6.0576
Antoine constant B	kPa°C	1198.97
Antoine constant C	kPa°C	227.206
Heat of combustion	kJ/kg mole	−4154746
UN number		1208
IMO classification		3.1/II
ADR/RID classification		3,3°(b)
UK exposure limits	mg/m^3 (ppm) 8hrTWA/MEL or OES	
USA exposure limits	mg/m^3 (ppm) TLV/TWA	
German exposure limits	mg/m^3 (ppm) MAK-TWA	
EU classification		F, Xn
EU risk phrases		48-20
EU safety phrases		9-16-24/25-29-51

[a] Ether = 1. [b] n-Butyl acetate = 1. [c] Nelson, Hemwall, Edwards. [d] Toluene. [e] Hydrocarbon solvent.

Solvent group		**HYDROCARBONS—ALIPHATIC/CYCLIC**
Chemical name		
Chemical formula		
Structural formula		
CAS No.		64742-49-0
Synonym/common name		
Trade names		Hexane Polymerisation
Suppliers		SHELL
Molecular mass		86
Boiling point/range	°C @ 760 mm Hg	66–69
Melting point	°C	
Density	kg/m^3 @ 20 °C	675
Evaporation time[a]		1.4
Evaporation time[b]		8.4
Vapour pressure	kPa @ 20 °C	19
Flash point	°C	−27 (IP 170)
Auto ignition temperature	°C	240
Upper flammability limit	%v/v in air	8.0
Lower flammability limit	%v/v in air	0.8
Viscosity	mPa.s (cP) @ 20 °C	0.31
Refractive index	n 20 D	1.379
Solubility in water	%w/w @ 20 °C	
Solubility of water	%w/w @ 20 °C	
Solubility parameter[c]	(J/cm^3)$^{1/2}$	14.9
Hydrogen bond index		0
Fractional polarity		0
Nitro cellulose dil ratio[d]		
Nitro cellulose dil ratio[e]		
Surface tension	mN/m @ 20 °C	18.5
Specific heat liquid	kJ/kg mole/°C @ 20 °C	189.2
Latent heat	kJ/kg mole	28810
Dielectric constant	@ 20 °C	1.9
Antoine constant A	kPa°C	6.0576
Antoine constant B	kPa°C	1198.97
Antoine constant C	kPa°C	227.206
Heat of combustion	kJ/kg mole	−4154746
UN number		1208
IMO classification		3.1/II
ADR/RID classification		3,3°(b)
UK exposure limits	mg/m^3 (ppm) 8hrTWA/MEL or OES	
USA exposure limits	mg/m^3 (ppm) TLV/TWA	
German exposure limits	mg/m^3 (ppm) MAK-TWA	
EU classification		F, Xn
EU risk phrases		48-20
EU safety phrases		9-16-24/25-29-51

[a] Ether = 1. [b] n-Butyl acetate = 1. [c] Nelson, Hemwall, Edwards. [d] Toluene. [e] Hydrocarbon solvent.

Solvent group		**HYDROCARBONS—ALIPHATIC/CYCLIC**
Chemical name		
Chemical formula		
Structural formula		
CAS No.		64742-49-0
Synonym/common name		
Trade names		SBP 62/78
Suppliers		SHELL
Molecular mass		89
Boiling point/range	°C @ 760 mm Hg	67–77
Melting point	°C	
Density	kg/m^3 @ 20 °C	680
Evaporation time[a]		1.6
Evaporation time[b]		7.8
Vapour pressure	kPa @ 20 °C	15.5
Flash point	°C	<0 (IP 170)
Auto ignition temperature	°C	240
Upper flammability limit	%v/v in air	8.0
Lower flammability limit	%v/v in air	0.6
Viscosity	mPa.s (cP) @ 20 °C	0.33
Refractive index	n 20 D	1.382
Solubility in water	%w/w @ 20 °C	
Solubility of water	%w/w @ 20 °C	
Solubility parameter[c]	(J/cm^3)$^{1/2}$	14.9
Hydrogen bond index		0
Fractional polarity		0
Nitro cellulose dil ratio[d]		
Nitro cellulose dil ratio[e]		
Surface tension	mN/m @ 20 °C	19
Specific heat liquid	kJ/kg mole/°C @ 20 °C	178.0
Latent heat	kJ/kg mole	28480
Dielectric constant	@ 20 °C	1.9
Antoine constant A	kPa°C	5.70103
Antoine constant B	kPa°C	1279.39
Antoine constant C	kPa°C	273.163
Heat of combustion	kJ/kg mole	−4307600
UN number		1271
IMO classification		3.1/II
ADR/RID classification		3,3°(b)
UK exposure limits	mg/m^3 (ppm) 8hrTWA/MEL or OES	
USA exposure limits	mg/m^3 (ppm) TLV/TWA	
German exposure limits	mg/m^3 (ppm) MAK-TWA	
EU classification		F, Xn
EU risk phrases		48-20
EU safety phrases		9-16-24/25-29-51

[a] Ether = 1. [b] n-Butyl acetate = 1. [c] Nelson, Hemwall, Edwards. [d] Toluene. [e] Hydrocarbon solvent.

Solvent group		HYDROCARBONS—ALIPHATIC/CYCLIC
Chemical name		
Chemical formula		
Structural formula		
CAS No.		64742-49-0
Synonym/common name		
Trade names		SBP 60/95 LNH
Suppliers		SHELL
Molecular mass		92
Boiling point/range	°C @ 760 mm Hg	60–90
Melting point	°C	
Density	kg/m^3 @ 20 °C	693
Evaporation time[a]		1.8
Evaporation time[b]		7.6
Vapour pressure	kPa @ 20 °C	19
Flash point	°C	−25 (IP 170)
Auto ignition temperature	°C	270
Upper flammability limit	%v/v in air	8.0
Lower flammability limit	%v/v in air	0.8
Viscosity	mPa.s (cP) @ 20 °C	0.31
Refractive index	n 20 D	1.388
Solubility in water	%w/w @ 20 °C	
Solubility of water	%w/w @ 20 °C	
Solubility parameter[c]	$(J/cm^3)^{1/2}$	15.1
Hydrogen bond index		0
Fractional polarity		0
Nitro cellulose dil ratio[d]		
Nitro cellulose dil ratio[e]		
Surface tension	mN/m @ 20 °C	19.5
Specific heat liquid	kJ/kg mole/°C @ 20 °C	
Latent heat	kJ/kg mole	29440
Dielectric constant	@ 20 °C	
Antoine constant A	kPa°C	6.15568
Antoine constant B	kPa°C	1278.38
Antoine constant C	kPa°C	244.419
Heat of combustion	kJ/kg mole	−4452800
UN number		1271
IMO classification		3.1/II
ADR/RID classification		3,3°(b)
UK exposure limits	mg/m^3 (ppm) 8hrTWA/MEL or OES	
USA exposure limits	mg/m^3 (ppm) TLV/TWA	
German exposure limits	mg/m^3 (ppm) MAK-TWA	
EU classification		F
EU risk phrases		
EU safety phrases		9-16-29-33

[a] Ether = 1. [b] n-Butyl acetate = 1. [c] Nelson, Hemwall, Edwards. [d] Toluene. [e] Hydrocarbon solvent.

Solvent group		**HYDROCARBONS—ALIPHATIC/CYCLIC**
Chemical name		
Chemical formula		
Structural formula		
CAS No.		64742-49-0
Synonym/common name		
Trade names		SBP 80/110 LNH
Suppliers		SHELL
Molecular mass		99
Boiling point/range	°C @ 760 mm Hg	87–104
Melting point	°C	
Density	kg/m^3 @ 20 °C	714
Evaporation time[a]		3.2
Evaporation time[b]		4.5
Vapour pressure	kPa @ 20 °C	8.5
Flash point	°C	−12 (IP 170)
Auto ignition temperature	°C	250
Upper flammability limit	%v/v in air	8.0
Lower flammability limit	%v/v in air	0.8
Viscosity	mPa.s (cP) @ 20 °C	0.41
Refractive index	n 20 D	1.397
Solubility in water	%w/w @ 20 °C	
Solubility of water	%w/w @ 20 °C	
Solubility parameter[c]	$(J/cm^3)^{1/2}$	15.1
Hydrogen bond index		0
Fractional polarity		0
Nitro cellulose dil ratio[d]		
Nitro cellulose dil ratio[e]		
Surface tension	mN/m @ 20 °C	20.5
Specific heat liquid	kJ/kg mole/°C @ 20 °C	198.0
Latent heat	kJ/kg mole	31680
Dielectric constant	@ 20 °C	2.0
Antoine constant A	kPa°C	6.47294
Antoine constant B	kPa°C	1600.81
Antoine constant C	kPa°C	273.149
Heat of combustion	kJ/kg mole	−4393620
UN number		1271
IMO classification		3.1/II
ADR/RID classification		3,3°(b)
UK exposure limits	mg/m^3 (ppm) 8hrTWA/MEL or OES	
USA exposure limits	mg/m^3 (ppm) TLV/TWA	
German exposure limits	mg/m^3 (ppm) MAK-TWA	
EU classification		F
EU risk phrases		
EU safety phrases		9-16-29-33

[a] Ether = 1. [b] n-Butyl acetate = 1. [c] Nelson, Hemwall, Edwards. [d] Toluene. [e] Hydrocarbon solvent.

Solvent group		HYDROCARBONS—ALIPHATIC/CYCLIC
Chemical name		
Chemical formula		
Structural formula		
CAS No.		64742-49-0
Synonym/common name		Heptane
Trade names		SBP 94/100
Suppliers		SHELL
Molecular mass		100
Boiling point/range	°C @ 760 mm Hg	94–99
Melting point	°C	
Density	kg/m^3 @ 20 °C	710
Evaporation time[a]		3.0
Evaporation time[b]		4.1
Vapour pressure	kPa @ 20 °C	8.5
Flash point	°C	−5 (IP 170)
Auto ignition temperature	°C	215
Upper flammability limit	%v/v in air	8.0
Lower flammability limit	%v/v in air	0.8
Viscosity	mPa.s (cP) @ 20 °C	0.44
Refractive index	n 20 D	1.397
Solubility in water	%w/w @ 20 °C	
Solubility of water	%w/w @ 20 °C	
Solubility parameter[c]	$(J/cm^3)^{1/2}$	15.3
Hydrogen bond index		0
Fractional polarity		0
Nitro cellulose dil ratio[d]		
Nitro cellulose dil ratio[e]		
Surface tension	mN/m @ 20 °C	20.5
Specific heat liquid	kJ/kg mole/°C @ 20 °C	210.0
Latent heat	kJ/kg mole	31500
Dielectric constant	@ 20 °C	1.9
Antoine constant A	kPa°C	6.09439
Antoine constant B	kPa°C	1303.88
Antoine constant C	kPa°C	220.525
Heat of combustion	kJ/kg mole	−4807000
UN number		1206
IMO classification		3.2/II
ADR/RID classification		3,3°(b)
UK exposure limits	mg/m^3 (ppm) 8hrTWA/MEL or OES	
USA exposure limits	mg/m^3 (ppm) TLV/TWA	
German exposure limits	mg/m^3 (ppm) MAK-TWA	
EU classification		F
EU risk phrases		
EU safety phrases		9-16-23-29-33

[a] Ether = 1. [b] n-Butyl acetate = 1. [c] Nelson, Hemwall, Edwards. [d] Toluene. [e] Hydrocarbon solvent.

Solvent group		**HYDROCARBONS—ALIPHATIC/CYCLIC**
Chemical name		
Chemical formula		
Structural formula		
CAS No.		64742-49-0
Synonym/common name		
Trade names		SBP 100/140
Suppliers		SHELL
Molecular mass		112
Boiling point/range	°C @ 760 mm Hg	105–137
Melting point	°C	
Density	kg/m^3 @ 20 °C	728
Evaporation time[a]		6.0
Evaporation time[b]		1.9
Vapour pressure	kPa @ 20 °C	3.5
Flash point	°C	1 (IP 170)
Auto ignition temperature	°C	220
Upper flammability limit	%v/v in air	8.0
Lower flammability limit	%v/v in air	0.8
Viscosity	mPa.s (cP) @ 20 °C	0.52
Refractive index	n 20 D	1.405
Solubility in water	%w/w @ 20 °C	
Solubility of water	%w/w @ 20 °C	
Solubility parameter[c]	$(J/cm^3)^{1/2}$	15.5
Hydrogen bond index		0
Fractional polarity		0
Nitro cellulose dil ratio[d]		
Nitro cellulose dil ratio[e]		
Surface tension	mN/m @ 20 °C	21.5
Specific heat liquid	kJ/kg mole/°C @ 20 °C	224.0
Latent heat	kJ/kg mole	
Dielectric constant	@ 20 °C	2.0
Antoine constant A	kPa°C	6.45962
Antoine constant B	kPa°C	1710.5
Antoine constant C	kPa°C	273.2
Heat of combustion	kJ/kg mole	−4970560
UN number		1300
IMO classification		3.2/II
ADR/RID classification		3,3°(b)
UK exposure limits	mg/m^3 (ppm) 8hrTWA/MEL or OES	
USA exposure limits	mg/m^3 (ppm) TLV/TWA	
German exposure limits	mg/m^3 (ppm) MAK-TWA	
EU classification		F
EU risk phrases		
EU safety phrases		9-16-29-33

[a] Ether = 1. [b] n-Butyl acetate = 1. [c] Nelson, Hemwall, Edwards. [d] Toluene. [e] Hydrocarbon solvent.

Solvent group		**HYDROCARBONS—ALIPHATIC/CYCLIC**
Chemical name		
Chemical formula		
Structural formula		
CAS No.		64742-49-0
Synonym/common name		
Trade names		SHELLSOL D25/SBP 140/165
Suppliers		SHELL
Molecular mass		132
Boiling point/range	°C @ 760 mm Hg	141–162
Melting point	°C	
Density	kg/m^3 @ 20 °C	750
Evaporation time[a]		20
Evaporation time[b]		0.6
Vapour pressure	kPa @ 20 °C	1
Flash point	°C	271 (IP 170)
Auto ignition temperature	°C	220
Upper flammability limit	%v/v in air	8.0
Lower flammability limit	%v/v in air	0.6
Viscosity	mPa.s (cP) @ 20 °C	0.68
Refractive index	n 20 D	1.415
Solubility in water	%w/w @ 20 °C	
Solubility of water	%w/w @ 20 °C	
Solubility parameter[c]	(J/cm^3)$^{1/2}$	15.5
Hydrogen bond index		0
Fractional polarity		0
Nitro cellulose dil ratio[d]		
Nitro cellulose dil ratio[e]		
Surface tension	mN/m @ 20 °C	23
Specific heat liquid	kJ/kg mole/°C @ 20 °C	
Latent heat	kJ/kg mole	
Dielectric constant	@ 20 °C	
Antoine constant A	kPa°C	9.5124
Antoine constant B	kPa°C	4640.78
Antoine constant C	kPa°C	461.254
Heat of combustion	kJ/kg mole	−5747280
UN number		1300
IMO classification		3.3/III
ADR/RID classification		3,31°(c)
UK exposure limits	mg/m^3 (ppm) 8hrTWA/MEL or OES	
USA exposure limits	mg/m^3 (ppm) TLV/TWA	
German exposure limits	mg/m^3 (ppm) MAK-TWA	
EU classification		
EU risk phrases		10
EU safety phrases		

[a] Ether = 1. [b] n-Butyl acetate = 1. [c] Nelson, Hemwall, Edwards. [d] Toluene. [e] Hydrocarbon solvent.

Solvent group		HYDROCARBONS—ALIPHATIC/CYCLIC
Chemical name		
Chemical formula		
Structural formula		
CAS No.		64742-48-9
Synonym/common name		
Trade names		SHELLSOL D40
Suppliers		SHELL
Molecular mass		141
Boiling point/range	°C @ 760 mm Hg	162–197
Melting point	°C	
Density	kg/m^3 @ 20 °C	765
Evaporation time[a]		85
Evaporation time[b]		0.19
Vapour pressure	kPa @ 20 °C	0.3
Flash point	°C	40 (IP 170)
Auto ignition temperature	°C	240
Upper flammability limit	%v/v in air	6.0
Lower flammability limit	%v/v in air	0.7
Viscosity	mPa.s (cP) @ 20 °C	0.87
Refractive index	n 20 D	1.424
Solubility in water	%w/w @ 20 °C	
Solubility of water	%w/w @ 20 °C	
Solubility parameter[c]	$(J/cm^3)^{1/2}$	15.5
Hydrogen bond index		0
Fractional polarity		0
Nitro cellulose dil ratio[d]		
Nitro cellulose dil ratio[e]		
Surface tension	mN/m @ 20 °C	24.5
Specific heat liquid	kJ/kg mole/°C @ 20 °C	
Latent heat	kJ/kg mole	
Dielectric constant	@ 20 °C	2.1
Antoine constant A	kPa°C	6.5971
Antoine constant B	kPa°C	2077.8
Antoine constant C	kPa°C	257.6
Heat of combustion	kJ/kg mole	−6415500
UN number		1300
IMO classification		3.3/III
ADR/RID classification		3,31°(c)
UK exposure limits	mg/m^3 (ppm) 8hrTWA/MEL or OES	
USA exposure limits	mg/m^3 (ppm) TLV/TWA	
German exposure limits	mg/m^3 (ppm) MAK-TWA	
EU classification		
EU risk phrases		10
EU safety phrases		

[a] Ether = 1. [b] n-Butyl acetate = 1. [c] Nelson, Hemwall, Edwards. [d] Toluene. [e] Hydrocarbon solvent.

Solvent group		**HYDROCARBONS—ALIPHATIC/CYCLIC**
Chemical name		
Chemical formula		
Structural formula		
CAS No.		64742-48-9
Synonym/common name		
Trade names		SHELLSOL D60
Suppliers		SHELL
Molecular mass		154
Boiling point/range	°C @ 760 mm Hg	187–211
Melting point	°C	
Density	kg/m^3 @ 20 °C	786
Evaporation time[a]		200
Evaporation time[b]		0.04
Vapour pressure	kPa @ 20 °C	0.1
Flash point	°C	66 (ASTM D93)
Auto ignition temperature	°C	225
Upper flammability limit	%v/v in air	6.0
Lower flammability limit	%v/v in air	0.76
Viscosity	mPa.s (cP) @ 20 °C	1.23
Refractive index	n 20 D	1.434
Solubility in water	%w/w @ 20 °C	
Solubility of water	%w/w @ 20 °C	
Solubility parameter[c]	$(J/cm^3)^{1/2}$	15.5
Hydrogen bond index		0
Fractional polarity		0
Nitro cellulose dil ratio[d]		
Nitro cellulose dil ratio[e]		
Surface tension	mN/m @ 20 °C	26
Specific heat liquid	kJ/kg mole/°C @ 20 °C	292.6
Latent heat	kJ/kg mole	
Dielectric constant	@ 20 °C	2.1
Antoine constant A	kPa°C	6.91546
Antoine constant B	kPa°C	2225.63
Antoine constant C	kPa°C	257.923
Heat of combustion	kJ/kg mole	−6976200
UN number		
IMO classification		
ADR/RID classification		3,32°(c)
UK exposure limits	mg/m^3 (ppm) 8hrTWA/MEL or OES	
USA exposure limits	mg/m^3 (ppm) TLV/TWA	
German exposure limits	mg/m^3 (ppm) MAK-TWA	
EU classification		
EU risk phrases		
EU safety phrases		

[a] Ether = 1. [b] n-Butyl acetate = 1. [c] Nelson, Hemwall, Edwards. [d] Toluene. [e] Hydrocarbon solvent.

Solvent group		**HYDROCARBONS—ALIPHATIC/CYCLIC**
Chemical name		
Chemical formula		
Structural formula		
CAS No.		64742-47-8
Synonym/common name		
Trade names		SHELLSOL D70
Suppliers		SHELL
Molecular mass		174
Boiling point/range	°C @ 760 mm Hg	185–245
Melting point	°C	
Density	kg/m^3 @ 20 °C	792
Evaporation time[a]		800
Evaporation time[b]		0.01
Vapour pressure	kPa @ 20 °C	0.06
Flash point	°C	73 (ASTM D93)
Auto ignition temperature	°C	225
Upper flammability limit	%v/v in air	5.5
Lower flammability limit	%v/v in air	0.6
Viscosity	mPa.s (cP) @ 20 °C	1.56
Refractive index	n 20 D	1.437
Solubility in water	%w/w @ 20 °C	
Solubility of water	%w/w @ 20 °C	
Solubility parameter[c]	$(J/cm^3)^{1/2}$	15.5
Hydrogen bond index		0
Fractional polarity		0
Nitro cellulose dil ratio[d]		
Nitro cellulose dil ratio[e]		
Surface tension	mN/m @ 20 °C	26
Specific heat liquid	kJ/kg mole/°C @ 20 °C	
Latent heat	kJ/kg mole	
Dielectric constant	@ 20 °C	2.15
Antoine constant A	kPa°C	5.9908
Antoine constant B	kPa°C	1753.0
Antoine constant C	kPa°C	221.03
Heat of combustion	kJ/kg mole	−7871238
UN number		
IMO classification		
ADR/RID classification		3,32°(c)
UK exposure limits	mg/m^3 (ppm) 8hrTWA/MEL or OES	
USA exposure limits	mg/m^3 (ppm) TLV/TWA	
German exposure limits	mg/m^3 (ppm) MAK-TWA	
EU classification		
EU risk phrases		
EU safety phrases		

[a] Ether = 1. [b] n-Butyl acetate = 1. [c] Nelson, Hemwall, Edwards. [d] Toluene. [e] Hydrocarbon solvent.

Solvent group		**HYDROCARBONS—ALIPHATIC/CYCLIC**
Chemical name		
Chemical formula		
Structural formula		
CAS No.		64742-46-7
Synonym/common name		
Trade names		SHELLSOL D120
Suppliers		SHELL
Molecular mass		218
Boiling point/range	°C @ 760 mm Hg	250–300
Melting point	°C	
Density	kg/m^3 @ 20 °C	822
Evaporation time[a]		>3900
Evaporation time[b]		<0.003
Vapour pressure	kPa @ 20 °C	
Flash point	°C	120 (DIN EN22719)
Auto ignition temperature	°C	215
Upper flammability limit	%v/v in air	5.5
Lower flammability limit	%v/v in air	0.5
Viscosity	mPa.s (cP) @ 20 °C	5.8
Refractive index	n 20 D	1.454
Solubility in water	%w/w @ 20 °C	
Solubility of water	%w/w @ 20 °C	
Solubility parameter[c]	(J/cm^3)$^{1/2}$	16.0
Hydrogen bond index		0
Fractional polarity		0
Nitro cellulose dil ratio[d]		
Nitro cellulose dil ratio[e]		
Surface tension	mN/m @ 20 °C	40
Specific heat liquid	kJ/kg mole/°C @ 20 °C	
Latent heat	kJ/kg mole	
Dielectric constant	@ 20 °C	
Antoine constant A	kPa°C	
Antoine constant B	kPa°C	
Antoine constant C	kPa°C	
Heat of combustion	kJ/kg mole	
UN number		
IMO classification		
ADR/RID classification		
UK exposure limits	mg/m^3 (ppm) 8hrTWA/MEL or OES	
USA exposure limits	mg/m^3 (ppm) TLV/TWA	
German exposure limits	mg/m^3 (ppm) MAK-TWA	
EU classification		
EU risk phrases		
EU safety phrases		

[a] Ether = 1. [b] n-Butyl acetate = 1. [c] Nelson, Hemwall, Edwards. [d] Toluene. [e] Hydrocarbon solvent.

Property	Unit	Value
Solvent group		**HYDROCARBONS—ALIPHATIC/CYCLIC/AROMATIC**
Chemical name		
Chemical formula		
Structural formula		
CAS No.		64742-80-9
Synonym/common name		
Trade names		INK SOLVENT 2629
Suppliers		SHELL
Molecular mass		212
Boiling point/range	°C @ 760 mm Hg	262–288
Melting point	°C	
Density	kg/m^3 @ 20 °C	831
Evaporation time[a]		>3900
Evaporation time[b]		<0.01
Vapour pressure	kPa @ 20 °C	<0.01
Flash point	°C	128 (ASTM D93)
Auto ignition temperature	°C	230
Upper flammability limit	%v/v in air	8.0
Lower flammability limit	%v/v in air	0.6
Viscosity	mPa.s (cP) @ 20 °C	3.47
Refractive index	n 20 D	1.465
Solubility in water	%w/w @ 20 °C	
Solubility of water	%w/w @ 20 °C	
Solubility parameter[c]	$(J/cm^3)^{1/2}$	16.6
Hydrogen bond index		0.9
Fractional polarity		0
Nitro cellulose dil ratio[d]		
Nitro cellulose dil ratio[e]		
Surface tension	mN/m @ 20 °C	
Specific heat liquid	kJ/kg mole/°C @ 20 °C	
Latent heat	kJ/kg mole	
Dielectric constant	@ 20 °C	
Antoine constant A	kPa°C	
Antoine constant B	kPa°C	
Antoine constant C	kPa°C	
Heat of combustion	kJ/kg mole	
UN number		
IMO classification		
ADR/RID classification		
UK exposure limits	mg/m^3 (ppm) 8hrTWA/MEL or OES	
USA exposure limits	mg/m^3 (ppm) TLV/TWA	
German exposure limits	mg/m^3 (ppm) MAK-TWA	
EU classification		
EU risk phrases		
EU safety phrases		

[a] Ether = 1. [b] n-Butyl acetate = 1. [c] Nelson, Hemwall, Edwards. [d] Toluene. [e] Hydrocarbon solvent.

Solvent group		**HYDROCARBONS—ALIPHATIC/CYCLIC/AROMATIC**
Chemical name		
Chemical formula		
Structural formula		
CAS No.		
Synonym/common name		HAWS
Trade names		HIGH AROMATIC WHITE SPIRIT
Suppliers		SHELL
Molecular mass		
Boiling point/range	°C @ 760 mm Hg	160–195
Melting point	°C	
Density	kg/m^3 @ 20 °C	815
Evaporation time[a]		
Evaporation time[b]		0.20
Vapour pressure	kPa @ 20 °C	0.3
Flash point	°C	40 (IP 170)
Auto ignition temperature	°C	265
Upper flammability limit	%v/v in air	6.5
Lower flammability limit	%v/v in air	0.7
Viscosity	mPa.s (cP) @ 20 °C	0.92
Refractive index	n 20 D	1.485
Solubility in water	%w/w @ 20 °C	
Solubility of water	%w/w @ 20 °C	
Solubility parameter[c]	(J/cm^3)$^{1/2}$	
Hydrogen bond index		
Fractional polarity		
Nitro cellulose dil ratio[d]		
Nitro cellulose dil ratio[e]		
Surface tension	mN/m @ 20 °C	
Specific heat liquid	kJ/kg mole/°C @ 20 °C	
Latent heat	kJ/kg mole	
Dielectric constant	@ 20 °C	
Antoine constant A	kPa°C	
Antoine constant B	kPa°C	
Antoine constant C	kPa°C	
Heat of combustion	kJ/kg mole	
UN number		1300
IMO classification		3.3/III
ADR/RID classification		3,31°(c)
UK exposure limits	mg/m^3 (ppm) 8hrTWA/MEL or OES	
USA exposure limits	mg/m^3 (ppm) TLV/TWA	
German exposure limits	mg/m^3 (ppm) MAK-TWA	
EU classification		
EU risk phrases		10
EU safety phrases		

[a] Ether = 1. [b] n-Butyl acetate = 1. [c] Nelson, Hemwall, Edwards. [d] Toluene. [e] Hydrocarbon solvent.

Solvent group		HYDROCARBONS—ALIPHATIC/CYCLIC/AROMATIC
Chemical name		
Chemical formula		
Structural formula		
CAS No.		
Synonym/common name		
Trade names		SHELLSOL E
Suppliers		SHELL
Molecular mass		125
Boiling point/range	°C @ 760 mm Hg	164–188
Melting point	°C	
Density	kg/m^3 @ 20 °C	856
Evaporation time[a]		79
Evaporation time[b]		0.20
Vapour pressure	kPa @ 20 °C	0.3
Flash point	°C	42 (IP 170)
Auto ignition temperature	°C	450
Upper flammability limit	%v/v in air	6.5
Lower flammability limit	%v/v in air	0.7
Viscosity	mPa.s (cP) @ 20 °C	0.79
Refractive index	n 20 D	1.487
Solubility in water	%w/w @ 20 °C	
Solubility of water	%w/w @ 20 °C	
Solubility parameter[c]	$(J/cm^3)^{1/2}$	17.2
Hydrogen bond index		4.5
Fractional polarity		0.001
Nitro cellulose dil ratio[d]		
Nitro cellulose dil ratio[e]		
Surface tension	mN/m @ 20 °C	28
Specific heat liquid	kJ/kg mole/°C @ 20 °C	
Latent heat	kJ/kg mole	
Dielectric constant	@ 20 °C	2.4
Antoine constant A	kPa°C	6.33229
Antoine constant B	kPa°C	1708.12
Antoine constant C	kPa°C	226.627
Heat of combustion	kJ/kg mole	−5186250
UN number		1300
IMO classification		3.3/III
ADR/RID classification		3,31°(c)
UK exposure limits	mg/m^3 (ppm) 8hrTWA/MEL or OES	
USA exposure limits	mg/m^3 (ppm) TLV/TWA	
German exposure limits	mg/m^3 (ppm) MAK-TWA	
EU classification		
EU risk phrases		10
EU safety phrases		

[a] Ether = 1. [b] n-Butyl acetate = 1. [c] Nelson, Hemwall, Edwards. [d] Toluene. [e] Hydrocarbon solvent.

Solvent group		**HYDROCARBONS—ALIPHATIC/CYCLIC/AROMATIC**
Chemical name		
Chemical formula		
Structural formula		
CAS No.		64742-82-1
Synonym/common name		LAWS
Trade names		LOW AROMATIC WHITE SPIRIT
Suppliers		SHELL
Molecular mass		140
Boiling point/range	°C @ 760 mm Hg	160–198
Melting point	°C	
Density	kg/m^3 @ 20 °C	775
Evaporation time[a]		85
Evaporation time[b]		0.19
Vapour pressure	kPa @ 20 °C	0.4
Flash point	°C	40 (IP 170)
Auto ignition temperature	°C	240
Upper flammability limit	%v/v in air	6.5
Lower flammability limit	%v/v in air	0.7
Viscosity	mPa.s (cP) @ 20 °C	0.84
Refractive index	n 20 D	1.434
Solubility in water	%w/w @ 20 °C	
Solubility of water	%w/w @ 20 °C	
Solubility parameter[c]	$(J/cm^3)^{1/2}$	16.2
Hydrogen bond index		0.5
Fractional polarity		0
Nitro cellulose dil ratio[d]		
Nitro cellulose dil ratio[e]		
Surface tension	mN/m @ 20 °C	26.4
Specific heat liquid	kJ/kg mole/°C @ 20 °C	
Latent heat	kJ/kg mole	
Dielectric constant	@ 20 °C	2.1
Antoine constant A	kPa°C	7.8400
Antoine constant B	kPa°C	2745.4
Antoine constant C	kPa°C	311.86
Heat of combustion	kJ/kg mole	−6090000
UN number		1300
IMO classification		3.3/III
ADR/RID classification		3,31°(c)
UK exposure limits	mg/m^3 (ppm) 8hrTWA/MEL or OES	
USA exposure limits	mg/m^3 (ppm) TLV/TWA	
German exposure limits	mg/m^3 (ppm) MAK-TWA	
EU classification		
EU risk phrases		10
EU safety phrases		

[a] Ether = 1. [b] n-Butyl acetate = 1. [c] Nelson, Hemwall, Edwards. [d] Toluene. [e] Hydrocarbon solvent.

Solvent group		HYDROCARBONS—ALIPHATIC/CYCLIC/AROMATIC
Chemical name		
Chemical formula		
Structural formula		
CAS No.		64742-80-9
Synonym/common name		
Trade names		SHELLSOL R
Suppliers		SHELL
Molecular mass		156
Boiling point/range	°C @ 760 mm Hg	205–270
Melting point	°C	
Density	kg/m^3 @ 20 °C	890
Evaporation time[a]		600
Evaporation time[b]		<0.01
Vapour pressure	kPa @ 20 °C	0.04
Flash point	°C	85 (ASTM D93)
Auto ignition temperature	°C	450
Upper flammability limit	%v/v in air	8.0
Lower flammability limit	%v/v in air	0.6
Viscosity	mPa.s (cP) @ 20 °C	
Refractive index	n 20 D	1.502
Solubility in water	%w/w @ 20 °C	
Solubility of water	%w/w @ 20 °C	
Solubility parameter[c]	(J/cm^3)$^{1/2}$	17.2
Hydrogen bond index		5.0
Fractional polarity		0.001
Nitro cellulose dil ratio[d]		
Nitro cellulose dil ratio[e]		
Surface tension	mN/m @ 20 °C	29
Specific heat liquid	kJ/kg mole/°C @ 20 °C	
Latent heat	kJ/kg mole	
Dielectric constant	@ 20 °C	2.5
Antoine constant A	kPa°C	6.98289
Antoine constant B	kPa°C	2304.84
Antoine constant C	kPa°C	251.488
Heat of combustion	kJ/kg mole	−6459960
UN number		3082
IMO classification		9/III
ADR/RID classification		3,32°(c)
UK exposure limits	mg/m^3 (ppm) 8hrTWA/MEL or OES	
USA exposure limits	mg/m^3 (ppm) TLV/TWA	
German exposure limits	mg/m^3 (ppm) MAK-TWA	
EU classification		
EU risk phrases		
EU safety phrases		

[a] Ether = 1. [b] n-Butyl acetate = 1. [c] Nelson, Hemwall, Edwards. [d] Toluene. [e] Hydrocarbon solvent.

Solvent group		HYDROCARBONS—ALIPHATIC/CYCLIC/AROMATIC
Chemical name		
Chemical formula		
Structural formula		
CAS No.		64742-82-1
Synonym/common name		
Trade names		SHELLSOL H
Suppliers		SHELL
Molecular mass		158
Boiling point/range	°C @ 760 mm Hg	180–210
Melting point	°C	
Density	kg/m^3 @ 20 °C	790
Evaporation time[a]		200
Evaporation time[b]		0.04
Vapour pressure	kPa @ 20 °C	0.15
Flash point	°C	63 (ASTM D93)
Auto ignition temperature	°C	230
Upper flammability limit	%v/v in air	6.5
Lower flammability limit	%v/v in air	0.7
Viscosity	mPa.s (cP) @ 20 °C	1.14
Refractive index	n 20 D	1.440
Solubility in water	%w/w @ 20 °C	
Solubility of water	%w/w @ 20 °C	
Solubility parameter[c]	(J/cm^3)$^{1/2}$	15.8
Hydrogen bond index		0.5
Fractional polarity		0
Nitro cellulose dil ratio[d]		
Nitro cellulose dil ratio[e]		
Surface tension	mN/m @ 20 °C	26.0
Specific heat liquid	kJ/kg mole/°C @ 20 °C	300.2
Latent heat	kJ/kg mole	
Dielectric constant	@ 20 °C	2.1
Antoine constant A	kPa°C	6.14569
Antoine constant B	kPa°C	1636.33
Antoine constant C	kPa°C	200.436
Heat of combustion	kJ/kg mole	−6873000
UN number		3082
IMO classification		9/III
ADR/RID classification		3,32°(c)
UK exposure limits	mg/m^3 (ppm) 8hrTWA/MEL or OES	
USA exposure limits	mg/m^3 (ppm) TLV/TWA	
German exposure limits	mg/m^3 (ppm) MAK-TWA	
EU classification		
EU risk phrases		
EU safety phrases		

[a] Ether = 1. [b] n-Butyl acetate = 1. [c] Nelson, Hemwall, Edwards. [d] Toluene. [e] Hydrocarbon solvent.

Solvent group		**HYDROCARBONS—ALIPHATIC/CYCLIC/AROMATIC**
Chemical name		
Chemical formula		
Structural formula		
CAS No.		64742-80-9
Synonym/common name		
Trade names		INK SOLVENT 2427
Suppliers		SHELL
Molecular mass		198
Boiling point/range	°C @ 760 mm Hg	243–267
Melting point	°C	
Density	kg/m^3 @ 20 °C	826
Evaporation time[a]		3900
Evaporation time[b]		<0.01
Vapour pressure	kPa @ 20 °C	<0.01
Flash point	°C	106 (ASTM D93)
Auto ignition temperature	°C	230
Upper flammability limit	%v/v in air	8.0
Lower flammability limit	%v/v in air	0.6
Viscosity	mPa.s (cP) @ 20 °C	2.63
Refractive index	n 20 D	1.462
Solubility in water	%w/w @ 20 °C	
Solubility of water	%w/w @ 20 °C	
Solubility parameter[c]	(J/cm^3)$^{1/2}$	16.4
Hydrogen bond index		0.7
Fractional polarity		0
Nitro cellulose dil ratio[d]		
Nitro cellulose dil ratio[e]		
Surface tension	mN/m @ 20 °C	
Specific heat liquid	kJ/kg mole/°C @ 20 °C	
Latent heat	kJ/kg mole	
Dielectric constant	@ 20 °C	
Antoine constant A	kPa°C	
Antoine constant B	kPa°C	
Antoine constant C	kPa°C	
Heat of combustion	kJ/kg mole	
UN number		
IMO classification		
ADR/RID classification		
UK exposure limits	mg/m^3 (ppm) 8hrTWA/MEL or OES	
USA exposure limits	mg/m^3 (ppm) TLV/TWA	
German exposure limits	mg/m^3 (ppm) MAK-TWA	
EU classification		
EU risk phrases		
EU safety phrases		

[a] Ether = 1. [b] n-Butyl acetate = 1. [c] Nelson, Hemwall, Edwards. [d] Toluene. [e] Hydrocarbon solvent.

Solvent group		HYDROCARBONS—ALIPHATIC/CYCLIC/AROMATIC
Chemical name		
Chemical formula		
Structural formula		
CAS No.		64742-80-9
Synonym/common name		
Trade names		INK SOLVENT 2831
Suppliers		SHELL
Molecular mass		226
Boiling point/range	°C @ 760 mm Hg	283–310
Melting point	°C	
Density	kg/m^3 @ 20 °C	835
Evaporation time[a]		>3900
Evaporation time[b]		<0.01
Vapour pressure	kPa @ 20 °C	<0.01
Flash point	°C	136 (ASTM D93)
Auto ignition temperature	°C	230
Upper flammability limit	%v/v in air	8.0
Lower flammability limit	%v/v in air	0.6
Viscosity	mPa.s (cP) @ 20 °C	4.17
Refractive index	n 20 D	1.464
Solubility in water	%w/w @ 20 °C	
Solubility of water	%w/w @ 20 °C	
Solubility parameter[c]	$(J/cm^3)^{1/2}$	16.6
Hydrogen bond index		0.9
Fractional polarity		0
Nitro cellulose dil ratio[d]		
Nitro cellulose dil ratio[e]		
Surface tension	mN/m @ 20 °C	
Specific heat liquid	kJ/kg mole/°C @ 20 °C	
Latent heat	kJ/kg mole	
Dielectric constant	@ 20 °C	
Antoine constant A	kPa°C	
Antoine constant B	kPa°C	
Antoine constant C	kPa°C	
Heat of combustion	kJ/kg mole	
UN number		
IMO classification		
ADR/RID classification		
UK exposure limits	mg/m^3 (ppm) 8hrTWA/MEL or OES	
USA exposure limits	mg/m^3 (ppm) TLV/TWA	
German exposure limits	mg/m^3 (ppm) MAK-TWA	
EU classification		
EU risk phrases		
EU safety phrases		

[a] Ether = 1. [b] n-Butyl acetate = 1. [c] Nelson, Hemwall, Edwards. [d] Toluene. [e] Hydrocarbon solvent.

Solvent group		HYDROCARBONS—AROMATIC
Chemical name		
Chemical formula		
Structural formula		
CAS No.		90989-39-2
Synonym/common name		
Trade names		SOLVAREX 10
Suppliers		TOTAL
Molecular mass		
Boiling point/range	°C @ 760 mm Hg	188–215
Melting point	°C	
Density	kg/m^3 @ 20 °C	890
Evaporation time[a]		130
Evaporation time[b]		
Vapour pressure	kPa @ 20 °C	
Flash point	°C	64
Auto ignition temperature	°C	
Upper flammability limit	%v/v in air	
Lower flammability limit	%v/v in air	
Viscosity	mPa.s (cP) @ 20 °C	
Refractive index	n 20 D	1.508
Solubility in water	%w/w @ 20 °C	
Solubility of water	%w/w @ 20 °C	
Solubility parameter[c]	$(J/cm^3)^{1/2}$	
Hydrogen bond index		
Fractional polarity		
Nitro cellulose dil ratio[d]		
Nitro cellulose dil ratio[e]		
Surface tension	mN/m @ 20 °C	
Specific heat liquid	kJ/kg mole/°C @ 20 °C	
Latent heat	kJ/kg mole	
Dielectric constant	@ 20 °C	
Antoine constant A	kPa°C	
Antoine constant B	kPa°C	
Antoine constant C	kPa°C	
Heat of combustion	kJ/kg mole	
UN number		1300
IMO classification		none
ADR/RID classification		3,32°(c)
UK exposure limits	mg/m^3 (ppm) 8hrTWA/MEL or OES	
USA exposure limits	mg/m^3 (ppm) TLV/TWA	
German exposure limits	mg/m^3 (ppm) MAK-TWA	
EU classification		
EU risk phrases		
EU safety phrases		

[a] Ether = 1. [b] n-Butyl acetate = 1. [c] Nelson, Hemwall, Edwards. [d] Toluene. [e] Hydrocarbon solvent.

Solvent group		**HYDROCARBONS—AROMATIC**
Chemical name		Solvent Naphtha–Heavy Aromatic
Chemical formula		
Structural formula		
CAS No.		64742-94-5
Synonym/common name		Heavy Solvent Naphtha
Trade names		C9+ II
Suppliers		PETROGAL
Molecular mass		
Boiling point/range	°C @ 760 mm Hg	185–209
Melting point	°C	
Density	kg/m^3 @ 20 °C	880–890
Evaporation time[a]		
Evaporation time[b]		
Vapour pressure	kPa @ 20 °C	
Flash point	°C	>60 (IP 170)
Auto ignition temperature	°C	>420
Upper flammability limit	%v/v in air	
Lower flammability limit	%v/v in air	
Viscosity	mPa.s (cP) @ 20 °C	
Refractive index	n 20 D	1.51301
Solubility in water	%w/w @ 20 °C	
Solubility of water	%w/w @ 20 °C	
Solubility parameter[c]	(J/cm^3)$^{1/2}$	
Hydrogen bond index		
Fractional polarity		
Nitro cellulose dil ratio[d]		
Nitro cellulose dil ratio[e]		
Surface tension	mN/m @ 20 °C	
Specific heat liquid	kJ/kg mole/°C @ 20 °C	
Latent heat	kJ/kg mole	
Dielectric constant	@ 20 °C	
Antoine constant A	kPa°C	
Antoine constant B	kPa°C	
Antoine constant C	kPa°C	
Heat of combustion	kJ/kg mole	
UN number		1256
IMO classification		3.3
ADR/RID classification		3,31°(c)
UK exposure limits	mg/m^3 (ppm) 8hrTWA/MEL or OES	
USA exposure limits	mg/m^3 (ppm) TLV/TWA	
German exposure limits	mg/m^3 (ppm) MAK-TWA	
EU classification		Xi
EU risk phrases		10-38-52/53
EU safety phrases		62-61

[a] Ether = 1. [b] n-Butyl acetate = 1. [c] Nelson, Hemwall, Edwards. [d] Toluene. [e] Hydrocarbon solvent.

Property	Unit	Value
Solvent group		**HYDROCARBONS—AROMATIC**
Chemical name		
Chemical formula		
Structural formula		
CAS No.		90989-39-2
Synonym/common name		
Trade names		SOLVAREX 90/160
Suppliers		TOTAL
Molecular mass		
Boiling point/range	°C @ 760 mm Hg	145–175
Melting point	°C	
Density	kg/m^3 @ 20 °C	870
Evaporation time[a]		25
Evaporation time[b]		
Vapour pressure	kPa @ 20 °C	
Flash point	°C	38
Auto ignition temperature	°C	
Upper flammability limit	%v/v in air	
Lower flammability limit	%v/v in air	
Viscosity	mPa.s (cP) @ 20 °C	
Refractive index	n 20 D	1.497
Solubility in water	%w/w @ 20 °C	
Solubility of water	%w/w @ 20 °C	
Solubility parameter[c]	$(J/cm^3)^{1/2}$	
Hydrogen bond index		
Fractional polarity		
Nitro cellulose dil ratio[d]		
Nitro cellulose dil ratio[e]		
Surface tension	mN/m @ 20 °C	
Specific heat liquid	kJ/kg mole/°C @ 20 °C	
Latent heat	kJ/kg mole	
Dielectric constant	@ 20 °C	
Antoine constant A	kPa°C	
Antoine constant B	kPa°C	
Antoine constant C	kPa°C	
Heat of combustion	kJ/kg mole	
UN number		1300
IMO classification		3.3
ADR/RID classification		3,31°(c)
UK exposure limits	mg/m^3 (ppm) 8hrTWA/MEL or OES	
USA exposure limits	mg/m^3 (ppm) TLV/TWA	
German exposure limits	mg/m^3 (ppm) MAK-TWA	
EU classification		
EU risk phrases		10
EU safety phrases		

[a] Ether = 1. [b] n-Butyl acetate = 1. [c] Nelson, Hemwall, Edwards. [d] Toluene. [e] Hydrocarbon solvent.

Solvent group		HYDROCARBONS—AROMATIC
Chemical name		
Chemical formula		
Structural formula		
CAS No.		64742-94-5
Synonym/common name		
Trade names		SOLVARO AFD
Suppliers		TOTAL
Molecular mass		
Boiling point/range	°C @ 760 mm Hg	190–225
Melting point	°C	
Density	kg/m^3 @ 20 °C	866
Evaporation time[a]		400
Evaporation time[b]		
Vapour pressure	kPa @ 20 °C	
Flash point	°C	73
Auto ignition temperature	°C	
Upper flammability limit	%v/v in air	
Lower flammability limit	%v/v in air	
Viscosity	mPa.s (cP) @ 20 °C	
Refractive index	n 20 D	1.491
Solubility in water	%w/w @ 20 °C	
Solubility of water	%w/w @ 20 °C	
Solubility parameter[c]	$(J/cm^3)^{1/2}$	
Hydrogen bond index		
Fractional polarity		
Nitro cellulose dil ratio[d]		
Nitro cellulose dil ratio[e]		
Surface tension	mN/m @ 20 °C	
Specific heat liquid	kJ/kg mole/°C @ 20 °C	
Latent heat	kJ/kg mole	
Dielectric constant	@ 20 °C	
Antoine constant A	kPa°C	
Antoine constant B	kPa°C	
Antoine constant C	kPa°C	
Heat of combustion	kJ/kg mole	
UN number		1223
IMO classification		none
ADR/RID classification		3,32°(c)
UK exposure limits	mg/m^3 (ppm) 8hrTWA/MEL or OES	
USA exposure limits	mg/m^3 (ppm) TLV/TWA	
German exposure limits	mg/m^3 (ppm) MAK-TWA	
EU classification		
EU risk phrases		
EU safety phrases		

[a] Ether = 1. [b] n-Butyl acetate = 1. [c] Nelson, Hemwall, Edwards. [d] Toluene. [e] Hydrocarbon solvent.

Solvent group		HYDROCARBONS—AROMATIC
Chemical name		
Chemical formula		
Structural formula		
CAS No.		90989-39-2
Synonym/common name		
Trade names		SOLVAREX 90/180
Suppliers		TOTAL
Molecular mass		
Boiling point/range	°C @ 760 mm Hg	165–195
Melting point	°C	
Density	kg/m^3 @ 20 °C	877
Evaporation time[a]		45
Evaporation time[b]		
Vapour pressure	kPa @ 20 °C	
Flash point	°C	45
Auto ignition temperature	°C	
Upper flammability limit	%v/v in air	
Lower flammability limit	%v/v in air	
Viscosity	mPa.s (cP) @ 20 °C	
Refractive index	n 20 D	1.500
Solubility in water	%w/w @ 20 °C	
Solubility of water	%w/w @ 20 °C	
Solubility parameter[c]	$(J/cm^3)^{1/2}$	
Hydrogen bond index		
Fractional polarity		
Nitro cellulose dil ratio[d]		
Nitro cellulose dil ratio[e]		
Surface tension	mN/m @ 20 °C	
Specific heat liquid	kJ/kg mole/°C @ 20 °C	
Latent heat	kJ/kg mole	
Dielectric constant	@ 20 °C	
Antoine constant A	kPa°C	
Antoine constant B	kPa°C	
Antoine constant C	kPa°C	
Heat of combustion	kJ/kg mole	
UN number		1300
IMO classification		3.3
ADR/RID classification		3,31°(c)
UK exposure limits	mg/m^3 (ppm) 8hrTWA/MEL or OES	
USA exposure limits	mg/m^3 (ppm) TLV/TWA	
German exposure limits	mg/m^3 (ppm) MAK-TWA	
EU classification		
EU risk phrases		10
EU safety phrases		

[a] Ether = 1. [b] n-Butyl acetate = 1. [c] Nelson, Hemwall, Edwards. [d] Toluene. [e] Hydrocarbon solvent.

Solvent group		**HYDROCARBONS—AROMATIC**
Chemical name		Mixture of xylenes
Chemical formula		
Structural formula		
CAS No.		
Synonym/common name		Xylenes/Xylol/Solvent xylenes
Trade names		
Suppliers		ICI
Molecular mass		
Boiling point/range	°C @ 760 mm Hg	137–143
Melting point	°C	< −25
Density	kg/m^3 @ 20 °C	
Evaporation time[a]		
Evaporation time[b]		
Vapour pressure	kPa @ 20 °C	0.8–2.0
Flash point	°C	25–31 (CLOSED CUP)
Auto ignition temperature	°C	480
Upper flammability limit	%v/v in air	7.0
Lower flammability limit	%v/v in air	1.0
Viscosity	mPa.s (cP) @ 20 °C	0.6
Refractive index	n 20 D	
Solubility in water	%w/w @ 20 °C	Insoluble
Solubility of water	%w/w @ 20 °C	
Solubility parameter[c]	(J/cm^3)$^{1/2}$	
Hydrogen bond index		
Fractional polarity		
Nitro cellulose dil ratio[d]		
Nitro cellulose dil ratio[e]		
Surface tension	mN/m @ 20 °C	
Specific heat liquid	kJ/kg mole/°C @ 20 °C	
Latent heat	kJ/kg mole	
Dielectric constant	@ 20 °C	
Antoine constant A	kPa°C	
Antoine constant B	kPa°C	
Antoine constant C	kPa°C	
Heat of combustion	kJ/kg mole	
UN number		1307
IMO classification		3.3/III
ADR/RID classification		3,31°(c)
UK exposure limits	mg/m^3 (ppm) 8hrTWA/MEL or OES	
USA exposure limits	mg/m^3 (ppm) TLV/TWA	
German exposure limits	mg/m^3 (ppm) MAK-TWA	
EU classification		
EU risk phrases		
EU safety phrases		

[a] Ether = 1. [b] n-Butyl acetate = 1. [c] Nelson, Hemwall, Edwards. [d] Toluene. [e] Hydrocarbon solvent.

Solvent group		HYDROCARBONS—AROMATIC
Chemical name		Mixture of aromatic hydrocarbons
Chemical formula		
Structural formula		
CAS No.		
Synonym/common name		
Trade names		PETRINEX T9
Suppliers		ICI
Molecular mass		
Boiling point/range	°C @ 760 mm Hg	180–290
Melting point	°C	0
Density	kg/m^3 @ 20 °C	
Evaporation time[a]		
Evaporation time[b]		
Vapour pressure	kPa @ 20 °C	0.010
Flash point	°C	70
Auto ignition temperature	°C	500
Upper flammability limit	%v/v in air	
Lower flammability limit	%v/v in air	
Viscosity	mPa.s (cP) @ 20 °C	
Refractive index	n 20 D	
Solubility in water	%w/w @ 20 °C	Insoluble
Solubility of water	%w/w @ 20 °C	
Solubility parameter[c]	$(J/cm^3)^{1/2}$	
Hydrogen bond index		
Fractional polarity		
Nitro cellulose dil ratio[d]		
Nitro cellulose dil ratio[e]		
Surface tension	mN/m @ 20 °C	
Specific heat liquid	kJ/kg mole/°C @ 20 °C	
Latent heat	kJ/kg mole	
Dielectric constant	@ 20 °C	
Antoine constant A	kPa°C	
Antoine constant B	kPa°C	
Antoine constant C	kPa°C	
Heat of combustion	kJ/kg mole	
UN number		3082
IMO classification		9/III
ADR/RID classification		3,32°(c)
UK exposure limits	mg/m^3 (ppm) 8hrTWA/MEL or OES	
USA exposure limits	mg/m^3 (ppm) TLV/TWA	
German exposure limits	mg/m^3 (ppm) MAK-TWA	
EU classification		
EU risk phrases		
EU safety phrases		

[a] Ether = 1. [b] n-Butyl acetate = 1. [c] Nelson, Hemwall, Edwards. [d] Toluene. [e] Hydrocarbon solvent.

Solvent group		**HYDROCARBONS—AROMATIC**
Chemical name		Mixture of aromatic hydrocarbons
Chemical formula		
Structural formula		
CAS No.		064742-95-6
Synonym/common name		
Trade names		AROMASOL H
Suppliers		ICI
Molecular mass		
Boiling point/range	°C @ 760 mm Hg	170–200
Melting point	°C	< −20
Density	kg/m^3 @ 20 °C	
Evaporation time[a]		
Evaporation time[b]		
Vapour pressure	kPa @ 20 °C	0.1
Flash point	°C	50
Auto ignition temperature	°C	500
Upper flammability limit	%v/v in air	8
Lower flammability limit	%v/v in air	1
Viscosity	mPa.s (cP) @ 20 °C	0.87
Refractive index	n 20 D	
Solubility in water	%w/w @ 20 °C	Insoluble
Solubility of water	%w/w @ 20 °C	
Solubility parameter[c]	(J/cm^3)$^{1/2}$	
Hydrogen bond index		
Fractional polarity		
Nitro cellulose dil ratio[d]		
Nitro cellulose dil ratio[e]		
Surface tension	mN/m @ 20 °C	
Specific heat liquid	kJ/kg mole/°C @ 20 °C	
Latent heat	kJ/kg mole	
Dielectric constant	@ 20 °C	
Antoine constant A	kPa°C	
Antoine constant B	kPa°C	
Antoine constant C	kPa°C	
Heat of combustion	kJ/kg mole	
UN number		1993
IMO classification		3.2/III
ADR/RID classification		3,31°(c)
UK exposure limits	mg/m^3 (ppm) 8hrTWA/MEL or OES	
USA exposure limits	mg/m^3 (ppm) TLV/TWA	
German exposure limits	mg/m^3 (ppm) MAK-TWA	
EU classification		Xn
EU risk phrases		10-20-36/37/38
EU safety phrases		2-26

[a] Ether = 1. [b] n-Butyl acetate = 1. [c] Nelson, Hemwall, Edwards. [d] Toluene. [e] Hydrocarbon solvent.

Solvent group		**HYDROCARBONS—AROMATIC**
Chemical name		Mixture of aromatic hydrocarbons
Chemical formula		
Structural formula		
CAS No.		064742-94-5
Synonym/common name		
Trade names		PETRINEX ASB
Suppliers		ICI
Molecular mass		
Boiling point/range	°C @ 760 mm Hg	167–210
Melting point	°C	< −50
Density	kg/m^3 @ 20 °C	
Evaporation time[a]		
Evaporation time[b]		
Vapour pressure	kPa @ 20 °C	0.010
Flash point	°C	24 (CLOSED CUP)
Auto ignition temperature	°C	550
Upper flammability limit	%v/v in air	8
Lower flammability limit	%v/v in air	1
Viscosity	mPa.s (cP) @ 20 °C	
Refractive index	n 20 D	
Solubility in water	%w/w @ 20 °C	Insoluble
Solubility of water	%w/w @ 20 °C	
Solubility parameter[c]	(J/cm^3)$^{1/2}$	
Hydrogen bond index		
Fractional polarity		
Nitro cellulose dil ratio[d]		
Nitro cellulose dil ratio[e]		
Surface tension	mN/m @ 20 °C	
Specific heat liquid	kJ/kg mole/°C @ 20 °C	
Latent heat	kJ/kg mole	
Dielectric constant	@ 20 °C	
Antoine constant A	kPa°C	
Antoine constant B	kPa°C	
Antoine constant C	kPa°C	
Heat of combustion	kJ/kg mole	
UN number		
IMO classification		
ADR/RID classification		3,32°(c)
UK exposure limits	mg/m^3 (ppm) 8hrTWA/MEL or OES	123(25)
USA exposure limits	mg/m^3 (ppm) TLV/TWA	
German exposure limits	mg/m^3 (ppm) MAK-TWA	
EU classification		
EU risk phrases		
EU safety phrases		

[a] Ether = 1. [b] n-Butyl acetate = 1. [c] Nelson, Hemwall, Edwards. [d] Toluene. [e] Hydrocarbon solvent.

Solvent group		**HYDROCARBONS—AROMATIC**
Chemical name		Mixture of toluene and C9/C_{10} aromatics
Chemical formula		
Structural formula		
CAS No.		68527-23-1
Synonym/common name		
Trade names		C7/C9 AROMATICS
Suppliers		ICI
Molecular mass		
Boiling point/range	°C @ 760 mm Hg	
Melting point	°C	
Density	kg/m^3 @ 20 °C	
Evaporation time[a]		
Evaporation time[b]		
Vapour pressure	kPa @ 20 °C	
Flash point	°C	4
Auto ignition temperature	°C	370
Upper flammability limit	%v/v in air	8
Lower flammability limit	%v/v in air	1
Viscosity	mPa.s (cP) @ 20 °C	
Refractive index	n 20 D	
Solubility in water	%w/w @ 20 °C	
Solubility of water	%w/w @ 20 °C	
Solubility parameter[c]	$(J/cm^3)^{1/2}$	
Hydrogen bond index		
Fractional polarity		
Nitro cellulose dil ratio[d]		
Nitro cellulose dil ratio[e]		
Surface tension	mN/m @ 20 °C	
Specific heat liquid	kJ/kg mole/°C @ 20 °C	
Latent heat	kJ/kg mole	
Dielectric constant	@ 20 °C	
Antoine constant A	kPa°C	
Antoine constant B	kPa°C	
Antoine constant C	kPa°C	
Heat of combustion	kJ/kg mole	
UN number		
IMO classification		
ADR/RID classification		
UK exposure limits	mg/m^3 (ppm) 8hrTWA/MEL or OES	
USA exposure limits	mg/m^3 (ppm) TLV/TWA	
German exposure limits	mg/m^3 (ppm) MAK-TWA	
EU classification		
EU risk phrases		
EU safety phrases		

[a] Ether = 1. [b] n-Butyl acetate = 1. [c] Nelson, Hemwall, Edwards. [d] Toluene. [e] Hydrocarbon solvent.

Solvent group		HYDROCARBONS—AROMATIC
Chemical name		Solvent naphtha
Chemical formula		
Structural formula		
CAS No.		64742-95-6
Synonym/common name		Light solvent naphtha
Trade names		C9 + I
Suppliers		PETROGAL
Molecular mass		
Boiling point/range	°C @ 760 mm Hg	160–183
Melting point	°C	
Density	kg/m^3 @ 20 °C	870–880
Evaporation time[a]		
Evaporation time[b]		
Vapour pressure	kPa @ 20 °C	
Flash point	°C	>450
Auto ignition temperature	°C	
Upper flammability limit	%v/v in air	
Lower flammability limit	%v/v in air	
Viscosity	mPa.s (cP) @ 20 °C	
Refractive index	n 20 D	1.50101
Solubility in water	%w/w @ 20 °C	
Solubility of water	%w/w @ 20 °C	
Solubility parameter[c]	$(J/cm^3)^{1/2}$	
Hydrogen bond index		
Fractional polarity		
Nitro cellulose dil ratio[d]		
Nitro cellulose dil ratio[e]		
Surface tension	mN/m @ 20 °C	
Specific heat liquid	kJ/kg mole/°C @ 20 °C	
Latent heat	kJ/kg mole	
Dielectric constant	@ 20 °C	
Antoine constant A	kPa°C	
Antoine constant B	kPa°C	
Antoine constant C	kPa°C	
Heat of combustion	kJ/kg mole	
UN number		1256
IMO classification		3.3
ADR/RID classification		3,31°(c)
UK exposure limits	mg/m^3 (ppm) 8hrTWA/MEL or OES	
USA exposure limits	mg/m^3 (ppm) TLV/TWA	
German exposure limits	mg/m^3 (ppm) MAK-TWA	
EU classification		
EU risk phrases		Xi
EU safety phrases		10-38-52/53
		62-61

[a] Ether = 1. [b] n-Butyl acetate = 1. [c] Nelson, Hemwall, Edwards. [d] Toluene. [e] Hydrocarbon solvent.

Solvent group		HYDROCARBONS—AROMATIC
Chemical name		
Chemical formula		
Structural formula		
CAS No.		90989-39-2
Synonym/common name		
Trade names		SOLVAREX 9
Suppliers		TOTAL
Molecular mass		
Boiling point/range	°C @ 760 mm Hg	165–180
Melting point	°C	
Density	kg/m^3 @ 20 °C	875
Evaporation time[a]		36
Evaporation time[b]		
Vapour pressure	kPa @ 20 °C	
Flash point	°C	44
Auto ignition temperature	°C	
Upper flammability limit	%v/v in air	
Lower flammability limit	%v/v in air	
Viscosity	mPa.s (cP) @ 20 °C	
Refractive index	n 20 D	1.500
Solubility in water	%w/w @ 20 °C	
Solubility of water	%w/w @ 20 °C	
Solubility parameter[c]	(J/cm^3)$^{1/2}$	
Hydrogen bond index		
Fractional polarity		
Nitro cellulose dil ratio[d]		
Nitro cellulose dil ratio[e]		
Surface tension	mN/m @ 20 °C	
Specific heat liquid	kJ/kg mole/°C @ 20 °C	
Latent heat	kJ/kg mole	
Dielectric constant	@ 20 °C	
Antoine constant A	kPa°C	
Antoine constant B	kPa°C	
Antoine constant C	kPa°C	
Heat of combustion	kJ/kg mole	
UN number		1300
IMO classification		3.3
ADR/RID classification		3,31°(c)
UK exposure limits	mg/m^3 (ppm) 8hrTWA/MEL or OES	
USA exposure limits	mg/m^3 (ppm) TLV/TWA	
German exposure limits	mg/m^3 (ppm) MAK-TWA	
EU classification		
EU risk phrases		10
EU safety phrases		

[a] Ether = 1. [b] n-Butyl acetate = 1. [c] Nelson, Hemwall, Edwards. [d] Toluene. [e] Hydrocarbon solvent.

Solvent group		HYDROCARBONS—AROMATIC
Chemical name		
Chemical formula		
Structural formula		
CAS No.		64742-94-5
Synonym/common name		
Trade names		SHELLSOL AD
Suppliers		SHELL
Molecular mass		
Boiling point/range	°C @ 760 mm Hg	180–202
Melting point	°C	
Density	kg/m^3 @ 20 °C	884
Evaporation time[a]		
Evaporation time[b]		
Vapour pressure	kPa @ 20 °C	
Flash point	°C	62 (ASTM D93)
Auto ignition temperature	°C	450
Upper flammability limit	%v/v in air	8.0
Lower flammability limit	%v/v in air	0.6
Viscosity	mPa.s (cP) @ 20 °C	
Refractive index	n 20 D	
Solubility in water	%w/w @ 20 °C	
Solubility of water	%w/w @ 20 °C	
Solubility parameter[c]	$(J/cm^3)^{1/2}$	17.8
Hydrogen bond index		5.3
Fractional polarity		0.001
Nitro cellulose dil ratio[d]		
Nitro cellulose dil ratio[e]		
Surface tension	mN/m @ 20 °C	30
Specific heat liquid	kJ/kg mole/°C @ 20 °C	
Latent heat	kJ/kg mole	
Dielectric constant	@ 20 °C	2.4
Antoine constant A	kPa°C	
Antoine constant B	kPa°C	
Antoine constant C	kPa°C	
Heat of combustion	kJ/kg mole	
UN number		3082
IMO classification		9/III
ADR/RID classification		3,32°(c)
UK exposure limits	mg/m^3 (ppm) 8hrTWA/MEL or OES	
USA exposure limits	mg/m^3 (ppm) TLV/TWA	
German exposure limits	mg/m^3 (ppm) MAK-TWA	
EU classification		
EU risk phrases		
EU safety phrases		

[a] Ether = 1. [b] n-Butyl acetate = 1. [c] Nelson, Hemwall, Edwards. [d] Toluene. [e] Hydrocarbon solvent.

Solvent group		HYDROCARBONS—AROMATIC
Chemical name		
Chemical formula		
Structural formula		
CAS No.		
Synonym/common name		
Trade names		HAN 8572
Suppliers		EXXON
Molecular mass		
Boiling point/range	°C @ 760 mm Hg	195–270
Melting point	°C	
Density	kg/m^3 @ 20 °C	
Evaporation time[a]		>1000
Evaporation time[b]		<0.01
Vapour pressure	kPa @ 20 °C	
Flash point	°C	74
Auto ignition temperature	°C	
Upper flammability limit	%v/v in air	
Lower flammability limit	%v/v in air	
Viscosity	mPa.s (cP) @ 20 °C	
Refractive index	n 20 D	1.525
Solubility in water	%w/w @ 20 °C	
Solubility of water	%w/w @ 20 °C	
Solubility parameter[c]	$(J/cm^3)^{1/2}$	
Hydrogen bond index		
Fractional polarity		
Nitro cellulose dil ratio[d]		
Nitro cellulose dil ratio[e]		
Surface tension	mN/m @ 20 °C	
Specific heat liquid	kJ/kg mole/°C @ 20 °C	
Latent heat	kJ/kg mole	
Dielectric constant	@ 20 °C	
Antoine constant A	kPa°C	
Antoine constant B	kPa°C	
Antoine constant C	kPa°C	
Heat of combustion	kJ/kg mole	
UN number		3082
IMO classification		9
ADR/RID classification		3,32°(c)
UK exposure limits	mg/m^3 (ppm) 8hrTWA/MEL or OES	
USA exposure limits	mg/m^3 (ppm) TLV/TWA	
German exposure limits	mg/m^3 (ppm) MAK-TWA	
EU classification		
EU risk phrases		
EU safety phrases		

[a] Ether = 1. [b] n-Butyl acetate = 1. [c] Nelson, Hemwall, Edwards. [d] Toluene. [e] Hydrocarbon solvent.

Property	Unit	Value
Solvent group		**HYDROCARBONS—AROMATIC**
Chemical name		
Chemical formula		
Structural formula		
CAS No.		
Synonym/common name		
Trade names		Solvesso 200
Suppliers		EXXON
Molecular mass		
Boiling point/range	°C @ 760 mm Hg	229–284
Melting point	°C	
Density	kg/m^3 @ 20 °C	
Evaporation time[a]		>1000
Evaporation time[b]		<0.01
Vapour pressure	kPa @ 20 °C	
Flash point	°C	97
Auto ignition temperature	°C	
Upper flammability limit	%v/v in air	
Lower flammability limit	%v/v in air	
Viscosity	mPa.s (cP) @ 20 °C	
Refractive index	n 20 D	1.593
Solubility in water	%w/w @ 20 °C	
Solubility of water	%w/w @ 20 °C	
Solubility parameter[c]	$(J/cm^3)^{1/2}$	
Hydrogen bond index		
Fractional polarity		
Nitro cellulose dil ratio[d]		
Nitro cellulose dil ratio[e]		
Surface tension	mN/m @ 20 °C	
Specific heat liquid	kJ/kg mole/°C @ 20 °C	
Latent heat	kJ/kg mole	
Dielectric constant	@ 20 °C	
Antoine constant A	kPa°C	
Antoine constant B	kPa°C	
Antoine constant C	kPa°C	
Heat of combustion	kJ/kg mole	
UN number		3082
IMO classification		9
ADR/RID classification		3,32°(c)
UK exposure limits	mg/m^3 (ppm) 8hrTWA/MEL or OES	
USA exposure limits	mg/m^3 (ppm) TLV/TWA	
German exposure limits	mg/m^3 (ppm) MAK-TWA	
EU classification		
EU risk phrases		
EU safety phrases		

[a] Ether = 1. [b] n-Butyl acetate = 1. [c] Nelson, Hemwall, Edwards. [d] Toluene. [e] Hydrocarbon solvent.

Solvent group		HYDROCARBONS—AROMATIC
Chemical name		
Chemical formula		
Structural formula		
CAS No.		
Synonym/common name		
Trade names		HAN 8070
Suppliers		EXXON
Molecular mass		
Boiling point/range	°C @ 760 mm Hg	185–230
Melting point	°C	
Density	kg/m^3 @ 20 °C	
Evaporation time[a]		500
Evaporation time[b]		<0.01
Vapour pressure	kPa @ 20 °C	
Flash point	°C	69
Auto ignition temperature	°C	
Upper flammability limit	%v/v in air	
Lower flammability limit	%v/v in air	
Viscosity	mPa.s (cP) @ 20 °C	
Refractive index	n 20 D	1.500
Solubility in water	%w/w @ 20 °C	
Solubility of water	%w/w @ 20 °C	
Solubility parameter[c]	(J/cm^3)$^{1/2}$	
Hydrogen bond index		
Fractional polarity		
Nitro cellulose dil ratio[d]		
Nitro cellulose dil ratio[e]		
Surface tension	mN/m @ 20 °C	
Specific heat liquid	kJ/kg mole/°C @ 20 °C	
Latent heat	kJ/kg mole	
Dielectric constant	@ 20 °C	
Antoine constant A	kPa°C	
Antoine constant B	kPa°C	
Antoine constant C	kPa°C	
Heat of combustion	kJ/kg mole	
UN number		3082
IMO classification		9
ADR/RID classification		3,32°(c)
UK exposure limits	mg/m^3 (ppm) 8hrTWA/MEL or OES	
USA exposure limits	mg/m^3 (ppm) TLV/TWA	
German exposure limits	mg/m^3 (ppm) MAK-TWA	
EU classification		
EU risk phrases		
EU safety phrases		

[a] Ether = 1. [b] n-Butyl acetate = 1. [c] Nelson, Hemwall, Edwards. [d] Toluene. [e] Hydrocarbon solvent.

Solvent group		HYDROCARBONS—AROMATIC
Chemical name		
Chemical formula		
Structural formula		
CAS No.		
Synonym/common name		
Trade names		Solvesso 100
Suppliers		EXXON
Molecular mass		
Boiling point/range	°C @ 760 mm Hg	161–178
Melting point	°C	
Density	kg/m^3 @ 20 °C	
Evaporation time[a]		35
Evaporation time[b]		0.2
Vapour pressure	kPa @ 20 °C	
Flash point	°C	49
Auto ignition temperature	°C	
Upper flammability limit	%v/v in air	
Lower flammability limit	%v/v in air	
Viscosity	mPa.s (cP) @ 20 °C	
Refractive index	n 20 D	1.501
Solubility in water	%w/w @ 20 °C	
Solubility of water	%w/w @ 20 °C	
Solubility parameter[c]	$(J/cm^3)^{1/2}$	
Hydrogen bond index		
Fractional polarity		
Nitro cellulose dil ratio[d]		
Nitro cellulose dil ratio[e]		
Surface tension	mN/m @ 20 °C	
Specific heat liquid	kJ/kg mole/°C @ 20 °C	
Latent heat	kJ/kg mole	
Dielectric constant	@ 20 °C	
Antoine constant A	kPa°C	
Antoine constant B	kPa°C	
Antoine constant C	kPa°C	
Heat of combustion	kJ/kg mole	
UN number		1993
IMO classification		3.3
ADR/RID classification		3,31°(c)
UK exposure limits	mg/m^3 (ppm) 8hrTWA/MEL or OES	
USA exposure limits	mg/m^3 (ppm) TLV/TWA	
German exposure limits	mg/m^3 (ppm) MAK-TWA	
EU classification		
EU risk phrases		10
EU safety phrases		43a

[a] Ether = 1. [b] n-Butyl acetate = 1. [c] Nelson, Hemwall, Edwards. [d] Toluene. [e] Hydrocarbon solvent.

Solvent group		HYDROCARBONS—AROMATIC
Chemical name		
Chemical formula		
Structural formula		
CAS No.		
Synonym/common name		
Trade names		HAN
Suppliers		EXXON
Molecular mass		
Boiling point/range	°C @ 760 mm Hg	183–277
Melting point	°C	
Density	kg/m³ @ 20 °C	
Evaporation time[a]		>1000
Evaporation time[b]		<0.01
Vapour pressure	kPa @ 20 °C	
Flash point	°C	72
Auto ignition temperature	°C	
Upper flammability limit	%v/v in air	
Lower flammability limit	%v/v in air	
Viscosity	mPa.s (cP) @ 20 °C	
Refractive index	n 20 D	1.545
Solubility in water	%w/w @ 20 °C	
Solubility of water	%w/w @ 20 °C	
Solubility parameter[c]	$(J/cm^3)^{1/2}$	
Hydrogen bond index		
Fractional polarity		
Nitro cellulose dil ratio[d]		
Nitro cellulose dil ratio[e]		
Surface tension	mN/m @ 20 °C	
Specific heat liquid	kJ/kg mole/°C @ 20 °C	
Latent heat	kJ/kg mole	
Dielectric constant	@ 20 °C	
Antoine constant A	kPa°C	
Antoine constant B	kPa°C	
Antoine constant C	kPa°C	
Heat of combustion	kJ/kg mole	
UN number		3082
IMO classification		9
ADR/RID classification		3,32°(c)
UK exposure limits	mg/m³ (ppm) 8hrTWA/MEL or OES	
USA exposure limits	mg/m³ (ppm) TLV/TWA	
German exposure limits	mg/m³ (ppm) MAK-TWA	
EU classification		
EU risk phrases		
EU safety phrases		

[a] Ether = 1. [b] n-Butyl acetate = 1. [c] Nelson, Hemwall, Edwards. [d] Toluene. [e] Hydrocarbon solvent.

Solvent group		HYDROCARBONS—AROMATIC
Chemical name		
Chemical formula		
Structural formula		
CAS No.		
Synonym/common name		
Trade names		VARSOL 140
Suppliers		EXXON
Molecular mass		
Boiling point/range	°C @ 760 mm Hg	290–315
Melting point	°C	
Density	kg/m^3 @ 20 °C	
Evaporation time[a]		>1000
Evaporation time[b]		<0.01
Vapour pressure	kPa @ 20 °C	
Flash point	°C	140
Auto ignition temperature	°C	
Upper flammability limit	%v/v in air	
Lower flammability limit	%v/v in air	
Viscosity	mPa.s (cP) @ 20 °C	
Refractive index	n 20 D	1.471
Solubility in water	%w/w @ 20 °C	
Solubility of water	%w/w @ 20 °C	
Solubility parameter[c]	$(J/cm^3)^{1/2}$	
Hydrogen bond index		
Fractional polarity		
Nitro cellulose dil ratio[d]		
Nitro cellulose dil ratio[e]		
Surface tension	mN/m @ 20 °C	
Specific heat liquid	kJ/kg mole/°C @ 20 °C	
Latent heat	kJ/kg mole	
Dielectric constant	@ 20 °C	
Antoine constant A	kPa°C	
Antoine constant B	kPa°C	
Antoine constant C	kPa°C	
Heat of combustion	kJ/kg mole	
UN number		
IMO classification		
ADR/RID classification		
UK exposure limits	mg/m^3 (ppm) 8hrTWA/MEL or OES	
USA exposure limits	mg/m^3 (ppm) TLV/TWA	
German exposure limits	mg/m^3 (ppm) MAK-TWA	
EU classification		
EU risk phrases		
EU safety phrases		

[a] Ether = 1. [b] n-Butyl acetate = 1. [c] Nelson, Hemwall, Edwards. [d] Toluene. [e] Hydrocarbon solvent.

Solvent group		**HYDROCARBONS—AROMATIC**
Chemical name		
Chemical formula		
Structural formula		
CAS No.		
Synonym/common name		
Trade names		Solvesso 200S
Suppliers		EXXON
Molecular mass		
Boiling point/range	°C @ 760 mm Hg	224–288
Melting point	°C	
Density	kg/m^3 @ 20 °C	
Evaporation time[a]		>1000
Evaporation time[b]		<0.01
Vapour pressure	kPa @ 20 °C	
Flash point	°C	99
Auto ignition temperature	°C	
Upper flammability limit	%v/v in air	
Lower flammability limit	%v/v in air	
Viscosity	mPa.s (cP) @ 20 °C	
Refractive index	n 20 D	1.594
Solubility in water	%w/w @ 20 °C	
Solubility of water	%w/w @ 20 °C	
Solubility parameter[c]	(J/cm^3)$^{1/2}$	
Hydrogen bond index		
Fractional polarity		
Nitro cellulose dil ratio[d]		
Nitro cellulose dil ratio[e]		
Surface tension	mN/m @ 20 °C	
Specific heat liquid	kJ/kg mole/°C @ 20 °C	
Latent heat	kJ/kg mole	
Dielectric constant	@ 20 °C	
Antoine constant A	kPa°C	
Antoine constant B	kPa°C	
Antoine constant C	kPa°C	
Heat of combustion	kJ/kg mole	
UN number		3082
IMO classification		9
ADR/RID classification		3,32°(c)
UK exposure limits	mg/m^3 (ppm) 8hrTWA/MEL or OES	
USA exposure limits	mg/m^3 (ppm) TLV/TWA	
German exposure limits	mg/m^3 (ppm) MAK-TWA	
EU classification		
EU risk phrases		
EU safety phrases		

[a] Ether = 1. [b] n-Butyl acetate = 1. [c] Nelson, Hemwall, Edwards. [d] Toluene. [e] Hydrocarbon solvent.

Solvent group		HYDROCARBONS—AROMATIC
Chemical name		
Chemical formula		
Structural formula		
CAS No.		
Synonym/common name		
Trade names		Solvesso 150
Suppliers		EXXON
Molecular mass		
Boiling point/range	°C @ 760 mm Hg	182–203
Melting point	°C	
Density	kg/m^3 @ 20 °C	
Evaporation time[a]		120
Evaporation time[b]		0.06
Vapour pressure	kPa @ 20 °C	
Flash point	°C	66
Auto ignition temperature	°C	
Upper flammability limit	%v/v in air	
Lower flammability limit	%v/v in air	
Viscosity	mPa.s (cP) @ 20 °C	
Refractive index	n 20 D	1.515
Solubility in water	%w/w @ 20 °C	
Solubility of water	%w/w @ 20 °C	
Solubility parameter[c]	$(J/cm^3)^{1/2}$	
Hydrogen bond index		
Fractional polarity		
Nitro cellulose dil ratio[d]		
Nitro cellulose dil ratio[e]		
Surface tension	mN/m @ 20 °C	
Specific heat liquid	kJ/kg mole/°C @ 20 °C	
Latent heat	kJ/kg mole	
Dielectric constant	@ 20 °C	
Antoine constant A	kPa°C	
Antoine constant B	kPa°C	
Antoine constant C	kPa°C	
Heat of combustion	kJ/kg mole	
UN number		3082
IMO classification		9
ADR/RID classification		3,32°(c)
UK exposure limits	mg/m^3 (ppm) 8hrTWA/MEL or OES	
USA exposure limits	mg/m^3 (ppm) TLV/TWA	
German exposure limits	mg/m^3 (ppm) MAK-TWA	
EU classification		
EU risk phrases		
EU safety phrases		

[a] Ether = 1. [b] n-Butyl acetate = 1. [c] Nelson, Hemwall, Edwards. [d] Toluene. [e] Hydrocarbon solvent.

Solvent group		**HYDROCARBONS—AROMATIC**
Chemical name		Benzene
Chemical formula		C_6H_6
Structural formula		C_6H_6
CAS No.		71-43-2
Synonym/common name		Benzene
Trade names		
Suppliers		ICI, PETROGAL, ENICHEM
Molecular mass		78.11
Boiling point/range	°C @ 760 mm Hg	79.8–80.2
Melting point	°C	5.35
Density	kg/m^3 @ 20 °C	882–886
Evaporation time[a]		
Evaporation time[b]		
Vapour pressure	kPa @ 20 °C	10.13
Flash point	°C	−11 (CLOSED CUP & IP 170)
Auto ignition temperature	°C	498 to >53
Upper flammability limit	%v/v in air	7.1–8.0
Lower flammability limit	%v/v in air	1.3–1.4
Viscosity	mPa.s (cP) @ 20 °C	
Refractive index	n 20 D	1.5011
Solubility in water	%w/w @ 20 °C	0.178
Solubility of water	%w/w @ 20 °C	
Solubility parameter[c]	(J/cm^3)$^{1/2}$	18.74
Hydrogen bond index		
Fractional polarity		
Nitro cellulose dil ratio[d]		
Nitro cellulose dil ratio[e]		
Surface tension	mN/m @ 20 °C	
Specific heat liquid	kJ/kg mole/°C @ 20 °C	
Latent heat	kJ/kg mole	30726
Dielectric constant	@ 20 °C	2.284
Antoine constant A	kPa°C	6.02232
Antoine constant B	kPa°C	1206.53
Antoine constant C	kPa°C	220.91
Heat of combustion	kJ/kg mole	−3301430
UN number		1114
IMO classification		3.2/II
ADR/RID classification		3,3°(b)
UK exposure limits	mg/m^3 (ppm) 8hrTWA/MEL or OES	16(5)
USA exposure limits	mg/m^3 (ppm) TLV/TWA	0.96(0.3)
German exposure limits	mg/m^3 (ppm) MAK-TWA	3.2(1)
EU classification		F, Carc 1, T, N
EU risk phrases		45-11-48/23/24/25-51/53
EU safety phrases		53-45-61

[a] Ether = 1. [b] n-Butyl acetate = 1. [c] Nelson, Hemwall, Edwards. [d] Toluene. [e] Hydrocarbon solvent.

Solvent group		**HYDROCARBONS—AROMATIC**
Chemical name		Toluene
Chemical formula		C_7H_8
Structural formula		$C_6H_5CH_3$
CAS No.		108-88-3
Synonym/common name		Toluene, Methylbenzene
Trade names		SOLVAREX Toluene
Suppliers		EXXON, ICI, PETROGAL, SHELL, TOTAL, ENICHEM
Molecular mass		92.14
Boiling point/range	°C @ 760 mm Hg	110–111
Melting point	°C	−95
Density	kg/m^3 @ 20 °C	866–871
Evaporation time[a]		6
Evaporation time[b]		1.72–2.0
Vapour pressure	kPa @ 20 °C	3
Flash point	°C	4 (IP 170) SHELL, 8 (IP 170) PETROGAL
Auto ignition temperature	°C	480–535
Upper flammability limit	%v/v in air	7.1–8.0
Lower flammability limit	%v/v in air	1.2
Viscosity	mPa.s (cP) @ 20 °C	0.58
Refractive index	n 20 D	1.497
Solubility in water	%w/w @ 20 °C	Insoluble
Solubility of water	%w/w @ 20 °C	
Solubility parameter[c]	$(J/cm^3)^{1/2}$	18.2
Hydrogen bond index		4.2
Fractional polarity		0.001
Nitro cellulose dil ratio[d]		
Nitro cellulose dil ratio[e]		
Surface tension	mN/m @ 20 °C	28.5
Specific heat liquid	kJ/kg mole/°C @ 20 °C	156.4
Latent heat	kJ/kg mole	33396
Dielectric constant	@ 20 °C	2.4
Antoine constant A	kPa°C	6.24335
Antoine constant B	kPa°C	1438.74
Antoine constant C	kPa°C	228.982
Heat of combustion	kJ/kg mole	−3904020
UN number		1294
IMO classification		3.2/II
ADR/RID classification		3,3°(b)
UK exposure limits	mg/m^3 (ppm) 8hrTWA/MEL or OES	188(50)
USA exposure limits	mg/m^3 (ppm) TLV/TWA	188(50)
German exposure limits	mg/m^3 (ppm) MAK-TWA	380(100)
EU classification		F, Xn
EU risk phrases		11-20
EU safety phrases		(2-)16-25-29-33

[a] Ether = 1. [b] n-Butyl acetate = 1. [c] Nelson, Hemwall, Edwards. [d] Toluene. [e] Hydrocarbon solvent.

Solvent group		**HYDROCARBONS—AROMATIC**
Chemical name		Styrene
Chemical formula		C_8H_8
Structural formula		$C_6H_5CH{=}CH_2$
CAS No.		100-42-5
Synonym/common name		Styrene
Trade names		
Suppliers		HULS
Molecular mass		104.2
Boiling point/range	°C @ 760 mm Hg	145.2
Melting point	°C	−31
Density	kg/m^3 @ 20 °C	907
Evaporation time[a]		19.5
Evaporation time[b]		
Vapour pressure	kPa @ 20 °C	0.71
Flash point	°C	31
Auto ignition temperature	°C	480
Upper flammability limit	%v/v in air	8.9
Lower flammability limit	%v/v in air	1.2
Viscosity	mPa.s (cP) @ 20 °C	0.76
Refractive index	n 20 D	1.546–1.547
Solubility in water	%w/w @ 20 °C	
Solubility of water	%w/w @ 20 °C	0.04
Solubility parameter[c]	$(J/cm^3)^{1/2}$	19.0
Hydrogen bond index		
Fractional polarity		
Nitro cellulose dil ratio[d]		
Nitro cellulose dil ratio[e]		
Surface tension	mN/m @ 20 °C	32.3
Specific heat liquid	kJ/kg mole/°C @ 20 °C	
Latent heat	kJ/kg mole	38700
Dielectric constant	@ 20 °C	2.426
Antoine constant A	kPa°C	6.34792
Antoine constant B	kPa°C	1629.2
Antoine constant C	kPa°C	230
Heat of combustion	kJ/kg mole	−4395630
UN number		2055
IMO classification		3.3/III
ADR/RID classification		3,31°(c)
UK exposure limits	mg/m^3 (ppm) 8hrTWA/MEL or OES	420(100)
USA exposure limits	mg/m^3 (ppm) TLV/TWA	213(50)
German exposure limits	mg/m^3 (ppm) MAK-TWA	85(20)
EU classification		Xn, Xi, R10
EU risk phrases		10-20-36/38
EU safety phrases		(2-)23

[a] Ether = 1. [b] n-Butyl acetate = 1. [c] Nelson, Hemwall, Edwards. [d] Toluene. [e] Hydrocarbon solvent.

Solvent group		**HYDROCARBONS—AROMATIC**
Chemical name		o-Xylene
Chemical formula		C_8H_{10}
Structural formula		$C_6H_4(CH_3)_2$
CAS No.		95-47-6
Synonym/common name		o-Xylene, 1,2-Dimethylbenzene
Trade names		
Suppliers		PETROGAL, ENICHEM
Molecular mass		106
Boiling point/range	°C @ 760 mm Hg	144.4
Melting point	°C	−25
Density	kg/m^3 @ 20 °C	880–884
Evaporation time[a]		
Evaporation time[b]		
Vapour pressure	kPa @ 20 °C	0.66
Flash point	°C	32 (CLOSED CUP) PETROGAL, 32 (CLOSED CUP) ENICHEM
Auto ignition temperature	°C	464
Upper flammability limit	%v/v in air	6.4–7
Lower flammability limit	%v/v in air	1.0
Viscosity	mPa.s (cP) @ 20 °C	0.81
Refractive index	n 20 D	1.5055
Solubility in water	%w/w @ 20 °C	0.0175
Solubility of water	%w/w @ 20 °C	
Solubility parameter[c]	(J/cm^3)$^{1/2}$	18.0
Hydrogen bond index		
Fractional polarity		
Nitro cellulose dil ratio[d]		
Nitro cellulose dil ratio[e]		
Surface tension	mN/m @ 20 °C	30.04
Specific heat liquid	kJ/kg mole/°C @ 20 °C	
Latent heat	kJ/kg mole	36820
Dielectric constant	@ 20 °C	2.568
Antoine constant A	kPa°C	6.13072
Antoine constant B	kPa°C	1479.82
Antoine constant C	kPa°C	214.315
Heat of combustion	kJ/kg mole	−4552860
UN number		1307
IMO classification		
ADR/RID classification		
UK exposure limits	mg/m^3 (ppm) 8hrTWA/MEL or OES	435(100)
USA exposure limits	mg/m^3 (ppm) TLV/TWA	434(100)
German exposure limits	mg/m^3 (ppm) MAK-TWA	440(100)
EU classification		Xn, Xi, R10
EU risk phrases		10-20/21-38
EU safety phrases		(2-)25

[a] Ether = 1. [b] n-Butyl acetate = 1. [c] Nelson, Hemwall, Edwards. [d] Toluene. [e] Hydrocarbon solvent.

Solvent group		**HYDROCARBONS—AROMATIC**
Chemical name		Xylene
Chemical formula		C_8H_{10}
Structural formula		$C_6H_4(CH_3)_2$
CAS No.		1330-20-7
Synonym/common name		Xylene, Dimethylbenzene
Trade names		SOLVAREX Xylene
Suppliers		EXXON, SHELL, TOTAL, ENICHEM
Molecular mass		106
Boiling point/range	°C @ 760 mm Hg	137.7–142
Melting point	°C	−25
Density	kg/m^3 @ 20 °C	870
Evaporation time[a]		13.5–15
Evaporation time[b]		0.76
Vapour pressure	kPa @ 20 °C	1
Flash point	°C	25 (IP 170) SHELL, 28 EXXON
Auto ignition temperature	°C	500
Upper flammability limit	%v/v in air	6.6
Lower flammability limit	%v/v in air	1.1
Viscosity	mPa.s (cP) @ 20 °C	
Refractive index	n 20 D	1.497
Solubility in water	%w/w @ 20 °C	
Solubility of water	%w/w @ 20 °C	
Solubility parameter[c]	(J/cm^3)$^{1/2}$	18.1
Hydrogen bond index		4.5
Fractional polarity		0.001
Nitro cellulose dil ratio[d]		
Nitro cellulose dil ratio[e]		
Surface tension	mN/m @ 20 °C	28.7
Specific heat liquid	kJ/kg mole/°C @ 20 °C	174.9
Latent heat	kJ/kg mole	36358
Dielectric constant	@ 20 °C	2.6
Antoine constant A	kPa°C	6.42712
Antoine constant B	kPa°C	1642.15
Antoine constant C	kPa°C	232.282
Heat of combustion	kJ/kg mole	−4545280
UN number		1307
IMO classification		3.3/III
ADR/RID classification		3,31°(c)
UK exposure limits	mg/m^3 (ppm) 8hrTWA/MEL or OES	435(100)
USA exposure limits	mg/m^3 (ppm) TLV/TWA	434(100)
German exposure limits	mg/m^3 (ppm) MAK-TWA	440(100)
EU classification		Xn, Xi, R10
EU risk phrases		10-20/21-38
EU safety phrases		(2-)25

[a] Ether = 1. [b] n-Butyl acetate = 1. [c] Nelson, Hemwall, Edwards. [d] Toluene. [e] Hydrocarbon solvent.

Solvent group		**HYDROCARBONS—AROMATIC**
Chemical name		Mixture of isomers and ethylbenzene
Chemical formula		C_8H_{10}
Structural formula		
CAS No.		1330-20-7
Synonym/common name		Xylene, Dimethylbenzene
Trade names		
Suppliers		PETROGAL, SA
Molecular mass		106
Boiling point/range	°C @ 760 mm Hg	137–143
Melting point	°C	
Density	kg/m^3 @ 20 °C	865–875
Evaporation time[a]		
Evaporation time[b]		
Vapour pressure	kPa @ 20 °C	
Flash point	°C	>26 (IP 170)
Auto ignition temperature	°C	>500
Upper flammability limit	%v/v in air	6.6
Lower flammability limit	%v/v in air	1
Viscosity	mPa.s (cP) @ 20 °C	
Refractive index	n 20 D	>1.495
Solubility in water	%w/w @ 20 °C	
Solubility of water	%w/w @ 20 °C	
Solubility parameter[c]	$(J/cm^3)^{1/2}$	
Hydrogen bond index		
Fractional polarity		
Nitro cellulose dil ratio[d]		
Nitro cellulose dil ratio[e]		
Surface tension	mN/m @ 20 °C	
Specific heat liquid	kJ/kg mole/°C @ 20 °C	
Latent heat	kJ/kg mole	
Dielectric constant	@ 20 °C	
Antoine constant A	kPa°C	
Antoine constant B	kPa°C	
Antoine constant C	kPa°C	
Heat of combustion	kJ/kg mole	
UN number		1307
IMO classification		3.3
ADR/RID classification		3,31°(c)
UK exposure limits	mg/m^3 (ppm) 8hrTWA/MEL or OES	435(100)
USA exposure limits	mg/m^3 (ppm) TLV/TWA	434(100)
German exposure limits	mg/m^3 (ppm) MAK-TWA	
EU classification		Xn
EU risk phrases		10-20/21-38
EU safety phrases		25

[a] Ether = 1. [b] n-Butyl acetate = 1. [c] Nelson, Hemwall, Edwards. [d] Toluene. [e] Hydrocarbon solvent.

Property	Unit	Value
Solvent group		**HYDROCARBONS—AROMATIC**
Chemical name		m-Xylene
Chemical formula		C_8H_{10}
Structural formula		$C_6H_4(CH_3)_2$
CAS No.		108-38-3
Synonym/common name		m-Xylene, 1,3-Dimethylbenzene
Trade names		
Suppliers		ENICHEM
Molecular mass		106.17
Boiling point/range	°C @ 760 mm Hg	139.7
Melting point	°C	−47.87
Density	kg/m^3 @ 20 °C	864
Evaporation time[a]		
Evaporation time[b]		
Vapour pressure	kPa @ 20 °C	0.8
Flash point	°C	27 (CLOSED CUP)
Auto ignition temperature	°C	527
Upper flammability limit	%v/v in air	7
Lower flammability limit	%v/v in air	1.1
Viscosity	mPa.s (cP) @ 20 °C	0.62
Refractive index	n 20 D	1.4972
Solubility in water	%w/w @ 20 °C	0.0146
Solubility of water	%w/w @ 20 °C	
Solubility parameter[c]	$(J/cm^3)^{1/2}$	18.0
Hydrogen bond index		
Fractional polarity		
Nitro cellulose dil ratio[d]		
Nitro cellulose dil ratio[e]		
Surface tension	mN/m @ 20 °C	28.66
Specific heat liquid	kJ/kg mole/°C @ 20 °C	
Latent heat	kJ/kg mole	36360
Dielectric constant	@ 20 °C	2.372
Antoine constant A	kPa°C	6.13785
Antoine constant B	kPa°C	1465.39
Antoine constant C	kPa°C	215.512
Heat of combustion	kJ/kg mole	−4551860
UN number		1307
IMO classification		
ADR/RID classification		
UK exposure limits	mg/m^3 (ppm) 8hrTWA/MEL or OES	
USA exposure limits	mg/m^3 (ppm) TLV/TWA	
German exposure limits	mg/m^3 (ppm) MAK-TWA	
EU classification		
EU risk phrases		
EU safety phrases		

[a] Ether = 1. [b] n-Butyl acetate = 1. [c] Nelson, Hemwall, Edwards. [d] Toluene. [e] Hydrocarbon solvent.

Solvent group		HYDROCARBONS—AROMATIC
Chemical name		Mixture of xylenes
Chemical formula		C_8H_{10}
Structural formula		
CAS No.		1330-20-7
Synonym/common name		Xylenes/Xylol/Solvent xylenes
Trade names		
Suppliers		ENICHEM
Molecular mass		106.17
Boiling point/range	°C @ 760 mm Hg	137–140
Melting point	°C	−25 to −50
Density	kg/m^3 @ 20 °C	864
Evaporation time[a]		
Evaporation time[b]		
Vapour pressure	kPa @ 20 °C	0.8
Flash point	°C	27–32 (CLOSED CUP)/ 27–46 (OPEN CUP)
Auto ignition temperature	°C	463
Upper flammability limit	%v/v in air	7
Lower flammability limit	%v/v in air	1.1
Viscosity	mPa.s (cP) @ 20 °C	0.7
Refractive index	n 20 D	
Solubility in water	%w/w @ 20 °C	0.0180
Solubility of water	%w/w @ 20 °C	
Solubility parameter[c]	(J/cm^3)$^{1/2}$	
Hydrogen bond index		
Fractional polarity		
Nitro cellulose dil ratio[d]		
Nitro cellulose dil ratio[e]		
Surface tension	mN/m @ 20 °C	
Specific heat liquid	kJ/kg mole/°C @ 20 °C	
Latent heat	kJ/kg mole	
Dielectric constant	@ 20 °C	2.24
Antoine constant A	kPa°C	
Antoine constant B	kPa°C	
Antoine constant C	kPa°C	
Heat of combustion	kJ/kg mole	
UN number		
IMO classification		
ADR/RID classification		
UK exposure limits	mg/m^3 (ppm) 8hrTWA/MEL or OES	
USA exposure limits	mg/m^3 (ppm) TLV/TWA	
German exposure limits	mg/m^3 (ppm) MAK-TWA	
EU classification		
EU risk phrases		
EU safety phrases		

[a] Ether = 1. [b] n-Butyl acetate = 1. [c] Nelson, Hemwall, Edwards. [d] Toluene. [e] Hydrocarbon solvent.

Solvent group		**HYDROCARBONS—AROMATIC**
Chemical name		p-Xylene
Chemical formula		C_8H_{10}
Structural formula		$C_6H_4(CH_3)_2$
CAS No.		106-42-3
Synonym/common name		p-Xylene, 1,4-dimethylbenzene
Trade names		
Suppliers		PETROGAL, ENICHEM
Molecular mass		106.17
Boiling point/range	°C @ 760 mm Hg	138.0–138.5
Melting point	°C	13.3
Density	kg/m^3 @ 20 °C	861–867
Evaporation time[a]		
Evaporation time[b]		
Vapour pressure	kPa @ 20 °C	0.86
Flash point	°C	27 (CLOSED CUP) ENICHEM
Auto ignition temperature	°C	528–530
Upper flammability limit	%v/v in air	7
Lower flammability limit	%v/v in air	1.1
Viscosity	mPa.s (cP) @ 20 °C	0.69
Refractive index	n 20 D	1.4958
Solubility in water	%w/w @ 20 °C	0.0185
Solubility of water	%w/w @ 20 °C	
Solubility parameter[c]	$(J/cm^3)^{1/2}$	17.94
Hydrogen bond index		
Fractional polarity		
Nitro cellulose dil ratio[d]		
Nitro cellulose dil ratio[e]		
Surface tension	mN/m @ 20 °C	28.31
Specific heat liquid	kJ/kg mole/°C @ 20 °C	
Latent heat	kJ/kg mole	35980
Dielectric constant	@ 20 °C	2.270
Antoine constant A	kPa°C	6.11140
Antoine constant B	kPa°C	1451.39
Antoine constant C	kPa°C	215.148
Heat of combustion	kJ/kg mole	−4552860
UN number		1307
IMO classification		3.3/III
ADR/RID classification		3,31°(c)
UK exposure limits	mg/m^3 (ppm) 8hrTWA/MEL or OES	435(100)
USA exposure limits	mg/m^3 (ppm) TLV/TWA	434(100)
German exposure limits	mg/m^3 (ppm) MAK-TWA	440(100)
EU classification		Xn, Xi, R10
EU risk phrases		10-20/21-38
EU safety phrases		(2-)25

[a] Ether = 1. [b] n-Butyl acetate = 1. [c] Nelson, Hemwall, Edwards. [d] Toluene. [e] Hydrocarbon solvent.

Solvent group		HYDROCARBONS—AROMATIC
Chemical name		1,4-Dimethylbenzene
Chemical formula		C_8H_{10}
Structural formula		$C_6H_4(CH_3)_2$
CAS No.		106-42-3
Synonym/common name		p-Xylene
Trade names		PARAXYLENE
Suppliers		ICI
Molecular mass		106.17
Boiling point/range	°C @ 760 mm Hg	138.3–138.5
Melting point	°C	13
Density	kg/m^3 @ 20 °C	
Evaporation time[a]		
Evaporation time[b]		
Vapour pressure	kPa @ 20 °C	1.150 @ 25 °C
Flash point	°C	24 (CLOSED CUP)
Auto ignition temperature	°C	530
Upper flammability limit	%v/v in air	6.6
Lower flammability limit	%v/v in air	1.08
Viscosity	mPa.s (cP) @ 20 °C	0.62
Refractive index	n 20 D	
Solubility in water	%w/w @ 20 °C	Marginally
Solubility of water	%w/w @ 20 °C	
Solubility parameter[c]	$(J/cm^3)^{1/2}$	17.94
Hydrogen bond index		
Fractional polarity		
Nitro cellulose dil ratio[d]		
Nitro cellulose dil ratio[e]		
Surface tension	mN/m @ 20 °C	
Specific heat liquid	kJ/kg mole/°C @ 20 °C	
Latent heat	kJ/kg mole	
Dielectric constant	@ 20 °C	
Antoine constant A	kPa°C	6.11140
Antoine constant B	kPa°C	1451.39
Antoine constant C	kPa°C	215.148
Heat of combustion	kJ/kg mole	
UN number		1307
IMO classification		3.3/III
ADR/RID classification		3,31°(c)
UK exposure limits	mg/m^3 (ppm) 8hrTWA/MEL or OES	435(100)
USA exposure limits	mg/m^3 (ppm) TLV/TWA	434(100)
German exposure limits	mg/m^3 (ppm) MAK-TWA	440(100)
EU classification		Xn, Xi, R10
EU risk phrases		10-20/21-38
EU safety phrases		(2-)25

[a] Ether = 1. [b] n-Butyl acetate = 1. [c] Nelson, Hemwall, Edwards. [d] Toluene. [e] Hydrocarbon solvent.

Solvent group		**HYDROCARBONS—AROMATIC**
Chemical name		Ethylbenzene
Chemical formula		C_8H_{10}
Structural formula		$C_6H_5CH_2CH_3$
CAS No.		100-41-4
Synonym/common name		Phenylethane
Trade names		
Suppliers		HULS, ICI
Molecular mass		106.2
Boiling point/range	°C @ 760 mm Hg	135–136
Melting point	°C	−94 to −95
Density	kg/m^3 @ 20 °C	867
Evaporation time[a]		
Evaporation time[b]		
Vapour pressure	kPa @ 20 °C	0.93
Flash point	°C	15 (CLOSED CUP) ICI, 23 HULS
Auto ignition temperature	°C	428–435
Upper flammability limit	%v/v in air	7.8–8.1
Lower flammability limit	%v/v in air	1.0
Viscosity	mPa.s (cP) @ 20 °C	0.7
Refractive index	n 20 D	
Solubility in water	%w/w @ 20 °C	Insoluble
Solubility of water	%w/w @ 20 °C	
Solubility parameter[c]	(J/cm^3)$^{1/2}$	17.98
Hydrogen bond index		
Fractional polarity		
Nitro cellulose dil ratio[d]		
Nitro cellulose dil ratio[e]		
Surface tension	mN/m @ 20 °C	29.05
Specific heat liquid	kJ/kg mole/°C @ 20 °C	
Latent heat	kJ/kg mole	35200
Dielectric constant	@ 20 °C	2.404
Antoine constant A	kPa°C	6.09280
Antoine constant B	kPa°C	1431.71
Antoine constant C	kPa°C	214.099
Heat of combustion	kJ/kg mole	−4564870
UN number		1175
IMO classification		3.2/II
ADR/RID classification		3,3°(b)
UK exposure limits	mg/m^3 (ppm) 8hrTWA/MEL or OES	435(100)
USA exposure limits	mg/m^3 (ppm) TLV/TWA	434(100)
German exposure limits	mg/m^3 (ppm) MAK-TWA	440(100)
EU classification		F, Xn, N
EU risk phrases		11-20-51/53
EU safety phrases		(2-)16-24/25/29-61

[a] Ether = 1. [b] n-Butyl acetate = 1. [c] Nelson, Hemwall, Edwards. [d] Toluene. [e] Hydrocarbon solvent.

Solvent group		HYDROCARBONS—AROMATIC
Chemical name		
Chemical formula		
Structural formula		
CAS No.		
Synonym/common name		
Trade names		SHELL CYCLO SOL 53 AROMATIC SOLVENT
Suppliers		SHELL
Molecular mass		120
Boiling point/range	°C @ 760 mm Hg	160–176
Melting point	°C	
Density	kg/m^3 @ 20 °C	
Evaporation time[a]		
Evaporation time[b]		0.21
Vapour pressure	kPa @ 20 °C	
Flash point	°C	
Auto ignition temperature	°C	
Upper flammability limit	%v/v in air	
Lower flammability limit	%v/v in air	
Viscosity	mPa.s (cP) @ 20 °C	
Refractive index	n 20 D	
Solubility in water	%w/w @ 20 °C	
Solubility of water	%w/w @ 20 °C	
Solubility parameter[c]	$(J/cm^3)^{1/2}$	
Hydrogen bond index		
Fractional polarity		
Nitro cellulose dil ratio[d]		
Nitro cellulose dil ratio[e]		
Surface tension	mN/m @ 20 °C	
Specific heat liquid	kJ/kg mole/°C @ 20 °C	
Latent heat	kJ/kg mole	
Dielectric constant	@ 20 °C	
Antoine constant A	kPa°C	
Antoine constant B	kPa°C	
Antoine constant C	kPa°C	
Heat of combustion	kJ/kg mole	
UN number		
IMO classification		
ADR/RID classification		
UK exposure limits	mg/m^3 (ppm) 8hrTWA/MEL or OES	
USA exposure limits	mg/m^3 (ppm) TLV/TWA	
German exposure limits	mg/m^3 (ppm) MAK-TWA	
EU classification		
EU risk phrases		
EU safety phrases		

[a] Ether = 1. [b] n-Butyl acetate = 1. [c] Nelson, Hemwall, Edwards. [d] Toluene. [e] Hydrocarbon solvent.

Solvent group		HYDROCARBONS—AROMATIC
Chemical name		
Chemical formula		
Structural formula		
CAS No.		64742-95-6
Synonym/common name		
Trade names		SHELLSOL A
Suppliers		SHELL
Molecular mass		120
Boiling point/range	°C @ 760 mm Hg	167-180
Melting point	°C	
Density	kg/m^3 @ 20 °C	876
Evaporation time[a]		46
Evaporation time[b]		0.20
Vapour pressure	kPa @ 20 °C	0.3
Flash point	°C	42 (IP 170)
Auto ignition temperature	°C	450
Upper flammability limit	%v/v in air	6.05
Lower flammability limit	%v/v in air	0.8
Viscosity	mPa.s (cP) @ 20 °C	0.79
Refractive index	n 20 D	1.503
Solubility in water	%w/w @ 20 °C	
Solubility of water	%w/w @ 20 °C	
Solubility parameter[c]	$(J/cm^3)^{1/2}$	18.0
Hydrogen bond index		5.0
Fractional polarity		0.001
Nitro cellulose dil ratio[d]		
Nitro cellulose dil ratio[e]		
Surface tension	mN/m @ 20 °C	29
Specific heat liquid	kJ/kg mole/°C @ 20 °C	216.0
Latent heat	kJ/kg mole	39000
Dielectric constant	@ 20 °C	2.4
Antoine constant A	kPa°C	6.7478
Antoine constant B	kPa°C	1912.9
Antoine constant C	kPa°C	240.33
Heat of combustion	kJ/kg mole	−4923600
UN number		1300
IMO classification		3.3/III
ADR/RID classification		3,31°(c)
UK exposure limits	mg/m^3 (ppm) 8hrTWA/MEL or OES	
USA exposure limits	mg/m^3 (ppm) TLV/TWA	
German exposure limits	mg/m^3 (ppm) MAK-TWA	
EU classification		
EU risk phrases		10
EU safety phrases		

[a] Ether = 1. [b] n-Butyl acetate = 1. [c] Nelson, Hemwall, Edwards. [d] Toluene. [e] Hydrocarbon solvent.

Solvent group		HYDROCARBONS—AROMATIC
Chemical name		1-Methylethylbenzene
Chemical formula		C_9H_{12}
Structural formula		$C_6H_5CH(CH_3)_2$
CAS No.		98-82-8
Synonym/common name		Cumene/Isopropylbenzene
Trade names		AROMASOL 12
Suppliers		ICI, ENICHEM
Molecular mass		120.19
Boiling point/range	°C @ 760 mm Hg	152–170
Melting point	°C	−96 to −100
Density	kg/m^3 @ 20 °C	861
Evaporation time[a]		
Evaporation time[b]		
Vapour pressure	kPa @ 20 °C	0.426
Flash point	°C	39 (CLOSED CUP) ICI,
Auto ignition temperature	°C	36 (CLOSED CUP) ENICHEM
Upper flammability limit	%v/v in air	400–424
Lower flammability limit	%v/v in air	6.5–88
Viscosity	mPa.s (cP) @ 20 °C	1
Refractive index	n 20 D	
Solubility in water	%w/w @ 20 °C	
Solubility of water	%w/w @ 20 °C	Immiscible
Solubility parameter[c]	$(J/cm^3)^{1/2}$	
Hydrogen bond index		17.6
Fractional polarity		
Nitro cellulose dil ratio[d]		
Nitro cellulose dil ratio[e]		
Surface tension	mN/m @ 20 °C	
Specific heat liquid	kJ/kg mole/°C @ 20 °C	28.20
Latent heat	kJ/kg mole	
Dielectric constant	@ 20 °C	37350
Antoine constant A	kPa°C	2.383
Antoine constant B	kPa°C	6.06588
Antoine constant C	kPa°C	1464.17
Heat of combustion	kJ/kg mole	208.207
UN number		−5215440
IMO classification		1918
ADR/RID classification		3.3/III
UK exposure limits	mg/m^3 (ppm) 8hrTWA/MEL or OES	3,31°(c)
USA exposure limits	mg/m^3 (ppm) TLV/TWA	120(25)
German exposure limits	mg/m^3 (ppm) MAK-TWA	246(50)
EU classification		245(50)
EU risk phrases		Xi, R10
EU safety phrases		10-37
		(2)

[a] Ether = 1. [b] n-Butyl acetate = 1. [c] Nelson, Hemwall, Edwards. [d] Toluene. [e] Hydrocarbon solvent.

Solvent group		HYDROCARBONS—AROMATIC
Chemical name		
Chemical formula		
Structural formula		
CAS No.		64742-94-5
Synonym/common name		
Trade names		SHELLSOL AB
Suppliers		SHELL
Molecular mass		130
Boiling point/range	°C @ 760 mm Hg	181–212
Melting point	°C	
Density	kg/m^3 @ 20 °C	895
Evaporation time[a]		148
Evaporation time[b]		0.07
Vapour pressure	kPa @ 20 °C	0.15
Flash point	°C	42 (ASTM D93)
Auto ignition temperature	°C	450
Upper flammability limit	%v/v in air	6.0
Lower flammability limit	%v/v in air	0.8
Viscosity	mPa.s (cP) @ 20 °C	1.07
Refractive index	n 20 D	1.512
Solubility in water	%w/w @ 20 °C	
Solubility of water	%w/w @ 20 °C	
Solubility parameter[c]	$(J/cm^3)^{1/2}$	17.8
Hydrogen bond index		5.3
Fractional polarity		0.001
Nitro cellulose dil ratio[d]		
Nitro cellulose dil ratio[e]		
Surface tension	mN/m @ 20 °C	30
Specific heat liquid	kJ/kg mole/°C @ 20 °C	234.0
Latent heat	kJ/kg mole	43420
Dielectric constant	@ 20 °C	2.4
Antoine constant A	kPa°C	7.45653
Antoine constant B	kPa°C	2459.76
Antoine constant C	kPa°C	269.313
Heat of combustion	kJ/kg mole	−5339750
UN number		3082
IMO classification		9/III
ADR/RID classification		3,32°(c)
UK exposure limits	mg/m^3 (ppm) 8hrTWA/MEL or OES	
USA exposure limits	mg/m^3 (ppm) TLV/TWA	
German exposure limits	mg/m^3 (ppm) MAK-TWA	
EU classification		
EU risk phrases		
EU safety phrases		

[a] Ether = 1. [b] n-Butyl acetate = 1. [c] Nelson, Hemwall, Edwards. [d] Toluene. [e] Hydrocarbon solvent.

Solvent group		**HYDROCARBONS—AROMATIC**
Chemical name		Tetrahydronaphthalene
Chemical formula		$C_{10}H_{12}$
Structural formula		
CAS No.		119-64-2
Synonym/common name		Tetralin
Trade names		
Suppliers		HULS
Molecular mass		132.21
Boiling point/range	°C @ 760 mm Hg	200–209
Melting point	°C	−31
Density	kg/m^3 @ 20 °C	969
Evaporation time[a]		200
Evaporation time[b]		
Vapour pressure	kPa @ 20 °C	0.024
Flash point	°C	71–77
Auto ignition temperature	°C	425
Upper flammability limit	%v/v in air	5.0
Lower flammability limit	%v/v in air	0.5
Viscosity	mPa.s (cP) @ 20 °C	2.2
Refractive index	n 20 D	1.5385–1.5395
Solubility in water	%w/w @ 20 °C	
Solubility of water	%w/w @ 20 °C	0.2
Solubility parameter[c]	(J/cm^3)$^{1/2}$	19.4
Hydrogen bond index		
Fractional polarity		
Nitro cellulose dil ratio[d]		
Nitro cellulose dil ratio[e]		
Surface tension	mN/m @ 20 °C	
Specific heat liquid	kJ/kg mole/°C @ 20 °C	
Latent heat	kJ/kg mole	43850
Dielectric constant	@ 20 °C	2.773
Antoine constant A	kPa°C	11.079
Antoine constant B	kPa°C	2797.9
Antoine constant C	kPa°C	1.187
Heat of combustion	kJ/kg mole	−5672710
UN number		
IMO classification		
ADR/RID classification		
UK exposure limits	mg/m^3 (ppm) 8hrTWA/MEL or OES	
USA exposure limits	mg/m^3 (ppm) TLV/TWA	
German exposure limits	mg/m^3 (ppm) MAK-TWA	
EU classification		Xi
EU risk phrases		19-36/38
EU safety phrases		(2-)26-28

[a] Ether = 1. [b] n-Butyl acetate = 1. [c] Nelson, Hemwall, Edwards. [d] Toluene. [e] Hydrocarbon solvent.

Solvent group		HYDROCARBONS—AROMATIC
Chemical name		
Chemical formula		
Structural formula		
CAS No.		
Synonym/common name		
Trade names		SHELL CYCLO SOL 63 AROMATIC SOLVENT
Suppliers		SHELL
Molecular mass		134
Boiling point/range	°C @ 760 mm Hg	173–208
Melting point	°C	
Density	kg/m^3 @ 20 °C	
Evaporation time[a]		
Evaporation time[b]		<0.1
Vapour pressure	kPa @ 20 °C	
Flash point	°C	
Auto ignition temperature	°C	
Upper flammability limit	%v/v in air	
Lower flammability limit	%v/v in air	
Viscosity	mPa.s (cP) @ 20 °C	
Refractive index	n 20 D	
Solubility in water	%w/w @ 20 °C	
Solubility of water	%w/w @ 20 °C	
Solubility parameter[c]	$(J/cm^3)^{1/2}$	
Hydrogen bond index		
Fractional polarity		
Nitro cellulose dil ratio[d]		
Nitro cellulose dil ratio[e]		
Surface tension	mN/m @ 20 °C	
Specific heat liquid	kJ/kg mole/°C @ 20 °C	
Latent heat	kJ/kg mole	
Dielectric constant	@ 20 °C	
Antoine constant A	kPa°C	
Antoine constant B	kPa°C	
Antoine constant C	kPa°C	
Heat of combustion	kJ/kg mole	
UN number		
IMO classification		
ADR/RID classification		
UK exposure limits	mg/m^3 (ppm) 8hrTWA/MEL or OES	
USA exposure limits	mg/m^3 (ppm) TLV/TWA	
German exposure limits	mg/m^3 (ppm) MAK-TWA	
EU classification		
EU risk phrases		
EU safety phrases		

[a] Ether = 1. [b] n-Butyl acetate = 1. [c] Nelson, Hemwall, Edwards. [d] Toluene. [e] Hydrocarbon solvent.

Solvent group		**HYDROCARBONS—AROMATIC**
Chemical name		Mixture based on aromatic hydrocarbons
Chemical formula		
Structural formula		
CAS No.		129813-62-3
Synonym/common name		Mixture of diphenyl alkanes, dialkylbenzene and tetralins
Trade names		VISTA 3050 SPECIALTY ALKYLATE
Suppliers		VISTA
Molecular mass		
Boiling point/range	°C @ 760 mm Hg	313
Melting point	°C	343–379
Density	kg/m^3 @ 20 °C	−57
Evaporation time[a]		890
Evaporation time[b]		
Vapour pressure	kPa @ 20 °C	
Flash point	°C	
Auto ignition temperature	°C	182 (PENSKY MARTENS)
Upper flammability limit	%v/v in air	379
Lower flammability limit	%v/v in air	
Viscosity	mPa.s (cP) @ 20 °C	
Refractive index	n 20 D	
Solubility in water	%w/w @ 20 °C	
Solubility of water	%w/w @ 20 °C	Insoluble
Solubility parameter[c]	$(J/cm^3)^{1/2}$	Insoluble
Hydrogen bond index		
Fractional polarity		
Nitro cellulose dil ratio[d]		
Nitro cellulose dil ratio[e]		
Surface tension	mN/m @ 20 °C	
Specific heat liquid	kJ/kg mole/°C @ 20 °C	
Latent heat	kJ/kg mole	
Dielectric constant	@ 20 °C	
Antoine constant A	kPa°C	
Antoine constant B	kPa°C	
Antoine constant C	kPa°C	
Heat of combustion	kJ/kg mole	
UN number		
IMO classification		
ADR/RID classification		
UK exposure limits	mg/m^3 (ppm) 8hrTWA/MEL or OES	
USA exposure limits	mg/m^3 (ppm) TLV/TWA	
German exposure limits	mg/m^3 (ppm) MAK-TWA	
EU classification		
EU risk phrases		
EU safety phrases		

[a] Ether = 1. [b] n-Butyl acetate = 1. [c] Nelson, Hemwall, Edwards. [d] Toluene. [e] Hydrocarbon solvent.

Property	Unit	Value
Solvent group		**HYDROCARBONS—AROMATIC**
Chemical name		Mixture based on aromatic hydrocarbons
Chemical formula		
Structural formula		
CAS No.		84961-70-6
Synonym/common name		Mixture of alkylated benzenes
Trade names		VISTA 150L SPECIALTY ALKYLATE
Suppliers		VISTA
Molecular mass		339
Boiling point/range	°C @ 760 mm Hg	323–389
Melting point	°C	−54
Density	kg/m^3 @ 20 °C	880
Evaporation time[a]		
Evaporation time[b]		
Vapour pressure	kPa @ 20 °C	
Flash point	°C	207 (PENSKY MARTENS)
Auto ignition temperature	°C	371
Upper flammability limit	%v/v in air	
Lower flammability limit	%v/v in air	
Viscosity	mPa.s (cP) @ 20 °C	
Refractive index	n 20 D	
Solubility in water	%w/w @ 20 °C	Insoluble
Solubility of water	%w/w @ 20 °C	Insoluble
Solubility parameter[c]	$(J/cm^3)^{1/2}$	
Hydrogen bond index		
Fractional polarity		
Nitro cellulose dil ratio[d]		
Nitro cellulose dil ratio[e]		
Surface tension	mN/m @ 20 °C	
Specific heat liquid	kJ/kg mole/°C @ 20 °C	
Latent heat	kJ/kg mole	
Dielectric constant	@ 20 °C	
Antoine constant A	kPa°C	
Antoine constant B	kPa°C	
Antoine constant C	kPa°C	
Heat of combustion	kJ/kg mole	
UN number		
IMO classification		
ADR/RID classification		
UK exposure limits	mg/m^3 (ppm) 8hrTWA/MEL or OES	
USA exposure limits	mg/m^3 (ppm) TLV/TWA	
German exposure limits	mg/m^3 (ppm) MAK-TWA	
EU classification		
EU risk phrases		
EU safety phrases		

[a] Ether = 1. [b] n-Butyl acetate = 1. [c] Nelson, Hemwall, Edwards. [d] Toluene. [e] Hydrocarbon solvent.

Property	Units	Value
Solvent group		**HYDROCARBONS—AROMATIC/ALIPHATIC**
Chemical name		Mixture based on saturated and aromatic hydrocarbons
Chemical formula		
Structural formula		
CAS No.		64741-85-1
Synonym/common name		MR SOLVENT
Trade names		VISTA MOLEX RAFFINATE
Suppliers		VISTA CHEMICAL CO.
Molecular mass		~167
Boiling point/range	°C @ 760 mm Hg	193–259
Melting point	°C	< −70
Density	kg/m^3 @ 20 °C	817
Evaporation time[a]		
Evaporation time[b]		0.021
Vapour pressure	kPa @ 20 °C	0.025
Flash point	°C	72 (PENSKY MARTENS)
Auto ignition temperature	°C	227
Upper flammability limit	%v/v in air	4.7
Lower flammability limit	%v/v in air	0.6
Viscosity	mPa.s (cP) @ 20 °C	1.7
Refractive index	n 20 D	
Solubility in water	%w/w @ 20 °C	Negligible
Solubility of water	%w/w @ 20 °C	Negligible
Solubility parameter[c]	$(J/cm^3)^{1/2}$	
Hydrogen bond index		
Fractional polarity		
Nitro cellulose dil ratio[d]		
Nitro cellulose dil ratio[e]		
Surface tension	mN/m @ 20 °C	
Specific heat liquid	kJ/kg mole/°C @ 20 °C	
Latent heat	kJ/kg mole	
Dielectric constant	@ 20 °C	
Antoine constant A	kPa°C	
Antoine constant B	kPa°C	
Antoine constant C	kPa°C	
Heat of combustion	kJ/kg mole	
UN number		1223
IMO classification		
ADR/RID classification		
UK exposure limits	mg/m^3 (ppm) 8hrTWA/MEL or OES	
USA exposure limits	mg/m^3 (ppm) TLV/TWA	525(100)*
German exposure limits	mg/m^3 (ppm) MAK-TWA	
EU classification		
EU risk phrases		
EU safety phrases		

[a] Ether = 1. [b] n-Butyl acetate = 1. [c] Nelson, Hemwall, Edwards. [d] Toluene. [e] Hydrocarbon solvent.

Solvent group		**HYDROCARBONS—CYCLIC**
Chemical name		Cyclopentane
Chemical formula		C_5H_{10}
Structural formula		
CAS No.		287-92-3
Synonym/common name		Cyclopentane
Trade names		
Suppliers		EC Erdolchemie
Molecular mass		70.5
Boiling point/range	°C @ 760 mm Hg	48–50
Melting point	°C	−94
Density	kg/m^3 @ 20 °C	
Evaporation time[a]		2
Evaporation time[b]		
Vapour pressure	kPa @ 20 °C	
Flash point	°C	−42
Auto ignition temperature	°C	380
Upper flammability limit	%v/v in air	8.7
Lower flammability limit	%v/v in air	1.1
Viscosity	mPa.s (cP) @ 20 °C	0.47
Refractive index	n 20 D	1.4065
Solubility in water	%w/w @ 20 °C	
Solubility of water	%w/w @ 20 °C	
Solubility parameter[c]	(J/cm^3)$^{1/2}$	
Hydrogen bond index		
Fractional polarity		
Nitro cellulose dil ratio[d]		
Nitro cellulose dil ratio[e]		
Surface tension	mN/m @ 20 °C	
Specific heat liquid	kJ/kg mole/°C @ 20 °C	
Latent heat	kJ/kg mole	
Dielectric constant	@ 20 °C	
Antoine constant A	kPa°C	
Antoine constant B	kPa°C	
Antoine constant C	kPa°C	
Heat of combustion	kJ/kg mole	
UN number		1146
IMO classification		3.1/II
ADR/RID classification		3,3°(b)
UK exposure limits	mg/m^3 (ppm) 8hrTWA/MEL or OES	
USA exposure limits	mg/m^3 (ppm) TLV/TWA	1720(600)
German exposure limits	mg/m^3 (ppm) MAK-TWA	
EU classification		F
EU risk phrases		11
EU safety phrases		(2-)9-16-29-33

[a] Ether = 1. [b] n-Butyl acetate = 1. [c] Nelson, Hemwall, Edwards. [d] Toluene. [e] Hydrocarbon solvent.

Solvent group		**HYDROCARBONS—CYCLIC**
Chemical name		Cyclohexane
Chemical formula		C_6H_{12}
Structural formula		
CAS No.		110-82-7
Synonym/common name		Cyclohexane
Trade names		
Suppliers		ICI
Molecular mass		84.16
Boiling point/range	°C @ 760 mm Hg	81
Melting point	°C	6.4
Density	kg/m^3 @ 20 °C	
Evaporation time[a]		
Evaporation time[b]		
Vapour pressure	kPa @ 20 °C	13.0 @ 25 °C
Flash point	°C	−20 (CLOSED CUP)
Auto ignition temperature	°C	260
Upper flammability limit	%v/v in air	8.4
Lower flammability limit	%v/v in air	1.3
Viscosity	mPa.s (cP) @ 20 °C	
Refractive index	n 20 D	
Solubility in water	%w/w @ 20 °C	Immiscible
Solubility of water	%w/w @ 20 °C	
Solubility parameter[c]	$(J/cm^3)^{1/2}$	16.78
Hydrogen bond index		
Fractional polarity		
Nitro cellulose dil ratio[d]		
Nitro cellulose dil ratio[e]		
Surface tension	mN/m @ 20 °C	25.24
Specific heat liquid	kJ/kg mole/°C @ 20 °C	
Latent heat	kJ/kg mole	30050
Dielectric constant	@ 20 °C	2.024
Antoine constant A	kPa°C	5.96407
Antoine constant B	kPa°C	1200.31
Antoine constant C	kPa°C	222.504
Heat of combustion	kJ/kg mole	−3919860
UN number		1145
IMO classification		3.1/II
ADR/RID classification		3,3°(b)
UK exposure limits	mg/m^3 (ppm) 8hrTWA/MEL or OES	340(100)
USA exposure limits	mg/m^3 (ppm) TLV/TWA	1030(300)
German exposure limits	mg/m^3 (ppm) MAK-TWA	1050(300)
EU classification		F
EU risk phrases		11
EU safety phrases		(2-)9-16-33

[a] Ether = 1. [b] n-Butyl acetate = 1. [c] Nelson, Hemwall, Edwards. [d] Toluene. [e] Hydrocarbon solvent.

Solvent group		HYDROCARBONS—CYCLIC
Chemical name		Decahydronaphthalene
Chemical formula		$C_{10}H_{18}$
Structural formula	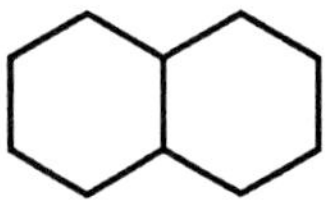	
CAS No.		91-17-8
Synonym/common name		Decalin
Trade names		
Suppliers		HULS
Molecular mass		138.2
Boiling point/range	°C @ 760 mm Hg	185–195
Melting point	°C	−30.4
Density	kg/m^3 @ 20 °C	888
Evaporation time[a]		100
Evaporation time[b]		
Vapour pressure	kPa @ 20 °C	0.29
Flash point	°C	54–61
Auto ignition temperature	°C	255
Upper flammability limit	%v/v in air	4.9
Lower flammability limit	%v/v in air	0.7
Viscosity	mPa.s (cP) @ 20 °C	3
Refractive index	n 20 D	1.474–1.475
Solubility in water	%w/w @ 20 °C	
Solubility of water	%w/w @ 20 °C	0.2
Solubility parameter[c]	(J/cm^3)$^{1/2}$	
Hydrogen bond index		
Fractional polarity		
Nitro cellulose dil ratio[d]		
Nitro cellulose dil ratio[e]		
Surface tension	mN/m @ 20 °C	
Specific heat liquid	kJ/kg mole/°C @ 20 °C	
Latent heat	kJ/kg mole	
Dielectric constant	@ 20 °C	
Antoine constant A	kPa°C	
Antoine constant B	kPa°C	
Antoine constant C	kPa°C	
Heat of combustion	kJ/kg mole	
UN number		
IMO classification		
ADR/RID classification		
UK exposure limits	mg/m^3 (ppm) 8hrTWA/MEL or OES	
USA exposure limits	mg/m^3 (ppm) TLV/TWA	
German exposure limits	mg/m^3 (ppm) MAK-TWA	
EU classification		
EU risk phrases		
EU safety phrases		

[a] Ether = 1. [b] n-Butyl acetate = 1. [c] Nelson, Hemwall, Edwards. [d] Toluene. [e] Hydrocarbon solvent.

Solvent group		HYDROCARBONS—UNSPECIFIED
Chemical name		
Chemical formula		C_{10}–C_{16} Range
Structural formula		
CAS No.		64742-47-8
Synonym/common name		Kerosene (Low odour)
Trade names		FINALAN 75
Suppliers		FINA
Molecular mass		
Boiling point/range	°C @ 760 mm Hg	200–260
Melting point	°C	
Density	kg/m^3 @ 20 °C	800 (15 °C)
Evaporation time[a]		
Evaporation time[b]		
Vapour pressure	kPa @ 20 °C	
Flash point	°C	77 (IP 34)
Auto ignition temperature	°C	
Upper flammability limit	%v/v in air	
Lower flammability limit	%v/v in air	
Viscosity	mPa.s (cP) @ 20 °C	2.07
Refractive index	n 20 D	
Solubility in water	%w/w @ 20 °C	
Solubility of water	%w/w @ 20 °C	
Solubility parameter[c]	(J/cm^3)$^{1/2}$	
Hydrogen bond index		
Fractional polarity		
Nitro cellulose dil ratio[d]		
Nitro cellulose dil ratio[e]		
Surface tension	mN/m @ 20 °C	
Specific heat liquid	kJ/kg mole/°C @ 20 °C	
Latent heat	kJ/kg mole	
Dielectric constant	@ 20 °C	
Antoine constant A	kPa°C	
Antoine constant B	kPa°C	
Antoine constant C	kPa°C	
Heat of combustion	kJ/kg mole	
UN number		
IMO classification		
ADR/RID classification		
UK exposure limits	mg/m^3 (ppm) 8hrTWA/MEL or OES	1500(200)
USA exposure limits	mg/m^3 (ppm) TLV/TWA	
German exposure limits	mg/m^3 (ppm) MAK-TWA	
EU classification		
EU risk phrases		22
EU safety phrases		16-24-25-43-62

[a] Ether = 1. [b] n-Butyl acetate = 1. [c] Nelson, Hemwall, Edwards. [d] Toluene. [e] Hydrocarbon solvent.

Solvent group		**HYDROCARBONS—UNSPECIFIED**
Chemical name		
Chemical formula		
Structural formula		
CAS No.		
Synonym/common name		
Trade names		Nappar 9
Suppliers		EXXON
Molecular mass		
Boiling point/range	°C @ 760 mm Hg	146–170
Melting point	°C	
Density	kg/m^3 @ 20 °C	
Evaporation time[a]		17
Evaporation time[b]		0.58
Vapour pressure	kPa @ 20 °C	
Flash point	°C	30
Auto ignition temperature	°C	
Upper flammability limit	%v/v in air	
Lower flammability limit	%v/v in air	
Viscosity	mPa.s (cP) @ 20 °C	
Refractive index	n 20 D	1.434
Solubility in water	%w/w @ 20 °C	
Solubility of water	%w/w @ 20 °C	
Solubility parameter[c]	(J/cm^3)$^{1/2}$	
Hydrogen bond index		
Fractional polarity		
Nitro cellulose dil ratio[d]		
Nitro cellulose dil ratio[e]		
Surface tension	mN/m @ 20 °C	
Specific heat liquid	kJ/kg mole/°C @ 20 °C	
Latent heat	kJ/kg mole	
Dielectric constant	@ 20 °C	
Antoine constant A	kPa°C	
Antoine constant B	kPa°C	
Antoine constant C	kPa°C	
Heat of combustion	kJ/kg mole	
UN number		1270
IMO classification		3.2
ADR/RID classification		3,31°(c)
UK exposure limits	mg/m^3 (ppm) 8hrTWA/MEL or OES	
USA exposure limits	mg/m^3 (ppm) TLV/TWA	
German exposure limits	mg/m^3 (ppm) MAK-TWA	
EU classification		Xi
EU risk phrases		10-38
EU safety phrases		24-43a

[a] Ether = 1. [b] n-Butyl acetate = 1. [c] Nelson, Hemwall, Edwards. [d] Toluene. [e] Hydrocarbon solvent.

Solvent group		HYDROCARBONS—UNSPECIFIED
Chemical name		
Chemical formula		C_{10}–C_{16} Range
Structural formula		
CAS No.		64742-47-8
Synonym/common name		Kerosene (Low odour)
Trade names		FINALAN 90
Suppliers		FINA
Molecular mass		
Boiling point/range	°C @ 760 mm Hg	222–254
Melting point	°C	
Density	kg/m^3 @ 20 °C	808 (15 °C)
Evaporation time[a]		
Evaporation time[b]		
Vapour pressure	kPa @ 20 °C	
Flash point	°C	92 (IP 34)
Auto ignition temperature	°C	
Upper flammability limit	%v/v in air	
Lower flammability limit	%v/v in air	
Viscosity	mPa.s (cP) @ 20 °C	1.96
Refractive index	n 20 D	
Solubility in water	%w/w @ 20 °C	
Solubility of water	%w/w @ 20 °C	
Solubility parameter[c]	(J/cm^3)$^{1/2}$	
Hydrogen bond index		
Fractional polarity		
Nitro cellulose dil ratio[d]		
Nitro cellulose dil ratio[e]		
Surface tension	mN/m @ 20 °C	
Specific heat liquid	kJ/kg mole/°C @ 20 °C	
Latent heat	kJ/kg mole	
Dielectric constant	@ 20 °C	
Antoine constant A	kPa°C	
Antoine constant B	kPa°C	
Antoine constant C	kPa°C	
Heat of combustion	kJ/kg mole	
UN number		
IMO classification		
ADR/RID classification		
UK exposure limits	mg/m^3 (ppm) 8hrTWA/MEL or OES	5 (OIL MIST)
USA exposure limits	mg/m^3 (ppm) TLV/TWA	
German exposure limits	mg/m^3 (ppm) MAK-TWA	
EU classification		
EU risk phrases		22
EU safety phrases		16-24-25-43-62

[a] Ether = 1. [b] n-Butyl acetate = 1. [c] Nelson, Hemwall, Edwards. [d] Toluene. [e] Hydrocarbon solvent.

Solvent group		HYDROCARBONS—UNSPECIFIED
Chemical name		
Chemical formula		C_{10}–C_{12} Range
Structural formula		
CAS No.		64742-47-8
Synonym/common name		Kerosene (Low odour)
Trade names		FINALAN 40
Suppliers		FINA
Molecular mass		
Boiling point/range	°C @ 760 mm Hg	162–197
Melting point	°C	
Density	kg/m^3 @ 20 °C	772 (15 °C)
Evaporation time[a]		
Evaporation time[b]		
Vapour pressure	kPa @ 20 °C	
Flash point	°C	44 (IP 34)
Auto ignition temperature	°C	
Upper flammability limit	%v/v in air	
Lower flammability limit	%v/v in air	
Viscosity	mPa.s (cP) @ 20 °C	0.96
Refractive index	n 20 D	
Solubility in water	%w/w @ 20 °C	
Solubility of water	%w/w @ 20 °C	
Solubility parameter[c]	(J/cm^3)$^{1/2}$	
Hydrogen bond index		
Fractional polarity		
Nitro cellulose dil ratio[d]		
Nitro cellulose dil ratio[e]		
Surface tension	mN/m @ 20 °C	
Specific heat liquid	kJ/kg mole/°C @ 20 °C	
Latent heat	kJ/kg mole	
Dielectric constant	@ 20 °C	
Antoine constant A	kPa°C	
Antoine constant B	kPa°C	
Antoine constant C	kPa°C	
Heat of combustion	kJ/kg mole	
UN number		1300
IMO classification		
ADR/RID classification		
UK exposure limits	mg/m^3 (ppm) 8hrTWA/MEL or OES	1740(300)
USA exposure limits	mg/m^3 (ppm) TLV/TWA	
German exposure limits	mg/m^3 (ppm) MAK-TWA	
EU classification		
EU risk phrases		10-22
EU safety phrases		2-16-24-25-43-62

[a] Ether = 1. [b] n-Butyl acetate = 1. [c] Nelson, Hemwall, Edwards. [d] Toluene. [e] Hydrocarbon solvent.

Solvent group		HYDROCARBONS—UNSPECIFIED
Chemical name		
Chemical formula		C_9–C_{16} Range
Structural formula		
CAS No.		64742-48-9
Synonym/common name		Kerosene (Low odour)
Trade names		FINALAN 60
Suppliers		FINA
Molecular mass		
Boiling point/range	°C @ 760 mm Hg	185–218
Melting point	°C	
Density	kg/m^3 @ 20 °C	798 (15 °C)
Evaporation time[a]		
Evaporation time[b]		
Vapour pressure	kPa @ 20 °C	
Flash point	°C	63 (IP 34)
Auto ignition temperature	°C	
Upper flammability limit	%v/v in air	
Lower flammability limit	%v/v in air	
Viscosity	mPa.s (cP) @ 20 °C	1.29
Refractive index	n 20 D	
Solubility in water	%w/w @ 20 °C	
Solubility of water	%w/w @ 20 °C	
Solubility parameter[c]	$(J/cm^3)^{1/2}$	
Hydrogen bond index		
Fractional polarity		
Nitro cellulose dil ratio[d]		
Nitro cellulose dil ratio[e]		
Surface tension	mN/m @ 20 °C	
Specific heat liquid	kJ/kg mole/°C @ 20 °C	
Latent heat	kJ/kg mole	
Dielectric constant	@ 20 °C	
Antoine constant A	kPa°C	
Antoine constant B	kPa°C	
Antoine constant C	kPa°C	
Heat of combustion	kJ/kg mole	
UN number		
IMO classification		
ADR/RID classification		
UK exposure limits	mg/m^3 (ppm) 8hrTWA/MEL or OES	1740(300)
USA exposure limits	mg/m^3 (ppm) TLV/TWA	
German exposure limits	mg/m^3 (ppm) MAK-TWA	
EU classification		
EU risk phrases		22
EU safety phrases		16-24-25-43-46

[a] Ether = 1. [b] n-Butyl acetate = 1. [c] Nelson, Hemwall, Edwards. [d] Toluene. [e] Hydrocarbon solvent.

Solvent group		**HYDROCARBONS—UNSPECIFIED**
Chemical name		
Chemical formula		C_9–C_{12} Range
Structural formula		
CAS No.		64742-47-8
Synonym/common name		Kerosene (Low odour)
Trade names		FINALAN 80
Suppliers		FINA
Molecular mass		
Boiling point/range	°C @ 760 mm Hg	211–239
Melting point	°C	
Density	kg/m^3 @ 20 °C	807 (15 °C)
Evaporation time[a]		
Evaporation time[b]		
Vapour pressure	kPa @ 20 °C	
Flash point	°C	
Auto ignition temperature	°C	
Upper flammability limit	%v/v in air	
Lower flammability limit	%v/v in air	
Viscosity	mPa.s (cP) @ 20 °C	2.97
Refractive index	n 20 D	
Solubility in water	%w/w @ 20 °C	
Solubility of water	%w/w @ 20 °C	
Solubility parameter[c]	$(J/cm^3)^{1/2}$	
Hydrogen bond index		
Fractional polarity		
Nitro cellulose dil ratio[d]		
Nitro cellulose dil ratio[e]		
Surface tension	mN/m @ 20 °C	
Specific heat liquid	kJ/kg mole/°C @ 20 °C	
Latent heat	kJ/kg mole	
Dielectric constant	@ 20 °C	
Antoine constant A	kPa°C	
Antoine constant B	kPa°C	
Antoine constant C	kPa°C	
Heat of combustion	kJ/kg mole	
UN number		
IMO classification		
ADR/RID classification		
UK exposure limits	mg/m^3 (ppm) 8hrTWA/MEL or OES	1500(200)
USA exposure limits	mg/m^3 (ppm) TLV/TWA	
German exposure limits	mg/m^3 (ppm) MAK-TWA	
EU classification		
EU risk phrases		22
EU safety phrases		16-24-25-43-62

[a] Ether = 1. [b] n-Butyl acetate = 1. [c] Nelson, Hemwall, Edwards. [d] Toluene. [e] Hydrocarbon solvent.

Solvent group		HYDROCARBONS—UNSPECIFIED
Chemical name		
Chemical formula		
Structural formula		
CAS No.		
Synonym/common name		
Trade names		VARSOL 60
Suppliers		EXXON
Molecular mass		
Boiling point/range	°C @ 760 mm Hg	186–216
Melting point	°C	
Density	kg/m^3 @ 20 °C	
Evaporation time[a]		150
Evaporation time[b]		0.04
Vapour pressure	kPa @ 20 °C	
Flash point	°C	64
Auto ignition temperature	°C	
Upper flammability limit	%v/v in air	
Lower flammability limit	%v/v in air	
Viscosity	mPa.s (cP) @ 20 °C	
Refractive index	n 20 D	1.446
Solubility in water	%w/w @ 20 °C	
Solubility of water	%w/w @ 20 °C	
Solubility parameter[c]	$(J/cm^3)^{1/2}$	
Hydrogen bond index		
Fractional polarity		
Nitro cellulose dil ratio[d]		
Nitro cellulose dil ratio[e]		
Surface tension	mN/m @ 20 °C	
Specific heat liquid	kJ/kg mole/°C @ 20 °C	
Latent heat	kJ/kg mole	
Dielectric constant	@ 20 °C	
Antoine constant A	kPa°C	
Antoine constant B	kPa°C	
Antoine constant C	kPa°C	
Heat of combustion	kJ/kg mole	
UN number		
IMO classification		
ADR/RID classification		3,32°(c)
UK exposure limits	mg/m^3 (ppm) 8hrTWA/MEL or OES	
USA exposure limits	mg/m^3 (ppm) TLV/TWA	
German exposure limits	mg/m^3 (ppm) MAK-TWA	
EU classification		
EU risk phrases		
EU safety phrases		

[a] Ether = 1. [b] n-Butyl acetate = 1. [c] Nelson, Hemwall, Edwards. [d] Toluene. [e] Hydrocarbon solvent.

Solvent group		HYDROCARBONS—UNSPECIFIED
Chemical name		
Chemical formula		C_{10}–C_{16} Range
Structural formula		
CAS No.		64742-47-8
Synonym/common name		Kerosene (Low odour)
Trade names		FINALAN 100
Suppliers		FINA
Molecular mass		
Boiling point/range	°C @ 760 mm Hg	236–272
Melting point	°C	
Density	kg/m^3 @ 20 °C	803
Evaporation time[a]		
Evaporation time[b]		
Vapour pressure	kPa @ 20 °C	
Flash point	°C	102 (IP 34)
Auto ignition temperature	°C	
Upper flammability limit	%v/v in air	
Lower flammability limit	%v/v in air	
Viscosity	mPa.s (cP) @ 20 °C	
Refractive index	n 20 D	
Solubility in water	%w/w @ 20 °C	
Solubility of water	%w/w @ 20 °C	
Solubility parameter[c]	$(J/cm^3)^{1/2}$	
Hydrogen bond index		
Fractional polarity		
Nitro cellulose dil ratio[d]		
Nitro cellulose dil ratio[e]		
Surface tension	mN/m @ 20 °C	
Specific heat liquid	kJ/kg mole/°C @ 20 °C	
Latent heat	kJ/kg mole	
Dielectric constant	@ 20 °C	
Antoine constant A	kPa°C	
Antoine constant B	kPa°C	
Antoine constant C	kPa°C	
Heat of combustion	kJ/kg mole	
UN number		
IMO classification		
ADR/RID classification		
UK exposure limits	mg/m^3 (ppm) 8hrTWA/MEL or OES	
USA exposure limits	mg/m^3 (ppm) TLV/TWA	
German exposure limits	mg/m^3 (ppm) MAK-TWA	
EU classification		
EU risk phrases		22
EU safety phrases		16-24-25-43-62

[a] Ether = 1. [b] n-Butyl acetate = 1. [c] Nelson, Hemwall, Edwards. [d] Toluene. [e] Hydrocarbon solvent.

Solvent group		HYDROCARBONS—UNSPECIFIED
Chemical name		
Chemical formula		
Structural formula		
CAS No.		
Synonym/common name		
Trade names		VARSOL 120
Suppliers		EXXON
Molecular mass		
Boiling point/range	°C @ 760 mm Hg	265–295∞
Melting point	°C	
Density	kg/m^3 @ 20 °C	
Evaporation time[a]		>1000
Evaporation time[b]		<0.01
Vapour pressure	kPa @ 20 °C	
Flash point	°C	122
Auto ignition temperature	°C	
Upper flammability limit	%v/v in air	
Lower flammability limit	%v/v in air	
Viscosity	mPa.s (cP) @ 20 °C	
Refractive index	n 20 D	1.468
Solubility in water	%w/w @ 20 °C	
Solubility of water	%w/w @ 20 °C	
Solubility parameter[c]	(J/cm^3)$^{1/2}$	
Hydrogen bond index		
Fractional polarity		
Nitro cellulose dil ratio[d]		
Nitro cellulose dil ratio[e]		
Surface tension	mN/m @ 20 °C	
Specific heat liquid	kJ/kg mole/°C @ 20 °C	
Latent heat	kJ/kg mole	
Dielectric constant	@ 20 °C	
Antoine constant A	kPa°C	
Antoine constant B	kPa°C	
Antoine constant C	kPa°C	
Heat of combustion	kJ/kg mole	
UN number		
IMO classification		
ADR/RID classification		
UK exposure limits	mg/m^3 (ppm) 8hrTWA/MEL or OES	
USA exposure limits	mg/m^3 (ppm) TLV/TWA	
German exposure limits	mg/m^3 (ppm) MAK-TWA	
EU classification		
EU risk phrases		
EU safety phrases		

[a] Ether = 1. [b] n-Butyl acetate = 1. [c] Nelson, Hemwall, Edwards. [d] Toluene. [e] Hydrocarbon solvent.

Solvent group		HYDROCARBONS—UNSPECIFIED
Chemical name		
Chemical formula		
Structural formula		
CAS No.		
Synonym/common name		
Trade names		Nappar 11
Suppliers		EXXON
Molecular mass		
Boiling point/range	°C @ 760 mm Hg	196–223∞
Melting point	°C	
Density	kg/m^3 @ 20 °C	
Evaporation time[a]		193
Evaporation time[b]		0.05
Vapour pressure	kPa @ 20 °C	
Flash point	°C	68
Auto ignition temperature	°C	
Upper flammability limit	%v/v in air	
Lower flammability limit	%v/v in air	
Viscosity	mPa.s (cP) @ 20 °C	
Refractive index	n 20 D	1.466
Solubility in water	%w/w @ 20 °C	
Solubility of water	%w/w @ 20 °C	
Solubility parameter[c]	(J/cm^3)$^{1/2}$	
Hydrogen bond index		
Fractional polarity		
Nitro cellulose dil ratio[d]		
Nitro cellulose dil ratio[e]		
Surface tension	mN/m @ 20 °C	
Specific heat liquid	kJ/kg mole/°C @ 20 °C	
Latent heat	kJ/kg mole	
Dielectric constant	@ 20 °C	
Antoine constant A	kPa°C	
Antoine constant B	kPa°C	
Antoine constant C	kPa°C	
Heat of combustion	kJ/kg mole	
UN number		1270
IMO classification		3.3
ADR/RID classification		3,32°(c)
UK exposure limits	mg/m^3 (ppm) 8hrTWA/MEL or OES	
USA exposure limits	mg/m^3 (ppm) TLV/TWA	
German exposure limits	mg/m^3 (ppm) MAK-TWA	
EU classification		Xi
EU risk phrases		38
EU safety phrases		24

[a] Ether = 1. [b] n-Butyl acetate = 1. [c] Nelson, Hemwall, Edwards. [d] Toluene. [e] Hydrocarbon solvent.

Solvent group		**HYDROCARBONS—UNSPECIFIED**
Chemical name		
Chemical formula		
Structural formula		
CAS No.		
Synonym/common name		
Trade names		Nappar 10
Suppliers		EXXON
Molecular mass		
Boiling point/range	°C @ 760 mm Hg	162–186
Melting point	°C	
Density	kg/m^3 @ 20 °C	
Evaporation time[a]		48
Evaporation time[b]		0.18
Vapour pressure	kPa @ 20 °C	
Flash point	°C	43
Auto ignition temperature	°C	
Upper flammability limit	%v/v in air	
Lower flammability limit	%v/v in air	
Viscosity	mPa.s (cP) @ 20 °C	
Refractive index	n 20 D	1.444
Solubility in water	%w/w @ 20 °C	
Solubility of water	%w/w @ 20 °C	
Solubility parameter[c]	(J/cm^3)$^{1/2}$	
Hydrogen bond index		
Fractional polarity		
Nitro cellulose dil ratio[d]		
Nitro cellulose dil ratio[e]		
Surface tension	mN/m @ 20 °C	
Specific heat liquid	kJ/kg mole/°C @ 20 °C	
Latent heat	kJ/kg mole	
Dielectric constant	@ 20 °C	
Antoine constant A	kPa°C	
Antoine constant B	kPa°C	
Antoine constant C	kPa°C	
Heat of combustion	kJ/kg mole	
UN number		1270
IMO classification		3.3
ADR/RID classification		3,31°(c)
UK exposure limits	mg/m^3 (ppm) 8hrTWA/MEL or OES	
USA exposure limits	mg/m^3 (ppm) TLV/TWA	
German exposure limits	mg/m^3 (ppm) MAK-TWA	
EU classification		Xi
EU risk phrases		10-38
EU safety phrases		24-43a

[a] Ether = 1. [b] n-Butyl acetate = 1. [c] Nelson, Hemwall, Edwards. [d] Toluene. [e] Hydrocarbon solvent.

Solvent group		HYDROCARBONS—UNSPECIFIED
Chemical name		
Chemical formula		
Structural formula		
CAS No.		
Synonym/common name		
Trade names		VARSOL 40
Suppliers		EXXON
Molecular mass		
Boiling point/range	°C @ 760 mm Hg	154–189
Melting point	°C	
Density	kg/m^3 @ 20 °C	
Evaporation time[a]		55
Evaporation time[b]		0.14
Vapour pressure	kPa @ 20 °C	
Flash point	°C	41
Auto ignition temperature	°C	
Upper flammability limit	%v/v in air	
Lower flammability limit	%v/v in air	
Viscosity	mPa.s (cP) @ 20 °C	
Refractive index	n 20 D	1.437
Solubility in water	%w/w @ 20 °C	
Solubility of water	%w/w @ 20 °C	
Solubility parameter[c]	$(J/cm^3)^{1/2}$	
Hydrogen bond index		
Fractional polarity		
Nitro cellulose dil ratio[d]		
Nitro cellulose dil ratio[e]		
Surface tension	mN/m @ 20 °C	
Specific heat liquid	kJ/kg mole/°C @ 20 °C	
Latent heat	kJ/kg mole	
Dielectric constant	@ 20 °C	
Antoine constant A	kPa°C	
Antoine constant B	kPa°C	
Antoine constant C	kPa°C	
Heat of combustion	kJ/kg mole	
UN number		1300
IMO classification		3.3
ADR/RID classification		3,31°(c)
UK exposure limits	mg/m^3 (ppm) 8hrTWA/MEL or OES	
USA exposure limits	mg/m^3 (ppm) TLV/TWA	
German exposure limits	mg/m^3 (ppm) MAK-TWA	
EU classification		
EU risk phrases		10
EU safety phrases		43a

[a] Ether = 1. [b] n-Butyl acetate = 1. [c] Nelson, Hemwall, Edwards. [d] Toluene. [e] Hydrocarbon solvent.

Solvent group		**HYDROCARBONS—UNSPECIFIED**
Chemical name		
Chemical formula		
Structural formula		
CAS No.		
Synonym/common name		
Trade names		VARSOL 80
Suppliers		EXXON
Molecular mass		
Boiling point/range	°C @ 760 mm Hg	202–239
Melting point	°C	
Density	kg/m^3 @ 20 °C	
Evaporation time[a]		600
Evaporation time[b]		<0.01
Vapour pressure	kPa @ 20 °C	
Flash point	°C	77
Auto ignition temperature	°C	
Upper flammability limit	%v/v in air	
Lower flammability limit	%v/v in air	
Viscosity	mPa.s (cP) @ 20 °C	
Refractive index	n 20 D	1.452
Solubility in water	%w/w @ 20 °C	
Solubility of water	%w/w @ 20 °C	
Solubility parameter[c]	$(J/cm^3)^{1/2}$	
Hydrogen bond index		
Fractional polarity		
Nitro cellulose dil ratio[d]		
Nitro cellulose dil ratio[e]		
Surface tension	mN/m @ 20 °C	
Specific heat liquid	kJ/kg mole/°C @ 20 °C	
Latent heat	kJ/kg mole	
Dielectric constant	@ 20 °C	
Antoine constant A	kPa°C	
Antoine constant B	kPa°C	
Antoine constant C	kPa°C	
Heat of combustion	kJ/kg mole	
UN number		
IMO classification		
ADR/RID classification		3,32°(c)
UK exposure limits	mg/m^3 (ppm) 8hrTWA/MEL or OES	
USA exposure limits	mg/m^3 (ppm) TLV/TWA	
German exposure limits	mg/m^3 (ppm) MAK-TWA	
EU classification		
EU risk phrases		
EU safety phrases		

[a] Ether = 1. [b] n-Butyl acetate = 1. [c] Nelson, Hemwall, Edwards. [d] Toluene. [e] Hydrocarbon solvent.

Solvent group		HYDROCARBONS—UNSPECIFIED
Chemical name		
Chemical formula		
Structural formula		
CAS No.		
Synonym/common name		
Trade names		VARSOL 30
Suppliers		EXXON
Molecular mass		
Boiling point/range	°C @ 760 mm Hg	137–164
Melting point	°C	
Density	kg/m^3 @ 20 °C	
Evaporation time[a]		26
Evaporation time[b]		0.44
Vapour pressure	kPa @ 20 °C	
Flash point	°C	29
Auto ignition temperature	°C	
Upper flammability limit	%v/v in air	
Lower flammability limit	%v/v in air	
Viscosity	mPa.s (cP) @ 20 °C	
Refractive index	n 20 D	1.430
Solubility in water	%w/w @ 20 °C	
Solubility of water	%w/w @ 20 °C	
Solubility parameter[c]	$(J/cm^3)^{1/2}$	
Hydrogen bond index		
Fractional polarity		
Nitro cellulose dil ratio[d]		
Nitro cellulose dil ratio[e]		
Surface tension	mN/m @ 20 °C	
Specific heat liquid	kJ/kg mole/°C @ 20 °C	
Latent heat	kJ/kg mole	
Dielectric constant	@ 20 °C	
Antoine constant A	kPa°C	
Antoine constant B	kPa°C	
Antoine constant C	kPa°C	
Heat of combustion	kJ/kg mole	
UN number		1300
IMO classification		3.3
ADR/RID classification		3,31°(c)
UK exposure limits	mg/m^3 (ppm) 8hrTWA/MEL or OES	
USA exposure limits	mg/m^3 (ppm) TLV/TWA	
German exposure limits	mg/m^3 (ppm) MAK-TWA	
EU classification		
EU risk phrases		10
EU safety phrases		43a

[a] Ether = 1. [b] n-Butyl acetate = 1. [c] Nelson, Hemwall, Edwards. [d] Toluene. [e] Hydrocarbon solvent.

Solvent group		HYDROCARBONS—UNSPECIFIED
Chemical name		
Chemical formula		
Structural formula		
CAS No.		
Synonym/common name		
Trade names		Isopar V
Suppliers		EXXON
Molecular mass		
Boiling point/range	°C @ 760 mm Hg	279–314∞
Melting point	°C	
Density	kg/m^3 @ 20 °C	
Evaporation time[a]		>1000
Evaporation time[b]		<0.01
Vapour pressure	kPa @ 20 °C	
Flash point	°C	134
Auto ignition temperature	°C	
Upper flammability limit	%v/v in air	
Lower flammability limit	%v/v in air	
Viscosity	mPa.s (cP) @ 20 °C	
Refractive index	n 20 D	1.452
Solubility in water	%w/w @ 20 °C	
Solubility of water	%w/w @ 20 °C	
Solubility parameter[c]	(J/cm^3)$^{1/2}$	
Hydrogen bond index		
Fractional polarity		
Nitro cellulose dil ratio[d]		
Nitro cellulose dil ratio[e]		
Surface tension	mN/m @ 20 °C	
Specific heat liquid	kJ/kg mole/°C @ 20 °C	
Latent heat	kJ/kg mole	
Dielectric constant	@ 20 °C	
Antoine constant A	kPa°C	
Antoine constant B	kPa°C	
Antoine constant C	kPa°C	
Heat of combustion	kJ/kg mole	
UN number		
IMO classification		
ADR/RID classification		
UK exposure limits	mg/m^3 (ppm) 8hrTWA/MEL or OES	
USA exposure limits	mg/m^3 (ppm) TLV/TWA	
German exposure limits	mg/m^3 (ppm) MAK-TWA	
EU classification		
EU risk phrases		
EU safety phrases		

[a] Ether = 1. [b] n-Butyl acetate = 1. [c] Nelson, Hemwall, Edwards. [d] Toluene. [e] Hydrocarbon solvent.

Solvent group		HYDROCARBONS—UNSPECIFIED
Chemical name		
Chemical formula		
Structural formula		
CAS No.		
Synonym/common name		
Trade names		VARSOL 110
Suppliers		EXXON
Molecular mass		
Boiling point/range	°C @ 760 mm Hg	246–275∞
Melting point	°C	
Density	kg/m^3 @ 20 °C	
Evaporation time[a]		>1000
Evaporation time[b]		<0.01
Vapour pressure	kPa @ 20 °C	
Flash point	°C	107
Auto ignition temperature	°C	
Upper flammability limit	%v/v in air	
Lower flammability limit	%v/v in air	
Viscosity	mPa.s (cP) @ 20 °C	
Refractive index	n 20 D	1.466
Solubility in water	%w/w @ 20 °C	
Solubility of water	%w/w @ 20 °C	
Solubility parameter[c]	(J/cm^3)$^{1/2}$	
Hydrogen bond index		
Fractional polarity		
Nitro cellulose dil ratio[d]		
Nitro cellulose dil ratio[e]		
Surface tension	mN/m @ 20 °C	
Specific heat liquid	kJ/kg mole/°C @ 20 °C	
Latent heat	kJ/kg mole	
Dielectric constant	@ 20 °C	
Antoine constant A	kPa°C	
Antoine constant B	kPa°C	
Antoine constant C	kPa°C	
Heat of combustion	kJ/kg mole	
UN number		
IMO classification		
ADR/RID classification		
UK exposure limits	mg/m^3 (ppm) 8hrTWA/MEL or OES	
USA exposure limits	mg/m^3 (ppm) TLV/TWA	
German exposure limits	mg/m^3 (ppm) MAK-TWA	
EU classification		
EU risk phrases		
EU safety phrases		

[a] Ether = 1. [b] n-Butyl acetate = 1. [c] Nelson, Hemwall, Edwards. [d] Toluene. [e] Hydrocarbon solvent.

Solvent group		HYDROCARBONS—UNSPECIFIED
Chemical name		
Chemical formula		
Structural formula		
CAS No.		
Synonym/common name		
Trade names		Nappar 7
Suppliers		EXXON
Molecular mass		
Boiling point/range	°C @ 760 mm Hg	100–105
Melting point	°C	
Density	kg/m^3 @ 20 °C	
Evaporation time[a]		3.8
Evaporation time[b]		3.92
Vapour pressure	kPa @ 20 °C	
Flash point	°C	<0
Auto ignition temperature	°C	
Upper flammability limit	%v/v in air	
Lower flammability limit	%v/v in air	
Viscosity	mPa.s (cP) @ 20 °C	
Refractive index	n 20 D	1.424
Solubility in water	%w/w @ 20 °C	
Solubility of water	%w/w @ 20 °C	
Solubility parameter[c]	(J/cm^3)$^{1/2}$	
Hydrogen bond index		
Fractional polarity		
Nitro cellulose dil ratio[d]		
Nitro cellulose dil ratio[e]		
Surface tension	mN/m @ 20 °C	
Specific heat liquid	kJ/kg mole/°C @ 20 °C	
Latent heat	kJ/kg mole	
Dielectric constant	@ 20 °C	
Antoine constant A	kPa°C	
Antoine constant B	kPa°C	
Antoine constant C	kPa°C	
Heat of combustion	kJ/kg mole	
UN number		2296
IMO classification		3.2
ADR/RID classification		3,3°(b)
UK exposure limits	mg/m^3 (ppm) 8hrTWA/MEL or OES	
USA exposure limits	mg/m^3 (ppm) TLV/TWA	
German exposure limits	mg/m^3 (ppm) MAK-TWA	
EU classification		F
EU risk phrases		11
EU safety phrases		9-16-29-33-43a

[a] Ether = 1. [b] n-Butyl acetate = 1. [c] Nelson, Hemwall, Edwards. [d] Toluene. [e] Hydrocarbon solvent.

Solvent group		**KETONES**
Chemical name		2-Propanone
Chemical formula		C_3H_6O
Structural formula		CH_3COCH_3
CAS No.		67-64-1
Synonym/common name		Acetone, dimethyl ketone
Trade names		DMK
Suppliers		UNION CARBIDE, BP, SHELL, ENICHEM
Molecular mass		58.08
Boiling point/range	°C @ 760 mm Hg	55.8–56.6
Melting point	°C	−94.7 to −95
Density	kg/m^3 @ 20 °C	790–792
Evaporation time[a]		2.0
Evaporation time[b]		5.6–9.5
Vapour pressure	kPa @ 20 °C	24.7
Flash point	°C	−18 (CLOSED CUP, −9.4 OPEN CUP) ENICHEM, −18 (IP 170) SHELL, −18 (OPEN CUP) BP
Auto ignition temperature	°C	465–540
Upper flammability limit	%v/v in air	13
Lower flammability limit	%v/v in air	2.1–2.6
Viscosity	mPa.s (cP) @ 20 °C	0.33
Refractive index	n 20 D	1.359
Solubility in water	%w/w @ 20 °C	Complete
Solubility of water	%w/w @ 20 °C	Complete
Solubility parameter[c]	$(J/cm^3)^{1/2}$	20.5
Hydrogen bond index		12.5
Fractional polarity		0.695
Nitro cellulose dil ratio[d]		4.2
Nitro cellulose dil ratio[e]		0.8
Surface tension	mN/m @ 20 °C	22.8
Specific heat liquid	kJ/kg mole/°C @ 20 °C	124.3
Latent heat	kJ/kg mole	30492
Dielectric constant	@ 20 °C	21.4
Antoine constant A	kPa°C	6.25478
Antoine constant B	kPa°C	1216.69
Antoine constant C	kPa°C	230.275
Heat of combustion	kJ/kg mole	−1690360
UN number		1090
IMO classification		3.1/II
ADR/RID classification		3,3°(b)
UK exposure limits	mg/m^3 (ppm) 8hrTWA/MEL or OES	1780(750)
USA exposure limits	mg/m^3 (ppm) TLV/TWA	1780(750)
German exposure limits	mg/m^3 (ppm) MAK-TWA	2400(1000)
EU classification		F
EU risk phrases		11
EU safety phrases		(2-)9-16-23-33

[a] Ether = 1. [b] n-Butyl acetate = 1. [c] Nelson, Hemwall, Edwards. [d] Toluene. [e] Hydrocarbon solvent.

Solvent group		**KETONES**
Chemical name		2-Butanone
Chemical formula		C_4H_8O
Structural formula		$CH_3CH_2COCH_3$
CAS No.		78-93-3
Synonym/common name		Methyl ethyl ketone/MEK
Trade names		
Suppliers		UNION CARBIDE, ELF ATOCHEM, EXXON, SHELL
Molecular mass		72.10
Boiling point/range	°C @ 760 mm Hg	79–80.5
Melting point	°C	−86
Density	kg/m^3 @ 20 °C	804–806
Evaporation time[a]		3.3
Evaporation time[b]		3.7
Vapour pressure	kPa @ 20 °C	9.5
Flash point	°C	−4 (IP 170) SHELL, −6 (CLOSED CUP) ELF ATOCHEM
Auto ignition temperature	°C	515
Upper flammability limit	%v/v in air	11.5
Lower flammability limit	%v/v in air	1.8
Viscosity	mPa.s (cP) @ 20 °C	0.4
Refractive index	n 20 D	1.379
Solubility in water	%w/w @ 20 °C	24–27
Solubility of water	%w/w @ 20 °C	10.0–12.5
Solubility parameter[c]	(J/cm^3)$^{1/2}$	19.0
Hydrogen bond index		10.5
Fractional polarity		0.510
Nitro cellulose dil ratio[d]		4.1
Nitro cellulose dil ratio[e]		0.9
Surface tension	mN/m @ 20 °C	24.8
Specific heat liquid	kJ/kg mole/°C @ 20 °C	162.2
Latent heat	kJ/kg mole	31728
Dielectric constant	@ 20 °C	18.5
Antoine constant A	kPa°C	6.18444
Antoine constant B	kPa°C	1259.22
Antoine constant C	kPa°C	221.758
Heat of combustion	kJ/kg mole	−2304563
UN number		1193
IMO classification		3.2/II
ADR/RID classification		3,3°(b)
UK exposure limits	mg/m^3 (ppm) 8hrTWA/MEL or OES	600(200)
USA exposure limits	mg/m^3 (ppm) TLV/TWA	590(200)
German exposure limits	mg/m^3 (ppm) MAK-TWA	590(200)
EU classification		F, Xi
EU risk phrases		11-36/37
EU safety phrases		(2-)9-16-25-33

[a] Ether = 1. [b] n-Butyl acetate = 1. [c] Nelson, Hemwall, Edwards. [d] Toluene. [e] Hydrocarbon solvent.

Solvent group		**KETONES**
Chemical name		Diethyl ketone
Chemical formula		$C_5H_{10}O$
Structural formula		$C_2H_5COC_2H_5$
CAS No.		96-22-0
Synonym/common name		Diethyl ketone
Trade names		
Suppliers		BASF
Molecular mass		86.1
Boiling point/range	°C @ 760 mm Hg	100.0–102.0
Melting point	°C	
Density	kg/m^3 @ 20 °C	815
Evaporation time[a]		6
Evaporation time[b]		
Vapour pressure	kPa @ 20 °C	
Flash point	°C	7 (DIN 51755 or 51758)
Auto ignition temperature	°C	445
Upper flammability limit	%v/v in air	7.7
Lower flammability limit	%v/v in air	1.6
Viscosity	mPa.s (cP) @ 20 °C	0.4
Refractive index	n 20 D	1.393
Solubility in water	%w/w @ 20 °C	
Solubility of water	%w/w @ 20 °C	
Solubility parameter[c]	(J/cm^3)$^{1/2}$	
Hydrogen bond index		
Fractional polarity		
Nitro cellulose dil ratio[d]		
Nitro cellulose dil ratio[e]		
Surface tension	mN/m @ 20 °C	
Specific heat liquid	kJ/kg mole/°C @ 20 °C	
Latent heat	kJ/kg mole	
Dielectric constant	@ 20 °C	
Antoine constant A	kPa°C	
Antoine constant B	kPa°C	
Antoine constant C	kPa°C	
Heat of combustion	kJ/kg mole	
UN number		
IMO classification		
ADR/RID classification		
UK exposure limits	mg/m^3 (ppm) 8hrTWA/MEL or OES	
USA exposure limits	mg/m^3 (ppm) TLV/TWA	
German exposure limits	mg/m^3 (ppm) MAK-TWA	
EU classification		
EU risk phrases		
EU safety phrases		

[a] Ether = 1. [b] n-Butyl acetate = 1. [c] Nelson, Hemwall, Edwards. [d] Toluene. [e] Hydrocarbon solvent.

Solvent group		**KETONES**
Chemical name		Cyclohexanone
Chemical formula		$C_6H_{10}O$
Structural formula	(ring structure: cyclohexane ring with C=O)	
CAS No.		108-94-1
Synonym/common name		Cyclohexanone
Trade names		CHK
Suppliers		UNION CARBIDE, SHELL, BASF
Molecular mass		98.14
Boiling point/range	°C @ 760 mm Hg	156
Melting point	°C	−16
Density	kg/m^3 @ 20 °C	946
Evaporation time[a]		40
Evaporation time[b]		0.3–0.4
Vapour pressure	kPa @ 20 °C	0.39
Flash point	°C	43.5 (IP 170) SHELL, 43 (DIN 51755 or 51758) BASF
Auto ignition temperature	°C	430
Upper flammability limit	%v/v in air	9.4
Lower flammability limit	%v/v in air	1.1
Viscosity	mPa.s (cP) @ 20 °C	2.20
Refractive index	n 20 D	1.455
Solubility in water	%w/w @ 20 °C	2.3–2.5
Solubility of water	%w/w @ 20 °C	8.0
Solubility parameter[c]	(J/cm^3)$^{1/2}$	20.3
Hydrogen bond index		13.7
Fractional polarity		0.38
Nitro cellulose dil ratio[d]		7.3
Nitro cellulose dil ratio[e]		1.3
Surface tension	mN/m @ 20 °C	34.9
Specific heat liquid	kJ/kg mole/°C @ 20 °C	161.9
Latent heat	kJ/kg mole	39648
Dielectric constant	@ 20 °C	
Antoine constant A	kPa°C	6.103304
Antoine constant B	kPa°C	1495.511
Antoine constant C	kPa°C	209.5517
Heat of combustion	kJ/kg mole	
UN number		1915
IMO classification		3.3/III
ADR/RID classification		3,31°(c)
UK exposure limits	mg/m^3 (ppm) 8hrTWA/MEL or OES	100(25)
USA exposure limits	mg/m^3 (ppm) TLV/TWA	100(25)
German exposure limits	mg/m^3 (ppm) MAK-TWA	200(50)
EU classification		Xn, R10
EU risk phrases		10–20
EU safety phrases		(2-)25

[a] Ether = 1. [b] n-Butyl acetate = 1. [c] Nelson, Hemwall, Edwards. [d] Toluene. [e] Hydrocarbon solvent.

Solvent group		**KETONES**
Chemical name		4-Methyl-3-pentene-2-one
Chemical formula		$C_6H_{10}O$
Structural formula		$(CH_3)_2C=CHCOCH_3$
CAS No.		141-79-7
Synonym/common name		Mesityl oxide
Trade names		MO
Suppliers		ELF ATOCHEM
Molecular mass		98.15
Boiling point/range	°C @ 760 mm Hg	129.8
Melting point	°C	−53
Density	kg/m^3 @ 20 °C	855
Evaporation time[a]		
Evaporation time[b]		1.06
Vapour pressure	kPa @ 20 °C	1.094
Flash point	°C	28 (CLOSED CUP)
Auto ignition temperature	°C	344
Upper flammability limit	%v/v in air	10.1
Lower flammability limit	%v/v in air	1.3
Viscosity	mPa.s (cP) @ 20 °C	0.6
Refractive index	n 20 D	1.4458
Solubility in water	%w/w @ 20 °C	2.8
Solubility of water	%w/w @ 20 °C	3.4
Solubility parameter[c]	$(J/cm^3)^{1/2}$	18.4
Hydrogen bond index		12.0
Fractional polarity		0.332
Nitro cellulose dil ratio[d]		3.6
Nitro cellulose dil ratio[e]		0.9
Surface tension	mN/m @ 20 °C	28.5
Specific heat liquid	kJ/kg mole/°C @ 20 °C	
Latent heat	kJ/kg mole	37500
Dielectric constant	@ 20 °C	
Antoine constant A	kPa°C	6.13756
Antoine constant B	kPa°C	1399.09
Antoine constant C	kPa°C	208.85
Heat of combustion	kJ/kg mole	
UN number		
IMO classification		
ADR/RID classification		
UK exposure limits	mg/m^3 (ppm) 8hrTWA/MEL or OES	60(15)
USA exposure limits	mg/m^3 (ppm) TLV/TWA	60(15)
German exposure limits	mg/m^3 (ppm) MAK-TWA	100(25)
EU classification		Xn, R10
EU risk phrases		10-20/21/22
EU safety phrases		(2-)25

[a] Ether = 1. [b] n-Butyl acetate = 1. [c] Nelson, Hemwall, Edwards. [d] Toluene. [e] Hydrocarbon solvent.

Solvent group		**KETONES**
Chemical name		4-Methyl-2-pentanone
Chemical formula		$C_6H_{12}O$
Structural formula		$(CH_3)_2CHCH_2COCH_3$
CAS No.		108-10-1
Synonym/common name		Methyl isobutyl ketone/MIBK
Trade names		
Suppliers		UNION CARBIDE, ELF ATOCHEM, HULS, SHELL
Molecular mass		100.16
Boiling point/range	°C @ 760 mm Hg	114–117
Melting point	°C	−84 to −85
Density	kg/m^3 @ 20 °C	799–802
Evaporation time[a]		7–7.2
Evaporation time[b]		1.60
Vapour pressure	kPa @ 20 °C	1.9
Flash point	°C	16 (CLOSED CUP) ELF ATOCHEM, 14 HULS, 14 (IP 170) SHELL
Auto ignition temperature	°C	460
Upper flammability limit	%v/v in air	7.5–9.0
Lower flammability limit	%v/v in air	1.3–1.7
Viscosity	mPa.s (cP) @ 20 °C	0.6
Refractive index	n 20 D	1.396
Solubility in water	%w/w @ 20 °C	2
Solubility of water	%w/w @ 20 °C	1–2.1
Solubility parameter[c]	(J/cm^3)$^{1/2}$	17.2
Hydrogen bond index		10.5
Fractional polarity		0.315
Nitro cellulose dil ratio[d]		3.6
Nitro cellulose dil ratio[e]		1.0
Surface tension	mN/m @ 20 °C	23.9
Specific heat liquid	kJ/kg mole/°C @ 20 °C	192.3
Latent heat	kJ/kg mole	36057
Dielectric constant	@ 20 °C	13.1
Antoine constant A	kPa°C	6.31286
Antoine constant B	kPa°C	1449.92
Antoine constant C	kPa°C	220.093
Heat of combustion	kJ/kg mole	−3530339
UN number		1245
IMO classification		3.1/II
ADR/RID classification		3,3°(b)
UK exposure limits	mg/m^3 (ppm) 8hrTWA/MEL or OES	205(50)
USA exposure limits	mg/m^3 (ppm) TLV/TWA	205(50)
German exposure limits	mg/m^3 (ppm) MAK-TWA	400(100)
EU classification		F
EU risk phrases		11
EU safety phrases		(2-)9-16-23-33

[a] Ether = 1. [b] n-Butyl acetate = 1. [c] Nelson, Hemwall, Edwards. [d] Toluene. [e] Hydrocarbon solvent.

Property	Unit	Value
Solvent group		**KETONES**
Chemical name		2-Heptanone
Chemical formula		$C_7H_{14}O$
Structural formula		$CH_3(CH_2)_4COCH_3$
CAS No.		110-43-0
Synonym/common name		Methyl n-amyl ketone
Trade names		
Suppliers		UNION CARBIDE
Molecular mass		114.19
Boiling point/range	°C @ 760 mm Hg	151.5
Melting point	°C	−35.0
Density	kg/m^3 @ 20 °C	817
Evaporation time[a]		
Evaporation time[b]		0.28
Vapour pressure	kPa @ 20 °C	
Flash point	°C	
Auto ignition temperature	°C	
Upper flammability limit	%v/v in air	
Lower flammability limit	%v/v in air	
Viscosity	mPa.s (cP) @ 20 °C	0.8
Refractive index	n 20 D	1.4087
Solubility in water	%w/w @ 20 °C	0.4
Solubility of water	%w/w @ 20 °C	1.3
Solubility parameter[c]	(J/cm^3)$^{1/2}$	17.4
Hydrogen bond index		
Fractional polarity		
Nitro cellulose dil ratio[d]		
Nitro cellulose dil ratio[e]		
Surface tension	mN/m @ 20 °C	
Specific heat liquid	kJ/kg mole/°C @ 20 °C	
Latent heat	kJ/kg mole	38300
Dielectric constant	@ 20 °C	11.98
Antoine constant A	kPa°C	6.15034
Antoine constant B	kPa°C	1462.981
Antoine constant C	kPa°C	201.929
Heat of combustion	kJ/kg mole	
UN number		
IMO classification		
ADR/RID classification		
UK exposure limits	mg/m^3 (ppm) 8hrTWA/MEL or OES	240(50)
USA exposure limits	mg/m^3 (ppm) TLV/TWA	233(50)
German exposure limits	mg/m^3 (ppm) MAK-TWA	
EU classification		Xn, R10
EU risk phrases		10-22
EU safety phrases		(2-)23

[a] Ether = 1. [b] n-Butyl acetate = 1. [c] Nelson, Hemwall, Edwards. [d] Toluene. [e] Hydrocarbon solvent.

Solvent group		**KETONES**
Chemical name		4-Hydroxy-4-methyl-2-pentanone
Chemical formula		$C_6H_{12}O_2$
Structural formula		$(CH_3)_2C(OH)CH_2COCH_3$
CAS No.		123-42-2
Synonym/common name		Diacetone alcohol
Trade names		
Suppliers		UNION CARBIDE
Molecular mass		116.16
Boiling point/range	°C @ 760 mm Hg	169.2
Melting point	°C	
Density	kg/m^3 @ 20 °C	940
Evaporation time[a]		
Evaporation time[b]		0.12
Vapour pressure	kPa @ 20 °C	
Flash point	°C	
Auto ignition temperature	°C	
Upper flammability limit	%v/v in air	
Lower flammability limit	%v/v in air	
Viscosity	mPa.s (cP) @ 20 °C	3.2
Refractive index	n 20 D	
Solubility in water	%w/w @ 20 °C	Complete
Solubility of water	%w/w @ 20 °C	Complete
Solubility parameter[c]	$(J/cm^3)^{1/2}$	18.9
Hydrogen bond index		6.5
Fractional polarity		0.312
Nitro cellulose dil ratio[d]		3.4
Nitro cellulose dil ratio[e]		0.6
Surface tension	mN/m @ 20 °C	
Specific heat liquid	kJ/kg mole/°C @ 20 °C	
Latent heat	kJ/kg mole	
Dielectric constant	@ 20 °C	
Antoine constant A	kPa°C	
Antoine constant B	kPa°C	
Antoine constant C	kPa°C	
Heat of combustion	kJ/kg mole	
UN number		
IMO classification		
ADR/RID classification		
UK exposure limits	mg/m^3 (ppm) 8hrTWA/MEL or OES	
USA exposure limits	mg/m^3 (ppm) TLV/TWA	
German exposure limits	mg/m^3 (ppm) MAK-TWA	
EU classification		
EU risk phrases		
EU safety phrases		

[a] Ether = 1. [b] n-Butyl acetate = 1. [c] Nelson, Hemwall, Edwards. [d] Toluene. [e] Hydrocarbon solvent.

Solvent group		**KETONES**
Chemical name		5-Methyl-3-heptanone
Chemical formula		$C_8H_{16}O$
Structural formula		$CH_3(CH_2)_4COCH_2CH_3$
CAS No.		541-85-5
Synonym/common name		Ethyl amyl ketone, ethyl isoamyl ketone
Trade names		EAK
Suppliers		ELF ATOCHEM
Molecular mass		128.21
Boiling point/range	°C @ 760 mm Hg	159.2
Melting point	°C	−46
Density	kg/m^3 @ 20 °C	820
Evaporation time[a]		
Evaporation time[b]		5
Vapour pressure	kPa @ 20 °C	
Flash point	°C	48 (CLOSED CUP)
Auto ignition temperature	°C	
Upper flammability limit	%v/v in air	
Lower flammability limit	%v/v in air	
Viscosity	mPa.s (cP) @ 20 °C	
Refractive index	n 20 D	1.4149
Solubility in water	%w/w @ 20 °C	0.3
Solubility of water	%w/w @ 20 °C	0.9
Solubility parameter[c]	(J/cm^3)$^{1/2}$	
Hydrogen bond index		
Fractional polarity		
Nitro cellulose dil ratio[d]		
Nitro cellulose dil ratio[e]		
Surface tension	mN/m @ 20 °C	
Specific heat liquid	kJ/kg mole/°C @ 20 °C	
Latent heat	kJ/kg mole	
Dielectric constant	@ 20 °C	
Antoine constant A	kPa°C	
Antoine constant B	kPa°C	
Antoine constant C	kPa°C	
Heat of combustion	kJ/kg mole	
UN number		
IMO classification		
ADR/RID classification		
UK exposure limits	mg/m^3 (ppm) 8hrTWA/MEL or OES	130(25)
USA exposure limits	mg/m^3 (ppm) TLV/TWA	131(25)
German exposure limits	mg/m^3 (ppm) MAK-TWA	
EU classification		Xi, R10
EU risk phrases		10-36/37
EU safety phrases		(2-)23

[a] Ether = 1. [b] n-Butyl acetate = 1. [c] Nelson, Hemwall, Edwards. [d] Toluene. [e] Hydrocarbon solvent.

Solvent group		**KETONES**
Chemical name		4-Methoxy-4-methyl-2-pentanone
Chemical formula		$C_7H_{14}O_2$
Structural formula		$(CH_3)_2C(OCH_3)CH_2COCH_3$
CAS No.		107-70-0
Synonym/common name		ME-6K
Trade names		ME-6K
Suppliers		SHELL
Molecular mass		130.19
Boiling point/range	°C @ 760 mm Hg	152.0–162.0
Melting point	°C	−35.4
Density	kg/m^3 @ 20 °C	902–912
Evaporation time[a]		46
Evaporation time[b]		0.27
Vapour pressure	kPa @ 20 °C	0.25
Flash point	°C	44 (IP 170)
Auto ignition temperature	°C	402
Upper flammability limit	%v/v in air	6.0
Lower flammability limit	%v/v in air	1.1
Viscosity	mPa.s (cP) @ 20 °C	1.25
Refractive index	n 20 D	1.418
Solubility in water	%w/w @ 20 °C	20.0
Solubility of water	%w/w @ 20 °C	9.0
Solubility parameter[c]	(J/cm^3)$^{1/2}$	17.4
Hydrogen bond index		12.5
Fractional polarity		0.19
Nitro cellulose dil ratio[d]		3.1
Nitro cellulose dil ratio[e]		0.8
Surface tension	mN/m @ 20 °C	27.7
Specific heat liquid	kJ/kg mole/°C @ 20 °C	278.6
Latent heat	kJ/kg mole	
Dielectric constant	@ 20 °C	
Antoine constant A	kPa°C	6.32318
Antoine constant B	kPa°C	1610.75
Antoine constant C	kPa°C	212.539
Heat of combustion	kJ/kg mole	−4008488
UN number		2293
IMO classification		3.3/III
ADR/RID classification		3,31°(c)
UK exposure limits	mg/m^3 (ppm) 8hrTWA/MEL or OES	
USA exposure limits	mg/m^3 (ppm) TLV/TWA	
German exposure limits	mg/m^3 (ppm) MAK-TWA	
EU classification		
EU risk phrases		10
EU safety phrases		23

[a] Ether = 1. [b] n-Butyl acetate = 1. [c] Nelson, Hemwall, Edwards. [d] Toluene. [e] Hydrocarbon solvent.

Solvent group		KETONES
Chemical name		3,5,5,Trimethyl-2-cyclohexene-1-one
Chemical formula		$C_9H_{14}O$
Structural formula		
CAS No.		78-59-1
Synonym/common name		Isophorone
Trade names		IPHO
Suppliers		UNION CARBIDE, BP, ELF ATOCHEM, HULS
Molecular mass		138.21
Boiling point/range	°C @ 760 mm Hg	210–216
Melting point	°C	−8
Density	kg/m^3 @ 20 °C	920–923
Evaporation time[a]		330
Evaporation time[b]		0.02
Vapour pressure	kPa @ 20 °C	0.04
Flash point	°C	96 (OPEN CUP) BP, 85 (CLOSED CUP) ELF ATOCHEM, 85 HULS
Auto ignition temperature	°C	460–462
Upper flammability limit	%v/v in air	3.8
Lower flammability limit	%v/v in air	0.8
Viscosity	mPa.s (cP) @ 20 °C	2.6
Refractive index	n 20 D	1.477
Solubility in water	%w/w @ 20 °C	<0.02–1.2
Solubility of water	%w/w @ 20 °C	2.3–4.3
Solubility parameter[c]	$(J/cm^3)^{1/2}$	19.2
Hydrogen bond index		14.9
Fractional polarity		0.190
Nitro cellulose dil ratio[d]		6.2
Nitro cellulose dil ratio[e]		
Surface tension	mN/m @ 20 °C	32.3
Specific heat liquid	kJ/kg mole/°C @ 20 °C	
Latent heat	kJ/kg mole	
Dielectric constant	@ 20 °C	
Antoine constant A	kPa°C	
Antoine constant B	kPa°C	
Antoine constant C	kPa°C	
Heat of combustion	kJ/kg mole	
UN number		
IMO classification		
ADR/RID classification		
UK exposure limits	mg/m^3 (ppm) 8hrTWA/MEL or OES	25(5)
USA exposure limits	mg/m^3 (ppm) TLV/TWA	28(5)
German exposure limits	mg/m^3 (ppm) MAK-TWA	28(5)
EU classification		Xi
EU risk phrases		36/37/38
EU safety phrases		(2-)26

Structural formula: O, $(CH_3)_2$, CH_3

[a] Ether = 1. [b] n-Butyl acetate = 1. [c] Nelson, Hemwall, Edwards. [d] Toluene. [e] Hydrocarbon solvent.

Solvent group		KETONES
Chemical name		Trimethylcyclohexanone
Chemical formula		$C_9H_{16}O$
Structural formula		
CAS No.		
Synonym/common name		Trimethylcyclohexanone
Trade names		
Suppliers		HULS
Molecular mass		140.2
Boiling point/range	°C @ 760 mm Hg	188.8
Melting point	°C	−10
Density	kg/m^3 @ 20 °C	888
Evaporation time[a]		109
Evaporation time[b]		
Vapour pressure	kPa @ 20 °C	0.1
Flash point	°C	64
Auto ignition temperature	°C	430
Upper flammability limit	%v/v in air	
Lower flammability limit	%v/v in air	1
Viscosity	mPa.s (cP) @ 20 °C	2.54
Refractive index	n 20 D	1.445
Solubility in water	%w/w @ 20 °C	
Solubility of water	%w/w @ 20 °C	1.4
Solubility parameter[c]	(J/cm^3)$^{1/2}$	
Hydrogen bond index		
Fractional polarity		
Nitro cellulose dil ratio[d]		
Nitro cellulose dil ratio[e]		
Surface tension	mN/m @ 20 °C	
Specific heat liquid	kJ/kg mole/°C @ 20 °C	
Latent heat	kJ/kg mole	
Dielectric constant	@ 20 °C	
Antoine constant A	kPa°C	
Antoine constant B	kPa°C	
Antoine constant C	kPa°C	
Heat of combustion	kJ/kg mole	
UN number		
IMO classification		
ADR/RID classification		3,32°(c)
UK exposure limits	mg/m^3 (ppm) 8hrTWA/MEL or OES	
USA exposure limits	mg/m^3 (ppm) TLV/TWA	
German exposure limits	mg/m^3 (ppm) MAK-TWA	
EU classification		
EU risk phrases		
EU safety phrases		

[a] Ether = 1. [b] n-Butyl acetate = 1. [c] Nelson, Hemwall, Edwards. [d] Toluene. [e] Hydrocarbon solvent.

Solvent group		**KETONES**
Chemical name		2,6-Dimethyl-4-heptanone/4,6-Dimethyl-2-heptanone (Isomer mixture)
Chemical formula		$C_9H_{18}O$
Structural formula		$(CH_3)_2CHCH_2COCH_2CH(CH_3)_2$
CAS No.		108-83-8
Synonym/common name		Diisobutyl ketone
Trade names		
Suppliers		SHELL, HULS, UNION CARBIDE
Molecular mass		142.24
Boiling point/range	°C @ 760 mm Hg	163.0–173.0
Melting point	°C	−42
Density	kg/m^3 @ 20 °C	806–812
Evaporation time[a]		48
Evaporation time[b]		0.15–0.2
Vapour pressure	kPa @ 20 °C	0.16–0.23
Flash point	°C	47 (IP 170) SHELL, 49 HULS
Auto ignition temperature	°C	345
Upper flammability limit	%v/v in air	6.2
Lower flammability limit	%v/v in air	0.8
Viscosity	mPa.s (cP) @ 20 °C	1.05
Refractive index	n 20 D	1.414
Solubility in water	%w/w @ 20 °C	0.05
Solubility of water	%w/w @ 20 °C	0.4–0.75
Solubility parameter[c]	(J/cm^3)$^{1/2}$	15.6–16.0
Hydrogen bond index		9.8
Fractional polarity		0.123
Nitro cellulose dil ratio[d]		1.5
Nitro cellulose dil ratio[e]		0.8
Surface tension	mN/m @ 20 °C	22.6
Specific heat liquid	kJ/kg mole/°C @ 20 °C	277.4
Latent heat	kJ/kg mole	39116
Dielectric constant	@ 20 °C	
Antoine constant A	kPa°C	6.07029
Antoine constant B	kPa°C	1476.4
Antoine constant C	kPa°C	195
Heat of combustion	kJ/kg mole	−5633016
UN number		1157
IMO classification		3.3/III
ADR/RID classification		3,31°(c)
UK exposure limits	mg/m^3 (ppm) 8hrTWA/MEL or OES	150(25)
USA exposure limits	mg/m^3 (ppm) TLV/TWA	145(25)
German exposure limits	mg/m^3 (ppm) MAK-TWA	290(50)
EU classification		Xi, R10
EU risk phrases		10-37
EU safety phrases		(2-)24

[a] Ether = 1. [b] n-Butyl acetate = 1. [c] Nelson, Hemwall, Edwards. [d] Toluene. [e] Hydrocarbon solvent.

Solvent group		NITRO COMPOUNDS
Chemical name		N-Methylpyrrolidone
Chemical formula		C_5H_9ON
Structural formula	(ring structure: N, O, CH_3)	
CAS No.		872-50-4
Synonym/common name		NMP
Trade names		
Suppliers		BASF
Molecular mass		99.1
Boiling point/range	°C @ 760 mm Hg	202.5–205.0
Melting point	°C	−24.4
Density	kg/m^3 @ 20 °C	1028
Evaporation time[a]		360
Evaporation time[b]		
Vapour pressure	kPa @ 20 °C	0.04
Flash point	°C	91 (DIN 51755 or 51758)
Auto ignition temperature	°C	270
Upper flammability limit	%v/v in air	9.5
Lower flammability limit	%v/v in air	1.3
Viscosity	mPa.s (cP) @ 20 °C	1.8
Refractive index	n 20 D	1.470
Solubility in water	%w/w @ 20 °C	
Solubility of water	%w/w @ 20 °C	
Solubility parameter[c]	$(J/cm^3)^{1/2}$	23.1
Hydrogen bond index		
Fractional polarity		
Nitro cellulose dil ratio[d]		
Nitro cellulose dil ratio[e]		
Surface tension	mN/m @ 20 °C	
Specific heat liquid	kJ/kg mole/°C @ 20 °C	
Latent heat	kJ/kg mole	
Dielectric constant	@ 20 °C	
Antoine constant A	kPa°C	
Antoine constant B	kPa°C	
Antoine constant C	kPa°C	
Heat of combustion	kJ/kg mole	
UN number		
IMO classification		
ADR/RID classification		
UK exposure limits	mg/m^3 (ppm) 8hrTWA/MEL or OES	
USA exposure limits	mg/m^3 (ppm) TLV/TWA	
German exposure limits	mg/m^3 (ppm) MAK-TWA	
EU classification		
EU risk phrases		
EU safety phrases		

[a] Ether = 1. [b] n-Butyl acetate = 1. [c] Nelson, Hemwall, Edwards. [d] Toluene. [e] Hydrocarbon solvent.

Solvent group		**PHENOLS**
Chemical name		Phenol
Chemical formula		C_6H_6O
Structural formula		C_6H_5OH
CAS No.		108-95-2
Synonym/common name		Phenol, hydroxybenzene, carbolic acid
Trade names		
Suppliers		ENICHEM
Molecular mass		94.11
Boiling point/range	°C @ 760 mm Hg	181.7
Melting point	°C	40.9
Density	kg/m^3 @ 20 °C	1057
Evaporation time[a]		
Evaporation time[b]		
Vapour pressure	kPa @ 20 °C	0.030
Flash point	°C	79 CC/85 OC
Auto ignition temperature	°C	715
Upper flammability limit	%v/v in air	9
Lower flammability limit	%v/v in air	1.5
Viscosity	mPa.s (cP) @ 20 °C	
Refractive index	n 20 D	1.5408
Solubility in water	%w/w @ 20 °C	8.7
Solubility of water	%w/w @ 20 °C	
Solubility parameter[c]	(J/cm^3)$^{1/2}$	23.1
Hydrogen bond index		
Fractional polarity		
Nitro cellulose dil ratio[d]		
Nitro cellulose dil ratio[e]		
Surface tension	mN/m @ 20 °C	
Specific heat liquid	kJ/kg mole/°C @ 20 °C	
Latent heat	kJ/kg mole	45689
Dielectric constant	@ 20 °C	
Antoine constant A	kPa°C	10.6887
Antoine constant B	kPa°C	273
Antoine constant C	kPa°C	−3053480
Heat of combustion	kJ/kg mole	1671
UN number		
IMO classification		
ADR/RID classification		
UK exposure limits	mg/m^3 (ppm) 8hrTWA/MEL or OES	
USA exposure limits	mg/m^3 (ppm) TLV/TWA	
German exposure limits	mg/m^3 (ppm) MAK-TWA	
EU classification		
EU risk phrases		
EU safety phrases		

[a] Ether = 1. [b] n-Butyl acetate = 1. [c] Nelson, Hemwall, Edwards. [d] Toluene. [e] Hydrocarbon solvent.

Solvent group		SULPHUR COMPOUNDS
Chemical name		Dimethyl sulphoxide
Chemical formula		C_2H_6OS
Structural formula		$(CH_3)_2SO$
CAS No.		67-68-5
Synonym/common name		DMSO
Trade names		
Suppliers		ELF ATOCHEM
Molecular mass		78.13
Boiling point/range	°C @ 760 mm Hg	189
Melting point	°C	18.5
Density	kg/m^3 @ 20 °C	1100
Evaporation time[a]		
Evaporation time[b]		
Vapour pressure	kPa @ 20 °C	
Flash point	°C	
Auto ignition temperature	°C	300-302
Upper flammability limit	%v/v in air	28.5
Lower flammability limit	%v/v in air	2.6
Viscosity	mPa.s (cP) @ 20 °C	2.14
Refractive index	n 20 D	1.478
Solubility in water	%w/w @ 20 °C	Complete
Solubility of water	%w/w @ 20 °C	
Solubility parameter[c]	$(J/cm^3)^{1/2}$	24.5
Hydrogen bond index		
Fractional polarity		
Nitro cellulose dil ratio[d]		
Nitro cellulose dil ratio[e]		
Surface tension	mN/m @ 20 °C	43.72
Specific heat liquid	kJ/kg mole/°C @ 20 °C	
Latent heat	kJ/kg mole	43140
Dielectric constant	@ 20 °C	
Antoine constant A	kPa°C	6.72165
Antoine constant B	kPa°C	1962.05
Antoine constant C	kPa°C	225.892
Heat of combustion	kJ/kg mole	−1977700
UN number		
IMO classification		
ADR/RID classification		3,32°(c)
UK exposure limits	mg/m^3 (ppm) 8hrTWA/MEL or OES	
USA exposure limits	mg/m^3 (ppm) TLV/TWA	
German exposure limits	mg/m^3 (ppm) MAK-TWA	
EU classification		
EU risk phrases		
EU safety phrases		

[a] Ether = 1. [b] n-Butyl acetate = 1. [c] Nelson, Hemwall, Edwards. [d] Toluene. [e] Hydrocarbon solvent.

Solvent group		**UNSPECIFIED**
Chemical name		
Chemical formula		
Structural formula		
CAS No.		
Synonym/common name		
Trade names		Exxate 1000
Suppliers		EXXON
Molecular mass		
Boiling point/range	°C @ 760 mm Hg	220–250
Melting point	°C	
Density	kg/m^3 @ 20 °C	
Evaporation time[a]		
Evaporation time[b]		<0.01
Vapour pressure	kPa @ 20 °C	
Flash point	°C	100
Auto ignition temperature	°C	
Upper flammability limit	%v/v in air	
Lower flammability limit	%v/v in air	
Viscosity	mPa.s (cP) @ 20 °C	
Refractive index	n 20 D	1.429
Solubility in water	%w/w @ 20 °C	
Solubility of water	%w/w @ 20 °C	
Solubility parameter[c]	$(J/cm^3)^{1/2}$	
Hydrogen bond index		
Fractional polarity		
Nitro cellulose dil ratio[d]		
Nitro cellulose dil ratio[e]		
Surface tension	mN/m @ 20 °C	
Specific heat liquid	kJ/kg mole/°C @ 20 °C	
Latent heat	kJ/kg mole	
Dielectric constant	@ 20 °C	
Antoine constant A	kPa°C	
Antoine constant B	kPa°C	
Antoine constant C	kPa°C	
Heat of combustion	kJ/kg mole	
UN number		
IMO classification		
ADR/RID classification		3,32°(c)
UK exposure limits	mg/m^3 (ppm) 8hrTWA/MEL or OES	
USA exposure limits	mg/m^3 (ppm) TLV/TWA	
German exposure limits	mg/m^3 (ppm) MAK-TWA	
EU classification		
EU risk phrases		
EU safety phrases		

[a] Ether = 1. [b] n-Butyl acetate = 1. [c] Nelson, Hemwall, Edwards. [d] Toluene. [e] Hydrocarbon solvent.

Solvent group		UNSPECIFIED
Chemical name		
Chemical formula		
Structural formula		
CAS No.		
Synonym/common name		
Trade names		Exxate 700
Suppliers		EXXON
Molecular mass		
Boiling point/range	°C @ 760 mm Hg	176–200
Melting point	°C	
Density	kg/m^3 @ 20 °C	
Evaporation time[a]		
Evaporation time[b]		0.08
Vapour pressure	kPa @ 20 °C	
Flash point	°C	66
Auto ignition temperature	°C	
Upper flammability limit	%v/v in air	
Lower flammability limit	%v/v in air	
Viscosity	mPa.s (cP) @ 20 °C	
Refractive index	n 20 D	1.417
Solubility in water	%w/w @ 20 °C	
Solubility of water	%w/w @ 20 °C	
Solubility parameter[c]	$(J/cm^3)^{1/2}$	
Hydrogen bond index		
Fractional polarity		
Nitro cellulose dil ratio[d]		
Nitro cellulose dil ratio[e]		
Surface tension	mN/m @ 20 °C	
Specific heat liquid	kJ/kg mole/°C @ 20 °C	
Latent heat	kJ/kg mole	
Dielectric constant	@ 20 °C	
Antoine constant A	kPa°C	
Antoine constant B	kPa°C	
Antoine constant C	kPa°C	
Heat of combustion	kJ/kg mole	
UN number		
IMO classification		
ADR/RID classification		3,3°(b)
UK exposure limits	mg/m^3 (ppm) 8hrTWA/MEL or OES	
USA exposure limits	mg/m^3 (ppm) TLV/TWA	
German exposure limits	mg/m^3 (ppm) MAK-TWA	
EU classification		
EU risk phrases		
EU safety phrases		

[a] Ether = 1. [b] n-Butyl acetate = 1. [c] Nelson, Hemwall, Edwards. [d] Toluene. [e] Hydrocarbon solvent.

Solvent group		**UNSPECIFIED**
Chemical name		
Chemical formula		
Structural formula		
CAS No.		
Synonym/common name		
Trade names		Exxate 1300
Suppliers		EXXON
Molecular mass		
Boiling point/range	°C @ 760 mm Hg	240–285
Melting point	°C	
Density	kg/m^3 @ 20 °C	
Evaporation time[a]		
Evaporation time[b]		<0.01
Vapour pressure	kPa @ 20 °C	
Flash point	°C	127
Auto ignition temperature	°C	
Upper flammability limit	%v/v in air	
Lower flammability limit	%v/v in air	
Viscosity	mPa.s (cP) @ 20 °C	
Refractive index	n 20 D	1.438
Solubility in water	%w/w @ 20 °C	
Solubility of water	%w/w @ 20 °C	
Solubility parameter[c]	$(J/cm^3)^{1/2}$	
Hydrogen bond index		
Fractional polarity		
Nitro cellulose dil ratio[d]		
Nitro cellulose dil ratio[e]		
Surface tension	mN/m @ 20 °C	
Specific heat liquid	kJ/kg mole/°C @ 20 °C	
Latent heat	kJ/kg mole	
Dielectric constant	@ 20 °C	
Antoine constant A	kPa°C	
Antoine constant B	kPa°C	
Antoine constant C	kPa°C	
Heat of combustion	kJ/kg mole	
UN number		
IMO classification		
ADR/RID classification		3,32°(c)
UK exposure limits	mg/m^3 (ppm) 8hrTWA/MEL or OES	
USA exposure limits	mg/m^3 (ppm) TLV/TWA	
German exposure limits	mg/m^3 (ppm) MAK-TWA	
EU classification		
EU risk phrases		
EU safety phrases		

[a] Ether = 1. [b] n-Butyl acetate = 1. [c] Nelson, Hemwall, Edwards. [d] Toluene. [e] Hydrocarbon solvent.

Solvent group		**UNSPECIFIED**
Chemical name		
Chemical formula		
Structural formula		
CAS No.		
Synonym/common name		
Trade names		Exxate 900
Suppliers		EXXON
Molecular mass		
Boiling point/range	°C @ 760 mm Hg	205–235
Melting point	°C	
Density	kg/m^3 @ 20 °C	
Evaporation time[a]		
Evaporation time[b]		0.012
Vapour pressure	kPa @ 20 °C	
Flash point	°C	90
Auto ignition temperature	°C	
Upper flammability limit	%v/v in air	
Lower flammability limit	%v/v in air	
Viscosity	mPa.s (cP) @ 20 °C	
Refractive index	n 20 D	1.425
Solubility in water	%w/w @ 20 °C	
Solubility of water	%w/w @ 20 °C	
Solubility parameter[c]	(J/cm^3)$^{1/2}$	
Hydrogen bond index		
Fractional polarity		
Nitro cellulose dil ratio[d]		
Nitro cellulose dil ratio[e]		
Surface tension	mN/m @ 20 °C	
Specific heat liquid	kJ/kg mole/°C @ 20 °C	
Latent heat	kJ/kg mole	
Dielectric constant	@ 20 °C	
Antoine constant A	kPa°C	
Antoine constant B	kPa°C	
Antoine constant C	kPa°C	
Heat of combustion	kJ/kg mole	
UN number		
IMO classification		
ADR/RID classification		3,32°(c)
UK exposure limits	mg/m^3 (ppm) 8hrTWA/MEL or OES	
USA exposure limits	mg/m^3 (ppm) TLV/TWA	
German exposure limits	mg/m^3 (ppm) MAK-TWA	
EU classification		
EU risk phrases		
EU safety phrases		

[a] Ether = 1. [b] n-Butyl acetate = 1. [c] Nelson, Hemwall, Edwards. [d] Toluene. [e] Hydrocarbon solvent.

Solvent group		**UNSPECIFIED**
Chemical name		Solvent water microemulsions
Chemical formula		
Structural formula		
CAS No.		
Synonym/common name		
Trade names		INVERT 1000/2000/5000
Suppliers		DOW
Molecular mass		
Boiling point/range	°C @ 760 mm Hg	
Melting point	°C	
Density	kg/m^3 @ 20 °C	
Evaporation time[a]		
Evaporation time[b]		
Vapour pressure	kPa @ 20 °C	
Flash point	°C	
Auto ignition temperature	°C	
Upper flammability limit	%v/v in air	
Lower flammability limit	%v/v in air	
Viscosity	mPa.s (cP) @ 20 °C	
Refractive index	n 20 D	
Solubility in water	%w/w @ 20 °C	
Solubility of water	%w/w @ 20 °C	
Solubility parameter[c]	(J/cm^3)$^{1/2}$	
Hydrogen bond index		
Fractional polarity		
Nitro cellulose dil ratio[d]		
Nitro cellulose dil ratio[e]		
Surface tension	mN/m @ 20 °C	
Specific heat liquid	kJ/kg mole/°C @ 20 °C	
Latent heat	kJ/kg mole	
Dielectric constant	@ 20 °C	
Antoine constant A	kPa°C	
Antoine constant B	kPa°C	
Antoine constant C	kPa°C	
Heat of combustion	kJ/kg mole	
UN number		
IMO classification		
ADR/RID classification		
UK exposure limits	mg/m^3 (ppm) 8hrTWA/MEL or OES	
USA exposure limits	mg/m^3 (ppm) TLV/TWA	
German exposure limits	mg/m^3 (ppm) MAK-TWA	
EU classification		
EU risk phrases		
EU safety phrases		

[a] Ether = 1. [b] n-Butyl acetate = 1. [c] Nelson, Hemwall, Edwards. [d] Toluene. [e] Hydrocarbon solvent.

Key parameter table: Boiling points

Chemical name	Trade name	Boiling point (°C)
2-Methylbutane	Exxol Isopentane	24–35
1,1-Dichloro-1-fluoroethane/methanol mixture	FORANE 14lb MGX	29.4
Mixture, 1,1-Dichloro-1-fluorethane/Methyl alcohol/Nitromethane	GENESOLV 2004	29.4
1,1-Dichloro-1-fluorethane (Stabilised)	FORANE, GENESOLV	32
Diethyl ether		34.5
n-Pentane	HYDROSOL Pentane	35–37
	Norpar 5 S	35–38
Dichloromethane	METHOKLONE, UKLENE, AEROTHENE, SOLVACLENE	39.8–40.2
	Exxol DSP 40/65	42–69
	SBP 40/65 LNH	44–62
	HYDROSOL Ess A	45–95
	HYDROSOL Ess G	45–65
Trichlorotrifluoroethane	Forane	47.6
Propionaldehyde		48.0
Cyclopentane		48–50
Cyclopentane	EXXOL Cyclopentane	49–56
Isohexane	HYDROSOL Hexane	50–62
	Exxol Isohexane	55–62
Methyl tert-butyl ether	DRIVERON S	55.3
2-Propanone	DMK	55.8–56.6
	Isohexane LNH	57–63
Methyl acetate		57.5
Trichloromethane		59.4–61.2
	HYDROSOL Ess 60/95	60–98
Mixture based on naphtha solvent refined light	SBP, SOLSOL A1 1	60–95
	SBP 60/95 LNH	60–90
	SHELL SOL B HT	61–77
Hexane	Exxol Hexane, HYDROSOL Hexane	64–70
Hexane (branched and linear)		64–70
	SHELL RUBBER SOLVENT	64–114
Isobutyraldehyde		64.1
Methanol		64.7
2,2-Oxybispropane	IPE	65–70
	Exxol DSP 60/95S	65–98
	Hexane Polymerisation	66–69
	Hexane Extraction	66–69
Tetrahydrofuran		66.5–66.5
	SBP 62/78	67–77
	Norpar 6	67–70
	Exxol DSP 65/100	68–99
	Exxol DHN50	72–76
	HYDROSOL Ess 70/80N	72–79
	HYDROSOL Ess C	73–98
1,1,1-Trichloroethane	BALTANE CF, CHLOROTHENE	74
Butyraldehyde		75.7
Ethyl acetate	ETHYL ACETATE, EtAc	76.0–78.0
Ethanol		78.2–80.0
2-Butanone		79–80.5
Benzene		79.8–80.2
	Nappar 6	80–83
Cyclohexane		81
2-Propanol	IPA, IPAC+	82.0–83.0
2-Methyl-2-propanol		82–83
	HYDROSOL Ess.80/120	83–108
1,2-Dichloroethane		83.5
	SBP 80/95 LNH	84–94
	Exxol DSP 80/110	86–108

Key parameter table: Boiling points *continued*

Chemical name	Trade name	Boiling point (°C)
2-Methyl-2-methoxy-butane	TAME	86.3
1,2,2-Trichloroethylene	TRIKLONE, NEU TRI, HI-TRI, TAVOXENE, ALTENE	87
	SBP 80/110 LNH	87–104
1-Methylethyl acetate		88.4
Isopropyl acetate		88.5
Heptane	Exxol Heptane, HYDROSOL Heptane	90–99
	SHELL TOLU-SOL A HT SOLVENT	90–97
	SHELL TOLU-SOL 19 EC SOLVENT	91–104
	SHELL TOLU-SOL 10 SOLVENT	91–105
	SHELL TOLU-SOL 3 SOLVENT	91–97
	SHELL TOLU-SOL 5 SOLVENT	91–98
	SHELL TOLU-SOL 25 SOLVENT	92–107
	SBP 94/100	94–99
1-Propanol		96.5–97.5
	HYDROSOL Ess 95/160N	98–153
	SHELL TOLU-SOL W HT SOLVENT	98–110
	Norpar 7	98–99.6
	SHELL TOLU-SOL 6 SOLVENT	99–108
2,2,4-Trimethyl pentane	Isooctane 100	99–104
2-Butanol	SBA, EXXON SBA	99.4–101.5
	Exxol DSP100/120	100–119
	Isopar C	100–105
	Nappar 7	100–105
Diethyl ketone		100.0–102.0
n-Propyl acetate		101.0–102.0
	Exxol DSP 100/140	101–137
	Exxol DSP 100/160	103–160
	HYDROSOL Ess <5	103–123
Valeraldehyde		103.7
	SBP 100/140	105–137
	HYDROSOL Ess F	105–160
	HYDROSOL Ess F <5	104–155
	HYDROSOL Ess 100/140	106–140
2-Methyl-1-propanol		106.0–109.0
2,4,4-Trimethylpentene-1 and 2		107–120
Toluene	SOLVAREX Toluene	110–111
4-Methyl-2-pentanone		114–117
2-Methylpropyl acetate		115.0–118.0
1-Butanol	NBA	116.0–118.0
1-Methoxy-propanol-2	DOWANOL PM, METHOXY PROPANOL, MePROX	117–125
	Isopar E	118–143
	Solvenon PM	119.0–122.0
	SHELL VM&P NAPHTHA HT	119–139
n-Butyl acetate	nBuAc	120–128
Propylene glycol methyl ether		121.0
	SHELL VM&P NAPHTHA EC	121–134
1,1,2,2-Tetrachloroethylene	DOWPER, Pertene, PERKLONE, SOLTENE	121
2-Methoxyethanol		123.5–125.0
	Norpar 8	125–127
2-Methyl-1-butanol		128.7
4-Methyl-3-pentene-2-one	MO	129.8
4-Methyl-2-pentanol		130.0–133.0
Ethoxy propanol	ETHOXY PROPANOL	132
Monochlorobenzene		132
2-Ethoxyethanol	CELLOSOLVE solvent	134–137

Key parameter table: Boiling points *continued*

Chemical name	Trade name	Boiling point (°C)
	SPIRDANE L2	135–160
Ethylbenzene		135–136
	SPIRDANE L1	135–160
1-Pentanol		137.0–139.0
Mixture of xylenes		137–143
	VARSOL 30	137–164
Mixture of xylenes		137–140
Mixture of isomers and ethylbenzene		137–143
Xylene	SOLVAREX Xylene	137.7–142
p-Xylene		138.0–138.0
1,4-Dimethylbenzene	PARAXYLENE	138.3–138.0
1-Chloro-4-(trifluoromethyl)benzene	OXSOL 100, PCBTF	139
2-(1-Methylethoxy)ethanol	iPrOX	139.5–144.0
m-Xylene		139.7
	Exxol D 30	141–164
	SHELLSOl D25/SBP 140/165	141–162
Acrylic acid		141.2
1-Methoxy-2-propanol acetate	MePROXAc	143.0–149.0
o-Xylene		144.4
n-Butyl propionate	UCAR n-Butyl Propionate	144.7–148
Solvent naphtha		145–210
	SPIRDANE D30	145–170
	SPIRDANE L30	145–170
	SOLVAREX 90/160	145–175
Styrene		145.2
1-Methoxy-2-acetoxy-propane	DOWANOL PMA	145.8
	Nappar 9	146–170
Methoxy propyl acetate		146
Amyl acetate		146.0–149.0
	Exxol DSP 145/160	146–159
Propylene glycol n-propyl ether	DOWANOL PnP	149.8
Mixture of aliphatic hydrocarbons	EVOLVE CH15	150–190
4-Hydroxy-4-methyl-2-pentanone	DA, DAAA	150–172
	Propyl CELLOSOLVE solvent	150.1
2-Heptanone		151.5
1-Methylethylbenzene	AROMASOL 12	152–170
N,N-Dimethylformamide		152.0–153.0
4-Methoxy-4-methyl-2-pentanone	ME-6K	152.0–162.0
2-Ethoxyethyl acetate	CELLOSOLVE acetate	153–159
	VARSOL 40	154–189
	SPIRDANE BT	155–200
	ISANE IP155	155–173
	SPIRDANE D40	155–200
	Exxol D 40	155–188
	SPIRDANE HT	155–200
Cyclohexanone	CHK	156
Ethoxy propyl acetate		158
	Isopar G	158–176
	SHELL SOL 340 HT	159–176
5-Methyl-3-heptanone	EAK	159.2
	HIGH AROMATIC WHITE SPIRIT	160–195
	SHELL CYCLO SOL 53 AROMATIC SOLVENT	160–176
	LOW AROMATIC WHITE SPIRIT	160–198
Solvent naphtha	C9+I	160–183
	Solvesso 100	161–178
Cyclohexanol		161.1
	SHELL MINERAL SPIRITS 150 EC	162–200
	FINALAN 40	162–197

Key parameter table: Boiling points *continued*

Chemical name	Trade name	Boiling point (°C)
	SHELL MINERAL SPIRITS 200 HT	162–206
(2-Ethoxy-methylethoxy)-propanol	ETHOXY PROPOXY PROPANOL	162–228
	SHELL MINERAL SPIRITS 145 EC	162–201
	SHELLSOL D40	162–197
	Nappar 10	162–186
2,6-Dimethyl-4-heptanone/4,6-Dimethyl-2-heptanone (Isomer mixture)		163.0–173.0
	SHELLSOL E	164–188
	SHELL MINERAL SPIRITS 135	164–202
n-Pentyl propionate	UCAR n-Pentyl Propionate	164.9
	ISANE IP165	165–195
	SOLVAREX 9	165–180
N,N-Dimethylacetamide		165.0–166.0
	SOLVAREX 90/180	165–195
	SHELLSOL A	167–180
Mixture of aromatic hydrocarbons	PETRINEX ASB	167–210
2-Butoxyethanol	nBuOX, Butyl CELLOSOLVE	168–173
Mixture of aliphatic hydrocarbons/ester blend	EVOLVE CH11	168–192
4-Hydroxy-4-methyl-2-pentanone		169.2
Alkoxypropanol formulation	PRIMACLEAN 1601	170–175
Mixture of aromatic hydrocarbons	AROMASOL H	170–200
Triisobutylene		170–183
Methyl acetoacetate		170–172
1-n-Butoxy-propanol-2	DOWANOL PnB	171
Ethylene glycol n-butyl ether	DOWANOL EB	171.1
	SHELLSOL TD	172–190
Para-menthadienes	TABS D5	173–185
	SHELL CYLCO SOL 63 AROMATIC SOLVENT	173–208
Mixture of aliphatic hydrocarbons	EVOLVE CH10	173–192
Isopentanoic acid		175.0
n-Heptanol		175
Blend of isomers	PROGLYDE DMM	175
2,6-Dimethyl-4-heptanol	DIBC	175–181
	Exxate 700	176–200
2,2,4,6,6-Pentamethyl heptane		176–192
Di-isobutyl carbinol		178.0
	SHELL SOL 71	179–204
	Isopar H	179–192
	SPIRDANE D60	180–215
	SHELLSOL H	180–210
Ethyl acetoacetate	ETHYL ACETOACETATE	180
	ISANE IP175	180–196
	SHELLSOL TK	180–199
	SHELLSOL AD	180–202
	SPIRDANE K1	180–215
	SPIRDANE K2	180–215
Mixture of aromatic hydrocarbons	PETRINEX T9	180–290
	Isopar K	181–204
	SHELLSOL AB	181–212
Phenol		181.7
Mixture of aliphatic hydrocarbons	EVOLVE CH12	182–212
	Isopar J	182–208
	Solvesso 150	182–203
	Exxol D 60	182–215
2-Ethylhexanol		183.0–185.0
	HAN	183–277
2(2-Methoxymethylethoxy)-propanol	MeDIPROX	184–196
(2-Methoxymethylethoxy)-propanol	METHOXY PROPOXY PROPANOL	184–197

Key parameter table: Boiling points *continued*

Chemical name	Trade name	Boiling point (°C)
Butyl glycol acetate		184.0–195.0
	ISANE IP185	185–210
Decahydronaphthalene		185–195
Mixture	Solvenon DPM	185.0–195.0
	SHELLSOL D70	185–245
Solvent naphtha-heavy aromatic	C9+ II	185–209
	FINALAN 60	185–218
	KETRUL HT	185–240
	KERDANE	185–240
	HAN 8070	185–230
	KETRUL BT	185–240
	KETRUL D70	185–240
Alkoxypropanol formulation	DOWANOL PX-16S	186
	VARSOL 60	186–216
1,2-Propylene glycol		186.0–190.0
	SHELLSOL D60	187–211
	SHELLSOL T	187–215
Mixture based on aliphatic hydrocarbons	VISTA LPA SOLVENT	187–259
	SOLVAREX 10	188–215
2-Butoxyethyl acetate	BUTYL GLYCOL ACETATE	188–194
Dipropylene glycol methyl ether		188.3
Trimethylcyclohexanone		188.8
Trimethylcyclohexanol		189–196
	Isopar L	189–210
Dimethyl sulphoxide		189
1-(2-Methoxy-propanol 2/1-(2-Methoxy-methylethoxy)-propanol 2	DOWANOL DPM	189
2-(2-Ethoxyethoxy)ethanol	CARBITOL solvent low gravity	190–203
	SOLVARO AFD	190–225
1n-Butoxy-propyl-acetate-2	DOWANOL PnBA	190
1,2-Diacetoxy-propane	DOWANOL PGDA	190
	SHELL SOL 142 HT	190–207
Mixture based on aliphatic hydrocarbons	VISTA LPA-150 SOLVENT	192–230
2-Ethylhexyl acetate		192.0–205.0
	Methyl CARBITOL solvent	192.0–195.0
	Butyl CELLOSOLVE acetate	192.3
	Norpar 12	193–220
1,2-Ethanediol		193–205
	Exxol D 70	193–213
Mixture based on saturated and aromatic hydrocarbons	VISTA MOLEX RAFFINATE	193–259
Diethylene glycol methyl ether	DOWANOL DM	194.1
	HAN 8572	195–270
2-Methyl-2,4-pentanediol	HG	195–200
	Nappar 11	196–223
	SOLVENT APV SPECIAL	198–203
Nonanol mixed isomers	Nonanol N	198.0–209.0
	KETRUL D80	198–240
	Exxol D 80	199–237
Methylbenzyl alcohol		200–205
n-Hexyl glycol	Hexyl CELLOSOLVE solvent	200.0–212.0
	FINALAN 75	200–260
Tetrahydronapththalene		200–209
Dimethyl esters of adipic, glutaric and succinic acids	ESTASOL	200–230
	RHODIASOLV RPDE	>200
Dichlorotoluene		202–208
	VARSOL 80	202–239
N-Methylpyrrolidone		202.5–205.0

Key parameter table: Boiling points *continued*

Chemical name	Trade name	Boiling point (°C)
	SHELLSOL R	205–270
	Exxate 900	205–235
Benzyl alcohol		205.4
	Isopar M	206–250
1,3-Butanediol		207–210
Dipropylene glycol methyl ether acetate	DOWANOL DPMA	209.3
3,5,5-Trimethyl-2-cyclohexene-1-one	IPHO	210–216
2,2,4,4,6,8,8-Heptamethyl nonane		210–250
2-(2-Ethoxyethoxy)ethyl acetate		210–222
	FINALAN 80	211–239
Dipropylene glycol n-propyl ether	DOWANOL DPnP	212.0
Mixture based on aliphatic hydrocarbons		213–237
Isodecanol mixed isomers		214.0–222.0
	KETRUL 211	215–240
	SHELLSOL TM	215–260
	KETRUL 210	215–240
	KETRUL 212	215–245
	SHELLSOL DMA	220–270
	SHELLSOL D90	220–270
	Exxate 1000	220–250
	Norpar 13	222–243
	FINALAN 90	222–254
	Solvesso 200S	224–288
2-(2-Butoxyethoxy)ethanol	nBuDIOX, BUTYL DIGLYCOL	225–234
	Exxol D 220/230	225–233
2-Ethylhexoic acid		227.0
Dipropylene glycol butylene ether		227.9
Dipropylene glycol (mixed isomers)		228.0–236.0
1,4-Butanediol		229
	Solvesso 200	229–284
1-(2-n-Butoxy-1-methyl-ethoxy)-propanol-2	DOWANOL DPnB	229
Diethylene glycol n-butyl ether	DOWANOL DB	230.0
Formulated fatty acid ester	PROTENE F	230
	Butyl CARBITOL solvent	230.6
Triethylene glycol methyl ether/Highers	DOWANOL TMH	232.0
2-(2-Butoxyethoxy)-ethanol acetate/Diethylene glycol n-butyl ether acetate	nBuDIOXAc	235.0–250.0
	Exxol D 100S	235–265
[2-[2-(ethoxyethoxy)ethoxy]ethanol		235–280
	ISANE IP235	235–265
[2-(2-Butoxyethoxy)ethyl acetate]		235–250
	FINALAN 100	236–272
	Exxol D 110	239–275
	Exxol D 240/260	239–266
	Isopar P	239–275
	Exxol D100	239–266
	SHELLSOL D100	240–262
	Solvenon PC	240.0–243.0
	Exxate 1300	240–285
Mixture based on saturated aliphatic hydrocarbons	VISTA 47 SOLVENT	240–277
Mixture based on aliphatic hydrocarbons	VISTA LPA-210 SOLVENT	240–282
Mixture	Solvenon PP	241.0–246.0
2-Phenoxy-propanol-2	DOWANOL PPh	242
Blend	DOWANOL TPM	243
2-Ethylhexane-1,3-diol		243–250
Tetradecane	VISTA C14 PARAFFIN	243–249
	INK SOLVENT 2427	243–267
Propylene carbonate		243.4

Key parameter table: Boiling points *continued*

Chemical name	Trade name	Boiling point (°C)
Diethylene glycol		243.8–246.0
Phenyl glycol		244.0–250.0
Aromatic based glycol ether	DOWANOL DALPAD A	245.6
Ethylene glycol phenyl ether	DOWANOL EPh	245.6
	VARSOL 110	246–275
	Butyl CARBITOL acetate	246.7
2-(Butoxyethoxy)ethyl acetate	BUTYL DIGLYCOL ACETATE	247
Tridecanol mixed isomers	Tridecanol A	248.0–264.0
Methoxytriglycol		249.0
	SHELLSOL D120	250–300
Tetradecane Pentadecane, Hexadecane	VISTA C14-C16 PARAFFINS	253–266
[2-[2-(butoxyethoxy)ethoxy]ethanol		255–295
Ethoxytriglycol		256.0
	Norpar 15	256–279
	Hexyl CARBITOL solvent	259.1
	Exxol D 120	260–294
	INK SOLVENT 2629	262–288
	VARSOL 120	265–295
	ISANE IP260	265–310
Mixture		265.0–350.0
Blend of isomers	DOWANOL TPnB	274
Di-isobutyl esters of adipic acid, glutaric acid and succinic acid	COASOL	274–289
	Isopar V	279–314
	INK SOLVENT 2831	283–310
Triethylene glycol butyl ether/Highers	DOWANOL TBH	283.0
2,2^-[1,2-Ethanediylbis(oxy)]bisethanol		287.4
Triethylene glycol		288.0
	Exxol D 140	288–310
	VARSOL 140	290–315
[2-[2-(butoxyethoxy)-ethoxy]ethanol		300–340
Mixture based on aromatic hydrocarbons	VISTA 150L SPECIALTY ALKYLATE	323–389
Mixture based on aromatic hydrocarbons	VISTA 3050 SPECIALTY ALKYLATE	343–379

Key parameter table: Melting points

Chemical name	Trade name	Melting point (°C)
n-Pentane	HYDROSOL Pentane	–129.7
1-Propanol		–126.2
Diethyl ether		–116
2-Butanol	SBA, EXXON SBA	–115
Ethanol		–114
Methyl tert-butyl ether	DRIVERON S	–109
2-Methoxy-2-methyl propane	MTBE+	–109
Tetrahydrofuran		–108
1,1-Dichloro-1-fluoroethane (Stabilised)	FORANE, GENESOLV	–103.5
2-Methyl-1-propanol		–103, –108
2,2,4-Trimethyl pentane	Isooctane 100	–103
2-Methylpropyl acetate		–98 to –99
Methyl acetate		–98
Methanol		–97.5 to –98
Dichloromethane	METHOKLONE, UKLENE, AEROTHENE, SOLVACLENE	–97
Butyraldehyde		–96.4
1-Methylethylbenzene	AROMASOL 12	–96 to –100
1-Methoxy-propanol-2	DOWANOL PM, METHOXY PROPANOL, MePROX	–96 to –100
Hexane	Exxol Hexane, HYDROSOL Hexane	–95.3
Hexane (branched and linear)		<–95
Toluene	SOLVAREX Toluene	–95
n-Propyl acetate		–95.0
2-Propanone	DMK	–94.7 to –95
Cyclopentane	EXXOL Cyclopentane	–94
Cyclopentane		–94
Ethylbenzene		–94 to –95
2,4,4,-Trimethylpentene-1 and 2		–93
Triisobutylene		–91
Heptane	Exxol Heptane, HYDROSOL Heptane	–90.6
Ethoxy propanol	ETHOXY PROPANOL	–90
2-(2-Ethoxyethoxy)ethanol	CARBITOL solvent low gravity	–90
4-Methyl-2-pentanol		–90
1-Butanol	NBA	–89, –90
Ethoxy propyl acetate		–89
Methoxy propyl acetate		–88
1,2,2-Trichloroethylene	TRIKLONE, NEU TRI, HI-TRI, TAVOXENE, ALTENE	–88
2-Propanol	IPA, IPAC+	–88
2-Butanone		–86
2,2-oxybispropane	IPE	–85
4-Methyl-2-pentanone		–84 to –85
2(2-Methoxymethylethoxy)-propanol	MeDIPROX	–83
1-(2-Methoxy-propoxy)-propanol	DOWANOL DPM	–83
2/1-(2-Methoxy-methylethoxy)-propanol 2 (2-Methoxymethylethoxy)-propanol	METHOXY PROPOXY PROPANOL	–83
Ethyl acetate	ETHYL ACETATE, EtAc	–82.4 to –84
2,2,4,6,6-Pentamethyl heptane		–81
2-Ethoxyethanol	CELLOSOLVE solvent	<–80
2-Ethylhexyl acetate		–80
Methyl acetoacetate		–80
Propionaldehyde		–80
2-Methyl-2-methoxy-butane	TAME	–80
1-Pentanol		–78.2
Blend	DOWANOL TPM	–78
n-Butyl acetate	nBuAc	–76.8 to –78
2-Ethylhexanol		–76
Blend of isomers	DOWANOL TPnB	<–75

Key parameter table: Melting points *continued*

Chemical name	Trade name	Melting point (°C)
1-n-Butoxy-propanol-2	DOWANOL PnB	<–75
1n-Butoxy-propyl-acetate-2	DOWANOL PnBA	<–75
1,2-Diacetoxy-propane	DOWANOL PGDA	<–75
1-(2-n-Butoxy-1-methyl-ethoxy)-propanol-2	DOWANOL DPnB	–75
1-Methylethyl acetate		–73.4
Isopropyl acetate		–73.4
Blend of isomers	PROGLYDE DMM	<–71
Mixture based on saturated and aromatic hydrocarbons	VISTA MOLEX RAFFINATE	<–70
Mixture of aliphatic hydrocarbons	EVOLVE CH12	<–70
Mixture of aliphatic hydrocarbons/ester blend	EVOLVE CH11	<–70
Mixture based on aliphatic hydrocarbons	VISTA LPA SOLVENT	<–70
Mixture based on aliphatic hydrocarbons	VISTA LPA-150 SOLVENT	<–70
Mixture based on aliphatic hydrocarbons		<–70
2-Butoxyethanol	nBuOX, Butyl CELLOSOLVE	–70
2,2,4,4,6,8,8-Heptamethyl nonane		–70
2-(2-Butoxyethoxy)ethanol	nBuDIOX, BUTYL DIGLYCOL	–68
1-Methoxy-2-acetoxy-propane	DOWANOL PMA	<–67
2,6-Dimethyl-4-heptanol	DIBC	–65
2-Butoxyethyl acetate	BUTYL GLYCOL ACETATE?	–65
Isobutyraldehyde		–65
1-Methoxy-2-propanol acetate	MePROXAc	–65
Trichloromethane		–63.5
2-Ethoxyethyl acetate	CELLOSOLVE acetate	–62
N,N-Dimethylformamide		–60.4
2-(1-Methylethoxy)ethanol	iPrOX	–60
Di-isobutyl esters of adipic acid, glutaric acid and succinic acid	COASOL	–60
Mixture based on aromatic hydrocarbons	VISTA 3050 SPECIALTY ALKYLATE	–57
Propylene carbonate		–54
Mixture based on aromatic hydrocarbons	VISTA 150L SPECIALTY ALKYLATE	–54
4-Methyl-3-pentene-2-one	MO	–53
Mixture based on aliphatic hydrocarbons	VISTA LPA-210 SOLVENT	<–51
(2-Ethoxy-methylethoxy)-propanol	ETHOXY PROPOXY PROPANOL	–51
Mixture of aromatic hydrocarbons	PETRINEX ASB	<–50
Mixture of aliphatic hydrocarbons	EVOLVE CH15	<–50
Mixture of aliphatic hydrocarbons	EVOLVE CH10	<–50
2-Methyl-2,4-pentanediol	HG	–50
m-Xylene		–47.87
5-Methyl-3-heptanone	EAK	–46
Monochlorobenzene		–45.2
Ethyl acetoacetate	ETHYL ACETOACETATE	–44
4-Hydroxy-4-methyl-2-pentanone	DA, DAA	–43
2,6-Dimethyl-4-heptanone/4,6-Dimethyl-2-heptanone (Isomer mixture)		–42
1,1,1-Trichloroethane	BALTANE CF, CHLOROTHENE	–37
1-Chloro,4-(trifluoromethyl)benzene	OXSOL 100, PCBTF	–36
1,2-Dichloroethane		–35.5
4-Methoxy-4-methyl-2-pentanone	ME-6K	–35.4
[2-[2-(butoxyethoxy)ethoxy]ethanol]		–35.1
Trichlorotrifluoroethane	Forane	–35
2-Heptanone		–35.0
n-Heptanol		–35
2-(2-Butoxyethoxy)-ethanol acetate/Diethylene glycol n-butyl ether acetate	nBuDIOXAc	–32
2-(Butoxyethoxy) ethyl acetate	BUTYL DIGLYCOL ACETATE	–32
[2-(2-Butoxyethoxy)ethyl acetate]		–32
Styrene		–31

Key parameter table: Melting points *continued*

Chemical name	Trade name	Melting point (°C)
Tetrahydronaphthalene		–31
Decahydronaphthalene		–30.4
Mixture of xylenes		<–25
Xylene	SOLVAREX Xylene	–25
Dimethyl esters of adipic, glutaric and succinic acids	ESTASOL	–25
Mixture of xylenes		–25 to –50
2-(2-ethoxyethoxy)ethyl acetate		–25
o-Xylene		–25
N-Methylpyrrolidone		–24.4
1,1,2,2-Tetrachloroethylene	DOWPER, Pertene, PERKLONE, SOLTENE	–22.3
Mixture of aromatic hydrocarbons	AROMASOL H	<–20
[2-[2-(ethoxyethoxy)ethoxy]ethanol		–19
Mixture based on saturated aliphatic hydrocarbons	VISTA 47 SOLVENT	–18
Cyclohexanone	CHK	–16
Benzyl alcohol		–15
1,2-Ethanediol		–11.5 to –12.7
Trimethylcyclohexanone		–10
3,5,5-Trimethyl-2-cyclohexene-1-one	IPHO	–8
Diethylene glycol		–7.8
2,2ˆ-[1,2-Ethanediylbis(oxy)]bisethanol		–7.2
Triethylene glycol		–4.3
Mixture of aromatic hydrocarbons	PETRINEX T9	0
Tetradecane	VISTA C14 PARAFFIN	2
Benzene		5.35
Cyclohexane		6.4
Tetradecane, Pentadecane, Hexadecane	VISTA C14-C16 PARAFFINS	8
1-Phenoxy-propanol-2	DOWANOL PPh	13
1,4-Dimethylbenzene	PARAXYLENE	13
p-Xylene		13.3
Acrylic acid		13.5
Dimethyl sulphoxide		18.5
1,4-Butanediol		19.7
Para-menthadienes	TABS D5	<20
	RHODIASOLV RPDE	20
Methylbenzyl alcohol		21.5
2-Methyl-2-propanol		24.3
Cyclohexanol		25.15
Trimethylcyclohexanol		(cis 37.3, trans 57.3)
Phenol		40.9

Key parameter table: Evaporation rates

Chemical name	Trade names	Evaporation rate (Diethylether = 1)
n-Pentane	HYDROSOL Pentane	<1.0
2-Methylbutane	Exxol Isopentane	<1.0
Diethyl ether		1
	SBO 40/65 LNH	1.0
	HYDROSOL Ess G	1
	Exxol DSP 40/65	1.1
	HYDROSOL Ess A	1.2
	Isohexane LNH	1.2
1,1-Dichloro-1-fluoroethane (Stabilised)	FORANE, GENESOLV	1.2
Isohexane	HYDROSOL Hexane	1.3
	Exxol Isohexane	1.3
2,2-Oxybispropane	IPE	1.4
	Hexane Polymerisation	1.4
	Hexane Extraction	1.4
Dichloromethane	METHOKLONE, UKLENE, AEROTHENE, SOLVACLENE	1.5–1.8
Hexane	Exxol Hexane, HYDROSOL Hexane	1.5
	Exxol DHN50	1.5
	Norpar 6	1.5
Methyl tert-butyl ether	DRIVERON S	1.6
	SBP 62/78	1.6
2-Methoxy-2-methyl propane	MTBE+	1.6
	SBP 60/95 LNH	1.8
	Nappar 6	2
	HYDROSOL Ess 60/95	2
Cyclopentane		2
	HYDROSOL Ess C	2
	Exxol DSP 65/100	2.0
2-Propanone	DMK	2.0
Tetrahydrofuran		2.0–2.3
	Exxol DSP 60/95S	2.1
Trichloromethane		2.5
	HYDROSOL Ess 70/80N	2.5
1,1,1-Trichloroethane	BALTANE CF, CHLOROTHENE	2.5
	Isopar C	2.8
Heptane	Exxol Heptane, HYDROSOL Heptane	2.8–3.2
	SBP 80/95 LNH	2.9
Ethyl acetate	ETHYL ACETATE, EtAc	2.9–3
2-Methyl-2-methoxy-butane	TAME	3
1,2,2-Trichloroethylene	TRILONE, NEU TRI, HI-TRI, TAVOXENE, ALTENE	3–3.8
	SBP 94/100	3.0
	HYDROSOL Ess.80/120	3.2
	SBP 80/110 LNH	3.2
	Exxol DSP 80/110	3.3
2-Butanone		3.3
	Norpar 7	3.8
	Nappar 7	3.8
	Norpar 5 S	<5
2,4,4,-Trimethylpentene-1 and 2		5
n-Propyl acetate		5
	Exxol DSP100/120	5.3
Toluene	SOLVAREX Toluene	6
	HYDROSOL Ess <5	6
	SBP 100/140	6.0
Diethyl ketone		6
Methanol		6.3
4-Methyl-2-pentanone		7–7.2
	Norpar 8	7

Key parameter table: Evaporation rates *continued*

Chemical name	Trade names	Evaporation rate (Diethylether = 1)
1,1,2,2-Tetrachloroethylene	DOWPER, Pertene, PERKLONE, SOLTENE	7.1–9.5
	HYDROSOL Ess100/140	7.5
	Isopar E	8
2-Methylpropyl acetate		8
Ethanol		8.3
	Exxol DSP 100/160	9.0
	Exxol DSP 100/140	9.0
	HYDROSOL Ess95/160N	10
n-Butyl acetate	nBuAc	11–12
2-Propanol	IPA, IPAC+	11
2-Methyl-2-propanol		11
	HYDROSOL Ess F	12
	HYDROSOL Ess F <(5	12
Xylene	SOLVAREX Xylene	13.5–15
2-Butanol	SBA, EXXON SBA	15
1-Propanol		16
	Nappar 9	17
Styrene		19.5
	SHELLSOL D25/SBP 140/165	20
	Solvenon PM	22
1-Methoxy-propanol-2	DOWANOL PM, METHOXY PROPANOL, MePROX	22–25
	SPIRDANE L2	22
	Exxol DSP 145/160	22
	SPIRDANE L1	22
	SOLVAREX 90/160	25
2-Methyl-1-propanol		25
	VARSOL 30	26
	Exxol D 30	27
	SPIRDANE D30	27
	SPIRDANE L30	27
Amyl acetate		28
1-Butanol	NBA	33
1-Methoxy-2-propanol acetate	MePROXAc	33
2-Methoxyethanol		34
	Solvesso 100	35
	SOLVAREX 9	36
	ISANE IP155	36
	Isopar G	40
Cyclohexanone	CHK	40
1-Methoxy-2-acetoxy-propane	DOWANOL PMA	43
2-Ethoxyethanol	CELLOSOLVE solvent	43
2,2,4,6,6-Pentamethyl heptane		43–46
	SOLVAREX 90/180	45
4-Methoxy-4-methyl-2-pentanone	ME-6K	46
	SHELLSOL A	46
2,6-Dimethyl-4-heptanone/4,6-Dimethyl-2-heptanone (Isomer mixture)		48
	Nappar 10	48
	ISANE IP165	50
2-(1-Methylethoxy)ethanol	iPrOX	52
	VARSOL 40	55
	Exxol D 40	55
Mixture of aliphatic hydrocarbons/ester blend	EVOLVE CH11	55
2-Ethoxyethyl acetate	CELLOSOLVE acetate	57
	SPIRDANE D40	65
	SPIRDANE HT	65
	SPIRDANE BT	65

Key parameter table: Evaporation rates *continued*

Chemical name	Trade names	Evaporation rate (Diethylether = 1)
	ISANE IP175	65
4-Methyl-2-pentanol		66
	SHELLSOL TD	70
Mixture of aliphatic hydrocarbons	EVOLVE CH10	75
	Isopar H	75
	SHELLSOL E	79
	LOW AROMATIC WHITE SPIRIT	85
	SHELLSOL D40	85
	Isopar K	93
1-Pentanol		96
N,N-Dimethylformamide		100
	Isopar J	100
Decahydronaphthalene		100
Mixture of aliphatic hydrocarbons	EVOLVE CH12	105
Trimethylcyclohexanone		109
Trimethylcyclohexanol		109
	SHELLSOL T	110
2-Butoxyethanol	nBuOX, Butyl CELLOSOLVE	119–160
	Solvesso 150	120
	Isopar L	125
	SOLVAREX 10	130
Blend of isomers	PROGLYDE DMM	131
	SHELLSOL AB	148
	ISANE IP185	150
	Exxol D 60	150
Alkoxypropanol formulation	PRIMACLEAN 1601	150
	VARSOL 60	150
4-Hydroxy-4-methyl-2-pentanone	DA, DAA	150
1-n-Butoxy-propanol-2	DOWANOL PnB	156
N,N-Dimethylacetamide		172
	Exxol D 70	175
	Norpar 12	185
Butyl glycol acetate		190
2-Butoxyethyl acetate	BUTYL GLYCOL ACETATE?	190
	Nappar 11	193
	SPIRDANE K2	200
	SPIRDANE K1	200
	SHELLSOL H	200
Tetrahydronaphthalane		200
	SPIRDANE D60	200
	SHELLSOL D60	200
1-nButoxy-propyl-acetate-2	DOWANOL PnBA	200
1,2-Diacetoxy-propane	DOWANOL PGDA	250
2,6-Dimethyl-4-heptanol	DIBC	300
2-Ethylhexyl acetate		320–500
3,5,5-Trimethyl-2-cyclohexene-1-one	IPHO	330
2(2-Methoxymethylethoxy)-propanol	MeDIPROX	360
N-Methylpyrrolidone		360
Mixture	Solvenon DPM	~380
	SOLVARD AFD	400
1-(2-Methoxy-propoxy)-propanol 2/1-(2-Methoxy-methylethoxy)-propanol 2	DOWANOL DPM	~400
	Exxol D 220/230	500
	HAN 8070	500
	KERDANE	550
	KETRUL HT	550
	KETRUL BT	550
	KETRUL D70	550
	Exxol D 80	600

Key parameter table: Evaporation rates *continued*

Chemical name	Trade names	Evaporation rate (Diethylether = 1)
	VARSOL 80	600
	SHELLSOL R	600
	Isopar M	680
2-Ethylhexanol		690
	SHELLSOL D70	800
2,2,4,6,8,8-Heptamethyl nonane		800
	KETRUL D80	800
	Norpar 13	870
1-(2-n-Butoxy-1-methyl-ethoxy)-propanol-2	DOWANOL DPnB	900
	Methyl CARBITOL solvent	$\sim$900
Methylbenzyl alcohol		920
1,2-Propylene glycol		1000
	KETRUL 212	1000
	KETRUL 210	1000
	KETRUL 211	1000
	Exxol D 100S	$>$1000
	Exxol D 120	$>$1000
	Exxol D 100	$>$1000
	Exxol D 240/260	$>$1000
	Exxol D 110	$>$1000
Diethylene glycol		$>$1000
	Norpar 15	$>$1000
	Exxol D 140	$>$1000
	ISANE IP260	$>$1000
	ISANE IP235	$>$1000
	VARSOL 120	$>$1000
	Isopar V	$>$1000
	Solvesso 200S	$>$1000
	HAN	$>$1000
	VARSOL 140	$>$1000
	Solvesso 200	$>$1000
	HAN 8572	$>$1000
Propylene carbonate		$>$1000
2-(2-Ethoxyethoxy)ethanol	CARBITOL solvent low gravity	1200
	SOLVENT APV SPECIAL	1200
1-Phenoxy-propanol-2	DOWANOL PPh	1200
n-Hexyl glycol	Hexyl CELLOSOLVE solvent	$\sim$1200
2-Ethylhexane-1,3-diol		$>$1200
1,4-Butanediol		$>$1200
2-[2-(Butoxyethoxy)-ethoxy]ethanol		$>$1200
2-[2-(Butoxyethoxy)ethoxy]ethanol		$>$1200
2-[2-(Ethoxyethoxy)ethoxy]ethanol		$>$1200
Blend	DOWANOL TPM	$>$1200
Blend of isomers	DOWANOL TPnB	$\gg$1200
	Solvenon PC	$>$1500
2-Methyl-2,4-pentanediol	HG	1680
2-(2-ethoxyethoxy)ethyl acetate		2400
Nonanol mixed isomers	Nonanol N	2500
1,2-Ethanediol		2500
[2-(2-Butoxyethoxy)ethyl acetate]		3000
2-(2-Butoxyethoxyl)ethanol	nBuDIOX, BUTYL DIGLYCOL	3750
	INK SOLVENT 2427	3900
	INK SOLVENT 2629	$>$3900
	INK SOLVENT 2831	$>$3900
	SHELLSOL DMA	$>$3900
	SHELLSOL D120	$>$3900
	SHELLSOL D90	$>$3900
	SHELLSOL D100	$>$3900
Isodecanol mixed isomers		4500

Key parameter table: Evaporation rates *continued*

Chemical name	Trade names	Evaporation rate (Diethylether = 1)
Mixture	Solvenon PP	~5000
Mixture		~8000
2-(Butoxyethoxy)-ethanol acetate/Diethylene glycol n-butyl ether acetate	nBuDIOXAc	8200
Dipropylene glycol (mixed isomers)		>10000
Tridecanol mixed isomers	Tridecanol A	>10000
Phenyl glycol		>10000

Key parameter table: Vapour pressures

Chemical name	Trade names	Vapour pressure kPa @ 20 °C
2-(2-Butoxyethoxy)-ethanol acetate/Diethylene glycol n-butyl ether acetate	nBuDIOXAc	0.0003
2-[2-(Ethoxyethoxy)ethoxy]ethanol		<0.001
2-[2-(Butoxyethoxy)ethoxy]ethanol		<0.001
2-[2-(butoxyethoxy)-ethoxy]ethanol		<0.001
2-Ethylhexane-1,3-diol		
	Isopar P	0.001
	BUTYL DIGLYCOL ACETATE	0.0013
2-(Butoxyethoxy)ethyl acetate		0.0013
[2-(2-Butoxyethoxy)ethyl acetate]	nBuDIOX, BUTYL DIGLYCOL	0.002–0.01
2-(2-Butoxyethoxy)ethanol		0.002
1,3-Butanediol	HG	0.003–0.00
2-Methyl-2,4-pentanediol	SHELLSOL D100	0.003
		0.004
Propylene carbonate	VISTA C14-C16 PARAFFINS	0.004
Tetradecane, Pentadecane, Hexadecane	VISTA LPA-210 SOLVENT	0.008
Mixture based on aliphatic hydrocarbons	VISTA C14 PARAFFIN	0.009
Tetradecane	PETRINEX T9	0.010
Mixture of aromatic hydrocarbons	PETRINEX ASB	0.010
Mixture of aromatic hydrocarbons	INK SOLVENT 2831	<0.01
	INK SOLVENT 2427	<0.01
	INK SOLVENT 2629	<0.01
1,4-Butanediol		<0.01
	SOLVENT APV SPECIAL	0.01
Mixture based on aliphatic hydrocarbons		0.012
1,2-Ethanediol		0.012
2-(2-ethoxyethoxy)ethyl acetate		0.013
Mixture based on saturated aliphatic hydrocarbons	VISTA 47 SOLVENT	0.013
Mixture based on aliphatic hydrocarbon	VISTA LPA SOLVENT	0.016
	SHELLSOL D90	0.02
	SHELLSOL DMA	0.02
2-(2-Ethoxyethoxy)ethanol	CARBITOL solvent low gravity	0.02
Tetrahydronaphthalane		0.024
Mixture based on saturated and aromatic hydrocarbons	VISTA MOLEX RAFFINATE	0.025
Mixture based on aliphatic hydrocarbons	VISTA LPA-150 SOLVENT	0.029
Methylbenzyl alcohol		0.03
Phenol		0.030
2-Butoxyethyl acetate	BUTYL GLYCOL ACETATE?	0.03–0.04
2,6-Dimethyl-4-heptanol	DIBC	0.03
	SHELLSOL R	0.04
3,5,5-Trimethyl-2-cyclohexene-1-one	IPHO	0.04
N-Methylpyrrolidone		0.04
2-Ethylhexanol		0.05–0.09
2(2-Methoxymethylethoxy)-propanol	MeDIPROX	0.052
2-Ethylhexyl acetate		0.06
	SHELLSOL D70	0.06
Mixture of aliphatic hydrocarbons	EVOLVE CH10	0.066
Alkoxypropanol formulation	DOWANOL PX-16S	0.07 (25 °C)
2-Butoxyethanol	nBuOX, Butyl CELLOSOLVE	0.08–0.13
	RHODIASOLV RPDE	0.08
Methyl acetoacetate		0.09
Mixture of aromatic hydrocarbons	AROMASOL H	0.1
2,2,4,6,6-Pentamethyl heptane		0.1
	SHELLSOL D60	0.1
Mixture of aliphatic hydrocarbons	EVOLVE CH12	0.1
Trimethylcyclohexanone		0.1
	SHELLSOL T	0.1

Key parameter table: Vapour pressures *continued*

Chemical name	Trade names	Vapour pressure kPa @ 20 °C
Ethyl acetoacetate	ETHYL ACETOACETATE	0.107
Alkoxypropanol formulation	PRIMACLEAN 1601	0.11
4-Hydroxy-4-methyl-2-pentanone	DA, DAA	0.12–0.164
(2-Ethoxy-methylethoxy)-propanol	ETHOXY PROPOXY PROPANOL	0.14
	SHELLSOL AB	0.15
	SHELLSOL H	0.15
2,6-Dimethyl-4-heptanone/4,6-Dimethyl-2-heptanone (Isomer mixture)		0.16–0.23
	SHELLSOL TD	0.2
Ethoxy propyl acetate		0.227
4-Methoxy-4-methyl-2-pentanone	ME-6K	0.25
Mixture of aliphatic hydrocarbons/ester blend	EVOLVE CH11	0.25
Decahydronaphthalene		0.29
	SHELLSOL A	0.3
	SHELLSOL E	0.3
	SHELLSOL D40	0.3
	HIGH AROMATIC WHITE SPIRIT	0.3
2-(1-Methylethoxy)ethanol	iPrOX	0.38
Cyclohexanone	CHK	0.39
Mixture of aliphatic hydrocarbons	EVOLVE CH15	<0.4
	LOW AROMATIC WHITE SPIRIT	0.4
1-Methoxy-2-propanol acetate	MePROXAc	0.42
4-Methyl-2-pentanol		0.42–0.49
1-Methylethylbenzene	AROMASOL 12	0.426
Methoxy propyl acetate		0.49
2-Ethoxyethanol	CELLOSOLVE solvent	0.5
1-Butanol	NBA	0.57–0.59
o-Xylene		0.66
Styrene		0.71
Mixture of xylenes		0.8
m-Xylene		0.8
Mixture of xylenes		0.8–2.0
P-xylene		0.86
Ethylbenzene		0.93
Ethoxy propanol	ETHOXY PROPANOL	<1
(2-Methoxymethylethoxy)-propanol	METHOXY PROPOXY PROPANOL	<1
1-Methoxy-propanol-2	DOWANOL PM, METHOXY PROPANOL, MePROX	1.0–1.17
	SHELLSOL D25/SBP 140/165	1
Xylene	SOLVAREX Xylene	1
Acrylic acid		1.03
2-Methyl-1-propanol		1.07–1.2
4-Methyl-3-pentene-2-one	MO	1.094
n-Butyl acetate	nBuAc	1.13–1.25
1,4-Dimethylbenzene	PARAXYLENE	1.150 (25 °C)
Cyclohexanol		1.33
2-Butanol	SBA, EXXON SBA	1.53–1.6
1,1,2,2-Tetrachloroethylene	DOWPER, Pertene, PERKLONE, SOLTENE	1.71–1.87
4-Methyl-2-Pentanone		1.9
2-Methylpropyl acetate		2.13
2-Ethoxyethyl acetate	CELLOSOLVE acetate	2.2
Toluene	SOLVAREX Toluene	3
	SBP 100/140	3.5
2-Methyl-2-propanol		4.1
2-Propanol	IPA, IPAC+	4.1
Ethanol		5.8–5.9
1,2,2-Trichloroethylene	TRIKLONE, NEU TRI, HI-TRI, TAVOXENE, ALTENE	7.23–7.72

Key parameter table: Vapour pressures *continued*

Chemical name	Trade names	Vapour pressure kPa @ 20 °C
1,2-Dichloroethane		8.13
	SBP 80/110 LNH	8.5
	SBP 80/95 LNH	8.5
	SPB 94/100	8.5
2-Butanone		9.5
Ethyl acetate	ETHYL ACETATE, EtAc	9.7–10.3
Benzene		10.13
Cyclohexane		13.0 (25 °C)
Methanol		13.1
1,1,1-Trichloroethane	BALTANE CF, CHLOROTHENE	13.3
	SBP 62/78	15.5
2,2-Oxybispropane	IPE	15.9
Tetrahydrofuran		17.3
	Hexane Extraction	19
	SBP 60/95 LNH	19
	Hexane Polymerisation	19
Trichloromethane		20.0
	Isohexane LNH	24
2-Propanone	DMK	24.7
2-Methoxy-2-methyl propane	MTBE+	26.9
Methyl tert-butyl ether	DRIVERON S	27.1
	SBP 40/65 LNH	33
Cyclopentane	EXXOL Cyclopentane	35.69
1-Chloro,4-(trifluoromethyl)benzene	OXSOL 100, PCBTF	35.8
Dichloromethane	METHOKLONE, UKLENE, AEROTHENE, SOLVACLENE	47
Diethyl ether		58.8
n-Pentane	HYDROSOL Pentane	65
Mixture 1,1-Dichloro-1-fluoroethane/Methyl alcohol/Nitromethane	GENESOLV 2004	65.5
1,1-Dichloro-1-fluoroethane (Stabilised)	FORANE, GENESOLV	65.855
2-Methylbutane	Exxol Isopentane	80

Key parameter table: Flash points

Chemical name	Trade name	Flash point °C
2-Methylbutane	Exxol Isopentane	–57(IP 170)SHELL
n-Pentane	HYDROSOL Pentane	–50(IP170)SHELL
	SBP 40/65 LNH	–43(IP 170)
Cyclopentane		–42
Diethyl ether		–40
2-Methoxy-2-methyl propane	MTBE+	–34(IP 170)
	Isohexane LNH	–33(IP 170)
Methyl tert-butyl ether	DRIVERON S	–28
	Hexane Polymerisation	–27(IP 170)
	Hexane Extraction	–27(IP 170)
	SBP 60/95 LNH	–25(IP 170)
2,2-oxybispropane	IPE	–24(IP 170)
Tetrahydrofuran		–22(DIN51755 or 51758) BASF, –21.5 HULS
Hexane (branched and linear)		<–20 (IP 170) PETROGAL
Cyclohexane		–20(CLOSED CUP)
2-Propanone	DMK	–18(CLOSED CUP)–9.4 (OPEN CUP)ENICHEM, –18(IP170)SHELL, –18(OPEN CUP)BP
Methyl acetate		–16
Mixture based on naphtha solvent refined light	SBP, SOLSOL A1 1	<–15
2,2,4-Trimethyl pentane	Isooctane 100	–14
	SBP 80/110 LNH	–12(IP 170)
2-Methyl-2-methoxy-butane	TAME	–11
Benzene		–11(CLOSED CUP &IP170)
	SBP 94/100	–5(IP 170)
2-Butanone		–4(IP 170)SHELL, –6(CLOSED CUP) ELF ATOCHEM
Ethyl acetate	ETHYL ACETATE, EtAc	–4(IP 170)SHELL
	HYDROSOL Ess95/160N	–3
	HYDROLSOL Ess100/140	<0
	HYDROSOL Ess <5	<0
	Exxol DSP 65/100	<0
	Exxol Isohexane	<0
	Isopar C	<0
	HYDROSOL Ess 60/95	<0
	HYDROSOL Ess C	<0
	HYDROSOL Ess G	<0
	HYDROSOL Ess 70/80N	<0
	Norpar 5 S	<0
	Exxol DSP100/120	<0

Key parameter table: Flash points *continued*

Chemical name	Trade name	Flash point °C
	Exxol DHN50	
	Exxol DSP 60/95S	<0
	Exxol DSP 40/65	<0
	HYDROSOL Ess A	<0
	HYDROSOL Ess F<5	<0
	HYDROSOL Ess.80/120	<0
Hexane	Exxol Hexane, HYDROSOL Hexane	<0
	Exxol DSP 100/140	<0
	SBP 62/78	<0
	SBP 80/95 LNH	<0(IP 170)
	Nappar 6	<0(IP 170)
	Nappar 7	<0
Isohexane	HYDROSOL Hexane	<0
	Exxol DSP 100/160	<0
Cyclopentane	EXXOL Cyclopentane	<0
	Norpar 6	<0
	Norpar 7	<0
Heptane	Exxo Heptane, HYDROSOL Heptane	<0
	Exxol DSP 80/110	<0
	SBP 100/140	1(IP 170)
1-Methylethyl acetate		2(CLOSED CUP)
2,4,4,-Trimethylpentene-1 and 2		2
Mixture of toluene and C9/C10 aromatics	C7/C9 AROMATICS	4
	HYDROSOL Ess F	4
Toluene	SOLVAREX Toluene	4(IP 170)SHELL, 8(IP 170)PETROGAL
Diethyl ketone		7(DIN 51755 or 51758)
	Isopar E	7
Methanol		10(IP 170)SHELL, 12(PENSKY MARTENS) BP
n-Propyl acetate		10(DIN 51755 or 51758) BASF
2-Propanol	IPA, IPAC+	12(OPEN CUP)BP, EXXON, 12(IP 170) SHELL, 13 HULS
Ethanol		12 to 13
2-Methyl-2-propanol		14
Ethylbenzene		15(CLOSED CUP)ICI, 23 HULS
	Norpar 8	16
4-Methyl-2-pentanone		16(CLOSED CUP)ELF ATOCHEM, 14 HULS, 14(IP 170) SHELL
1,2-Dichloroethane		18.3(OPEN CUP) 13 (CLOSED CUP) ENICHEM
2-Methylpropyl acetate		19(DIN 51755 or 51758) BASF, 17 BP

Key parameter table: Flash points *continued*

Chemical name	Trade name	Flash point °C
	SPIRDANE L2	22
	SPIRDANE L1	22
1-Propanol		23 BASF
2-Butanol	SBA, EXXON SBA	23.9(CLOSED CUP)ELF, 24(IP 170)SHELL, 25 EXXON
Mixture of aromatic hydrocarbons	PETRINEX ASB	24(CLOSED CUP)
1,4-Dimethylbenzene	PARAXYLENE	24(CLOSED CUP)
n-Butyl acetate	nBuAc	25(IP 170)SHELL, 27(DIN 51755 or 51758)BASF, 29 BP, 23(CLOSED CUP) 38 (OPEN CUP)
Mixture of xylenes		25–31(CLOSED CUP)
Xylene	SOLVAREX Xylene	25(IP170)SHELL, 28 EXXON
Mixture of isomers and ethylbenzene		>26(IP 170)
2-Methyl-1-propanol		27(DIN51755 or 51758)BASF, 28BP/HULS, 28(IP 170)SHELL
	SPIRDANE L30	27
	SPIRDANE D30	27
p-Xylene		27(CLOSED CUP)ENICHEM
Mixture of xylenes		27–32 (CLOSED CUP)/27–46 (OPEN CUP)
m-Xylene		27 (CLOSED CUP)
	Exxol D30	28
4-Methyl-3-pentene-2-one	MO	28 (CLOSED CUP)
	VARSOL 30	29
	Nappar 9	30
1-Methoxy-propanol-2	DOWANOL PM, METHOXY PROPANOL, MePROX	30(IP 170)SHELL,35 BP,31(SETA)DOW
Styrene		31
	Solvenon PM	32(DIN 51755 or 51758)
	Exxol DSP 145/160	32
o-Xylene		32(CLOSED CUP)PETROGAL, 32(CLOSED CUP)ENICHEM
Solvent naphtha		>33
1-Butanol	NBA	34(DIN 51755 or 51758)BASF, 43(IP 170)SHELL, 37 BP
	SPIRDANE D40	38
	SOLVAREX 90/160	38
	SPIRDANE HT	38
	SPIRDANE BT	38
2-Methoxyethanol		38(DIN 51755 or 51758)
1-Methylethylbenzene	AROMASOL 12	39(CLOSED CUP)ICI, 36(CLOSED CUP)ENICHEM

Key parameter table: Flash points *continued*

Chemical name	Trade name	Flash point °C
Ethoxy propanol	ETHOXY PROPANOL	40
	LOW AROMATIC WHITE SPIRIT	40(IP 170)
	Exxol D40	40
Amyl acetate		40(DIN 51755 or 51758)BASF
	HIGH AROMATIC WHITE SPIRIT	40(IP 170)
	SHELLSOL D40	40(IP 170)
	ISANE IP155	41
	VARSOL 40	41
4-Methyl-2-pentanol		41(CLOSED CUP & IP170)
	Isopar G	41
Triisobutylene		42
	SHELLSOL E	42(IP 170)
	SHELLSOL AB	42(ASTM D93)
Mixture of aliphatic hydrocarbons	EVOLVE CH15	42(SETA CLOSED CUP)
	SHELLSOL A	42(IP 170)
1-Methoxy-2-acetoxy-propane	DOWANOL PMA	42
	Nappar 10	43
1-Chloro,4-(trifluoromethyl)benzene	OXSOL 100, PCBTF	43(TAG CLOSED CUP)
2-Ethoxyethanol	CELLOSOLVE solvent	43
Cyclohexanone	CHK	43.5(IP 170)SHELL, 43(DIN 51755 or 51758)BASF
4-Methoxy-4-methyl-2-pentanone	ME-6K	44(IP 170)
	SHELLSOL TD	44(IP 170)
	SOLVAREX 9	44
	FINALAN 40	44(IP 34)
	SOLVAREX 90/180	45
1-Methoxy-2-propanol acetate	MePROXAc	45(DIN 51755 or 51758)BASF, 45(IP 170)SHELL
2,2,4,6,6-Pentamethyl heptane		46
2-(1-Methylethoxy)ethanol	iPrOX	46(IP 170)
2,6-Dimethyl-4-heptanone/4,6-Dimethyl-2-heptanone (Isomer mixture)		47(IP 170)SHELL, 49 HULS
Mixture of aliphatic hydrocarbons/ester blend	EVOLVE CH11	48(SETA CLOSED CUP)
5-Methyl-3-heptanone	EAK	48(CLOSED CUP)
	ISANE IP165	48
1-Pentanol		49(DIN 51755 or 51758)BASF
	Solvesso 100	49
Mixture of aromatic hydrocarbons	AROMASOL H	50
Methoxy propyl acetate		50
Para-menthadienes	TABS D5	51

Key parameter table: Flash points *continued*

Chemical name	Trade name	Flash point °C
2-Ethoxyethyl acetate	CELLOSOLVE acetate	52
Ethoxy propyl acetate		54
Decahydronaphthalane		54–61
Ethyl acetoacetate	ETHYL ACETOACETATE	57
4-Hydroxy-4-methyl-2-pentanone	DA, DAA	58(ASTM D93 & CLOSED CUP)
N,N-Dimethylformamide		58(DIN 51755 or 51758)
Cyclohexanol		58(OPEN CUP), 63 (CLOSED CUP)ENICHEM
Mixture of aliphatic hydrocarbons	EVOLVE CH10	58(SETA CLOSED CUP)
Mixture of aliphatic hydrocarbons	EVOLVE CH12	58(SETA CLOSED CUP)
	Isopar H	58
	SPIRDANE K1	59
	SHELLSOL T	59(ASIM D93)
	Isopar K	59
	SHELLSOL TK	59(ASTM D93)
	SPIRDANE D60	60
	SPIRDANE K2	60
Solvent naphtha-heavy aromatic	C9+ II	>60(IP 170)
	Isopar J	61
	ISANE IP175	61
	Exxol D 60	62
	SHELLSOL AD	62(ASTM D93)
	SHELLSOL H	63(ASTM D93)
Alkoxypropanol formulation	PRIMACLEAN 1601	63
1-n-Butoxy-propanol-2	DOWANOL PnB	63(PENSKY MARTENS CLOSED CUP)
	FINALAN 60	63(IP 34)
Trimethylcyclohexanone		64
	SOLVAREX 10	64
	VARSOL 60	64
	Isopar L	65
	KETRUL BT	65
	KERDANE	65
	KETRUL D70	65
	KETRUL HT	65
	ISANE IP185	65
Blend of isomers	PROGLYDE DMM	65(PENSKY MARTENS CLOSED CUP)
	SHELLSOL D60	66(ASTM D93)
	Exxate 700	66
	Solvesso 150	66

Key parameter table: Flash points *continued*

Chemical name	Trade name	Flash point °C
Mixture based on aliphatic hydrocarbons	VISTA LPA-150 SOLVENT	66(TAG CLOSED CUP)
2-Butoxyethanol	nBuOX, Butyl CELLOSOLVE	67(DIN51755 or 51758)BASF, 67(ASTM D93)SHELL, 69(OPEN CUP)BP
	Nappar 11	68
	HAN 8070	69
Mixture of aromatic hydrocarbons	PETRINEX T9	70
N,N-Dimethylacetamide		70(DIN 51755 or 51758)
Tetrahydronaphthalene		71–77
Mixture based on aliphatic hydrocarbons	VISTA LPA SOLVENT	71(PENSKY MARTENS)
Mixture based on saturated and aromatic hydrocarbons	VISTA MOLEX RAFFINATE	72(PENSKY MARTENS)
	Exxol D70	72
	HAN	72
	SHELLSOL D70	73(ASTM D93)
	SOLVARO AFD	73
2,6-Dimethyl-4-heptanol	DIBC	74(ASTM D93)
	Norpar 12	74
	HAN 8572	74
	KETRUL D80	75
	Exxol D80	75
1-(2-Methoxy-propoxy)-propanol 2/1-(2-Methoxy-methylethoxy)-propanol 2	DOWANOL DPM	75(SETA)
2-Ethylhexanol		76(DIN 51755 or 51758)BASF, 77 BP, 82 HULS
Trimethylcyclohexanol		76
	FINALAN 75	77(IP 34)
Methyl acetoacetate		77(PENSKY MARTENS CLOSED CUP)
	VARSOL 80	77
1-n-Butoxy-propyl-acetate-2	DOWANOL PnBA	77(PENSKY MARTENS CLOSED CUP)
2-Ethylhexyl acetate		77(DIN 51755 or 51758)BASF
	Isopar M	78
2-Butoxyethyl acetate	BUTYL GLYCOL ACETATE?	78 HULS, 88(OPEN CUP)BP
Butyl glycol acetate		78(DIN 51755 or 51758)
(2-Methoxymethylethoxy)-propanol	METHOXY PROPOXY PROPANOL	79
Phenol		79(CLOSED CUP), 85(OPEN CUP)
Alkoxypropanol formulation	DOWANOL PX-16S	79
2-(2-Methoxymethylethoxy)-propanol)	MeDIPROX	79(ASTM D93)
Mixture	Solvenon DPM	80(DIN 51755 or 51758)
Mixture based on aliphatic hydrocarbons		82(TAG CLOSED CUP)
	KETRUL 211	85

Key parameter table: Flash points *continued*

Chemical name	Trade name	Flash point °C
	KETRUL 212	85
	KETRUL 210	85
	SHELLSOL R	85(ASTM D93)
1,2-Diacetoxy-propane	DOWANOL PGDA	86
	SHELLSOL TM	86(ASTM D93)
(2-Ethoxy-methylethoxy)-propanol	ETHOXY PROPOXY PROPANOL	86.5
	Exxate 900	90
	Exxol D 220/230	90
Methylbenzyl alcohol		90
N-Methylpyrrolidone		91(DIN 51755 or 51758)
	Methyl CARBITOL solvent	91(DIN 51755 or 51758)BASF
	FINALAN 90	92(IP 34)
2-Methyl-2,4-pentanediol	HG	93(ASTM D93)SHELL, 97.9(CLOSED CUP)ELF ATOCHEM
n-Hexyl glycol	Hexyl CELLOSOLVE solvent	94(DIN 51755 or 51758)BASF
	SOLVENT APV SPECIAL	94
2-(2-Ethoxyethoxy)ethanol	CARBITOL solvent low gravity	94
	Norpar 13	95
	SHELLSOL D90	95(ASTM D93)
	SHELLSOL DMA	95(ASTM D93)
Nonanol mixed isomers	Nonanol N	96(DIN 51755 or 51758)BASF
3,5,5-Trimethyl-2-cyclohexene-1-one	IPHO	96(OPEN CUP)BP, 85(CLOSED CUP)ELF ATOCHEM, 85 HULS
	Solvesso 200	97
	Solvesso 200S	99
	Exxate 1000	100
	RHODIASOLV RPDE	>100
Isodecanol mixed isomers		102(DIN 51755 or 51758)BASF
2,2,4,6,8,8-Heptamethyl nonane		102
	FINALAN 100	102(IP 34)
	ISANE IP235	102
	Exxol D 100S	103
1,2-Propylene glycol		103(DIN 51755 or 51758)BASF, 103 HULS
	Exxol D 100	103
	Exxol D 240/260	103
	Isopar P	104

Key parameter table: Flash points *continued*

Chemical name	Trade name	Flash point °C
	SHELLSOL D100	105(ASTM D93)
2-(2-Butoxyethoxy)ethanol	nBuDIOX,BUTYL DIGLYCOL	105(DIN 15755 or 15758)BASF, 105(ASTM)SHELL, 105 HULS, 116(OPEN CUP)BP
Mixture based on aliphatic hydrocarbons	VISTA LPA-210 SOLVENT	105(PENSKY MARTENS)
	Exxol D 110	106
	INK SOLVENT 2427	106(ASTM D93)
2-(2-Ethoxyethoxy)ethyl acetate		107
	VARSOL 110	107
Mixture based on saturated aliphatic hydrocarbons	VISTA 47 SOLVENT	107(PENSKY MARTENS)
[2-(2-Butoxyethoxy)ethyl acetate]		108
Dimethyl esters of adipic, glutaric and succinic acids	ESTASOL	108 OPEN CUP
1,3-Butanediol		109
Formulated fatty acid ester	PROTENE F	>110 CLOSED CUP ASTM3828 or IP 303(SETA FLASH)
Ethoxylated fatty acid ester	PROTENE R	>110 CLOSED CUP ASTM3828 or IP 303(SETA FLASH)
1,2-Ethanediol		111(CLOSED CUP), 119(OPEN CUP) ENICHEM, 111(DIN 51755 or 51758)BASF
1-(2-n-Butoxy-1-methyl-ethoxy)-propanol-2	DOWANOL DPnB	111(PENSKY MARTENS)
Tetradecane	VISTA C14 PARAFFIN	112(PENSKY MARTENS)
2-(Butoxyethoxy)ethyl acetate	BUTYL DIGLYCOL ACETATE	115(OPEN CUP)
Blend	DOWANOL TPM	116(PENSKY MARTENS CLOSED CUP)
2-(2-Butoxyethoxy)-ethanol acetate/Diethylene glycol n-butyl ether acetate	nBuDIOXAc	116(ASTM D92-COC)SHELL, 102(DIN 51755 or 51758)BASF
	Norpar 15	118
	Exxol D 120	120
2-[2-(ethoxyethoxy)ethoxy]ethanol		120
Mixture	Solvenon PP	120(DIN 51755 or 51758)
	SHELLSOL D120	120(DIN EN22719)
Tetradecane, Pentadecane, Hexadecane	VISTA C14-C16 PARAFFINS	121(PENSKY MARTENS)
Phenyl glycol		121(DIN51755 or 51758)
	VARSOL 120	122
	ISANE IP260	122
	Solvenon PC	123(DIN 51755 or 51758)BASF
Tridecanol mixed isomers	Tridecanol A	123(DIN 51755 or 51758)BASF
Blend of isomers	DOWANOL TPnB	124(PENSKY MARTENS CLOSED CUP)
	Exxate 1300	127
1-Phenoxy-propanol-2	DOWANOL PPh	127
	INK SOLVENT 2629	128(ASTM D93)
2-Ethylhexane-1,3-diol		128

Key parameter table: Flash points *continued*

Chemical name	Trade name	Flash point °C
Propylene carbonate		130
2-[2-(butoxyethoxy)ethoxy]ethanol	COASOL	130
Di-isobutyl esters of adipic acid, glutaric acid and succinic acid	Isopar V	131
	Exxol D 140	134
		135
1,4-Butanediol	INK SOLVENT 2831	135
		136(ASTM D93)
Dipropylene glycol (mixed isomers)	VARSOL 140	138(DIN 51755 or 51758)BASF, 120 HULS
		140
Diethylene glycol		140(DIN 51755 or 51758)BASF
Mixture		144(DIN15755 or 15758)
2-[2-(butoxyethoxy)-ethoxy]ethanol	VISTA 3050 SPECIALTY ALKYLATE	170
Mixture based on aromatic hydrocarbons	VISTA 150L SPECIALTY ALKYLATE	182(PENSKY MARTENS)
Mixture based on aromatic hydrocarbons	SHELLSOL D25/SBP 140/165	207(PENSKY MARTENS)
	C9 + I	271(IP 170)
Solvent naphtha	FORANE 141b MGX	>450
1,1-Dichloro-1-fluoroethane/methanol mixture	FORANE, GENESOLV	NONE
1,1-Dichloro-1-fluoroethane (Stabilised)		NONE
Trichloromethane		NONE
1,1,1-Trichloroethane	BALTANE CF, CHLOROTHENE	NONE
1,2,2-Trichloroethylene	TRIKLONE, NEU TRI, HI-TRI, TAVOXENE, ALTENE	NONE
Dichloromethane	METHOKLONE, UKLENE, AEROTHENE, SOLVACLENE	NONE
Mixture 1,1-Dichloro-1-fluoroethane/Methyl alcohol/ Nitromethane	GENESOLV 2004	NONE
1,1,2,2-Tetrachloroethylene	DOWPER, Pertene, PERKLONE, SOLTENE	NONE

Key parameter table: Solubility parameters

Chemical name	Trade name	Solubility parameter $(J/cm^3)^{1/2}$
2-Methylbutane	Exxol Isopentane	13.9
2,2,4-Trimethyl pentane	Isooctane 100	14.01
	SBP 40/65 LNH	14.1
2-Methoxy-2-methyl propane	MTBE+	14.3
2,2-Oxybispropane	IPE	14.3
n-Pentane	HYDROSOL Pentane	14.3
	Isohexane LNH	14.7
Hexane	Exxol Hexane, HYDROSOL Hexane	14.87
	SBP 62/78	14.9
	Hexane Polymerisation	14.9
	Hexane Extraction	14.9
	SHELLSOL TD	14.9
	SHELLSOL T	15.0
	SBP 80/95 LNH	15.1
	SBP 60/95 LNH	15.1
	SBP 80/110 LNH	15.1
Heptane	Exxol Heptane, HYDROSOL Heptane	15.20
	SHELLSOL D100	15.3
	SBP 94/100	15.3
	SHELLSOL D25/SBP 140/165	15.5
	SHELLSOL D40	15.5
	SHELLSOL D60	15.5
	SHELLSOL D70	15.5
	SBP 100/140	15.5
Diethyl ether		15.6
2,6-Dimethyl-4-heptanone/4,6-Dimethyl-2-heptanone (Isomer mixture)		15.6–16
	SHELLSOL H	15.8
	SHELLSOL D120	16.0
	LOW AROMATIC WHITE SPIRIT	16.2
	INK SOLVENT 2427	16.4
Cyclopentane	EXXOL Cyclopentane	16.57
	INK SOLVENT 2831	16.6
	INK SOLVENT 2629	16.6
Cyclohexane		16.78
2-Methylpropyl acetate		17.0
	SHELLSOL R	17.2
	SHELLSOL E	17.2
Isopropyl acetate		17.2
4-Methyl-2-pentanone		17.2
1-Methoxy-2-propanol acetate	MePROXAc	17.4
2-(2-Butoxyethoxy)-ethanol acetate/Diethylene glycol n-butyl ether acetate	nBuDIOXAc	17.4
Amyl acetate		17.4
2-Heptanone		17.4
4-Methoxy-4-methyl-2-pentanone	ME-6K	17.4
1,1,1-Trichloroethane	BALTANE CF, CHLOROTHENE	17.4
1-Methylethylbenzene	AROMASOL 12	17.6
n-Butyl acetate	nBuAc	17.6
	SHELLSOL AB	17.8
	SHELLSOL AD	17.8
2-Ethoxyethyl acetate	CELLOSOLVE acetate	17.8
2(2-Methoxymethylethoxy)-propanol	MeDIPROX	17.8
Isobutyraldehyde		17.9
1,4-Dimethylbenzene	PARAXYLENE	17.94
p-Xylene		17.94
Ethylbenzene		17.98
	SHELLSOL A	18.0

Key parameter table: Solubility parameters *continued*

Chemical name	Trade name	Solubility parameter $(J/cm^3)^{1/2}$
m-Xylene		18.0
n-Propyl acetate		18.0
o-Xylene		18.0
Xylene	SOLVAREX Xylene	18.1
Toluene	SOLVAREX Toluene	18.2
2-Butoxyethanol	nBuOX, Butyl CELLOSOLVE	18.2–18.6
2-(2-Butoxyethoxy)ethanol	nBuDIOX, BUTYL DIGLYCOL	18.2
Butyraldehyde		18.4
4-Methyl-3-pentene-2-one	MO	18.4
[2-(2-Butoxyethoxy)ethyl acetate)		18.6
Ethyl acetate	ETHYL ACETATE, EtAc	18.6
Benzene		18.74
4-Hydroxy-4-methyl-2-pentanone	DA, DAA	18.8
2-(1-Methylethoxy)ethanol	iPrOX	18.8
4-Hydroxy-4-methyl-2-pentanone		18.9
2-Butanone		19.0
1,2,2-Trichloroethylene	TRIKLONE, NEU TRI, HI-TRI, TAVOXENE, ALTENE	19.0
1,1,2,2-Tetrachloroethylene	DOWPER, Pertene, PERKLONE, SOLTENE	19.0
2-(2-ethoxyethoxy)ethyl acetate		19.0
Styrene		19.0
2,6-Dimethyl-4-heptanol	DIBC	19.2
3,5,5-Trimethyl-2-cyclohexene-1-one	IPHO	19.2
Tetrahydrofuran		19.2
Di-isobutyl carbinol		19.3
Propionaldehyde		19.3
2-Ethylhexanol		19.4
Tetrahydronaphthalene		19.4
1-Methoxy-propanol-2	DOWANOL PM, METHOXY PROPANOL, MeProx	19.4
Trichloromethane		19.42
Methyl acetate		19.6
2-(2-Ethoxyethoxy)ethanol	CARBITOL solvent low gravity	19.6
Dichloromethane	METHOKLONE, UKLENE, AEROTHENE, SOLVACLENE	20.21
2-Methoxyethanol		20.3
Cyclohexanone	CHK	20.3
2-Ethoxyethanol	CELLOSOLVE solvent	20.3
4-Methyl-2-pentanol		20.5
2-Propanone	DMK	20.5
2-Methyl-2-propanol		21.7
2-Methyl-1-propanol		21.9
Triethylene glycol		21.9
2-Butanol	SBA, EXXON SBA	22.1
N,N-Dimethylacetamide		22.1
1-Pentanol		22.3
2-Methyl-2,4-pentanediol	HG	23.1
N-Methylpyrrolidone		23.1
Phenol		23.1
1-Butanol	NBA	23.3
2-Propanol	IPA, IPAC+	23.5
1,3-Butanediol		23.7
Dimethyl sulphoxide		24.5
Acrylic acid		24.5
1,4-Butanediol		24.7
Benzyl alcohol		24.8
1-Propanol		24.91

Key parameter table: Solubility parameters *continued*

Chemical name	Trade name	Solubility parameter $(J/cm^3)^{1/2}$
Ethanol		26.0
Propylene carbonate		27.2
1,2-Ethanediol		29.1
Diethylene glycol		29.13
Methanol		29.7

Key parameter table: Auto-ignition temperatures

Chemical name	Trade name	Auto-ignition temperature °C
1-Phenoxy-propanol-2	DOWANOL PPh	135
Blend of isomers	PROGLYDE DMM	165
Diethyl ether		170
Blend of isomers	DOWANOL TPnB	186
1-(2-n-Butoxy-1-methyl-ethoxy)-propanol-2	DOWANOL DPnB	189
(2-Ethoxy-methylethoxy)-propanol	ETHOXY PROPOXY PROPANOL	190
Mixture		203
2(2-Methoxymethylethoxy)-propanol	MeDIPROX	205
Mixture of aliphatic hydrocarbons	EVOLVE CH10	205
Mixture	Solvenon DPM	205
2-(2-Butoxyethoxy)ethanol	nBuDIOX, BUTYL DIGLYCOL	210–230
Tetrahydrofuran		212
	SBP 94/100	215
	Methyl CARBITOL solvent	215
	SHELLSOL D120	215
Mixture based on aliphatic hydrocarbons	VISTA LPA-210 SOLVENT	218
	SHELLSOL D90	220
	SHELLSOL DMA	220
	SHELLSOL D100	220
n-Hexyl glycol	Hexyl CELLOSOLVE solvent	220
2-[2-(ethoxyethoxy)ethoxy]ethanol		220
Mixture based on naphtha solvent refined light	SBP, SOLSOL A1 1	220
	SBP 100/140	220
2-(2-Ethoxyethoxy)ethanol	CARBITOL solvent low gravity	220
	SHELLSOL D25/SBP 140/165	220
	SOLVENT APV SPECIAL	220
	SHELLSOL D70	225
	SHELLSOL D60	225
Mixture based on saturated and aromatic hydrocarbon	VISTA MOLEX RAFFINATE	227
Mixture based on aliphatic hydrocarbons	VISTA LPA-150 SOLVENT	228
Mixture based on aliphatic hydrocarbons		228
	INK SOLVENT 2831	230
Solvent naphtha		230
	INK SOLVENT 2629	230
2-Butoxyethanol	nBuOX, Butyl CELLOSOLVE	230–245
	SHELLSOL H	230
	INK SOLVENT 2427	230
2-[2-(butoxyethoxy)ethoxy]ethanol		230
Mixture based on aliphatic hydrocarbons	VISTA LPA SOLVENT	231
2-Ethylhexanol		231–330
2-(1-Methylethoxy)ethanol	iPrOX	240
	SHELLSOL D40	240
	Hexane Extraction	240
	Hexane Polymerisation	240
	LOW AROMATIC WHITE SPIRIT	240
	SBP 62/78	240
Hexane (branched and linear)		>240
Alkoxypropanol formulation	PRIMACLEAN 1601	242
	SBP 40/65 LNH	245
	SBP 80/110 LNH	250
2-[2-(butoxyethoxy)-ethoxy]ethanol		250
2-Ethoxyethanol	CELLOSOLVE solvent	250
Decahydronaphthalene		255
Ethoxy propanol	ETHOXY PROPANOL	255
	Isohexane LNH	260
Cyclohexane		260

Key parameter table: Auto-ignition temperatures *continued*

Chemical name	Trade name	Auto-ignition temperature °C
1-n-Butoxy-propanol-2	DOWANOL PnB	260
2-(2-Butoxyethoxy)-ethanol acetate/Diethylene glycol n-butyl ether acetate	nBuDIOXAc	265–290
	HIGH AROMATIC WHITE SPIRIT	265
Tridecanol mixed isomers	Tridecanol A	265
	Solvenon PM	270
1-Methoxy-propanol-2	DOWANOL PM, METHOXY PROPANOL, MePROX	270–287
Blend	DOWANOL TPM	270
2-Ethylhexyl acetate		270–295
	SBP 60/95 LNH	270
1-(2-Methoxy-propoxy)-propanol 2/1-(2-Methoxy-methylethoxy)-propanol 2	DOWANOL DPM	270
(2-Methoxymethylethoxy)-propanol	METHOXY PROPOXY PROPANOL	270
N-Methylpyrrolidone		270
Nonanol mixed isomers	Nonanol N	270
Isodecanol mixed isomers		275
	SBP 80/95 LNH	275
1-n-Butoxy-propyl-acetate-2	DOWANOL PnBA	279
Methyl acetoacetate		280
Butyl glycol acetate		280
n-Pentane	HYDROSOL Pentane	285
2,6-Dimethyl-4-heptanol	DIBC	290
[2-(2-Butoxyethoxy)ethyl acetate]		290
Mixture of aliphatic hydrocarbons	EVOLVE CH15	292
Ethyl acetoacetate	ETHYL ACETOACETATE	295
2-(Butoxyethoxy)ethyl acetate	BUTYL DIGLYCOL ACETATE	299
2,4,4-Trimethylpentene-1 and 2		300
Cyclohexanol		300
1-Pentanol		300
Dimethyl sulphoxide		300–302
4-Methyl-2-pentanol		305
2-Methoxyethanol		310
2-(2-Ethoxyethoxy)ethyl acetate		310
Dipropylene glycol (mixed isomers)		310–350
1-Methoxy-2-propanol acetate	MePROXAc	315
Ethoxy propyl acetate		325
1-Methoxy-2-acetoxy-propane	DOWANOL PMA	333
2-Butoxyethyl acetate	BUTYL GLYCOL ACETATE?	340–375
1-Butanol	NBA	340–372
2-Ethylhexane-1,3-diol		340
4-Methyl-3-pentene-2-one	MO	344
2,6-Dimethyl-4-heptanone/4,6-Dimethyl-2-heptanone (Isomer mixture)		345
Amyl acetate		345
Diethylene glycol		345
n-Heptanol		350
Methoxy propyl acetate		354
Triisobutylene		355
1-Propanol		360
Ethanol		365–425
n-Butyl acetate	nBuAc	370–421
Dimethyl esters of adipic, glutaric and succinic acids	ESTASOL	370
Mixture of aliphatic hydrocarbons/ester blend	EVOLVE CH11	370
Mixture of toluene and C9/C10 aromatics	C7/C9 AROMATICS	370
Mixture based on aromatic hydrocarbons	VISTA 150L SPECIALTY ALKYLATE	371

Key parameter table: Auto-ignition temperatures *continued*

Chemical name	Trade name	Auto-ignition temperature °C
Trimethylcyclohexanol		375
Mixture of aliphatic hydrocarbons	Evolve CH12	375
Mixture based on aromatic hydrocarbons	VISTA 3050 SPECIALTY ALKYLATE	379
Cyclopentane		380
2-Butanol	SBA, EXXON SBA	390–406
2-Methyl-1-propanol		390–427
2-Propanol	IPA, IPAC+	399–425
Di-isobutyl esters of adipic acid, glutaric acid and succinic acid	COASOL	400
1-Methylethylbenzene	AROMASOL 12	400–424
2-Ethoxyethyl acetate	CELLOSOLVE acetate	400
N,N-Dimethylacetamide		400
4-Methoxy-4-methyl-2-pentanone	ME-6K	402
2-Methylpropyl acetate		405–423
	SHELLSOL TD	405
	SHELLSOL T	405
2,2-Oxybispropane	IPE	405
1,2-Ethanediol		410
N,N-Dimethylformamide		410
1,2,2-Trichloroethylene	TRIKLONE, NEU TRI, HI-TRI, TAVOXENE, ALTENE	410
2,2,4,6,6-Pentamethyl heptane		410
1,2-Propylene glycol		410
1,2-Dichloroethane		413
2-Methyl-2-methoxy-butane	TAME	415
2,2,4-Trimethyl pentane	Isooctane 100	420
1,4-Butanediol		420
2-Methylbutane	Exxol Isopentane	420
2,2,4,4,6,8,8-Heptamethyl nonane		420
Solvents naphtha-heavy aromatic	C9+ II	>420
Ethyl acetate	ETHYL ACETATE, EtAc	425–530
2-Methyl-2,4-pentanediol	HG	425
Tetrahydronaphthalene		425
Ethylbenzene		428–435
n-Propyl acetate		430
Cyclohexanone	CHK	430
Trimethylcyclohexanone		430
1,2-Diacetoxy-propane	DOWANOL PGDA	431
Benzyl alcohol		436
1,3-Butanediol		440
Diethyl ketone		445
	SHELLSOL R	450
	SHELLSOL E	450
	SHELLSOL AD	450
	SHELLSOL AB	450
	SHELLSOL A	450
Methanol		455–464
Methyl tert-butyl ether	DRIVERON S	460
1-Methylethyl acetate		460
3,5,5-Trimethyl-2-cyclohexene-1-one	IPHO	460–462
4-Methyl-2-pentanone		460
2-Methoxy-2-methyl-propane	MTBE+	460
Mixture of xylenes		463
o-Xylene		464
2-Propanone	DMK	465–540
	Solvenon PC	~465
Mixture of xylenes		480

Key parameter table: Auto-ignition temperatures *continued*

Chemical name	Trade name	Auto-ignition temperature °C
Styrene		480
Methylbenzyl alcohol		480
Mixture	Solvenon PP	480
Toluene	Solvarex Toluene	480–535
2-Methyl-2-propanol		490
Benzene		498–535
Mixture of aromatic hydrocarbons	PETRINEX T9	500
Mixture of aromatic hydrocarbons	AROMASOL H	500
Xylene	SOLVAREX Xylene	500
Mixture of isomers and ethylbenzene		>500
1-Chloro,4-(trifluoromethyl)benzene	OXSOL 100, PCBTF	>500
Dichlorotoluene		>500
Propylene carbonate		510
2-Butanone		515
m-Xylene		527
p-Xylene		528–530
1,4-Dimethylbenzene	PARAXYLENE	530
1,1-Dichloro-1-fluoroethane/methanol mixture	FORANE 141b MGX	532
1,1-Dichloro-1-fluoroethane (Stabilised)	FORANE, GENESOLV	532–550
Phenyl glycol		535
1,1,1-Trichloroethane	BALTANE CF, CHLOROTHENE	537
Mixture 1,1-Dichloro-1-fluoroethane/Methyl alcohol/Nitromethane	GENESOLV 2004	550
Mixture of aromatic hydrocarbons	PETRINEX ASB	550
4-Hydroxy-4-methyl-2-pentanone	DA, DAA	603–620
Trichloromethane		624
Dichloromethane	METHOKLONE, UKLENE, AEROTHENE, SOLVACLENE	624–662
Monochlorobenzene		637
Phenol		715
1,1,2,2-Tetrachloroethylene	DOPWER, Pertene, PERKLONE, SOLTENE	NONE

Key parameter table: Chemical names/suppliers

Chemical name	Suppliers
(2-Ethoxy-methylethoxy)-propanol	BP
(2-Methoxymethylethoxy)-propanol	BP
1,1,1-Trichloroethane	ELF ATOCHEM, DOW, VULCAN
1,1,2,2-Tetrachloroethylene	DOW, ELF ATOCHEM, ICI, ENICHEM, VULCAN, SOLVAY
1,1-Dichloro-1-fluoroethane (Stabilised)	ELF ATOCHEM, ALLIED SIGNAL
1,1-Dichloro-1-fluoroethane/methanol mixture	ELF ATOCHEM
1,2,2-Trichloroethylene	ICI, ELF ATOCHEM, DOW, ENICHEM, SOLVAY
1,2-Diacetoxy-propane	DOW
1,2-Dichloroethane	VULCAN, ENICHEM
1,2-Ethanediol	SHELL, UNION CARBIDE, BASF, ENICHEM
1,2-Propylene glycol	BASF, HULS
1,3-Butanediol	HULS
1,4-Butanediol	HULS
1,4-Dimethylbenzene	ICI
1-(2-Methoxy-propoxy)-propanol 2/ 1-(2-Methoxy-methylethoxy)-propanol 2	DOW
1-(2-n-Butoxy-1-methyl-ethoxy)-propanol-2	DOW
1-Butanol	BP, ELF ATOCHEM, HULS, SHELL, UNION CARBIDE, BASF
1-Chloro,4-(trifluoromethyl)benzene	OCCIDENTAL
1-Methoxy-2-acetoxy-propane	DOW
1-Methoxy-2-propanol acetate	BASF, SHELL
1-Methoxy-propanol-2	SHELL, BP, DOW
1-Methylethyl acetate	RHÔNE POULENC, PARDIE ACETIQUES
1-Methylethylbenzene	ICI, ENICHEM
1-n-Butoxy-propanol-2	DOW
1-Pentanol	UNION CARBIDE, BASF
1-Phenoxy-propanol-2	DOW
1-Propanol	UNION CARBIDE, BASF
1-n-Butoxy-propyl-acetate-2	DOW
2-(2-Methoxymethylethoxy)-propanol	SHELL
2,2,4,4,6,8,8,-Heptamethyl nonane	EC Erdolchemie
2,2,4,6,6-Pentamethyl heptane	BP, EC Erdolchemie
2,2,4-Trimethyl pentane	EC Erdolchemie
2,2-oxybispropane	SHELL
2,2ˆ-[1,2-Ethanediylbis(oxy)]bisethanol	SHELL
2,4,4,-Trimethylpentene-1 and 2	EC Erdolchemie
2,6-Dimethyl-4-heptanol	SHELL
2,6-Dimethyl-4-heptanone/4,6-Dimethyl-2-heptanone (Isomer mixture)	SHELL, HULS, UNION CARBIDE
2-(1-Methylethoxy)ethanol	SHELL
2-(2-Butoxyethoxy)-ethanol acetate/Diethylene glycol n-butyl ether acetate	BASF, SHELL
2-(2-Butoxyethoxy)-ethanol	BP, HULS, SHELL, BASF
2-(2-Ethoxyethoxy)ethanol	UNION CARBIDE, HULS
2-(2-ethoxyethoxy)ethyl acetate	HULS
2-(Butoxyethoxy)ethyl acetate	BP
2-Butanol	ELF ATOCHEM, EXXON, SHELL
2-Butanone	UNION CARBIDE, ELF ATOCHEM, EXXON, SHELL
2-Butoxyethanol	BP, HULS, SHELL, UNION CARBIDE, BASF
2-Butoxyethyl acetate	BP, HULS
2-Ethoxyethanol	UNION CARBIDE, HULS
2-Ethoxyethyl acetate	UNION CARBIDE, HULS
2-Ethylhexane-1,3-diol	HULS
2-Ethylhexanol	BP, ELF ATOCHEM, HULS, SHELL, UNION CARBIDE, BASF
2-Ethylhexoic Acid	UNION CARBIDE
2-Ethylhexyl acetate	HULS, BASF
2-Heptanone	UNION CARBIDE

Key parameter table: Chemical names/suppliers *continued*

Chemical name	Suppliers
2-Methoxy-2-methyl propane	SHELL
2-Methoxyethanol	BASF
2-Methyl-1-butanol	UNION CARBIDE
2-Methyl-1-propanol	BP, ELF ATOCHEM, SHELL, BASF
2-Methyl-2,4-pentanediol	ELF ATOCHEM, SHELL, UNION CARBIDE
2-Methyl-2-methoxy-butane	EC Erdolchemie
2-Methyl-2-propanol	HULS
2-Methylbutane	SHELL, EXXON
2-Methylpropyl acetate	UNION CARBIDE, BASF, BP
2-Propanol	BP, EXXON, SHELL, UNION CARBIDE, HULS
2-Propanone	UNION CARBIDE, BP, SHELL, ENICHEM
3,5,5-Trimethyl-2-cyclohexene-1-one	UNION CARBIDE, BP, ELF ATOCHEM, HULS
4-Hydroxy-4-methyl-2-pentanone	ELF ATOCHEM, SHELL
4-Hydroxy-4-methyl-2-pentanone	UNION CARBIDE
4-Methoxy-4-methyl-2-pentanone	SHELL
4-Methyl-2-pentanol	ELF ATOCHEM, SHELL, UNION CARBIDE
4-Methyl-2-pentanone	UNION CARBIDE, ELF ATOCHEM, HULS, SHELL
4-Methyl-3-pentene-2-one	ELF ATOCHEM
5-Methyl-3-heptanone	ELF ATOCHEM
Acrylic acid	UNION CARBIDE
Alkoxypropanol formulation	DOW
Alkoxypropanol formulation	DOW
Amyl acetate	UNION CARBIDE, BASF
Aromatic based glycol ether	DOW
Benzene	ICI, PETROGAL, ENICHEM
Benzyl alcohol	ELF ATOCHEM
Blend	DOW
Blend of isomers	DOW
Blend of isomers	DOW
Butyl glycol acetate	BASF
Butyraldehyde	UNION CARBIDE
Cyclohexane	ICI
Cyclohexanol	ENICHEM
Cyclohexanone	UNION CARBIDE, SHELL, BASF
Cyclopentane	EC Erdolchemie
Decahydronaphthalene	HULS
Di-isobutyl carbinol	UNION CARBIDE
Di-isobutyl esters of adipic acid, gluatric acid and succinic acid	CHEMOXY
Dichloromethane	ICI, DOW, ELF ATOCHEM, AKZO, VULCAN, SOLVAY
Dichlorotoluene	ELF ATOCHEM
Diethyl ether	HULS
Diethyl ketone	BASF
Diethylene glycol	SHELL, UNION CARBIDE, BASF, ENICHEM
Diethylene glycol methyl ether	DOW
Diethylene glycol n-butyl ether	DOW
Dimethyl esters of adipic, glutaric and succinic acids	CHEMOXY
Dimethyl suphoxide	ELF ATOCHEM
Dipropylene glycol (mixed isomers)	BASF, HULS
Dipropylene glycol butylene ether	UNION CARBIDE
Dipropylene glycol methyl ether	UNION CARBIDE
Dipropylene glycol methyl ether acetate	DOW
Dipropylene glycol n-propyl ether	DOW
Ethanol	BP, HAYMAN, HULS, UNION CARBIDE
Ethoxy propanol	BP
Ethoxy propyl acetate	BP
Ethoxylated fatty acid ester	ELF ATOCHEM
Ethoxytriglycol	UNION CARBIDE

Key parameter table: Chemical names/suppliers *continued*

Chemical name	Suppliers
Ethyl acetate	UNION CARBIDE, BP, HULS, SHELL
Ethyl acetoacetate	BP
Ethylbenzene	HULS, ICI
Ethylene glycol n-butyl ether	DOW
Ethylene glycol phenyl ether	DOW
Formulated fatty acid ester	ELF
Heptane	EXXON, TOTAL
Hexane	EXXON, TOTAL
Hexane (branched and linear)	PETROGAL, SA
Isobutyraldehyde	UNION CARBIDE
Isodecanol mixed isomers	BASF
Isohexane	TOTAL
Isopentanoic acid	UNION CARBIDE
Isopropyl acetate	UNION CARBIDE
m-Xylene	ENICHEM
Methanol	BP, ELF ATOCHEM, SHELL
Methoxy propyl acetate	BP
Methoxytriglycol	UNION CARBIDE
Methyl acetate	CHEMOXY
Methyl acetoacetate	BP
Methyl tert-butyl ether	HULS
Methylbenzyl alcohol	HULS
Mixture	BASF
Mixture	BASF
Mixture	BASF
Mixture 1,1-Dichloro-1-fluoroethane/Methyl alcohol/Nitromethane	ALLIED SIGNAL INC
Mixture based on aliphatic hydrocarbons	VISTA
Mixture based on aliphatic hydrocarbons	VISTA
Mixture based on aliphatic hydrocarbons	VISTA
Mixture based on aromatic hydrocarbons	VISTA
Mixture based on aromatic hydrocarbons	VISTA
Mixture based on naphtha solvent refined light	PETROGAL, SA
Mixture based on saturated aliphatic hydrocarbons	VISTA
Mixture based on saturated and aromatic hydrocarbons	VISTA CHEMICAL CO.
Mixture of aliphatic hydrocarbons	ICI
Mixture of aliphatic hydrocarbons	ICI
Mixture of aliphatic hydrocarbons	ICI
Mixture of aliphatic hydrocarbons/ester blend	ICI
Mixture of aromatic hydrocarbons	ICI
Mixture of aromatic hydrocarbons	ICI
Mixture of aromatic hydrocarbons	ICI
Mixture of isomers and ethylbenzene	PETROGAL, SA
Mixture of xylenes	ENICHEM
Mixture of xylenes	ICI
Mixture of toluene and C9/C10 aromatics	ICI
Monochlorobenzene	ELF ATOCHEM
N,N-Dimethylacetamide	BASF
N,N-Dimethylformamide	BASF
n-Butyl acetate	BP, SHELL, BASF, UNION CARBIDE
n-Butyl propionate	CHEMOXY, UNION CARBIDE
n-Heptanol	ELF ATOCHEM
n-Hexyl glycol	UNION CARBIDE, BASF
N-Methylpyrrolidone	BASF
n-Pentane	SHELL, TOTAL
n-Pentyl propionate	UNION CARBIDE
n-Propyl acetate	UNION CARBIDE, BASF
Nonanol mixed isomers	BASF

Key parameter table: Chemical names/suppliers *continued*

Chemical name	Suppliers
o-Xylene	PETROGAL, ENICHEM
P-xylene	PETROGAL, ENICHEM
Para-menthadienes	BUSH BOAKE ALLEN LTD
Phenol	ENICHEM
Phenyl glycol	BASF
Propionaldehyde	UNION CARBIDE
Propylene carbonate	HULS
Propylene glycol methyl ether	UNION CARBIDE
Propylene glycol n-propyl ether	DOW, UNION CARBIDE
Solvent naphtha	PETROGAL
Solvent naphtha-heavy aromatic	PETROGAL
Solvent water microemulsions	DOW
Styrene	HULS
Tetradecane	VISTA
Tetradecane, Pentadecane, Hexadecane	VISTA
Tetrahydrofuran	HULS, BASF
Tetrahydronaphthalene	HULS
Toluene	EXXON, ICI, PETROGAL, SHELL, TOTAL, ENICHEM
Trichloromethane	ELF ATOCHEM, AKZO, VULCAN
Trichlorotrifluoroethane	ELF ATOCHEM
Tridecanol mixed isomers	BASF
Triethylene glycol	UNION CARBIDE
Triethylene glycol butyl ether/Highers	DOW
Triethylene glycol methyl ether/Highers	DOW
Triisobutylene	EC Erdolchemie
Trimethylcyclohexanol	HULS
Trimethylcyclohexanone	HULS
Valeraldehyde	UNION CARBIDE
Xylene	EXXON, SHELL, TOTAL, ENICHEM
[2-(2-Butoxyethoxy)ethyl acetate]	HULS
[2-[2-(Butoxyethoxy)-ethoxy]ethanol]	HULS
[2-[2-(Ethoxyethoxy)ethoxy]ethanol]	HULS

Key parameter table: Trade names/suppliers

Trade name	Suppliers
ALTENE D6, TRICHLOROETHYLENE D1	ELF ATOCHEM
AROMASOL 12	ICI
BALTANE CF	ELF ATOCHEM
Butyl CARBITOL acetate	UNION CARBIDE
Butyl CARBITOL solvent	UNION CARBIDE
Butyl CELLOSOLVE acetate	UNION CARBIDE
BUTYL DIGLYCOL ACETATE	BP
BUTYL DIGLYCOL ETHER	BP
C7/C9 AROMATICS	ICI
C9+ II	PETROGAL
C9+ I	PETROGAL
CARBITOL solvent low gravity	UNION CARBIDE
CELLOSOLVE acetate	UNION CARBIDE
COASOL	CHEMOXY
DOWANOL DALPAD A	DOW
DOWANOL DB	DOW
DOWPER	DOW
DRIVERON S	HULS
ESTASOL	CHEMOXY
ETHOXY PROPANOL	BP
ETHOXY PROPOXY PROPANOL	BP
ETHOXY PROPYL ACETATE	BP
ETHYL ACETATE	BP
ETHYL ACETOACETATE	BP
EVOLVE CH10	ICI
Exxate 1000	EXXON
EXXOL Cyclopentane	EXXON
Exxol D 100	EXXON
EXXON MEK	EXXON
EXXON SBA	EXXON
FINALAN 100	FINA
Forane	ELF ATOCHEM
FORANE 141b DGX	ELF ATOCHEM
GENESOLV 2000	ALLIED SIGNAL INC
HAN	EXXON
Hexane Extraction	SHELL
Hexane Polymerisation	SHELL
Hexyl CARBITOL solvent	UNION CARBIDE
Hexyl CELLOSOLVE solvent	UNION CARBIDE
HIGH AROMATIC WHITE SPIRIT	SHELL
HYDROSOL Ess 60/95	TOTAL
INK SOLVENT 2427	SHELL
INVERT 1000/2000/5000	DOW
iPrOX	SHELL
ISANE IP155	TOTAL
Isohexane LNH	SHELL
Isooctane 100	EC Erdolchemie
Isopar C	EXXON
KERDANE	TOTAL
KETRUL 210	TOTAL
LOW AROMATIC WHITE SPIRIT	SHELL
ME-6K	SHELL
MeDIPROX	SHELL
MePROX	SHELL
METHOKLONE	ICI
Methyl CARBITOL solvent	UNION CARBIDE
Nappar 10	EXXON
nBuDIOX	SHELL
nBuDIOXAc	SHELL
nBuOX	SHELL

Key parameter table: Trade names/suppliers *continued*

Trade name	Suppliers
NEU TRI	DOW
Nonanol N	BASF
Norpar 12	EXXON
OXSOL 100, PCBTF	OCCIDENTAL CHEMICAL CORP.
Perchloroethylene SMD	ENICHEM
PERKLONE	ICI
Pertene	ELF ATOCHEM
PETRINEX ASB	ICI
PRIMACLEAN 1601	DOW
PROGLYDE DMM	DOW
Propyl CELLOSOLVE solvent	UNION CARBIDE
PROTENE F	ELF
RHODIASOLV RPDE	RHÔNE POULENC
SBP 100/140	SHELL
SHELL CYCLO SOL 53 AROMATIC SOLVENT	SHELL
SHELL MINERAL SPIRITS 135	SHELL
SHELL RUBBER SOLVENT	SHELL
SHELL SOL 142 HT	SHELL
SHELL TOLU-SOL 10 SOLVENT	SHELL
SHELL VM&P NAPHTHA EC	SHELL
SHELL VM&P NAPHTHA HT	SHELL
SHELLSOL A	SHELL
SOLTENE	SOLVAY
SOLVACLENE	SOLVAY
SOLVAREX 10	TOTAL
SOLVARO AFD	TOTAL
Solvenon DPM	BASF
SOLVENT APV SPECIAL	HULS
Solvesso 100	EXXON
SPIRDANE BT	TOTAL
TABS D5	BUSH BOAKE ALLEN
TAME	EC Erdolchemie
TAVOXENE	SOLVAY
Trichloroethylene BD	ENICHEM
Tridecanol A	BASF
TRIKLONE	ICI
UCAR n-Pentyl Propionate	UNION CARBIDE
UKLENE	ELF ATOCHEM
VARSOL 110	EXXON
VISTA 150L SPECIALTY ALKYLATE	VISTA
VISTA 47 SOLVENT	VISTA
VISTA C14 PARAFFIN	VISTA
VISTA MOLEX RAFFINATE	VISTA

Supplier information

Company name: **AKZO NOBEL CHEMICALS bv**
Address: Stations Plein 4, PO Box 247, 3800 AE Amersfoort, The Netherlands

Telephone No.: +31 33 67 6910
Fax No.: +31 33 67 6134

Contact: R.H.L. de LEEUW, Desk Manager, Chlorine and Derivatives

---

Company name: **ALLIED SIGNAL INC.**
Address: Fluorocarbons, PO Box 1053, Morristown, New Jersey 07962-1053, USA

Telephone No.: +1 201 455 2000
Fax No.: +1 201 455 2615

Contact:

Company name: **BASF AKTIENGESELLSCHAFT**
Address: Industrial Chemical Division, Marketing Solvents, 67056 Ludwigshafen, Germany

Telephone No.: +49 621 60 43085
Fax No.: +49 621 60 42525

Contact: R. DOEWELER

Company name: **BP CHEMICALS LIMITED**
Address: Hull Research and Technology Centre, Salt End, Hull, HU12 8DS, UK

Telephone No.: +44 1482 896251
Fax No.: +44 1482 894868

Contact: C.F. HILL

Company name: **BUSH BOAKE ALLEN LTD**
Address: Dans Road, Widnes, Cheshire, WA8 0RF, UK

Telephone No.: +44 151 424 3131
Fax No.: +44 151 424 3268

Contact: DR. M. VEAL, Product Sales Manager

Company name: **CHEMOXY INTERNATIONAL plc**
Address: All Saints Refinery, Cargo Fleet Road, Middlesbrough, Cleveland, TS3 6AF, UK

Telephone No.: +44 1642 248555
Fax No.: +44 1642 244340

Contact:

Company name: **DOW EUROPE SA**
Address: Bachtobel Str. 3, 8810, Horgen, Switzerland

Telephone No.: +41 1728 2771
Fax No.: +41 1728 2786

Contact: R. BATTAGELLO, TS & D Engineer
G. MERKOFER, TS & D Engineer
J. DONLEAVY, Business Director
W. BAGGENSTOSS

Company name: **THE DOW CHEMICAL COMPANY**
Address: Performance Products Dept., Midland, Michigan 48674, USA

Telephone No.: +1 800 447 4369
Fax No.:

Contact:

Company name: **EC ERDOLCHEMIE GMBH**
Address: Alte Strasse 201, 50769 Köln, Postfach 750212, Germany

Telephone No.: +49 2133 55 6116
Fax No.: +49 2133 55 6199

Contact: H. TSCHORN, Dipl. Ing.

Company name: **ELF ATOCHEM SA**
Address: 4–8 Cours Michelet, La Defense 10, Cedex 42, 92091, Paris, France

Telephone No.: +33 49 00 80 80
Fax No.: +33 49 00 74 47

Contact: M. GUERARD, Product Manager

Company name: **ELF ATOCHEM SA**
Address: 12 Place de L'Iris, Cedex, 54-92062, Paris La Defense 2, France

Telephone No.: +33 49 00 80 80
Fax No.: +33 47 96 97 05

Contact: A.S. GUERARD, Product Manager Oxygenated Solvents

Company name: **ELF ATOCHEM UK LTD**
Address: Colthrop Way, Thatcham, Newbury, Berkshire, RG13 4LW, UK

Telephone No.: +44 1635 870000
Fax No.: +44 1635 861212

Contact:

Company name: **ENICHEM S.p.A**
Address: Via Toranelli 26, 20124 Milano, Italy

Telephone No.: +39 6977 8332
Fax No.:

Contact: Ing. E. LOMBARDI, Quality Assurance Manager

Company name: **EXXON CHEMICAL EUROPE INC**
Address: Machelen Chemical Technology Center, Hermeslaan 2, B-1831, Machelen, Belgium

Telephone No.: +32 2 722 21 11
Fax No.: +32 2 722 27 80

Contact:

Company name: **EXXON CHEMICAL LIMITED**
Address: PO Box 122, 4600 Parkway, Fareham, Hampshire, PO15 7AP, UK

Telephone No.: +44 1489 884400
Fax No.:

Contact:

Company name: **FINA PLC**
Address: Fina House, Ashley Avenue, Epsom, Surrey, KT18 5AD, UK

Telephone No.: +44 1372 726226
Fax No.: +44 1372 744520

Contact: S. WESTLAKE, Technical Sales Manager

Company name: **HAYMAN LTD**
Address: 70 Eastways Park, Withan, Essex, CM8 3YE, UK

Telephone No.: +44 1376 517517
Fax No.: +44 1376 510709

Contact:

Company name: **HULS AKTIENGESELLSCHAFT**
Address: Postfach 1320, D-4370, Marl, Germany

Telephone No.: +49 2365 481
Fax No.: +49 2365 49-2000

Contact:

Company name: **ICI CHEMICALS AND POLYMERS LTD**
Address: Petrochemicals Business, PO Box 90, Wilton Centre, Middlesbrough, Cleveland, TS90 8JE, UK

Telephone No.: +44 1642 454144
Fax No.:

Contact: P. MACKENZIE

Company name: **ICI CHEMICALS AND POLYMERS LTD**
Address: Chlor Chemicals Business, PO Box 14, The Heath, Runcorn, Cheshire, WA7 4QG, UK

Telephone No.: +44 1928 511050
Fax No.: +44 1928 581072

Contact: GRAHAM P. HOWE/R. TWEDDLE/C. McKENZIE

Company name: **OCCIDENTAL CHEMICAL CORP**
Address: Basic Chemicals Group, Occidental Tower, 5005 LBJ Freeway, Dallas, TX 75244, USA

Telephone No.: +1 800 752 5151
Fax No.:

Contact: DIANE TRAMONTANA, Business Director

Company name: **PARDIES ACETIQUES**
Address: 25 Quai Paul Doumer, 92408 Courbevoie, France

Telephone No.: +33 1 47 68 04 84
Fax No.: +33 1 47 68 05 00

Contact: P. EATON, Marketing Director

Company name: **PETROGAL SA**
Address: Av. Fontes Pereira de Melo, 6-2 Dt, 1000-Lisbon, Portugal

Telephone No.: +351 1 3102656
Fax No.: +351 1 3102953

Contact: P. SPOHR, Export Manager

Company name: **RHÔNE POULENC**
Address: 25, Quai Paul Doumer, 92408 Corbevoie Cedex, France

Telephone No.: +1 47 68 12 34
Fax No.:

Contact:

Company name: **SHELL CHEMICAL COMPANY**
Address: 1 Shell Plaza, PO Box 2463, Houston, TX 77252, USA

Telephone No.: +1 713 241 5793
Fax No.:

Contact: R.W.F. CHOUFFOT

Company name: **SHELL CHEMICALS EUROPE LTD**
Address: Shell International Chemical Company Ltd, Shell Centre, London, SE1 7NA, UK

Telephone No.: +44 171 934 1234
Fax No.: +44 171 934 4549

Contact:

Company name: **SHELL CHIMIE**
Address: 89 Boulevard Franklin Roosevelt, 82564 Rueil-Malmaison, Cedex, France

Telephone No.: +33 1 47 14 71 00
Fax No.: +33 1 47 14 82 99

Contact:

Company name: **SHELL NEDERLAND CHEMIE B.V.**
Address: Postbus 3030, 3190 GH Hoogvliet, Rotterdam, The Netherlands

Telephone No.: +31 10 231 7000
Fax No.: +31 10 231 7180

Contact:

Company name: **DEUTSCHE SHELL CHEMIE GMBH**
Address: Postfach 5220, 65727 Eschborn, Germany

Telephone No.: +49 6196 4740
Fax No.: +49 6196 4745 02

Contact:

Company name: **SOLVAY SA**
Address: 33 Rue de Prince Albert, B-1050, Brussels, Belgium

Telephone No.: +32 2 509 6111
Fax No.: +32 2 509 6505

Contact: M. SERVAIS, Product Manager

Company name: **TOTAL PETROLEUM INC**
Address: East Superior St., Alma, MI 48802, USA

Telephone No.: +1 517 463 7630
Fax No.: +1 517 463 9623

Contact: K. BATTLE, Manager Specialty Products

Company name: **TOTAL SOLVENTS**
Address: 51 Esplanade du General de Gaulle, La Defense 10. 92800 Puteaux, France

Telephone No.: +33 1 41 35 33 64
Fax No.: +33 1 41 35 47 74

Contact:

Company name: **UNION CARBIDE CORPORATION**
Address: Customer Information Centre, 10235, West Little York Road, Houston, TX 77040, USA

Telephone No.: +1 800 568 4000
Fax No.:

Contact:

Company name: **UNION CARBIDE CORPORATION**
Address: 39 Old Ridgebury Road, Danbury, CT 06817-001, USA

Telephone No.: +1 203 794 2000
Fax No.:

Contact:

Company name: **VISTA CHEMICAL CO**
Address: 900 Threadneedle, PO Box 19029, Houston, TX 77224, USA

Telephone No.: +1 713 788 3000
Fax No.: +1 713 588 3456

Contact: W.L. SORENSEN, Snr Research Engineer

Company name: **VULCAN CHEMICALS**
Address: Division of Vulcan Materials Company, PO Box 530390, Birmingham, AL 35253-0390, USA

Telephone No.: +1 205 877 3021
Fax No.: +1 205 877 3448

Contact:

Index